AF736104

Computational Nondestructive Evaluation Handbook

Computational Nondestructive Evaluation Handbook

Ultrasound Modeling Techniques

Authored by Sourav Banerjee and Cara A.C. Leckey

CRC Press
Taylor & Francis Group
Boca Raton London New York

CRC Press is an imprint of the
Taylor & Francis Group, an **informa** business

Published [2020]
by CRC Press
6000 Broken Sound Parkway NW, Suite 300, Boca Raton, FL 33487-2742

and by CRC Press
2 Park Square, Milton Park, Abingdon, Oxon, OX14 4RN

© 2020 Taylor & Francis Group, LLC

CRC Press is an imprint of Taylor & Francis Group, LLC

Reasonable efforts have been made to publish reliable data and information, but the author and publisher cannot assume responsibility for the validity of all materials or the consequences of their use. The authors and publishers have attempted to trace the copyright holders of all material reproduced in this publication and apologize to copyright holders if permission to publish in this form has not been obtained. If any copyright material has not been acknowledged please write and let us know so we may rectify in any future reprint.

Except as permitted under U.S. Copyright Law, no part of this book may be reprinted, reproduced, transmitted, or utilized in any form by any electronic, mechanical, or other means, now known or hereafter invented, including photocopying, microfilming, and recording, or in any information storage or retrieval system, without written permission from the publishers.

For permission to photocopy or use material electronically from this work, access www.copyright.com or contact the Copyright Clearance Center, Inc. (CCC), 222 Rosewood Drive, Danvers, MA 01923, 978-750-8400. For works that are not available on CCC please contact mpkbookspermissions@tandf.co.uk

Trademark notice: Product or corporate names may be trademarks or registered trademarks, and are used only for identification and explanation without intent to infringe.

ISBN: 978-1-138-31454-2 (hbk)

Typeset in Times
by Cenveo® Publisher Services

Dedication

This book is dedicated to science that are yet to be discovered

And specially dedicated to my

Father, Mr. Som Nath Banerjee

&

Mother, Mrs. Malabika Banerjee

- Sourav Banerjee

This book is dedicated to physics—yes, physics—
thanks for always keeping us on our toes!

(And also, to my family for their endless support)

- Cara A.C. Leckey

Contents

Preface

Nondestructive evaluation (NDE) has impact across industries that require the verification of material, component, and/or structural quality. Whether applied in a manufacturing setting, or in-service throughout the lifetime of a part, NDE plays a key role for ensuring component durability and safety. Since NDE is cross-cutting with application to civil infrastructure and power, automotive, and aerospace industries, among others; there exist a wide variety of inspection techniques for the corresponding variety of materials, geometries, defect types, etc. Some of the most common inspection methods are ultrasound, thermography, eddy current, and X-ray—with ultrasound currently being the most widely used method in many fields.

The rise of advanced materials and manufacturing methods, such as tailored composites and additive manufacturing, has led to numerous applications where heritage inspection methods may not perform as well as they do on traditional materials/components. Further, for complex structures solely relying on iterative experimental methods for the development of new inspection approaches may not lead to an optimized method with the required capabilities. Computational NDE has an important role to fill these gaps.

While computational NDE can broadly be defined to include both NDE modeling and computational-based data analysis; the focus of this book is methods in NDE modeling and simulation. Current computational hardware (such as manycore hardware) has made it feasible to run realistic NDE simulations in a timely fashion. NDE modeling and simulation tools create a cost-effective route to explore a large variety of inspection approaches and defect scenarios for a given inspection challenge. Computational methods can allow for a more thorough and rigorous exploration of a problem space than would be feasible, or practical, with only experimental methods. Further, rapid (and validated) simulation tools can enable method optimization, model-based inversion, and physics-informed analysis. These in turn can lead to improved inspection results and improved defect characterization. Additionally, computational NDE methods can enable a path to inspectability-informed design and model-assisted probability of inspection.

The objective of this book is to outline a variety of methods for NDE modeling and simulation. Many of the described methods, such as finite difference-based methods, are applicable to numerous NDE techniques (e.g., thermography, electromagnetic-based methods, ultrasound). However, since ultrasound is currently the most widely used NDE technique across numerous industries, this book primarily focuses on the mathematical details for elastodynamic-based NDE (e.g., ultrasound) and give examples of ultrasonic NDE modeling and simulation.

It is our hope that readers find this handbook to contain the basics needed for computational implementation of the methods that are highlighted. As new materials, structures, and manufacturing techniques arise, we expect an increased reliance computational NDE to achieve inspection requirements.

We would like to thank a number of individuals for useful discussions, writing computer codes, plotting figures, and typing a few equations in regard to the writing of this book, including Erik Frankforter with NASA Langley Research Center, Sajan Shrestha, Hossain Ahmed, Subir Patra, Fariha Mir, Vahid Tavaf and Mustahseen M Indaleeb with the University of South Carolina.

Sourav Banerjee & Cara Leckey

About the Authors

Dr. Sourav Banerjee is currently an Associate Professor in the Department of Mechanical Engineering at the University of South Carolina (UofSC), Columbia, South Carolina, USA. He is currently serving as a Director of the Integrated Material Assessment and Predictive Simulation Laboratory (i-MAPS) at UofSC since 2012. Before joining academia, Dr. Banerjee worked in the industries, performed NDE inspection, design and matured many technologies as project engineer, Senior Research Scientist, and as Director of Product Development, respectively. Dr. Banerjee's current research is focused on Ultrasonic and Acoustic waves while catering to multiple fields including experimental and computational NDE, structural health monitoring (SHM), design and analysis of multifunctional metamaterials, energy harvesting from acoustic/ultrasonic waves, noise barriers and ultrasonic wave-based biomedical device applications. Dr. Banerjee has published more than ~120 research articles of which more than ~56 are in peer-reviewed international journals. He has authored three book chapters and one book related to his research area on ultrasonics. Dr. Banerjee is very interested in computational NDE that he recognizes has potential to solve many autonomous inspection challenges in the future generation NDE inspections. Dr. Banerjee has given academic and professional lectures at multiple countries including Germany, Italy, China, India, Russia, to name a few. He works with many industries and government agencies around the world. He is the recipient of structural health monitoring (SHM) Person of the year award 2019 awarded at the Stanford University during IWSHM 2019 and the Breakthrough Star award awarded by the Office of Vice President of Research at the UofSC in 2019. Not only research, Dr. Banerjee is a passionate about teaching. His innovative teaching methods for new smartphone-generation students have made them more engaged during the class and helped them learn through practical engineering problems. For his innovative teaching method for large classes, he received Michael J. Mungo Teaching Award, the higher teaching award at UofSC in 2017.

Dr. Banerjee is serving as an executive committee member for the ASME NDPD division, committee member for the SPIE Smart Structure and NDE conference, etc. He is serving as a symposium organizer on the topic related to NDE at the ASME IMECE conference since 2016. He serves on the editorial board of Scientific Reports published by Nature Publishing Group, *International Aeronautics Journal*, *International Journal of Aeronautics and Aerospace Engineering*, etc. Dr. Banerjee also serves as an advisory board member of the *Journal of Nondestructive Evaluation, Diagnostics and Prognostics of Engineering Systems*, published by ASME.

Dr. Banerjee received Ph.D. in Engineering Mechanics and Applied Mathematics from University of Arizona, Tucson, USA, M.Tech in Structural Engineering from Indian Institute of Technology (IIT), Bombay, India, and the B.E. in Civil Engineering from the Indian Institute of Engineering Science and Technology (IIEST), Shibpur, formerly known as Shibpur BE College, India.

Dr. Cara A.C. Leckey is currently the Deputy Branch Chief for the Nondestructive Evaluation Sciences Branch at NASA Langley Research Center. She has been a physicist in the Nondestructive Evaluation group at NASA Langley since 2010, with a research focus on computational NDE. During her time at NASA, Dr. Leckey has led a variety of computational NDE-focused projects, including research on the development and optimization of NDE simulation codes and the use of NDE simulation to solve aerospace inspection challenges, as well as multidisciplinary work focused on the optimization of high-performance computing code for several of NASA's scientific/engineering software tools. Her work often involves collaboration with academia, industry, and other government agencies. Dr. Leckey is the recipient of the NASA Early Career Achievement Medal and the NASA Aeronautics Research Mission Directorate Associate Administrator's High Potentials Award. She previously served as a subtopic manager for NASA's Small Business Innovation Research (SBIR) program for a topic on focused on computational NDE. She currently serves as an Associate Technical Editor for the journals *Materials Evaluation* and *Research in NDE.* Dr. Leckey received Ph.D. and M.S. degrees in physics from the College of William and Mary, and a B.S. in physics from University of Mary Washington.

1 Computational Nondestructive Evaluation (CNDE)

1.1 INTRODUCTION

Across numerous industries, structural components are designed for a targeted service lifetime. Rapid advancements in manufacturing (such as additive manufacturing), new materials (such as composites), and complex structural designs (that are enabled by model-based engineering) are occurring across civil, energy, transportation, defense, and space industries. These advancements have an impact on the current and future design, manufacturing, and resulting service life of infrastructure and structural components (e.g., highways, bridges, buildings, pipelines, ground vehicles, military equipment, defense vehicles, ships, marine structures, commercial and military aircraft, space vehicles, power plants, nuclear waste structures). While designed for a specific service life, unexpected structural failures can occur due to off-nominal usage. Additionally, failure can occur due to prolonged service. Structural failures can lead to economic losses, ecological impacts, and loss of human life. The preservation and maintenance of existing infrastructure is a concern that affects social and economic development in a global manner. The smart management of structures as a whole, or management of the structural components, is a key to protecting economic investments and ensuring safety. Therefore, it is essential to guarantee that, as per its design, a structure or structural component will safely perform over a designed period of time. This challenge has motivated scientists and engineers to investigate different techniques that enable them to understand the state of the materials and structures and the risk factor involved with the state-of-health of the structure.

Nondestructive evaluation (NDE), also known as nondestructive testing (NDT), [1] is a field that is focused on the health diagnosis of materials and structures. In the past two decades, a subset of NDE called in situ NDE has also emerged, commonly called structural health monitoring (SHM) [2]. NDE/SHM is a field of applied physics and engineering. NDE/SHM is concerned with assessing the structural integrity of load-bearing and critical components without causing damage to the structure. NDE/SHM is a field of study that is of interest to the scientific and engineering research community, but which also has a large impact on industrial practices and the quality assurance of high-valued manufacturing products, structures, infrastructure, etc.

During the World War II, to fulfill the demand of quality assurance of mechanical systems, machine components, and the defense equipment, the field of NDE emerged.

However, after the war, lack of business cases in the defense sectors pushed the majority of the NDE industries to focus on consumer products due to their enormous industrial growth. Simultaneously, quality assurance of the aerospace systems became a new interest for government agencies. In the last few decades, there has been a tremendous rise in the use of NDE techniques applied to numerous facets of civil, mechanical, aerospace, and nuclear engineering. This is primarily because, most of the NDE methods work in a noninvasive manner, i.e., without direct contact or cutting into the material, as well as without altering the state and properties of the material. Even methods that require contact with the structures like contact-based methods and SHM with bonded or embedded sensors, do not change the functionality of the materials and structures. Additionally, advances with noninvasively built models help to inspect the nonvisible areas of the structures in such cases.

NDE/SHM, in a general sense, refers to methods of evaluation (inspection) of structural materials that neither destroy the material, compromise the strength of the structure nor cause any damage to the material or structure during or after inspection. In a general sense, irrespective of the specific NDE method, active NDE entails the generation of external energy (such as ultrasound, heat, electromagnetic waves, etc.) that is sent into the material or structure, which then interacts with the material and any existing defects. The energy that has interacted with the material/structure/defect is then recorded by a sensing method specific to the respective NDE/SHM method used for inspection (e.g., ultrasonic transducer, infrared camera, etc.). Passive NDE entails the monitoring (e.g., with an infrared camera) of energy (such as heat) that is generated within the specimen, such as due to material friction from loading conditions. There have been several NDE methods developed during the past century (such as eddy current, terahertz, ultrasound, thermography, vibroacoustic, among other methods). "Ultrasonic NDE" is currently the most commonly used method. In ultrasonic NDE, the probing energy is in the form of a stress wave that propagates into the material. The wave can be generated and sensed via electromechanical means. The term "ultrasonic wave" refers to an acoustic stress wave that encompasses the frequency ranges above the hearing frequencies of the human ear (~20 kHz). Hence, the stress wave sent into the material through transmitter probe or sensed from the material by sensing probe cannot be heard by the human ear. Ultrasound encompasses a wide frequency range, from 20 kHz and above, up to 2 GHz. But most of the engineering applications do not need frequencies above ~100 MHz. Majority of NDE applications use frequencies between ~1 MHz and ~10 MHz. On the other hand, ultrasound-based in situ NDE (i.e., guided wave SHM) applications are commonly limited within ~750 kHz. Before going into much detail, let's first see why NDE is necessary and what other NDE techniques are available.

1.1.1 Various NDE Methods

Materials and structures under operation are always subjected to fatigue loading. Under continuous fatigue, materials will fail at a certain point. Hence, accurate prediction of material failure is imperative for their safe operation during their lifetime. Accurate prediction of material failure is the bottleneck for any critical mission. Missions could be sophisticated space missions to the moon or mars, or could be safe travel in commercial aerospace vehicles, or travel through safe structures, underground or above ground on the earth. Immaterial of the specific material or structure we target; it requires two things (1) intermittent or continuous NDE/SHM that can inspect and estimate the state of the health of the material or structure and (2) a very good failure prediction tool to predict the remaining useful life of the material based on the current state of health. In order to achieve the first requirement, condition-based maintenance (CBM) is proposed [3–7]. Under the concept of CBM, integrated sensor data can be continuously analyzed to assess the health of the structure. And if necessary, maintenance is performed based on that assessment. If any damages are found, and if found to be severe, components or structural parts may require immediate removal and replacement. But if the damage is not significant and component could survive further, a remaining useful life is estimated using virtual digital models. However, this step is very risky. It requires confidence not only on the predictive models, but also requires confidence on the characterization of the damage from the analyses based on the NDE/SHM sensor data.

Questions that must be answered when selecting and assessing an appropriate NDE/SHM method are: what type of defects or damage needs to be detected/characterized? Which NDE/SHM method(s) should be used? And/or what type of damage should are the methods applicable for? There are plenty of choices. A wide spectrum of inspection methods is available to use. These methods use their respective physics of sensing and sensor materials for NDE/SHM under the umbrella of CBM. NDE/SHM encompasses a wide range of sensing techniques from thermography to radar, to ultrasonic, to X-ray [8], etc. Numerous techniques have been developed and utilized in the NDE/SHM field. Some of them are briefly discussed in this section. A few common methods and their associated sensing technologies are listed in Table 1.1 for quick reference.

Visual inspection or visual testing [9] is the most basic type of NDE technique and uses human interpretation. The technique is widely adopted (particularly for in-service parts) due to its fast applicability and affordability. This is a very basic NDE method and may have significant error from human interpretation.

Ultrasonic testing [10] or ultrasonic NDE employs high frequency stress waves to detect the existence, orientation, and characteristics of defects in the material with fast scans and resolution up to half the wavelength of the ultrasonic wave (often an adequate resolution for many applications). There are two approaches of ultrasonic NDE that are most widely used; pulse echo and through transmission approaches (discussed in detail in later sections).

Thermography testing or thermal imaging [11] inspects components by detecting the change in thermal conductivity of the material. Although, the technique can inspect a large surface area, it can require sensitive and expensive instrumentation (such as specialized infrared cameras). Infrared thermography testing (IRT) is based on the recording of the thermal radiation emitted by a surface of a specimen. The technique can involve active inspection, where a heat source is applied to the specimen under inspection and the diffusion of heat in the component is measured using an IR camera. Alternatively, IRT methods can be used in passive mode, where IR camera can be employed to record the heat sources generated inside the specimen (e.g., due to the application of external loading).

TABLE 1.1
Different NDE/SHM Methods Used for Material State Awareness

Inspection Method	Basic Physics	Sensor/Sensing Technologies	Advantages	Limitations
Visual inspection	Optical physics: images created out of the light reflected, refracted, and transmitted from the object of inspection	Human eye, optical camera, etc.	Inexpensive	Only surface condition is evaluated, depends on human interpretation, incipient states and internal damage are beyond the reach of this method and thus is not reliable
Thermal imaging	Physics of heat conduction: detects the change in thermal conductivity of the material	Infrared camera, IRT	Can cover a large surface area of the structure	Expensive instrumentation and hefty equipment. Not easy for hidden areas of the structure or complex geometry components
Radiography	Physics of X-ray and/or gamma ray	X-ray source/ detector, gamma ray source/detector	Can accurately image small sized defects, voids, inclusions, cracks, fiber defects, and delaminations	Equipment is expensive and generally not movable. For inspecting large structures, the equipment size can become unmanageable
Electromagnetic testing	Physics of electrical current in conductive materials. Induced electric currents and/or magnetic fields inside a material	Eddy current probe, ECT, RFT, MFL, ACFM	Electromagnetic response to detect fractures, faults, corrosion, etc. in the materials	Works in conductive materials and surface to near surface defects can be measured

TABLE 1.1 *(Continued)*
Different NDE/SHM Methods Used for Material State Awareness

Inspection Method	Basic Physics	Sensor/Sensing Technologies	Advantages	Limitations
Acoustic emission	Physics of wave propagation in materials. It detects the energy released during the defect initiation and stress wave propagation in the material	Piezoelectric sensors that work under mechanoelectrical transduction	Can detect crack initiation and propagation in metallic structures, matrix microcracking, fiber-matrix debonding, localized delamination, and fiber breakage in composite materials	Very sensitive to damage, appropriate physics-based knowledge is required to analyze acoustic emission data
Shearography	Physics of laser optics, uses coherent light or coherent sound waves	Similar to electronic speckle pattern interferometry, uses laser light emitter and detector, optical cameras	Measures stress concentration, detects delaminations and other defects, large area inspection	
Ultrasonic	Physics of wave propagation in elastic solids, longitudinal wave, shear wave, Rayleigh wave, Scholte wave, leaky Lamb waves, guided Lamb wave, etc.	Piezoelectric transducer/receiver, piezoelectric wafers, piezoelectric crystals, piezoelectric polymers, laser Doppler vibrometer as sensors	Ultrasonic testing is versatile, can be applied for most common defects mentioned above, can cover large areas, can inspect nonvisible hidden areas, can work with conductive or nonconductive materials	Often requires a coupling medium, e.g., water, to propagate into the material under inspection. In air coupled ultrasonic applications, attenuation is the biggest challenge, requires high voltage actuation

Radiographic testing [12] makes use of X-rays (for thin structures) or gamma rays (for thick structures) to visualize the state of the structure. It is useful for detecting voids, inclusions, cracks, fiber defects, and delaminations in material or structural components.

Electromagnetic testing [13] induces electric currents, and/or magnetic fields inside a material and measures the electromagnetic response in order to detect cracks, fractures, faults, corrosion, and other defects in the materials. Electromagnetic methods include Eddy current testing [14], remote field testing (RFT) [14], magnetic flux leakage (MFL) [14], and alternating current field measurement (ACFM) [15], among others.

Acoustic emission (AE) [16] is a dynamic process that detects initiation and propagation of defects with the help of piezoelectric sensors. These AE sensors sense mechanical perturbations, such as a stress wave propagating and spreading out and away from the origin of the defect. AE can be used to detect crack initiation and propagation in metallic structures, matrix microcracking, fiber-matrix debonding, localized delamination growth, and fiber breakage in composite materials. Although, AE is very sensitive to damage, appropriate knowledge is required to analyze AE data and without being paired with other NDE methods, may not yield a quantitative characterization of damage size.

Acousto-ultrasonic testing [17] is a method that combines acoustic and ultrasonic testing of materials under stress to assess noncritical flaws. However, it is not used while detecting individual large flaws such as delamination or voids, which may adversely affect the strength of the material.

Shearography testing [18] is a laser-based optical method useful in characterization of delamination defects by detecting stress concentrations.

In addition to the above methods, Table 1.1 summarizes different NDE methods and their respective physics.

1.1.2 Computational Ultrasonic NDE

Different inspection methods use different physics and generate different types of data. This means that the methods enable diagnosis of different properties. Hence, understanding of the NDE data from the respective inspection method is very crucial for interpreting the NDE result to yield information about the state-of-health of the component under inspection. Although it is important to understand respective data formats, it is not practical to review and analyze all existing NDE methods and their respective data formats in one concise book. Hence, in this book only the ultrasonic NDE [19] will be discussed. In fact, out of all the NDE methods, nondestructive inspection with ultrasound is the most popular and widely applicable NDE method for various engineering and biomedical applications. Further, in this book we focus on ultrasonic NDE/SHM sensors and ultrasonic signals.

For accurately identifying the material state and to yield material state awareness, an extensive knowledge of wave interaction with defects and both pristine and degraded material states is crucial. The ultrasonic waves are introduced as a probing energy into the material with the help of piezoelectric transducer. The wave energy travels into the material and interacts with the current material state. Diagnosis of the material state is then carried out by sensing the propagated (scattered, reflected, and/or transmitted) energy from a distant location. The fundamental notion of ultrasonic NDT is to investigate/analyze the ultrasonic sensor/receiver signals to acquire internal information about the structure under interrogation. A detailed study of ultrasonic wave propagation in different structures is essential for optimal understanding of the ultrasonic NDE sensor data. The sensitivity of the ultrasonic probing energy to certain shapes and orientations of defects is also very crucial. For this purpose, there is an absolute need to understand the wave propagation behavior in materials and structures.

Improved understanding of sensor signals, and more precisely, the nature of wave propagation in different types of materials (such as isotropic and anisotropic materials), has been a topic of interest in NDE research for the past few decades. Due to the widespread application of engineered anisotropic materials, most prevalently composites, in many different fields of engineering and technology (e.g., mechanical and aerospace industries), tools to aid the understanding of wave propagation in these complex materials are of great importance across numerous industries relying on NDE. Wave propagation behavior is studied by investigating wave interaction with the material in the form of scattering, reflection, and transmission of the wave, giving rise to geometric dispersion. These interactions of waves depend on many factors such as the part geometry, mechanical properties, number and nature of the interfacial conditions, loading conditions, and environmental conditions the part is subjected to [20]. The mechanical properties (number of independent material constants) vary from the simple case of isotropic materials to the most general anisotropic case for triclinic materials.

Simulation of energy interaction with different materials and structures is the heart of the field of computational nondestructive evaluation (CNDE). CNDE is comparatively a new word coined in last few years in the field of ultrasonic NDE, and generally is used to refer both computational physics-based modeling and simulation of NDE, as well as computational means for NDE data analysis. Computational NDE methods can allow for a more thorough and rigorous exploration of a problem space than would be feasible, or practical, with only experimental methods. Further, rapid (and validated) simulation tools can enable inspection optimization, model-based inversion, and physics-informed analysis. These in turn can lead to improved inspection results and improved defect characterization. Additionally, computational NDE methods can enable a path to inspectability-informed design and model-assisted probability of inspection. As stated in the preface, this book focuses on NDE modeling and simulation for ultrasonic NDE.

Analytical and numerical computation of acoustic or ultrasonic wave propagation in fluid and solid media has been a field of research for more than a century [21, 22]. Until the beginning of 1950s, efforts were mostly limited to analytical solutions of acoustic and ultrasonic wave propagation. After the 1950s the numerical computation of wave propagation started gaining popularity. Immediately, it was realized that while the wave propagation problems can be solved using numerical methods, the available computational power was not enough to solve complex problems. Since the 1970s, increasing computational capability has made the numerical computation of wave propagation a reality. Increasing computational power not only boosted the numerical efforts, but also created an opportunity to solve advanced analytical wave propagation problems (e.g., to find generalized wave dispersion behavior at higher frequencies). Despite significant research activities on computational wave field modeling under CNDE, there remain wave propagation behaviors in fluid, isotropic, and especially in anisotropic media that are still not fully understood.

Currently, the broad concepts of "Digital NDE pipeline" and "Digital Twin" have been promoted by government agencies in their respective countries. These drivers, along with the availability of low-cost multicore and manycore computational hardware, have generated a renewed focus on CNDE. CNDE could be used for predicting a system's response in scenarios that are either impossible, nonpractical, or too costly to be implemented experimentally. Ultrasonic inspection and accurate data interpretation are keys to structural certification across numerous engineering fields, and can be the determining factor on whether a part/structure is accepted or rejected for in-service use. If the signals are interpreted inaccurately, the complete certification process is jeopardized. Further, ultrasonic inspection of components that are already in-service is a key to ensuring structural safety. Hence, it is important to establish confidence in the interpretation of the ultrasonic signals.

Ultrasonic signals carry signatures of the material state and are altered by degraded material properties. When ultrasonic energy interacts with the internal damages, e.g., cracks, delamination, debonding, dents, etc., the energy reflects, refracts, scatters, or attenuates/decays. These physical interactions create the features in the ultrasonic signal that can be correlated with the damage state. Further, the external environment that the structure is subjected to can lead to changes in the ultrasonic signal. For example, metallic structures in a corrosive environment or materials such as composites under conditions with high humidity or excessive heat can lead to change in the ultrasonic signal. Different sources can cause different types of changes in ultrasonic signals. Using the ultrasonic NDE experiments, researchers have been able to quantify and understand the rudimentary damage states in the materials. For example, when ultrasonic waves interact with a crack, much of the wave energy may reflect (depending on the ultrasonic wavelength compared to the crack size and depending on the crack orientation with respect to the wave incidence). In such a case, the transmitted wave signal in a "pitch-catch" type set may correspondingly have a drop-in amplitude due to some of the energy being reflected. As another example, when ultrasonic energy interacts with a delamination in a composite, the coda part of the signal can contain a prolonged vibration due to energy trapping and leakage from the delamination region.

Although these are a few examples of specific wave damage interactions, there are plethora of possibilities. The effects of complex damage states, such as hidden delaminations, kissing bonds, and trapped cracks, are not easily understood when relying solely on experiment-based NDE methods. CNDE can provide a path for developing insight into these complex damage scenarios. However, currently CNDE is not widely used for shining light onto real-world inspection challenges, and the field would benefit from increased reliance on CNDE. Here, it is also important to note, that throughout this introduction – the discussion is on the use of validated NDE models. Validation is an important step that one must take to ensure that a model/simulation produces results that are a realistic representation of the NDE scenario. Whether one is using custom developed code, or commercial packages, models must be validated (in a quantitative fashion) before they are relied on to provide insight and guidance for an inspection case. Model validation can be a significant undertaking, requiring the setup of specific experiments to acquire validation comparisons. Yet, validation should be the first step when using a modeling code/tool for an NDE scenario that is outside the realm of what was previously validated (in terms of the physics involved).

This book creates an opportunity to explore the methods of ultrasonic CNDE modeling, with a focus on the application of methods to different NDE problems. It is expected that in the not-to-distant future CNDE will be matured enough to enable inspectability-informed design, computational-based NDE optimization, inversion-based defect characterization, and may lead to the ability to create a digital nondestructive evaluator twin (DigiNDTwin) of a structure. DigiNDTwin would be a tool with a "twin" digital structure and uses CNDE to determine the appropriate inspection for a structure (which may be at a distant location), receives the resulting real NDE data, processes that data to enable rapid assessment of structural health. Here we will bring our attention back to the fundamentals and will talk about the mathematical and computational basics that will lead us to the above-mentioned exciting future. We fundamentally need to understand ultrasonic signals and wave characteristics in various materials.

It is neither feasible nor cost-effective to perform all the possible experiments with various damage scenarios [23–27] and geometries. Hence, it is not possible to understand the ultrasonic wave interactions in the structure with all possible damages and material degradation using experiments. Waves in anisotropic media with complicated geometries are difficult to understand using the traditional ray tracing methods [19, 25, 28], which worked very well in isotropic or metallic structures. Then a question arises. What is the best approach to understand a large number of possible scenarios for ultrasonic inspection? And how do we classify the different signals based on ultrasonic energy and damage interactions? The most cost-effective possibility is through "virtual experiment" using physics-based simulations. A reliable simulation environment that could generate a library of multiple possible wave-damage interaction scenarios, will be useful for understanding the ultrasonic signals in detail. Consequently, such task can only be possible by implementing efficient numerical simulations of the wave behavior in the system.

Due to the necessity of simulating elastodynamic phenomena in various materials, CNDE is maturing rapidly with renewed effort. Enhanced computational capability, unique algorithms and parallel computing facilities have accelerated the field of CNDE with wave simulation in the most cost-effective and least computationally taxing fashion. Wave simulation is a field of research that began with the development of computer technology and numerical algorithms for solving multivariable differential and integral equations [29]. In the field of computational wave modeling, these algorithms are important tools providing insight into wave propagation for a variety of applications. Additionally, computational modeling offers a way to visualize physics phenomena that is difficult to visualize through empirical efforts. With the help of CNDE, researchers can compute the wave field for different materials with numerous wave-damage interaction scenarios and study the wave interaction behavior for those various situations. Hence, CNDE allows researchers to evaluate the behavior of ultrasonic probes under various conditions, visualize and characterize generated wave fields in the material, study wave propagation characteristics and scattered energy from reflectors, and predict the probability of detecting different types of defects and orientations of the material. Moreover, it is suitable for making predictions for complex geometries and materials. The knowledge gained through simulation can help to identify the quantitative changes in signals; which in turn will help us in the reliable interpretation of the real-life damage and material degradation scenarios, including the extent of internal damage and degree of material degradation [19, 24, 28, 30–33]. All of the simulated scenarios could then form a library of ultrasonic signals resulting from different damage states to enable data-driven analysis and machine learning-based analysis. When a real-life NDE/SHM signal is measured from a real NDE/SHM experiment or scheduled inspection, material state and damage state in the material could be accessed through an automated algorithm developed from the library. Automated analysis is an enabler for in situ NDE/SHM inspections requiring intermittent or continuous structural health assessment and CBM.

In summary a "virtual experiment" through physics-based simulation is a component of CNDE, which broadly is defined to include both computational-based modeling/simulation and data analysis. Many of the modeling/simulation methods (such as finite-difference, finite element) described in this book are applicable across NDE techniques, but here we focus on the mathematical specifics of ultrasonic CNDE methods. Computation-based study of the interaction of ultrasonic energy with the materials and structures has primarily arisen over the past three decades, while NDE has been around for more than a century. It was immediately realized that having access to a tool that can virtually simulate ultrasonic NDE inspections prior to performing them in the laboratories or in the field can add tremendous value across numerous industries that require NDE. In nutshell CNDE could have following five major positive impact

- **Designing NDE inspection methods in the laboratory:** Performing virtual NDE experiments using physics-based models will facilitate pre-evaluation of the NDE setup for feasibility.
- **Design, optimization, and assessment of SHM systems for real time on-board or off-board monitoring:** Performing virtual experiments will facilitate evaluation of an SHM setup before implementing SHM for real world inspection of materials and structures under operation. On-board SHM systems monitor the structure during operation, and an off-board system inspect the structures when they are idle. CNDE can aid optimization of the setup of SHM systems (e.g., layout and number of sensors for the application). CNDE can also enable efficient assessment of SHM system capabilities and limitations.
- **Optimized NDE inspection with CNDE:** Digital NDE models can be used to design NDE inspection methods with optimized locations and numbers of the actuators/transducers and sensor/receivers on the structures to yield the most effective damage detection and characterization. Optimized inspections can increase the probability of detection.
- **Reduction of NDE/SHM big-data:** NDE/SHM data often falls into the category of "big-data" and data processing can be costly in time and money. With the help of CNDE, NDE signal features can be linked to different types of damage states (whether using inverse methods or data-driven or machine learning methods). Once identified, during a real-world inspection only the important features of the data set could be saved, reducing the data set requiring further analysis and overall time required to process the data. Another option to reduce 'big-data' will be the compressive sensing. Compressive sensing similarly stores only few data points sampled below Nyquist frequency. The hypothesis of compressive sensing starts from the assumption that not all basis functions and their respective coefficients are necessary to express a signal completely. The system is most of the time over defined.

That means only few orthogonal basis functions and respective coefficients may dominantly contribute to the primary image or to the important features of the signals. Being said, there are verities of features that could be necessary for the signal processing and its correct analysis depending on their method of inspection. However, they are not available to our disposal as priori knowledge. In such case it is not only important to know the reduced order basis functions that suffice to reconstruct a signal, but it is equally important to tailor the reconstruction that is focused on specific features of the signal. If a SHM or NDE signal is stored under sampled, i.e. sampled below Nyquist frequency, during the inspection, signal features and right basis functions should be noted. CNDE may provide such opportunity.

- **DigiNDTwin for mission control from a distant location:** When NDE/SHM sensor data are sent back to the virtual model from a distant location, a DigiNDTwin model will help accurately diagnose the material and structural state. For example, a Mars mission structure equipped with NDE/SHM systems the digital twin would be used to inform what inspection should be performed on the real structure (likely to be performed as an automated inspection). Real NDE data would then be sent to the virtual digital twin to be automatically evaluated. This concept is shown in Fig. 1.1.
- **Supporting progressive failure models**: CNDE can aid determination of the damage state, including determination of the location of the damage on the structure and characterization of the damage (e.g., damage type, size, etc.). Once damage is identified and quantified, damage geometry with degraded material properties could be incorporated in the progressive failure models for estimation of remaining useful life of the materials and structures. The importance of CNDE and its usefulness for aiding the design of NDE experiments and inspections, and the link to progressive failure models is schematically presented in Fig. 1.2.

As it is now clear that why CNDE is important for NDE/SHM, the following section reviews different ultrasonic NDE/SHM scenarios that are frequently used in NDE/SHM experiments and inspection.

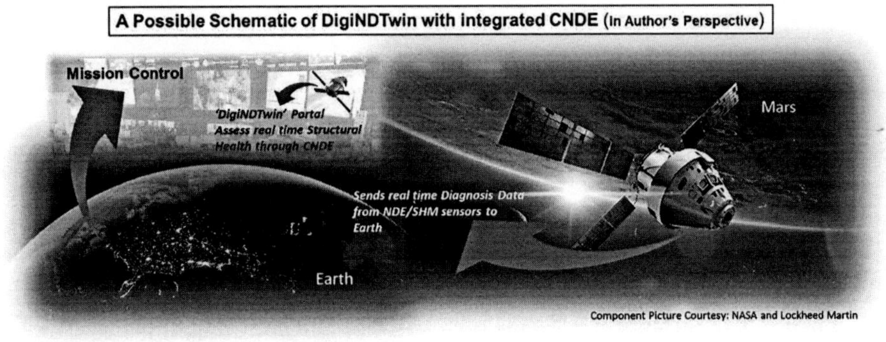

FIGURE 1.1 A possible schematic of DigiNDTwin with integrated vision for CNDE (an author's perspective).

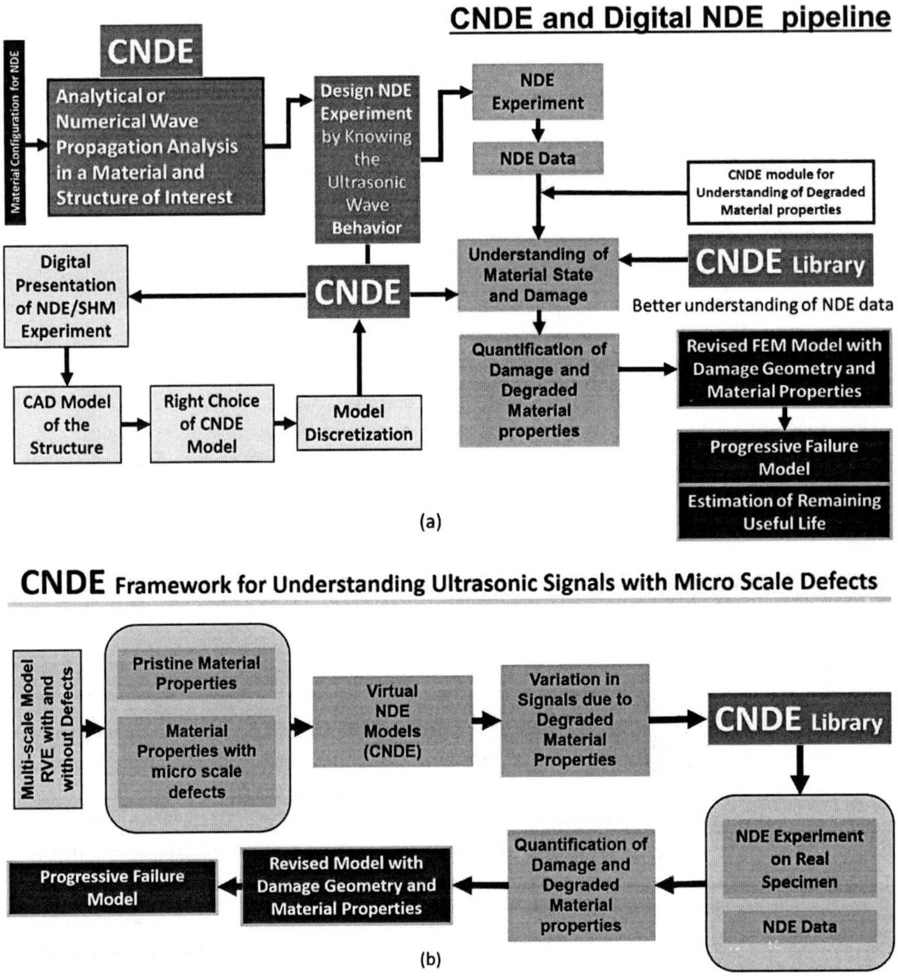

FIGURE 1.2 (a) Integrated architecture for CNDE and digital NDE pipeline. (b) A CNDE framework for incorporating microscale defects in the models.

1.2 PHYSICS AND APPARATUS FOR ULTRASONIC TECHNIQUE

1.2.1 Ultrasonic NDE

In ultrasonic NDE, an ultrasonic transducer imparts energy into the material being inspected. Many ultrasonic transducers contain piezoelectric crystals which perform electro-mechanical transduction [34] to convert electrical energy into ultrasonic energy. Ultrasonic transducers can behave as a transmitter (converting electrical energy into mechanical energy), a receiver (converting mechanical energy into electrical energy), or a transreceiver (operating in pulse-echo mode).

Other non-piezoelectric ultrasonic transducers also exist, which use magnetostriction [35]. For magnetorestrictive transducers the transducer material generates stress wave when exposed to magnetic field. Irrespective of mode of actuation, stress waves generated by the transducer propagate into and interact with the material being inspected. The wave energy is propagates, reflects, and scatters in the material and returns to the transreceiver or to the receiver, where the mechanical energy is converted back to the electrical signal. These signals are analyzed to determine the material state and/or the presence of defects or damage in the material. Figs. 1.3 and 1.4 show general schematics of typical ultrasonic NDE methods, apparatus, and arrangements. The ultrasonic energy requires a coupling media to propagate into the material under inspection and hence, needs water or another couplant as an intermediary between the transducer and material. Couplant and water-based ultrasonic methods have no restrictions on the frequency of application and can go as high as 2 GHz. Most general NDE applications using contact type transducers with couplant are between 1 MHz and 10 MHz. In some applications, a series of piezoelectric crystals are arranged to make one transducer where each crystal element can be excited individually with a prescribed time delay. This approach creates a phased-array ultrasonic transducer that can generate angled incidence oblique waves or be used for total waveform capture and paired with methods such as total focusing method. Phased arrays are useful for many applications in metals and composites. To enhance the inspection resolution, concave or focused ultrasonic transducers have also been developed. These transducers are predominantly used immersed in water, i.e., in the water tank or in scanning acoustic microscopes (SAM). Some materials, however, are sensitive to water. To avoid using water, air-coupled ultrasonic transducers have been developed but must be placed very close to the material surface to get enough energy into the part (since there is no ideal couplant, but air between the transducer and the part), while simultaneously avoiding the direct contact with the part. Air coupled applications are restricted to around 400 kHz. Recently polymer-based dry-contact ultrasonic transducers have also been developed, however, they require very high amplification and work with specially designed hardware. Both air-coupled and dry-contact ultrasonic NDE need significant amplification of the signals and primarily are used for low frequency (<400 kHz) applications.

1.2.2 Ultrasonic in situ NDE or SHM Method

Ultrasonic in situ NDE method or SHM, can be divided into two modes of operation: passive and active. For both passive and active SHM, embedded or surface mounted sensors can be used. A common sensor type is ceramic or polymer-based piezoelectric wafers. Fiber optic-based sensors have also been investigated for ultrasonic SHM. In passive SHM, if the material or structure is impacted by external sources, the resulting wave energy can be detected by the piezoelectric sensors. In active mode of SHM, piezoelectric transducers are excited to generate ultrasonic waves in the structure and the propagated ultrasonic waves are detected by piezoelectric sensors (which may be the same as the same sensors as the excitation transducers, or different sensors). Excitation of the piezoelectric actuators is performed by sending electric signals to the actuators and electro-mechanical

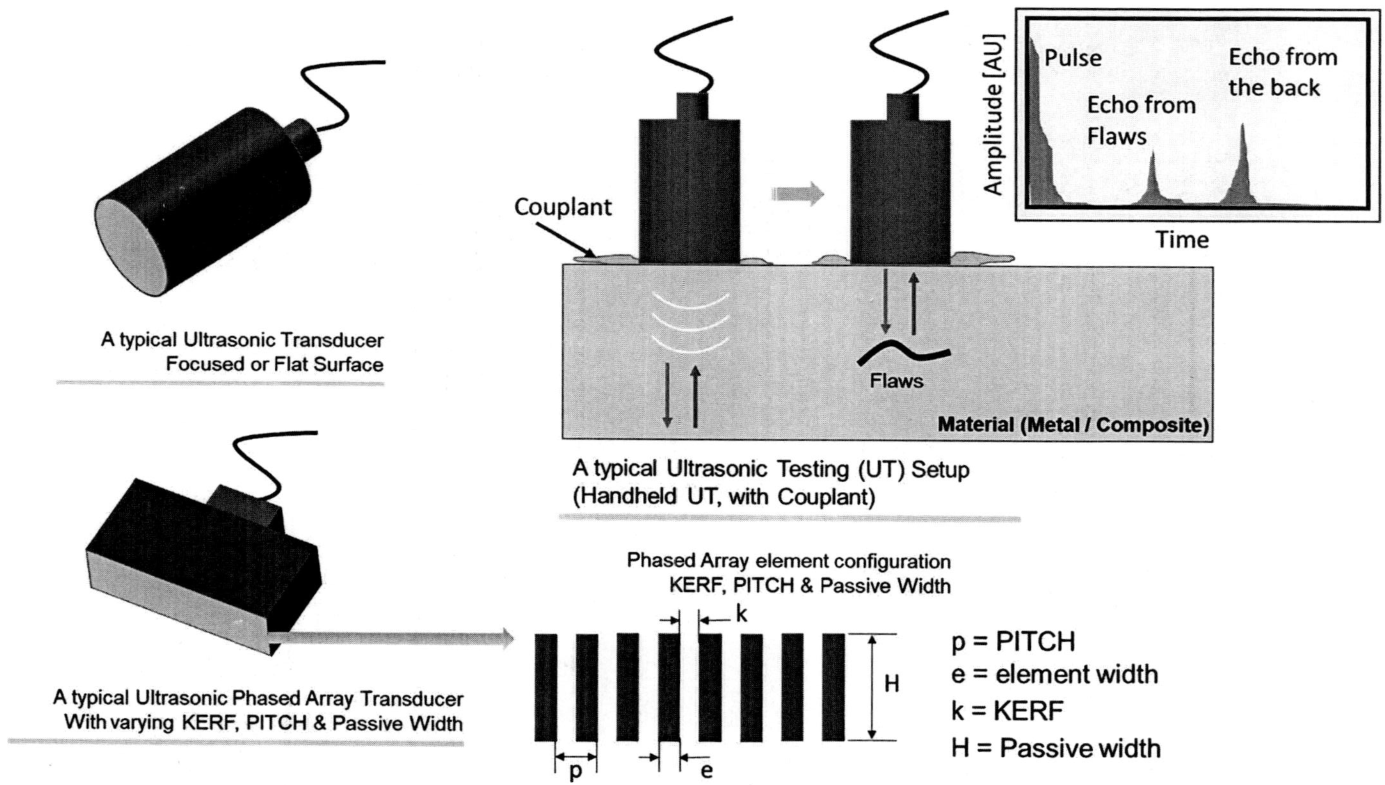

FIGURE 1.3 Typical ultrasonic transducers used in NDE of metal and composites. These transducer arrangements are transreceiver type, operates in pulse-echo mode, and used with couplants to send the ultrasonic energy to the test specimen. Typical ultrasonic traditional and phased array transducers and their typical test setup.

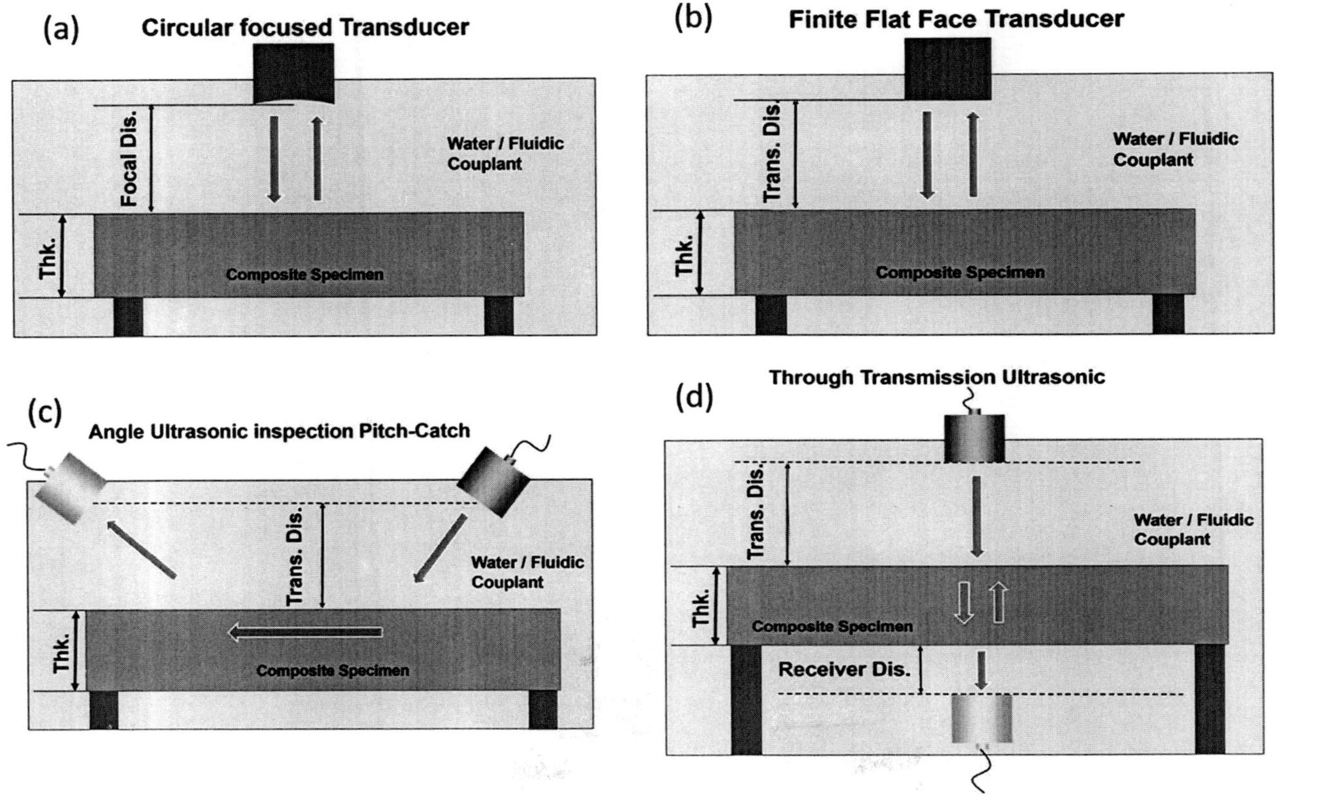

FIGURE 1.4 Typical arrangements of ultrasonic transducers in a water tank: (a) focused transducer in pulse-echo mode, enhanced resolution, focusing of ultrasonic beam is possible, (b) flat surface transducer in pulse-echo mode, (c) guided wave inspection under water in pitch-catch mode, (d) through transmission arrangement in pitch-patch mode.

transduction causes the actuators to mechanically vibrate and send energy into the material or structure in the form of a stress wave. For reception of the resulting waves, the mechanical ultrasonic vibration causes the mechano-electric transduction to convert the mechanical vibration back into an electrical signal [36]. Piezoelectric wafer sensors are similar to traditional ultrasonic transducers used for NDE applications, where piezoelectric crystals perform electro-mechanical transduction [2]. The sensors used for ultrasonic SHM are commonly attached (e.g., adhesively bonded) or embedded at select point locations and would require sensor destruction for sensor removal. This differs from traditional ultrasonic NDE which entails the use of re-usable, nonattached transducers (which may be scanned across the part surface, for example). Similar wafer type sensors can also be used as AE sensors in passive SHM mode. Schematics of typical ultrasonic SHM systems are shown in Fig. 1.5.

In an alternate SHM setup a small number of piezoelectric sensors may be attached to the component, but it is desirable to collect data of the resulting wave propagation at a larger number of locations. In such a case laser Doppler vibrometers (LDVs) can be used to record the wave field in a noncontact fashion [37]. This setup is more suitable for an off-board SHM type setup (i.e., taking measurements when a component is not in use). A schematic of typical laser UT system is shown in Fig. 1.6. If no embedded or surface mounted sensors are allowed on the structure, conventional NDE discussed in Section 2.1 will be the only options.

1.2.3 Ultrasonic NDE/SHM of Metals vs. Composites

More details of different wave types can be found in later chapters, specifically in Chapter 4, but in this section a few basic concepts are presented. In bulk metallic isotropic solid materials, waves propagate with spherical wave fronts, i.e., the wave velocities of the three fundamental stress waves (longitudinal or P-wave, shear-horizontal or SH-wave, shear-vertical or SV-wave) are equal in all directions. SH-wave and SV-wave velocities are equal and are close to half of the P-wave velocity in the material. Due to the spherical nature of the waves in isotropic materials, application of ultrasonic NDE in metallic structures is less complicated than for anisotropic composites. In anisotropic composites, complications arise because the wavefronts do not propagate in a spherical fashion.

Conventional signal amplitude, frequency peaks, time of flight alone are enough to detect damage and defects in metallic structures but might be misleading or erroneous in anisotropic materials like composites. This is due to the very nature of wave energy propagation in composites, where anisotropic propagation of scattered, reflected, and refracted waves in the material complicate the wave field. Thus, a common solution of increasing voltage input or amplification of the input signal may not be a viable solution to achieve better signal-to-noise ratio (SNR) which is commonly adopted in attenuative isotropic materials. Although composites tend to be attenuative, they are simultaneously aberrative in terms of the direction of propagation of wave energy in the material. Based on this difficulty, the following physical properties are compared in metal and composites (Table 1.2). A detailed systematic description of each of these properties can be found in a later chapter of this book.

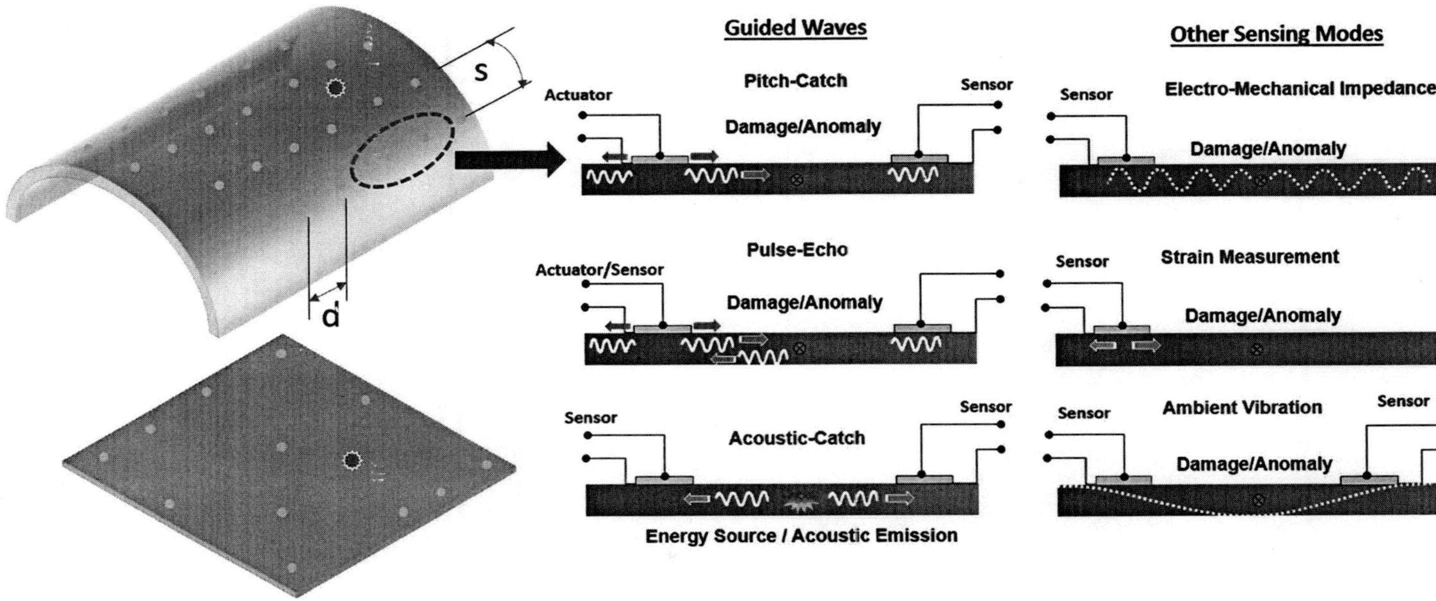

FIGURE 1.5 Guided wave (pitch-catch, pulse-echo, acoustic emission) and other sensing modes (e.g., electromechanical impedance) with surface mounted or embedded piezoelectric actuator/sensors.

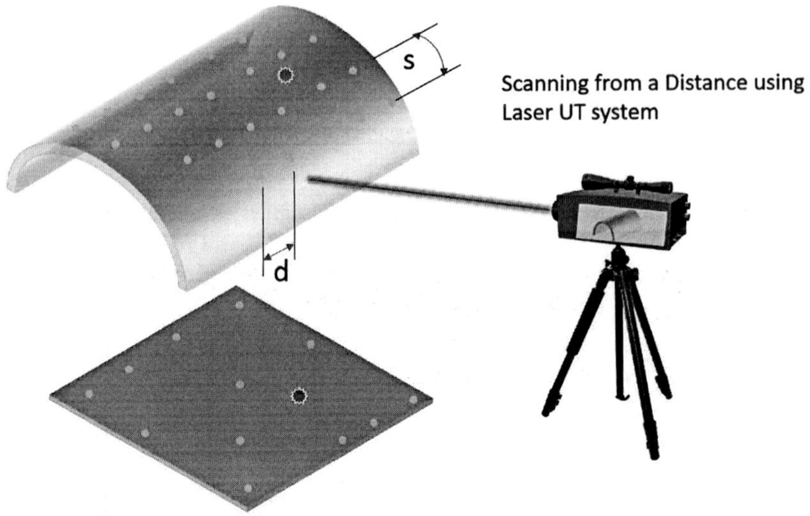

FIGURE 1.6 Laser ultrasonic: long distance noncontact guided wave sensing system using laser Doppler vibrometer.

TABLE 1.2
Difference of Wave Phenomena in Metallic and Composite Materials

Property	Metal (Isotropic)	Composite (Anisotropic)
Velocity slowness	Same in all direction	Different in different directions and depends on the stacking sequence and orientation of the layups
Attenuation	Low and frequency dependent	High and frequency dependent
Aberration	Not an issue	Dominates most of the practical NDE scenarios
Phase velocity and group velocity	Are in the same direction	They make an angle between them depending on the stacking sequence and orientation of the layups
Signal amplitude in transmission	Transmitted signal amplitude along a propagation path in a pulse-echo mode is always unique and maximum	This is not true, there is a possibility of another path where signal amplitude could be higher due to skewed energy propagation (e.g., Fig. 1.7)
Signal amplitude in reflection	Like reflected signal amplitude along a propagation path in a pulse-echo mode is always unique and maximum	This is not true, there is a possibility of another path where signal amplitude could be higher and always lower at the transmission point due to skewed energy propagation (e.g., Fig. 1.7)

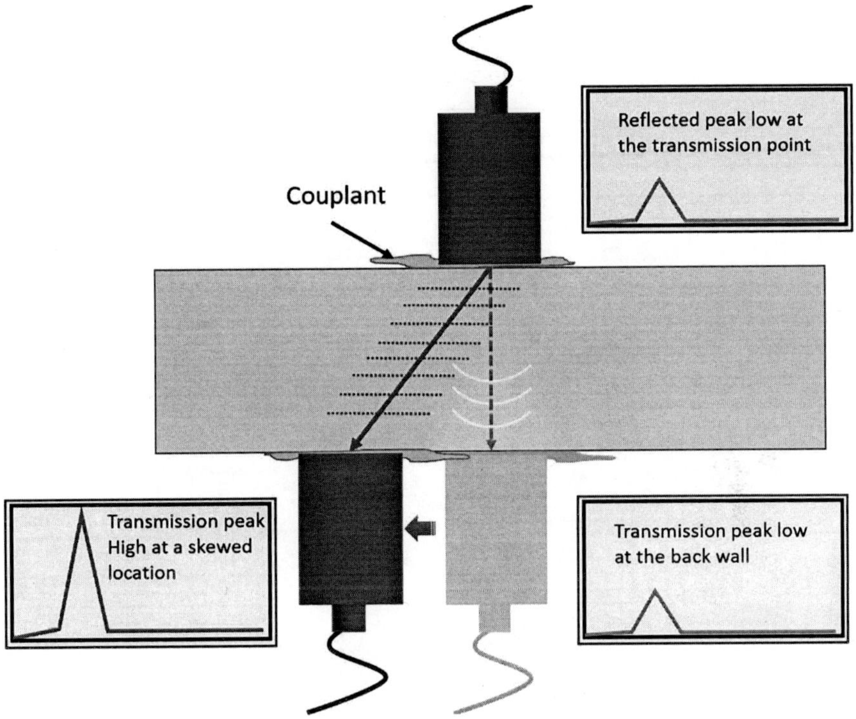

FIGURE 1.7 Typical ultrasonic NDE situation for composite, the fundamental aberration from NDE of metallic structure.

Hence, it is necessary to exploit the wave propagation in different directions. Collectively measuring the signals from reflected, refracted, and scattered energy in composite material would be an effective approach in ultrasonic NDE of composites. In this light phased-array ultrasonic, transducers are well suited for composite inspections. As mentioned previously, phased-array ultrasonic NDE transducers (Fig. 1.3) contain several piezoelectric elements arranged in an array, which may be linear in shape or some other shape. Each element can be excited at a different time, with a slight offset from each other as shown in Fig. 1.8. This arrangement can provide the opportunity to change the ultrasonic wave front shape in the material and enhance volumetric inspection of the material. Hence, for effective NDE of composites, p, e, k, and H in Fig. 1.3 are the valuable parameters including the time delay. It is also apparent that instead of a point measurement, area inspection may be more valuable for the composite NDE. Thus C-scan images generated by phased-array ultrasonic transducer a good method for ultrasonic NDE of composites (Table 1.3).

1.3 HISTORICAL BACKGROUND OF CNDE

To accomplish our goal to precisely and quickly compute the wave field and virtually generate ultrasonic signals for materials and structures with complex scenarios, efficient methodologies, and appropriate computational resources are required.

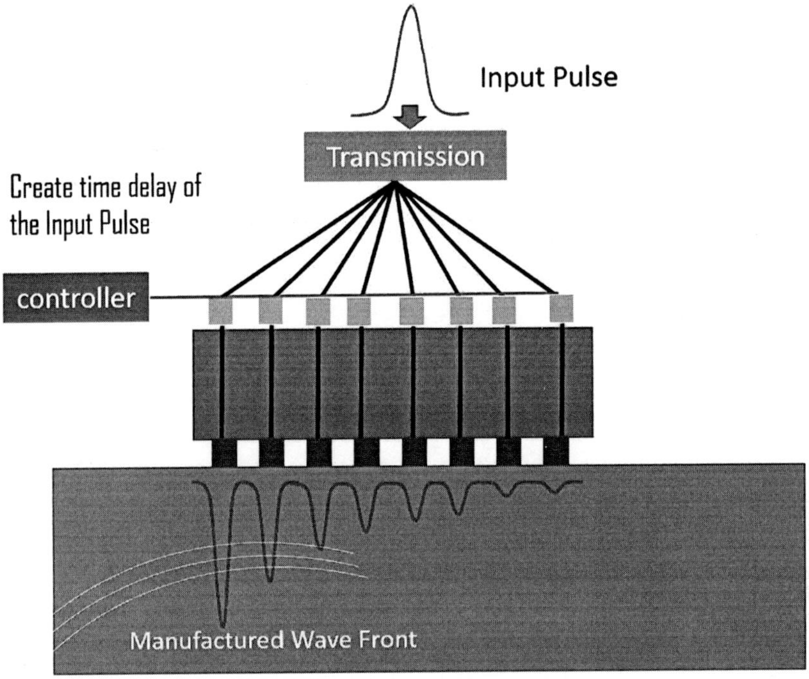

FIGURE 1.8 Typical ultrasonic NDE setup using phased-array transducer.

TABLE 1.3
Ultrasonic NDE Methods for Metallic and Composite Structures

Different Ultrasonic NDE Methods	Metal (Isotropic)	Composite (Anisotropic)
Pulse-echo mode without phased array	Very effective, for thickness detection, defect identification	Can be effective for through-thickness bulk wave-based measurements, may be affected by attenuation and wave energy skewing
Phased array ultrasonic	Effective for volumetric inspection at complex joints, welds, etc.	Effective for composite, as all possible wave vectors and energy directions are comprehended in the image generated by the apparatus
Phased array ultrasonic frequency	Typical frequencies 1–10 MHz	Typical frequency 0.5–10 MHz depending on ply thickness

Some problems may only require basic analysis in the frequency domain and whereas others require detailed time domain analysis. Traditional NDE models should include solid-fluid interactions (Fig. 1.4). Whereas, in situ NDE or SHM models have sensors that are integrated with the structure (Fig. 1.5). Further for accurate inspection processes, effective design of the transducers and piezoelectric wafers is required. Additionally, the interpretation of the ultrasonic signals requires detailed understanding of the bounded ultrasonic wave field generated in front of the transducers. The ultrasonic bounded beam generated by the transducer (Fig. 1.9) governs the energy, shape, spread, aberration, and focus of the ultrasonic waves. Depending on the energy variation within the ultrasonic beam, interactions with the materials and interfaces can change rapidly. Understanding of the interaction between the bounded ultrasonic beam and materials is necessary and creates new types of wave propagation problems that are not generally solved in traditional wave propagation analysis and are different from the vibration, bulk wave propagation and, dispersion of guided wave propagation in the wires, rods, plates and various simple structural components that were extensively studied even before World War II. Numerical modeling techniques can be useful in understanding the ultrasonic beam. Additionally, models of guided waves-based SHM can require different considerations than when modeling traditional bulk wave NDE methods.

Almost all numerical methods suffer from inaccuracy if the model is not properly created. For numerical methods, accuracy is governed by the resolution of discretization of the problem (in time, space, and/or frequency domains). In a general

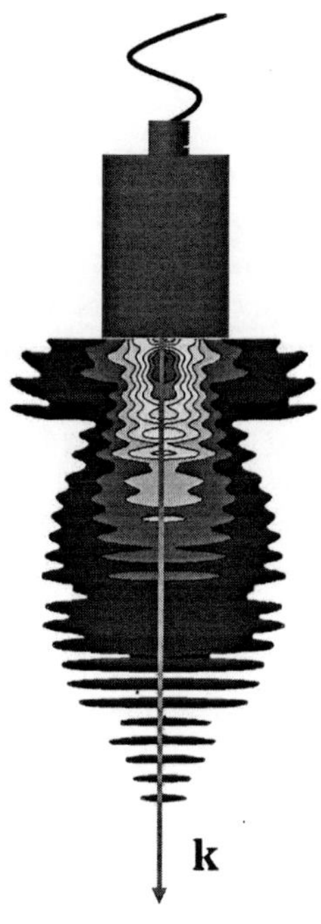

FIGURE 1.9 Ultrasonic wave field in water in front of a typical ultrasonic transducer.

sense, higher discretization gives a better and more accurate computation of the wave field. But too fine of a discretization can lead to increased numerical dispersion and also comes with a high computational cost (requiring more computation time and more computer memory). Hence, without compromising the accuracy of the results, researchers attempted to make the wave field simulation methods more cost effective. With this same goal, over time, many numerical methods have been developed to simulate wave propagation in solids and fluids. The different methods tend to be fitting for various situations. For example, some methods may be computationally fast but may not contain the required level of real-world detail for specific cases, while others may be able to include real-world detail but may be computationally expensive for certain problems. It can be difficult to decide which wave simulation method is the most useful and appropriate for a specific problem. There are no such guidelines available from the research community. It is common that experts in one method have attempted to solve NDE problems with their respective expertise, without exploring other methods. Different methods have been attempted by different research groups or individuals, and there has been no overarching communication on their respective outcomes.

Based on authors experience in the field of NDE and SHM, it is their opinion that all numerical methods have their respective pros and cons and no individual method should be wholly discarded. Hence, it is wise to explore and understand all the available methods at our disposal in order to understand their respective advantages and disadvantages. A problem driven selection process would be most fitting when selecting a computation method that is most appropriate for a specific scenario. This book is specifically dedicated to achieving this goal. Before going into further details on specific methods, an overview of available methods is provided to create the groundwork for later chapters in this book.

Studies of the elastic wave propagation in materials and structures have been of great interest to the research community in the field of NDE since the 1950s. This interest has grown by the recent expansion in the use of anisotropic materials such as composites, super alloys, and crystal structures for a wide variety of applications. In the post-war era, NDE was taken to new heights with the invention and formulation of novel configurations of ultrasonic transducers for various applications. It became essential to efficiently design transducers as well as to accurately interpret the resulting ultrasonic signals. Even though the wave propagation behavior in wires, rods, plates, and various simple structural components had been studied extensively, this was not the case for ultrasonic transducers. Hence, there arose a need for understanding wave interaction phenomena in finite dimensional transducers where application of plane wave assumptions is not appropriate.

This need resulted in mathematical modeling of the ultrasonic wave using the Rayleigh-Sommerfeld integral technique [38]. The technique is commonly known as the ray tracing method. The ray tracing method simplifies the ultrasonic wave problem by representing the waves as geometric rays. The mathematical analysis involves studying the incident waves and providing a reliable and intuitive understanding of the wave energy through the structure. This technique relies on the high-frequency assumption of the ray theory. With the evolution of the technique, it was later known as dynamic ray tracing which required the solution of the Eikonal equation [39]. Dynamic ray tracing is an improvement over the previously proposed models with a paraxial approximation for plane wave incidence. Since it is known that the incident wavefronts are neither plane nor spherical in the real experiments using transducers, this modeling led to an understanding which was incomplete. Also, the technique involved significant information loss due to failure to extract exact amplitude information and track all the wave modes in a complex media. To enhance the accuracy of modeling, various computational techniques have been developed and utilized by researchers. Some of the popular computational simulation techniques used in CNDE are finite element method (FEM) [40], boundary element method (BEM) [41, 42], indirect boundary integral equation (IBIE) [43–46], multi-Gaussian beam model (M-GBM) [47–49], charge simulation technique (CST) [50], multiple multi-pole method (MMP) [51–53], spectral element method (SEM) [54], elastodynamic finite integration technique (EFIT) [55], local interaction simulation approach (LISA) [56–59], distributed point source method (DPSM) [60–70], Gaussian distributed point source method (G-DPSM) [71], among others. Among all the techniques mentioned above, FEM has been the most commonly used approach because of the availability of many commercial packages. Unfortunately, spurious reflections at the element boundaries can exist in FEM simulations. Moreover, it has also been found to be less accurate at higher frequencies than DPSM. It was reported that an alternate method DPSM is at least two times faster than the FEM [72], however it requires elastodynamic Green's functions to be computed numerically. DPSM is a frequency domain method where the wave field response at each individual frequency can be visualized. SAFE method was successfully used in computation of time domain wave fields in metallic and composite structures. SEM was also used for calculation of dispersion curves, transmission strength, and Green's functions [73–76]. LISA and EFIT are promising numerical methods to compute wave fields in time domain and have been used for a variety of materials [55, 77]. All of the above methods have their pros and cons. A method that could overcome the ubiquitous limitations of: respective techniques such as incorrect modeling of critical reflection phenomena, spurious reflections at higher frequencies at the element interfaces, inability to adapt to changes in interface curvature, necessity of far field approximation, ability to accurately model waves in any material (including highly anisotropic), and high computational expense, would be a versatile method for CNDE. However, at this point in time, no single method is the obvious choice to wholly overcome these limitations.

1.4 OVERVIEW OF THE CHAPTERS

This book is written to provide a fundamental understanding of the tools that are available to perform CNDE modeling and simulation with available computational technology. Ultrasonic NDE ranges between a wide range of frequencies and all methods may not be applicable for all frequency ranges based on their computational demand. In this book primarily, DPSM, SEM, LISA, EFIT and the recently developed Perielastodynamic method are presented as available and versatile CNDE tools. All the above methods demand the basic understanding of mechanics of materials, wave propagation in isotropic to generic anisotropic materials, numerical methods and computational tools for integration, etc. This book covers the necessary details for each method to be implemented by graduate students, new researchers in the field of NDE and/or computer programmers who contribute their programming skills

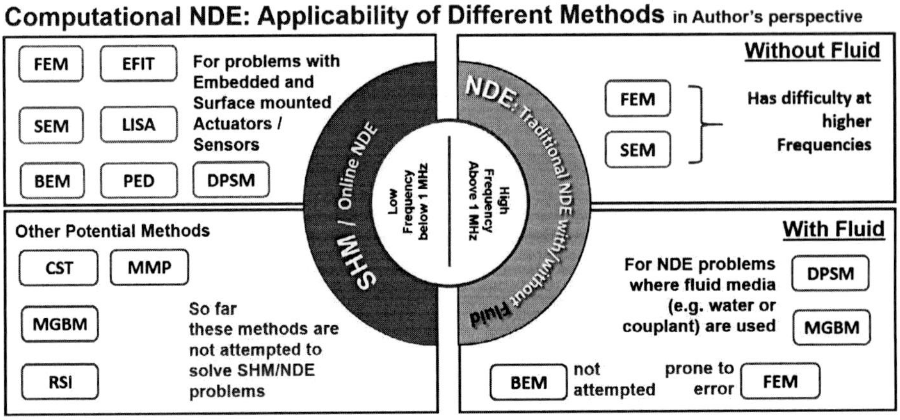

FIGURE 1.10 A graphical presentation of applicability of the CNDE methods based on the frequency range and type of transducer/actuator and sensors used. Abbreviations are as follows: FEM, finite element method; SEM, spectral element method; BEM, boundary element method; EFIT, elastodynamic finite integration technique; LISA, local interaction simulation approach; PED, Peri-elastodynamic approach; DPSM, distributed point source method; CST, charge simulation technique; MGBM, multi-Gaussian beam model; RSI, Rayleigh-Sommerfeld integral method; MMP, multiple multipole technique.

in NDE. Further, a few basics examples of method implementation are presented in this book. Chapter 2 covers the basics including necessary mathematical treatments for vectors, vector fields, and their respective transformations between coordinate systems. Chapter 3 covers the basics of the mechanics of materials, where, details of stress, strain, and governing elastodynamic equations are discussed. Chapter 4 is dedicated to introducing the basic physics of wave propagation in bulk solid isotropic and anisotropic materials. Chapter 5 covers the physics of guided wave propagation in isotropic and anisotropic wave guides. Before going into the CNDE methods individually, Chapter 6 reviews the basics of errors, numerical integration, numerical solutions of dynamic equations, and an overview of parallel computing. Chapter 7 presents the DPSM method starting from the fundamentals. Chapter 8 presents the EFIT method, Chapter 9 covers the LISA, Chapter 10 presents SEM, and Chapter 11 explains the Perielastodynamic (PED) method in detail. In various chapters, there are a few sample problems that frequently occur in ultrasonic NDE. Fig. 1.10 shows the applicability and attempted CNDE methods in author's perspective in a concise form based on the NDE and SHM applications below and above 1 MHz.

1.5 SUMMARY

This chapter first introduced the field of NDE. Various NDE methods were reviewed and their respective advantages and limitations were discussed. Selected ultrasonic NDE methods were further explained in detail. The need for CNDE was explained through several schematics, and a brief historical overview of CNDE was presented. More details on the CNDE modeling and simulation methods listed in this chapter can be found in the following chapters.

REFERENCES

1. Department of Defense Handbook, Appendix C, in *Nondestructive Evaluation System Reliability Assessment: Ultrasonic Test System*. 2009 April 7, Department of Defense.
2. Giurgiutiu, V., *Structural Health Monitoring with Piezoelectric Wafer Active Sensors*, 2nd ed. 2014, Amsterdam: Elsevier Academic Press.
3. Andrew, K.S., J. Daming Lin, and D. Banjevic, A review on machinery diagnostics and prognostics implementing condition-based maintenance. *Mechanical Systems and Signal Processing*, 2006. **20**(7): pp. 1483–1510.
4. Lindgren, E.A., J.S. Knoop, J.C. Aldrin, G.J. Steffes, C.F. Buynak, Aging aircraft NDE: capabilities, challenges, and opportunities, in *AIP Conference Proceedings*, 2007.
5. Fitzwater, L.T., Y. Torng, K. Halbert, *Air Vehicle Integration and Technology Research (AVIATR). Task Order 0003: Condition-Based Maintenance Plus Structural Integrity (CBM+SI) Demonstration (April 2011 to August 2011)*, 2011.
6. Guillén López, A.J., G.F.J. Fernández, A.C. Márquez, A framework for effective management of CBM programs, in *Advanced Maintenance Modelling for Asset Management*, Crespo Márquez A., González-Prida Díaz V., Gómez Fernández J., Editors. 2018, Cham: Springer.
7. Yam, R.C.M., et al., Intelligent predictive decision support system for condition-based maintenance. *The International Journal of Advanced Manufacturing Technology*, 2001. **17**(5): pp. 383–391.
8. Halmshaw, R., *Non-destructive Testing*, 1987, London: Edward Arnold.
9. Bossi, R. and V. Giurgiutiu, Nondestructive testing of damage in aerospace composites, in *Polymer Composites in the Aerospace Industry*, 2015, Amsterdam: Elsevier, pp. 413–448.
10. Peng, W., et al., A brief review of the application and problems in ultrasonic fatigue testing. *AASRI Procedia*, 2012. **2**: pp. 127–133.
11. Kroeger, T., Thermographic inspection of composites. *Reinforced Plastics*, 2014. **58**(4): pp. 42–43.
12. Tan, K.T., N. Watanabe, and Y. Iwahori, X-ray radiography and micro-computed tomography examination of damage characteristics in stitched composites subjected to impact loading. *Composites Part B: Engineering*, 2011. **42**(4): pp. 874–884.
13. Yang, S.-H., et al., Non-contact detection of impact damage in CFRP composites using millimeter-wave reflection and considering carbon fiber direction. *NDT & E International*, 2013. **57**: pp. 45–51.
14. *Nondestructive Testing Handbook*, 3rd ed, ed. E.T. (ET). Vol. **5**. The American Society for Nondestructive Testing, 0145WCD.
15. Mix, P.E., *Introduction to Nondestructive Testing: A Training Guide*. 2005, New York: John Wiley and Sons.
16. Sarasini, F. and C. Santulli, Non-destructive testing (NDT) of natural fibre composites: acoustic emission technique, in *Natural Fibre Composites*. 2014, Amsterdam: Elsevier, pp. 273–302.
17. Su, Z., et al., Acousto-ultrasonics-based fatigue damage characterization: linear versus nonlinear signal features. *Mechanical Systems and Signal Processing*, 2014. **45**(1): pp. 225–239.
18. Hung, Y., L. Yang, and Y. Huang, Non-destructive evaluation (NDE) of composites: digital shearography, in *Non-Destructive Evaluation (NDE) of Polymer Matrix Composites*. 2013, Amsterdam: Elsevier, pp. 84–115.
19. Sachse, W., et al., Recent developments in quantitative ultrasonic NDE of composites. *Ultrasonics*, 1990. **28**(2): pp. 97–104.
20. Nayfeh, A.H., *Wave Propagation in Layered Anisotropic Media: With Application to Composites*. Vol. **39**. 1995, Amsterdam: Elsevier.

21. Achenbach, J.D., *Wave Propagation in Elastic Solids.* 1999, New York: Elsevier.
22. Graff, K.F., *Wave Mothion in Elastic Solids.* 1975, New York: Dover Publication.
23. Mal, A.K., C.C. Yin, and Y. Bar-Cohen, Ultrasonic nondestructive evaluation of cracked composite laminates. *Composites Engineering*, 1991. **1**(2): pp. 85–101.
24. Mouchtachi, A., et al., Ultrasonic Study of Elastic Anisotropy of Material Composite. *Applied Composite Materials*, 2004. **11**(6): pp. 341–351.
25. Moser, B., L. Weber, and A. Mortensen, Damage accumulation during cyclic loading of a continuous alumina fibre reinforced aluminium composite. *Scripta Materialia*, 2005. **53**(10): pp. 1111–1115.
26. Li, B., et al., Damage localization in composite laminates based on a quantitative expression of anisotropic wavefront. *Smart Materials and Structures*, 2013. **22**(6): pp. 065005.
27. Marguères, P. and F. Meraghni, Damage induced anisotropy and stiffness reduction evaluation in composite materials using ultrasonic wave transmission. *Composites Part A: Applied Science and Manufacturing*, 2013. **45**(0): pp. 134–144.
28. Pradhan, B., N. Venu Kumar, and N.S. Rao, Stiffness degradation resulting from 90° ply cracking in angle-ply composite laminates. *Composites Science and Technology*, 1999. **59**(10): pp. 1543–1552.
29. Carcione, J.M., *Wave Fields in Real Media: Wave Propagation in Anisotropic, Anelastic, Porous and Electromagnetic Media.* Vol. **38**. 2007, Amsterdam: Elsevier.
30. Sahay, S.K., R.A. Kline, and R. Mignogna, Phase and group velocity considerations for dynamic modulus measurement in anisotropic media. *Ultrasonics*, 1992. **30**(6): pp. 373–382.
31. Hood, J.A. and R.B. Mignogna, Determination of elastic moduli in anisotropic media from ultrasonic contact measurements. *Ultrasonics*, 1995. **33**(1): pp. 45–54.
32. Every, A.G. and K.Y. Kim, Determination of elastic constants of anisotropic solids from elastodynamic Green's functions. *Ultrasonics*, 1996. **34**(2–5): pp. 471–472.
33. Sharma, M.D., Group velocity along general direction in a general anisotropic medium. *International Journal of Solids and Structures*, 2002. **39**(12): pp. 3277–3288.
34. Vijaya, M.S., *Piezoelectric Materials and Devices: Applications in Engineering and Medical Sciences.* 2017, Boca Raton: CRC Press.
35. Vinogradov, S., A. Cobb, and J. Fisher, New magnetostrictive transducer designs for emerging application areas of NDE. *Materials (Basel)*, 2018. **11(5)**: p. 755.
36. Yu, L., V. Giurgiutiu, In situ 2-D piezoelectric wafer active sensors arrays for guided wave damage detection. *Ultrasonics*, 2008. **48**(2): pp. 117–134.
37. Tian, Z., S. Howden, Z. Ma, W. Xiao, L. Yu, Pulsed laser-scanning laser Doppler vibrometer (PL-SLDV) phased arrays for damage detection in aluminum plates. *Mechanical Systems and Signal Processing*, 2019. **121**: pp. 158–170.
38. Franco, E., Andrade, M. A. B., Adamowski, J.C., Biuochi, F., Acoustic beam modeling of ultrasonic transducers and arrays using the impulse response and the discrete representation methods. *Journal of the Brazilian Society of Mechanical Sciences and Engineering*, 2011. **XXXIII**(4): pp. 408–416.
39. Biondi, B., Solving the frequency-dependent Eikonal equation. *62nd Ann. Internat. Mtg.*, Soc. Expl. Geophys., 1992. **Expanded Abstracts**: p. 1351.
40. Software, P., *Finite Element Analysis of Propagating Elastic Waves.* 2009, Weilinger Associates Inc.
41. Shaw, R.P., *Ch.6 Boundary Integral Equation Methods Applied to Wave Problems*, in *Developments in Boundary Element Methods - 1*, P.K. Banerjee and R. Butterfield Editors. 1979, Applied Science Publishers. pp. 121–153.
42. X. Zhao., J.L. Rose., Boundary element modeling for defect characterization potential in a wave guide. *International Journal of Solids and Structures*, 2003. **40**(11): p. 2645.

43. F.J. Sannchez-Sesma, M. Campillo, Diffraction of P, SV and Rayleigh waves by topographic features: a boundary integral formulation. *Bulletin of the Seismological Society of America*, 1991. **81**: pp. 2234–2253.
44. F.J. Sannchez-Sesma, M. Campillo, Topographic effects for incident P, SV and Rayleigh waves. *Tectonophysics*, 1993. **218**: pp. 113–125.
45. T. Pointer, E. Liu, J.A. Hudson, Numerical modeling of seismic waves scattered by hydrofractures: application of the indirect boundary element method. *Geophysical Journal International*, 1998. **135**: pp. 289–303.
46. Bouchon, M., Sannchez-Sesma, F.J., Boundary integral equations and boundary elements methods in elastodynamics, in *Advances in Wave Propagation in Heterogeneous Earth, Advances in Geophysicis*, V. Maupin, R. Dmowska and R.S. Wu, Editor. 2007, New York, Boston: Elsevier-Academic Press.
47. J. J. Wen, M.A. Breazeale., A diffraction beam field expressed as the superposition of Gaussian beams. *Journal of the Acoustical Society of America*, 1988. **83**: p. 1752.
48. B. P. Newberry, R.B.Thompson., A paraxial theory for the propagation of ultrasonic beams in anisotropic solids. *Journal of the Acoustical Society of America*, 1989. **85**: p. 2290.
49. Spies, M., Transducer field modeling in anisotropic media by superposition of Gaussian base functions. *Journal of the Acoustical Society of America*, 1999. **105**: p. 633.
50. C. Rajamohan, J. Raamachandran, Bending of anisotropic plates by charge simulation method. *Advances in Engineering Software*, 1999. **30**(5): p. 369.
51. Ballisti, C.H., The multiple multipole method (MMP) in electro and magnetostatic problems. *IEEE Transactions on Magnetics*, 1983. **19**(6): p. 2367.
52. Hafner, C., MMP calculations of guided waves. *IEEE Transactions on Magnetics*, 1985. **21**: p. 2310.
53. Imhof, M.G., Computing the elastic scattering from inclusions using the multiple multipoles method in three dimensions. *Geophysical Journal International*, 2004. **156**: p. 287.
54. J Moll, C Rezk-Salama, R T Schulte, T Klinkert, C -P Fritzen, A Kolb, Modelling of wave-based SHM systems using the spectral element method, under interactive simulation and visualization of lamb wave propagation in isotropic and anisotropic structures. *Journal of Physics: Conference Series*, 2011. **305**: p. 012095.
55. Cara A.C. Leckey, Matthew D. Rogge, F. Raymond Parker, Guided waves in anisotropic and quasi-isotropic aerospace composites: Three-dimensional simulation and experiment. *Ultrasonics*, 2014. **54**: pp. 385–394.
56. Delsanto, P.P., Schechter, R.S., Mignogna, R.B., Connection machine simulation of ultrasonic wave propagation in materials III: The three-dimensional case. *Wave Motion*, 1997. **26**(4): pp. 329–339.
57. Nadella, K.S.a.C., Carlos ES, Local interaction simulation approach for modeling wave propagation in composite structures. *CEAS Aeronautical Journal*, 2013. **4**(1): pp. 35–48.
58. Shen, Y.a.C., Carlos ES, Local interaction simulation approach for efficient modeling of linear and nonlinear ultrasonic guided wave active sensing of complex structures. *Journal of Nondestructive Evaluation, Diagnostics and Prognostics of Engineering Systems*, 2018. **1**(1): p. 011008.
59. Sundararaman, S.a.A., Douglas E, Modeling guided waves for damage identification in isotropic and orthotropic plates using a local interaction simulation approach. *Journal of Vibration and Acoustics*, 2008. **130**(4): p. 041009.
60. Placko, D. and T. Kundu, Theoretical study of magnetic and ultrasonic sensors: dependence of magnetic potential and acoustic pressure on the sensor geometry. Advanced NDE for structural and biological health monitoring, in *T. Kundu (Ed.), Proceedings of SPIE, SPIE's 6th Annual International Symposium on NDE for Health Monitoring and Diagnostics*. 2001: Newport Beach, California. pp. 52–62.

61. Placko, D., T. Kundu, and R. Ahmad, Theoretical computation of acoustic pressure generated by ultrasonic sensors in the presence of an interface. Smart NDE and health monitoring of structural and biological systems. in *SPIE's 7th Annual International Symposium in NDE and Health Monitoring and Diagnostics*. 2002: San Diego, CA. pp. 157–168.
62. Banerjee, S., Kundu, T., Symmetric and anti-symmetric Rayleigh-Lamb modes in sinusoidally corrugated waveguides: An Analytical approach. *International Journal of Solids and Structures*, 2006. **43**(6551–6567).
63. Banerjee, S. and T. Kundu, Elastic wave propagation in sinusoidally corrugated waveguides. *Journal of the Acoustical Society of America*, 2006. **119**(4): pp. 2006–2017.
64. Banerjee, S., Kundu, T., Ultrasonic field modeling in plates immersed in fluids. *International Journal of Solids and Structures*, 2007. **44**(18–19): pp. 6013–6029.
65. Banerjee, S., Alnuaimi, N., Kundu, T., DPSM technique for ultrasonic field modelling near fluid-solid interface. *Ultrasonics*, 2007. **46**(3): pp. 235–250.
66. Banerjee, S. and T. Kundu, Advanced application of distributed point source method – ultrasonic field modeling in solid media, in *DPSM for Modeling Engineering Problems*, T. Kundu and D. Placko, Editors. 2007, Hoboken, New Jersey: John & Willey Publication.
67. Placko, D. and T. Kundu, *DPSM for Modeling Engineering Problems*. 2007, New York: John Wiley & Sons.
68. Banerjee, S., Kundu, T., Semi-analytical modeling of ultrasonic fields in solids with internal anomalies immersed in a fluid. *Wave Motion*, 2008. **45**(5): pp. 581–595.
69. Banerjee, S., Kundu, T., Elastic wave field computation in multilayered non-planar solid structures: a mesh-free semi-analytical approach. *Journal of Acoustical Society of America*, 2008. **123**(3): pp. 1371–1382.
70. Banerjee, S., Das, S., Kundu, T., Placko, D., Controlled space radiation concept for mesh-free semi-analytical techniquc to model wave fields in complex geometries. *Ultrasonics*, 2009. **48**(8): pp. 615–622.
71. Rahani, E.K., and Kundu, T., Gaussian-DPSM (G-DPSM) and element source method (ESM) modifications to DPSM for ultrasonic field modeling. *Ultrasonics* 2011. **51**(5): pp. 625–631.
72. Wada, Y., Kundu, T., Nakamura, K, Mesh-free distributed point source method for modeling viscous fluid motion between disks vibrating at ultrasonic frequency. *The Journal of the Acoustical Society of America*, 2014. **136**(466).
73. Hayashi, T., Guided wave simulation and visualization by a semianalytical finite element method. *Materials Evaluation*, 2003. **61**(1).
74. Lanza Discalea, F., et al., Health monitoring of UAV wing skin-to-spar joints using guided waves and macro fiber composite transducers. *Journal of intelligent material systems and structures*, 2007. **18**(4): pp. 373–388.
75. Marzani, A., et al., A semi-analytical finite element formulation for modeling stress wave propagation in axisymmetric damped waveguides. *Journal of Sound and Vibration*, 2008. **318**(3): pp. 488–505.
76. Zhuang, W., A. Shah, and S. Dong, Elastodynamic Green's function for laminated anisotropic circular cylinders. *Journal of Applied Mechanics*, 1999. **66**(3): pp. 665–674.
77. Shen, Y., Cesnik, C. E. S., Modeling of nonlinear interactions between guided waves and fatigue cracks using local interaction simulation approach. *Ultrasonics*, 2017. **74**: pp. 106–123.

2 Vector Fields and Tensor Analysis

2.1 UNDERSTANDING VECTORS

In mathematics a vector is defined as a quantity that both has magnitude and direction. To describe a vector in most simple words, we can say that a vector is used to describe the position of a point in a mathematically viable space relative to another point in the same space. If the vectors pointed to the objects in the space are static, i.e., nonmoving object, the vectors are called the static vectors. Let's consider a fancy modern room with three orthogonal axes aligned along the intersection of two walls and the floor as shown in Fig. 2.1a. The three axes x, y, and z are pointed away from each other following the right-hand coordinate rule in a Cartesian coordinate system. Let's consider three points in this coordinate system to carry on our discussion of vectors. The three objects that we considered are the accent light (center of the accent light is designated as point A), a fancy bowl on the table (the base of the bowl is designated as point B), and the reflection of a camera on the mirror (the base of the camera on the stand is designated as point C). The coordinates of the point A are A_1, A_2, A_3 are the linear shortest distances of the point A from the origin O along x, y, and z directions, respectively. Similarly, the coordinates of the point B are B_1, B_2, B_3 are the linear shortest distances of the point B from the origin O along x, y, and z directions, respectively. Also, the coordinates of the point C are C_1, C_2, C_3 are the linear shortest distances of the point C from the origin O along x, y, and z directions, respectively, where C_2 is negative. In Fig. 2.1a, x, y, and z axes are alternatively named as x_1, x_2, and x_3, respectively. From now on, throughout this book, we will use x_1, x_2, and x_3 or simply 1, 2, and 3 to designate the x, y, and z coordinates without the loss of generality. Locations of the three points, A, B, and C can be designated by three vectors $\mathbf{V_A}$, $\mathbf{V_B}$, and $\mathbf{V_C}$, respectively. In our selected coordinate system in the fancy room as shown in Fig. 2.1a these pointing vectors can be expanded as written in Eqs. (2.1)–(2.3).

$$\mathbf{V_A} = A_1\widehat{\mathbf{e_1}} + A_2\widehat{\mathbf{e_2}} + A_3\widehat{\mathbf{e_3}} \tag{2.1}$$

$$\mathbf{V_B} = B_1\widehat{\mathbf{e_1}} + B_2\widehat{\mathbf{e_2}} + B_3\widehat{\mathbf{e_3}} \tag{2.2}$$

$$\mathbf{V_C} = C_1\widehat{\mathbf{e_1}} + C_2\widehat{\mathbf{e_2}} + C_3\widehat{\mathbf{e_3}} \tag{2.3}$$

where $\widehat{\mathbf{e_1}}$, $\widehat{\mathbf{e_2}}$, $\widehat{\mathbf{e_3}}$ are the unit vectors along 1, 2, and 3 directions, respectively. The vectors in Eqs. (2.1)–(2.3) are the static vectors, because they are pointed to the objects that are in specific location in the space. If we consider a different coordinate system (with unit vectors, $\widehat{\mathbf{b_1}}$, $\widehat{\mathbf{b_2}}$, $\widehat{\mathbf{b_3}}$) in the same room as shown in Fig. 2.1b,

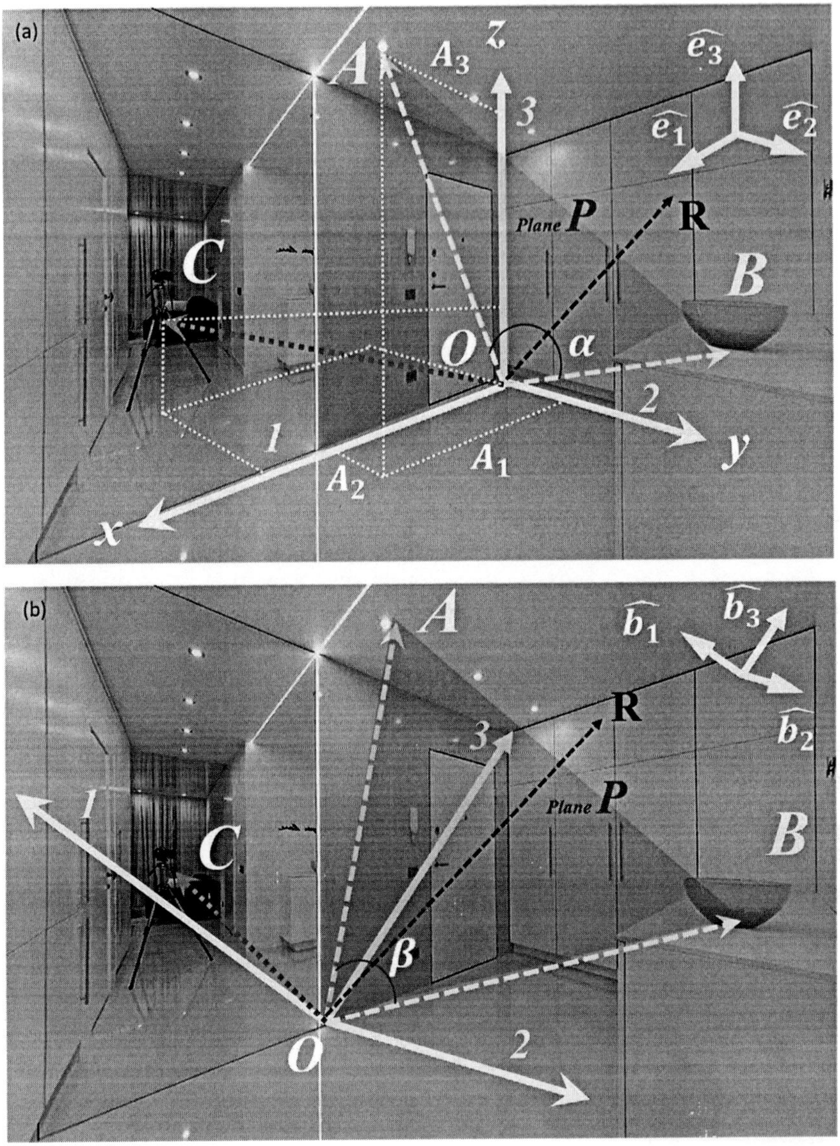

FIGURE 2.1 (a) Representation of vectors pointing to specific points in space in a $x\ y\ z$ Cartesian coordinate system. (b) Representation of the modified vectors pointing to the same points in an arbitrary orthogonal coordinate system in the same space.

it is to be noted that the objects A, B, and C does not change their locations. But the vectors pointing to the points A, B, and C are changed with different magnitude due to composition of different unit vectors. Please note that the choice of the coordinate systems with unit vectors $\widehat{\mathbf{e}_1}$, $\widehat{\mathbf{e}_2}$, $\widehat{\mathbf{e}_3}$ or $\widehat{\mathbf{b}_1}$, $\widehat{\mathbf{b}_2}$, $\widehat{\mathbf{b}_3}$ are up to us to decide, so, we can establish a unique relationship between these two coordinate systems without worrying about the objects in the room. Let's assume we know such relation and the relationship can be written as follows

$$\widehat{e_1} = \psi_{11}\widehat{\mathbf{b_1}} + \psi_{12}\widehat{\mathbf{b_2}} + \psi_{13}\widehat{\mathbf{b_3}} \tag{2.4}$$

$$\widehat{e_2} = \psi_{21}\widehat{\mathbf{b_1}} + \psi_{22}\widehat{\mathbf{b_2}} + \psi_{23}\widehat{\mathbf{b_3}} \tag{2.5}$$

$$\widehat{e_3} = \psi_{31}\widehat{\mathbf{b_1}} + \psi_{32}\widehat{\mathbf{b_2}} + \psi_{33}\widehat{\mathbf{b_3}} \tag{2.6}$$

Further we will bring the matrix formulation in our discussion and we can write

$$\begin{Bmatrix} \mathbf{V}_A \\ \mathbf{V}_B \\ \mathbf{V}_C \end{Bmatrix} = \begin{bmatrix} A_1 & A_2 & A_3 \\ B_1 & B_2 & B_3 \\ C_1 & C_2 & C_3 \end{bmatrix} \begin{Bmatrix} \widehat{\mathbf{e_1}} \\ \widehat{\mathbf{e_2}} \\ \widehat{\mathbf{e_3}} \end{Bmatrix} = \mathbf{\Phi}.\hat{\mathbf{e}} \tag{2.7}$$

$$\hat{\mathbf{e}} = \begin{Bmatrix} \widehat{e_1} \\ \widehat{e_2} \\ \widehat{e_3} \end{Bmatrix} = \begin{bmatrix} \psi_{11} & \psi_{12} & \psi_{13} \\ \psi_{21} & \psi_{22} & \psi_{23} \\ \psi_{31} & \psi_{32} & \psi_{33} \end{bmatrix} \begin{Bmatrix} \widehat{\mathbf{b_1}} \\ \widehat{\mathbf{b_2}} \\ \widehat{\mathbf{b_3}} \end{Bmatrix} = \Psi.\hat{\mathbf{b}} \tag{2.8}$$

Hence, the positions vectors of the points A, B, and C (i.e., $\mathbf{V_A}$, $\mathbf{V_B}$, $\mathbf{V_C}$) in the new coordinate system shown in Fig. 2.1b can be written as

$$\begin{Bmatrix} \mathbf{V}_A \\ \mathbf{V}_B \\ \mathbf{V}_C \end{Bmatrix} = \begin{bmatrix} A_1 & A_2 & A_3 \\ B_1 & B_2 & B_3 \\ C_1 & C_2 & C_3 \end{bmatrix} \begin{bmatrix} \psi_{11} & \psi_{12} & \psi_{13} \\ \psi_{21} & \psi_{22} & \psi_{23} \\ \psi_{31} & \psi_{32} & \psi_{33} \end{bmatrix} \begin{Bmatrix} \widehat{\mathbf{b_1}} \\ \widehat{\mathbf{b_2}} \\ \widehat{\mathbf{b_3}} \end{Bmatrix} = \mathbf{\Phi\Psi}.\hat{\mathbf{b}} \tag{2.9}$$

Comments & Summary

- Vectors [1] are nothing but an array of numbers that describes the position of a point in a space (we will discuss about time later) defined by a specific coordinate system.
- Although when the positions of the points are unaltered in a space, it is feasible to choose a coordinate system differently, which may change the numbers in the array that describes the vectors.
- Any coordinate system much have its unique orthogonal basis vectors.
- Any vector in a space with defined coordinate system must have components along the basis vectors with which the coordinate system is defined, e.g., $\hat{\mathbf{e}}$ or $\hat{\mathbf{b}}$.
- Any "n" dimensional coordinate system must have "n" basis vectors, i.e., equal to the number of directions in the system. For example, a typical Cartesian coordinate system shown in Fig. 2.1a has three directions.

Therefore, in Cartesian coordinate system we can say we have three basis vectors, or three-unit vectors, or three eigen vectors.

- There are infinite possibilities of having infinite coordinate systems or basis system in a space.

2.2 A BRIEF REVIEW OF INDEX NOTATION

In this discussion we will bring Eqs. (2.1)–(2.3) of the position vectors $\mathbf{V_A}$, $\mathbf{V_B}$, $\mathbf{V_C}$ for the points A, B, and C, respectively. Instead of elaborately writing the components of the vectors multiplied with the basis vectors, a short index notation (or Einstein notation) can be used to describe the same equation in a short form as written bellow.

$$\mathbf{V_A} = A_i\widehat{\mathbf{e_i}} \tag{2.10}$$

$$\mathbf{V_B} = B_i\widehat{\mathbf{e_i}} \tag{2.11}$$

$$\mathbf{V_C} = C_i\widehat{\mathbf{e_i}} \tag{2.12}$$

where i takes the values 1, 2, and 3 in three-dimensional Cartesian coordinate system or can take any numbers from 1 to "n" in a "n" dimensional coordinate system. The repeated occurrence of i, with the parameters on the same side of the equations, means the summation of the parameters over the index with varying numbers, carrying the identification of each coordinates in the respective coordinate system. Comparing Eqs. (2.1) and (2.10) it can be realized that $A_1\widehat{\mathbf{e_1}} + A_2\widehat{\mathbf{e_2}} + A_3\widehat{\mathbf{e_3}}$ can be easily shorten by $A_i\widehat{\mathbf{e_i}}$. Similarly, the logic is true for any vectors.

2.2.1 DOT PRODUCT OF TWO VECTORS

Let's consider a dot product between two vectors $\mathbf{V_A}$ and $\mathbf{V_B}$. Without the loss of any generality and not to confuse the two vectors, we will use two indices i and j that behaves in a same manner described in the section above. We will write $\mathbf{V_A} = A_i\widehat{\mathbf{e_i}}$ and $\mathbf{V_B} = B_j\widehat{\mathbf{e_j}}$. Now the dot product will be as follows in Eq. (2.13), where we introduced a new parameter δ_{ij}.

$$\mathbf{V_A} \,.\, \mathbf{V_B} = A_i\widehat{\mathbf{e_i}} \,.\, B_j\widehat{\mathbf{e_j}} = A_iB_j\left(\widehat{\mathbf{e_i}}.\widehat{\mathbf{e_j}}\right) = A_iB_j\left(\delta_{ij}\right) = A_iB_i \tag{2.13}$$

δ_{ij} is called Kronecker delta which represents the dot product of two basis vectors in a coordinate system with the respective indices i and j. Hence, the dot product of $\widehat{e_i}$ and $\widehat{e_j}$ can be written as δ_{ij}. When i and j take same value ($\widehat{\mathbf{e_i}}.\widehat{\mathbf{e_i}}$), it represents the dot product of a basic (or unit vector) to itself. If the index values are different for i and j, it represents the dot product of two basic or unit vectors in two orthogonal directions. And we know that the dot product of two unit vectors along same direction is unity and dot products of two unit vectors in orthogonal directions are always zero. Hence, the factor $B_j\left(\delta_{ij}\right)$ in Eq. (2.13) will result B_i only. Now we can write the dot product between $\mathbf{V_A}$ and $\mathbf{V_B}$ as

$$\mathbf{V_A} \,.\, \mathbf{V_B} = A_iB_i \tag{2.14}$$

Here we used an identity with a property related to the Kronecker delta $B_j\left(\delta_{ij}\right)=B_i$. The Kronecher delta can be described as an identity matrix in any "n" dimensional coordinate system. Following equation is in three-dimensional Cartesian coordinate system will be helpful to visualize this fact.

$$\delta_{ij}=\begin{bmatrix}\widehat{\mathbf{e_1}}.\widehat{\mathbf{e_1}} & \widehat{\mathbf{e_1}}.\widehat{\mathbf{e_2}} & \widehat{\mathbf{e_1}}.\widehat{\mathbf{e_3}}\\ \widehat{\mathbf{e_2}}.\widehat{\mathbf{e_1}} & \widehat{\mathbf{e_2}}.\widehat{\mathbf{e_2}} & \widehat{\mathbf{e_2}}.\widehat{\mathbf{e_3}}\\ \widehat{\mathbf{e_3}}.\widehat{\mathbf{e_1}} & \widehat{\mathbf{e_3}}.\widehat{\mathbf{e_2}} & \widehat{\mathbf{e_3}}.\widehat{\mathbf{e_3}}\end{bmatrix}=\begin{bmatrix}1.1.\cos(0^\circ) & 1.1.\cos(90^\circ) & 1.1.\cos(90^\circ)\\ 1.1.\cos(90^\circ) & 1.1.\cos(0^\circ) & 1.1.\cos(90^\circ)\\ 1.1.\cos(90^\circ) & 1.1.\cos(90^\circ) & 1.1.\cos(0^\circ)\end{bmatrix}$$
$$=\begin{bmatrix}1 & 0 & 0\\ 0 & 1 & 0\\ 0 & 0 & 1\end{bmatrix} \tag{2.15}$$

To understand the physical meaning of the dot product, we can say that the dot product between two vectors represents the magnitude of the component of one vector on other, when the calculation happens inherently on a plane created by the two chosen vectors, e.g., $\mathbf{V_A}$ and $\mathbf{V_B}$ created the plane P shown in Fig. 2.1a and b. Please note that the orientations of the planes P are different in two different coordinate systems. The scalar quantity A_iB_i (as no $\hat{\mathbf{e}}$ or $\hat{\mathbf{b}}$ basis vectors are associated with it) in Eq. (2.14) is nothing but the component of $\mathbf{V_A}$ on $\mathbf{V_B}$ or the component of $\mathbf{V_B}$ on $\mathbf{V_A}$. $A_iB_i=|\mathbf{V_A}|.|\mathbf{V_B}|\cos(\alpha)$ in Fig. 2.1a and $A_iB_i=|\mathbf{V_A}|.|\mathbf{V_B}|\cos(\beta)$ in Fig. 2.1b.

2.2.2 Cross Product of Two Vectors

Let's consider a cross product between two vectors $\mathbf{V_A}$ and $\mathbf{V_B}$. Without the loss of any generality and not to confuse the two vectors, we will use two indices i and j that behave in a same manner described in the section above. We will write $\mathbf{V_A}=A_i\widehat{\mathbf{e_i}}$ and $\mathbf{V_B}=B_j\widehat{\mathbf{e_j}}$. Now the cross product will be as follows in Eq. (2.16), where we introduced a new parameter ϵ_{kij}.

$$\mathbf{V_A}\times\mathbf{V_B}=A_i\widehat{\mathbf{e_i}}\times B_j\widehat{\mathbf{e_j}}=A_iB_j\left(\widehat{\mathbf{e_i}}\times\widehat{\mathbf{e_j}}\right)=\epsilon_{kij}A_iB_j\ \widehat{\mathbf{e_k}} \tag{2.16}$$

ϵ_{kij} is called the permutation symbol and the index k takes the values 1, 2, and 3 like the indices i and j. Now to explain the permutation symbol, how it works, let's expand the $\epsilon_{kij}A_iB_j$ expression in the following equation

$$\begin{aligned}\epsilon_{kij}A_iB_j&=\widehat{\mathbf{e_1}}\left(\epsilon_{111}A_1B_1+\epsilon_{112}A_1B_2+\epsilon_{113}A_1B_3+\epsilon_{121}A_2B_1+\epsilon_{122}A_2B_2+\epsilon_{123}A_2B_3\right.\\&\left.+\epsilon_{131}A_3B_1+\epsilon_{132}A_3B_2+\epsilon_{133}A_3B_3\right)+\widehat{\mathbf{e_2}}\left(\epsilon_{211}A_1B_1+\epsilon_{212}A_1B_2+\epsilon_{213}A_1B_3+\epsilon_{221}A_2B_1\right.\\&\left.+\epsilon_{222}A_2B_2+\epsilon_{223}A_2B_3+\epsilon_{231}A_3B_1+\epsilon_{232}A_3B_2+\epsilon_{233}A_3B_3\right)+\widehat{\mathbf{e_3}}\left(\epsilon_{311}A_1B_1+\epsilon_{312}A_1B_2\right.\\&\left.+\epsilon_{313}A_1B_3+\epsilon_{321}A_2B_1+\epsilon_{322}A_2B_2+\epsilon_{323}A_2B_3+\epsilon_{331}A_3B_1+\epsilon_{332}A_3B_2+\epsilon_{333}A_3B_3\right)\end{aligned} \tag{2.17}$$

This is a humongous equation with only few valuable elements. When the permutation symbol has all the indices 1, 2, and 3 in a sequence in clockwise or counterclockwise, they survive with values +1 and −1, respectively. Else the value of the permutation symbol is zero. This can be mathematically written as

TABLE 2.1
Permutation Symbol

$\epsilon_{kij} = 0$	if $k = i$ or $i = j$ or $j = k$ or $k = i = j$
$\epsilon_{kij} = +1$	if $k \rightarrow i \rightarrow j$ are in clockwise sequence, e.g., 1-2-3 or 2-3-1 or 3-1-2
$\epsilon_{kij} = -1$	if $k \leftarrow i \leftarrow j$ are in counterclockwise sequence, e.g., 3-2-1, 2-1-3, 1-3-2

Based on Table 2.1, Eq. (2.17) can be rewritten with fewer nonzero terms where the basis vectors survived in the expression. This means that the cross product of two vectors is a vector quantity.

$$\epsilon_{kij} A_i B_j = \widehat{\mathbf{e_1}}\left(A_2B_3 - A_3B_2\right) + \widehat{\mathbf{e_2}}\left(-A_1B_3 + A_3B_1\right) + \widehat{\mathbf{e_3}}\left(A_1B_2 - A_2B_1\right) \quad (2.18)$$

The vector (**R**) we obtain from the cross products of $\mathbf{V_A}$ and $\mathbf{V_B}$ in two different coordinate systems is depicted in Fig. 2.1a and b, respectively. The black dotted arrows represent the direction of the vector perpendicular to the planes created by the vectors $\mathbf{V_A}$ and $\mathbf{V_B}$. The cross product of two vectors essentially represents the capability of two vectors to rotate the system about the axis perpendicular to the imaginary plane that is created by them. A new vector **R** with magnitudes along the basis vectors in the respective coordinate system is written in Eq. (2.19). Further, the unit vector along **R** (that represents the direction normal to the plane created by the $\mathbf{V_A}$ and $\mathbf{V_B}$ vectors) is written in Eq. (2.10).

$$\mathbf{R} = \mathbf{V_A} \times \mathbf{V_B} = \left(A_2B_3 - A_3B_2\right)\widehat{\mathbf{e_1}} + \left(A_3B_1 - A_1B_3\right)\widehat{\mathbf{e_2}} + \left(A_1B_2 - A_2B_1\right)\widehat{\mathbf{e_3}} \quad (2.19)$$

$$\hat{\mathbf{r}} = \frac{\left(A_2B_3 - A_3B_2\right)\widehat{\mathbf{e_1}} + \left(A_3B_1 - A_1B_3\right)\widehat{\mathbf{e_2}} + \left(A_1B_2 - A_2B_1\right)\widehat{\mathbf{e_3}}}{\sqrt{\left(A_2B_3 - A_3B_2\right)^2 + \left(A_3B_1 - A_1B_3\right)^2 + \left(A_1B_2 - A_2B_1\right)^2}} \quad (2.20)$$

2.3 UNDERSTANDING THE VECTOR FIELD

We will start our discussion with the same figures in Fig. 2.1a and b. Let's talk about the vectors $\mathbf{V_A}$, $\mathbf{V_B}$, and $\mathbf{V_C}$. The position vector of the object A, which is a recess light in the room is not movable. So, the vector $\mathbf{V_A}$ is a static vector and its component along the basis vectors (A_i) is frozen in space and time. However, the components of the position vectors of the objects B and C, B_i and C_i, respectively are not frozen in space and time, because the objects are movable. Let's consider, at time $t = 0$, the bowl was at the location shown in Fig. 2.1a. However, a household object like a bowl can be used for many purposes and we have considered moving it from the table to the floor after any arbitrary time τ. What happens to the vector, $\mathbf{V_B}$? The vector pointing to the base of the bowl will move with the bowl by changing its magnitude. Please imagine a retractable rod connecting the base of the bowl and the origin O. This will change the components of the vector $\mathbf{V_B}$ along the basis vectors.

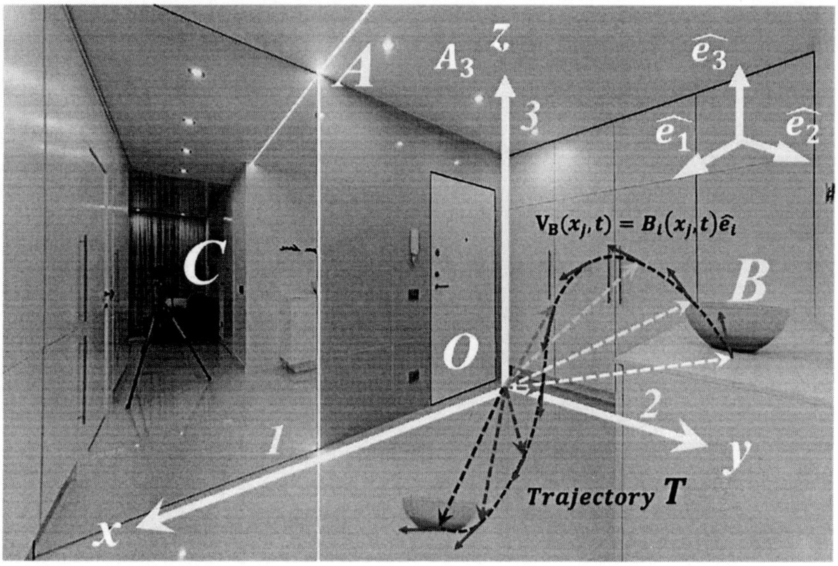

FIGURE 2.2 Description of a changing vector pointing to a movable object and introduction to the vector field.

Imagine the process when the motion was taking place (during time $t = 0 \rightarrow \tau$) and we were moving the bowl from the table to the floor. During that time the values of the components B_i were changing in space and time. Hence, we can say that the components are the functions of the spatial coordinates and time, i.e., $B_i(x, y, z, t)$ or $B_i(x_1, x_2, x_3, t)$ or using the help of index notation $B_i(x_j, t)$. Similarly, the camera C is also a movable object and the components of the vector pointing to the object C are the functions of the spatial coordinates and time. Thus, we can say

$$\mathbf{V_B}(x_j,t) = B_i(x_j,t)\widehat{\mathbf{e_i}} \text{ and } \mathbf{V_C}(x_j,t) = C_i(x_j,t)\widehat{\mathbf{e_i}} \tag{2.21}$$

The above expression actually creates a vector field [1]. Now let's focus on Fig. 2.2 a derivative of Fig 2.1a. In Fig. 2.2 a trajectory or a path T is shown along which the base of the bowl was moved from the table to the floor. At each point along the trajectory T, there are tangential vectors (small arrows in the figure along the trajectory) representing the direction of the motion of the object in the respective coordinate system, which is a three-dimensional Cartesian coordinate system in Fig. 2.2. Now if there are several trajectories like T are in the system representing the motion of all the particles in a system, the small tangential vectors or the arrows along all the trajectories may create a map of a playground where all the particles are playing in the field. Such a field is called a vector field. As the motion of the particles, deformations, velocities, and the forces are the vector quantity, they all create a vector field due to their nonstationary or nonstatic operations. Fig. 2.3 shows few examples of the vector field that we see in science and engineering. Wind velocity around an airfoil shape used in aerospace engineering, fluid flow inside a venturi meter, magnetic field

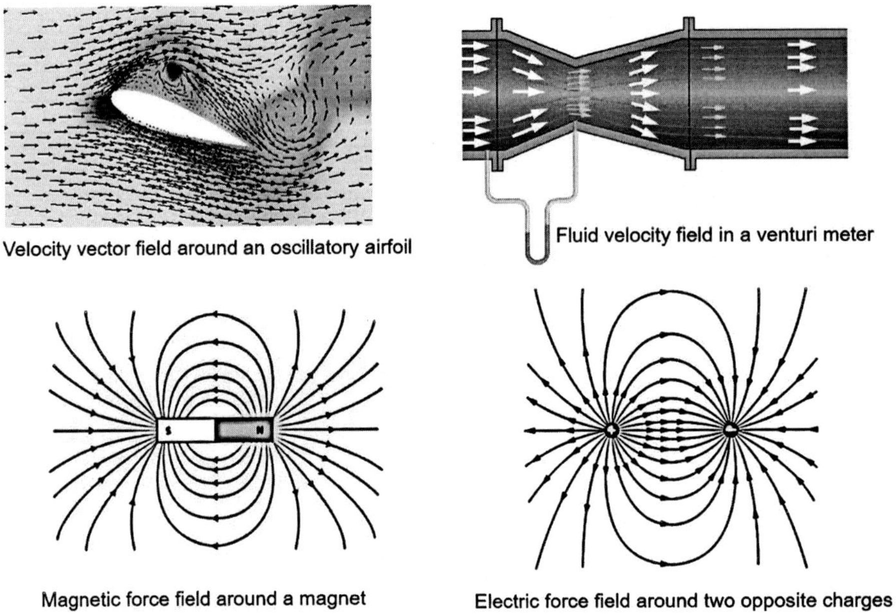

FIGURE 2.3 Examples of different types of vector fields in aerodynamics, fluid flow, magnetic and electric force field.

due to a magnet, electric field due to charges, etc. are the examples of vector field. Vector field is manifested by the local motion of a particle or response of a local particle due to a local force or the spatial or spatio-temporal variation of the direction of the force itself. Irrespective of any of the above, vector field is a map of local vectors with direction and magnitudes as shown in Fig. 2.3.

2.3.1 Gradient Operator

Gradient of any arbitrary scalar function $\varphi(x_j,t)$ along an axis x_1 can be written as $\partial\varphi / \partial x_1$. Similarly, the gradients along x_2 and x_3 are $\partial\varphi / \partial x_2$ and $\partial\varphi / \partial x_3$, respectively. As three gradients along three different directions result three different values, together they can be written in a vector form as follows

$$Grad\left(\varphi\left(x_j,t\right)\right)=\frac{\partial\varphi}{\partial x_1}\widehat{\mathbf{e_1}}+\frac{\partial\varphi}{\partial x_2}\widehat{\mathbf{e_2}}+\frac{\partial\varphi}{\partial x_3}\widehat{\mathbf{e_3}} \tag{2.22}$$

So, it is apparent that the gradient of a scalar function is a vector quantity where the basis vectors are associated in the expression. Eq. (2.22) can be seen in a little bit different form as written in Eq. (2.23)

$$Grad\left(\varphi\left(x_j,t\right)\right)=\left(\frac{\partial}{\partial x_1}\widehat{\mathbf{e_1}}+\frac{\partial}{\partial x_2}\widehat{\mathbf{e_2}}+\frac{\partial}{\partial x_3}\widehat{\mathbf{e_3}}\right)\varphi=\nabla\varphi=\varphi_{,i}\widehat{\mathbf{e_i}} \tag{2.23}$$

where ∇ is introduced as a vector operator and $\varphi_{,i}$ in index notation means the vector quantity has elements that are partial derivatives of φ with respect to x_i and i can take values 1, 2, and 3. We will call ∇ a gradient operator.

Now, when we operate the ∇, a gradient operator on a vector-valued function of position and time like an arbitrary vector field $\mathbf{V} = V_i(x_j,t)\widehat{e_i}$ with components $V_i(x_j,t)$ along the basis vectors $\widehat{e_i}$, the operation can be done in one of the two ways. Because ∇ and $\mathbf{V}$ are both the vectors, one is a vector type operator, and another is a vector-valued function. As we discussed two types of operations with vectors in Section 2.2, we can have a dot product $(\nabla.\mathbf{V})$ or can have a cross product $(\nabla \times \mathbf{V})$, which are called divergence and curl of a vector field, respectively.

2.3.2 Divergence of a Vector Field

The dot product between the gradient operator and a vector-valued function is called the divergence of the vector field. We can write the expressions of the dot product as follows

$$\begin{aligned}\nabla.\mathbf{V} &= \left(\frac{\partial}{\partial x_1}\widehat{\mathbf{e_1}} + \frac{\partial}{\partial x_2}\widehat{\mathbf{e_2}} + \frac{\partial}{\partial x_3}\widehat{\mathbf{e_3}}\right).\mathbf{V} = \frac{\partial}{\partial x_k}\widehat{\mathbf{e_k}}.V_i(x_j,t)\widehat{\mathbf{e_i}} \\ &= \frac{\partial V_i(x_j,t)}{\partial x_k}\widehat{\mathbf{e_k}}.\widehat{\mathbf{e_i}} = \frac{\partial V_i(x_j,t)}{\partial x_k}\delta_{ik} = \frac{\partial V_i(x_j,t)}{\partial x_i} = V_{i,i}(x_j,t)\end{aligned} \tag{2.24}$$

We used the property of the Kronecker delta in Eq. (2.24) similar to the property used in Eq. (2.13). It is apparent that the gradient of a vector filed is a scalar quantity and no basis vectors survived in the final expression in Eq. (2.24). Eq. (2.24) can be expanded using the index notation

$$V_{i,i}(x_j,t) = \frac{\partial V_1(x_j,t)}{\partial x_1} + \frac{\partial V_2(x_j,t)}{\partial x_2} + \frac{\partial V_3(x_j,t)}{\partial x_3} \tag{2.25}$$

The above quantity in Eq. (2.25), a scalar valued function measures the rate at which the vector filed diverges in space time fabric. As there is no direction in the divergence of the vector field, we can assume that the value of the divergence is a radius of a circular vector front, which can only increase or decrease, independent of any direction. If we draw a set of vectors pointing outwards or inwards around the perimeter of a circle, they are said to be divergent vectors, as they will never intersect each other. Hence, for any arbitrary vector field $\mathbf{V}$, if we calculate the divergence, $\nabla.\mathbf{V}$, it measures the rate at which the vector field is diverging or going away from each other at every point in the space and time. Fig. 2.4 shows this concept in a schematic diagram. If the divergence of a vector field is positive at any spatial point at any particular time, it means that the vector field is diverging at that point or going away from each other, but if the divergence of the field is negative at a point at any particular time, it means that the vector filed is converging at that spatial point at that time, respectively. If the divergence of a vector filed at a point is equal to zero ($\nabla.\mathbf{V} = 0$), then it means that the vector field does not have any component along the radial direction of the imaginary circle we described above. It means that all the

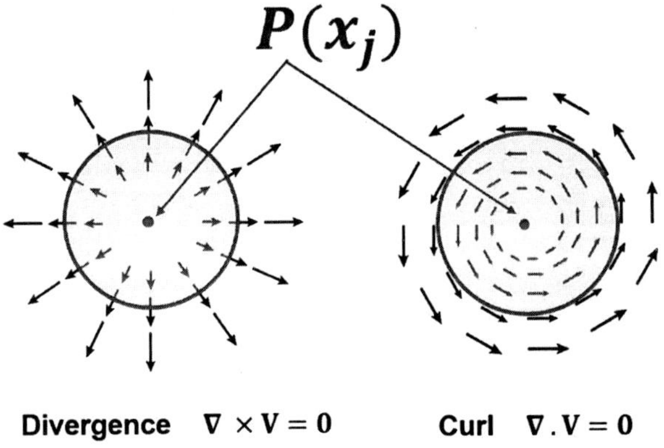

FIGURE 2.4 Pictorial representation of divergence and curl of a vector field.

components of the vector field at that point are along the tangential direction along the imaginary circle described above.

2.3.3 Curl of a Vector Field

Cross product of gradient operator and any arbitrary vector valued function can be written and expanded as follows

$$\nabla \times \mathbf{V} = \frac{\partial}{\partial x_i}\widehat{\mathbf{e_i}} \times V_j\widehat{\mathbf{e_j}} = \frac{\partial}{\partial x_i} V_j\left(\widehat{\mathbf{e_i}} \times \widehat{\mathbf{e_j}}\right) = \epsilon_{kij}\frac{\partial V_j}{\partial x_i}\widehat{\mathbf{e_k}} \tag{2.26}$$

Using the description of permutation symbol in Table 2.1, in Cartesian coordinate system the curl of a vector field can be expanded as

$$\epsilon_{kij}\frac{\partial V_j}{\partial x_i}\widehat{\mathbf{e_k}} = \widehat{\mathbf{e_1}}\left(\frac{\partial V_3}{\partial x_2} - \frac{\partial V_2}{\partial x_3}\right) + \widehat{\mathbf{e_2}}\left(-\frac{\partial V_3}{\partial x_1} + \frac{\partial V_1}{\partial x_3}\right) + \widehat{\mathbf{e_3}}\left(\frac{\partial V_2}{\partial x_1} - \frac{\partial V_1}{\partial x_2}\right) \tag{2.27}$$

It is apparent that the curl of a vector field is a vector quantity and has three components along three basis vectors $\widehat{\mathbf{e}}_1$, $\widehat{\mathbf{e}}_2$, $\widehat{\mathbf{e}}_3$, respectively. It is to be noted that the curl of the vector filed is actually a vector valued function and changing at ever point in the space. Thus the unit normal vector $\hat{\mathbf{r}}$ which is a function of (x_j,t), pointing to the direction represented by the curl of the vector filed and can be written as

$$\hat{\mathbf{r}} = \frac{\left(\frac{\partial V_3}{\partial x_2} - \frac{\partial V_2}{\partial x_3}\right)\widehat{\mathbf{e_1}} + \left(-\frac{\partial V_3}{\partial x_1} + \frac{\partial V_3}{\partial x_1}\right)\widehat{\mathbf{e_2}} + \left(\frac{\partial V_2}{\partial x_1} - \frac{\partial V_1}{\partial x_2}\right)\widehat{\mathbf{e_3}}}{\sqrt{\left(\frac{\partial V_3}{\partial x_2} - \frac{\partial V_2}{\partial x_3}\right)^2 + \left(-\frac{\partial V_3}{\partial x_1} + \frac{\partial V_3}{\partial x_1}\right)^2 + \left(\frac{\partial V_2}{\partial x_1} - \frac{\partial V_1}{\partial x_2}\right)^2}} \tag{2.28}$$

The $\hat{\mathbf{r}}$ vector represents the normal direction of a plane in which the vector field is rotating. Hence, the curl measures the rotation of the vector field at any point in space and time. Fig. 2.4 shows this concept in a schematic diagram. As discussed above, if the divergence of a vector filed at a point is equal to zero ($\nabla.\mathbf{V} = \mathbf{0}$), then it means that the vector field does not have any component along the radial direction of the imaginary circle we described in the previous section. It means that all the components of the vector field are along the tangential direction along the imaginary circle described above at that point. It is intuitive that the vector components only along the tangential direction can only rotate the field at that point. This means the vector field is curled at that point in space. This signifies the inherent meaning of the curl of a vector field. In most general cases, $\nabla.\mathbf{V} \neq \mathbf{0}$ and $\nabla \times \mathbf{V} \neq \mathbf{0}$ at a spatial point means the vector filed has partial tendency of rotation while the field has tendency of diverging outward or inward at that point.

Later in the upcoming chapters, while describing the wave propagation in materials we will use this concept multiple times.

2.4 CONCEPT OF TENSOR AND TENSOR ANALYSIS IN BRIEF

Almost every time as graduate student, we cannot recognize the necessity of the chapter in a textbook on tensors. First of all, why do we need the tensor analysis in engineering, is not clear. Hence, here in this chapter we will cover the definition of the tensors [2] and the fundamental reason behind the use of tensors in our analysis. We will review the tensor analysis in brief with the fundamental knowledge required in almost all chapters in this book.

First, tensor is nothing, but a generalized term used to describe a set of numbers in a system related in a certain way in a multidimensional system. Let's assume an array of numbers or a two- or three-dimensional matrix of few numbers which appears numerous times in almost all mathematical description that we use. We know that these arrays and matrices actually describe the geometrical objects in a space. However, do we know, how these numbers are related to the coordinate system? If the coordinate system is transformed, what happens to the numbers that describe the object? To answer these questions, we need tensor. Again, in slightly different words, tensor is nothing but a multidimensional geometrical object that describes the relationship between the objects in the same space. It might be confusing, and question may arise in our mind, what is an object? No worries, we will refer to Fig. 2.1 again. Objects are like the tangible objects we just described in Fig. 2.1. To describe an object in a three-dimensional Cartesian coordinate system we used the vectors in Section 2.1. So alternatively, we may call the vectors itself the objects, because they point to the objects in the real space. Similarly, to describe multiple objects in the same coordinate system, we can have multiple vectors and that can construct a matrix. N numbers of vectors in a three-dimensional space will construct a $3 \times N$ matrix. Also, a surface of an object like a spherical ball can be described by multiple vectors pointing to the discrete points distributed on the surface of the object.

2.4.1 First-Order and Second-Order Tensors

Vectors and matrices are in general called tensors. Tensors are categorized with their "rank" or "order" with respect to their dimensions. A scalar quantity does not have any direction, and it is just a number which cannot form an array of numbers; thus a scalar quantity is called a zeroth-order tensor. The vectors describing the objects A, B, and C in Section 2.1 (Fig. 2.1), $\mathbf{V_A}$, $\mathbf{V_B}$, and $\mathbf{V_C}$ are called the first-order tensors, because they are just made of one-dimensional array of numbers, namely A_i, B_i, and C_i, respectively. So a cluster of numbers made of A_i, B_i, and C_i, where each takes three numbers in three-dimensional coordinate system, will result a 3×3 matrix $\mathbf{\Phi}$, used in Eq. (2.7). To access all the numbers in the matrix, we would need two indices (one for row and one for column) and we call the matrix a two-dimensional tensor or a second-order tensor. So the first-order and the second-order tensors can be described mathematically in the following equations.

$$A_i = \mathbf{V_A}.\widehat{\mathbf{e_i}} \tag{2.29}$$

$$B_i = \mathbf{V_B}.\widehat{\mathbf{e_i}} \tag{2.30}$$

$$C_i = \mathbf{V_C}.\widehat{\mathbf{e_i}} \tag{2.31}$$

$$\mathbf{\Phi} = \begin{bmatrix} A_1 & A_2 & A_3 \\ B_1 & B_2 & B_3 \\ C_1 & C_2 & C_3 \end{bmatrix}_{3\times 3} = \begin{bmatrix} \Phi_{11} & \Phi_{12} & \Phi_{13} \\ \Phi_{21} & \Phi_{22} & \Phi_{23} \\ \Phi_{31} & \Phi_{32} & \Phi_{33} \end{bmatrix} \tag{2.32}$$

Using index notation we can write the $\mathbf{\Phi}$ matrix as $\mathbf{\Phi}_{ij}$, where i and j take values 1, 2, and 3, in a three-dimensional coordinate system. A_i, B_i, and C_i are the first-order tensors, and $\mathbf{\Phi}_{ij}$ is second-order tensor.

Next, we will advance our understanding with the tensors. Tensors are not only describing objects in the space but also (a) describe the relationships between the objects in the same space, (d) may describe its relation to the object in a transformed coordinate system, and/or (c) may describe the relationship between two independent coordinate systems. For example, the Ψ matrix used in Eq. (2.8) is also a tensor Ψ_{ij} as per the above definition, which is describing the relation between the $\hat{\mathbf{e}}$ and the $\hat{\mathbf{b}}$ coordinate systems. Although will be discussed in Chapter 3, here we can introduce the concept where are we going to use the concept of tensor. Let's say we want to describe the stresses at a point in space inside a material body at a specific time due to the propagation of the wave through the material. At any arbitrary point, x_j at an arbitrary time t there are nine stress values that we use, where there are only six independent stresses. Now, comparing Eq. (2.32), where we also have nine values, we can say that a stress matrix is a second-order tensor σ_{ij} and can be written as

$$\sum = \begin{bmatrix} \sigma_{11} & \sigma_{12} & \sigma_{13} \\ \sigma_{21} & \sigma_{22} & \sigma_{23} \\ \sigma_{31} & \sigma_{32} & \sigma_{33} \end{bmatrix} \tag{2.33}$$

Like stress, strain is also a second-order tensor. However, displacements, velocities, and accelerations at a point in the body are the first-order tensors.

2.4.2 Transformation Laws of Tensors

We will again refer back to Fig. 2.1a and b. We have two coordinate systems with two set of unit vectors $\widehat{\mathbf{e_1}}, \widehat{\mathbf{e_2}}, \widehat{\mathbf{e_3}}$ and $\widehat{\mathbf{b_1}}, \widehat{\mathbf{b_2}}, \widehat{\mathbf{b_3}}$. We have their relationship written in Eq. (2.8). So using index notation, we will write

$$\widehat{\mathbf{e_i}} = \Psi_{ij}\widehat{\mathbf{b_j}} \tag{2.34}$$

The vector, $\mathbf{V_A} = A_i\widehat{\mathbf{e_i}}$, transformed to the new coordinate system with unit vectors $\widehat{\mathbf{b_1}}, \widehat{\mathbf{b_2}}, \widehat{\mathbf{b_3}}$ will take a form substituting Eq. (2.34) viz.

$$\mathbf{V_A} = A_i\widehat{\mathbf{e_i}} = A_i\Psi_{ij}\widehat{\mathbf{b_j}} \tag{2.35}$$

and

$$\begin{aligned} A_i\Psi_{ij}\widehat{\mathbf{b_j}} = {} & A_1\Psi_{11}\widehat{\mathbf{b_1}} + A_1\Psi_{12}\widehat{\mathbf{b_2}} + A_1\Psi_{13}\widehat{\mathbf{b_3}} + A_2\Psi_{21}\widehat{\mathbf{b_1}} + A_2\Psi_{22}\widehat{\mathbf{b_2}} \\ & + A_2\Psi_{23}\widehat{\mathbf{b_3}} + A_3\Psi_{31}\widehat{\mathbf{b_1}} + A_1\Psi_{32}\widehat{\mathbf{b_2}} + A_1\Psi_{33}\widehat{\mathbf{b_3}} \end{aligned} \tag{2.36}$$

Here in Eq. (2.36), we summed the index i and j, over 1, 2, and 3, as they appear more than once on the same side of the equation (ride side). They are called the dummy indices. If the indices appear only once on the same side of the equation and are same on the either side of the equation, they are called the free indices. Dummy indices must carry the summation over the number of indices available, for example, here we had 1, 2, and 3, which is equal to the number of dimensions.

Next, we will take a second-order tensor and will transform from any arbitrary coordinate system with unit vectors $\widehat{\mathbf{e_1}}, \widehat{\mathbf{e_2}}, \widehat{\mathbf{e_3}}$ to another arbitrary coordinate system with unit vectors $\widehat{\mathbf{b_1}}, \widehat{\mathbf{b_2}}, \widehat{\mathbf{b_3}}$. Let's assume that the σ_{ij} in Eq. (2.33) describes the stress state at an arbitrary point at the bottom of the bowl in Fig. 2.1a in the $\hat{\mathbf{e}}$ coordinate system. Then, what would be the stress state (σ'_{ij}) at the same point in the bowl described by the $\hat{\mathbf{b}}$ coordinate system? We will assume each row of the Σ matrix as a vector. Assuming i and j describe the index for row and column, respectively, we can write the transformation equation for each row as written for the vector $\mathbf{V_A}$ in Eq. (2.35). However, here we will play a trick and introduce two more new indices m and n such that the dummy indices are not confused with the free indices. We can expand the stress state at the point specified in $\hat{\mathbf{b}}$ coordinate system as follows.

$$\sigma'_{ij} = \Psi_{im}\Psi_{nj}\sigma_{mn} \tag{2.37}$$

Conversely, we can write

$$\sigma_{ij} = \Psi_{im}\Psi_{jn}\sigma'_{mn} \tag{2.38}$$

Here we will expand our discussion bit further and introduce a fourth-order tensor as a product of two second-order tensors. We know that at any specific point the stress-strain relation can be written as

$$\sigma_{ij} = C_{ijmn}\varepsilon_{mn} \tag{2.39}$$

which is no different from Eq. (2.37) or (2.38). So, the C_{ijmn}, which is a fourth-order tensor as we have four indices, is product of two second-order tensors like $\Psi_{im}\Psi_{jn}$. Hence, we can say, to establish a relation between two first-order tensors, we would need a second-order tensor, to establish a relation between two second-order tensors, we would need a fourth-order tensor, and to establish a relation between two nth-order tensors, we would need a n^2-order tensor. We can also say that the total number of elements in a tensor is equal to the (No. of indices for the Tensor)^ No. of dimension of the coordinate system used.

These understandings will be further used in the following chapters to discuss the wave propagation in material body.

2.5 COVARIANT, CONTRAVARIANT TENSORS, AND JACOBIAN MATRIX

In the previous sections, we used the basis vectors or the unit vectors of specific coordinate system, for example, $\hat{\mathbf{e}}$ and $\hat{\mathbf{b}}$, to discuss the vectors and the vector fields. It is convenient to realize that the basis vectors are orthogonal to each other, as described in Fig. 2.1a and b. Previously, we tailored our discussion to a three-dimensional coordinate system. However, we did not discuss the basis vectors in a generic form in N-dimensional coordinate system. Let's assume a set of orthogonal basis vectors $\langle \widehat{\mathbf{e_1}}\ \widehat{\mathbf{e_2}}\ \widehat{\mathbf{e_3}} \ldots \widehat{\mathbf{e_N}} \rangle$ in an N-dimensional coordinate system with axes $x_1\ x_2\ x_3\ \ldots\ x_N$, we write x_j, where j takes the values 1, 2, 3 to N. Let's assume another set of orthogonal basis vectors $\langle \widehat{\mathbf{b_1}}\ \widehat{\mathbf{b_2}}\ \widehat{\mathbf{b_3}} \ldots \widehat{\mathbf{b_M}} \rangle$ in an M-dimensional coordinate system with axes $y_1\ y_2\ y_3\ \ldots\ y_M$, we write y_k, where k takes the values 1, 2, 3 to M as shown in Fig. 2.5.

2.5.1 Transformation of Scalar and Vector Objects and Covariant Vectors

Our objective in this section is to transform the gradient of a scalar object described in x_j coordinate to y_k coordinate system. Any arbitrary potential function ϕ, like, energy flux, heat flux, wave potentials are generally called the scalar functions described in a space, and we will call scalar object. From our previous discussions, we saw that an object remains an object unaltered, when the coordinate is transformed. But it is to be realized that the derivatives of the scalar object along a particular direction, or the overall gradient and the components of the gradient along the basis vectors change in the newly described coordinate system. Let's assume that a scalar object ϕ is known in x_j coordinate system and its derivatives (derivatives of a scalar object are a vector quantity and are called gradient) are completely described in x_j, which is $\phi_{,i}\widehat{\mathbf{e_i}}$ in index notation referring Eq. (2.23). Next following our objective, we need $\phi_{,k}\widehat{\mathbf{b_k}}$. We will expand the $\phi_{,k}$ in the following equation using the conventional chain rule.

$$\frac{\partial \phi}{\partial y_k} = \frac{\partial \phi}{\partial x_1}\frac{\partial x_1}{\partial y_k} + \frac{\partial \phi}{\partial x_2}\frac{\partial x_2}{\partial y_k} + \frac{\partial \phi}{\partial x_3}\frac{\partial x_3}{\partial y_k} + \cdots + \frac{\partial \phi}{\partial x_N}\frac{\partial x_N}{\partial y_k} = \frac{\partial \phi}{\partial x_j}\frac{\partial x_j}{\partial y_k} \tag{2.40}$$

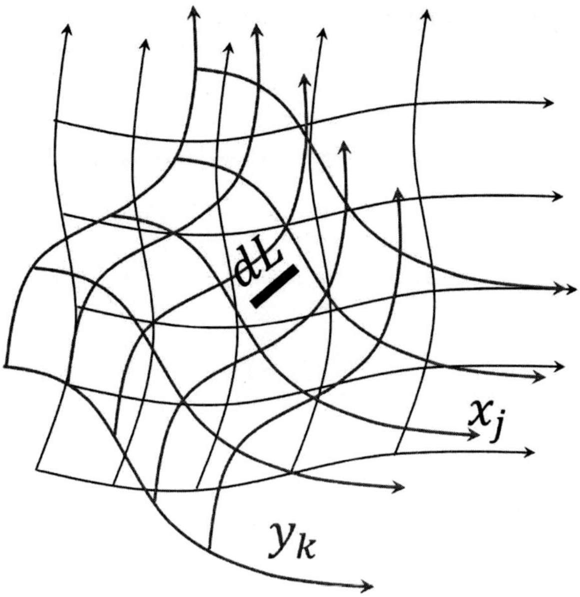

FIGURE 2.5 Transformation of coordinate system from x_j basis system to y_k basis system.

Expanding the above expression for all the derivatives in a matrix form we can write,

$$\begin{Bmatrix} \dfrac{\partial \phi}{\partial y_1} \\ \dfrac{\partial \phi}{\partial y_2} \\ \dfrac{\partial \phi}{\partial y_3} \\ \dfrac{\partial \phi}{\partial y_4} \\ \vdots \\ \vdots \\ \vdots \\ \vdots \\ \dfrac{\partial \phi}{\partial y_M} \end{Bmatrix} = \begin{bmatrix} \dfrac{\partial x_1}{\partial y_1} & \dfrac{\partial x_2}{\partial y_1} & \dfrac{\partial x_3}{\partial y_1} & \dfrac{\partial x_4}{\partial y_1} & \dfrac{\partial x_5}{\partial y_1} & \cdots & \cdots & \cdots & \dfrac{\partial x_N}{\partial y_1} \\ \dfrac{\partial x_1}{\partial y_2} & \dfrac{\partial x_2}{\partial y_2} & \dfrac{\partial x_3}{\partial y_2} & \dfrac{\partial x_4}{\partial y_2} & \dfrac{\partial x_5}{\partial y_2} & \cdots & \cdots & \cdots & \dfrac{\partial x_N}{\partial y_1} \\ \dfrac{\partial x_1}{\partial y_3} & \dfrac{\partial x_2}{\partial y_3} & \dfrac{\partial x_3}{\partial y_3} & \dfrac{\partial x_4}{\partial y_3} & \dfrac{\partial x_5}{\partial y_3} & \cdots & \cdots & \cdots & \dfrac{\partial x_N}{\partial y_1} \\ \dfrac{\partial x_1}{\partial y_4} & \dfrac{\partial x_2}{\partial y_4} & \dfrac{\partial x_3}{\partial y_4} & \dfrac{\partial x_4}{\partial y_4} & \dfrac{\partial x_5}{\partial y_4} & \cdots & \cdots & \cdots & \dfrac{\partial x_N}{\partial y_1} \\ \vdots & \vdots & \vdots & \vdots & \vdots & \ddots & \vdots & \vdots & \vdots \\ \vdots & \vdots & \vdots & \vdots & \vdots & \cdots & \ddots & \vdots & \vdots \\ \vdots & \vdots & \vdots & \vdots & \vdots & \cdots & \cdots & \vdots & \vdots \\ \vdots & \vdots & \vdots & \vdots & \vdots & \cdots & \cdots & \ddots & \vdots \\ \dfrac{\partial x_1}{\partial y_M} & \dfrac{\partial x_2}{\partial y_M} & \dfrac{\partial x_3}{\partial y_M} & \dfrac{\partial x_4}{\partial y_M} & \dfrac{\partial x_5}{\partial y_M} & \cdots & \cdots & \cdots & \dfrac{\partial x_N}{\partial y_1} \end{bmatrix} \begin{Bmatrix} \dfrac{\partial \phi}{\partial x_1} \\ \dfrac{\partial \phi}{\partial x_2} \\ \dfrac{\partial \phi}{\partial x_3} \\ \dfrac{\partial \phi}{\partial x_4} \\ \vdots \\ \vdots \\ \vdots \\ \vdots \\ \dfrac{\partial \phi}{\partial x_N} \end{Bmatrix} \quad (2.41)$$

The matrix in Eq. (2.141) is of size, $M \times N$. The gradient vectors on either side of Eq. (2.41) are with dimensions $M \times 1$ and $N \times 1$, respectively are called the covariant vector. The relation of a covariant vector can be further written in index notation as

$$\frac{\partial\phi}{\partial y_k} = \frac{\partial\phi}{\partial x_j}\left(\frac{\partial x_j}{\partial y_k}\right) \; or \; \phi_{,k}(y) = \left(\frac{\partial x_j}{\partial y_k}\right)\phi_{,j}(x) \tag{2.42}$$

where j is the dummy index and k is the free index. Here, the gradient of the scalar object is a first-order tensor. In previous sections we described that a second-order tensor can be expressed as a product of two first-order tensors. Hence, let's assume that two covariant vectors form a second-order covariant tensor $T_{ji}(x)$, where the index i takes the values 1, 2, 3 to N in the x_j coordinate system. Thus, we can write the transformation of the covariant tensor $T_{ji}(x)$ from x_j coordinate to y_k coordinate system as follows

$$\phi_{,k}(y)\phi_{,s}(y) = \left(\frac{\partial x_j}{\partial y_k}\right)\left(\frac{\partial x_i}{\partial y_s}\right)\phi_{,j}(x)\phi_{,i}(x) \quad \text{or}$$
$$T_{ks}(y) = \left(\frac{\partial x_j}{\partial y_k}\right)\left(\frac{\partial x_i}{\partial y_s}\right)T_{ji}(x) \tag{2.43}$$

where k and s take values 1, 2, 3 to M.

2.5.2 Transformation of Basis, Contravariant Vectors, and Jacobian

Let's assume a small object of length dL (Fig. 2.5) in the x_j coordinate system. The magnitude of dL in x_j can be written as

$$dL = \sqrt{dx_1^2 + dx_2^2 + dx_3^2 + \cdots + dx_N^2} \tag{2.44}$$

where $dx_1, dx_2, dx_3 \ldots dx_N$ are the element length of the object along the basis vectors, $\langle \widehat{\mathbf{e}_1}\ \widehat{\mathbf{e}_2}\ \widehat{\mathbf{e}_3} \ldots \widehat{\mathbf{e}_N} \rangle$. Hence, we can also write the $\overline{\mathbf{dL}}$ vector as

$$\overline{\mathbf{dL}} = dx_1\widehat{\mathbf{e}_1} + dx_2\widehat{\mathbf{e}_2} + dx_3\widehat{\mathbf{e}_3} + \cdots + dx_N\widehat{\mathbf{e}_N} \tag{2.45}$$

Similarly, the same object can also be written in the M-dimensional y_k coordinate system as follows,

$$\overline{\mathbf{dL}} = dy_1\widehat{\mathbf{b}_1} + dy_2\widehat{\mathbf{b}_2} + dy_3\widehat{\mathbf{b}_3} + \cdots + dy_M\widehat{\mathbf{b}_M} \tag{2.46}$$

where $dy_1, dy_2, dy_3 \ldots dy_M$ are the element length of the object along the basis vectors $\langle \widehat{\mathbf{b}_1}\ \widehat{\mathbf{b}_2}\ \widehat{\mathbf{b}_3} \ldots \widehat{\mathbf{b}_M} \rangle$.

However, how the components dx_j are related to the components dy_k, are not described yet. But it is evident that $dy_k = f\left(dx_j, \frac{\partial y_k}{\partial x_j}\right)$. So we can write

$$dy_k = \frac{\partial y_k}{\partial x_j}dx_j \tag{2.47}$$

where j is the dummy index and signifies summation over the N dimensions. We will expand Eq. (2.47) and write in a matrix form.

$$\begin{Bmatrix} dy_1 \\ dy_2 \\ dy_3 \\ dy_4 \\ \vdots \\ \vdots \\ \vdots \\ dy_M \end{Bmatrix} = \begin{bmatrix} \frac{\partial y_1}{\partial x_1} & \frac{\partial y_1}{\partial x_2} & \frac{\partial y_1}{\partial x_3} & \frac{\partial y_1}{\partial x_4} & \frac{\partial y_1}{\partial x_5} & \cdots\cdots\cdots & \frac{\partial y_1}{\partial x_N} \\ \frac{\partial y_2}{\partial x_1} & \frac{\partial y_2}{\partial x_2} & \frac{\partial y_2}{\partial x_3} & \frac{\partial y_2}{\partial x_4} & \frac{\partial y_2}{\partial x_5} & \cdots\cdots\cdots & \frac{\partial y_2}{\partial x_N} \\ \frac{\partial y_3}{\partial x_1} & \frac{\partial y_3}{\partial x_2} & \frac{\partial y_3}{\partial x_3} & \frac{\partial y_3}{\partial x_4} & \frac{\partial y_3}{\partial x_5} & \cdots\cdots\cdots & \frac{\partial y_3}{\partial x_N} \\ \frac{\partial y_4}{\partial x_1} & \frac{\partial y_4}{\partial x_2} & \frac{\partial y_4}{\partial x_3} & \frac{\partial y_4}{\partial x_4} & \frac{\partial y_4}{\partial x_5} & \cdots\cdots\cdots & \frac{\partial y_4}{\partial x_N} \\ \vdots & \vdots & \vdots & \vdots & \vdots & \ddots \; \vdots \; \vdots & \vdots \\ \vdots & \vdots & \vdots & \vdots & \vdots & \cdots \ddots \; \vdots & \vdots \\ \vdots & \vdots & \vdots & \vdots & \vdots & \cdots\cdots \; \vdots & \vdots \\ \vdots & \vdots & \vdots & \vdots & \vdots & \cdots\cdots \ddots & \vdots \\ \frac{\partial y_M}{\partial x_1} & \frac{\partial y_M}{\partial x_2} & \frac{\partial y_M}{\partial x_3} & \frac{\partial y_M}{\partial x_4} & \frac{\partial y_M}{\partial x_5} & \cdots\cdots\cdots & \frac{\partial y_M}{\partial x_N} \end{bmatrix} \begin{Bmatrix} dx_1 \\ dx_2 \\ dx_3 \\ dx_4 \\ dx_5 \\ \vdots \\ \vdots \\ \vdots \\ dx_M \end{Bmatrix} \tag{2.48}$$

Please, note the difference between the matrices in Eq. (2.48) and Eq. (2.41). They are not the same. The vector composed of the components of the dL elements in both x_j and y_k coordinate systems in Eq. (2.48) is called the contravariant vectors and their relation written in the matrix form in Eq. (2.48) is called the Jacobian matrix. Whereas, the matrix in Eq. (2.41) is not the Jacobian matrix that are used for coordinate transformation only. Not to confuse the covariant vectors and tensors with the contravariant, we will use the upper indices to describe the contravariant vectors and in index notation Eq. (247) can be modified as follows

$$Y^k(y) = \frac{\partial y_k}{\partial x_j} X^j(x) \tag{2.49}$$

where $Y^k(y) = dy_k$ and $X^j(x) = dx_j$. Here, contravariant vector is a first-order tensor. In previous sections we described that a second-order tensor can be expressed as a product of two first-order tensors. Hence, let's assume that two contravariant vectors form a second-order contravariant tensor $T^{ji}(x)$, where the index i also takes the values 1, 2, 3 to N in the x_j coordinate system. Thus, we can write the transformation of the covariant tensor $T^{ji}(x)$, from x_j coordinate to y_k coordinate system as follows

$$Y^k(y)Y^s(y) = \left(\frac{\partial y_k}{\partial x_j}\right)\left(\frac{\partial y_s}{\partial x_i}\right) X^j(x) X^i(x) \quad \text{or}$$

$$T^{rs}(y) = \left(\frac{\partial y_k}{\partial x_j}\right)\left(\frac{\partial y_s}{\partial x_i}\right) T^{ji}(x) \tag{2.43}$$

where k and s take values 1, 2, 3 to M.

2.6 EXAMPLES ON INDEX NOTATIONS

PROBLEM 1

Determine the simplest form of the following expressions when, $i, j, k = 1, 2$, and 3

1. $\in_{3jk}\ a_j a_k$
2. $\in_{ijk}\ \delta_{jk}$
3. $\in_{1jk}\ a_2 T_{kj}$
4. $\in_{1jk}\ \delta_{3j} v_k$

ANSWER

Problem 1.1 Expanding $\in_{3jk}\ a_j\ a_k$, one can write

$$\in_{3jk}\ a_j\ a_k = \in_{312}\ a_1\ a_2 + \in_{313}\ a_1\ a_3 + \in_{321}\ a_2\ a_1 + \in_{322}\ a_2\ a_2 + \in_{323}\ a_2\ a_3 + \in_{311}\ a_1\ a_1 + \in_{331}\ a_3\ a_1 + \in_{332}\ a_3\ a_2 + \in_{333}\ a_3\ a_3$$

As we know when i, j, and k are in clockwise; $\in_{ijk} = 1$

i, j, and k are in counterclockwise; $\in_{ijk} = -1$

other combination of i, j, and k results $\in_{ijk} = 0$

So, $\in_{3jk}\ a_j\ a_k = a_1\ a_2 - a_2\ a_1 + 0 = 0$

Problem 1.2 $\in_{ijk}\ \delta_{jk} = \in_{ijj}$, as multiplication with Kronecker delta will switch the parent index with its own index and will result unity. However, as stated in Problem 1.1 any arbitrary combination of i, j, and k results $\in_{ijk} = 0$, $\in_{ijj} = 0$. Hence, $\in_{ijk}\ \delta_{jk} = 0$.

Problem 1.3 Expanding $\in_{1jk}\ a_2\ T_{kj}$ one can write

$$\begin{aligned}\in_{1jk}\ a_2\ T_{kj} &= \in_{111}\ a_2\ T_{11} + \in_{112}\ a_2\ T_{21} + \in_{113}\ a_2\ T_{31} + \in_{121}\ a_2\ T_{12} + \in_{122}\ a_2\ T_{22} \\ &\quad + \in_{123}\ a_2\ T_{32} + \in_{131}\ a_2\ T_{13} + \in_{132}\ a_2\ T_{23} + \in_{133}\ a_2\ T_{33} \\ &= a_2\ T_{32} - a_2\ T_{23} \\ &= a_2\left(T_{32} - T_{23}\right)\end{aligned}$$

If $T_{32} = T_{23}$, then $\in_{1jk}\ a_2\ T_{kj} = 0$.

Problem 1.4 Expanding $\in_{1jk}\ \delta_{3j}\ v_k$ one can get

$$\begin{aligned}\in_{1jk}\ \delta_{3j}\ v_k &= \in_{11k}\ \delta_{31}\ v_k + \in_{12k}\ \delta_{32}\ v_k + \in_{13k}\ \delta_{33}\ v_k \\ &= \in_{123}\ \delta_{32}\ v_3 + \in_{132}\ \delta_{33}\ v_2, \text{ since } \delta_{32} = 0 \\ &= (-1)(1)\ v_2 \\ &= -v_2\end{aligned}$$

PROBLEM 2

Prove that the following identities are true when $i, j, k, p, q, r = 1,2,3$

1. $\epsilon_{pqr}\epsilon_{pqr} = 6$
2. $\epsilon_{pqi}\epsilon_{pqj} = 2\delta_{ij}$
3. $\epsilon_{pqi}\epsilon_{pkl} = \delta_{ik}\delta_{jl} - \delta_{il}\delta_{jk}$

ANSWER

Problem 2.1 First of all the following equality holds true when the first index p is equal for both the permutation symbols ϵ_{pij} *and* ϵ_{pkl} , hence,

$$\epsilon_{pij}\ \epsilon_{pkl} = \epsilon_{jip}\ \epsilon_{lkp}$$

We note that the quantity $\epsilon_{jip}\ \epsilon_{lkp}$ is nonzero if and only if, j, i, and p are all different. Similarly, l, k, and p are all different. Therefore, either $j = l$ and $i = k$ or $j = k, i = l$.

For the first case ($j = l$ and $i = k$), $\epsilon_{jip} = \epsilon_{lkp}$ and therefore $\epsilon_{jip}\ \epsilon_{lkp} = 1$

For the second case ($j = k, i = l$), $\epsilon_{jip} = -\epsilon_{lkp}$ because interchanges of two indices change the sign of the value of ϵ_{lkp}

Consequently, $\epsilon_{jip}\epsilon_{lkp} = -1$

Now if j, i, l, and k are fixed and p run through the values 1, 2, 3, the only one term will be nonzero is that summation. So, the whole summation will either be +1 or −1. So, this fact can be written as: $\delta_{jl}\ \delta_{ik} - \delta_{jk}\delta_{il}$ because this expression equal to +1 for $j = l$ and $i = k$, $j \neq i$ and equal to −1 for $j = k$ and $i = l$ $i \neq j$. If neither, it will be equals to zero. Hence, one can write the following identity as

$$\epsilon_{pij}\ \epsilon_{pkl} = \delta_{jl}\ \delta_{ik} - \delta_{jk}\delta_{il}$$

again for $\epsilon_{pqr}\ \epsilon_{pqr}$, plugging $p = r$ and $j = l, i = k$ in the above identity we get

$$\epsilon_{lkr}\ \epsilon_{lkr} = \delta_{ll}\ \delta_{kk} - \delta_{lk}\ \delta_{kl} = 2(3) - 0 \quad (l \neq k) = 6$$

Problem 2.2 Similar to the problem above, $\epsilon_{pqi}\ \epsilon_{pqj} = \epsilon_{piq}\ \epsilon_{pjq}$

Plugging, $j = p = l$, $k = q$, $i = i$, $l = j$ in the previous equation one can write

$$\begin{aligned}\epsilon_{piq}\ \epsilon_{pjq} &= \delta_{pp}\ \delta_{ij} - \delta_{pj}\ \delta_{ip} \\ &= \delta_{pp}\ \delta_{ij} - \delta_{ij} \\ &= 3\delta_{ij} - \delta_{ij} \\ &= 2\delta_{ij}\end{aligned}$$

Problem 2.3 Redundant, proved under Problem 2.1

PROBLEM 3
Use the indicial notation to prove the following results for vectors u, v, w, x

1. $\mathbf{w}\times(\mathbf{u}+\mathbf{v})=\mathbf{w}\times\mathbf{u}+\mathbf{w}\times\mathbf{v}$
2. $\mathbf{u}\times\mathbf{v}=-(\mathbf{v}\times\mathbf{u})$
3. $\mathbf{w}\times(\mathbf{u}\times\mathbf{v})=\mathbf{u}(\mathbf{w}.\mathbf{v})-\mathbf{v}(\mathbf{w}.\mathbf{u})$
4. $(\mathbf{x}\times\mathbf{w}).(\mathbf{u}\times\mathbf{v})=(\mathbf{x}.\mathbf{u})(\mathbf{w}.\mathbf{v})-(\mathbf{x}.\mathbf{v})(\mathbf{w}.\mathbf{u})$
5. $(\mathbf{x}\times\mathbf{w})\times(\mathbf{u}\times\mathbf{v})=\mathbf{w}\left[(\mathbf{x}.(\mathbf{u}\times\mathbf{v}))\right]-\mathbf{x}[\mathbf{w}.(\mathbf{u}\times\mathbf{v})]$
6. $\mathbf{div}(\mathbf{u}\times\mathbf{v})=\mathbf{v}\ \mathit{curl}\ \mathbf{u}-\mathbf{u}\ \mathit{curl}\ \mathbf{v}$
7. $\mathbf{div}(\mathit{curl}\ (\mathbf{v}))=\mathbf{0}$

ANSWER

Problem 3.1 $\mathbf{w}\times(\mathbf{u}+\mathbf{v})$ can be expanded using index notation as follows

$$
\begin{aligned}
&= w_i\hat{\mathbf{e}}_i \times \left(u_j\ \hat{\mathbf{e}}_j + v_k\hat{\mathbf{e}}_k\right) \\
&= \epsilon_{ijk}\ w_j\ \left(u_k + v_k\right)\hat{\mathbf{e}}_i \\
&= \epsilon_{ijk}\ \left(w_j u_k + w_j\ v_k\right)\hat{\mathbf{e}}_i \\
&= \epsilon_{ijk}\ \left(w_j u_k + w_j\ v_k\right)\hat{\mathbf{e}}_i \\
&= \epsilon_{ijk}\ w_j u_k\hat{\mathbf{e}}_i + \epsilon_{ijk}\ w_j v_k\hat{\mathbf{e}}_i \\
&= (\mathbf{w}\times\mathbf{u})+(\mathbf{w}\times\mathbf{v})
\end{aligned}
$$

Problem 3.2 $(\mathbf{u}\times\mathbf{v})$ can be expanded as

$$
\begin{aligned}
&= \epsilon_{ijk}\ u_j\ v_k\hat{\mathbf{e}}_i \\
&= -\epsilon_{ikj}\ u_j\ v_k\hat{\mathbf{e}}_i \\
&= -\epsilon_{ikj}\ u_j\ v_k\hat{\mathbf{e}}_i \\
&= -\epsilon_{ikj}\ v_k\ u_j\ \hat{\mathbf{e}}_i \\
&= -(\mathbf{v}\times\mathbf{u})
\end{aligned}
$$

Problem 3.3 $\mathbf{w}\times(\mathbf{u}\times\mathbf{v})$ can be expanded as

$$
\begin{aligned}
&= \epsilon_{ijk}\ w_j\ \left(\epsilon_{klm}\ u_l\ v_m\right) \\
&= \epsilon_{ijk}\ \epsilon_{klm}\ w_j\ u_l\ v_m \\
&= \left(\delta_{il}\ \delta_{jm}\ -\ \delta_{im}\ \delta_{jl}\right)\ w_j\ u_l\ v_m \\
&= \left(\delta_{il}\ \delta_{jm}\ w_j\ u_l\ v_m - \delta_{im}\ \delta_{jl}\ \ w_j\ u_l\ v_m\right) \\
&= \left(u_i\ \ w_m\ v_m -\ v_i\ \ w_l\ v_l\right) \\
&= \mathbf{u}(\mathbf{w}.\mathbf{v})-\mathbf{v}(\mathbf{w}.\mathbf{u})
\end{aligned}
$$

Problem 3.4 $(\mathbf{x}\times\mathbf{w}).(\mathbf{u}\times\mathbf{v})$ can be expanded as

$$= \epsilon_{ijk}\ x_j\ w_k\ \epsilon_{ilm}\ u_l\ v_m$$

$$= \left(\epsilon_{i\underline{jk}}\ \ \epsilon_{i\underline{lm}}\right) x_j\ w_k\ u_l\ v_m$$

$$= \left(\delta_{jl}\ \delta_{km}\ -\ \delta_{jm}\ \delta_{kl}\right)\ x_j\ w_k\ u_l\ v_m$$

$$= \delta_{jl}\ \delta_{km}\ x_j\ w_k\ u_l\ v_m\ -\ \delta_{jm}\ \delta_{kl}\ x_j\ w_k\ u_l\ v_m$$

$$= x_l\ v_k\ w_k\ u_l\ -\ x_m\ w_l\ u_l\ v_m$$

$$= x_l\ u_l\ v_k\ w_k - x_m\ \ v_m\ w_l\ u_l$$

$$= (\mathbf{x}.\mathbf{u})\ (\mathbf{w}.\mathbf{v}) - (\mathbf{x}.\mathbf{v})\ (\mathbf{w}.\mathbf{u})$$

Problem 3.5 $(\mathbf{x}\times\mathbf{w})\times(\mathbf{u}\times\mathbf{v})$ can be expanded as

$$= \epsilon_{ijk}\ \left[\epsilon_{jlm}\ \ x_l\ w_m\right]\left[\epsilon_{knp}\ \ u_n\ v_p\right]$$

$$= \epsilon_{ijk}\ \ c_{jlm}\ \epsilon_{knp}\ \ x_l\ w_m\ u_n\ v_p$$

$$= -\epsilon_{jik}\ \ \epsilon_{jlm}\ \epsilon_{knp}\ \ x_l\ w_m\ u_n\ v_p$$

$$= \left(-\delta_{il}\ \delta_{km}\ +\ \delta_{im}\ \delta_{kl}\right)\ \epsilon_{knp}\ x_l\ w_m\ u_n\ v_p$$

$$= \epsilon_{knp}\ \delta_{im}\ \delta_{kl}\ x_l\ w_m\ u_n\ v_p\ -\epsilon_{knp}\ \ \delta_{il}\ \delta_{km}\ \ x_l\ w_m\ u_n\ v_p$$

$$= \epsilon_{knp}\ \ w_i\ x_k\ u_n\ v_p\ -\epsilon_{knp}\ \ \ x_i\ w_k\ u_n\ v_p$$

$$= w_i\ \left[\epsilon_{knp}\ u_n\ v_p\ x_k\right] - x_i\ \left[\epsilon_{knp}\ u_n\ v_p\ w_k\right]$$

$$= \mathbf{w}\left[\left(\mathbf{x}\ .(\mathbf{u}\times\mathbf{v})\right)\right] - \mathbf{x}\left[\left(\mathbf{w}\ .(\mathbf{u}\times\mathbf{v})\right)\right]$$

Problem 3.6 div $(\mathbf{u}\times\mathbf{v})$ can be expanded as

$$= \nabla.\ (\mathbf{u}\times\mathbf{v})$$

$$= \frac{\partial}{\partial x_i}\left(\epsilon_{ijk}\ \ u_j\ v_k\right)$$

$$= \epsilon_{ijk}\ \ u_j\ \frac{\partial v_k}{\partial x_i} + \epsilon_{ijk}\ \frac{\partial u_j}{\partial x_i}\ \ v_k$$

$$= -\epsilon_{jik}\ \ \frac{\partial v_k}{\partial x_i}\ u_j + \epsilon_{ijk}\ \frac{\partial u_j}{\partial x_i}\ \ v_k$$

$$= \mathbf{v}.\ (\nabla\times\mathbf{u}) - \mathbf{u}.\ (\nabla\times\mathbf{v})$$

Which is equivalent to, **v** (curl **u**) – **u**(curl **v**)

Problem 3.7 div $(curl\ (\mathbf{v}))$ can be expanded as

$$= \nabla . (\nabla \times \mathbf{v})$$

$$= \frac{\partial}{\partial x_i}\left(\epsilon_{ijk} \ \frac{\partial}{\partial x_j} \ v_k \right)$$

$$= \epsilon_{ijk} \ \frac{\partial^2 v_k}{\partial x_j \ \partial x_i}$$

$$= \epsilon_{ijk} \ \ v_{k,ji}$$

If we swap the i, j, the result will be the same and there will be no change in sign. However, if the indices are swapped again then the operation will result $-\epsilon_{ijk} \ \ v_{k,ij}$. Hence,

$$2 \ \epsilon_{ijk} \ \ v_{k,ij} = 0$$

$$\epsilon_{ijk} \ \ v_{k,ij} = 0$$

$$\nabla \ . \left(\nabla \ \times \mathbf{v}\right) = 0 \text{ or div } (\text{curl } \mathbf{v}) = 0$$

Problem 3.8 $\text{curl}(\text{grad}(\varphi))$ can be expanded as

$$= \nabla \times (\nabla . \ \varphi)$$

$$= \epsilon_{ijk} \ \ \frac{\partial}{\partial x_j} \frac{\partial \varphi}{\partial x_k}$$

$$= \epsilon_{ijk} \ \ \frac{\partial^2 \varphi}{\partial x_j \ \partial x_k}$$

$$= \epsilon_{ijk} \ \ \varphi,_{ji}$$

If we swap the i, j, the result will be the same and there will be no change in sign. However, if the indices are swapped again then the operation will result $-\epsilon_{ijk}\ \varphi,_{ij}$. Hence,

$$2 \ \epsilon_{ijk} \ \ \varphi,_{ij} = 0$$

$$\epsilon_{ijk} \ \ \varphi,_{ij} = 0$$

$$\nabla \times (\nabla \ . \ \varphi) = 0 \quad \text{or} \quad \text{curl } (\text{div } \varphi) = 0$$

PROBLEM 4

Prove for vector A B C using vector calculus and index notation

1. $\nabla.(\nabla\times\mathbf{A})=0$
2. $\nabla(\mathbf{A}.\mathbf{B})=(\mathbf{A}.\nabla)\mathbf{B}+(\mathbf{B}.\nabla)\mathbf{A}+\mathbf{A}\times(\nabla\times\mathbf{B})+\mathbf{B}\times(\nabla\times\mathbf{A})$
3. $\nabla\times(\nabla\times\mathbf{C})=\nabla(\nabla.\mathbf{C})-\nabla^2\mathbf{C}$

ANSWER

Problem 4.1 $\nabla.\,(\nabla\times\mathbf{A})$ can be written as

$$=\frac{\partial}{\partial x_i}\left(\epsilon_{ijk}\ \frac{\partial}{\partial x_j}\ A_k\right)$$

$$=\epsilon_{ijk}\ \frac{\partial^2 A_k}{\partial x_j\ \partial x_i}$$

$$=\epsilon_{ijk}\ \ A_{k,ji}$$

If we swap the i, j, the result will be the same and there will be no change in sign. However, if the indices are swapped again then the operation will result $-\epsilon_{ijk}\ \ A_{k,ij}$. Hence,

$$2\ \epsilon_{ijk}\ \ A_{k,ij}=0$$

$$\epsilon_{ijk}\ \ A_{k,ij}=0$$

$$\nabla\,.\,(\nabla\times\mathbf{A})=0$$

Problem 4.2 $(\mathbf{A}.\nabla)\ \mathbf{B}+\ (\mathbf{B}.\nabla)\ \mathbf{A}+\mathbf{A}\times(\nabla\times\mathbf{B})+\mathbf{B}\times(\nabla\times\mathbf{A})$ can be written as

$$=\left(A_k\ \frac{\partial}{\partial x_k}\right)B_i+\left(B_k\ \frac{\partial}{\partial x_k}\right)A_i+\epsilon_{ijk}\ A_j\left(\epsilon_{klm}\ \frac{\partial B_m}{\partial x_k}\right)+\epsilon_{ijk}\ B_j\left(\epsilon_{klm}\ \frac{\partial A_m}{\partial x_k}\right)$$

$$=\left(A_k\ \frac{\partial B_i}{\partial x_k}\right)+\left(B_k\ \frac{\partial A_i}{\partial x_k}\right)+\epsilon_{ijk}\ \epsilon_{klm}\left(A_j\ \frac{\partial B_m}{\partial x_k}+B_j\ \frac{\partial A_m}{\partial x_k}\right)$$

$$=A_k\ \frac{\partial B_i}{\partial x_k}+B_k\ \frac{\partial A_i}{\partial x_k}+\left(\delta_{il}\ \delta_{jm}-\delta_{im}\delta_{jl}\right)\left(A_j\ \frac{\partial B_m}{\partial x_k}+B_j\ \frac{\partial A_m}{\partial x_k}\right)$$

$$=A_k\ \frac{\partial B_i}{\partial x_k}+B_k\ \frac{\partial A_i}{\partial x_k}+\delta_{il}\ \delta_{jm}\left(A_j\ \frac{\partial B_m}{\partial x_k}+B_j\ \frac{\partial A_m}{\partial x_k}\right)-\delta_{im}\delta_{jl}\left(A_j\ \frac{\partial B_m}{\partial x_k}+B_j\ \frac{\partial A_m}{\partial x_k}\right)$$

$$= A_k \frac{\partial B_i}{\partial x_k} + B_k \frac{\partial A_i}{\partial x_k} + A_j \frac{\partial B_j}{\partial x_i} + B_j \frac{\partial A_j}{\partial x_i} - A_j \frac{\partial B_i}{\partial x_j} - B_j \frac{\partial A_i}{\partial x_j}$$

$$= A_k \frac{\partial B_i}{\partial x_k} + B_k \frac{\partial A_i}{\partial x_k} + A_j \frac{\partial B_j}{\partial x_i} + B_j \frac{\partial A_j}{\partial x_i} - A_k \frac{\partial B_i}{\partial x_k} - B_k \frac{\partial A_i}{\partial x_k}$$

$$= A_j \frac{\partial B_j}{\partial x_i} + B_j \frac{\partial A_j}{\partial x_i} = \frac{\partial}{\partial x_i} \left(A_j \; B_j \right) = \nabla \left(\mathbf{A.B} \right)$$

Problem 4.3 $\nabla \times (\nabla \times \mathbf{C})$ can be written as

$$= \epsilon_{ijk} \left(\frac{\partial}{\partial x_j} \left(\epsilon_{klm} \frac{\partial}{\partial x_j} C_j \right) \right)$$

$$= \epsilon_{ijk} \; \epsilon_{klm} \left(\frac{\partial^2 C_j}{\partial x_j \; \partial x_l} \right)$$

$$= \epsilon_{kij} \; \epsilon_{klm} \left(\frac{\partial^2 C_j}{\partial x_j \; \partial x_l} \right)$$

$$= \left(\delta_{il} \; \delta_{jm} - \delta_{im} \delta_{jl} \right) \left(\frac{\partial^2 C_j}{\partial x_j \; \partial x_l} \right)$$

$$= \delta_{il} \; \delta_{jm} \left(\frac{\partial^2 C_j}{\partial x_j \; \partial x_l} \right) - \delta_{im} \delta_{jl} \left(\frac{\partial^2 C_j}{\partial x_j \; \partial x_l} \right)$$

$$= \frac{\partial^2 C_j}{\partial x_j \; \partial x_i} - \frac{\partial^2 C_i}{\partial x_j \; \partial x_j} = \frac{\partial}{\partial x_j} \left(\frac{\partial C_j}{\partial x_j} \right) - \frac{\partial}{\partial x_j} \left(\frac{\partial C_i}{\partial x_i} \right)$$

$$= \nabla \left(\nabla . \mathbf{C} \right) - \nabla^2 \mathbf{C}$$

2.7 SUMMARY

Chapter 2 briefly reviewed the concept of vectors and vector fields. Mapping of vectors and objects from a specific coordinate system (with orthogonal basis) to any arbitrary coordinate system (with different orthogonal basis), when the relation between the basis systems are known is discussed. Differential operations like, divergence and curl of a vector fields are also discussed and explained with their physical meaning. All mathematical expressions are then presented introducing the index notation, where importance of dummy and free index is discussed. Extending the discussion of vectors and vector fields, concept of tensor of different rank is introduced. Covariant and contravariant vectors and tensors are discussed thereafter, while the concept of Jacobian matrix is introduced. The topics discussed in this chapter will be immensely valuable in the following chapters, where discussions are based on the fundamental concepts presented in this chapter.

2.8 APPENDIX

2.8.1 Divergence Theorem

In any arbitrary vector field discussed in Section 2.3, an imaginary surface (e.g., spherical, cuboidal, ellipsoidal, cylindrical, etc.) could be assumed. Sometimes it is imperative in science and engineering to find how much flux has passed through that surface. The surface integral of a vector field is called the flux integral. Name comes from the fact that it computes the flux of fluid of density ρ with velocity field $\mathbf{V}(x_j,t)$ flowing through a surface Γ. The surface is shown in Fig. A.2.1 with an example of a spherical surface and an arbitrary velocity field $\mathbf{V}(x_j) = x_1x_2\hat{\mathbf{e}}_1 + x_2x_3\hat{\mathbf{e}}_2 + x_3x_1\hat{\mathbf{e}}_3$. The vector field also passes through an incremental surface element $d\vec{\Gamma}$ taken from the spherical surface. To find the flux flowing through that incremental surface, one should take a dot product between the vector field and the direction normal to that incremental surface. Once the dot product is taken which signifies the component of the vector field along the normal to the incremental local surface area, integral of the product over the entire surface will give the total flux flowing away or going in through the surface Γ. Mathematically, if the velocity field is known then the total mass of fluid passing through the surface Γ can be written as

$$m = \rho\mathbf{V} \quad ; \quad M = \oiint \rho\mathbf{V}.d\vec{\Gamma} = \oiint \rho\mathbf{V}.\,\hat{\mathbf{n}}\,d\Gamma \tag{A.2.1}$$

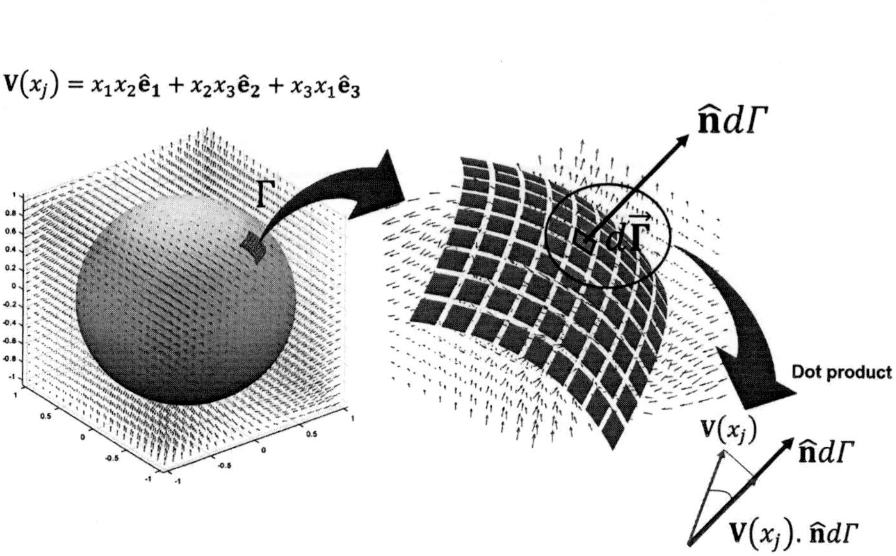

FIGURE A.2.1 Schematic representation of a vector field passing through a surface, projection of the vector field on the surface normal at each point on the surface.

Similarly, the net charge enclosed by a closed surface in an electric field $\mathbf{E}(x_j, t)$ can be written as

$$q = \epsilon_0 \mathbf{E}; \qquad Q = \epsilon_0 \oiint \mathbf{E}.d\vec{\Gamma} = \epsilon_0 \oiint \mathbf{E}.\hat{\mathbf{n}}\, d\Gamma \tag{A.2.2}$$

where ϵ_0 is the permittivity of the free surface.

To carry this discussion forward often the surface integrals in Eqs. (A.2.1) and (A.2.2) are difficult to perform. The surface through which the flux passes through, needs to be parameterized and $\hat{\mathbf{n}}d\Gamma$ must be properly calculated. The unit normal to a surface at any arbitrary point ($\hat{\mathbf{n}}$) can be calculated by taking the cross product of the respective tangential vectors toward the change of the parameters that defines the curvature of the surface. For example, a spherical surface discussed in Chapter 3 is defined by its radius, in plane horizontal angle and the azimuthal angle. Here, the detailed calculation of the unit normal is not discussed and can be found elsewhere. Gauss-divergence theorem in fact omits this requirement of the calculation of the unit normal for specific surfaces. Gauss-divergence theorem is stated as follows. The total flux of a vector field passes through a surface supposedly calculated through a surface integral can be simple calculated by the volume integral of the divergence of the vector field over the entire volume enclosed by the surface. Mathematically, Eqs. (A.2.1) and (A.2.2) can then be written using divergence theorem as follows

$$\oiint \rho \mathbf{V}.\hat{\mathbf{n}}\, d\Gamma = \int_{\Omega} \nabla.\mathbf{V}\; d\Omega \tag{A.2.3}$$

Or

$$\epsilon_0 \oiint \mathbf{E}.\hat{\mathbf{n}}\, d\Gamma = \int_{\Omega} \nabla.\mathbf{E}\; d\Omega \tag{A.2.4}$$

where $\nabla.\mathbf{V}$ is the divergence of a vector field ($\mathbf{V}$) and is true for any vector fields. Divergence of a vector field is explained in Section 2.3 in detail.

2.8.2 Stokes Theorem

In any arbitrary vector field discussed in Section 2.3, an imaginary surface (e.g., spherical, cuboidal, ellipsoidal, cylindrical, etc.) could be assumed to be cut by a plane as shown in Fig. A.2.2. Sometimes it is imperative in science and engineering to find how much flux is flowing along the boundary of an arbitrary surface if the surface is cut by a plane. The contour integral of a vector field is called the contour integral along a boundary of an object. The surface is shown in Fig. A.2.2 is an arbitrary surface and an arbitrary velocity field $\mathbf{V}(x_j) = x_1x_2\hat{\mathbf{e}}_1 + x_2x_3\hat{\mathbf{e}}_2 + x_3x_1\hat{\mathbf{e}}_3$. The vector field also passes through an incremental contour element $d\vec{\mathbf{c}}$ taken along the cut boundary of the surface Γ. To find the total field along the incremental contour one should take a dot product between the vector field and the direction normal to that incremental contour vector as shown in Fig. A.2.2. Once the dot product is taken

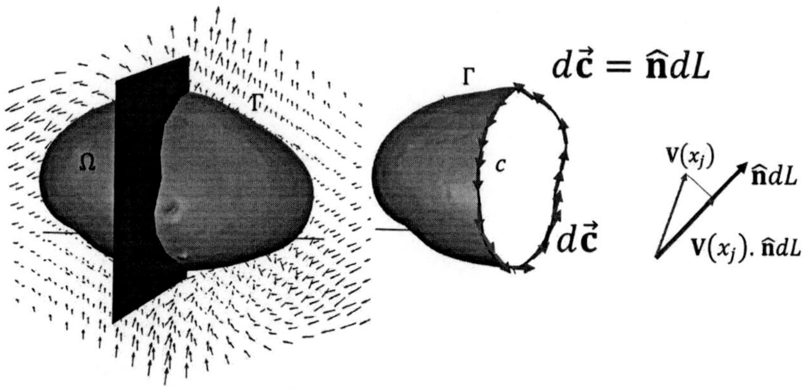

FIGURE A.2.2 Schematic representation of a vector field passing through a surface, projection of the vector field on the tangent of a curve which is a contour enclosing the surface.

which signifies the component of the vector field along the normal to the contour vector, integral of the product over the entire contour line will give the total flux flowing along the boundary if the surface Γ when cut by a plane. Mathematically, if the velocity field is known then the total flow can be written as

$$F = \oint \mathbf{V}.d\vec{\mathbf{c}} = \oint \mathbf{V}.\hat{\mathbf{n}}\, dL \tag{A.2.5}$$

where $\hat{\mathbf{n}}$ is the unit normal vector pointing along the tangent to the contour of the surface. Further, Stokes theorem says that the vector field along a contour line enclosing a boundary of a surface can be found by taking curl of the vector field over the entire surface. Hence, mathematically Stokes theorem can be written as

$$\oint \mathbf{V}.\hat{\mathbf{n}}\, dL = \int_{\Gamma} (\nabla \times \mathbf{V}).d\vec{\Gamma} \tag{A.2.6}$$

REFERENCES

1. Shercliff, J.A., *Vector Fields: Vector Analysis Developed Through its Applications to Engineering and Physics*. 1977, Cambridge: Cambridge University Press.
2. Itskov, M., *Tensor Algebra and Tensor Analysis for Engineers*. 2009, Berlin, Heidelberg: Springer.

3 Mechanics of Continua

3.1 COORDINATE SYSTEM

Conventionally, discussion on continuum mechanics starts with the concept of stress and strain. Here, however, instead of starting our discussion from stress and strain, we will start our introduction to continuum mechanics [1–5] with the description of the coordinate systems. Appropriate use of the state variables in appropriate coordinate systems is the most important concept required in continuum mechanics. From Chapter 2, we understood, how an object or a tensor can be described in a particular coordinate system, and how are they transformed to an alternate coordinate system, if the relation between the original and the target coordinate system is known. It is timely to mention that the "relation" mentioned above is inherently between the bases of the two coordinate systems (refer Eq. (2.8)). In other words, the "relation" is in between the contravariant vectors in the original and the target coordinate systems (refer Eq. (2.48)) discussed in the previous chapter. In this chapter we will restrict our discussion to a three-dimensional coordinate system.

Let's assume a continuous object consists of numerous material particles interconnected by imaginary vectors shown in Fig. 3.1. A sample material point (P_1) is assumed to occupy a region in space which consists of the material volume ϑ_1. Let's assume the vector pointing to the material point P_1 can be written as $\mathbf{P_1} = X_{11}\widehat{\mathbf{E_1}} + X_{21}\widehat{\mathbf{E_2}} + X_{31}\widehat{\mathbf{E_3}}$. Similarly another material point (P_2) is assumed to occupy a region in space which consists of the material volume ϑ_2. Let's assume the vector pointing to the material point P_2 can be written as $\mathbf{P_2} = X_{12}\widehat{\mathbf{E_1}} + X_{22}\widehat{\mathbf{E_2}} + X_{32}\widehat{\mathbf{E_3}}$. If the material body in Fig. 3.1 has N numbers of material points, the system will have N numbers of vectors that point to all the material points. We can call any one of them by $\mathbf{P}_N$. Please recollect the following lines from Chapter 2: "In mathematics a vector is defined as a quantity that has both magnitude and direction. To describe a vector, in most simple words, we can say that a vector is used to describe the position of a point in a mathematically viable space relative to another point in the same space. If the vectors pointed to the objects in the space are static, i.e., nonmoving objects, the vectors are called the static vectors." But, the $\mathbf{P}_N$ here is not a static vector because the material that occupies the space is subjected to deform and the material points should move under the external forces. The external boundary forces and the internal stresses in a material body are generated due to pure mechanical, coupled with thermomechanical, electromechanical, mechanochemical, or biomechanical actions, which will be discussed later. Thus an arbitrary point P_N can move in space relative to the other points in the material space and, as a result, the components of the vector $\mathbf{P}_N$ change. Sometimes, these motions are small and both the deformed and the undeformed states can be referred by the same parent coordinate system fixed at the origin O (Fig. 3.1). In other words, it's like watching the event of

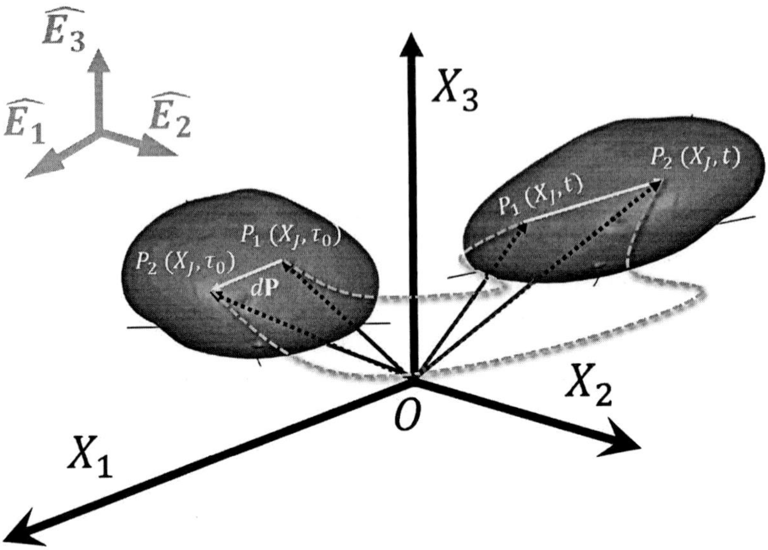

FIGURE 3.1 Deformation of a deformable body: deformed position of two particles and a vector connecting them in Cartesian coordinate system is compared with their initial position and the vector before deformation.

the motion standing at a specific point in space. It means that the coordinate system, its axes, its origin, and the bases are static and are frozen in space and time. This also means that the observer (or the coordinate system) is fixed in space and time. Such frozen coordinate systems are described individually in Fig. 2.1 in Chapter 2 and in Fig. 3.1. However, another situation also exists.

When the motion is large, what if the observer is moving during the motion to track the particles? Like bowl B was moved from the table to the ground in Chapter 2. This situation will arise if the coordinate system $\hat{\mathbf{e}}$ in Fig. 2.1a in Chapter 2 will follow bowl B and move with the bowl. It is assumed that when the motion is complete, the observer comes to its final stop. Then it is obvious that the observer sees the final state of the material from its current location. But, now to describe the current state, can the observer refer the current position of the points with respect to the original state from where both the object and the observer started in the first place? That means, in that description, we have to consider both the deformation of the body and the motion of the observer. Hence, the word "both" here carries a significant mathematical meaning that many tend to confuse. In such description, an additional velocity term appears which resembles the motion of the coordinate system itself. The first case where the coordinate system or the observer is static is called the *Lagrangian coordinate* or the *material coordinate system*. The second case where the observer is moving with the motion is called the *Eulerian coordinate* or the *spatial coordinate system*. In the first case, we will designate the basis vectors of a material coordinate system with a capital $\hat{\mathbf{E}}$ and the points in space with capital X_1, X_2, and X_3 or simply the X_j. In the second case, we will designate the basis vectors in a spatial coordinate system with a small $\hat{\mathbf{e}}$ and the points in space with small x_1, x_2, and x_3 or simply the x_j. In following subsections, we will mathematically discuss these systems in detail.

3.1.1 Lagrangian Coordinate or Material Coordinate System

Let's assume a material body at time $t = \tau_0$ in Fig. 3.1. The material consists of material points and occupies a region in space. If the material body has N numbers of material points, the position vectors of N material particles can be described by their respective position using X_1, X_2, and X_3 variables in a fixed $\hat{\mathbf{E}}$ coordinate system, which has a fixed origin at O. These N position vectors can be written as

$$\mathbf{X}_N(X_J, t) = X_{IN}(t)\widehat{\mathbf{E_I}} \tag{3.1}$$

where I and J are the capital letter index used for Lagrangian or material description, takes values, 1, 2, and 3 similar to the indices described in the previous chapter with lower cases. Here, I is the dummy index, and N is the free index. Next, extending the idea, an infinitesimal vector $d\mathbf{P}$ (see Fig. 3.1) in the material body can be similarly expressed as $d\mathbf{P} = dX_I\widehat{\mathbf{E_I}}$.

Recalling Eq. (2.44), we can write the infinitesimal length of an element in the material body described in Lagrangian coordinate system [4] as

$$dL^2 = d\mathbf{P}.d\mathbf{P} = dX_I\widehat{\mathbf{E_I}}.dX_J\widehat{\mathbf{E_J}} = dX_I.dX_J\widehat{\mathbf{E_I}}.\widehat{\mathbf{E_J}} = dX_I.dX_I \tag{3.2}$$

where, recollecting Eq. (2.13) from Chapter 2, we can say $\widehat{\mathbf{E_I}}.\widehat{\mathbf{E_J}} = \delta_{IJ}$.

Similarly, any tensor in a material coordinate system will be a function of the variables, X_1, X_2, and X_3 and time t. For example, stress – a second-order tensor at any point in the material body that has ($2^3 = 9$) nine elements in a three-dimensional coordinate system is a function of X_1, X_2, and X_3 and time t and can be expressed as $\sigma_{IJ}(X_j, t)$.

3.1.2 Eulerian Coordinate or Spatial Coordinate System

Starting at time $t = \tau_0$ a material body has deformed and moved to a new position after $t = t$. The material particles can now be described by a new coordinate system $\hat{\mathbf{e}}$ as shown in Fig. 3.2. The spatial coordinates of the particles can be described by x_1, x_2, and x_3 or simply the x_j. It is obvious that the new coordinates of any particle in the new coordinate system $\hat{\mathbf{e}}$ are inherently related to the original position of the particle in the old coordinate system $\hat{\mathbf{E}}$. Hence, we can say

$$x_k = x_k(X_J, t) \tag{3.3}$$

However, it is important to know the relation between the $\hat{\mathbf{E}}$ and $\hat{\mathbf{e}}$ coordinate system at all the instances during the motion. Please note that $\hat{\mathbf{E}}$ and $\hat{\mathbf{e}}$ vectors are the contravariant vectors and their relation gives us the Jacobian matrix (Eq. (2.48)).

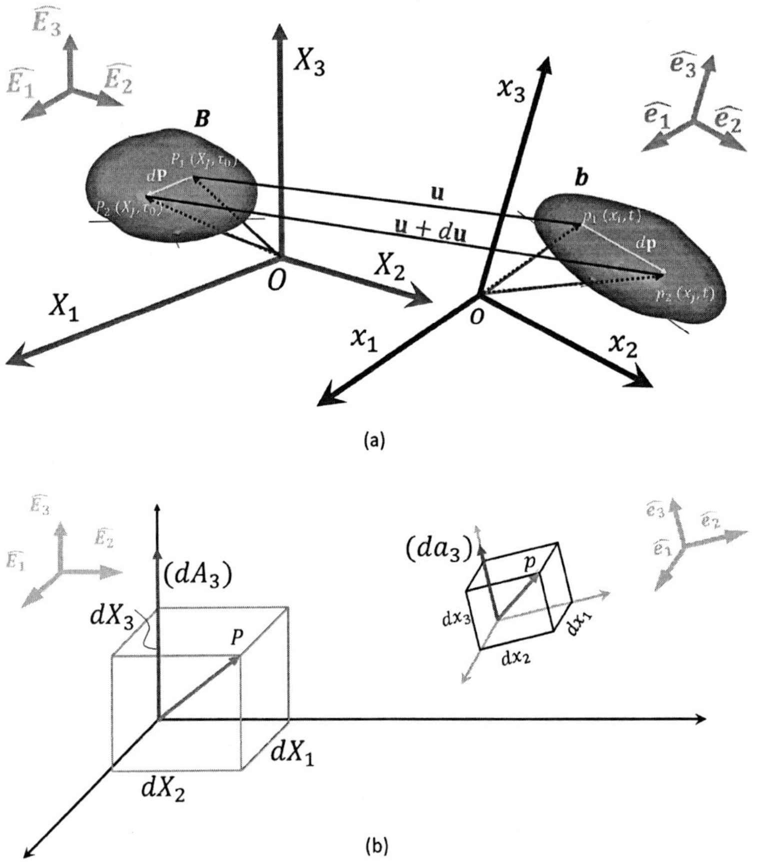

FIGURE 3.2 Schematic presentation of the original state of the material in original coordinate system and deformed state in a new coordinate system that moves with the deformable body. (a) Each point in the body is traced before and after deformation and shows the displacement of the respective points. (b) Change in length of a vector in undeformed and deformed state with respect to their local coordinate system.

The Jacobian matrix that establishes the relation between the bases of $\hat{\mathbf{E}}$ and $\hat{\mathbf{e}}$ vectors transforming $\hat{\mathbf{E}}$ to $\hat{\mathbf{e}}$ can be written as

$$\mathbf{J} = \text{Jacobian} = \left|\frac{\partial x_i}{\partial X_I}\right| = \begin{vmatrix} \dfrac{\partial x_1}{\partial X_1} & \dfrac{\partial x_1}{\partial X_2} & \dfrac{\partial x_1}{\partial X_3} \\ \dfrac{\partial x_2}{\partial X_1} & \dfrac{\partial x_2}{\partial X_2} & \dfrac{\partial x_2}{\partial X_3} \\ \dfrac{\partial x_3}{\partial X_1} & \dfrac{\partial x_3}{\partial X_2} & \dfrac{\partial x_3}{\partial X_3} \end{vmatrix} \tag{3.4}$$

which is similar to Eq. (2.48) mentioned in Chapter 2.

Let's now consider Fig. 3.2 in detail. The material points P_1 and P_2 in the body ***B*** is located at points p_1 and p_2 in the body ***b***, respectively after the deformation. The body ***B*** was located in $\hat{\mathbf{E}}$ coordinate system, and the body ***b*** is now located in the new $\hat{\mathbf{e}}$ coordinate system. In this forward motion, we would need the definition of the points written in Eq. (3.3). However, there is a possibility of the inverse motion in space, and the points p_1 and p_2 in the body ***b*** at time t can be traced back to the points P_1 and P_2 in the body ***B***. Hence, it is feasible that the original coordinates of the points in $\hat{\mathbf{E}}$, can be represented by the new coordinate system $\hat{\mathbf{e}}$ as follows:

$$X_K = X_K\left(x_j, t\right) \tag{3.5}$$

With this definition, we must state our assumption about a continuous body (***B*** or ***b***) which is fundamental to the continuum mechanics called *axiom of continuity*. It is assumed that the equations above for the particle mapping are single valued, but they must have the continuous partial derivatives with respect to the arguments. There are total six equations in Eqs. (3.3) and (3.5) combined, and it is assumed that each member of these equation set is the unique inverse of the other in a neighborhood of any material point like P_1 or P_2. This axiom of continuity restricts any region of positive volume in the material body to collapse to a zero volume or inflate to an infinite volume during the deformation. Because matter is *indestructible*. It also assumed that the matter is *impenetrable* and ensures that an infinitesimal material volume will never penetrate the other neighboring material volume. Hence, we can say that the material particles are infinitesimal deformable volumes that follow the above axiom to make a continuous material body.

Recalling Eq. (2.44), we can write the infinitesimal length of the element between the points p_1 and p_2 in the body ***b*** at time t described in Eulerian coordinate system as

$$dl^2 = d\mathbf{p}.d\mathbf{p} = dx_i\widehat{\mathbf{e}_\mathbf{i}}.dx_j\widehat{\mathbf{e}_\mathbf{j}} = dx_i.dx_j\widehat{e_i}.\widehat{e_j} = dx_i.dx_i, \text{ where } \widehat{\mathbf{e}_\mathbf{i}}.\widehat{e_j} = \delta_{ij}.$$

With this understating, next we will discuss the deformation of the material body in brief.

3.2 MOTION OF A DEFORMABLE BODY

Motion of a body talks about the global kinetics and kinematics of a continuous material body. Motion of a continuous media can be seen as a movie of the deformation of the media over time. In order to describe the motion of a body, we must know the following definitions of the material derivatives, path lines and stream lines. Path lines and stream lines frequently occur in fluid mechanics but carry some significance in wave propagation and thus discussed in this chapter.

3.2.1 Material Derivatives

In this section we will discuss the concepts of material derivatives. Material derivatives are only considered with respect to time. The material time rate of change of any vector or tensor is our concern herein. Let's assume an arbitrary tensor **T**. With our discussion in Chapter 2, we know that the tensors could be of rank one, which results a vector, like a displacement vector, and of rank two, which results a matrix like a stress matrix. Now to find the material time rate of change of **T**, when the material axis (i.e., the $\hat{\mathbf{E}}$ coordinate system) is fixed for the original material points (keeping **X** constant), we can write the expression for a vector as

$$\frac{d\mathbf{T}}{dt} = \left.\frac{\partial \mathbf{T}}{\partial t}\right|_{\mathbf{X}} \; ; \; \mathbf{T} = \mathbf{T}(\mathbf{X},\mathrm{t}) = \mathrm{T}_K(X_J,t)\hat{\mathbf{E}}_K \; ; \; \frac{d\mathbf{T}}{dt} = \frac{\partial T_K}{\partial t}\hat{\mathbf{E}}_K \tag{3.6}$$

where $\hat{\mathbf{E}}_K$ are the constants unit vectors and the tensor is defined in *material coordinate* or *Lagrangian coordinate system* as discussed in Section 3.1.1. Next if the vector is a *spatial function* described in Section 3.1.2, or defined in *Eulerian coordinate system*, then the material derivatives will take a different form and can be written as

$$\mathbf{T} = \mathbf{T}(x_j,t) = \mathbf{T}_k(x_j,t)\widehat{\mathbf{e}_\mathbf{k}} \; ; \; \frac{d\mathbf{T}}{dt} = \left(\left.\frac{\partial T_k}{\partial t}\right|_{\mathbf{x}} + \frac{\partial T_k}{\partial x_j}\frac{\partial x_j}{\partial t}\right)\hat{\mathbf{e}}_k \tag{3.7}$$

Please recollect that we are using, capital indices for *Lagrangian* and small indices for *Eulerian* coordinate systems. Here, the tensor is the function of spatial coordinate system **x** with unit vectors $\hat{\mathbf{e}}$, and the spatial coordinates **x** are function of material coordinates **X** with unit vector $\hat{\mathbf{E}}$ (refer Eq. (3.3)). Hence, we can write the material derivatives in a general form as follows:

$$\frac{d\mathbf{T}}{dt} = \dot{\mathbf{T}} = \frac{DT_k}{Dt}\hat{\mathbf{e}}_k \; ; \; \text{where, } \frac{DT_k}{Dt} = \frac{\partial T_k}{\partial t} + T_{k,j}\frac{\partial x_j}{\partial t} \tag{3.8}$$

where the generalized material derivative $\frac{DT_k}{Dt}$ is a summation of nonstationarly rate and the convective time rate, respectively. If the tensor is defined in Lagrangian system, the convective time rate would be zero, else if the tensor is defined in Eulerian coordinate system, both the terms will survive. With this understanding, we can write the expression for acceleration (a_k) as rate of change of velocity (v_k) in the Eulerian coordinate system as follows:

$$\frac{Dv_k}{Dt} = a_k(x,t) = \frac{\partial v_k}{\partial t} + v_{k,l}v_l \tag{3.9}$$

With this similar understanding, we present the material derivatives of few common terms frequently used in continuum mechanics and we will be using in this chapter. Detailed derivations are skipped herein but it can be found in Ref. [2].

3.2.1.1 Material Derivative of Displacement Gradient

Using Eqs. (3.3) and (3.5), their derivatives can be described by $dX_K = X_{K,k}dx_k$ and $dx_k = x_{k,K}dX_K$, where $X_{K,k}$ and $x_{k,K}$ are the displacement gradients in Lagrangian and Eulerian coordinate systems, respectively. In the following equations the material derivatives of the displacement gradients are derived. Please note that here onwards, most of the mathematical steps are shown in calculations but are not explicitly explained, because a prior knowledge of the readers on index notation is expected.

$$\frac{D}{Dt}(dx_k)=\frac{D}{Dt}\left(\frac{\partial x_k}{\partial X_K}dX_K\right)=\frac{\partial}{\partial X_K}\left(\frac{Dx_k}{Dt}\right)dX_K=\frac{\partial v_k}{\partial X_K}dX_K=\frac{\partial v_k}{\partial x_l}\frac{\partial x_l}{\partial X_K}dX_K \tag{3.10}$$

So we can write the material derivative of the differential displacement as

$$\frac{D}{Dt}(dx_k)=v_{k,l}dx_l \tag{3.11}$$

And so, using Eqs. (3.10) and (3.11), the material derivative of the displacement gradient in Eulerian coordinate system can be written as

$$\frac{D}{Dt}(x_{k,K})=v_{k,l}x_{l,K} \tag{3.12}$$

And in Lagrangian system can be written as

$$\frac{D}{Dt}(X_{K,k})=-v_{m,k}\frac{\partial X_K}{\partial x_m} \tag{3.13}$$

3.2.1.2 Material Derivative of Jacobian

The Jacobian is mentioned in Eq. (3.4) and can be expressed as a determinant of the matrix $x_{i,I}$, i.e., $|x_{i,I}|$. The material derivative of the Jacobian using Eq. (3.12) is derived herein as follows:

$$\frac{D}{Dt}(J)=\frac{D}{Dt}\left(|x_{i,I}|\right)=\frac{\partial J}{\partial x_{k,K}}\frac{D}{Dt}(x_{k,K})=(\text{cofactor } x_{k,K})\; v_{k,l}x_{l,K} \tag{3.14}$$

after further simplification and employing the Kronecker delta function, we can write

$$\frac{D}{Dt}(J)=JX_{k,K}v_{k,l}x_{l,K}=Jv_{k,k} \tag{3.15}$$

Please note that the velocity term is always present in the material derivatives, which says that the velocity of the reference frame moving with the particle motion is an important factor in continuum mechanics. Using the material derivatives of the Jacobian, the material derivative of the differential velocity can be found very easily as follows:

$$\frac{D}{Dt}(dv)=\frac{D}{Dt}(JdV)=v_{k,k}JdV=v_{k,k}dv \tag{3.16}$$

3.2.1.3 Material Derivative of Square of an Arc Length

As discussed in Section 3.1, the square of an arc length of a line segment in Lagrangian and Eulerian system can be expressed as $dl^2 = dx_k dx_k$ and $dL^2 = dX_K dX_K$, respectively. Now the material derivative of the square of the arc length in *Lagrangian system* can be derived as follows:

$$\frac{D}{Dt}\left(dl^2\right) = \frac{D}{Dt}\left(dx_k dx_k\right) = dx_k \frac{D\left(dx_k\right)}{Dt} + dx_k \frac{D\left(dx_k\right)}{Dt} \tag{3.17}$$

Utilizing Eq. (3.11), we can further write

$$\frac{D}{Dt}\left(dl^2\right) = dx_k v_{k,l} dx_l + v_{k,l} dx_l dx_k = 2\ v_{k,l} dx_k dx_l$$

or

$$\frac{D}{Dt}\left(dl^2\right) = \left(v_{k,l} + v_{l,k}\right) dx_k dx_l \tag{3.18}$$

This is because Eq. (3.17) is symmetric with indices k and l, which takes the values like all the indices used in this book (Chapters 1–3). Next the material derivative of the square of an arc length of a line segment in *Eulerian system* can be derived using the expression $dx_k = x_{k,K} dX_K$ as follows

$$\begin{aligned}\frac{D}{Dt}\left(dL^2\right) &= dX_K \frac{D\left(dX_K\right)}{Dt} + dX_K \frac{D\left(dX_K\right)}{Dt} = \left(v_{k,l} + v_{l,k}\right) x_{k,K} x_{l,L} dX_K dX_L \\ &= \dot{\mathbb{E}}_{KL} dX_K dX_L\end{aligned} \tag{3.19}$$

where $\dot{\mathbb{E}}_{KL}$ is the Lagrangian strain rate tensor that will be discussed later when we will discuss strains in deformable bodies.

3.2.1.4 Material Derivative of Element of an Area

With the above expressions, now we will take some big leaps. Instead of writing the step-by-step derivations, we will just write the material derivatives of few common parameters. One can easily derive these expressions using the processes presented above in the previous sections. Material derivative of an elemental area (please also refer Section 3.3) can be found as

$$\frac{D}{Dt}\left(da_m\right) = v_{i,i} da_m - v_{i,m} da_i \tag{3.20}$$

3.2.1.5 Material Derivatives of Line (ℓ) and Surface ($\mathfrak{s}$) Integral of a Scalar Field φ

Material derivative of the line integral of a scalar field φ over a material line ℓ can be written as follows:

$$\frac{D}{Dt}\int_{\ell} \varphi\, dx_k = \int_{\ell} (\dot{\varphi}\, dx_k + \varphi\ v_{k,l}\ dx_l) \tag{3.21}$$

However, if the line is fixed in the material body, the second part in the expression in Eq. (3.21) will disappear. Next, the material derivative of the surface integral of any scalar field φ over a material surface $\mathcal{S}$ can be found as

$$\frac{D}{Dt}\int_{\mathcal{S}} \varphi\, da_m = \int_{\mathcal{S}} \dot{\varphi}\, da_m + \varphi\left(-v_{i,m}\ da_i + v_{i,i}\ da_m\right) \tag{3.22}$$

3.2.1.6 Material Derivatives of Surface ($\mathcal{S}$) Integral of a Vector Field

Assuming an arbitrary vector field $q_k\left(x_j, t\right)$ in the material body, the material derivative of $\mathbf{q}$ can be expressed as

$$\frac{D}{Dt}\int_{\mathcal{S}} q_k\ da_k = \int \left(\dot{q}_k - q_l v_{l,k} + q_k v_{l,l}\right) da_k \tag{3.23}$$

or alternatively

$$\frac{D}{Dt}\int_{\mathcal{S}} \mathbf{q}.d\mathbf{a} = \int_{\mathcal{S}} d\mathbf{a}.\left(\frac{\partial \mathbf{q}}{\partial t} + \nabla\times\left(\mathbf{q}\times\mathbf{v}\right) + \mathbf{v}.\nabla\mathbf{q}\right) \tag{3.24}$$

where $\nabla\times\left(\mathbf{q}\times\mathbf{v}\right)$ is the curl of the vector field obtained from $\left(\mathbf{q}\times\mathbf{v}\right)$, which is a cross product of the vector fields, $\mathbf{q}$ and $\mathbf{v}$, $\nabla\mathbf{q}$ is the gradient of the vector field, and $\mathbf{v}.\nabla\mathbf{q}$ is the dot product of the vector fields, $\mathbf{v}$ and $\nabla\mathbf{q}$. Please refer Chapter 2 for the concepts of dot and cross products of the vector fields.

3.2.2 Path Lines and Stream Lines

Here it is apparent that path lines and stream lines in a moving body are essentially related to the velocity filed of the material particles as they appear in the material derivatives. Now imagine all the material particles that are in motion in a continuous body are to be traced (each particle) with reference to a fixed coordinate origin. It's like following a bunch of racing cars after they started from the start line, with respect to the start line itself, which is known as $\mathbf{X}$. As the material particles are followed, the local coordinate changes continuously in Eulerian coordinate system. The trace of these particles originally located at $\mathbf{X}$ as time passes is called the *path line*. We can write the equation of a path line as follows

$$x_k = x_k\left(X_J, t\right) \text{ when } \mathbf{X} \text{ is fixed} \tag{3.25}$$

The path lines can also be achieved as an integral curve of $dx_k = v_k dt$ that passes through $\mathbf{X}$ at $t = 0$ [4]. That is joining the dots along which the particles have moved visualized through a long exposure time.

Alternatively, stream lines are the curve that represent the tangent to the velocity filed of the material particles at an instance of time. Stream lines could be achieved

FIGURE 3.3 Difference between Lagrangian representation and Eulerian representation of a waterfall.

from the integral curves to $v_k = Cdx_k$, where C is any arbitrary constant. This also means that along a stream line the ratio of the incremental distance traveled by the material particle at any instance along any direction to the velocity of the particle along that same direction is always constant and can be presented as

$$\frac{dx_1}{v_1} = \frac{dx_2}{v_2} = \frac{dx_3}{v_3} = \frac{1}{C} \tag{3.26}$$

The mathematical treatment above can be explained using the following example. Imagine a photograph of a waterfall. If we take the photograph of a beautiful waterfall using our smartphone, we will capture only a small instance of the flow of the water. This picture from the smartphone shows the location of the water particles at that instance when the capture button was pressed. If we join the lines with similar velocity, we will get the *stream line*. However, if a professional photographer takes the picture of the waterfall with his expensive lens with a long exposure time, we will see a beautiful picture of the waterfall. In this picture we see the trace of the particles over a time and see the white *path lines* that are nicely aligned. So the smartphone picture is a Lagrangian device, but the expensive camera with long exposure time is an Eulerian device, just to understand the difference. Fig. 3.3 shows the difference between these two pictures.

3.3 DEFORMATION AND STRAIN IN A DEFORMABLE BODY

In this section we will refer to Figs. 3.2, 3.4a, and 3.4b to briefly understand the deformation of a deformable body. It is mentioned in the previous sections that any arbitrary points P_1 and P_2 in the original body has moved to the new points p_1 and p_2, respectively in the body $\boldsymbol{b}$ after deformation. The point P_1 is displaced by $\mathbf{u}$, where $\mathbf{u}$ is the displacement vector and has three components. Similarly, the point P_2 at a $d\mathbf{P}$ distance from P_1 is displaced by $\mathbf{u}+d\mathbf{u}$. The new location of the points p_1 and p_2 is at $d\mathbf{p}$ distance. With this basic understanding, next we will present the deformation of an arc length, deformation of an elemental area and volume to link them to the deformation and strain tensors in Lagrangian and Eulerian coordinate systems, respectively. Please refer Fig. 3.4 which is a more elaborative description

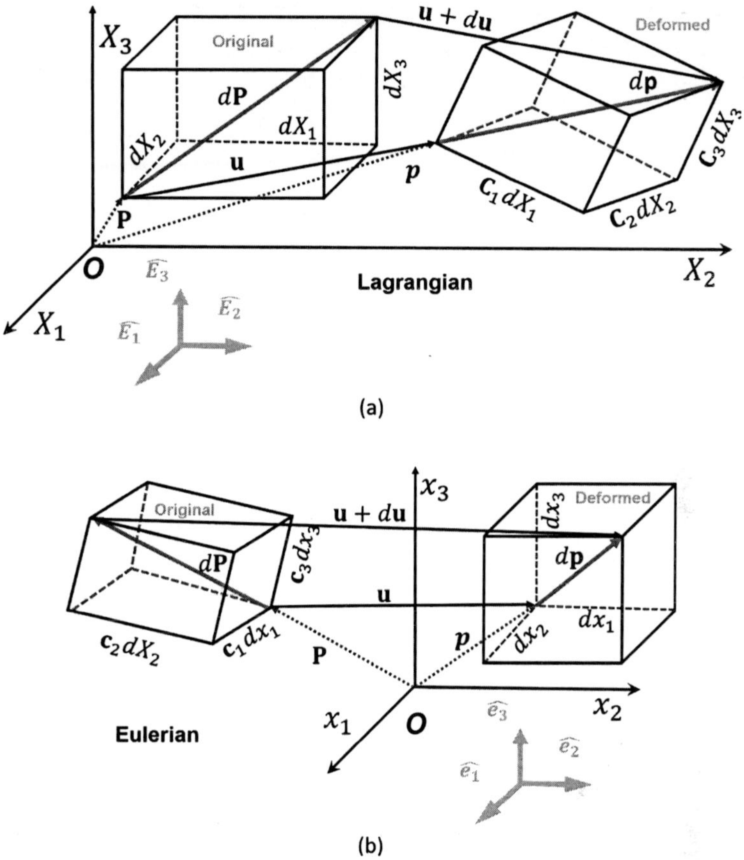

FIGURE 3.4 (a) Lagrangian coordinate system showing how the deformation is measured in reference coordinate system. (b) Eulerian coordinate system showing how the deformation is measured in deformed coordinate system.

of the material element inside the body. In Lagrangian system (Fig. 3.4a) the deformation is observed with respect to a fixed coordinate system from where the deformation was started. However, in Eulerian system (Fig. 3.4b) the material points are described with respect to its current position and are traced back to the original state from where the deformation was started. Please note that in Fig. 3.4 the $d\mathbf{P}$ vector is deformed to a $d\mathbf{p}$ vector, with the displacement of two opposite corners of the parallelepiped being $\mathbf{u}$ and $\mathbf{u}+d\mathbf{u}$, respectively. Original components of the $d\mathbf{P}$ vector before deformation were dX_K. However, after deformation, every sides of the parallelepiped are changed due to the deformation and can be expressed by multiplying the deformation vectors to the respective components. In the following section the original and deformed states are mathematically explained in both the Lagrangian and Eulerian systems in a tabular form (Table 3.1). For understanding the left column, please refer Fig. 3.4a and while understanding the right column, please refer Fig. 3.4b.

TABLE 3.1
Cauchy's and Green's Deformation Tensors [2]

	Lagrangian *Description of the Deformed State*	**Eulerian** *Description of the Original State*
T1.1	$d\mathbf{p} = dx_k\hat{\mathbf{e}}_k$	$d\mathbf{P} = dX_K\hat{\mathbf{E}}_K$
T1.2	$dl^2 = d\mathbf{p}.d\mathbf{p}$ $= dx_k dx_l\hat{\mathbf{e}}_k.\,\hat{\mathbf{e}}_l$ $= \delta_{kl}dx_k dx_l$ $dl^2 = dx_k dx_k$	$dL^2 = d\mathbf{P}.d\mathbf{P}$ $= dX_K dX_L\hat{\mathbf{E}}_K.\hat{\mathbf{E}}_L$ $= \delta_{KL}dX_K dX_L$ $dL^2 = dX_K dX_K$
T1.3	$dx_k = \frac{\partial x_k}{\partial X_K}dX_K$	$dX_K = \frac{\partial X_K}{\partial x_k}dx_k$
T1.4	$dl^2 = \frac{\partial x_k}{\partial X_K}\frac{\partial x_k}{\partial X_K}dX_K dX_K$	$dL^2 = \frac{\partial X_K}{\partial x_k}\frac{\partial X_K}{\partial x_k}dx_k dx_k$
T1.5	$d\mathbf{p} = \frac{\partial x_k}{\partial X_K}dX_K\hat{\mathbf{e}}_k$ $= \left(\frac{\partial x_k}{\partial X_K}\hat{\mathbf{e}}_k\right)dX_K$ $d\mathbf{p} = \mathbf{C}_K dX_K$	$d\mathbf{P} = \frac{\partial X_K}{\partial x_k}dx_k\hat{\mathbf{E}}_K$ $= \left(\frac{\partial X_K}{\partial x_k}\hat{\mathbf{E}}_K\right)dx_k$ $d\mathbf{P} = \mathbf{c}_k dx_k$
T1.6	$dl^2 = d\mathbf{p}.\,d\mathbf{p}$ $= \mathbf{C}_K dX_K.\mathbf{C}_L dX_L$ $= (\mathbf{C}_K.\mathbf{C}_L)\,dX_K dX_L$	$dL^2 = d\mathbf{P}.\,d\mathbf{P}$ $= \mathbf{c}_k dx_k.\mathbf{c}_l dx_l$ $= (\mathbf{c}_k.\mathbf{c}_l)dx_k dx_l$
T1.7	$\mathbf{C}_K\,.\,\mathbf{C}_L = \frac{\partial x_k}{\partial X_K}\cdot\frac{\partial x_l}{\partial X_L}\delta_{kl}$ $= \frac{\partial x_k}{\partial X_K}\cdot\frac{\partial x_k}{\partial X_L}$ $\mathbf{C}_K\,.\,\mathbf{C}_L = C_{KL}(\mathbf{X},t)$ $C_{KL}(\mathbf{X},t)$ is **Green's deformation tensor**	$\mathbf{c}_k\,.\,\mathbf{c}_l = \frac{\partial X_K}{\partial x_k}\cdot\frac{\partial X_L}{\partial x_l}\delta_{KL}$ $= \frac{\partial X_K}{\partial x_k}\cdot\frac{\partial X_K}{\partial x_l}$ $\mathbf{c}_k\,.\,\mathbf{c}_l = c_{kl}(\mathbf{x},t)$ $c_{kl}(\mathbf{x},t)$ is **Cauchy's deformation tensor**
T1.8	Inverse $\mathbf{C}_K^{-1} = \frac{\partial X_K}{\partial x_k}\hat{\mathbf{e}}_K$ $\mathbf{C}_K^{-1}.\,\mathbf{C}_L^{-1} = \frac{\partial X_K}{\partial x_k}\cdot\frac{\partial X_L}{\partial x_l}\delta_{kl}$ $C_{KL}^{-1} = B_{KL}(\mathbf{X},t)$ $C_{KL}^{-1}C_{LM} = \delta_{KM}$ *$C_{KL}^{-1} = B_{KL}$ is Piola deformation tensor* **Right Cauchy-Green tensor**	Inverse $\mathbf{c}_k^{-1} = \frac{\partial x_k}{\partial X_K}\hat{\mathbf{E}}_K$ $\mathbf{c}_k^{-1}.\,\mathbf{c}_l^{-1} = b_{kl}(\mathbf{x},t)$ $c_{kl}^{-1} = \frac{\partial x_k}{\partial X_K}\cdot\frac{\partial x_l}{\partial X_L}\delta_{KL}$ $c_{kl}^{-1}\,c_{lm} = \delta_{km}$ *$c_{kl}^{-1} = b_{kl}$ is Finger deformation tensor* **Left Cauchy-Green tensor**

Please note that the bold parameters are the vectors and the tensors are expressed with their indices.

3.3.1 Cauchy's and Green's Deformation Tensor

Please refer the definition of an arc length of an element line in a deformable body described above. Table 3.1 presents the step-by-step derivation of the deformation tensors in both the coordinate systems.

3.3.2 Description of Strain in a Deformable body

Please refer Fig. 3.2 where the new square of the arc length of the element between the two material points p_1 and p_2 is dl^2. This new element length has been modified from the original length dL^2 between the two original material points P_1 and P_2. Hence the change of length can be measured in two coordinate systems, as shown in Table 3.2.

3.3.3 Strain in terms of Displacement

Please refer Figs. 3.2 and 3.4 where the displacements of the particles P_1 and P_2 from its original state to the new state located at p_1 and p_2 are $\mathbf{u}$ and $\mathbf{u} + d\mathbf{u}$, respectively. Based on the discussions in Chapter 2, it is necessary to visualize that both the displacement vectors have components in $\hat{\mathbf{E}}$ and $\hat{\mathbf{e}}$ coordinate systems. Again, using the rule of finding vector resultant (not discussed in this book and can be found in [2]), we can say

$$d\mathbf{P} = d\mathbf{p} - d\mathbf{u} \quad \text{or} \quad d\mathbf{p} = d\mathbf{P} + d\mathbf{u} \tag{3.27}$$

TABLE 3.2
Lagrangian and Eulerian Strain Tensor and Strain Rate [1–3]

	Lagrangian (Fig. 3.4a)	**Eulerian (Fig. 3.4b)**
T2.1	Change in length $dl^2 - dL^2 = dx_k dx_k - dX_k dX_k$ $= C_{KL}(\mathbf{X},t) dX_k dX_L - dX_k dX_k$ $= dX_K dX_L \left(C_{KL}(\mathbf{X},t) - \delta_{KL}\right)$	Change in length $dl^2 - dL^2 = dx_k dx_k - dX_k dX_k$ $= dx_k dx_k - c_{kl}(\mathbf{x},t) dx_k dx_l$ $= dx_k dx_l \left(\delta_{kl} - c_{kl}(\mathbf{x},t)\right)$
T2.2	***Lagrangian Strain Tensor*** $2\mathbb{E}_{KL} = C_{KL}(\mathbf{X},t) - \delta_{KL}$ With respect to undeformed state (Fig. 3.4a)	***Eulerian Strain Tensor*** $2e_{kl} = \delta_{kl} - c_{kl}(\mathbf{x},t)$ With respect to deformed state (Fig. 3.4b)
T2.3	***Lagrangian to Eulerian Strain Transformation*** $\dot{e}_{kl} = \mathbb{E}_{KL} X_{K,k} X_{L,l}$	***Eulerian to Lagrangian Strain Transformation*** $\mathbb{E}_{KL} = e_{kl} x_{k,K} x_{l,L}$
T2.4	***Lagrangian Strain Rate Tensor*** $2\dot{\mathbb{E}}_{KL} = \dot{C}_{KL}(\mathbf{X},t)$ $\dot{C}_{KL}(\mathbf{X},t) = \frac{D}{Dt}\left(\frac{\partial x_k}{\partial X_K} \cdot \frac{\partial x_k}{\partial X_L}\right)$ using Eq. (3.12) we get $\dot{\mathbb{E}}_{KL} = \frac{1}{2}\left(v_{k,l} + v_{l,k}\right)\frac{\partial x_k}{\partial X_K} \cdot \frac{\partial x_l}{\partial X_L}$ Or $\dot{\mathbb{E}}_{KL} = d_{kl} x_{k,K} x_{l,L}$ Where, $d_{kl} = \frac{1}{2}\left(v_{k,l} + v_{l,k}\right)$ $\dot{C}_{KL}(\mathbf{X},t) = 2 d_{kl} x_{k,K} x_{l,L}$ Initial condition when the medium is not strained $\dot{C}_{KL}(\mathbf{X},0) = 2 d_{kl} \delta_{kK} \delta_{lL}$	***Eulerian Strain Rate Tensor*** $2\dot{e}_{kl} = -\dot{c}_{kl}(\mathbf{x},t)$ $\dot{c}_{kl}(\mathbf{x},t) = \frac{D}{Dt}\left(\frac{\partial X_K}{\partial x_k} \cdot \frac{\partial X_L}{\partial x_l}\right)$ using T2.3 and Eq. (3.13) we get $\dot{e}_{kl} = \frac{D}{Dt}\left(\mathbb{E}_{KL} X_{K,k} X_{L,l}\right)$ as $\dot{\mathbb{E}}_{KL} = d_{kl} x_{k,K} x_{l,L}$ $\dot{e}_{kl} = d_{kl} - e_{pk} v_{p,l} - e_{pl} v_{p,k}$ $\dot{c}_{kl}(\mathbf{x},t) = -c_{pk} v_{p,l} - c_{pl} v_{p,k}$ Initial condition when the medium is not strained $\dot{c}_{kl}(\mathbf{x},0) = -2 d_{kl}$

TABLE 3.3
Strain-Displacement Relation

	Lagrangian	Eulerian
T3.1	$\mathbf{C}_K = \frac{\partial \mathbf{p}}{\partial X_K} = \frac{\partial \mathbf{P}}{\partial X_K} + \frac{\partial \mathbf{u}}{\partial X_K}$ *Displacement vector* $d\mathbf{u} = dU_M \hat{\mathbf{E}}_M$ M is an arbitrary index takes values 1, 2, and 3	$\mathbf{c}_k = \frac{\partial \mathbf{P}}{\partial x_k} = \frac{\partial \mathbf{p}}{\partial x_k} - \frac{\partial \mathbf{u}}{\partial x_k}$ *Displacement vector* $d\mathbf{u} = du_m \hat{\mathbf{e}}_m$ m is an arbitrary index takes values 1, 2, and 3
T3.2	$\mathbf{C}_K = \hat{\mathbf{E}}_K + U_{M,K} \hat{\mathbf{E}}_M$	$\mathbf{c}_k = \hat{\mathbf{e}}_k - u_{m,k} \hat{\mathbf{e}}_m$
T3.3	**Green's deformation tensor** $C_{KL} = \mathbf{C}_K . \mathbf{C}_L$ $= \left(\hat{\mathbf{E}}_K + U_{M,K}\hat{\mathbf{E}}_M\right).\left(\hat{\mathbf{E}}_L + U_{N,L}\hat{\mathbf{E}}_N\right)$ $= \delta_{KL} + U_{L,K} + U_{K,L} + \delta_{NM} U_{M,K} U_{N,L}$	**Cauchy's deformation tensor** $c_{kl} = \mathbf{c}_k . \mathbf{c}_l$ $= \left(\hat{\mathbf{e}}_k - u_{m,k}\hat{\mathbf{e}}_m\right).\left(\hat{e}_l - u_{n,l}\hat{\mathbf{e}}_n\right)$ $= \delta_{kl} - u_{l,k} - u_{k,l} + \delta_{nm} u_{m,k} u_{n,l}$
T3.4	Substituting the expression for deformation tensor in the **Lagrangian Strain Tensor** expression in T2.2 we get $2\mathbb{E}_{KL} = U_{L,K} + U_{K,L} + \delta_{NM} U_{M,K} U_{N,L}$	Substituting the expression for deformation tensor in the **Eulerian Strain Tensor** expression in T2.2 we get $2\mathrm{e}_{kl} = u_{l,k} + u_{k,l} - \delta_{nm} u_{m,k} u_{n,l}$
T3.5	With the study of small deformation, the nonlinear terms can be approximated to zero, **Lagrangian Strain Tensor** $\mathbb{E}_{KL} = \frac{1}{2}\left(U_{K,L} + U_{L,K}\right)$	With the study of small deformation, the nonlinear terms can be approximated to zero, **Eulerian Strain Tensor** $\mathrm{e}_{kl} = \frac{1}{2}\left(u_{l,k} + u_{k,l}\right)$

Using the expression in T1.5, we can write the following equation that connects the deformation tensor and the displacement vectors (Table 3.3).

3.4 MASS, MOMENTUM, AND ENERGY

Here in this section also we will follow a nonstandard approach. Instead of discussing stresses in a material body, right after discussing strains, we will discuss few axioms of continuum mechanics which will be used in the stress hypothesis. In order to describe the axioms of continuum mechanics mathematically, it is necessary to define the following terms in a mathematical language.

3.4.1 Mass of a Body

Mass is considered continuous in a material body and expressed using the local density of the body $\rho(\mathbf{x})$ per unit volume. The total mass of the body can be expressed as

$$M = \int_{V} \rho(\mathbf{x})\, d\mathrm{v} \tag{3.28}$$

However, the total mass with discrete lumped masses distributed in a continuous body will give the following equation for the total mass as

$$M = \int_{V} \rho \, d\mathrm{v} + \sum_{\alpha} M_{\alpha} \tag{3.29}$$

where M_α is the lumped mass at the αth point in the body.

3.4.2 Momentum of a Deformable Body

Momentum of a discrete particle is defined as the product of the mass of the particle and the velocity of the particle in a discrete sense. However, in a continuous media, the density of the body can be different at different points. Similarly, the velocities can also be different at different material particles. Hence, the volume integral of momentum **P** can be expressed as

$$\mathbf{P} = \int_{V} \rho(\mathbf{x})\, \vec{\mathbf{v}}\, d\mathrm{v} \;\; or \;\; P_i = \int_{V} \rho\, v_i d\mathrm{v} \quad \text{where } \vec{\mathbf{v}} = \dot{\mathbf{x}} \tag{3.30}$$

3.4.3 Angular Momentum of a Deformable Body

Moment of momentum is called the angular moment which is a cross product between the momentum of a particle and the distance of the particle from the axis about which the moment is taken. Hence, in a continuous media the total angular momentum can be expressed by the volume integral of the local angular momentum and can be expressed as follows:

$$\mathbf{H} = \int_{V} \vec{\mathbf{p}} \times \rho(\mathbf{x})\, \vec{\mathbf{v}}\, d\mathrm{v} \;\; or \;\; H_i = \int_{V} \rho(x_m) \in_{ijk} x_j v_k \; d\mathrm{v} \tag{3.31}$$

where $\vec{\mathbf{p}}$ is the respective distance of the material particles in a continuous body from a specific point about which the total angular momentum is to be calculated. Hence, the vector $\vec{p}$ inherently depends on the position vectors of the particles. Here, the fixed point was considered the point O in Fig. 3.2.

3.4.4 Kinetic Energy Stored in a Deformable Body

Kinetic energy of a material particle is defined as the half of the product of the mass of the particle and the square of the velocity of the particle. Whereas in a continuous body, as density and the velocity of the material particles are changing across the continuous body, the total kinetic energy can be written as the volume integral of the local kinetic energy. The volume integral of the local kinetic energy stored in a deformable body can be expressed as

$$K = \frac{1}{2}\int \rho(\mathbf{x})\left(\vec{\mathrm{v}}\,.\,\vec{\mathrm{v}}\right) d\mathrm{v} \; or \; K = \frac{1}{2}\int_{\mathbb{V}} \rho\, v_k v_k \; d\mathrm{v} \tag{3.32}$$

where k is the index notation taking values 1, 2, and 3, and v_k is the particle velocity along the kth direction.

3.5 FUNDAMENTAL AXIOM OF CONTINUUM MECHANICS

In this section we will present the fundamental axiom of continuum mechanics that will use the definitions presented in the previous sections and will be further used in the following sections. First question we must ask, what is axiom. Axioms are statements that are considered universally true and should be used as fundamental assumption which are regarded as being established, accepted, or self-evidently true. We may find the following axioms are reasonable for our further discussion, which will provide key relationship, without which the mechanics of continuum is impossible.

3.5.1 Axiom 1: Principle of Conservation of Mass

As mass cannot be created or destroyed, it is reasonable to assume that the total mass in a continuous body is unchanged during the motion. If the statement is valid for a small region around a material particle, it can be said that the mass is conserved locally. But, if the mass is locally conserved, then it is obvious that the total mass of a continuous body is also conserved globally. To ensure this condition, we can say that the mass before and after deformation is equal. Alternatively, the rate of change of mass of a continuous body is equal to zero during the motion, which gives the following equations:

$$\int_{\mathrm{v}} \rho(\mathbf{x})\, d\mathrm{v} = \int_{\mathbb{V}} \rho_0(\mathbf{X}) d\mathbb{V} \quad \text{or} \quad \frac{D}{Dt}\int_{\mathrm{v}} \rho\, d\mathrm{v} = \int_{\mathrm{v}} \frac{D}{Dt}(\rho d\mathrm{v}) = 0 \tag{3.33}$$

where the local density of the body is expressed as $\rho_0(\mathbf{X})$ and $\rho(\mathbf{x})$, before and after deformation, respectively. The above integral can be expressed both in material and spatial coordinate systems, as shown in Table 3.4.

TABLE 3.4
Conservation of Mass in Lagrangian and Eulerian System

	Lagrangian (Fig. 3.4a)	Eulerian (Fig. 3.4b)
T4.1	$\int_{\mathbb{V}} (\rho_0 - \rho\mathbf{J})\, d\mathbb{V} = 0$	$\int_{\mathrm{v}} (\rho - \rho_0\mathbf{J}^{-1}) d\mathrm{v} = 0$

3.5.2 Axiom 2: Principle of Balance of Momentum

This principle is fundamentally similar to Newton's second law which says that the rate of change of velocity of a particle is directly proportional to the force applied to the particle along the direction of the motion. However, in this axiom it is said that the rate of change of momentum (***P***) of a continuous body as an integral of the local momentum in a particular direction is equal to the resultant force (***F***) acting on the body along the same direction.

$$\frac{d\boldsymbol{P}}{dt} = \boldsymbol{F}, \qquad \frac{D}{Dt}\int_{\mathrm{v}} \rho v_k \, d\mathrm{v} = F_k \tag{3.34}$$

3.5.3 Axiom 3: Principle of Balance of Angular Momentum

It is established that a material body under deformation and motion should be under balance, when the total forces in all three directions and the moments with respect to all three axes are conserved. That means that the internal forces and moments in the body are equal to the applied forces and moments on the body. The balances of forces are explained in Eq. (3.34), but the balance of moments can be explained by the following equations. The rate of change of integral of the total local moment of momentum (***H***) or the angular momentum should be equal to the total applied global moment (***M***) as written in Eq. (3.35)

$$\frac{d\boldsymbol{H}}{dt} = \boldsymbol{M} \;\; or \;\; \frac{D}{Dt}\int_{\mathrm{v}} \rho \in_{klm} x_l v_m \, d\mathrm{v} = M_k \tag{3.35}$$

3.5.4 Axiom 4: Principle of Conservation of Energy

During and after the deformation it is necessary that the total energy of the system should be conserved. Because the energy cannot be destroyed or created during an enclosed process, but could be transformed from one form to another. Hence, the rate of change of addition of the total kinetic energy and the total internal energy is equal to the total work done by the external forces plus other external or internal energy (U_α) that enters or leave the system, respectively. Based on this axiom, the equation for continuum mechanics can be written as follows:

$$\frac{D}{Dt}(K + E) = \mathcal{W} + \sum_{\alpha} U_\alpha \tag{3.36}$$

where K and E are the total kinetic energy and total internal energy, respectively. $\mathcal{W}$ is the work done on the system by the external forces per unit time. The other external energy that can enter or leave the system per unit time can be classified into heat energy, electrical energy, and chemical energy. According to this axiom, the energies are additive. If the internal energy density per unit mass is defined as ε, then the total internal energy can be written as follows:

$$E = \int_{v} \rho\varepsilon \, dv \tag{3.37}$$

The internal energy in thermodynamic sense is a state function which is independent of any process. In continuum mechanics, the total internal energy is the summation of the total internal strain energy, the total internal heat energy, and the total dissipative stress power which is equal to the product of dissipative stress and the symmetric part of the velocity gradient during motion.

3.6 INTERNAL STRESS STATE IN A DEFORMABLE BODY

In this section the stress state in a deformable body is briefly reviewed. A deformable body subjected to external and internal forces under motion must satisfy the global laws of motion, as a result of the axioms of continuum mechanics discussed in Section 3.5. During deformation or motion like it happens during wave propagation in a media, the internal forces are the result of action and reaction between two neighboring particles. These internal forces act along a line joining the two particles, following Newton's third law. Hence, the resultant force between the pair of particles is zero but the internal stresses are nonzero. Augustin-Louis Cauchy [2] first explained the internal stress state in a deformable body based on a hypothesis called stress hypothesis. The orthogonal surfaces cutting through the line joining any two neighboring particles interact with each other through traction forces to keep two neighboring elements in balance. In Fig. 3.5, a three-dimensional

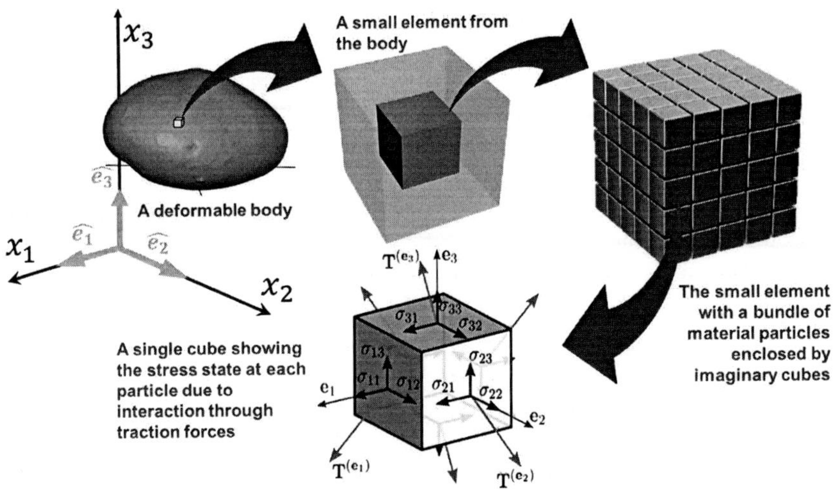

FIGURE 3.5 A 3D arbitrary object, its smallest part enclosing a material particle and a cube on which stress state is expressed.

arbitrary object is considered, and an arbitrary white particle block is identified to be interrogated. The particle block identified might have multiple particles inside, and each of them can be further imagined to be surrounded by six imaginary planes making a three-dimensional array of imaginary cubes enclosing each particle. Taking out an arbitrary cube from that system, the stress state is explained. The traction ($\mathbf{T}^{(\hat{e}_j)}$ or T_j in Fig. 3.5) acting on the surfaces along the line joining the neighboring particles, must have three components in a three-dimensional Cartesian coordinate system. The components on each plane have the normal stresses perpendicular to the plane and the shear stresses along the surface of the planes as shown in Fig 3.5. In a three-dimensional coordinate system, there are six such planes through which the neighboring particles are interacting with each other. Hence, there are six traction forces, out of which only three are unique to keep the element (the imaginary cube enclosing a single particle) in equilibrium. In other word, all planes have one resultant direction of stress, but have three components with one normal stress and two shear stresses according to Cauchy as shown in Fig. 3.5. $\mathbf{T}^{(\hat{e}_j)}$ or T_j or the three values T_1, T_2, and T_3, are sufficient to describe the stress state at a point. However, each traction has three components, thus nine values of stresses are the unique to a point. These nine values are described in a two-dimensional tensor form as σ_{ij}, where, i and j takes values 1, 2, and 3. It is now obvious that the traction and stresses at a point are related to each other, in terms of the normal direction of the six planes that are imagined surrounding the material particle. If the normal direction of each planes are expressed with respect to their direction cosine with n_1, n_2, and n_3, then the traction and the stresses are related using the following equation:

$$T_i = \sigma_{ij} n_j \tag{3.38}$$

The stress tensor σ_{ij} can be written in a matrix form as follows:

$$\sum = \sigma_{ij} = \begin{bmatrix} \sigma_{11} & \sigma_{12} & \sigma_{13} \\ \sigma_{21} & \sigma_{22} & \sigma_{23} \\ \sigma_{31} & \sigma_{32} & \sigma_{33} \end{bmatrix} \tag{3.39}$$

The normal stresses here in Eq. (3.39) occupy the diagonal elements and the off-diagonal elements are the shear stresses. The matrix is a symmetric, because the shear stresses are in balance and the local concentrated moment at the point is assumed to be zero. Hence, there are six independent stress components in a two-dimensional tensor.

Sometimes it is necessary to know the stress state on another plane, particularly to know the principal direction of the stress state in which the normal stresses are maximum but no shear stresses act on the planes. Transformation of second-order tensor is discussed in detail in Section 2.4.2 in Chapter 2. Please refer Eqs. (2.34) through (2.39) where it is shown how a second-order tensor could be transformed from a coordinate system with basis vector $\hat{\mathbf{e}}$ to a different coordinate system with basis vector $\hat{\mathbf{b}}$. Here we repeat the discussion and write the stress state on a coordinate

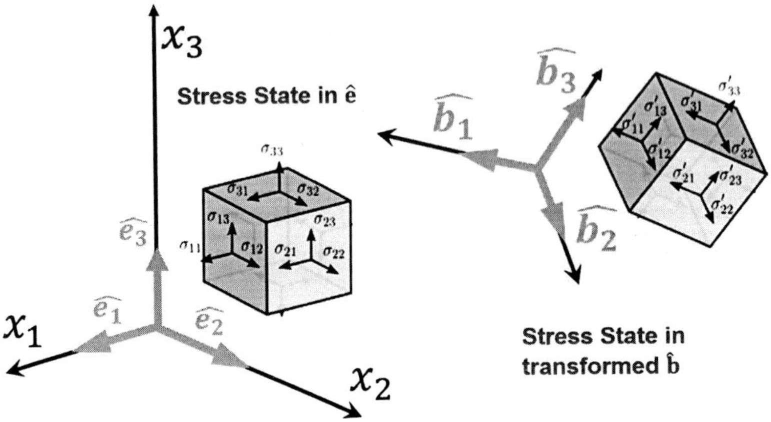

FIGURE 3.6 Stress state of a 3D stress element in an undeformed and deformed coordinate system, the deformed stress element shows the transformed stress states.

system with basis vector $\hat{\mathbf{b}}$ as shown in Fig. 3.6. The stress state (σ'_{ij}) in the coordinate system with basis vector $\hat{\mathbf{b}}$ can be written as

$$\sigma'_{ij} = \Psi_{im}\Psi_{nj}\sigma_{mn} \tag{3.40}$$

where Ψ_{im} and Ψ_{nj} are the second-order tensor called transformation matrix that holds the relation between the $\hat{\mathbf{e}}$ and $\hat{\mathbf{b}}$ coordinate systems (Eq. (2.8)) discussed in Section 2.4.2 in Chapter 2.

3.7 EXTERNAL AND INTERNAL LOAD ON A DEFORMABLE BODY

A deformable body is subjected to deformation when the body is exposed to different types of loads as mentioned in the previous sections. These loads are external and internal as shown in Fig. 3.7 on the surface Γ and in Ω, respectively. They can be categorized into three types. (a) Extrinsic body loads, which are the forces and couples that act externally on the body. Generally, these forces act through the center of mass of the body and the load density per unit mass is assumed under this category. Gravity and electrostatics forces are generally considered as the extrinsic body forces. Let's $\mathbf{f}$ be the body force per unit mass. (b) Extrinsic surface loads, which are the forces and couples that act externally on the surface of the body. Extrinsic surface forces generally arise from the interaction between two bodies (sloid or fluid) through surface contacts. Surface force per unit area is called the surface traction and the couple per unit area is called the surface couple. Hydrostatic forces, surface traction by electrostatic field, piezoelectric effects are considered in this category. Boundary condition in continuum mechanics are greatly influenced by this surface tractions and surface couples. Let's $\mathbf{t}^{(n)}$ be the traction force per unit surface area on the surface with direction normal ($\mathbf{n}$). (c) Intrinsic loads, which arises from the mutual interaction between the surrounded neighboring particles as shown in Fig. 3.5. These forces as

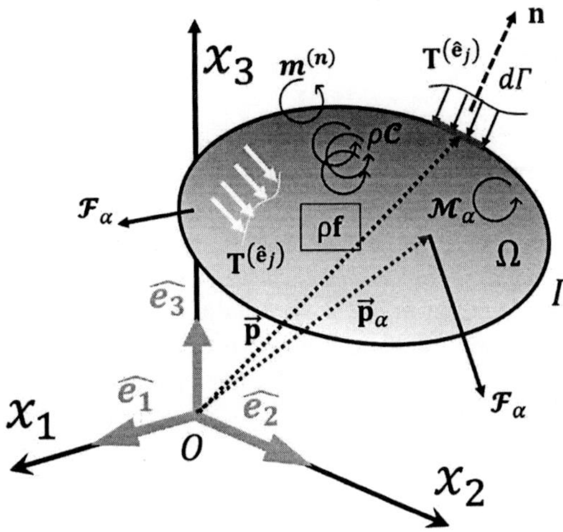

FIGURE 3.7 A deformable body with external and internal loads on the surface Γ and in Ω.

(as traction on the surfaces on the cubes, Figs. 3.5 and 3.6) shown act along the lines connecting the points but act in the opposite directions, causing the resultant internal force equal to zero. Assuming concentrated force ($\mathcal{F}_\alpha$) and concentrated couple ($\mathcal{M}_\alpha$) acting on the body at the points $\vec{\mathbf{p}}_\alpha$ (Fig. 3.7), we can write the total external force $\boldsymbol{F}$ and the total moment $\boldsymbol{M}$ as follows: [2]

$$\mathbf{F} = \oint_{\mathcal{S}} \mathbf{T}^{(\hat{e}_j)} \,.\, d\mathbf{a} + \int_{v} \rho \,\mathbf{f}\, dv + \sum_{\alpha} \mathcal{F}_\alpha \tag{3.41}$$

$$\mathbf{M} = \oint_{\mathcal{S}} \left(m^{(n)} + \vec{\mathbf{p}} \times \mathbf{T}^{(\hat{e}_j)} \right) d\mathbf{a} + \int_{v} \rho (\mathcal{C} + \vec{\mathbf{p}} \times \mathbf{f}) dv + \sum_{\alpha} (\mathcal{M}_\alpha + \vec{\mathbf{p}}_\alpha \times \mathcal{F}_\alpha) \tag{3.42}$$

where $\mathcal{C}$ is the couple per unit mass, $m^{(n)}$ is the surface couple per unit area on the surface with direction normal ($\boldsymbol{n}$), $\vec{\mathbf{p}}$ is the vector which signifies the directional distance between the traction and the point about which the moment is calculated in the body (Fig. 3.7).

3.8 FUNDAMENTAL ELASTODYNAMIC EQUATION

Applying the Axioms 2 and 3 presented in sections 3.5.2 and 3.5.3, respectively Eqs. (3.34) and (3.35) can be revised to Euler equation of motion, viz

$$\frac{D}{Dt} \int_{v} \rho \mathbf{v} \, dv = \oint_{\mathcal{S}} \mathbf{T}^{(\hat{e}_j)} \,.\, d\mathbf{a} + \int_{v} \rho \,\mathbf{f}\, dv + \sum_{\alpha} \mathcal{F}_\alpha \tag{3.43}$$

$$\frac{D}{Dt}\int_{\mathrm{v}}\rho(\vec{\mathbf{p}}\times\mathbf{v})\,d\mathrm{v}=\oint_{\mathcal{S}}\left(\boldsymbol{m}^{(n)}+\vec{\mathbf{p}}\times\mathbf{T}^{(\hat{\mathbf{e}}_j)}\right)d\mathbf{a}+\int_{\mathrm{v}}\rho(\mathcal{C}+\vec{\mathbf{p}}\times\mathbf{f})\,d\mathrm{v}+\sum_{\alpha}(\mathcal{M}_{\alpha}+\vec{\mathbf{p}}_{\alpha}\times\mathcal{F}_{\alpha}) \tag{3.44}$$

These are the equations for global balance of momenta of the entire body subjected to the external loads and moments. In case of a nonpolar material body, we can ignore the $\mathcal{C}$ and $\boldsymbol{m}^{(n)}$. Additionally, if the material body is not subjected to any concentrated forces and the concentrated local couples at discrete points, we can ignore the end terms in Eqs. (3.43) and (3.44). Performing the material derivatives in Eqs. (3.43) and (3.44) and applying the result from Axiom 1 (Eq. (3.33)) which says $\frac{D}{Dt}\int_{\mathrm{v}}\rho d\mathrm{v}=0$, we get

$$\int_{\mathrm{v}}\rho\frac{d\mathbf{v}}{dt}\,d\mathrm{v}=\oint_{\mathcal{S}}\mathbf{T}^{(\hat{\mathbf{e}}_j)}\,.\,d\mathbf{a}+\int_{\mathrm{v}}\rho\,\mathbf{f}\,d\mathrm{v} \tag{3.45}$$

$$\int_{\mathrm{v}}\rho\left(\vec{\mathbf{p}}\times\frac{d\mathbf{v}}{dt}\right)d\mathrm{v}=\oint_{\mathcal{S}}\left(\vec{\mathbf{p}}\times\mathbf{T}^{(\hat{\mathbf{e}}_j)}\right)d\mathbf{a}+\int_{\mathrm{v}}\rho(\vec{\mathbf{p}}\times\mathbf{f})\,d\mathrm{v} \tag{3.46}$$

The above equations can be written in index notations as follows:

$$\int_{\mathrm{v}}\rho\frac{dv_k}{dt}\,d\mathrm{v}=\oint_{\mathcal{S}}\sigma_{kj}\,n_j\,d\Gamma+\int_{\mathrm{v}}\rho\,f_k\,d\mathrm{v} \tag{3.47}$$

$$\int_{\mathrm{v}}\rho\in_{klm}x_l\frac{dv_m}{dt}\,d\mathrm{v}=\oint_{\mathcal{S}}\in_{klm}x_l\sigma_{mj}n_j\,d\Gamma+\int_{\mathrm{v}}\rho\in_{klm}x_lf_m\,d\mathrm{v} \tag{3.48}$$

where $d\Gamma$ is the elementary area on the surface of the body with direction cosine n_j. Now, applying divergence theorem, we could transfer the surface integral to the volume integral and the above equations modify to a local balance equation ignoring the volume integral of all the terms we get

$$\sigma_{kj,j}+\rho f_k=\rho\ddot{u}_k \tag{3.49}$$

and

$$\sigma_{kj}=\sigma_{jk} \tag{3.50}$$

The above two equations are the two fundamental equations of continuum mechanics, one of which describes the elastodynamic behavior in the material (Eq. (3.49)), and another says that at any point in the material, in the absence of the local moment the local stress matrix is always symmetric. These two equations are the fundamental to the wave propagation in solids. Hence, we will be solving these equations and will be discussing them in detail in this book in the following chapters.

3.9 THERMODYNAMICS OF CONTINUA

Wave propagation in a continuum media must account for the conservation of energy in the body. During the wave propagation in solid or fluid media, the body must satisfy the equation of equilibrium derived in Section 3.7 that leads to Eqs. (3.49) and (3.50). In addition to these equations, global conservation of energy has to be satisfied during the propagation. As the particles in the body moves with certain velocity during the wave propagation, the material body will be subjected to continuously changing kinetic energy ($\mathcal{K}$). In addition to the kinetic energy the body also has the continuously changing internal energy ($\mathcal{E}$) that arises from the change in the strain energy due to the deformation of the body. Dissipation of the energy is due to the material's damping characteristics. It is obvious that the wave propagation in a body can only initiate due to probing of an external energy to the body. The work done ($\mathcal{W}$) by the external forces should contribute to the changes of the kinetic and internal energies mentioned above. Considering this fact, we can write the general thermodynamic equation of the body imposing the principle of conservation of global energy which says that the time rate of change of the sum of the kinetic energy and the internal energy is equal to the work done on the body and all the other energies supplied to the body such as heat energy, electrical energy, electromagnetic, or chemical energy designated as U_α, when α signifies the identification of different nonmechanical external energies. Hence, the conservation of global energy will viz.

$$\dot{\mathcal{K}} + \dot{\mathcal{E}} = \mathcal{W} + \sum_{\alpha} U_\alpha \tag{3.51}$$

Ignoring all other different energies except heat energy ($\mathcal{Q}$), Eq. (3.51) will revise to

$$\dot{\mathcal{K}} + \dot{\mathcal{E}} = \mathcal{W} + \mathcal{Q} \tag{3.51}$$

3.9.1 Conservation of Local Energy

Now it is time to expand the global energy equation in terms of the energies at the local points. At any point if the velocity of the particle with density ρ is $\mathbf{v} = v_j\hat{e}_j$ due to the wave propagation, then the global kinetic energy can be written as the volume integral of the local kinetic energy as follows:

$$\mathcal{K} = \frac{1}{2}\int_{\mathrm{v}} \rho\, v_j v_j \, d\mathrm{v} \qquad v_j = \dot{x}_j \tag{3.52}$$

Similarly, the global internal energy can be expressed as volume integral of local internal energy density per unit mass.

$$\mathcal{E} = \int_{v} \rho\, \varepsilon\, dv \quad ; \quad \varepsilon = \text{internal energy density} \tag{3.53}$$

As shown in Fig. 3.7, the exterior surface of the body is subjected to the traction forces. The surface integral of the time rate of displacement; i.e., the velocity of a particle on the surface multiplied with the traction force will be the measure of the work done on the body due to the surface loads. Similarly, the volume integral of the body force multiplied with the particle velocity will measure the work done on the body by the body forces. Hence, the total work done on the material body by the external and internal forces is summarized as follows:

$$\mathcal{W} = \oint_{\mathcal{S}} \sigma_{ji} v_i n_j \, d\Gamma + \int_{v} \rho\, f_i v_i \, dv \tag{3.54}$$

If the material body is subjected to the thermal loads on the surface and the body, assuming the heat vector $\mathfrak{q}$ acting on unit surface area and the distributed heat source $\mathfrak{h}$ per unit mass, the heat energy $\mathcal{Q}$ can be written as

$$\mathcal{Q} = \oint_{\mathcal{S}} \mathfrak{q}.\mathbf{n} \, d\Gamma + \int_{v} \rho \mathfrak{h} \, dv \tag{3.55}$$

Next utilizing the definition and expressions presented in Section 3.2.1 on the material derivatives and applying the divergence theorem to transfer the surface integral to volume integral, we can write the energy terms in Eq. (3.51) more elaborately

$$\frac{D}{Dt}\left(\frac{1}{2}\int_{v} \rho\, v_k v_k \, dv\right) + \frac{D}{Dt}\left(\int_{v} \rho\varepsilon \, dv\right) = \int_{v} \left(\sigma_{ji} v_i\right)_{,j} dv + \int_{v} \rho f_i v_i \, dv$$
$$+ \int_{v} \left(\mathfrak{q}_{k,k} + \rho \mathfrak{h}\right) dv \tag{3.56}$$

Further the material derivatives of kinetic and internal energies are shown in a step-by-step derivation of the equations as follows:

Material Derivative of Kinetic Energy:

$$\dot{\mathcal{K}} = \frac{D}{Dt}\left(\frac{1}{2}\int_{v} \rho\, v_k v_k \, dv\right) = \frac{1}{2}\int \left[(\rho\, dv)\frac{D}{Dt}(v_k.v_k) + v_k v_k \frac{D}{Dt}(\rho dv)\right]$$
$$= \int_{v} \left(\rho\, v_k \dot{v}_k dv + \frac{1}{2} v_k v_k \frac{D}{Dt}(\rho dv)\right) \tag{3.57.1}$$

$$\dot{\mathcal{K}} = \int_{v} \left(\rho v_k \dot{v}_k dv + \frac{1}{2} v_k v_k \rho \frac{D}{Dt}(dv) + \frac{1}{2} v_k v_k \dot{\rho}\, dv \right)$$

$$= \int_{v} \left(\rho v_k \dot{v}_k + \frac{1}{2} \rho v_k v_k v_{l,l} + \frac{1}{2} \dot{\rho} v_k v_k \right) dv \tag{3.57.2}$$

Material Derivative of Internal Energy:

$$\dot{\mathcal{E}} = \frac{D}{Dt} \int_{v} \rho \varepsilon \, dv = \int_{v} (\rho dv) \dot{\varepsilon} + \varepsilon \frac{D}{Dt} (\rho dv) \tag{3.58.1}$$

$$\dot{\mathcal{E}} = \int_{v} \rho \, \dot{\varepsilon} dv + \varepsilon \, \dot{\rho} dv + \rho \varepsilon v_{k,k} \, dv = \int_{v} (\rho \dot{\varepsilon} + \varepsilon \dot{\rho} + \rho \varepsilon v_{k,k}) dv \tag{3.58.2}$$

Now let's rearrange Eq. (3.56) while we utilize Eqs. (3.57.1) and (3.58.1)

$$\left(\varepsilon + \frac{1}{2} v_k v_k \right) \frac{D}{Dt} \left(\int_{v} \rho dv \right) + \int_{v} \rho \dot{\varepsilon} dv - \int_{v} \rho v_k \dot{v}_k \, dv$$

$$= \int_{v} (\sigma_{ji} v_{i,j} + \sigma_{ji,j} v_i) dv + \int_{v} \rho f_i v_i \, dv + \int_{v} (\mathfrak{q}_{k,k} + \rho \mathfrak{h}) dv \tag{3.59}$$

Further simplification and rearrangement in the light of Eqs. (3.33) and (3.49) that are derived from the two axioms of continuum mechanics.

$$\left(\varepsilon + \frac{1}{2} v_k v_k \right) \frac{D}{Dt} \left(\int_{v} \rho dv \right) + \int_{v} (\rho \dot{\varepsilon} - \sigma_{ji} v_{i,j} - \mathfrak{q}_{k,k} - \rho \mathfrak{h}) \, dv$$

$$= \int_{v} v_i (\sigma_{ji,j} v_i + \rho f_i - \rho \dot{v}_i) dv \tag{3.60}$$

Eq. (3.33) is derived from the conservation of mass and Eq. (3.49) is derived from the conservation of linear momentum. If we apply those equations in Eq. (3.60) for the global conservation of energy and ignore the volume integral, we get the equation for the *local conservation of energy* which reads

$$\rho \dot{\varepsilon} = \sigma_{ji} v_{i,j} + \mathfrak{q}_{k,k} + \rho \mathfrak{h} \tag{3.61}$$

3.9.2 Conservation of Mechanical Energy (Kinetic, Internal, and Potential Energy)

Referring back to Eq. (3.54) we see that the body force vector (f_i) is something should be described in more detail. The body force vector (f_i) on a local particle can be described as a function of the position vector of the local particle, because a force field in a deformable body may be developed due to an arbitrary potential $\mathbb{U}$ and can be expressed as the spatial derivatives of that potential function. For example, electrical and magnetic force fields contributed to the body force are the derivative of the electrical and magnetic potential functions, respectively. Similarly, we can write, $f_i = -\mathbb{U}_{,i}$ and the work done in Eq. (3.54) can be revised to

$$\mathcal{W} = \oint_{\mathcal{S}} \sigma_{ji} v_i n_j \, d\Gamma - \int_{\mathrm{v}} \rho \, \mathbb{U}_{,i} v_i \, d\mathrm{v} = \oint_{\mathcal{S}} \sigma_{ji} v_i n_j \, d\Gamma - \dot{\mathcal{U}} + \int_{\mathrm{v}} \mathbb{U} \frac{D}{Dt} \left(\int_{\mathrm{v}} \rho \, d\mathrm{v} \right) \tag{3.62}$$

where $\mathcal{U}$ is the potential density function or the potential energy. Substituting Eq. (3.62) in Eq. (3.51), we can write the expression for the summation of kinetic energy, internal energy, and potential energy as

$$\dot{\mathcal{K}} + \dot{\mathcal{E}} + \dot{\mathcal{U}} = \oint_{\mathcal{S}} \sigma_{ji} v_i n_j \, d\Gamma + \mathcal{Q} \tag{3.63}$$

Where, Eq. (3.33) is used, which is a result from the Axiom 1 in continuum mechanics enumerates that during a deformation process the total potential carried by the locally created or annihilated mass is equal to zero.

Now if the deformable body during wave propagation is totally insulated and the energy of the surface traction is zero, we get the conservation of mechanical energy, which states when the local mass is conserved in an insulated body with vanishing surface integral of traction forces, the summation of kinetic energy, internal energy, and potential energy is constant. Mathematically we can write,

$$\mathcal{K} + \mathcal{E} + \mathcal{U} = \text{constant} \tag{3.64}$$

3.9.3 Internal Energy and Strain Energy

There has been a persistent misconception among students that concerns the internal energy and strain energy in a deformable body. In most occasions, they are considered same, but that is not true. Here in this section we explain the meaning of the both in the light of Sections 3.8.1 and 3.8.2. In the previous sections we introduced the potential function to explain the body force that gave us the conservation of mechanical energy in Eq. (3.64). However, question may arise, why only the body force? Why don't we also define the potential function for the surface traction? Answer is yes, we can also define the potential for the surface traction σ_{ji}. However, the nature of surface traction is bit different from the body force. The body force is always recoverable. But the surface stresses can be divided into two parts: recoverable or reversible and non-recoverable or irreversible or dissipative. Potential functions can be defined for the recoverable variables only. We define a potential function $\mathcal{T}$ to capture the recoverable part of the stress tensor multiplied with the velocity gradient, i.e., $\sigma_{ji} v_{i,j}$. Please recollect Table 3.3 where we showed that the strain is the gradient of the displacements. In general strain energy is the product of stress and the displacement gradient and strains are considered unknows in the governing equations. Time rate of change of this strain energy has partially appeared on the right side of Eq. (3.63). Let's also recollect the volume integral of Eq. (3.61) to describe the global balance of energy and rewrite the equation using the new functions we described in this section

$$\int_{v} \rho\dot{\varepsilon}\, dv = \int_{v} \rho\dot{\mathcal{T}}\, dv + \int_{v} [\sigma]_{ji}^{D}\, v_{i,j}\, dv + \int_{v} \left(q_{k,k} + \rho h\right) dv \tag{3.65}$$

where $[\sigma]_{ji}^{D}$ is the dissipative or irreversible part of the stress tensor [2]. Using Eqs. (3.58.1) and (3.55), when the conservation of mass is satisfied, we can express the above equation in the global form as follows:

$$\dot{\mathcal{E}} = \dot{\mathcal{T}} + \mathcal{D} + \mathcal{Q} \tag{3.66}$$

where $\mathcal{T} = \int_{v} \rho\mathcal{T} dv$ is the total strain energy and $\mathcal{D} = \int_{v} [\sigma]_{ji}^{D}\, v_{i,j} dv$ is the total dissipative power. $\dot{\mathcal{T}}$ is the rate of change of strain energy. Hence, the total internal energy and the strain energy are not the same. However, in case of two scenarios described below, they are equal; i.e., internal energy is equal to the strain energy if and only if

- A deformable body is totally insulated, and the dissipated stress tensor is equal to zero, $\mathcal{E} = \mathcal{T}$, if $\mathcal{D} = \mathcal{Q} = 0$
  or
- In a deformable body, the total dissipated energy is totally converted into heat energy, $\mathcal{E} = \mathcal{T}$ if, $\mathcal{D} + \mathcal{Q} = 0$

Hence, the above results (internal energy is equal to the strain energy) are true only under the condition of adiabatic process. That is, no heat source may present in the body and no heat is lost or gained, to and from the environment. Hence, in thermodynamically admissible elastic solid, recollecting Eq. (3.61), without the loss of generality we can rewrite the equation of internal energy density as follows:

$$\rho\dot{\varepsilon} = \sigma_{kl} v_{k,l} \tag{3.67}$$

where different name of the indices are used but they still take values 1, 2, and 3. As we just proved, under adiabatic condition the internal energy is equivalent to the strain energy, we can express the internal energy density as a function of deformation gradient tensor or more inclusively, Green's deformation tensor C_{KL}. Let's assume $\rho_0\varepsilon = \Sigma(C_{KL})$, where ρ_0 is the density of the medium at the initial state. The Σ is the potential function is in fact closely related to the free energy of the system and can be proved applying Clausius-Duhem inequality [2], not described in this book in detail. The potential Σ is referred with many names, such as strain energy density function, internal energy density function, or work function. On the other hand, the Lagrangian strain tensor is the function of Green's deformation tensor, and thus we can rewrite the above expression as $\rho_0\varepsilon = \Sigma(\mathbb{E}_{KL})$. Next, substituting the expression in Eq. (3.67) with much mathematical steps omitted herein (refer [2]), we can write the first form of constitutive relation as follows:

$$\sigma_{kl}v_{k,l} = \frac{\rho}{\rho_0}\frac{\partial\Sigma}{\partial\mathbb{E}_{KL}}\dot{\mathbb{E}}_{KL} \tag{3.68}$$

Further exploiting the expressions T2.4 in Table 3.2, we can write

$$\sigma_{kl} = 2\frac{\rho}{\rho_0}\frac{\partial\Sigma}{\partial\mathbb{E}_{KL}}x_{k,K}x_{l,L} \tag{3.69}$$

If the material is incompressible during deformation, we can say $\rho_0 = \rho$. Additionally, if we assume, during deformation the displacement gradient is very small, referring T2.3, we can further say

$$\sigma_{kl} = \frac{\partial\Sigma(e_{kl})}{\partial e_{kl}} \tag{3.70.1}$$

In this book, in other chapters we will use the following indices i,j,k,l,m,n,p,q interchangeably where all take values 1, 2, and 3. Please note it is immaterial if we change the indices as follows without the meaning of the above equation and Eq. (3.70) can be rewritten as

$$\sigma_{ij} = \frac{\partial\Sigma(e_{ij})}{\partial e_{ij}} \tag{3.70.2}$$

3.10 CONSTITUTIVE LAW OF CONTINUA

Based on the discussion above during wave propagation in solid (under the external force field), it is apparent that at any point in a continuum, six stresses and three displacements are the unknowns and must be found from the balance equations to completely describe the wave field. Axioms of continuum mechanics says that in addition to globally satisfying boundary conditions, it is necessary for local mass to be conserved. Also, the linear and angular momentums must be conserved at the local points in the body. These four conditions gave use Eqs. (3.33), (3.49), and (3.50) reiterated herein in Eq. (3.67)

$$\left.\begin{aligned} \frac{\partial\rho}{\partial t} + (\rho v_i)_{,i} &= 0 \\ \sigma_{ij,j} + \rho f_i &= \rho\ddot{u}_i \\ \sigma_{ij} &= \sigma_{ji} \end{aligned}\right\} \tag{3.71}$$

There are total four independent equations with six stresses and three displacements and density of the media that are the unknowns in the equations. System is not deterministic at this moment. For given displacements, strains at a point can be calculated and will be automatically compatible, but in cases of strain being unknown, not any displacement pattern will satisfy because, strains are the function of displacement gradients and must obey certain rules. Thus, we need to bring six new strain variables into the equations (Eq. (3.72)). Strain displacement equations from Table 3.3 are reiterated here

$$\mathbf{e}_{kl} = \frac{1}{2}\left(u_{k,l} + u_{l,k}\right) \tag{3.72}$$

This will give us ten equations but sixteen unknown. Still the system remains indeterministic. Above we just mentioned about the compatibility of the strains at a local point for a given displacement field. Hence, the compatibility conditions will be additional six equations that will bind the deformation of the system in a geometrically admissible form. Although not derived, the compatibility equations are given below

$$\mathbf{e}_{ji,lm} + \mathbf{e}_{lm,ji} - \mathbf{e}_{jm,il} - \mathbf{e}_{il,jm} = 0 \tag{3.73}$$

Where all indices take values 1, 2, and 3, without the loss of generality. In case of displacement formulation, we will have Eqs. (3.71) and (3.72), because the displacements will automatically make the strains compatible and Eq. (3.73) is automatically satisfied. But in case of stress formulation, strain-displacement relation (Eq. (3.72)) cannot be exploited, but the compatibility condition (Eq. (3.73)) must be imposed. In either case we have ten equations and sixteen unknows. There is always a deficit of six equations.

These six equations are supplied by the constitutive law of thermodynamically admissible elastic materials where stress and strains are related using Eq. (3.70). The work function $\Sigma(\mathbf{e}_{kl})$ can be expressed as a polynomial of the strain components in a general nonlinear elastic solid as follows:

$$\sum = \sum_{o} + \sum_{kl} \mathbf{e}_{kl} + \frac{1}{2}\sum_{klij} \mathbf{e}_{kl}\mathbf{e}_{ij} + \ldots \tag{3.74}$$

where Σ_o, Σ_{kl}, and Σ_{klij} are the function of position vector $\mathbf{x}$ in an inhomogeneous solid and are constants in homogeneous solids. In a linear material we could retain up to the quadratic terms.

Applying Eq. (3.70) or (3.71) and much mathematical and logical steps we get the following constitutive law:

$$\sigma_{ij} = \mathbb{C}_{ijkl}\mathbf{e}_{kl} \tag{3.75}$$

where $\mathbb{C}_{ijkl} = \frac{1}{2}\left(\Sigma_{klij} + \Sigma_{ijkl}\right)$ is the constitutive matrix (a fourth-order tensor) and to be experimentally determined. However, $\mathbb{C}_{ijkl}$ is subjected to the restrictions as follows:

$$\mathbb{C}_{ijkl} = \mathbb{C}_{jikl} = \mathbb{C}_{jilk} \;;\; \mathbb{C}_{ijkl} = \mathbb{C}_{klij} \tag{3.76}$$

For the sake of the brevity, based on earlier discussion, we could further write a polynomial expansion of stress in general nonlinear materials as follows:

$$\sigma_{ij} = \mathbb{C}_{ij} + \mathbb{C}_{ijkl}e_{kl} + \mathbb{C}_{ijklmn}e_{kl}e_{mn} \tag{3.77}$$

Please note that σ_{ij} is a second-order tensor with nine elements. However, Eq. (3.39) describes the symmetric nature of the stress tensor, which resulted six independent elements in the stress matrix. Similarly, the strain tensor (T3.5) will also result six independent strain elements in the strain matrix. Instead of writing them in a matrix form it is found to be convenient to write their relation in a vector form. Hence, the stress matrix will become a vector of six stress elements, and so does the strain matrix. These six stresses and six strains are related by the constitutive law, resulted fourth-order constitutive tensor $\mathbb{C}_{ijkl}$. However, it needs only 36 elements to be in the constitutive matrix to relate six stresses to six strains. Using first three equalities in Eq. (3.76) we could expand the stress-strain relation through constitutive matrix as follows:

$$\begin{Bmatrix} \sigma_{11} \\ \sigma_{22} \\ \sigma_{33} \\ \sigma_{23} \\ \sigma_{31} \\ \sigma_{12} \end{Bmatrix} = \begin{bmatrix} \mathbb{C}_{1111} & \mathbb{C}_{1122} & \mathbb{C}_{1133} & \mathbb{C}_{1123} & \mathbb{C}_{1131} & \mathbb{C}_{1112} \\ \mathbb{C}_{2211} & \mathbb{C}_{2222} & \mathbb{C}_{2233} & \mathbb{C}_{2223} & \mathbb{C}_{2231} & \mathbb{C}_{2212} \\ \mathbb{C}_{3311} & \mathbb{C}_{3322} & \mathbb{C}_{3333} & \mathbb{C}_{3323} & \mathbb{C}_{3331} & \mathbb{C}_{3312} \\ \mathbb{C}_{2311} & \mathbb{C}_{2322} & \mathbb{C}_{2333} & \mathbb{C}_{2323} & \mathbb{C}_{2331} & \mathbb{C}_{2312} \\ \mathbb{C}_{3111} & \mathbb{C}_{3122} & \mathbb{C}_{3133} & \mathbb{C}_{3123} & \mathbb{C}_{3131} & \mathbb{C}_{3112} \\ \mathbb{C}_{1211} & \mathbb{C}_{1222} & \mathbb{C}_{1233} & \mathbb{C}_{1223} & \mathbb{C}_{1231} & \mathbb{C}_{1212} \end{bmatrix} \begin{Bmatrix} e_{11} \\ e_{22} \\ e_{33} \\ 2e_{23} \\ 2e_{31} \\ 2e_{12} \end{Bmatrix} \tag{3.78}$$

Further using the fourth equality in Eq. (3.76) obtained from consistent strain energy density function or the work function, we have fewer independent coefficients. Because of this thermodynamically admissible state of linear elastic material there are 21 independent constitutive coefficients in the $\mathbb{C}_{ijkl}$ matrix. Matrix in Eq. (3.78) should be a symmetric matrix. Hence, we can redefine Eq. (3.78) in the following form:

$$\begin{Bmatrix} \sigma_{11} \\ \sigma_{22} \\ \sigma_{33} \\ \sigma_{23} \\ \sigma_{31} \\ \sigma_{12} \end{Bmatrix} = \begin{bmatrix} \mathbb{C}_{11} & \mathbb{C}_{12} & \mathbb{C}_{13} & \mathbb{C}_{14} & \mathbb{C}_{15} & \mathbb{C}_{16} \\ \mathbb{C}_{21} & \mathbb{C}_{22} & \mathbb{C}_{23} & \mathbb{C}_{24} & \mathbb{C}_{25} & \mathbb{C}_{26} \\ \mathbb{C}_{31} & \mathbb{C}_{32} & \mathbb{C}_{33} & \mathbb{C}_{34} & \mathbb{C}_{35} & \mathbb{C}_{36} \\ \mathbb{C}_{41} & \mathbb{C}_{42} & \mathbb{C}_{43} & \mathbb{C}_{44} & \mathbb{C}_{45} & \mathbb{C}_{46} \\ \mathbb{C}_{51} & \mathbb{C}_{52} & \mathbb{C}_{53} & \mathbb{C}_{54} & \mathbb{C}_{55} & \mathbb{C}_{56} \\ \mathbb{C}_{61} & \mathbb{C}_{62} & \mathbb{C}_{63} & \mathbb{C}_{64} & \mathbb{C}_{65} & \mathbb{C}_{66} \end{bmatrix} \begin{Bmatrix} e_{11} \\ e_{22} \\ e_{33} \\ 2e_{23} \\ 2e_{31} \\ 2e_{12} \end{Bmatrix} \tag{3.79}$$

In index notation, $\sigma_m = \mathbb{C}_{mn}e_n$

How the indices of the fourth-order tensor $\mathbb{C}_{ijkl}$ are mapped to the indices of a second-order tensor $\mathbb{C}_{mn}$ or the constitutive matrix with 21 independent coefficients is described in Table 3.5.

Further due to material symmetry the number of independent constitutive coefficients reduce as described below. Without any symmetry but with 21 independent coefficients a material is called triclinic material.

TABLE 3.5
Constitutive Mapping from Fourth-Order Tensor to a Second-Order Tensor

$\sigma_{ij} \xrightarrow{yields}$	σ_m	$e_{kl} \xrightarrow{yields}$	e_n	*ij or ji*	*m or n*
σ_{11}	σ_1	e_{11}	e_1	11	1
σ_{22}	σ_2	e_{22}	e_2	22	2
σ_{33}	σ_3	e_{33}	e_3	33	3
σ_{23}	σ_4	e_{23}	e_4	23, 32	4
σ_{31}	σ_5	e_{31}	e_5	31, 13	5
σ_{12}	σ_6	e_{12}	e_6	12, 21	6

3.10.1 Materials with One Plane of Symmetry: Monoclinic Materials

Referring cartesian coordinate system (x_j) described in Chapter 2, if the material is symmetric with respect to a plane described by an arbitrary axis x_1 (i.e., x_1 is normal to the x_2x_3 plane) then stress and strain states must have the mirror symmetry with respect to the x_1 plane of symmetry. This will cause only 13 independent constitutive coefficients to survive [4, 5] as follows:

$$\begin{Bmatrix} \sigma_{11} \\ \sigma_{22} \\ \sigma_{33} \\ \sigma_{23} \\ \sigma_{31} \\ \sigma_{12} \end{Bmatrix} = \begin{bmatrix} \mathbb{C}_{11} & \mathbb{C}_{12} & \mathbb{C}_{13} & \mathbb{C}_{14} & 0 & 0 \\ & \mathbb{C}_{22} & \mathbb{C}_{23} & \mathbb{C}_{24} & 0 & 0 \\ & & \mathbb{C}_{33} & \mathbb{C}_{34} & 0 & 0 \\ & & & \mathbb{C}_{44} & 0 & 0 \\ & Sym & & & \mathbb{C}_{55} & \mathbb{C}_{56} \\ & & & & & \mathbb{C}_{66} \end{bmatrix} \begin{Bmatrix} e_{11} \\ e_{22} \\ e_{33} \\ 2e_{23} \\ 2e_{31} \\ 2e_{12} \end{Bmatrix} \tag{3.80}$$

If the plane of symmetry changes to x_2 plane, then the symmetric stress and strain states will result the constitutive relation as follows in Eq. (3.81). Please note that Eqs. (3.80) and (3.81) both have the same number of independent 13 coefficients, but the zeros are placed at different locations in the matrix depending on the plane of symmetry.

$$\begin{Bmatrix} \sigma_{11} \\ \sigma_{22} \\ \sigma_{33} \\ \sigma_{23} \\ \sigma_{31} \\ \sigma_{12} \end{Bmatrix} = \begin{bmatrix} \mathbb{C}_{11} & \mathbb{C}_{12} & \mathbb{C}_{13} & 0 & \mathbb{C}_{15} & 0 \\ & \mathbb{C}_{22} & \mathbb{C}_{23} & 0 & \mathbb{C}_{25} & 0 \\ & & \mathbb{C}_{33} & 0 & \mathbb{C}_{35} & 0 \\ & & & \mathbb{C}_{44} & \mathbb{C}_{45} & 0 \\ & Sym & & & \mathbb{C}_{55} & 0 \\ & & & & & \mathbb{C}_{66} \end{bmatrix} \begin{Bmatrix} e_{11} \\ e_{22} \\ e_{33} \\ 2e_{23} \\ 2e_{31} \\ 2e_{12} \end{Bmatrix} \tag{3.81}$$

3.10.2 Materials with Two Planes of Symmetry: Orthotropic Materials

Drawing conclusion from the above discussion, if the material possesses both x_1 plane and x_2 plane of symmetry then the constitutive equation must be inclusive of both Eq. (3.80) and Eq. (3.81). When the material is symmetric about two orthogonal planes then it automatically makes the material symmetric about the third plane, i.e., symmetric about the x_3 plane. The constitutive matrix of such material will have only nine independent coefficients as follows:

$$\begin{Bmatrix} \sigma_{11} \\ \sigma_{22} \\ \sigma_{33} \\ \sigma_{23} \\ \sigma_{31} \\ \sigma_{12} \end{Bmatrix} = \begin{bmatrix} \mathbb{C}_{11} & \mathbb{C}_{12} & \mathbb{C}_{13} & 0 & 0 & 0 \\ & \mathbb{C}_{22} & \mathbb{C}_{23} & 0 & 0 & 0 \\ & & \mathbb{C}_{33} & 0 & 0 & 0 \\ & & & \mathbb{C}_{44} & 0 & 0 \\ & Sym & & & \mathbb{C}_{55} & 0 \\ & & & & & \mathbb{C}_{66} \end{bmatrix} \begin{Bmatrix} e_{11} \\ e_{22} \\ e_{33} \\ 2e_{23} \\ 2e_{31} \\ 2e_{12} \end{Bmatrix} \tag{3.82}$$

These materials are called orthotropic or orthorhombic materials. Most layered composites that has different layers with of angle fiber orientations belong to this category.

3.10.3 Materials with Three Planes of Symmetry and One Plane of Isotropy: Transversely Isotropic Materials

In addition to the three planes of symmetry described in the previous category, if the material has one additional plane of isotropy, i.e., the material behaves like an isotropic material about an axis then the plane perpendicular to axis or the transverse plane has infinite plane of symmetry. In addition to all plane of symmetry, if the material is isotropic about the x_3 axis, then the constitutive matrix will have only six coefficients, but five independent coefficients as follows:

$$\begin{Bmatrix} \sigma_{11} \\ \sigma_{22} \\ \sigma_{33} \\ \sigma_{23} \\ \sigma_{31} \\ \sigma_{12} \end{Bmatrix} = \begin{bmatrix} \mathbb{C}_{11} & \mathbb{C}_{12} & \mathbb{C}_{13} & 0 & 0 & 0 \\ & \mathbb{C}_{11} & \mathbb{C}_{13} & 0 & 0 & 0 \\ & & \mathbb{C}_{33} & 0 & 0 & 0 \\ & & & \mathbb{C}_{44} & 0 & 0 \\ & Sym & & & \mathbb{C}_{44} & 0 \\ & & & & & \dfrac{\mathbb{C}_{11} - \mathbb{C}_{12}}{2} \end{bmatrix} \begin{Bmatrix} e_{11} \\ e_{22} \\ e_{33} \\ 2e_{23} \\ 2e_{31} \\ 2e_{12} \end{Bmatrix} \tag{3.83}$$

These materials are called transversely isotropic materials. They have same modulus along the x_1 and x_2 axis and same shear modulus along x_3 axis, irrespective of the plane on which it acts.

3.10.4 Materials with Three Planes and Three Axes of Symmetry: Isotropic Materials

Drawing conclusion from the above, adding one more axis of symmetry will cause one more plane of isotropy. When two orthogonal planes are isotropic then the third plane must also be isotropic. We can write the constitutive matrix of an isotropic materials as follows. In this case there are only two independent material constants in the constitutive equation.

$$\begin{Bmatrix} \sigma_{11} \\ \sigma_{22} \\ \sigma_{33} \\ \sigma_{23} \\ \sigma_{31} \\ \sigma_{12} \end{Bmatrix} = \begin{bmatrix} \mathbb{C}_{11} & \mathbb{C}_{12} & \mathbb{C}_{12} & 0 & 0 & 0 \\ & \mathbb{C}_{11} & \mathbb{C}_{12} & 0 & 0 & 0 \\ & & \mathbb{C}_{11} & 0 & 0 & 0 \\ & & & \frac{\mathbb{C}_{11}-\mathbb{C}_{12}}{2} & 0 & 0 \\ & Sym & & & \frac{\mathbb{C}_{11}-\mathbb{C}_{12}}{2} & 0 \\ & & & & & \frac{\mathbb{C}_{11}-\mathbb{C}_{12}}{2} \end{bmatrix} \begin{Bmatrix} e_{11} \\ e_{22} \\ e_{33} \\ 2e_{23} \\ 2e_{31} \\ 2e_{12} \end{Bmatrix} \tag{3.84}$$

The above coefficients can be further expressed in terms of Lame constants $\mathbb{L}$ and $\mathbb{m}$, where,

$$\mathbb{C}_{11} = \mathbb{L} + 2\mathbb{m} \quad and \quad \mathbb{C}_{12} = \mathbb{L} \tag{3.85}$$

Substituting Eq. (3.85), Eq. (3.84) can be further written in index notation as follows:

$$\sigma_{ij} = \mathbb{L}\delta_{ij}e_{kk} + 2\mathbb{m}e_{ij} \tag{3.86}$$

Conversely, strains in terms of stresses (Hook's Law) in isotropic material can be written as

$$e_{ij} = \frac{\sigma_{ij}}{2\mathbb{m}} + \delta_{ij}\frac{\mathbb{L}\sigma_{kk}}{2\mathbb{m}(3\mathbb{L}+2\mathbb{m})} \tag{3.87}$$

3.11 APPENDIX

3.11.1 Important Equations in Cartesian Coordinate System

In this chapter most of the equations are presented using index notations however, few equations will be extensively used in the following chapters. Hence, for reference they are expanded in this appendix. Primarily Eqs. (3.71)–(3.73) are expanded in the following set of equations, respectively.

Governing Differential Equations of Motion

$$\frac{\partial \sigma_{11}}{\partial x_1}+\frac{\partial \sigma_{12}}{\partial x_2}+\frac{\partial \sigma_{13}}{\partial x_3}+f_1=\rho\frac{\partial^2 u_1}{\partial t^2} \quad \text{(A.3.1.1)}$$

$$\frac{\partial \sigma_{21}}{\partial x_1}+\frac{\partial \sigma_{22}}{\partial x_2}+\frac{\partial \sigma_{23}}{\partial x_3}+f_2=\rho\frac{\partial^2 u_2}{\partial t^2} \quad \text{(A.3.1.2)}$$

$$\frac{\partial \sigma_{31}}{\partial x_1}+\frac{\partial \sigma_{32}}{\partial x_2}+\frac{\partial \sigma_{33}}{\partial x_3}+f_3=\rho\frac{\partial^2 u_3}{\partial t^2} \quad \text{(A.3.1.3)}$$

Strain-Displacement Relations

$$e_{11}=\left(u_{1,1}\right) \quad \text{(A.3.2.1)}$$

$$e_{22}=\left(u_{2,2}\right) \quad \text{(A.3.2.2)}$$

$$e_{33}=\left(u_{3,3}\right) \quad \text{(A.3.2.3)}$$

$$e_{12}=\frac{1}{2}\left(u_{1,2}+u_{2,1}\right) \quad \text{(A.3.2.4)}$$

$$e_{23}=\frac{1}{2}\left(u_{2,3}+u_{3,2}\right) \quad \text{(A.3.2.5)}$$

$$e_{31}=\frac{1}{2}\left(u_{3,1}+u_{1,3}\right) \quad \text{(A.3.2.6)}$$

Strain Compatibility Equations

$$\frac{\partial^2 e_{11}}{\partial x_2{}^2}+\frac{\partial^2 e_{22}}{\partial x_1{}^2}=2\frac{\partial^2 e_{12}}{\partial x_1 x_2} \quad \text{(A.3.3.1)}$$

$$\frac{\partial^2 e_{22}}{\partial x_3{}^2}+\frac{\partial^2 e_{33}}{\partial x_2{}^2}=2\frac{\partial^2 e_{23}}{\partial x_2 x_3} \quad \text{(A.3.3.2)}$$

$$\frac{\partial^2 e_{33}}{\partial x_1{}^2}+\frac{\partial^2 e_{11}}{\partial x_3{}^2}=2\frac{\partial^2 e_{31}}{\partial x_3 x_1} \quad \text{(A.3.3.3)}$$

$$\frac{\partial^2 e_{11}}{\partial x_2 x_3}=\frac{\partial}{\partial x_1}\left(-\frac{\partial e_{23}}{\partial x_1}+\frac{\partial e_{31}}{\partial x_2}+\frac{\partial e_{12}}{\partial x_3}\right) \quad \text{(A.3.3.4)}$$

$$\frac{\partial^2 e_{22}}{\partial x_3 x_1}=\frac{\partial}{\partial x_2}\left(+\frac{\partial e_{23}}{\partial x_1}-\frac{\partial e_{31}}{\partial x_2}+\frac{\partial e_{12}}{\partial x_3}\right) \quad \text{(A.3.3.5)}$$

$$\frac{\partial^2 e_{33}}{\partial x_1 x_2}=\frac{\partial}{\partial x_3}\left(+\frac{\partial e_{23}}{\partial x_1}+\frac{\partial e_{31}}{\partial x_2}-\frac{\partial e_{12}}{\partial x_3}\right) \quad \text{(A.3.3.6)}$$

Stress-Strain ($\sigma_{ij} - e_{kl}$) Relation for Triclinic Material

$$\begin{Bmatrix} \sigma_{11} \\ \sigma_{22} \\ \sigma_{33} \\ \sigma_{23} \\ \sigma_{31} \\ \sigma_{12} \end{Bmatrix} = \begin{bmatrix} \mathbb{C}_{11} & \mathbb{C}_{12} & \mathbb{C}_{13} & \mathbb{C}_{14} & \mathbb{C}_{15} & \mathbb{C}_{16} \\ \mathbb{C}_{21} & \mathbb{C}_{22} & \mathbb{C}_{23} & \mathbb{C}_{24} & \mathbb{C}_{25} & \mathbb{C}_{26} \\ \mathbb{C}_{31} & \mathbb{C}_{32} & \mathbb{C}_{33} & \mathbb{C}_{34} & \mathbb{C}_{35} & \mathbb{C}_{36} \\ \mathbb{C}_{41} & \mathbb{C}_{42} & \mathbb{C}_{43} & \mathbb{C}_{44} & \mathbb{C}_{45} & \mathbb{C}_{46} \\ \mathbb{C}_{51} & \mathbb{C}_{52} & \mathbb{C}_{53} & \mathbb{C}_{54} & \mathbb{C}_{55} & \mathbb{C}_{56} \\ \mathbb{C}_{61} & \mathbb{C}_{62} & \mathbb{C}_{63} & \mathbb{C}_{64} & \mathbb{C}_{65} & \mathbb{C}_{66} \end{bmatrix} \begin{Bmatrix} e_{11} \\ e_{22} \\ e_{33} \\ 2e_{23} \\ 2e_{31} \\ 2e_{12} \end{Bmatrix} \quad \text{(A.3.4)}$$

Stress-Strain ($\sigma_{ij} - e_{kl}$) Relation for Isotropic Material

$$\begin{Bmatrix} \sigma_{11} \\ \sigma_{22} \\ \sigma_{33} \\ \sigma_{23} \\ \sigma_{31} \\ \sigma_{12} \end{Bmatrix} = \begin{bmatrix} \mathbb{L}+2\mathbb{m} & \mathbb{L} & \mathbb{L} & 0 & 0 & 0 \\ & \mathbb{L}+2\mathbb{m} & \mathbb{L} & 0 & 0 & 0 \\ & & \mathbb{L}+2\mathbb{m} & 0 & 0 & 0 \\ & & & \mathbb{m} & 0 & 0 \\ & Sym & & & \mathbb{m} & 0 \\ & & & & & \mathbb{m} \end{bmatrix} \begin{Bmatrix} e_{11} \\ e_{22} \\ e_{33} \\ 2e_{23} \\ 2e_{31} \\ 2e_{12} \end{Bmatrix} \quad \text{(A.3.5)}$$

3.11.2 Important Equations in Cylindrical Coordinate System

3.11.2.1 Transformation to Cylindrical Coordinate System

Transformation of a tensor is discussed in sections 2.4 and 2.5 in Chapter 2. In that section universal rule for any generalized coordinate transformation is discussed. However, here specifically the transformation of the basis vectors in Cartesian coordinate system to the basis vectors of polar coordinate system is discussed in detail. This is to find the Ψ_{ij} matrix discussed in Eq. (2.34) to establish the relation between most two popular coordinate systems. Here, the polar coordinate system and cylindrical coordinate system are synonymous. To initiate the discussion it is necessary to refer Fig. A.3.1. First a right hand coordinate system is shown at the corner of the figure to visualize the transformation. In cylindrical coordinate system the x_3 axis remain same. Hence, the basis vector along x_3 axis remains same which is $\widehat{\mathbf{e}_3}$. But other two basis vectors are one along the radial direction (r) which is $\hat{\mathbf{e}}_r$ and along angular direction (θ) which is $\hat{\mathbf{e}}_\theta$ as shown in Fig. A.3.1. The x_3 axis is rotated by an angle θ. Let's consider any arbitrary point P which is now located at x_1, x_2, x_3. After coordinate transformation, location of the point P does not change but the representation of the point changes to say R, Θ, x_3. OA and OC in Fig. A.3.1 are x_1 and x_2, respectively. Similarly, OB and OD in Fig. A.3.1 are R and Θ, respectively. Based on geometry, OB can be seen as a summation of OF and FB. Based on these geometrical relations, following expressions can be written as

$$OA = CP = x_1 \ ; OC = AP = x_2 \ ; OB = OF + FB = x_1 \cos\theta + x_2 \sin\theta \quad \text{(A.3.6)}$$

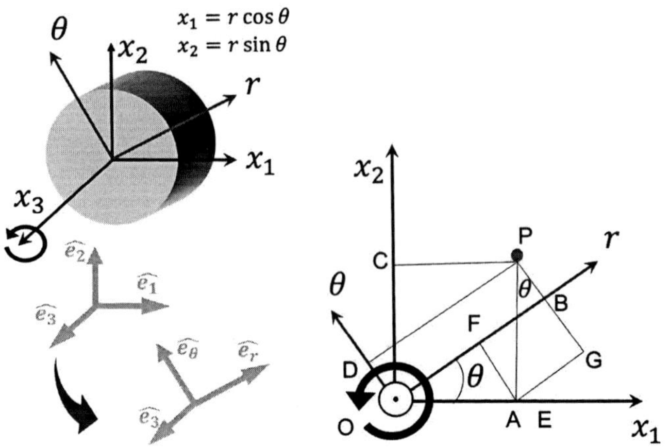

FIGURE A.3.1 Cylindrical coordinate system: description of a particle in transformed coordinate system.

Similarly,

$$OD = BP = GP - GB = GP - AF = x_2 \cos\theta - x_1 \sin\theta \tag{A.3.7}$$

Since, $OB = R$ and $OD = \Theta$, the relation between R, Θ, x_3 and x_1, x_2, x_3 can be written as

$$\begin{Bmatrix} R \\ \Theta \\ x_3 \end{Bmatrix} = \begin{bmatrix} \cos\theta & \sin\theta & 0 \\ -\sin\theta & \cos\theta & 0 \\ 0 & 0 & 1 \end{bmatrix} \begin{Bmatrix} x_1 \\ x_2 \\ x_3 \end{Bmatrix} \tag{A.3.8}$$

Hence, the basis vectors for the respective coordinate systems are also related by

$$\begin{Bmatrix} \hat{\mathbf{e}}_r \\ \hat{\mathbf{e}}_\theta \\ \hat{\mathbf{e}}_3 \end{Bmatrix} = \begin{bmatrix} \cos\theta & \sin\theta & 0 \\ -\sin\theta & \cos\theta & 0 \\ 0 & 0 & 1 \end{bmatrix} \begin{Bmatrix} \hat{\mathbf{e}}_1 \\ \hat{\mathbf{e}}_2 \\ \hat{\mathbf{e}}_3 \end{Bmatrix} \tag{A.3.9}$$

The transformation matrix Ψ_{ij} in Eq. (2.34) for transforming the coordinates or tensors from Cartesian to cylindrical coordinate will be

$$\Psi = \begin{bmatrix} \cos\theta & \sin\theta & 0 \\ -\sin\theta & \cos\theta & 0 \\ 0 & 0 & 1 \end{bmatrix} \tag{A.3.10}$$

Using this rule the displacement vector u_i in Cartesian coordinate system can be transformed to polar coordinate as follows:

$$\begin{Bmatrix} u_r \\ u_\theta \\ u_3 \end{Bmatrix} = \begin{bmatrix} \cos\theta & \sin\theta & 0 \\ -\sin\theta & \cos\theta & 0 \\ 0 & 0 & 1 \end{bmatrix} \begin{Bmatrix} u_1 \\ u_2 \\ u_3 \end{Bmatrix} \tag{A.3.11}$$

3.11.2.2 Gradient Operator in Cylindrical Coordinate System

The gradient operator ∇ discussed in Section 2.3.1 in Cartesian coordinate system can be expressed by

$$\nabla = \left(\frac{\partial}{\partial x_1} \hat{\mathbf{e}}_1 + \frac{\partial}{\partial x_2} \hat{\mathbf{e}}_2 + \frac{\partial}{\partial x_3} \hat{\mathbf{e}}_3 \right) \tag{A.3.12}$$

Which is similarly written in cylindrical coordinate system as follows:

$$\nabla = \left(\frac{\partial}{\partial r} \hat{\mathbf{e}}_r + \frac{1}{r}\frac{\partial}{\partial \theta} \hat{\mathbf{e}}_\theta + \frac{\partial}{\partial x_3} \hat{\mathbf{e}}_3 \right) \tag{A.3.13}$$

- Gradient of a scalar field $\varphi(x_j,t)$ in Cartesian coordinate system is written in Eq. (2.22) and similarly, gradient of a scalar field $\varphi(r,\theta,x_3,t)$ in cylindrical coordinate system can be written as

$$Grad\left(\varphi(r,\theta,x_3,t)\right) = \nabla\varphi = \left(\frac{\partial \varphi}{\partial r} \hat{\mathbf{e}}_r + \frac{1}{r}\frac{\partial \varphi}{\partial \theta} \hat{\mathbf{e}}_\theta + \frac{\partial \varphi}{\partial x_3} \hat{\mathbf{e}}_3 \right) \tag{A.3.14}$$

- Divergence of a vector field $\mathbf{V} = V_i(x_j,t)\hat{\mathbf{e}}_i$ in Cartesian coordinate system is written in Eq. (2.24) and similarly, divergence of the following vector field $\mathbf{V} = V_r(r,\theta,x_3,t)\hat{\mathbf{e}}_r + V_\theta(r,\theta,x_3,t)\hat{\mathbf{e}}_\theta + V_3(r,\theta,x_3,t)\hat{\mathbf{e}}_3$ in cylindrical coordinate system can be written as

$$\nabla.\mathbf{V} = \left(\frac{\partial V_r}{\partial r} + \frac{1}{r}\frac{\partial V_\theta}{\partial \theta} + \frac{\partial V_3}{\partial x_3} \right) \tag{A.3.15}$$

- Curl of a vector field $\mathbf{V} = V_i(x_j,t)\hat{\mathbf{e}}_i$ in Cartesian coordinate system is written in Eq. (2.27) and similarly, curl of the following vector field $\mathbf{V} = V_r(r,\theta,x_3,t)\hat{\mathbf{e}}_r + V_\theta(r,\theta,x_3,t)\hat{\mathbf{e}}_\theta + V_3(r,\theta,x_3,t)\hat{\mathbf{e}}_3$ in cylindrical coordinate system can be written as

$$\nabla \times \mathbf{V} = \hat{\mathbf{e}}_r \left(\frac{1}{r}\frac{\partial V_3}{\partial \theta} - \frac{\partial V_\theta}{\partial x_3} \right) + \hat{\mathbf{e}}_\theta \left(-\frac{\partial V_3}{\partial r} + \frac{\partial V_r}{\partial x_3} \right) + \hat{\mathbf{e}}_3 \left(\frac{1}{r}\frac{\partial (rV_\theta)}{\partial r} - \frac{1}{r}\frac{\partial V_r}{\partial \theta} \right) \tag{A.3.16}$$

3.11.2.3 Strain-Displacement Relation in Cylindrical Coordinate System

Strain-displacement relations in Cartesian coordinate system are written in Eqs. (A.72.1) through (A.72.6). Similarly the strain-displacement relations in cylindrical coordinate system can be written as

$$e_{rr} = \left(\frac{\partial u_r}{\partial r}\right) \tag{A.3.17.1}$$

$$e_{\theta\theta} = \left(\frac{u_r}{r} + \frac{1}{r}\frac{\partial u_\theta}{\partial \theta}\right) \tag{A.3.17.2}$$

$$e_{33} = \left(\frac{\partial u_3}{\partial x_3}\right) \tag{A.3.17.3}$$

$$e_{r\theta} = \frac{1}{2}\left(\frac{1}{r}\frac{\partial u_r}{\partial \theta} - \frac{u_\theta}{r} + \frac{\partial u_\theta}{\partial r}\right) \tag{A.3.17.4}$$

$$e_{\theta 3} = \frac{1}{2}\left(\frac{1}{r}\frac{\partial u_3}{\partial \theta} + \frac{\partial u_\theta}{\partial x_3}\right) \tag{A.3.17.5}$$

$$e_{3r} = \frac{1}{2}\left(\frac{\partial u_r}{\partial x_3} + \frac{\partial u_3}{\partial r}\right) \tag{A.3.17.6}$$

3.11.2.4 Governing Differential Equations of Motion in Cylindrical Coordinate System

Governing differential equation of motion in Cartesian coordinate system is written in Eqs. (A.71.1) through (A.71.3). Similarly the governing equation of motion in cylindrical coordinate system can be written as

$$\frac{\partial \sigma_{rr}}{\partial r} + \frac{1}{r}\frac{\partial \sigma_{r\theta}}{\partial \theta} + \frac{\partial \sigma_{r3}}{\partial x_3} + \frac{1}{r}(\sigma_{rr} - \sigma_{\theta\theta}) + f_r = \rho\frac{\partial^2 u_r}{\partial t^2} \tag{A.3.18.1}$$

$$\frac{\partial \sigma_{\theta r}}{\partial r} + \frac{1}{r}\frac{\partial \sigma_{\theta\theta}}{\partial \theta} + \frac{\partial \sigma_{\theta 3}}{\partial x_3} + \frac{2}{r}(\sigma_{r\theta}) + f_\theta = \rho\frac{\partial^2 u_\theta}{\partial t^2} \tag{A.3.18.2}$$

$$\frac{\partial \sigma_{3r}}{\partial r} + \frac{1}{r}\frac{\partial \sigma_{3\theta}}{\partial \theta} + \frac{\partial \sigma_{33}}{\partial x_3} + \frac{1}{r}(\sigma_{3r}) + f_3 = \rho\frac{\partial^2 u_3}{\partial t^2} \tag{A.3.18.3}$$

3.11.3 Important Equations in Spherical Coordinate System

In spherical coordinate system the unit vectors are $\hat{\mathbf{e}}_r$, $\hat{\mathbf{e}}_\theta$, and $\hat{\mathbf{e}}_\phi$ as shown in Fig. A.3.2. The displacement field in transformed coordinate system is

$$\mathbf{u} = u_r(r,\theta,\phi,t)\hat{\mathbf{e}}_r + u_\theta(r,\theta,\phi,t)\hat{\mathbf{e}}_\theta + u_\phi(r,\theta,\phi,t)\hat{\mathbf{e}}_\phi \tag{A.3.19}$$

3.11.3.1 Gradient Operator in Spherical Coordinate System

The gradient operator ∇ discussed in Section 2.3.1 in Cartesian coordinate system, which is similarly written in cylindrical spherical coordinate system as follows:

$$\nabla = \left(\frac{\partial}{\partial r}\hat{\mathbf{e}}_r + \frac{1}{r\sin\phi}\frac{\partial}{\partial\theta}\hat{\mathbf{e}}_\theta + \frac{1}{r}\frac{\partial}{\partial\phi}\hat{\mathbf{e}}_\phi \right) \tag{A.3.20}$$

- Gradient of a scalar field $\varphi(r,\theta,\phi,t)$ in spherical coordinate system can be written as

$$Grad\,(\varphi(r,\theta,x_3,t)) = \nabla\varphi = \left(\frac{\partial\varphi}{\partial r}\hat{\mathbf{e}}_r + \frac{1}{r\sin\phi}\frac{\partial\varphi}{\partial\theta}\hat{\mathbf{e}}_\theta + \frac{1}{r}\frac{\partial\varphi}{\partial\phi}\hat{\mathbf{e}}_\phi \right) \tag{A.3.21}$$

- Divergence of the following vector field $\mathbf{V} = V_r(r,\theta,\phi,t)\hat{\mathbf{e}}_r + V_\theta(r,\theta,\phi,t)\hat{\mathbf{e}}_\theta + V_\phi(r,\theta,\phi,t)\hat{\mathbf{e}}_\phi$ in spherical coordinate system can be written as

$$\nabla.\mathbf{V} = \left(\frac{\partial V_r}{\partial r} + \frac{1}{r\sin\phi}\frac{\partial V_\theta}{\partial\theta} + \frac{1}{r}\frac{\partial V_\phi}{\partial\phi} \right) \tag{A.3.22}$$

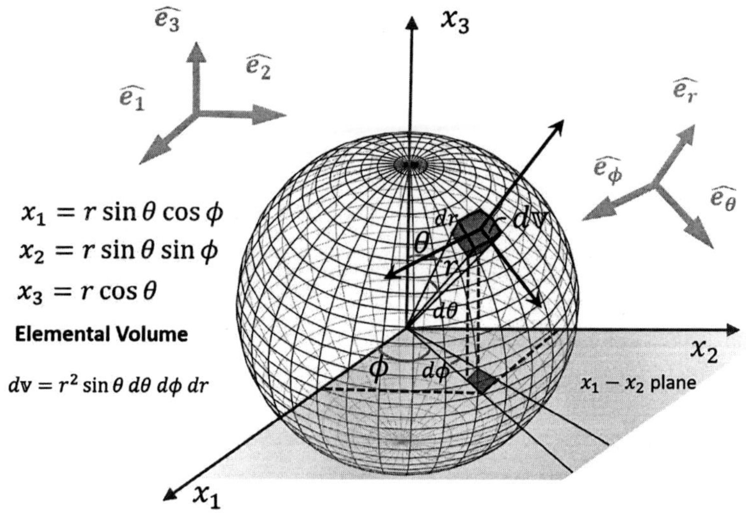

FIGURE A.3.2 Spherical coordinate system: description of a particle and a volume element in transformed coordinate system.

- Curl of the following vector field $\mathbf{V} = V_r(r,\theta,\phi,t)\hat{\mathbf{e}}_r + V_\theta(r,\theta,\phi,t)\hat{\mathbf{e}}_\theta + V_\phi(r,\theta,\phi,t)\hat{\mathbf{e}}_\phi$ in spherical coordinate system can be written as

$$\nabla \times \mathbf{V} = \frac{1}{r^2 \sin\phi}\left[\hat{\mathbf{e}}_r\left(\frac{\partial(rV_\phi)}{\partial\theta} - \frac{\partial(r\sin\phi V_\theta)}{\partial\phi}\right) + \hat{\mathbf{e}}_\theta r\sin\phi\left(-\frac{\partial(rV_\phi)}{\partial r} + \frac{\partial V_r}{\partial\phi}\right) + \hat{\mathbf{e}}_\phi r\left(\frac{\partial(r\sin\phi V_\theta)}{\partial r} - \frac{\partial V_r}{\partial\theta}\right)\right] \tag{A.3.23}$$

3.11.3.2 Strain-Displacement Relation in Spherical Coordinate System

The strain-displacement relations in spherical coordinate system can be written as

$$e_{rr} = \left(\frac{\partial u_r}{\partial r}\right) \tag{A.3.24.1}$$

$$e_{\theta\theta} = \left(\frac{u_r}{r} + \frac{1}{r\sin\phi}\frac{\partial u_\theta}{\partial\theta} + \frac{u_\phi}{r}\cot\phi\right) \tag{A.3.24.2}$$

$$e_{\phi\phi} = \left(\frac{u_r}{r} + \frac{1}{r}\frac{\partial u_\phi}{\partial\phi}\right) \tag{A.3.24.3}$$

$$e_{r\theta} = \frac{1}{2}\left(\frac{1}{r\sin\phi}\frac{\partial u_r}{\partial\theta} - \frac{u_\theta}{r} + \frac{\partial u_\theta}{\partial r}\right) \tag{A.3.24.4}$$

$$e_{\theta\phi} = \frac{1}{2}\left(\frac{1}{r\sin\phi}\frac{\partial u_\phi}{\partial\theta} + \frac{1}{r}\frac{\partial u_\theta}{\partial\phi} - \frac{u_\theta}{r}\cot\phi\right) \tag{A.3.24.5}$$

$$e_{\phi r} = \frac{1}{2}\left(\frac{1}{r}\frac{\partial u_r}{\partial\phi} + \frac{\partial u_\phi}{\partial r} - \frac{u_\phi}{r}\right) \tag{A.3.24.6}$$

3.11.3.3 Governing Differential Equations of Motion in Spherical Coordinate System

Governing differential equation of motion in Cartesian coordinate system is written in Eqs. (A.71.1) through (A.71.3). Similarly the governing equation of motion in spherical coordinate system can be written as

$$\frac{\partial\sigma_{rr}}{\partial r} + \frac{1}{r\sin\phi}\frac{\partial\sigma_{r\theta}}{\partial\theta} + \frac{1}{r}\frac{\partial\sigma_{r\phi}}{\partial\phi} + \frac{1}{r}\left(2\sigma_{rr} - \sigma_{\theta\theta} + \sigma_{r\phi}\cot\phi - \sigma_{\phi\phi}\right) + f_r = \rho\frac{\partial^2 u_r}{\partial t^2} \tag{A.3.25.1}$$

$$\frac{\partial \sigma_{\theta r}}{\partial r}+\frac{1}{r\sin\phi}\frac{\partial \sigma_{\theta\theta}}{\partial \theta}+\frac{1}{r}\frac{\partial \sigma_{\theta\phi}}{\partial \phi}+\frac{2}{r}\left(\sigma_{\theta\phi}\right)\cot\phi+\frac{3}{r}\left(\sigma_{\theta r}\right)+f_\theta=\rho\frac{\partial^2 u_\theta}{\partial t^2} \quad \text{(A.3.25.2)}$$

$$\frac{\partial \sigma_{\phi r}}{\partial r}+\frac{1}{r\sin\phi}\frac{\partial \sigma_{\phi\theta}}{\partial \theta}+\frac{1}{r}\frac{\partial \sigma_{\phi\phi}}{\partial \phi}+\frac{1}{r}\left(3\sigma_{\phi r}+\cot\theta\left(\sigma_{\phi\phi}-\sigma_{\theta\theta}\right)\right)$$
$$+f_\phi=\rho\frac{\partial^2 u_\phi}{\partial t^2} \quad \text{(A.3.25.3)}$$

3.11.4 Fundamental Concept of Classical Mechanics

Wave propagation in a medium is called elastodynamic problem. It means that the system is dynamic and continuously evolving over time. According to classical mechanics [6] a continuously evolving system can be described in Fig. A.3.3. Let's assume that a system was at state $\mathbb{S}_1$ at time t_1 and has evolved to a state $\mathbb{S}_2$ after time t_2. The path that it takes to evolve say $\boldsymbol{q}(t)$ as shown in Fig. A.3.3. Now a classical question can be asked for this system. Is it necessary for the system to take this particular path $\boldsymbol{q}(t)$? Or any other path was possible? This can be answered by perturbing the path $\boldsymbol{q}(t)$ slightly by $\delta\boldsymbol{q}$. According to the principle of least action in classical mechanics it is said that the action ($\mathfrak{L}$) which is the spatio-temporal integration of a scalar function called Lagrangian ($\mathcal{L}$) defined for a system should be minimum during an evolution. In plain words, a system is a lazy individual who tries to do minimum work by performing minimum action during the evolution. This small statement is universal for any evolution of physical system to biological system to our Universe. The Lagrangian scalar function of a system is nothing but the difference between the kinetic energy and the potential energy. Later in the light of Chapter 4, a refined definition is presented in Section 4.5.2. Hence, based in this

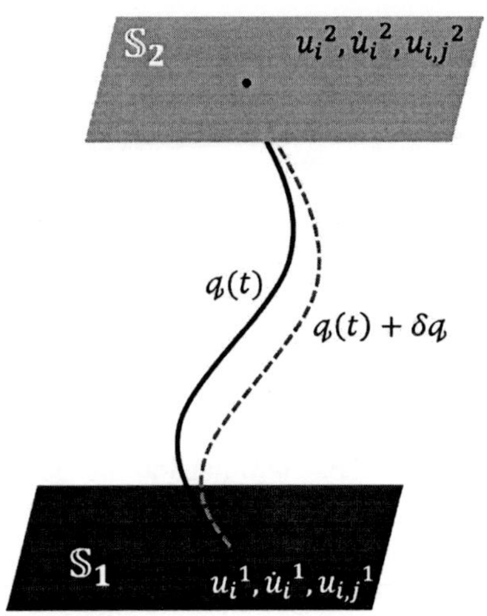

FIGURE A.3.3 A continuously evolving system: evolution of the system from state $\mathbb{S}\_\mathbf{1}$ to $\mathbb{S}\_\mathbf{2}$ through a most optimized path $\boldsymbol{q}(t)$.

principle of least action, if the path is slightly perturbed, the perturbed action ($\delta\mathfrak{L}$) must also be equal to zero. This means that at state $\mathbb{S}_1$, all the state variables (which defines the system explicitly, e.g., in a deformable body displacements u_i, velocities $\dot{u}_i$, and spatial derivatives $u_{i,j}$ are the state variables) of the system, has evolved from one set of values $(u_i^1, \dot{u}_i^1, u_{i,j}^1)$ to another set of values $(u_i^2, \dot{u}_i^2, u_{i,j}^2)$ either through a path $q(t)$ or slightly deviated path $q(t) \pm \delta q$, the variation of the action $\delta\mathfrak{L}$ must vanish. This principle is called the Hamilton's principle in classical mechanics [6, 7]. Mathematically the variation of action can be written as

$$\delta\mathfrak{L} = \delta\int_{t_1}^{t_2}\int_{v} \mathcal{L}\left(u_i, \dot{u}_i, u_{i,j}\right) dv\, dt = \delta\int_{t_1}^{t_2}\int_{v} \mathcal{L}\left(u_i, \dot{u}_i, u_{i,j}\right) dx_j\, dt = 0 \qquad \text{(A.3.26)}$$

This same principle is used in Section 4.5.2 in Chapter 4 to derive the governing differential equation for wave propagation using integral approach.

Although not used in this book, a brief introduction to the Hamiltonian equations are introduced herein. Please note Hamilton's principle and Hamiltonian equations are not the same. If a system is described by two generalized coordinates, for example, position x_i and momentum $p_i = \rho\dot{u}_i$ (the trajectory of evolution is $q(\mathbf{x}, \mathbf{p})$) where, both position and momentum are vectors, $\mathbf{x}$ and $\mathbf{p}$), then the Hamiltonian equation of evolution can be written as

$$\frac{d\mathbf{p}}{dt} = -\frac{\partial H}{\partial \mathbf{x}} \qquad \text{(A.3.27.1)}$$

$$\frac{d\mathbf{x}}{dt} = +\frac{\partial H}{\partial \mathbf{p}} \qquad \text{(A.3.27.2)}$$

where H is called the Hamiltonian of the system. In contrast to the subtraction in Lagrangian, Hamiltonian is described by the summation of the kinetic energy and the potential energy of the system. The trajectory or the path of evolution $q(\mathbf{x}, \mathbf{p})$ is the solution of the Hamiltonian equation with prescribed initial conditions.

3.12 SUMMARY

Fundamental of continuum mechanics or mechanics of deformable bodies is reviewed in this chapter tailored to the understanding of wave propagation in deformable bodies. Eulerian and Lagrangian coordinate systems are reviewed to give better understanding what coordinate system should be used in explaining the wave propagation. Fundamental elastodynamic equation is derived from the fundamentals and later introducing the thermodynamics of continua internal energies and strain energy densities are explained. With different plane of symmetry, different materials are introduced with their respective necessary material coefficients. Wave propagation in different material types requires understanding of required material coefficients if a particular material type is anisotropic, depending on their degree of anisotropy. Briefly in Appendix, elastodynamic equations, stain-displacement relations are introduced in cylindrical and spherical coordinate systems.

REFERENCES

1. Capaldi, F., *Continuum Mechanics: Constitutive Modeling of Structural and Biological Materials*. 2012, Cambridge: Cambridge University Press.
2. Eringen, A.C., *Mechanics of Continua*. 1980, Huntington, New York: Robert E. Krieger Publishing Company.
3. Thomas Mase, G., Smelser, R.E., Rossmann, J.S., *Continuum Mechanics for Engineers*, 3rd ed. 2009, Boca Raton: CRC Press.
4. Lai, W.M., Rubin, D., Krempl, E., *Introduction to Continuum Mechanics*. 1974, New York: Pergamon Press, Inc.
5. Shabana, A.A., *Computational Continuum Mechanics*, 3rd ed. 2018, Hoboken, New Jersey: John Wiley & Sons, Ltd.
6. Morin, D., *Introduction to Classical Mechanics*. 2007, Cambridge: Cambridge University Press.
7. Torres del Castillo, G., *An Introduction to Hamiltonian Mechanics*. 2018, Switzerland: Springer.

4 Acoustic and Ultrasonic Waves in Elastic Media

4.1 BASIC TERMINOLOGIES IN WAVE PROPAGATION

In order to discuss wave propagation in fluid or solid media [1–7], clear understanding of the following terminologies is important and is frequently used in this book. A wave could be a physical quantity like, stress, displacement, or velocity components in the host media. A host media is a material or a structure of which nondestructive evaluation is to be performed. Energy is introduced in the material using an ultrasonic transducer which creates the disturbance in the medium and then the disturbance is transmitted from one point to the next. This disturbance is essentially the wave which progress through the medium and is called wave propagation and essential physical phenomena needs deeper physical understanding for their computational analysis. The most important terminologies are

4.1.1 Wave Fronts, Rays, and Plane Waves

A wave front is an equi-disturbance surface in three dimensions and a line in two dimensions. For example, considering the particle displacements in a media, a wave front is a surface passing through all the material points that have equal displacements. The normal to the wave front defines the direction of wave propagation is called the rays and generally in isotropic and fluid media the rays are direction of the energy propagation. However, please note that does not mean that the wave energy will always propagates along the direction of the wave propagation. This particular abnormality will be discussed in Section 4.3. Let's consider an isotropic media in Fig. 4.1. A point source is excited in the media (Fig. 4.1a) and the energy is isotopically propagated along all directions. This creates a wave field and a snapshot in time is presented in Fig. 4.1b. The red lines in Fig. 4.1b are having equal displacement and these lines advance in time making the wave front. The red lines are very small as they are taken from the small square box on the wave front in Fig. 4.1a. Collectively joining such many lines will result the shape of the wave fronts. In fluid and in isotropic solid materials, the wave fronts are circular in two-dimensional geometries and spherical in three-dimensional geometries. However, circular or spherical wave fronts are not inevitable. Waves having circular cylindrical surface as the wave front are called circular crested waves. Similarly, waves having flat surface or a straight-line surface as the wave front are called straight crested waves. In anisotropic materials wave fronts may take different shapes based on different wave velocities at different directions. Wave fronts are inherently related with wave velocities.

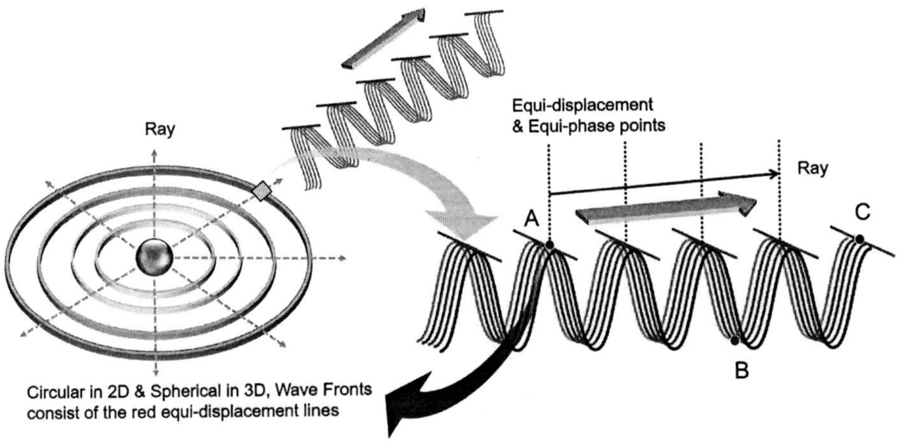

FIGURE 4.1 a) Propagation of waves due to a point source in a media creating circular wave fronts b) schematic representation of a Ray, equi-displacement and equi-phase points.

4.1.2 Phase Wave Velocity

Intuitively, it is apparent from Fig. 4.1a and b that the wave fronts will be circular, if the wave energy in the media moves at constant velocity in all the possible directions. From our background knowledge we know that the wave propagation in a media is resulted from transferring of the wave energy (introduced by a source, here ultrasonic transducer) from one particle to the next through their local displacements, but the particles do not sway along with the wave itself. Particle at point A and B in Fig. 4.1b will remain at their respective positions but the energy at point A will transfer to the point B after certain period based on the wave velocity in the media. After certain period in time, point B will have the similar displacement pattern like now at A and C, and A and C will similar displacement pattern like B. Point C and A are in same phase and have same displacement but four wavelengths away. Hence, between A and C there are total five wave fronts. According to the definition of velocity, we know that distance travel by unit time is called velocity. Hence, for wave same definition is applicable. Wave velocity is defined by the distance travel by the wave energy per unit time. For example, if the wave travels a distance of x in time t, and the wave velocity is c then the propagation of the wave can be defined by a function $f(x-ct)$ of the physical quantities. The wave function has to be constant over the planes (red lines in Fig. 4.1b) which have their normal along the direction of the ray at any arbitrary distance $x+\Delta x$ after a time period of $t+\Delta t$, we can write

$$f(x-ct)=f\left(x+\Delta x-c(t+\Delta t)\right) \tag{4.1}$$

The argument of the function must be constant to hold the equation to be true. Hence, equating the arguments we get

$$c=\frac{\Delta x}{\Delta t}=\frac{dx}{dt} \tag{4.2}$$

To be accurate, c is actually the phase velocity of the medium. Here let's introduce two more terms. Frequency (f) and wave number (w). Wave number is defined as the number of waves per unit length, thus the SI unit of wave number is per meter $(1/m)$. Frequency is the number of wave cycles per unit time and thus the SI unit is radian per second (1/sec). Hence, to get the unit of velocity $\left(\frac{m}{\text{sec}}\right)$ it is clear that we can write

$$c = \frac{f}{w} \tag{4.3}$$

Physically this equation is obvious. Please refer Fig. 4.1b. Let's consider the points A and C and assume the wave energy took time t to travel from A to C. Between A and C there are four wave lengths with each wavelength equal to say λ. Hence, 4λ is traveled by the wave in t sec and simple the phase wave velocity is $c = \frac{4\lambda}{t}$. Now using the definition of wave number, we can write $w = \frac{1}{\lambda}$. Frequency ω by definition is the number of cycles traveled in unit time. From point A to point C the wave energy traveled 4 cycles in time t. Thus $f = \frac{4}{t}$. Using Eq. (4.3) we still get the same phase wave velocity $c = \frac{4\lambda}{t}$. Please note that the number 4 was just an example. The numeric number 4 can be replaced by any arbitrary rational number n and $c = \frac{n\lambda}{t}$, because A and C could be arbitrarily chosen at any distance apart and can be written as any fraction of the wave length λ.

4.1.3 Plane Harmonic Wave

Discussion of wave number and frequency is not complete, unless we discuss the plane harmonic waves in a media. Here we will restrict our discussion to a one-dimensional problem. A plane harmonic wave is described by the time variation of the form $e^{-i\omega t}$ called the time harmonic part of the wave, where ω is the angular frequency. On the contrary spatial distribution of any physical parameters (e.g., stress, displacements, or velocities) of the wave field is predominantly dependent on their phase part of the harmonic wave form. The phase part of the plane harmonic wave is described as e^{ikx}, where k is the wave number. For a plane harmonic wave, any arbitrary wave parameter can be expressed as a potential function with amplitude (A), for example, the potential function for the displacement in a media at any spatial and temporal point can be written as $Ae^{i(kx-\omega t)}$. Wave parameters expressed in terms of plane harmonic waves can be expressed as a potential function as follows

$$\phi = Ae^{i(kx-\omega t)} = Ae^{ik\left(x-\frac{\omega}{k}t\right)} \tag{4.4}$$

Fig. 4.2 shows one-dimensional wave propagation phenomena on a spatio-temporal plane. Two lines from the figure are extracted, one at a constant time another at a constant spatial point. The wave forms along those two lines are also presented in the figure. At constant time, the wave form shows the spatial variation of the wave parameter with wavelength λ. At a fixed spatial point, the wave form shows the temporal variation with time period T. One cycle of wave is considered 2π radians.

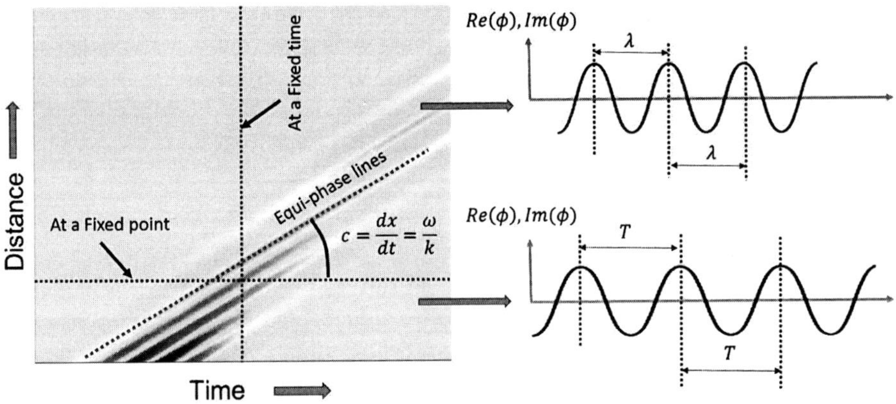

FIGURE 4.2 One dimensional wave propagation on a spatio-temporal plane, it shows during wave propagation, how the crests and troughs of a wave signal changes their location in time when evaluated along a distance.

As depicted in Fig. 4.2, if the fixed point is moved to a new spatial point by $\frac{2\pi}{k}$ or the temporal point is moved to a new point by $\frac{2\pi}{\omega}$, numerical value of the potential function (ϕ) will retain the same value. Hence, by definition wavelength and time period can be expressed as

$$\lambda = \frac{2\pi}{k} \quad ; \quad k = 2\pi w \quad ; \quad T = \frac{2\pi}{\omega} \tag{4.5}$$

Wave number k is the number of wavelengths over a dimensionless distance 2π. Similarly, the angular frequency ω is the number of cycles (full time period) over a dimensionless period of 2π. There is a different between just frequency and angular frequency. Frequency f introduced before for a wave is the number of cycles propagates per unit time. As one cycle is 2π radian, the angular frequency ω is defines as

$$\omega = 2\pi f \tag{4.6}$$

Recollecting Eq. (4.3), and substitute the respective expressions for f and w, the phase wave velocity can be written as

$$c = \frac{\omega}{k} \tag{4.7}$$

Hence, the harmonic wave potential in Eq. (4.4) can be further modified to

$$\phi = Ae^{ik(x-ct)} = f(x-ct) \tag{4.8}$$

Here, the term $x - ct$ is the phase and remain constant to retain the constant value of the plane wave potential ϕ. It is apparent that the phase wave velocity c is the slope of the equi-phase lines as shown in Fig. 4.2. It is timely to introduce another parameter that will be valuable later is the slowness. Slowness is the inverse of the phase wave velocity. Hence, phase slowness can be defined as $s = 1/c = k/\omega$ and its unit is sec/m in SI unit.

4.1.4 Wave Groups and Group Wave Velocity

As the name of the section indicates, this situation only occurs when multiple different plane waves or a group of plane waves with closely matching phase wave velocities and wave numbers (i.e., closely matching frequencies) interact with the each other and propagate with additional features in the medium. In fact, this case we must handle most of the time in wave propagation. Let's assume displacement (u) in one-dimensional medium is the superposition of two plane waves with close but not the same frequencies and the superposition of the plane waves may result

$$u = Ae^{i(k_1 x - \omega_1 t)} + Ae^{i(k_2 x - \omega_2 t)} \tag{4.9}$$

where k_1, k_2 and ω_1, ω_2 are two closely spaced wave numbers and angular frequencies deviated from the central k and ω as follows.

$$k_1 = k - \Delta k \;\; ; \;\; k_2 = k + \Delta k \;\; ; \;\; \omega_1 = \omega - \Delta\omega \;\; ; \;\; \omega_2 = \omega + \Delta\omega \tag{4.10}$$

Considering the real part of the displacement function we can write

$$Re(u) = A\cos\left((k - \Delta k)x - (\omega - \Delta\omega)t\right) + A\cos\left((k + \Delta k)x - (\omega + \Delta\omega)t\right) \tag{4.11}$$

Applying standard trigonometric identities and simplifying Eq. (4.11) we get

$$Re(u) = 2A\cos(kx - \omega t)\cos(\Delta kx - \Delta\omega t) \tag{4.12}$$

The wave in Eq. (4.12) has two components, one is with velocity $c = \frac{\omega}{k}$ and another with velocity $c_g = \frac{\Delta\omega}{\Delta k} = \frac{d\omega}{dk}$. The former, we discussed before, is called the phase wave velocity and later is called the group wave velocity. In a medium due to superposition of multiple plane wave potentials with different frequencies and wave numbers causes the wave to form wave packets and not all wave packets move with the same velocities. Fig. 4.3 shows a typical case of simple group waves derived in Eq. (4.12). Here, the particles are experiencing the phase velocity $c = \frac{\omega}{k}$ called the carrier waves. However, there is another wave with larger wavelength called the modulation wave is caused by the superposition of the plane waves. This wave travels with a velocity $c_g = \frac{d\omega}{dk}$, which is not the same as the phase velocity. This is the velocity of the group wave or the wave packets. Group velocity is also called the velocity of the wave energy and referred as c_E. Thus, direction of group wave velocity and the direction of wave energy velocity are synonymous. A small mathematical calculation to justify this synonymity is presented in the Appendix of this chapter. In Fig. 4.3, the blue wave is the carrier wave travels with phase velocity and the red wave is the modulation of the wave packet travels with group velocity. In a medium it is possible to have the phase velocity greater, equal, and less than the group wave velocity and

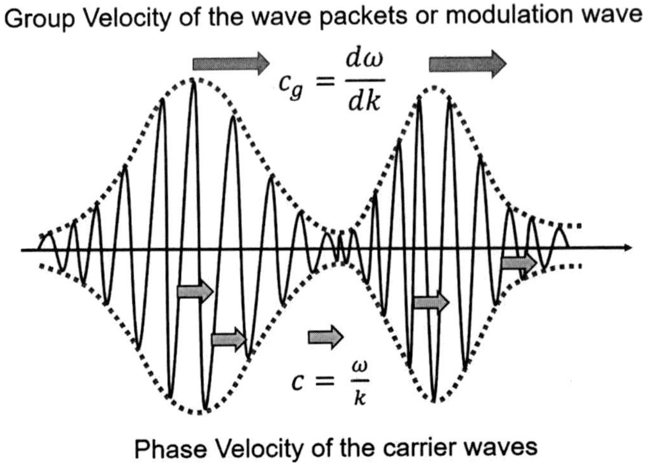

FIGURE 4.3 Schematic representation of phase wave and group wave velocity.

the resulted phenomena can only be visualized through animation, but it is attempted to describe the phenomena in words below.

- $c > c_g$: when phase velocity is greater than the group velocity, wave will cross the wave packets faster than the packet itself. It will appear that the wave is originating at the back of the wave packet and travelling toward the front of the packet.
- $c = c_g$: when phase velocity is equal to the group velocity, the carrier waves will appear to be stand still with respect to the envelop or the wave packets. They will travel together and there will be no distortion of the shape of the wave packets.
- $c < c_g$: when phase velocity is less than the group velocity, carrier waves will fall back and will not be able to keep up with the wave packets which move faster. It will appear that the wave is originating at the front of the wave packet and travelling toward the back of the packet.

4.1.5 Wave Dispersion

Group wave velocity is $c_g = \frac{d\omega}{dk}$ and the angular frequency is $\omega = kc$. Hence, substituting the expression for the angular frequency in the expression for group velocity we get

$$c_g = c + k\frac{dc}{dk} \tag{4.13}$$

In general, a system is said to be dispersive when its phase velocity depends on the angular frequency or in other words on the wavelength. Thus, in a specific

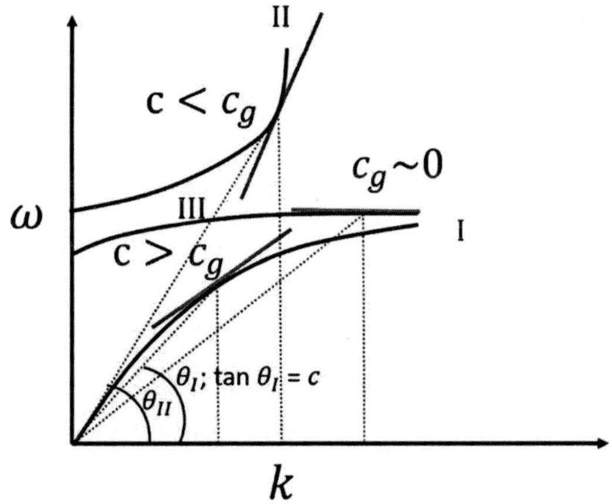

FIGURE 4.4 A hypothetical frequency spectrum in wave-number and frequency domain ($k - \omega$) to explain the situations with the group velocities.

media the angular frequencies and the wave numbers follow a specific dependency called frequency spectrum or dispersion relation. Similar situations described in Section 4.1.4 are discussed again in the light of Eq. (4.13). Fig. 4.4 also describes the situation in the light of a hypothetical frequency spectrum (i.e., a dispersion function describes the relation of angular frequency and wave number in a media). Three (I, II, III) different hypothetical curves are described.

- $c > c_g$, i.e., $\frac{dc}{dk} < 0$: It will appear that the wave is originating at the back of the wave packet and travelling toward the front of the packet and continuously changes the shape of the packet along the length of the travel of the wave. Wave velocity changes with the input frequency. This situation is called positive dispersion. Convex curvature of a dispersion curve generally indicates this situation. For example, curve I in Fig. 4.4 shows that the red line has lower slope than the black dotted line at angle θ_I.
- $c = c_g$, i.e., $\frac{dc}{dk} = 0$ when phase velocity is equal to the group velocity, the carrier waves will appear to be stand still with respect to the envelop or the wave packets. This is a no dispersion situation. Wave packets retain its shape. At very specific cases, i.e., at very specific k and ω this may occur. When the slope of the dispersion curve or the frequency spectrum matches with the slope of the triangles generated by the dotted lines in Fig. 4.4, this situation occurs.
- Additionally, a flat dispersion curve over the wave numbers (curve III in Fig. 4.4) results a unique situation which also appears to be nondispersive, because the group velocity of the wave packet becomes zero. It means that the wave packet does not move. This situation occurs when phase velocity is directly proportional to dc/dk. This signifies that in any specific frequency multiple wave numbers can be present. A straight-line dispersion due to local resonance sometimes results such nondispersive.

- $c < c_g$, i.e., $\frac{dc}{dk} > 0$ when phase velocity is less than the group velocity, carrier waves will fall back and will not be able to keep up with the wave packets which moves faster. Wave velocity changes with the input frequency. This situation is called negative dispersion. Concave curvature of a dispersion curve generally indicates this situation as described through curve II in Fig. 4.4. Here, the slope of the red line is higher than the tan of the angle θ_{II}.

4.2 WAVE PROPAGATION IN FLUID MEDIA

A typical ultrasonic nondestructive evaluation setup always concerns fluid and solid media with fluid-solid interfaces. Here predominantly the incompressible fluids are considered. When ultrasonic energy is excited using a transducer in water or in any incompressible fluid media, the energy propagates in all possible direction resulting spherical wave fronts. The wave fronts in fluid media propagate with a constant velocity irrespective of the frequency of the wave actuated in the media. Hence, a perfect fluid is a nondispersive media which means that the angular frequency (ω) vs. wave number (k) relationship is linear with a constant slope, like it is also hypothetically presented in Fig. 4.4. Perfect incompressible fluid cannot support any shear stresses, and pressure at a point is equal and compressive from all possible directions. This gives us a relation that the normal stresses $\sigma_{11} = \sigma_{22} = \sigma_{33} = -\mathrm{p}$, where $\mathrm{p}(\mathbf{x},t)$ is pressure at a point at an instance in a fluid media. The negative sign is to designate compressive stress that always prevails in fluid. By virtue of its material characteristics, fluid cannot take any tensile stresses. Recalling the wave propagation equation or the governing partial differential equations of motion in any elastic media described in Chapter 3, and eliminating the shear stresses from Eq. (A.3.71) we can write

$$\frac{\partial \sigma_{11}}{\partial x_1} + f_1 = \rho \frac{\partial^2 u_1}{\partial t^2} \quad or \quad -\frac{\partial \mathrm{p}}{\partial x_1} + f_1 = \rho \frac{\partial^2 u_1}{\partial t^2} \tag{4.14.1}$$

$$\frac{\partial \sigma_{22}}{\partial x_2} + f_2 = \rho \frac{\partial^2 u_2}{\partial t^2} \quad or \quad -\frac{\partial \mathrm{p}}{\partial x_2} + f_2 = \rho \frac{\partial^2 u_2}{\partial t^2} \tag{4.14.2}$$

$$\frac{\partial \sigma_{33}}{\partial x_3} + f_3 = \rho \frac{\partial^2 u_3}{\partial t^2} \quad or \quad -\frac{\partial \mathrm{p}}{\partial x_3} + f_3 = \rho \frac{\partial^2 u_3}{\partial t^2} \tag{4.14.3}$$

Referring the gradient of a scalar field (here pressure field) described in Chapter 2 in Eq. (2.23), we can write the above equations in the following from

$$-\nabla \mathrm{p} + \mathbf{f} = \rho \ddot{\mathbf{u}} \quad \text{or} \quad \langle -\mathrm{p}_{,i} + f_i = \rho \ddot{u}_i \rangle \widehat{\mathbf{e}_i} \quad \text{or simply} \quad -\mathrm{p}_{,i} + f_i = \rho \ddot{u}_i \tag{4.15}$$

This shows (middle expression in Eq. (4.15)) that the above equations for three independent directions are analogous to a vector field which has three orthogonal independent components in three directions. $\nabla \mathrm{p}$ the gradient of the scalar pressure field is indeed representing a vector field. Hence, to understand how the wave propagates in fluid it is inevitable to understand how the wave fronts propagate in the media. Wave propagation is conceptually related to the divergence of a vector field, because it seems we are in quest of the same phenomena discussed in Chapter 2 in Section 2.3.2. To find how the radius of the vector front changes over time we need to find the divergence of the field $\nabla \mathrm{p}$, because gradient of the scalar pressure field is a vector field. Hence, we apply the divergence operator on Eq. (4.15) and we get

$$-\nabla.\nabla \mathrm{p}+\nabla.\mathbf{f}=\rho\nabla.\ddot{\mathbf{u}} \quad \text{or} \quad -\mathrm{p}_{,ii}+f_{i,i}=\rho\ddot{u}_{i,i} \tag{4.16}$$

The isotropic constitutive relation presented in Chapter 3, Eq. (3. 86) can be modified for fluid and can be written as $\sigma_{ij}=\mathbb{L}\delta_{ij}e_{kk}$. The constitutive equations for fluid can be further modified considering the discussions in Chapters 2 and 3.

$$-\mathrm{p}=\mathbb{L}\left(e_{11}+e_{22}+e_{33}\right)=\mathbb{L}\left(u_{1,1}+u_{2,2}+u_{3,3}\right)=\mathbb{L}u_{i,i}=\mathbb{L}\ \nabla.\mathbf{u} \tag{4.17}$$

Substituting Eq. (4.17) into Eq. (4.16) we get

$$-\mathrm{p}_{,ii}-\frac{\rho}{\mathbb{L}}\ddot{\mathrm{p}}+f_{i,i}=0 \quad or \quad -\mathrm{p}_{,ii}-\frac{1}{c_f^2}\ddot{\mathrm{p}}+f_{i,i}=0 \tag{4.18}$$

where $c_f=\sqrt{\mathbb{L}/\rho}$ is the wave velocity in the fluid media, which is the square root of the ratio of the modulus ($\mathbb{L}$) and the density (ρ) of the fluid media. Further neglecting body force, we can write the final form of the equation of wave propagation in fluid in the following form

$$\nabla^2\mathrm{p}-\frac{1}{c_f^2}\frac{\partial^2\mathrm{p}}{\partial t^2}=0 \tag{4.19.1}$$

$$\text{or} \quad \frac{\partial^2\mathrm{p}}{\partial x_1^{\ 2}}+\frac{\partial^2\mathrm{p}}{\partial x_2^{\ 2}}+\frac{\partial^2\mathrm{p}}{\partial x_3^{\ 2}}-\frac{1}{c_f^2}\frac{\partial^2\mathrm{p}}{\partial t^2}=0 \tag{4.19.2}$$

where $\nabla^2=\nabla.\nabla$ signifies the divergence of the gradient of a scalar field.

4.2.1 Pressure Potential in Fluid

Now let's refer the discussion in Section 4.1.3 and wave potential for the plane harmonic waves in Eq. (4.8). Please note that the wave potential discussed above is one dimensional along a specific axis x. Thus, we have only one wave number along the x direction. However, for a wave propagating in a three-dimensional homogeneous isotropic media like fluid, wave number along any direction of wave propagation (i.e., along the ray of the wave) orthogonal to the wave front will be the primary wave number that will follow the plane harmonic wave equations in the media, such as $k=c\omega=\frac{2\pi}{\lambda}$. Let's term this wave number k_f in fluid.

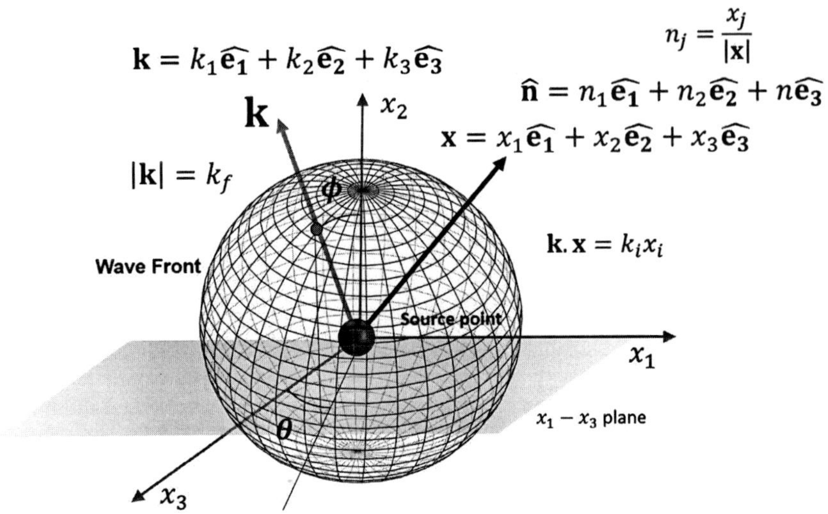

FIGURE 4.5 A spherical wave front in fluid and necessary mathematical terms to calculate phase $k_i x_i$.

Please refer Fig. 4.5, where, a spherical wave front is shown. In the figure, $\mathbf{k} = k_i\hat{e}_I$ is a vector pointing to the direction of wave propagating with magnitude of the wave number $= k_f$. k_i is the three components of the wave number in Cartesian coordinate system. $\mathbf{x} = x_i\hat{e}_I$ is a vector pointing to any arbitrary direction passing through a point x_i where, the pressure wave field is to be investigated. According to Section 4.1.3, phase is multiplication of wave numbers and distance of the wave travels. In same light, phase could be described as a dot product of **k** and **x** here, because both are the vectors. Hence, referring Chapter 2 with discussions on dot products and vectors, the pressure plane wave potential function in Eq. (4.8) in three dimensions will be modified to

$$\mathrm{p}(x_j, t) = Ae^{i(\mathbf{k}.\mathbf{x} - \omega t)} = Ae^{i(k_i x_i - \omega t)} \tag{4.20}$$

Alternatively, Eq. (4.20) can be written by separating the spatial and temporal part of the harmonic wave as follows

$$\mathrm{p}(x_j, t) = Ae^{i(k_i x_i)}e^{-i\omega t} = \mathrm{p}(x_j)e^{-i\omega t} \tag{4.21}$$

Performing second-order derivative of Eq. (4.21) with respect to time and eliminating the temporal harmonic part of the wave potential, Eq. (4.19.1) can be rewritten as

$$\nabla^2 \mathrm{p} + k_f^2\ \mathrm{p} = 0 \tag{4.22}$$

where $k_f = \omega / c_f$ and the equation is called a typical homogeneous Helmholtz equation [8].

4.2.2 Generalized Wave Potential in Fluid

Alternatively, to solve Eq. (4.22) omitting the time harmonic part let's assume an arbitrary spatial wave potential function $\phi(x_j)$ and gradient of which can be described as the displacement function in the fluid media. We can write

$$\phi = Ae^{i(k_i x_i)} \quad \text{and} \quad \mathbf{u} = \nabla\phi \tag{4.23}$$

Referring Eq. (4.17), substituting Eq. (4.23) and the expression for c_f we can directly write the expression for pressure as

$$\mathrm{p} = -\mathbb{L}\nabla.(\nabla\phi) = \mathbb{L}k_f^2\phi = \mathbb{L}\frac{\omega^2}{c_f^2}\phi = \rho\omega^2\phi \tag{4.24}$$

This will automatically satisfy Eq. (4.22) and is the solution of Eq. (4.19.1) or (4.19.2). Please note Eq. (4.24) is the homogeneous solution of the homogeneous governing differential equation without any active source or body force in the fluid media. The solution will be modified for Eq. (4.16) based on the type of body force considered in the media. Considering a point source in fluid, we could still solve the pressure field analytically discussed in the later chapters, but acoustic and ultrasonic field generated by finite size transducer cannot be solve analytically. For such cases, a typical NDE situation has to rely on the numerical computation and will be the topic of our later chapters.

4.3 WAVE PROPAGATION IN BULK ISOTROPIC SOLID MEDIA

4.3.1 Navier's Equation of Motion

The governing differential equation of motion which is also known as the equilibrium equation in a solid medium described in Chapter 3, Eq. (3.71) can be modified without body force as written in Eq. (4.25). Elimination of the body force from Eq. (3.71) may not work if there is an active wave source in the solid media. However, for general plane wave incidence without any active wave source propagation of the acoustic or ultrasonic wave will follow the following equation of equilibrium.

$$\sigma_{ij,j} - \rho\ddot{u}_i = 0 \tag{4.25}$$

where σ_{ij} is the stress and u_i is the displacement at a point ($\mathbf{x} = x_i\hat{\mathrm{e}}_i$) in a bulk isotropic solid. For homogeneous solid media the density (ρ) remains constant; however, stress (σ_{ij}) and displacement (u_i) are in general functions of both space (x_i) and time (t).

Recollecting Eq. (3.75), the constitutive law for any linear material can be written as

$$\sigma_{ij} = \mathbb{C}_{ijkl}\mathrm{e}_{kl} \tag{4.26}$$

where $\mathbb{e}_{kl}$ is the strain in the isotropic solid. For any generalized linear elastic isotropic or anisotropic material $\mathbb{C}_{ijkl}$ is the matrix of elastic constants written in Eqs. (3.78) and (3.79). For isotropic materials the stress-strain relation modifies to the following form,

$$\sigma_{ij} = \mathbb{L}\delta_{ij}\mathbb{e}_{kk} + 2\mathbb{m}\mathbb{e}_{ij} \tag{4.27}$$

where $\mathbb{L}$ and $\mathbb{m}$ are the two Lamé constants for isotropic material and δ_{ij} is the Kronecker delta described in Chapter 3. After substituting the expression of strain $\mathbb{e}_{kl} = \frac{1}{2}\left(u_{k,l} + u_{l,k}\right)$ written in Eq. (3.72), into Eq. (4.27) one can get

$$\sigma_{ij} = \mathbb{L}\delta_{ij}u_{k,k} + \mathbb{m}\left(u_{i,j} + u_{j,i}\right) \tag{4.28}$$

Performing the derivatives of σ_{ij} with respect to x_j and substituting in Eq. (4.25) we get

$$\mathbb{L}\delta_{ij}u_{k,kj} + \mathbb{m}\left(u_{i,jj} + u_{j,ij}\right) = \rho\ddot{u}_i \tag{4.29}$$

Based on the discussions in Chapter 2, after successive index operation of the above equation, we get

$$\mathbb{L}u_{k,ki} + \mathbb{m}\left(u_{i,jj} + u_{j,ij}\right) = \rho\ddot{u}_i \tag{4.30}$$

or

$$\left(\mathbb{L} + \mathbb{m}\right)u_{j,ji} + \mathbb{m}\left(u_{i,jj}\right) = \rho\ddot{u}_i \tag{4.31}$$

Eq. (4.31) can be written in vector form

$$\left(\mathbb{L} + \mathbb{m}\right)\nabla\left(\nabla.\mathbf{u}\right) + \mathbb{m}\nabla^2\mathbf{u} = \rho\ddot{\mathbf{u}} \tag{4.32}$$

Displacement at a point in a solid media consists of three components (u_1, u_2, u_3) in three orthogonal directions (x_1, x_2, x_3) and the components of the displacement changes from point to point $(u_i(x_j, t))$ due to the wave propagation. Hence, by definition, displacement is a vector field. Thus, identities of a vector field proved in Section 2.7 in Chapter 2 will be valid entirely for the displacement field. The identity in Problem 2.4.3 in Chapter 2 will be very useful to further simplifying Eq. (4.32). Hence, applying

$$\nabla^2\mathbf{u} = \nabla\left(\nabla.\mathbf{u}\right) - \nabla\times\left(\nabla\times\mathbf{u}\right) \tag{4.33}$$

We get

$$\left(\mathbb{L} + \mathbb{m}\right)\nabla\left(\nabla.\mathbf{u}\right) + \mathbb{m}\nabla\left(\nabla.\mathbf{u}\right) - \mathbb{m}\nabla\times\left(\nabla\times\mathbf{u}\right) = \rho\ddot{\mathbf{u}}$$

$$\text{or } \left(\mathbb{L} + 2\mathbb{m}\right)\nabla\left(\nabla.\mathbf{u}\right) - \mathbb{m}\nabla\times\left(\nabla\times\mathbf{u}\right) = \rho\ddot{\mathbf{u}} \tag{4.34}$$

The above equation is called the Navier's equation of motion [9, 10] without body force. Applying permutation symbol discussed in Sections 2.2.2 and 2.3.3 in Chapter 2, Navier's equation of motion without body force in index notation can be further written as follows

$$(\mathbb{L}+2\mathbb{m})u_{j,ji}-\mathbb{m}\epsilon_{ijk}\epsilon_{kpq}u_{q,pj}=\rho\ddot{u}_i \tag{4.35}$$

If we have kept the body force in Eq. (4.25) keeping Eq. (3.71) intact, the Navier's equation of motion in Eq. (4.35) would take the following form in vector and index notations, respectively

$$(\mathbb{L}+2\mathbb{m})\nabla(\nabla.\mathbf{u})-\mathbb{m}\nabla\times(\nabla\times\mathbf{u})+\mathbf{f}=\rho\ddot{\mathbf{u}} \tag{4.36}$$

$$(\mathbb{L}+2\mathbb{m})u_{j,ji}-\mathbb{m}\epsilon_{ijk}\epsilon_{kpq}u_{q,pj}+f_i=\rho\ddot{u}_i \tag{4.37}$$

4.3.2 Solving Navier's Equation of Motion: Solution of Wave Propagation in Isotropic Solids

4.3.2.1 Helmholtz Decomposition

If we are just interested in knowing how acoustic and ultrasonic wave propagates in a bulk isotropic solid media, we do not need to consider any active source or the body force in the equation and Eq. (4.34) or (4.35) is sufficient. In NDE there is always a finite size transducer which emits and transmits the energy to the solid media. Their contribution supposed to be considered under the body force in Eq. (4.36) or (4.37). However, excitation energy might be of different forms and due to the finite size of the emitters, it is extremely difficult to have an analytical solution of the wave field. Thus, numerical or computational analysis is mandatory, which is the topic of our following chapters. Here in this chapter to analytically understand the basic wave phenomena in fluid and solid media, without body force, plane wave solution method is discussed for Eq. (4.34) or (4.35). Pressure was the primary unknown in Eq. (4.19) and thus the plane wave solution was assumed for pressure in Eq. (4.20). Here in Eq. (4.34) the primary unknown is the displacement filed ($\mathbf{u}=u_i\hat{\mathrm{e}}_\mathrm{i}$) and the plane wave solution should be assumed for the displacement field. However, unlike pressure in Eq. (4.20), which is a scalar field, the displacement field is a vector field. In this light we can say that the plane wave potentials (scalar or vector potentials) must have components to represent each component of the displacement filed. In other words, each component of the displacement field must be expressed by respective plane wave potentials. This complicates the situation and shrouds our convenient decision to suitably select a wave potential for the displacement field as we did before. Based on our discussion in Chapter 2, we know that gradient of a scalar field (here a scalar wave potential) produces a vector field, which could represent the displacement field we need. However, according to Chapter 2, also the curl of a vector field (here a vector wave potential with three individual components) can also produce a vector field. This new vector field could also potentially represent the displacement field we need. Hence, the question is which is more accurate. It is inevitable that a vector field must have both orthogonal components and we could take the most suitable method, the method of superposition. We will assume that the displacement wave field in isotropic solid is a superposition of gradient of a scalar wave potential (ϕ) and curl of a vector wave potential (ψ). Further could be said that any sufficiently smooth vector filed (**u**) could be resolved into the sum of an irrotational (i.e., curl free – refer Sections 2.3.2 and 2.3.3 in Chapter 2) vector field that could be obtained from the gradient of a scalar potential and a solenoidal (i.e., divergence free – refer Section 2.3.3 in Chapter 2) vector filed that could be obtained by taking curl of a vector potential. Hence, the displacement wave field is decomposed into both scalar and vector potentials. Such decomposition was first proposed by Hermann Von Helmholtz (1821–1894) and called the Helmholtz decomposition [9, 10]. However, this decomposition may not result a unique relationship between u_i, ϕ, and ψ_j. To achieve unique relationship, an auxiliary condition was imposed and is called gauge condition. It is said that the decomposition of a vector field into one scalar and three components of one vector potential will result unique relationship only if the divergence of the vector potential is zero (i.e., $\nabla.\psi = 0$). This means that the vector potential must be non-divergent and should only produce vortex field of displacements. Recently this argument has be questioned in few publications [11] but the discussion is omitted as it is not relevant at this stage.

After Helmholtz decomposition, we can write the displacement wave field as

$$\mathbf{u} = \nabla\phi + \nabla\times\psi \tag{4.38.1}$$

$$u_i = \phi_{,i} + \epsilon_{ijk}\psi_{k,j} \tag{4.38.2}$$

4.3.2.2 Navier's Equation of Motion to Helmholtz Equation

The Navier's equation of motion can be simplified further by implementing the Helmholtz decomposition, i.e., substituting Eq. (4.38) into Eq. (4.34). Eq. (4.34) will be modified to

$$(\mathbb{L}+2\mathrm{m})\nabla\left(\nabla^2\phi+\nabla.(\nabla\times\psi)\right)-\mathrm{m}\nabla\times\left(\nabla\times(\nabla\phi+\nabla\times\psi)\right)=\rho\left(\nabla\ddot{\phi}+\nabla\times\ddot{\psi}\right) \tag{4.39}$$

Please refer the identities (Problems 2.3.7 and 2.3.8) proved in Chapter 2, where we mentioned that divergence of the curl of a vector field in zero and equivalently, curl of the gradient of a scalar field is also zero. Additionally, the identity in Problem 2.4.3 in Chapter 2, reiterated for the displacement field in Eq. (4.33), equivalent for the vector potential ψ is also valuable herein. Applying the following identities

$$\nabla.(\nabla\times\psi)=0 \;\; ; \;\; \nabla\times(\nabla\phi)=0 \;\; ; \;\; \nabla\times\nabla\times\psi=\nabla(\nabla.\psi)-\nabla^2\psi \tag{4.40}$$

Eq. (4.39) could be modified to

$$(\mathbb{L}+2\mathrm{m})\nabla\left(\nabla^2\phi)\right)-\mathrm{m}\nabla\times\left(\nabla(\nabla.\psi)-\nabla^2\psi\right)=\rho\left(\nabla\ddot{\phi}+\nabla\times\ddot{\psi}\right) \tag{4.41}$$

$$(\mathbb{L}+2\mathrm{m})\nabla\left(\nabla^2\phi)\right)-\mathrm{m}\nabla\times\left(\nabla(\nabla.\psi)-\nabla^2\psi\right)=\rho\left(\nabla\ddot{\phi}+\nabla\times\ddot{\psi}\right) \tag{4.42}$$

and the gauge condition ($\nabla.\psi = 0$) we get

$$(\mathbb{L}+2\mathrm{m})\nabla(\nabla^2\phi) - \mathrm{m}\nabla\times(-\nabla^2\psi) = \rho(\nabla\ddot{\phi} + \nabla\times\ddot{\psi}) \tag{4.43}$$

Applying the commutability of the del operator

$$\nabla\left[(\mathbb{L}+2\mathrm{m})(\nabla^2\phi) - \rho\ddot{\phi}\right] + \nabla\times\left[\mathrm{m}(\nabla^2\psi) - \rho\ddot{\psi}\right] = 0 \tag{4.44}$$

Please note that in Eq. (4.44) it is saying that the superposition of the gradient of a scalar field and the curl of a vector field is equal to zero. However, this is only possible when the scalar and vector fields are nonexistent, independently. Hence, the sufficient conditions for Eq. (4.44) to be valid are

$$(\mathbb{L}+2\mathrm{m})(\nabla^2\phi) - \rho\ddot{\phi} = 0 \tag{4.45}$$

$$\mathrm{m}(\nabla^2\psi) - \rho\ddot{\psi} = 0 \tag{4.46}$$

Eqs. (4.45) and (4.46) can be written in the form of wave equations as follows

$$\nabla^2\phi - \frac{1}{c_p^2}\ddot{\phi} = 0 \tag{4.47}$$

$$\nabla^2\psi - \frac{1}{c_s^2}\ddot{\psi} = 0 \tag{4.48}$$

where $c_p = \sqrt{\frac{(\mathbb{L}+2\mathrm{m})}{\rho}}$ and $c_s = \sqrt{\frac{\mathrm{m}}{\rho}}$ are the P-wave or longitudinal and S-wave or shear wave velocities in the isotropic solid media.

As we have found before that the solution of a typical wave equation (e.g., Eq. (4.19.1)) in linear media can be described as a superposition of several plane harmonic waves, similar we could assume to solve Eqs. (4.47) and (4.48) herein. A wave potential is composed of a spatial phase part and a time harmonic temporal part. Thus, we could assume $\phi(x_j,t) = \varphi(x_j)e^{-i\omega t}$ and $\psi_i(x_j,t) = \Psi_i(x_j)e^{-i\omega t}$ or $\psi(x_j,t) = \Psi(x_j)e^{-i\omega t}$ and substitute in Eqs. (4.47) and (4.48) the equations transform into Helmholtz equations as follows

$$\nabla^2\varphi + k_p^2\varphi = 0 \tag{4.49}$$

$$\nabla^2\Psi + k_s^2\Psi = 0 \tag{4.50}$$

where $k_p^2 = \omega^2/c_p^2$ and $k_s^2 = \omega^2/c_s^2$ are the square of the longitudinal wave number and the shear wave number, respectively.

4.3.2.3 Generalized Wave Potentials in Isotropic Solids

Following the discussion in Section 4.2.2 where, we found, how to write a generalized wave potential for Helmholtz type equations (Eqs. (4.22), (4.49), and (4.50)), we will further express the spatial phase part of the wave potentials in Eqs. (4.49) and (4.50) as follows

$$\varphi = Ae^{i(\mathbf{k^p}.\mathbf{x})} = Ae^{i(k_j^p x_j)} \quad \text{and} \quad \boldsymbol{\Psi} = \mathbf{B}e^{i(\mathbf{k^s}.\mathbf{x})} = \mathbf{B}e^{i(k_j^s x_j)} \tag{4.51}$$

where $\mathbf{k^p}$ and $\mathbf{k^s}$ describes the wave vectors pointed toward the propagation of the P-wave and the S-wave respectively. As they are assumed different, it implies that the propagation direction of both the P-wave and S-wave will not be same in the isotropic bulk media. Please note $\mathbf{k^p}.\mathbf{x} = k_j^p x_j$, $|\mathbf{k^p}| = k_p$ and $\mathbf{k^s}.\mathbf{x} = k_j^s x_j$, $|\mathbf{k^s}| = k_s$. Please keep a note of the superscript and the subscript p and s. Superscripts are of the wave vectors and subscripts are for the magnitude of the respective wave numbers. A and $\mathbf{B}$ in Eq. (4.51) are the amplitude of the scalar and vector potentials, respectively. Additionally, please note that the amplitude vector $\mathbf{B}$ has three different amplitudes (B_1, B_2, B_3 designated as B_k) for three corresponding orthogonal vector potentials. Please refer Fig. 4.6 where, two different wave directions, one for P-wave ($\mathbf{k^p}$) and another for S-wave ($\mathbf{k^s}$) are drawn. If we are interested in investigating the wave field at an arbitrary spatial point (x_1, x_2, x_3) then dot product between the wave direction vector and the position vector $\mathbf{x} = x_j\hat{e}_j$ will describe the phases of the P-wave and the S-wave, respectively, which is actually written in Eq. (4.51). One very important aspect, we have overlooked so far, which is the physical nature of the P-waves and S-waves described by the scalar potential φ and the vector potential $\boldsymbol{\Psi}$. To describe this, first, we need to understand what these P-waves and the S-waves in isotropic solids are.

4.3.2.4 Longitudinal Waves and Shear Waves in Isotropic Solids

Helmholtz decomposition of the displacement wave field gives scalar and vector potentials in isotropic solids which have very specific displacement patterns.

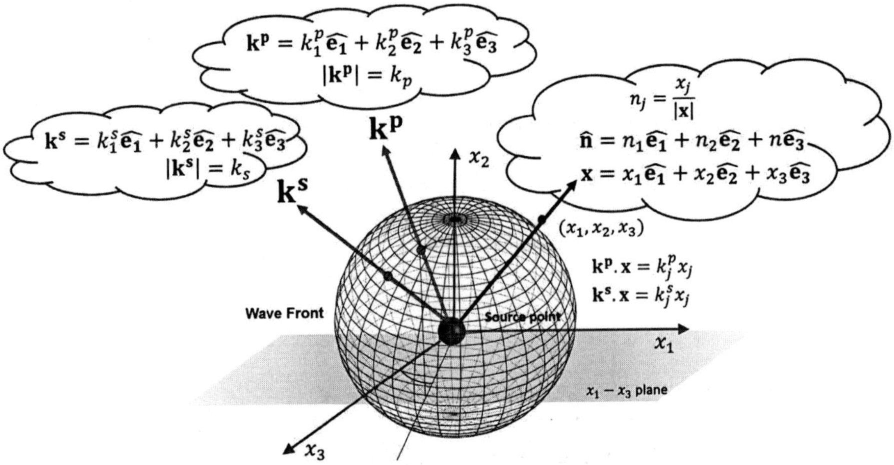

FIGURE 4.6 Spherical wave fronts in isotropic solid: two wave directions one for P-wave ($\boldsymbol{k^p}$) and another for S-wave ($\boldsymbol{k^s}$) are shown to describe the mathematical equations for their respective phase $k_j^p x_j$ and $k_j^s x_j$.

To understand their pattern, let's consider one potential at a time. If a wave field is developed such a way that it has no rotational filed, i.e., the field is a curl free wave field, then we can assume $\psi = 0$. Expanding Eq. (4.38) with generalized wave potentials expressed in Eq. (4.51) we can write

$$\mathbf{u} = \nabla\varphi e^{-i\omega t} \quad \text{or} \quad \mathbf{u} = \mathbf{k}^{\mathbf{P}}.Ae^{i\left(\mathbf{k}^{\mathbf{P}}.\mathbf{x}\right)}e^{-i\omega t} \tag{4.52.1}$$

$$u_i = \varphi_{,i}e^{-i\omega t} \quad \text{or} \quad u_k = ik_i^P Ae^{i\left(k_j^P x_j\right)}e^{-i\omega t} \tag{4.52.2}$$

Here, in Eqs. (4.52.1) and (4.52.2) we can see that the direction of the particle displacement (u_j) in the wave field is along the direction of the wave propagation (k_j^P). That means the particle of the media will move along the direction of the wave propagation as shown in Fig. 4.7. This is called the P-wave or the longitudinal wave which is mathematically represented by a scalar potential.

Next, if a wave field is developed such a way that it has no divergence, i.e., the field is a divergence free wave field, then we can assume $\phi = 0$. Expanding Eq. (4.38) with generalized wave potentials expressed in Eq. (4.51) we can write

$$\mathbf{u} = \nabla \times \Psi \; e^{-i\omega t} \tag{4.53.1}$$

Applying the rule described in Eq. (2.26) and assuming the amplitudes (B_k) as constants we can express Eq. (4.53.1) in index notations as follows

$$u_k = \hat{\mathbf{e}}_{\mathbf{k}}\epsilon_{kij}\frac{\partial\left(\mathrm{B}_j e^{i\left(k_l^s x_l\right)}\right)}{\partial x_i}e^{-i\omega t} = i\;\hat{\mathbf{e}}_{\mathbf{k}}\epsilon_{kij}\mathrm{B}_j k_p^S \delta_{pi} e^{i\left(k_l^s x_l\right)}e^{-i\omega t} \tag{4.53.2}$$

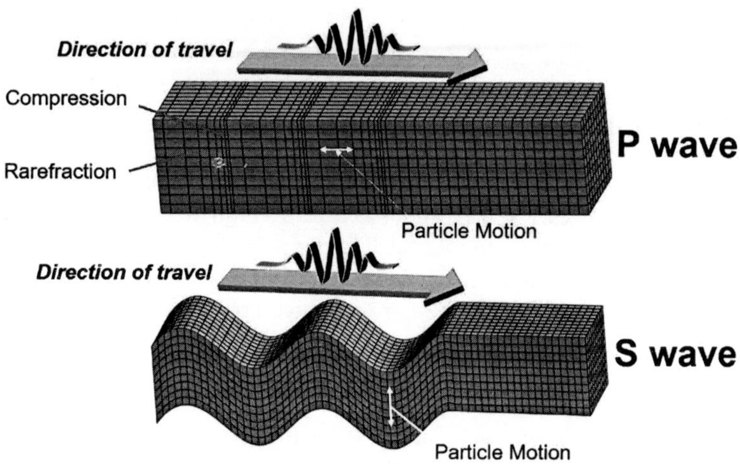

FIGURE 4.7 Schematic description of P-wave and S-wave.

and further based on the discussion presented in Chapter 2 right after Eq. (2.13) we apply the Kronecker delta rule and get

$$u_k = \hat{\mathbf{e}}_{\mathbf{k}} \epsilon_{kij} C_j k_i^s e^{i\left(k_l^s x_l\right)} e^{-i\omega t} \tag{4.53.3}$$

where $C_j = iB_j$. Expanding the displacements in three orthogonal directions following the method described in Eq. (2.27) but omitting the unit vectors which are implied, we can write

$$\begin{aligned} u_1 &= \left[C_3 k_2^s e^{i\left(k_l^s x_l\right)} - C_2 k_3^s e^{i\left(k_l^s x_l\right)} \right] e^{-i\omega t} \\ u_2 &= \left[C_1 k_3^s e^{i\left(k_l^s x_l\right)} - C_3 k_1^s e^{i\left(k_l^s x_l\right)} \right] e^{-i\omega t} \\ u_3 &= \left[C_2 k_1^s e^{i\left(k_l^s x_l\right)} - C_1 k_2^s e^{i\left(k_l^s x_l\right)} \right] e^{-i\omega t} \end{aligned} \tag{4.54}$$

Next let's take a dot product of the above displacement field $\mathbf{u}$ with the wave propagation direction $\mathbf{k}^s$. We can expand the dot product using the description in Eq. (2.14) as follows

$$\mathbf{u}.\mathbf{k}^s = u_i k_i^s = u_1 k_1^s + u_2 k_2^s + u_3 k_3^s \tag{4.55}$$

Substituting Eq. (4.54) into Eq. (4.55) one can easily show that the sum in Eq. (4.55) will result zero. Hence, we can conclude that the displacement field of a divergence free wave field is orthogonal to the direction of wave propagation. None of the particle displacements will be along the direction of the wave vector $\mathbf{k}^s$ as shown in Fig. 4.7. This is called the S-wave or the transverse wave or shear wave or S-wave which is mathematically represented by a vector potential with three orthogonal components. The vector potential will have three components. In case of two-dimensional in-plane problems the survivability of the components of the vector potential will depends on the choice of the axes in the problem. For example, if $x_1 - x_2$ plane is considered, then out of plane displacement u_3 should be enforced to zero. In order to do that Eq. (4.54) is telling that C_1 and C_2 should be zero and hence, only Ψ_3 component will survive. On the contrary if $x_1 - x_3$ plane is considered, then out of plane displacement u_2 should be enforced to zero. In order to do that Eq. (4.54) is telling that C_1 and C_3 should be zero and hence, only Ψ_2 component will survive. Hence, it is to note that the components of the vector potential in the Helmholtz decomposition of displacement wave field written in Eq. (4.38.2) should be judicially selected and appropriate derivate should be used based on the selection of the coordinate system.

4.3.2.5 In Plane and Out of Plane Shear Waves in Isotropic Solids

From the above discussion it is evident that P-wave with wave velocity c_p has particle motion along the principal direction of the wave vector $\mathbf{k}^\mathbf{P}$ or along the wave propagation direction. Similarly, it is also clear that S-waves with wave velocity c_s have

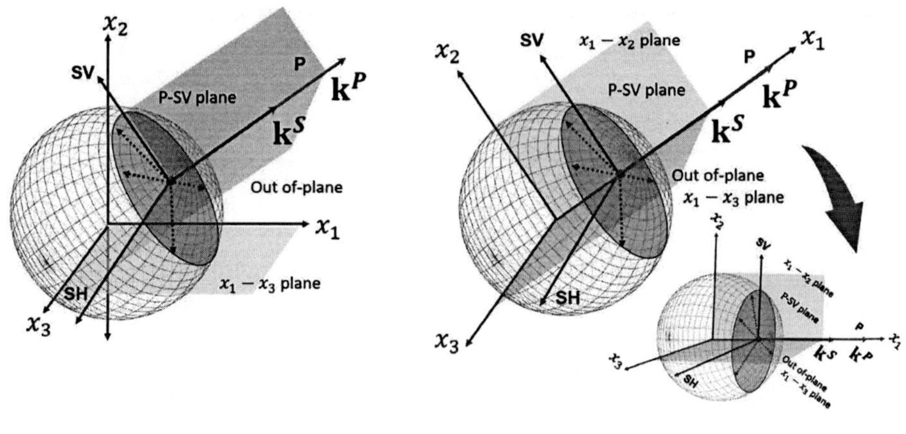

FIGURE 4.8 a) Showing possible particle motions when the wave vectors for P-wave and S-wave are along the same arbitrary direction in cartesian coordinate system, b) Showing possible particle motions when the wave vectors for P-wave and S-wave are along the x_1 axis.

particle motions orthogonal to the direction of wave vector $\mathbf{k}^s$or the wave propagation direction. Referring to Fig. 4.8, where $\mathbf{k}^P$ and $\mathbf{k}^s$ are in same arbitrary direction, it can be visualized that the particles have several options in choosing the direction of motion to carry the S-wave, as long as they are orthogonal to the wave vector $\mathbf{k}^P$ and $\mathbf{k}^s$. In Fig. 4.8, a Cartesian coordinate system is used. Instead of considering infinite possibilities, which are redundant, we will consider only two possible orthogonal directions. One is called in-plane motion, and another is called the out of plane motion. One of them is coupled with P-wave and another is independent and decoupled from the P-wave. In Fig. 4.8, $\mathbf{k}^P$ and $\mathbf{k}^s$ wave vectors are pointed to an arbitrary direction. A circular plane orthogonal to the wave vectors can be imagined as shown in Fig. 4.8. The particle displacements of the P-wave are along the wave vector $\mathbf{k}^P$ designated as P, in Fig. 4.8. However, the particle displacements of the S-waves have multiple possibilities (dotted arrows in Fig. 4.8), out of which only two directions are marked, along SV and SH, while the wave vectors are pointing to the direction of $\mathbf{k}^P$ and $\mathbf{k}^s$ which signifies the wave propagation direction. Particle displacements along SV are orthogonal to the P-wave vector and considered in-plane with the P-wave, which are coupled. Considering the same P-wave vector, particle motion along SH is also orthogonal to the P-wave but out of plane to the P-SV plane as indicated in Fig. 4.8. The particle motion in the SV direction is in-plane with the P-wave but results a shear wave with vertical polarization (please note that the vertical direction does not always refer the x_3 axis or the direction of the gravity) and called shear vertical wave or SV-wave which has the wave vector $\mathbf{k}^s$ and the wave velocity c_s. Similarly, the particle motion in the SH direction is out of plane with respect to the P-SV plane, but results a shear wave with horizontal polarization (please note that the horizontal direction does not always refer to the x_2 or x_1 axis or orthogonal to the gravity) and called shear horizontal wave or SH-wave which has the wave vector $\mathbf{k}^s$ and the wave velocity c_s. Here, vertical and horizontal polarization is merely the terms coined with respect to the direction of the wave vector.

In most common cases wave propagation direction is aligned with the x_1 axis. Thus, x_2 axis, which is vertical to the x_1 axis designates the polarization of the SV waves in right-hand coordinate system. Alternatively, x_3 axis which is horizontal and orthogonal to the x_1 axis designates the polarization of the SH waves. Fig. 4.8b shows that same situation presented in Fig. 4.8a with the x_1 axis intentinally aligned to the wave vector for our convenience. Without loss of the generality, in most of our analysis this convenience is availed all most all the time. With x_1 axis pointed along the $\mathbf{k}^{\mathrm{P}}$ and $\mathbf{k}^{\mathrm{s}}$ vector, x_3 axis is the direction of particle motion for the SV waves and x_2 axis is the direction of the particle motion for the SH waves. From this section it is clear that in isotropic infinite media, three wave modes, P-wave, SV-wave, and SH-wave propagate with three orthogonal polarization. It is also necessary to mention herein that the SH-wave is independent from P and SV wave because it is orthogonal to the P-SV plane and does not carry any components of the P and SV waves.

4.3.2.6 Wave Potentials for P, SV, and SH Waves and Their Relation

It is evident that P-wave or the dilatational wave is expressed by a scalar potential φ. Similarly, S-waves are expressed by vector potential $\boldsymbol{\Psi}$. But as we consider SV and SH separately based on the particle motions in two mutually orthogonal directions, each could be expressed by their own potentials similar to a scalar potential. As the particle motion for the SH wave is orthogonal to the SV and P wave, it does not have any component on the P-SV plane and demands separate treatment. Considering the coordinate alignment used in Fig. 4.8b let's consider a case where there is no in plane motion of the particles on the P-SV plane, i.e., u_1 and u_2 are zero. Only u_3 motion survives. Let's consider the wave equation in Eq. (4.37) which could be expanded with all displacement components before applying the Helmholtz decomposition. Further plugging $u_1 = u_2 = 0$ Eq. (4.37) could be rewritten only for the u_3 displacement as follows

$$\mathrm{m}\left(u_{3,11} + u_{3,22}\right) - \rho \ddot{u}_3 = 0 \tag{4.56.1}$$

$$\nabla^2 u_3 - \frac{1}{c_s^2} \ddot{u}_3 = 0 \tag{4.56.2}$$

which is similar to Eq. (4.47) or (4.48). Next, substituting plane wave potential for SH wave

$$u_3\left(x_j, t\right) = \Psi_{SH}\left(x_j\right) e^{-i\omega t} \tag{4.57}$$

Eqs. (4.56.2) revised to

$$\nabla^2 \Psi_{SH} + k_s^2 \Psi_{SH} = 0 \tag{4.58}$$

where $k_s^2 = \omega^2 / c_s^2$ is the square of the shear wave number discussed previously. Based on generalized plane wave potential discussed in Section 4.2.2.3, the $\Psi_{SH}\left(x_j\right)$ potential can be further assumed to be

$$\Psi_{SH}\left(x_j\right) = B_{SH} e^{i\left(k_j^s x_j\right)} \tag{4.59}$$

where B_{SH} is the amplitude of the SH wave along the x_3 axis with wave propagating along the x_1 axis with velocity c_s and wave number k_s at radial frequency ω.

Keeping our discussion along the same line, we can express the wave potentials for the P-wave and SV-wave as follows

$$\varphi_P\left(x_j\right) = A_P e^{i\left(k_j^P x_j\right)} \tag{4.60}$$

$$\Psi_{SV}\left(x_j\right) = B_{SV} e^{i\left(k_j^S x_j\right)} \tag{4.61}$$

SV-wave and P-wave propagate on a same plane and hence they are coupled. Thus, when P-wave is incident on an isotropic interface, it will break down into both P-wave and SV-wave in the reflection and also break down in the transmission to another isotropic media beyond the interface. However, when SH-wave is incident on an isotropic interface, it will not break down into P-wave or SV-wave but will remain SH-wave in the reflection and remain SH-wave in transmission to another isotropic media beyond the interface. To express these multiple waves generated due to the interaction of the waves at the interface, different wave potentials with different wave amplitude should be described.

4.3.3 Wave Interactions at the Bulk Isotropic Interfaces

In ultrasonic NDE, wave interactions at the interfaces are the most crucial to understand, because they cause reflection, refraction, transmission, and scattering of the waves. Interfaces between two different isotropic materials, bonded joints between two metals, interface between coating and parent material, boundary of a semi-infinite media, and interfaces between layered materials are few examples of isotropic interfaces. There are several possibilities of wave interaction at different types of interfaces and at free boundaries. Most of the cases are omitted in this book but the detail analysis of various cases can be found in Refs. [5]. Here in this section only the wave interaction at an interface between two isotropic materials is discussed using the wave potential derived in the previous section. Similar approach is applicable for all other different types of interfaces and boundaries. Let's refer Fig. 4.9a and b where materials 1 and 2 are joined at the interface. In Fig. 4.9a, an incident P wave is assumed. In Fig. 4.9b, an incident SH-wave is assumed. As pointed out previously, at the interface, due to incident of a P-wave, the wave will both transmit and reflect at the interface. However, there will be two reflected wave modes and two transmitted wave modes. Two reflected wave modes will be composed of one P-wave and another SV-wave, which will be back to the material 1. Two transmitted wave modes will be composed of one P-wave and another SV-wave, which will go into the material 2. Generation of P and SV wave due to the interface is inevitable because they are coupled and inseparable. Solution using only one wave potential at the interface will not be enough to satisfy all the boundary or interface conditions. On the contrary, at the interface, due to incident of an SH-wave, the wave will both transmit and reflect at the interface maintain the SH-wave mode. That means there will be only one reflected wave mode back to material 1 and one transmitted wave mode into

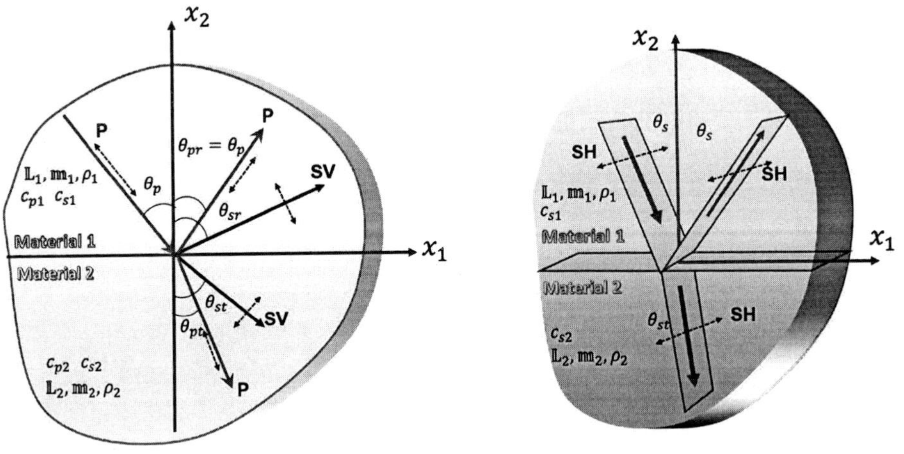

FIGURE 4.9 a) Reflection and refraction of P-wave at the interface of two materials; b) Reflection and refraction of SH-wave at the interface of two materials.

the material 2, which are both SH-waves. As SH-wave is decoupled, an anti-plane mode with orthogonal particle displacement to the direction of wave propagation will be independent from the generation of P and SV wave at the interface. Thus, only SH-wave potentials will be sufficient to satisfy all the conditions at the boundary or the interface. Next let's take one case at a time.

4.3.3.1 P-Wave Incident at the Interface

The discussion above is pictorially presented in Fig. 4.9a. It clearly shows that to solve wave interaction problem at an isotropic interface, there will be five generalized wave potentials to consider. In the following wave potentials, the time harmonic part $e^{-i\omega t}$ is omitted without the loss of generality.

incident P-wave $$\varphi_{PI}\left(x_j\right) = A_{PI}e^{i\left(k_j^P x_j\right)} \tag{4.62.1}$$

reflected P-wave $$\varphi_{Pr}\left(x_j\right) = A_{Pr}e^{i\left(k_j^P x_j\right)} \tag{4.62.2}$$

transmitted P-wave $$\varphi_{Pt}\left(x_j\right) = A_{Pt}e^{i\left(k_j^P x_j\right)} \tag{4.62.3}$$

reflected SV-wave $$\Psi_{SVr}\left(x_j\right) = B_{SVr}e^{i\left(k_j^S x_j\right)} \tag{4.62.4}$$

transmitted SV-wave $$\Psi_{SVt}\left(x_j\right) = B_{SVt}e^{i\left(k_j^S x_j\right)} \tag{4.62.5}$$

The case with P-SV wave interaction presented in Fig. 4.9a is an in-plane problem and thus it is a two-dimensional problem. There will be no out of plane wave vectors. Thus, both the wave vectors $\mathbf{k}^{\mathbf{p}}$ and $\mathbf{k}^{\mathbf{s}}$ individually will have two components along x_1 and x_2 axis, respectively. As shown in Fig. 4.9a the incident P-wave makes an angle θ_p with x_2 axis. Reflected and transmitted P-waves makes θ_{pr} and θ_{pt} angles with x_2 axis respectively. Similarly, reflected and transmitted SV-waves makes θ_{sr} and θ_{st} angles with x_2 axis, respectively. To uniquely satisfy the interface conditions it is necessary that θ_{pr} should be equal to θ_p. Proof of this requirement is omitted in this book but can be found in many basic books on wave propagation [3–7, 9, 10]. Applying Snell's law [5, 6] a relation between the reflected and transmitted angles and the incident angle can be found (please refer Appendix in Section 4.5.3). They are $\theta_{sr} = \sin^{-1}\left(\frac{c_{s1}}{c_{p1}}\sin\theta_p\right)$, $\theta_{st} = \sin^{-1}\left(\frac{c_{s2}}{c_{p1}}\sin\theta_p\right)$, and $\theta_{pt} = \sin^{-1}\left(\frac{c_{p2}}{c_{p1}}\sin\theta_p\right)$. Based on this arrangement, reflected waves are called the up going waves and have both positive wave number components. However, the transmitted waves are called the down going waves and x_2 components of the wave number vectors are negative. Additionally, there are two materials in the problem. Hence, both the magnitudes of P-wave and S-wave numbers in material 1 and material 2, respectively will be different based on the material properties indicated in Fig. 4.9a. Recalling the previous discussion on wave number vectors in Section 4.3.2.3, let's assume the magnitudes of the P-wave and SV-wave numbers in two materials are k_{p1} k_{p2} k_{s1} and k_{s2}, respectively.

$$k_{p1} = \frac{\omega}{c_{p1}} \;;\; c_{p1} = \sqrt{\frac{(\mathbb{L}_1 + 2\mathbb{m}_1)}{\rho_1}} \;;\; k_{p2} = \frac{\omega}{c_{p2}} \;;\; c_{p2} = \sqrt{\frac{(\mathbb{L}_2 + 2\mathbb{m}_2)}{\rho_2}} \tag{4.63.1}$$

$$k_{s1} = \frac{\omega}{c_{s1}} \;;\; c_{s1} = \sqrt{\frac{\mathbb{m}_1}{\rho_1}} \;;\; k_{s2} = \frac{\omega}{c_{s2}} \;;\; c_{s2} = \sqrt{\frac{\mathbb{m}_2}{\rho_2}} \tag{4.63.2}$$

Next the components of the wave numbers along x_1 axes are

In Material 1, wave numbers along x_1 axis:

$$k_1^p = k_{p1}\sin\theta_p \;;\; k_1^s = k_{s1}\sin\theta_{sr} \tag{4.64.1}$$

In Material 2, wave number along x_1 axis:

$$k_1^p = k_{p2}\sin\theta_{pt} \;;\; k_1^s = k_{s2}\sin\theta_{st} \tag{4.64.2}$$

As the wave propagates as a union and superposed together it is not possible to have different wave numbers along the same x_1 axis. Hence, let's assume a common wave number k along the x_1 axis and rewrite Eqs. (4.64.1) and (4.64.2) as follows

$$k = k_{p1}\sin\theta_p = k_{s1}\sin\theta_{sr} = k_{p2}\sin\theta_{pt} = k_{s2}\sin\theta_{st} \tag{4.65}$$

The components of the wave numbers along x_2 axes are however, not the same and can be independent. This independency causes the different direction of wave vectors pointing to their respective wave propagation direction as shown in Fig. 4.9a. Components of the wave numbers along x_2 axis are

In Material 1, wave numbers along x_2 axis:

$$\left[k_2^p\right]_1 = k_{p1}\cos\theta_p \quad ; \quad \left[k_2^s\right]_1 = k_{s1}\cos\theta_{sr} \tag{4.66.1}$$

In Material 2, wave number along x_2 axis:

$$\left[k_2^p\right]_2 = k_{p2}\cos\theta_{pt} \quad ; \quad \left[k_2^s\right]_2 = k_{s1}\cos\theta_{sr} \tag{4.66.2}$$

where subscripts with the bracketed parameter indicate the material. For convenience it is better to use different parameters for the wave numbers along x_2 axis in two different material bodies. Introducing new parameter, the above expression can be presented as follows

In Material 1, wave numbers along x_2 axis:

$$\alpha_1 = \left[k_2^p\right]_1 = k_{p1}\cos\theta_p \quad ; \quad \beta_1 = \left[k_2^s\right]_1 = k_{s1}\cos\theta_{sr} \tag{4.67.1}$$

In Material 2, wave number along x_2 axis:

$$\alpha_2 = \left[k_2^p\right]_2 = k_{p2}\cos\theta_{pt} \quad ; \quad \beta_2 = \left[k_2^s\right]_2 = k_{s1}\cos\theta_{sr} \tag{4.67.2}$$

where α is the P-wave number along x_2 axis and β is the SV-wave number along x_2 axis. Subscripts 1 and 2 are to indicate in which material they propagates and $\theta_{sr} = \sin^{-1}\left(\frac{c_{s1}}{c_{p1}}\sin\theta_p\right)$, $\theta_{st} = \sin^{-1}\left(\frac{c_{s2}}{c_{p1}}\sin\theta_p\right)$, and $\theta_{pt} = \sin^{-1}\left(\frac{c_{p2}}{c_{p1}}\sin\theta_p\right)$ obtained applying the Snell's law (please refer Appendix in Section 4.5.3).

With the above notations the wave vector components can be expressed as follows

$$\text{incident P-wave} \quad : \quad \mathbf{k^p} = k\hat{\mathbf{e}}_1 - \alpha_1\hat{\mathbf{e}}_2 \tag{4.68.1}$$

$$\text{reflected P-wave} \quad : \quad \mathbf{k^{pr}} = k\hat{\mathbf{e}}_1 + \alpha_1\hat{\mathbf{e}}_2 \tag{4.68.2}$$

$$\text{transmitted P-wave} \quad : \quad \mathbf{k^{pt}} = k\hat{\mathbf{e}}_1 - \alpha_2\hat{\mathbf{e}}_2 \tag{4.68.3}$$

$$\text{reflected SV-wave} \quad : \quad \mathbf{k^{sr}} = k\hat{\mathbf{e}}_1 + \beta_1\hat{\mathbf{e}}_2 \tag{4.68.4}$$

$$\text{transmitted SV-wave} \quad : \quad \mathbf{k^{st}} = k\hat{\mathbf{e}}_1 - \beta_2\hat{\mathbf{e}}_2 \tag{4.68.5}$$

Substituting the wave vectors in Eq. (4.68.i) into Eq. (4.62.i) and performing the dot product between the respective wave vectors and **x** the wave potentials can be rewritten as

$$\text{incident P-wave} \qquad \varphi_{P1}\left(x_j\right) = A_{P1}e^{i(kx_1 - \alpha_1 x_2)} \tag{4.69.1}$$

$$\text{reflected P-wave} \qquad \varphi_{Pr}\left(x_j\right) = A_{Pr}e^{i(kx_1 + \alpha_1 x_2)} \tag{4.69.2}$$

transmitted P-wave $$\varphi_{Pt}\left(x_j\right) = A_{Pt} e^{i(kx_1 - \alpha_2 x_2)} \tag{4.69.3}$$

reflected SV-wave $$\Psi_{SVr}\left(x_j\right) = B_{SVr} e^{i(kx_1 + \beta_1 x_2)} \tag{4.69.4}$$

transmitted SV-wave $$\Psi_{SVt}\left(x_j\right) = B_{SVt} e^{i(kx_1 - \beta_2 x_2)} \tag{4.69.5}$$

Normalizing all the wave potentials with respect to the incident amplitude A_{P1} the total P-wave and SV potentials in materials 1 and 2 can be written with respect to their reflection and transmission coefficients, respectively

Material 1:

$$\varphi_1 = e^{i(kx_1 - \alpha_1 x_2)} + R_p e^{i(kx_1 + \alpha_1 x_2)} \quad ; \quad \Psi_1 = R_s e^{i(kx_1 + \beta_1 x_2)} \tag{4.70.1}$$

Material 2:

$$\varphi_2 = T_p e^{i(kx_1 - \alpha_2 x_2)} \quad ; \quad \Psi_2 = T_s e^{i(kx_1 - \beta_2 x_2)} \tag{4.70.2}$$

where $R_p = A_{Pr}/A_{P1}$; $R_s = B_{SVr}/A_{P1}$, $T_p = A_{Pt}/A_{P1}$, $T_s = B_{SVt}/A_{P1}$. The problem depicted here is an in-plane two-dimensional problem. Hence, the Helmholtz decomposition of displacements will carry only two potentials in it. Following Eq. (4.38.2) and omitting the time harmonic part from the generalized plane wave potentials, the displacement wave fields in materials 1 and 2 can be presented as

In Material 1, displacements along x_1 and x_2 axis:

$$\left[u_1\right]_1 = \frac{\partial \varphi_1}{\partial x_1} + \frac{\partial \Psi_1}{\partial x_2} \tag{4.71.1}$$

$$\left[u_2\right]_1 = \frac{\partial \varphi_1}{\partial x_2} - \frac{\partial \Psi_1}{\partial x_1} \tag{4.71.2}$$

In Material 2, displacements along x_1 and x_2 axis:

$$\left[u_1\right]_2 = \frac{\partial \varphi_2}{\partial x_1} + \frac{\partial \Psi_2}{\partial x_2} \tag{4.72.1}$$

$$\left[u_2\right]_2 = \frac{\partial \varphi_2}{\partial x_2} - \frac{\partial \Psi_2}{\partial x_1} \tag{4.72.2}$$

Recalling the stress equation (Eq. (4.28)) discussed in Section 4.3.1 and substituting the displacement field equations in Eqs. (4.71) and (4.72) for materials 1 and 2 the stress components will viz.

In Material 1, stress components σ_{11}, σ_{12}, and σ_{22} are

$$\left[\sigma_{11}\right]_1 = m\left[\frac{c_{p1}^2}{c_{s1}^2}\left(\frac{\partial^2 \varphi_1}{\partial x_1^2} + \frac{\partial^2 \varphi_1}{\partial x_2^2}\right) + 2\left(\frac{\partial^2 \Psi_1}{\partial x_1 x_2} - \frac{\partial^2 \varphi_1}{\partial x_2^2}\right)\right] \tag{4.73.1}$$

$$[\sigma_{22}]_1 = \mathrm{m}\left[\frac{c_{p1}^2}{c_{s1}^2}\left(\frac{\partial^2\varphi_1}{\partial x_1^2}+\frac{\partial^2\varphi_1}{\partial x_2^2}\right)-2\left(\frac{\partial^2\varphi_1}{\partial x_1^2}+\frac{\partial^2\Psi_1}{\partial x_1 x_2}\right)\right] \tag{4.73.2}$$

$$[\sigma_{12}]_1 = \mathrm{m}\left[2\frac{\partial^2\varphi_1}{\partial x_1 x_2}+\left(\frac{\partial^2\Psi_1}{\partial x_2^2}-\frac{\partial^2\Psi_1}{\partial x_1^2}\right)\right] \tag{4.73.3}$$

In Material 2, stress components σ_{11}, σ_{12}, and σ_{22} are

$$[\sigma_{11}]_2 = \mathrm{m}\left[\frac{c_{p2}^2}{c_{s2}^2}\left(\frac{\partial^2\varphi_2}{\partial x_1^2}+\frac{\partial^2\varphi_2}{\partial x_2^2}\right)+2\left(\frac{\partial^2\Psi_2}{\partial x_1 x_2}-\frac{\partial^2\varphi_2}{\partial x_2^2}\right)\right] \tag{4.73.4}$$

$$[\sigma_{22}]_2 = \mathrm{m}\left[\frac{c_{p2}^2}{c_{s2}^2}\left(\frac{\partial^2\varphi_2}{\partial x_1^2}+\frac{\partial^2\varphi_2}{\partial x_2^2}\right)-2\left(\frac{\partial^2\varphi_2}{\partial x_1^2}+\frac{\partial^2\Psi_2}{\partial x_1 x_2}\right)\right] \tag{4.73.5}$$

$$[\sigma_{12}]_2 = \mathrm{m}\left[2\frac{\partial^2\varphi_2}{\partial x_1 x_2}+\left(\frac{\partial^2\Psi_2}{\partial x_2^2}-\frac{\partial^2\Psi_2}{\partial x_1^2}\right)\right] \tag{4.73.6}$$

At the interface it is necessary that the displacements u_1 and u_2 and stresses σ_{22} and σ_{12} should be continuous in order to satisfy the wave propagation equation both in material 1 and material 2. So, the interface conditions could be written as follows

$$[u_1]_1 = [u_1]_2 \quad ; \quad [u_2]_1 = [u_2]_2 \quad ; \quad [\sigma_{22}]_1 = [\sigma_{22}]_2 \quad ; \quad [\sigma_{12}]_1 = [\sigma_{12}]_2 \tag{4.74}$$

Enforcing these interface conditions and substituting Eqs. (4.70.1) and (4.70.2) into Eq. (4.74), four independent equations are obtained and can be written in a matrix form as follows

$$\begin{bmatrix} k & \beta_1 & -k & \beta_2 \\ \alpha_1 & -k & \alpha_2 & k \\ 2k\alpha_1 & -2k^2+k_{s1}^2 & 2k\alpha_2\dfrac{\mathrm{m}_2}{\mathrm{m}_1} & \left(2k^2-k_{s2}^2\right)\dfrac{\mathrm{m}_2}{\mathrm{m}_1} \\ \left(2k^2-k_{s1}^2\right) & 2k\beta_1 & -\left(2k^2-k_{s2}^2\right)\dfrac{\mathrm{m}_2}{\mathrm{m}_1} & 2k\beta_2\dfrac{\mathrm{m}_2}{\mathrm{m}_1} \end{bmatrix} \begin{Bmatrix} R_p \\ R_s \\ T_p \\ T_s \end{Bmatrix} = \begin{Bmatrix} -k \\ \alpha_1 \\ 2k\alpha_1 \\ -\left(2k^2-k_{s1}^2\right) \end{Bmatrix} \tag{4.75}$$

Eq. (4.75) presents four simultaneous linear equations which has a rank four matrix and can be uniquely solved. Hence, the reflection and transmission coefficients for the P-wave and SV-waves at the interface can be calculated. Once the coefficients are calculated, they can be plugged into Eq. (4.70.i) with a coordinate location (x_j) and then into Eq. (4.72.i) to calculate the displacement wave field anywhere in the materials. Similarly, further the displacement field or the wave potentials with reflection coefficients can be plugged into Eq. (4.73.i) to calculate the stress field anywhere in the materials.

4.3.3.2 SH-Wave Incident at the Interface

Discussion on SH-wave is pictorially presented in Fig. 4.9b. SH-wave is decoupled from P and SV waves. Thus, it clearly shown that to solve SH-wave interaction problem at an isotropic interface, there will be only three generalized wave potentials to consider. In the following wave potentials, the time harmonic part $e^{-i\omega t}$ is omitted without the loss of generality.

$$\text{incident SH-wave} \qquad \Psi_{SH}\left(x_j\right) = B_{SH} e^{i\left(k_j^s x_j\right)} \tag{4.76.1}$$

$$\text{reflected SH-wave} \qquad \Psi_{SHr}\left(x_j\right) = B_{SHr} e^{i\left(k_j^s x_j\right)} \tag{4.76.2}$$

$$\text{transmitted SH-wave} \qquad \Psi_{SHt}\left(x_j\right) = B_{SHt} e^{i\left(k_j^s x_j\right)} \tag{4.76.3}$$

The case with SH wave interaction presented in Fig. 4.9b is an anti-plane problem but still it is a two-dimensional problem. Whereas, the particle displacements are orthogonal to the wave propagation direction. There will be no out of plane wave vectors but out of plane displacements. The wave vector $\mathbf{k}^s$ will have two components along x_1 and x_2 axis, respectively. As shown in Fig. 4.9b the incident SH-wave makes an angle θ_s with x_2 axis. Reflected and transmitted SH-waves makes θ_{sr} and θ_{st} angles with x_2 axis respectively. To uniquely satisfy the interface conditions it is necessary that θ_{sr} should be equal to θ_s ($\theta_{sr} = \theta_s$). Applying Snell's law [5, 6] a relation between the transmitted angles and the incident angle can be found which is $\theta_{st} = \sin^{-1}\left(\frac{c_{s2}}{c_{s1}}\sin\theta_s\right)$ (please refer Appendix in Section 4.5.3). Based on this arrangement, reflected wave is called the up going wave and has both positive wave number components. However, the transmitted wave is called the down going wave and x_2 component of the wave number vector is negative. Additionally, there are two materials in the problem. Hence, the magnitudes of SH-wave numbers in material 1 and material 2 will be different based on the material properties indicated in Fig. 4.9b. Recalling the previous discussion on wave number vectors in Section 4.3.2.3, let's assume the magnitudes of the SH-wave numbers in two materials are k_{s1} and k_{s2}, respectively.

$$k_{s1} = \frac{\omega}{c_{s1}} \quad ; \quad c_{s1} = \sqrt{\frac{\mathrm{m}_1}{\rho_1}} \quad ; \quad k_{s2} = \frac{\omega}{c_{s2}} \quad ; \quad c_{s2} = \sqrt{\frac{\mathrm{m}_2}{\rho_2}} \tag{4.77}$$

Next the components of the wave numbers along x_1 axes are

<u>*In Material 1, wave numbers along x_1 axis:*</u>

$$k_1^s = k_{s1}\sin\theta_s \tag{4.78.1}$$

<u>*In Material 2, wave number along x_1 axis:*</u>

$$k_1^s = k_{s2}\sin\theta_{st} \tag{4.78.2}$$

As the wave propagates as a union and superposed together it is not possible to have different wave numbers along the same x_1 axis. Hence, let's assume a common wave number k along the x_1 axis and rewrite Eqs. (4.78.1) and (4.78.2) as follows

$$k = k_{s1} \sin\theta_s = k_{s2} \sin\theta_{st} \tag{4.79}$$

The components of the wave numbers along x_2 axes are however, not the same and can be independent. This independency causes the different direction of wave vectors pointing to their respective wave propagation direction as shown in Fig. 4.9b. Components of the wave numbers along x_2 axis are

In Material 1, wave numbers along x_2 axis:

$$\left[k_2^s\right]_1 = k_{s1} \cos\theta_s \tag{4.80.1}$$

In Material 2, wave number along x_2 axis:

$$\left[k_2^s\right]_2 = k_{s1} \cos\theta_{st} \tag{4.80.2}$$

where subscripts with the bracketed parameter indicate the material. For convenience it is better to use different parameters for the wave numbers along x_2 axis in two different material bodies. Introducing new parameter, the above expression can be presented as follows

In Material 1, wave numbers along x_2 axis:

$$\beta_1 = \left[k_2^s\right]_1 = k_{s1} \cos\theta_s \tag{4.81.1}$$

In Material 2, wave number along x_2 axis:

$$\beta_2 = \left[k_2^s\right]_2 = k_{s1} \cos\theta_{st} \tag{4.81.2}$$

where β is the SH-wave number along x_2 axis. Subscripts 1 and 2 are to indicate in which material they propagate and $\theta_{st} = \sin^{-1}\left(\frac{c_{s2}}{c_{s1}} \sin\theta_s\right)$ obtained applying the Snell's law (please refer Appendix in Section 5).

With the above notations the wave vector components can be expressed as follows

$$\text{incident P-wave} \quad : \quad \mathbf{k}^s = k\hat{\mathbf{e}}_1 - \beta_1\hat{\mathbf{e}}_2 \tag{4.82.1}$$

$$\text{reflected SH-wave} \quad : \quad \mathbf{k}^{sr} = k\hat{\mathbf{e}}_1 + \beta_1\hat{\mathbf{e}}_2 \tag{4.82.2}$$

$$\text{transmitted SH-wave} \quad : \quad \mathbf{k}^{st} = k\hat{\mathbf{e}}_1 - \beta_2\hat{\mathbf{e}}_2 \tag{4.83.3}$$

Substituting the wave vectors in Eq. (4.82.i) into Eq. (4.76.i) and performing the dot product between the respective wave vectors and **x** the wave potentials can be rewritten as

$$\text{incident P-wave} \qquad \Psi_{SH}\left(x_j\right) = B_{SH} e^{i(kx_1 - \beta_1 x_2)} \tag{4.83.1}$$

reflected SV-wave $$\Psi_{SHr}\left(x_j\right) = B_{SHr}e^{i(kx_1+\beta_1 x_2)} \quad (4.83.4)$$

transmitted SV-wave $$\Psi_{SHt}\left(x_j\right) = B_{SHt}e^{i(kx_1-\beta_2 x_2)} \quad (4.83.5)$$

Normalizing all the wave potentials with respect to the incident amplitude B_{SH} the total SH-wave potentials in materials 1 and 2 can be written with respect their reflection and transmission coefficients, respectively

Material 1:

$$\Psi_1 = e^{i(kx_1-\beta_1 x_2)} + R_s e^{i(kx_1+\beta_1 x_2)} \quad (4.84.1)$$

Material 2:

$$\Psi_2 = T_s e^{i(kx_1-\beta_2 x_2)} \quad (4.84.2)$$

where $R_s = B_{SHr}/B_{SH}$; $T_s = B_{SHt}/B_{SH}$. The problem depicted here is an anti-plane two-dimensional problem. That means both u_1 and u_2 displacement fields are equal to zero. Only u_3 displacement field will survive. As there is no other displacement components present for SH-wave interaction with the interface, Helmholtz decomposition of the resultant displacement field is redundant and hence, the potential function assumed in Eqs. (4.84.1) and (4.84.2) can be directly assumed for the u_3 displacement. Thus u_3 displacement potentials in two materials can be written as

Material 1:

$$[u_3]_1 = e^{i(kx_1-\beta_1 x_2)} + R_s e^{i(kx_1+\beta_1 x_2)} \quad (4.85.1)$$

Material 2:

$$[u_3]_2 = T_s e^{i(kx_1-\beta_2 x_2)} \quad (4.85.2)$$

Recalling the stress equation (Eq. (4.28)) discussed in Section 4.3.1 and substituting the displacement field equations in Eq. (4.85) for materials 1 and 2 the stress components will viz.

In Material 1, stress components ($\sigma_{11} = \sigma_{22} = \sigma_{33} = \sigma_{12} = 0$) but σ_{13} and σ_{23} are

$$[\sigma_{13}]_1 = \mathrm{m}\left[\frac{\partial[u_3]_1}{\partial x_1}\right] \quad (4.86.1)$$

$$[\sigma_{23}]_1 = \mathrm{m}\left[\frac{\partial[u_3]_1}{\partial x_2}\right] \quad (4.86.2)$$

In Material 2, stress components ($\sigma_{11} = \sigma_{22} = \sigma_{33} = \sigma_{12} = 0$) but σ_{23} and σ_{13} are

$$[\sigma_{13}]_2 = \mathrm{m}\left[\frac{\partial[u_3]_2}{\partial x_1}\right] \quad (4.87.1)$$

$$\left[\sigma_{23}\right]_2 = \mathrm{m}\left[\frac{\partial\left[u_3\right]_2}{\partial x_2}\right] \tag{4.87.2}$$

At the interface it is not necessary that the displacements u_3 to be continuous but it is necessary that the stresses σ_{13} and σ_{23} be continuous in order to satisfy the wave propagation equation both in material 1 and material 2. So, the interface conditions could be written as follows

$$\left[\sigma_{13}\right]_1 = \left[\sigma_{13}\right]_2 \quad ; \quad \left[\sigma_{23}\right]_1 = \left[\sigma_{23}\right]_2 \tag{4.88}$$

Enforcing these interface conditions and substituting Eqs. (4.85.1) and (4.85.2) into Eq. (4.88), two independent equations are obtained and can be written in a matrix form as follows

$$\begin{bmatrix} 1 & -1 \\ \mathrm{m}_1\beta_1 & \mathrm{m}_2\beta_2 \end{bmatrix} \begin{Bmatrix} R_s \\ T_s \end{Bmatrix} = \begin{Bmatrix} -1 \\ \mathrm{m}_1\beta_1 \end{Bmatrix} \tag{4.89}$$

Eq. (4.89) presents two simultaneous linear equations which can be easily solved and the unknowns R_s and T_s can be found as

$$R_s = \frac{1-Q}{1+Q} \quad ; \quad T_s = \frac{2}{1+Q} \tag{4.90}$$

where $Q = \mathrm{m}_2\beta_2/\mathrm{m}_1\beta_1$ is a factor derived from the material properties of the two materials that created the interface. Once the coefficients are calculated, they can be plugged into Eq. (4.85.1) with a coordinate location (x_j) to calculate the displacement wave field anywhere in the materials. Similarly, further the displacement field or the wave potentials with reflection coefficients can be plugged into Eq. (4.86.1) to calculate the stress field anywhere in the materials.

4.4 WAVE PROPAGATION IN BULK ANISOTROPIC SOLID MEDIA

In the previous Section 4.2, wave propagation in bulk isotropic solid is discussed with the necessary understanding. It is clear that wave fronts in fluid and isotropic material are spherical because wave velocity in all directions is equal. Although P-waves and S-waves have different wave velocities, their wave fronts are still spherical as shown in Fig. 4.10a. However, waves in anisotropic solid media [12] does not propagate with spherical wave front as it does in isotropic materials. This situation is shown in Fig. 4.10b for an anisotropic material GaAs. Considering a source at the origin, wave velocity of the shear wave in aluminum, an isotropic material is constant irrespective of the direction of propagation or irrespective of the pointing wave vector shown in Fig. 4.8a. Due to constant amplitude in all directions, wave velocity plot in three-dimension generates a spherical (Fig. 4.10a) plot of the wave front. Similarly, considering a source at the origin in GaAs, an anisotropic material, quasi shear wave velocities (will be discussed later) are different

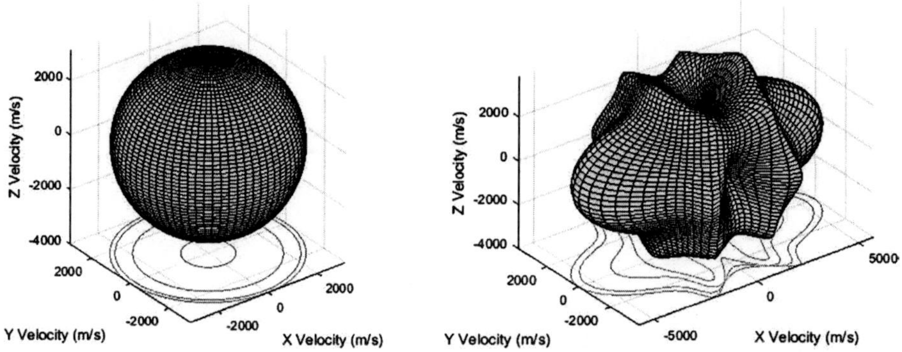

FIGURE 4.10 a) Spherical wave front in isotropic media (Aluminum) showing wave velocity profile in 3D b) 3D wave velocity profile in anisotropic media (GaAs), showing nonspherical wave front.

in different directions of propagation or the pointing wave vectors shown. Due to different amplitudes of velocities along different directions, wave velocity plot in three dimensions (3D) generates specific architectures (like in Fig. 4.10b) of the wave fronts and these architectures are uniquely dependent on the material property matrix or the $\mathbb{C}_{mn}$ matrix in Eq. (3.79). Comparing Fig. 4.10a and b it is clear that analyzing wave propagation in an anisotropic material is way challenging than that of an isotropic material and needs special treatments. Composite materials are good example of anisotropic solid material with different degrees. For example, triclinic, monoclinic, orthotropic, transversely isotropic materials are the example of generalized anisotropic material with different degrees of anisotropy or with different plane and/or axis of symmetry discussed in Section 3.9 in Chapter 3. Although composites will get more attention in this book in later chapters related to NDE problems, waves in anisotropic bulk media is also significant in geophysical problems for underground exploration, understanding plate tectonics, exploring or finding soil and rock strata, etc.

In this section, wave propagation in bulk anisotropic solids with different degrees of anisotropy will be discussed first. Ultrasonic NDE of composite materials are much complex than an NDE of isotropic solids. As discussed before, ultrasonic transducer emits ultrasonic energy in the forward direction from the surface of the transducer. For example, if an ultrasonic transducer is placed in an anisotropic media (or rather on the surface of the anisotropic media) and emits the ultrasonic energy in the forward direction, say along the $\mathbf{k} = k_i\hat{\mathbf{e}}_i$ vector as shown in Fig. 4.11a, the propagation of the energy of the wave will not be along the direction of the $\mathbf{k}$ vector, as it will be in isotropic media. This case is quite evident from a simple comparison of wave front presented in Fig. 4.10. The unit vector $\hat{\mathbf{n}} = n_i\hat{\mathbf{e}}_i$ in Fig. 4.11a is along the $\mathbf{k}$ vector, where n_i is the direction cosines of the unit vector $\hat{\mathbf{n}}$, i.e., cosines of the respective angles made by the $\mathbf{k}$ vector with the coordinate axes x_1 x_2 and x_3 respectively According to Section 4.3, inside an isotropic material the wave energy will propagate along the direction of the $\mathbf{k}$ vector. However, inside an anisotropic media, wave energy does not propagate along the $\mathbf{k}$ vector but creates different wave front for the

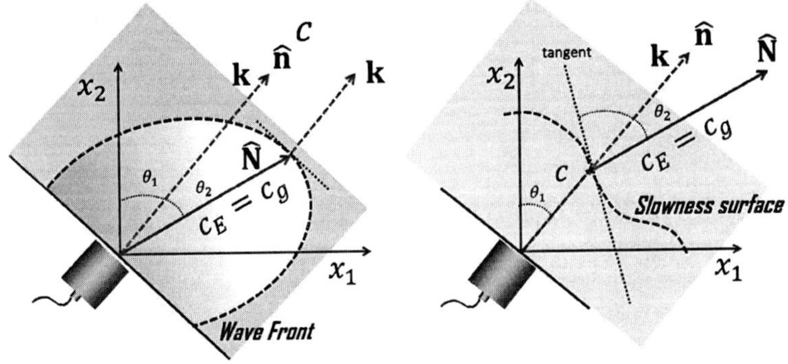

FIGURE 4.11 An artistic view showing the schematics of the wave field inside a bulk anisotropic media a) showing the wave front b) showing the slowness surface.

wave energy to propagated. It can be seen that the velocity of the wave energy c_E is at an angle θ_2 with respect to the wave vector $\mathbf{k}$. This is along a different direction than the phase velocity c along the direction $\hat{\mathbf{n}}$. Please note that the plane of energy propagation may not be on the same plane as the incident wave which is evident from an example Fig. 4.10b. This means that the unit vector $\hat{\mathbf{N}} = N_i\hat{\mathbf{e}}_i$ in Fig. 4.11a may not be on the x_1-x_2 plane. It can be along any direction in the 3D co-ordinate system based on the material properties and the direction of the $\mathbf{k}$ vector. Here, N_i is the direction cosines of the unit vector $\hat{\mathbf{N}}$, i.e., cosines of the respective angles made by the unit vector $\hat{\mathbf{N}}$ with the coordinate axes x_1 x_2 and x_3 respectively. Next, Fig. 4.11b shows a hypothetical velocity slowness surface in an anisotropic material. Slowness surface is most vital component to discuss wave propagation in anisotropic material, which was not necessary to introduce for isotropic material, however, similar could be defined and still the following discussions could be equally valid for isotropic material. Please note that the slowness surface is a generalized concept for all material types, but inevitable to realize wave propagation in anisotropic materials. Slowness surface is a surface in 3D created by plotting the inverse of the phase velocities along the different directions of the wave propagation. Alternatively saying, along the $\mathbf{k}$ vector or along the direction of the wave propagation inverse of the wave velocities are plotted. In isotropic material slowness surfaces are spherical due to its isotropy. But in anisotropic material nature or the architecture of the slowness surface could be very different depending on the architecture of the wave velocity surface obtained by directly plotting the phase wave velocities along the different directions of the wave propagation (i.e., along the $\mathbf{k}$ vector). A sample wave velocity surface and a sample slowness surface are shown in Fig. 4.12, for GaAs anisotropic material, where these architectures are realized for a quasi shear wave. Referring back to Fig. 4.11b, a normal to the slowness surface $\hat{\mathbf{N}}$ at the intersection of wave vector $\mathbf{k}$ and the slowness surface points shows the direction of the propagation of wave energy. $\hat{\mathbf{N}}$ can be defined as wave energy vector. As discussed in Section 4.14, wave energy velocity and wave group velocity are synonyms. Hence, the unit or normal vector $\hat{\mathbf{N}}$ also points toward the internal wave group velocity direction, when

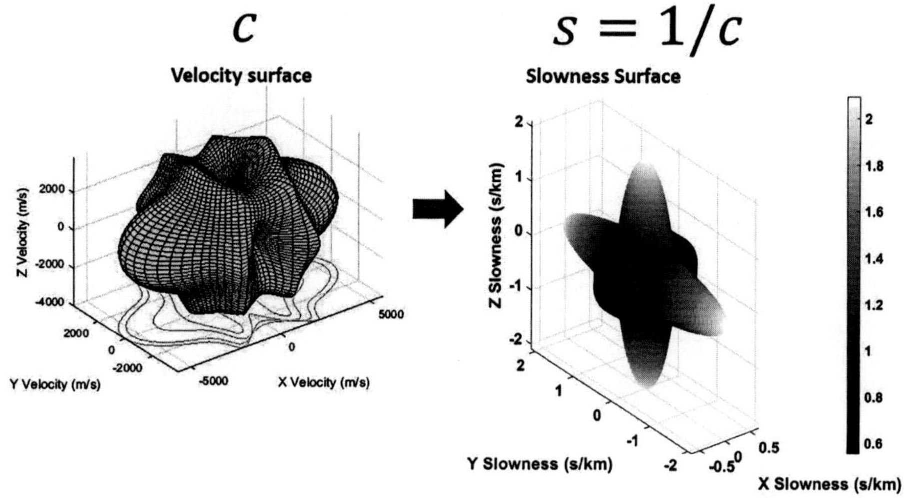

FIGURE 4.12 A sample wave velocity surface and a sample slowness surface is shown for GaAs anisotropic material.

the **k** vector points toward the unit vector $\hat{\mathbf{n}}$ along the given wave propagation direction. In isotropic material **k** vector and the vector $\hat{\mathbf{N}}$ are pointing toward the same direction but in anisotropic they are not. Conversely, we can also make a comment that in anisotropic material the wave front surfaces and the slowness surfaces are not concentric as it is the case for the isotropic materials.

Next it is necessary to summarize three different surfaces that are very important in anisotropic media. Although existed for isotropic media, they did not play much importance in analyzing the wave propagation and the analysis presented in Section 4.3 was possible without discussing them explicitly. This is because they were all concentric and never intersected each other in isotropic material. In anisotropic media these three surfaces are not necessarily concentric. They play much importance in understanding the wave propagation in anisotropic material. Let's talk about these three surfaces, wave front surface, wave velocity surface, and wave slowness surface. Wave front surfaces are created by joining the particles that are in same phase in the material when the wave propagates. Wave velocity surfaces are created by joining the tip of the vectors with magnitude that represents the respective amplitude of the phase wave velocities in the direction of the pointing vector like it is created in Fig. 4.10a for shear wave phase velocity in isotropic and in Fig. 4.10b for quasi shear wave phase velocity in anisotropic media. Slowness surfaces are created by joining the tip of the vectors with magnitude that represents the respective amplitude of the inverse of the phase wave velocities in the direction of the pointing vector (or wave vector or wave propagation direction) as it is created in Fig. 4.12 from the inverse of the magnitudes used to create the wave velocity surface presented for GaAs in Fig. 4.10b.

To represents the relationship between these surfaces, we can direct our attention to the Fig. 4.13a, where three different but very important possible intersecting

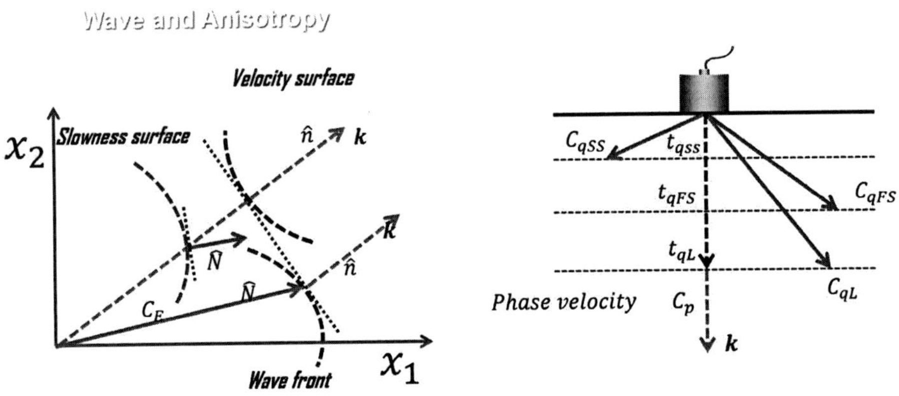

FIGURE 4.13 a) Schematics of the wave surfaces and their relation b) relation of the phase velocity with the fundamental wave modes in anisotropic media.

surfaces are shown. Velocity surface is the surface constructed with the values of the modal wave velocities around the direction of the wave vector **k** along the unit vector with direction cosines $\hat{\mathbf{n}}$. But that does not mean that the wave packet is propagated along the same direction with that same velocity. Now by taking the inverse of the wave velocity magnitudes plotted around the same direction of wave vector **k** along the unit vector with direction cosines $\hat{\mathbf{n}}$, another similar surface emerged, which is called the slowness surface as indicated in Fig. 4.12. Interestingly normal to that slowness surface represents the direction of the energy propagation. Next, if the equal energy lines with equal phase are plotted in the media, one will get the wave fronts. As depicted in Fig. 4.11a, the wave front is propagated along the direction of the energy propagation. A normal to the slowness surface represents the direction of the wave front as well as shown in Fig. 4.13a.

P-wave, SH-wave, and SV-wave are the three fundamental wave modes in isotropic bulk media. We should expect similar in anisotropic media, but they are not quite similar. Thus, the modes in anisotropic media are called the quasi modes. For example, when the wave energy is introduced into an anisotropic media, the wave breaks down into three fundamental wave modes [12, 13], quasi longitudinal (qL), quasi shear 1 or quasi slow shear (qSS), and quasi shear 2 or quasi fast shear (qFS). While, the wave vector **k** represents the intended direction of the wave propagation in an anisotropic media, the wave energy propagates in a different direction due to the above mentioned three wave modes with their three respective wave velocities along three different directions, which creates three different phase wave velocity surfaces. However, interference of these phase wave velocity surfaces is very complex to realize and causing the variation of wave velocities with direction, producing a unique wave front with unique direction of maximum energy propagation with maximum energy velocity or group wave velocity. Nonetheless, some form of projection of these wave modes along the direction of the **k** vector will contribute to the phase velocity of the wave packet along the **k** direction as shown in Fig. 4.13b. In this figure the distances are illustrated to times which are proportional to their respective phase velocities along the direction of the wave vector **k**. The distances travelled by each mode are also proportional to the group velocities of respective wave types. This could be further understood using the Snell's law discussed in Section 4.5.3. Hence, by now, we have realized that it is just not difficult but impossible to track these wave traces using the conventional ray tracing methods in the composites. Indeed, we need a sophisticated mathematical method for modeling the wave in anisotropic media where all the above-mentioned situations can be incorporated by virtue of their mathematical formulations, automatically. But before any of the method is introduced it is better that one should be able to generate the modal wave velocity surfaces and the slowness surfaces that are presented in Figs. 4.10 and 4.12 for different modes in different material types with different degrees of anisotropy. It seems they are the most crucial to understand the waves in bulk anisotropic media in the first place.

4.4.1 Governing Elastodynamic Equation in Anisotropic Media

In this section the physics of wave propagation in anisotropic media is discussed and will be shown how the physical discussion in the previous section makes sense, mathematically. It is assumed that readers are familiar with the index notations and familiar with the basics of continuum mechanics discussed in Chapters 2 and 3, respectively. Let's recall the fundamental elastodynamic equation or the equation of motion in a solid body derived in Eq. (3.71) in Chapter 3.

$$\sigma_{ij,j} + F_i = \rho \ddot{u}_i \tag{4.91}$$

where $F_i = \rho f_i$ is the body force, σ_{ij} is the stress, and u_i is the displacement at a point ($\mathbf{x} = x_i \hat{\mathrm{e}}_\mathrm{i}$) in a bulk anisotropic solid. For homogeneous anisotropic solid media the density (ρ) remains constant; however, stress (σ_{ij}) and displacement (u_i) are in general functions of both space (x_i) and time (t). Eq. (4.91) was derived using the differential approach however, the same equation could be derived using integral principle described briefly in Appendix. Recollecting Eq. (3.75), the constitutive law for any linear material and the strain displacement relation in linear elastic material written in Eq. (3.72) can be written as

$$\sigma_{ij} = \mathbb{C}_{ijkl}\mathrm{e}_{kl} \quad ; \quad \mathrm{e}_{kl} = \frac{1}{2}\left(u_{k,l} + u_{l,k}\right) \tag{4.92}$$

where e_{kl} is the strain in the anisotropic solid. For any generalized linear elastic anisotropic material $\mathbb{C}_{ijkl}$ is the matrix of elastic constants written in Eqs. (3.78) and (3.79). Assuming the material is linear during the NDE inspection due to the very short exposure time of the wave compared to the loading history of the material, all the above equations are valid in anisotropic material. Substituting the geometrically linear strain-displacement relation and the constitutive law in Eq. (4.92) into the elastodynamic equation in Eq. (4.91), the equation takes the form in terms of displacement [14] as shown in Eq. (4.93).

$$\mathbb{C}_{ijml}\frac{\partial^2 u_m}{\partial x_j \partial x_l} + F_i = \rho \ddot{u}_i \tag{4.93}$$

Next it would be necessary to solve this complicated equation to realize the facts about the anisotropic waves discussed in the previous section. In NDE experiments a transducer is used with a central frequency of actuation, which means that the transducer is able to generate the maximum amplitude of the displacement at that frequency. However, if not impossible, it is extremely difficult to make a monochromatic ultrasonic transducer. So, when the transducer is actuated, there will be other proximal frequencies generating displacements with gradually lower amplitudes. Assuming linear materials where superposition theorem is valid in the Fourier domain (discussed in Chapter 5) and the reciprocity theorem is valid, we will proceed to solve the above equation using a monochromatic harmonic displacement function and such solutions can be commutable to the other neighboring frequencies generated by the transducer. Here it is necessary to bring the discussion of plane harmonic waves again that are mentioned in Sections 4.3.2.2 and 4.3.2.3. For isotropic media scalar and vector potentials were assumed to decompose the displacement field. They were $\phi(x_j,t)=\varphi(x_j)e^{-i\omega t}$ and $\psi_i(x_j,t)=\Psi_i(x_j)e^{-i\omega t}$ or $\psi(x_j,t)=\Psi(x_j)e^{-i\omega t}$. Please note that the wave propagation problem is solvable for one unique frequency at a time because the time harmonic part of the potential functions $e^{-i\omega t}$ that goes into the equation is separable and are omitted most of the time from the equation due to its appearance on either side of the equation. Many such cases are depicted in Section 4.3.2. However, the wave numbers ($k=\omega/c$) in the equations are calculated using a specific angular frequency ω, which is the same frequency appeared in this time harmonic part $e^{-i\omega t}$. Hence, although the $e^{-i\omega t}$ part is taken out from the equation, it is actually not separated, but contributed monochromatic wave form for the solution. Thus, these potentials were also monochromatic. In anisotropic media, Helmholtz decomposition will not be suitable and thus there is no point in assuming those potentials separately. This is because the three wave surfaces are not concentric, and three wave modes may not be orthogonal to each other. Additionally, they may not create exact dilatational or distortional wave field as it does in isotropic material discussed in Section 4.3.2.4. Thus, to solve the wave propagation problem monochromatically, let's assume monochromatic displacement function like those potentials in isotropic media, but directly imposed on each components of the displacement wave field in anisotropic solid.

$$u_m = Ag_m e^{i(\mathbf{k}.\mathbf{x}-\omega t)} \tag{4.94}$$

where A is the scalar amplitude of the wave, ω is the monochromatic wave frequency, $\mathbf{k}$ is the wave vector, $\mathbf{x}$ is the position vector where, $\mathbf{k}.\mathbf{x}$ is the dot product between $\mathbf{k}$ and $\mathbf{x}$ represents the phase component of the wave as discussed in Section 4.3. Only one very important thing is different in Eq. (4.94) which is the parameter g_m. Here, $\hat{\mathbf{g}} = g_m\hat{\mathbf{e}}_\mathbf{m}$ is called the polarization direction or the unit vectors pointing to the direction of propagating wave modes that contribute to the particle displacement along three orthogonal directions. After substituting Eq. (4.94) into Eq. (4.93) and following few mathematical simplifications [14] we get the following equation.

$$\left[\mathbb{C}_{ijml}k_jk_l - \rho\delta_{im}\omega^2\right]Ag_m = -F_i \tag{4.95}$$

The above system of equations is subjected to external force or the body force and depends on the quality and magnitude of the force field. However, to extract the nature of the wave field and the wave modes in an unperturbed system, it is necessary to solve the homogeneous system of equations. Hence, without body force the Navier's equation in anisotropic media and the nontrivial solution of the equation can be written as

$$\left[\mathbb{C}_{ijml}n_jn_l - \rho\delta_{im}c^2\right]g_m = 0 \tag{4.96}$$

The above equation is an eigen value problem, where $c^2 = \omega^2/k^2$ is the square of the phase velocity of the wave along the direction of **k** vector, n_j are the direction cosines of the wave propagation direction, i.e., again along the **k** vector (Fig. 4.11a). Solution of this equation gives the wave modes that are propagating in the material with material constants $\mathbb{C}_{ijml}$. Eq. (4.96) is the well-known equation called Christoffel's equation [12]. The solution of this equation will provide three eigen modes with wave velocities c_{qL}, c_{qFS}, and c_{qSS}, respectively, along the directions found from the eigen vector matrix $\mathbf{E_v}$ in Fig. 4.14 composed on g_m, the polarization vectors of each modes. The figure explains the meaning of the solution of Eq. (4.96) graphically and the understanding of the eigen vector solution in a pictorial form. The solution is performed for a specific direction of wave propagation along the **k** vector with direction normal $\hat{\mathbf{n}}$. In a 3D Cartesian right-hand coordinate system, the solution provided three eigen wave modes that are the wave velocities in three different directions. These are the direction represented by the eigen vectors $\mathbf{g_L}, \mathbf{g_{FS}}, \mathbf{g_{SS}}$, respectively obtained from the eigen value problem in Eq. (4.96). Immediately now it can be realized that the superposition of the projections of the phase velocities of the wave modes (c_{qL}, c_{qFS}, and c_{qSS}) on the **k** vector gives the phase velocity of the wave along the **k** or the wave propagation direction. However, direction of the wave energy will be different as discussed above

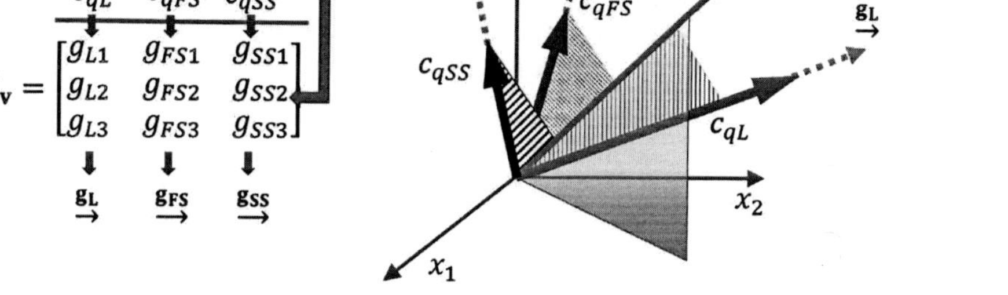

FIGURE 4.14 Graphical understanding of Christoffel Solution that gives phase slowness of different wave modes in all directions.

which will be at the resultant direction with the group velocity or the velocity of the wave energy. There are three wave modes quasi longitudinal (qL), quasi fast shear (qFS) or quasi shear 2 and quasi slow shear (qSS) or quasi shear 1 wave modes with velocities c_{qL}, c_{qFS}, and c_{qSS}, respectively. These three wave modes are propagating along the directions $\mathbf{g_L}$, $\mathbf{g_{FS}}$, $\mathbf{g_{SS}}$, respectively. The $\mathbf{g_L}$, $\mathbf{g_{FS}}$, $\mathbf{g_{SS}}$ are the eigen vectors with components along x_1, x_2, and x_3 axes are written as

$$\mathbf{g_L} = g_{L1}\hat{\mathbf{e}}_1 + g_{L2}\hat{\mathbf{e}}_2 + g_{L3}\hat{\mathbf{e}}_3 \tag{4.97.1}$$

$$\mathbf{g_{FS}} = g_{FS1}\hat{\mathbf{e}}_1 + g_{FS2}\hat{\mathbf{e}}_2 + g_{FS3}\hat{\mathbf{e}}_3 \tag{4.97.2}$$

$$\mathbf{g_{SS}} = g_{SS1}\hat{\mathbf{e}}_1 + g_{SS2}\hat{\mathbf{e}}_2 + g_{SS3}\hat{\mathbf{e}}_3 \tag{4.97.3}$$

These above equations are graphically explained in Fig. 4.14. It is clear that for each wave propagation direction **k** in three dimension (3D) with varying angles, c_{qL}, c_{qFS}, and c_{qSS} and the elements in $\mathbf{E_v}$ matrix will be different. If the magnitude of the wave velocities c_{qL}, c_{qFS}, and c_{qSS} are plotted for all possible wave propagation direction (**k** along $\hat{\mathbf{n}}$) in three dimension by discretizing a sphere (unit normals $\hat{\mathbf{n}}$ at each point on the surface of the sphere are calculated from the basic equations used in spherical coordinate system), the velocity surfaces are generated for all three modes separately. Such wave velocity surfaces can be generated with different material properties $\mathbb{C}_{ijml}$ (Eq. (3.78)) or $\mathbb{C}_{mn}$ (Eq. (3.79)) produce different wave velocity surfaces. By taking the inverse of the velocity surfaces slowness surfaces can be created like it is shown in Fig. (4.12). Few examples are presented herein.

4.4.2 Wave Modes in all Possible Directions of Wave Propagation in 3D

4.4.2.1 Comparison between Isotropic and Anisotropic Slowness Profiles

Let's assume an anisotropic material GaAs with the following material constants

$$\begin{bmatrix} C_{11} & C_{12} & C_{13} & C_{14} & C_{15} & C_{16} \\ C_{21} & C_{22} & C_{23} & C_{24} & C_{25} & C_{26} \\ C_{31} & C_{32} & C_{33} & C_{34} & C_{35} & C_{36} \\ C_{41} & C_{42} & C_{43} & C_{44} & C_{45} & C_{46} \\ C_{51} & C_{52} & C_{53} & C_{54} & C_{55} & C_{56} \\ C_{61} & C_{62} & C_{63} & C_{64} & C_{65} & C_{66} \end{bmatrix} = \begin{bmatrix} 72.3 & 10.2 & 10.3 & 0.00 & 0.00 & 0.00 \\ 10.2 & 78.3 & 24.4 & 0.00 & 0.00 & 0.00 \\ 10.3 & 24.4 & 78.3 & 0.00 & 0.00 & 0.00 \\ 0.00 & 0.00 & 0.00 & 5.00 & 0.00 & 0.00 \\ 0.00 & 0.00 & 0.00 & 0.00 & 175.2 & 0.00 \\ 0.00 & 0.00 & 0.00 & 0.00 & 0.00 & 175.2 \end{bmatrix} GPa \tag{4.98}$$

Eq. (4.98) was plugged into Eq. (4.96) to obtain the 3D wave velocity surfaces for each quasi wave mode. Next for each direction the inverse of the wave velocities or the slowness were calculated which resulted the slowness surfaces. Fig. 4.15a–c shows the wave velocity slowness for the qL, qFS, qSS modes, respectively. Similarly, Fig. 4.16 shows the wave velocity slowness of the longitudinal and two shear wave modes in aluminum, which is

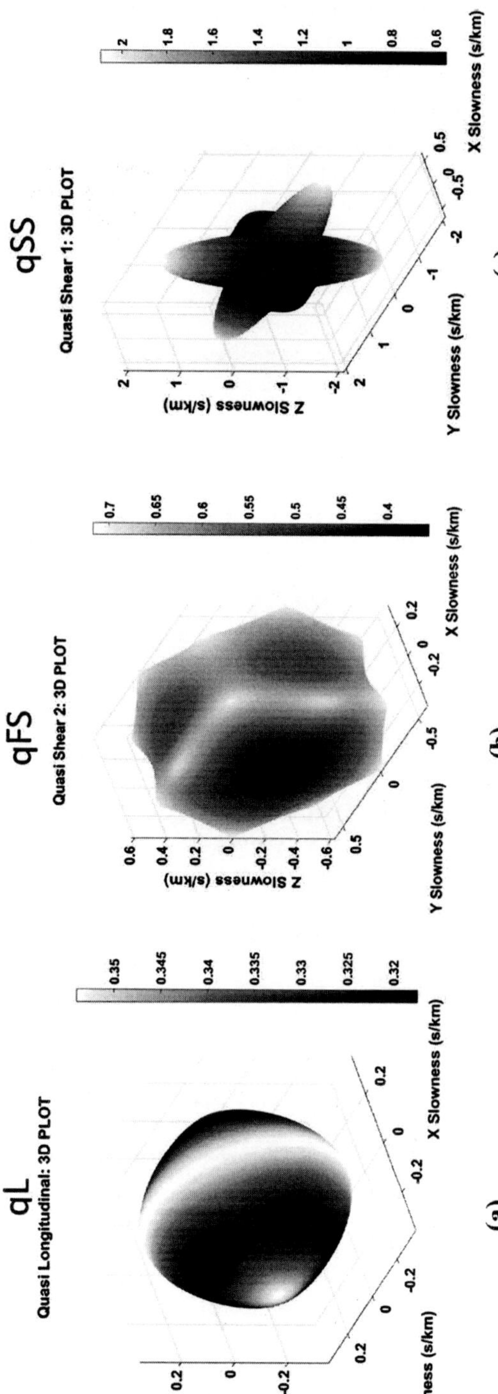

FIGURE 4.15 3D slowness plot for GaAs a) quasi longitudinal mode b) quasi fast shear (qS2) mode c) quasi slow shear (qS1) mode.

an isotropic material. Constitutive property matrix for aluminum with Lamé constants ($\mathbb{L} = 6.13$ Gpa; $\mathbb{m} = 2.64$ GPa) and density 2700 kg/m^3 is

$$\mathbb{C}_{mn} = \begin{bmatrix} 11.41 & 6.13 & 6.13 & 0 & 0 & 0 \\ & 11.41 & 6.13 & 0 & 0 & 0 \\ & & 11.41 & 0 & 0 & 0 \\ & & & 2.64 & 0 & 0 \\ & Sym & & & 2.64 & 0 \\ & & & & & 2.64 \end{bmatrix} GPa \tag{4.99}$$

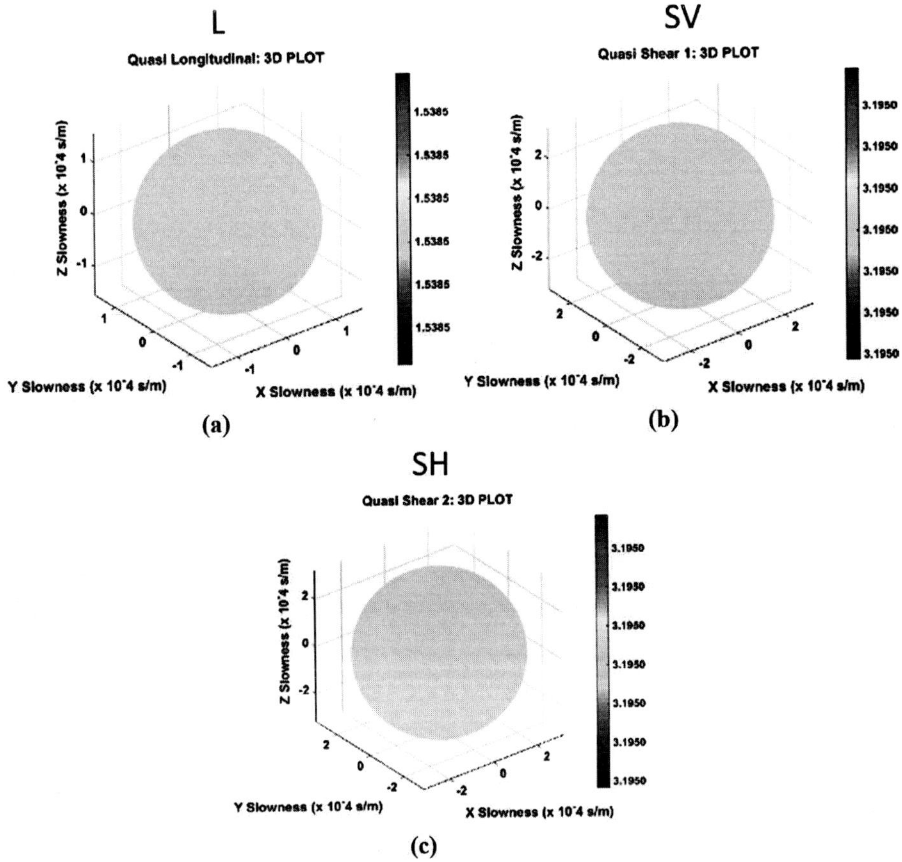

FIGURE 4.16 Slowness plots for Isotropic material. The unit of the slowness is on the order of 10^{-4} sec/m. The 3D surface plots have X-slowness in x-axis, Y-slowness in y-axis, and Z-slowness in z-axis. The contour plots have slowness plots for three planes: X-Y, Y-Z and X-Z planes with slowness parameters in respective axes as required. (a) 3D surface plot for L mode, (b) 3D surface plot for SV mode, (c) 3D surface plot for SH mode, (d): X-Y Contour plot for L mode, (e) Y-Z Contour plot for L mode, (f) X-Z Contour plot for L mode, (g) X-Y Contour plot for SV mode, (h) Y-Z Contour plot for SV mode, (i) X-Z Contour plot for SV mode, (j) X-Y Contour plot for SH mode, (k) Y-Z Contour plot for SH mode, (l) X-Z Contour plot for SH mode.

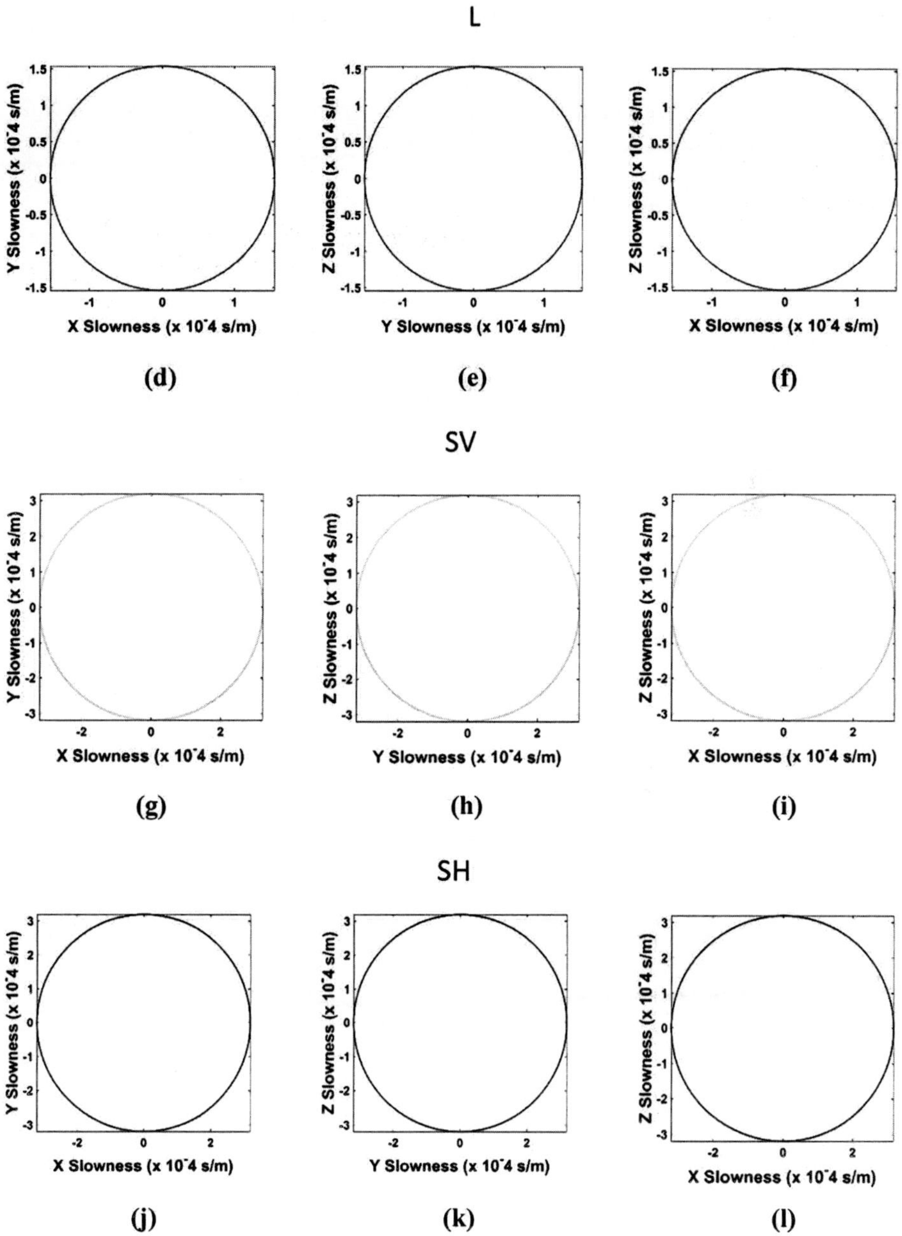

FIGURE 4.16 *(Continued)*

As it is known that isotropic material has the infinite planes of symmetry, it is possible to prove so through the plots in Fig. 4.16a–l. It can be seen with the help of the 3D plots for L, SH, and SV modes that the phase slowness is constant along all the directions. Also, the magnitude of the phase slowness is lower for longitudinal mode than the shear modes. It is also possible to further verify both the arguments with the help of contour plots for X-Y, Y-Z, and X-Z planes as it can be seen that the phase slowness is indeed constant in those planes as well as the magnitude of L being lower than that of SH and SV modes, which is obvious as per our previous discussions on isotropic materials in Section 4.3. Please note that X, Y, and Z planes in Fig. 4.16 are synonymous to the x_1, x_2, and x_3 axis used through out this book. The slowness surfaces are spherical in isotropic material are nonspherical in anisotropic material. Following the similar notation in the figures, next three different anisotropic materials are considered. Three different materials are (a) monoclinic material, (b) fully orthotropic material, and (c) orthotropic, transversely isotropic material.

4.4.2.2 Slowness Profiles for Monoclinic Material

Material properties of a typical monoclinic material with one plane of symmetry (refer Section 3.9.1) in matrix form can be written as

$$\mathbb{C}_{mn} = \begin{bmatrix} 102.6 & 24.1 & 6.3 & 40 & 0 & 0 \\ & 18.7 & 6.4 & 10 & 0 & 0 \\ & & 13.3 & -0.1 & 0 & 0 \\ & & & 23.6 & 0 & 0 \\ & Sym & & & 3.8 & 0.9 \\ & & & & & 5.3 \end{bmatrix} GPa \quad \& \ \rho = 1500\ kg/m^3 \quad (4.100)$$

Monoclinic material has only one plane of symmetry. By looking at the 3D surface plots (Fig. 4.17a–c), it can be seen that the phase slowness behavior is different for each wave modes. While, with the analysis of the contour plots (Fig. 4.17d–l) and the slowness behavior is symmetric in Y-Z and X-Z plane but is nonsymmetric in X-Y plane. Hence, it is possible to deduce that the material has the symmetry along the X-Y plane.

4.4.2.3 Slowness Profiles for Fully Orthotropic Material

Material properties of a typical orthotropic material with two/three plane of symmetry (refer Section 3.9.2) in matrix form can be written as

$$\mathbb{C}_{mn} = \begin{bmatrix} 70 & 23.9 & 6.2 & 0 & 0 & 0 \\ & 33 & 6.8 & 0 & 0 & 0 \\ & & 14.7 & 0 & 0 & 0 \\ & & & 21.9 & 0 & 0 \\ & Sym & & & 4.7 & 0 \\ & & & & & 4.2 \end{bmatrix} Gpa \quad \& \ \rho = 1500\ kg/m^3 \quad (4.101)$$

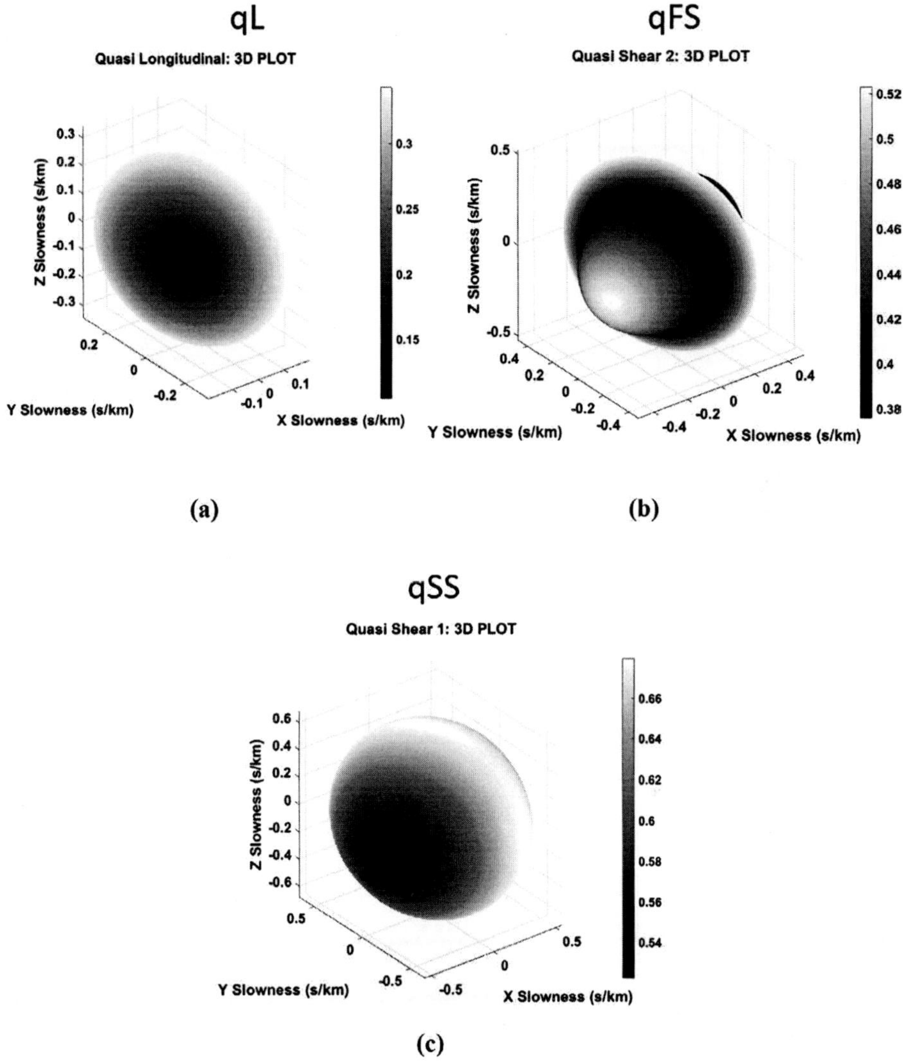

FIGURE 4.17 Slowness plots for Monoclinic material. The unit of the slowness is on the order of 10^{-4} sec/m. The 3D surface plots have X-slowness in x-axis, Y-slowness in y-axis, and Z-slowness in z-axis. The contour plots have slowness plots for three planes: X-Y, Y-Z and X-Z planes with slowness parameters in respective axes as required. a) 3D surface plot for qL mode, b) 3D surface plot for qFS mode, c) 3D surface plot for qSS mode, d) X-Y Contour plot for qL mode, e) Y-Z Contour plot for qL mode, f) X-Z Contour plot for qL mode, g) X-Y Contour plot for qFS mode, h) Y-Z Contour plot for qFS mode, i) X-Z Contour plot for qFS mode, j) X-Y Contour plot for qSS mode, k) Y-Z Contour plot for qSS mode, l) X-Z Contour plot for qSS mode.

(Continued)

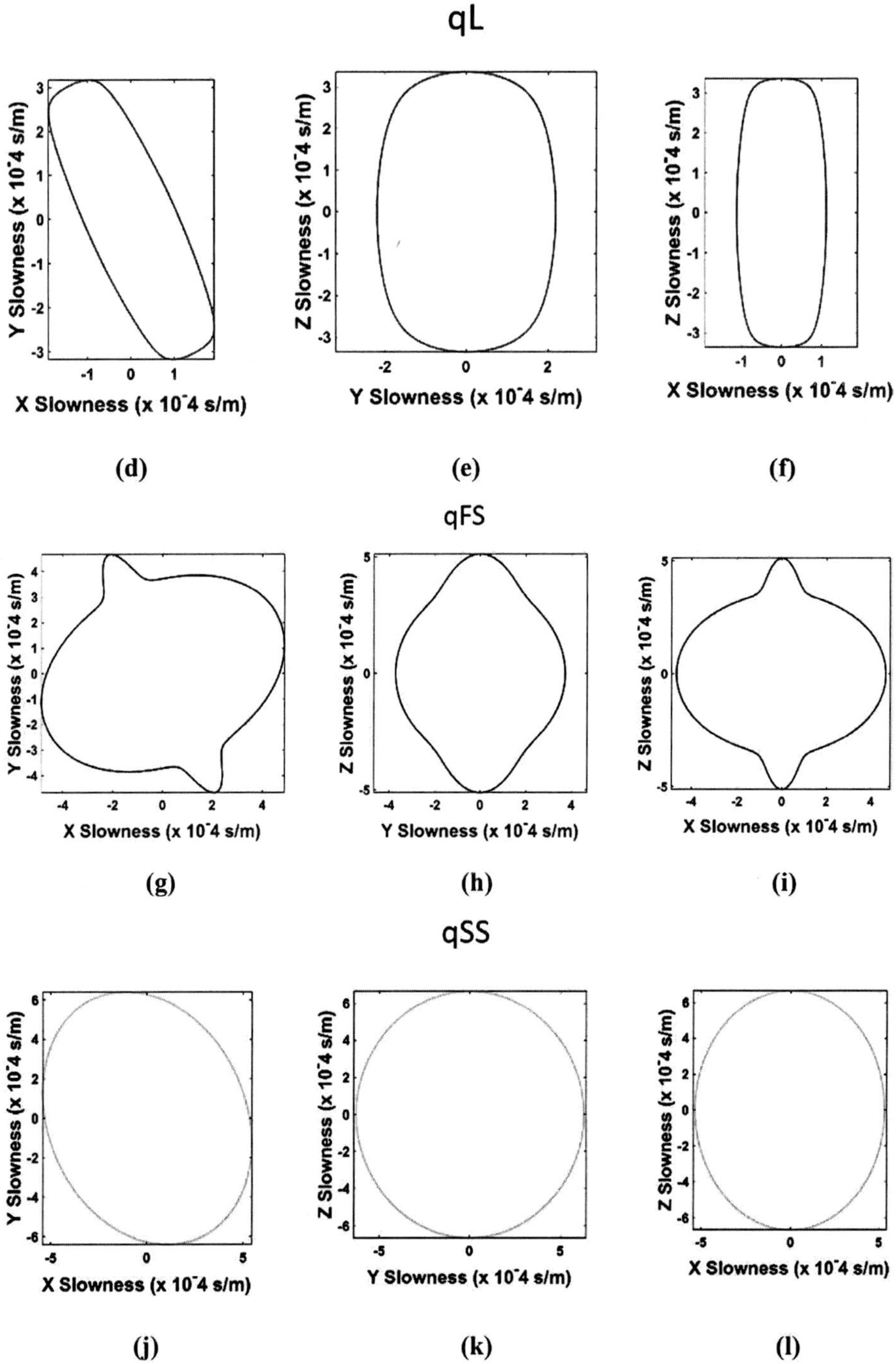

FIGURE 4.17 *(Continued)*

Fully orthotropic material has the three planes of symmetry. By looking at the 3D surface plots (Fig 4.18a–c), it can be seen that the phase slowness behavior is different for each wave modes. The symmetry of the phase slowness in each of the three X-Y, Y-Z, and X-Z planes are evident through the contour plots (Fig 4.18d–l). For each wave mode, the slowness differs in three planes which are in agreement with

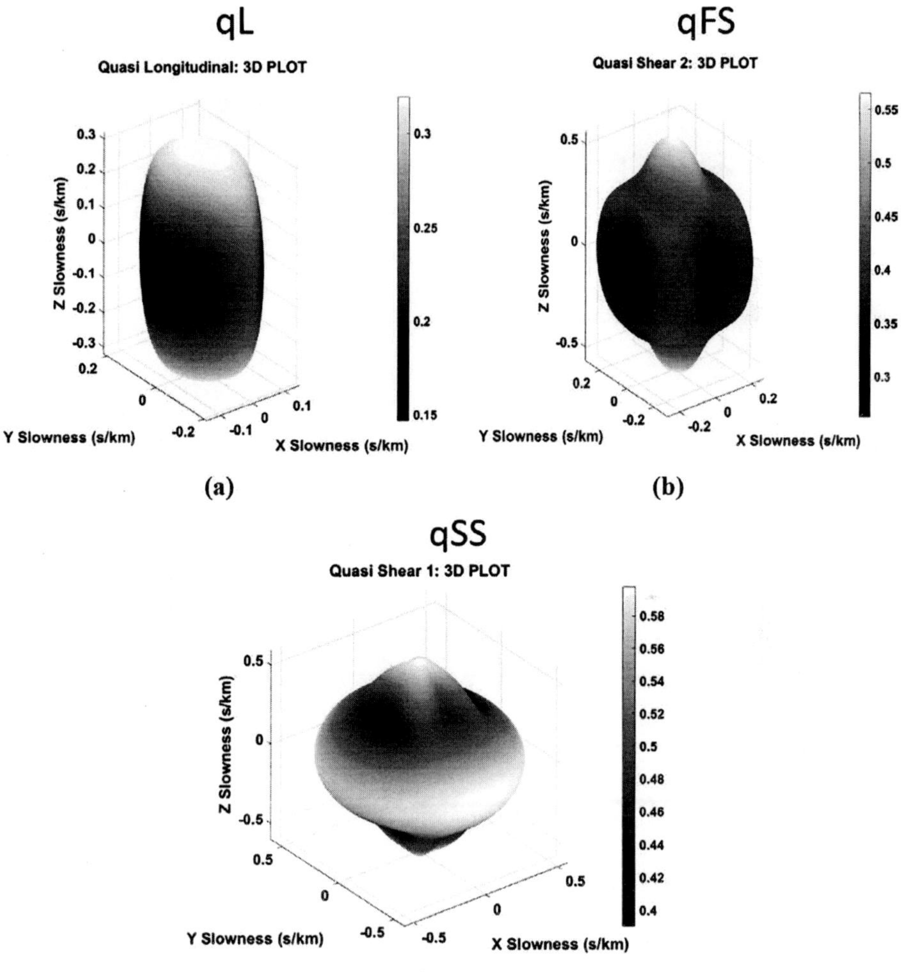

FIGURE 4.18 Slowness plots for Fully Orthotropic material. The unit of the slowness is on the order of 10^{-4} sec/m. The 3D surface plots have X-slowness in x-axis, Y-slowness in y-axis, and Z-slowness in z-axis. The contour plots have slowness plots for three planes: X-Y, Y-Z and X-Z planes with slowness parameters in respective axes as required. a) 3D surface plot for qL mode, b) 3D surface plot for qFS mode, c) 3D surface plot for qSS mode, d) X-Y Contour plot for qL mode, e) Y-Z Contour plot for qL mode, f) X-Z Contour plot for qL mode, g) X-Y Contour plot for qFS mode, h) Y-Z Contour plot for qFS mode, i) X-Z Contour plot for qFS mode, j) X-Y Contour plot for qSS mode, k) Y-Z Contour plot for qSS mode, l) X-Z Contour plot for qSS mode.

(Continued)

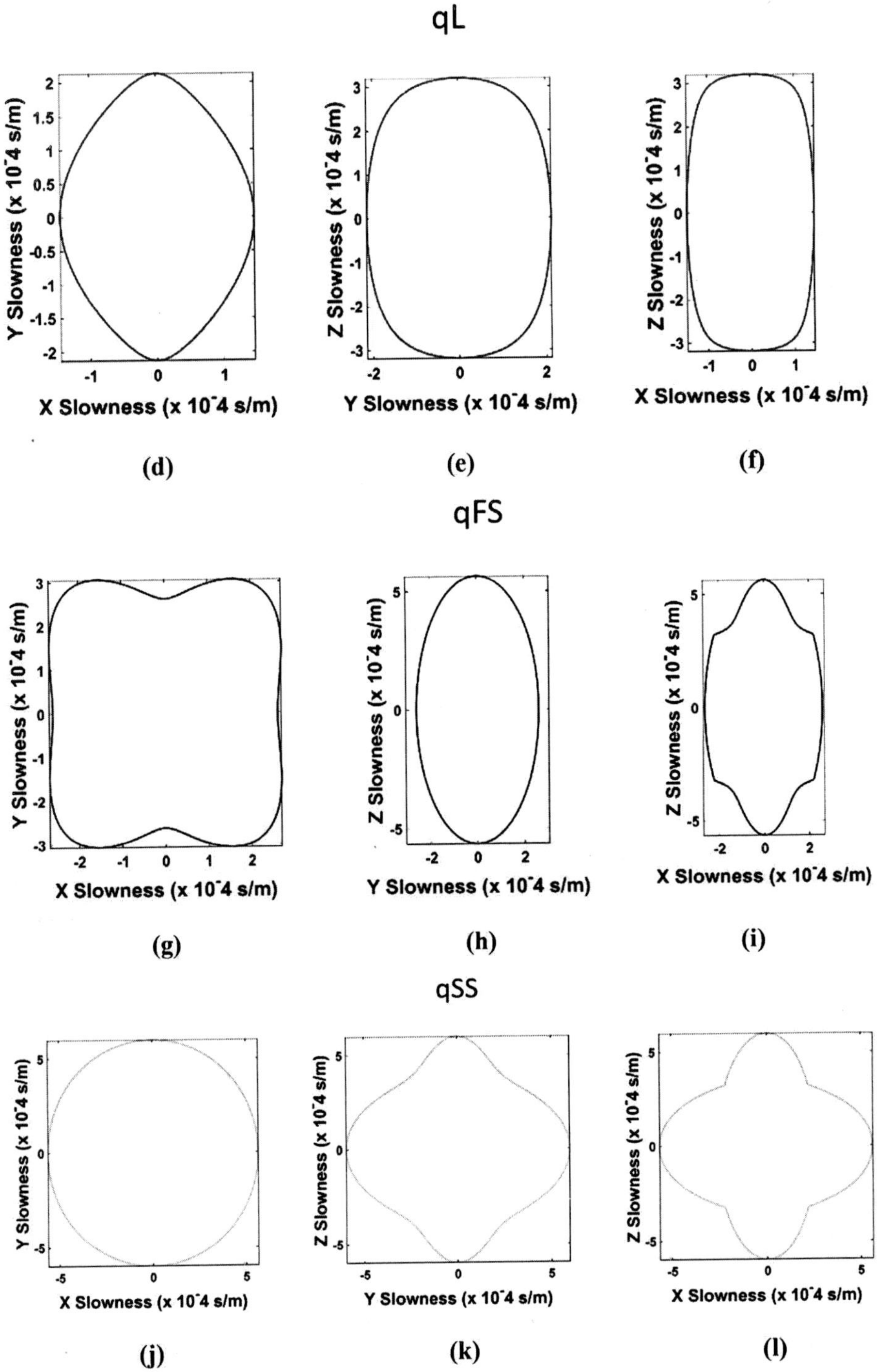

FIGURE 4.18 *(Continued)*

the characteristics of the material properties along three planes. Also, the slowness behavior is different for all three modes.

4.4.2.4 Slowness Profiles for Transversely Isotropic

Material properties of a typical transversely isotropic material with two/three plane of symmetry and one axis of symmetry (refer Section 3.9.3) in matrix form can be written as

$$\mathbb{C}_{mn} = \begin{bmatrix} 143.7 & 6.22 & 6.22 & 0 & 0 & 0 \\ & 13.4 & 6.5 & 0 & 0 & 0 \\ & & 13.4 & 0 & 0 & 0 \\ & & & 3.45 & 0 & 0 \\ & Sym & & & 5.7 & 0 \\ & & & & & 5.7 \end{bmatrix} \text{Gpa} \quad \text{and} \quad \rho = 1560 \text{ kg/m}^3 \qquad (4.102)$$

Orthotropic transversely isotropic has one additional axis of symmetry in addition to three planes of symmetry and the material properties are symmetric along the axis perpendicular to the plane of isotropy. Please note that here the x_1 axis is the axis of symmetry, whereas in Eq. (3.83) x_3 axis was considered axis of symmetry. Here, it is necessary to note that the major axis with higher material constant (i.e., fiber direction in composite material) was along the x_3 axis in Eq. (3.83) in Section 3.9.3. However, here for simplicity and later application in CNDE problems, the major axis with higher material constant (i.e., fiber direction in composite material) is presented along the x_1 axis. Thus, there is a change in equality of two constants. Instead of $\mathbb{C}_{31} = \mathbb{C}_{32}$ in Eq. (3.83), in Eq. (4.102) $\mathbb{C}_{12} = \mathbb{C}_{13}$. Also instead of $\mathbb{C}_{66} = \frac{\mathbb{C}_{11}-\mathbb{C}_{12}}{2}$ in Eq. (3.83), $\mathbb{C}_{44} = \frac{\mathbb{C}_{33}-\mathbb{C}_{13}}{2}$ in Eq. (4.102). While utilizing these equations, one should consider their respective definition of the coordinate system. By looking at the 3D surface plots (Fig. 4.19a–c), it can be seen that the phase slowness behavior is different for each wave modes. From Fig. 4.19e, h, and k, it is possible to clearly identify that the slowness behavior is constant in Y-Z plane, hence the material property is isotropic in Y-Z plane. Also, with the help of the contour plots in Figs. 4.19d, f, g, h, j, and l, symmetry of the phase slowness about X-axis can be identified. Other than the plane of isotropy, i.e., Y-Z plane, the slowness behavior is different in other planes as shown by the X-Y and X-Z contour plots. This section equipped the book with the necessary tools for wave propagation modeling in anisotropic solid plates for computational NDE discussed in the subsequent chapters.

4.4.3 Wave Interactions at the Bulk Anisotropic Interfaces

4.4.3.1 Geometrical Understanding of Reflection and Refraction in Anisotropic Solid

In Section 4.4.2 slowness surfaces are plotted and described in different types of anisotropic materials. Here it is important to clarify the understanding of wave vectors (simply known as the slowness vectors) and the wave energy vectors one more time.

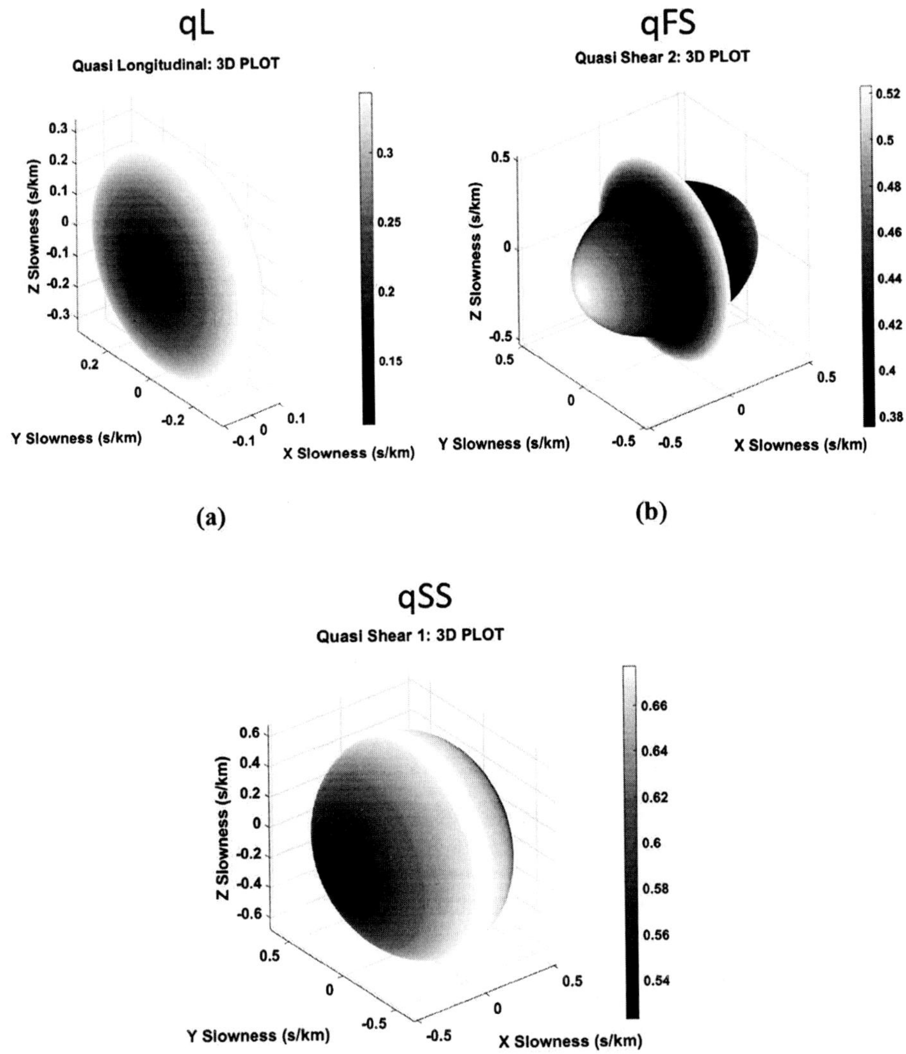

FIGURE 4.19 Slowness plots for Orthotropic Transversely Isotropic material. The unit of the slowness is on the order of 10^{-4} sec/m. The 3D surface plots have X-slowness in x-axis, Y-slowness in y-axis, and Z-slowness in z-axis. The contour plots have slowness plots for three planes: X-Y, Y-Z and X-Z planes with slowness parameters in respective axes as required. a) 3D surface plot for qL mode, b) 3D surface plot for qFS mode, c) 3D surface plot for qSS mode, d) X-Y Contour plot for qL mode, e) Y-Z Contour plot for qL mode, f) X-Z Contour plot for qL mode, g) X-Y Contour plot for qFS mode, h) Y-Z Contour plot for qFS mode, i) X-Z Contour plot for qFS mode, j) X-Y Contour plot for qSS mode, k) Y-Z Contour plot for qSS mode, l) X-Z Contour plot for qSS mode.

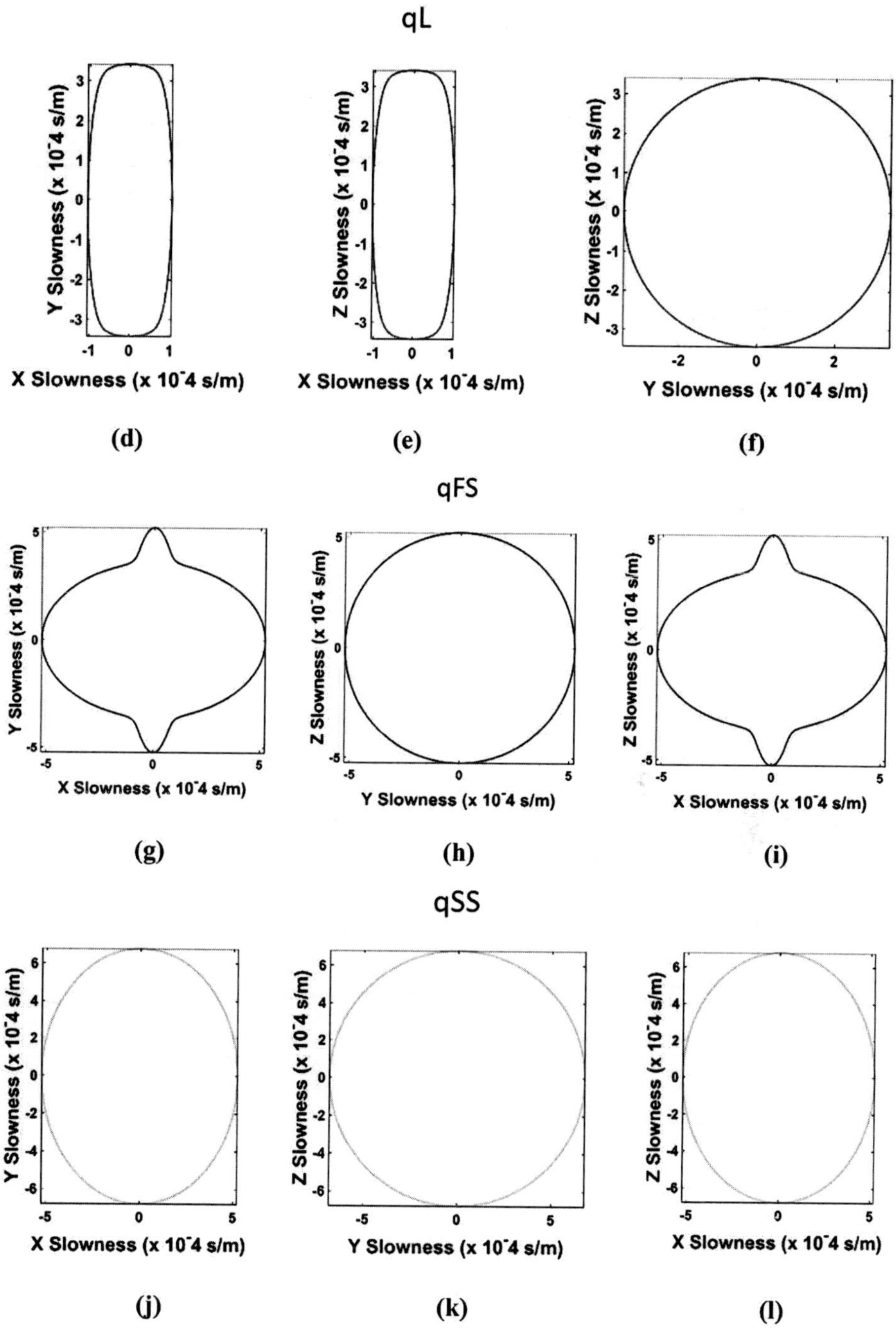

FIGURE 4.19 *(Continued)*

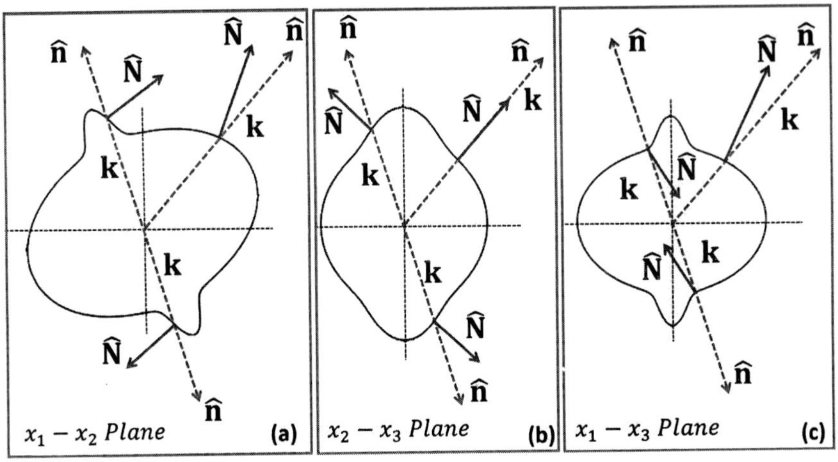

FIGURE 4.20 Wave propagation direction $\boldsymbol{k}$ is not aligned with the wave energy propagation direction: quasi Fast Shear (qFS) wave slowness profiles for a monoclinic material on a) $x_1 - x_2$ plane b) $x_2 - x_3$ plane and c) $x_1 - x_3$ plane.

Wave propagation direction $\mathbf{k}$ is not aligned with the wave energy propagation direction. To demonstrate this fact more clearly, quasi fast shear (qFS) wave slowness profiles on $x_1 - x_2$ plane, $x_2 - x_3$ plane, and $x_1 - x_3$ plane for a monoclinic material are shown in Fig. 4.20a, b, and c directly taken from the Fig. 4.17g, h, and i, respectively. In these figures three $\mathbf{k}$ vectors on each plane, i.e., total nine wave directions or the slowness vector directions are selected and marked. Depending on where these $\mathbf{k}$ vectors (unit vector is $\hat{\mathbf{n}}$) are intersected with the slowness curve in two dimension and a slowness surface in three dimension, the normal to the local surface is going to govern the direction of the wave energy propagation along respective normal unit vector $\hat{\mathbf{N}}$ as shown in Fig. 4.20. In some cases, or at some points, coincidentally, the slowness vector and the wave energy vector are perfectly aligned or parallel like it is shown in Fig. 4.20b for the qFS mode on $x_2 - x_3$ plane. This understanding is equivalent for all the wave modes in anisotropic solids.

Next, when two different anisotropic materials are joined or welded or bonded at an interface, wave reflection and refraction of an incident wave should abide by the physics described above. Further, due to the requirements from Snell's law in anisotropic materials, it is known (please refer Appendix for more details) that the wave number along the interface on the incident plane has to be a unique value and should be same for all the wave modes, irrespective of material properties on the either side of the interface. This very requirement gives the opportunity to analyze the actual direction of wave energy of each mode in anisotropic material due to an interface as described in the following paragraphs and Fig. 4.21.

In Fig. 4.21 an interface ($x_1 - x_3$ plane) is assumed between two different materials, one is monoclinic, and another one is orthotropic. The direction of the orthogonal to the interface is along the x_2 axis ($\hat{\mathbf{e}}_2$). Let's assume the top material is monoclinic and the bottom materials is orthotropic. With this assumption the slowness profiles on $x_1 - x_2$ plane and $x_2 - x_3$ plane will be important to analyze each mode in both

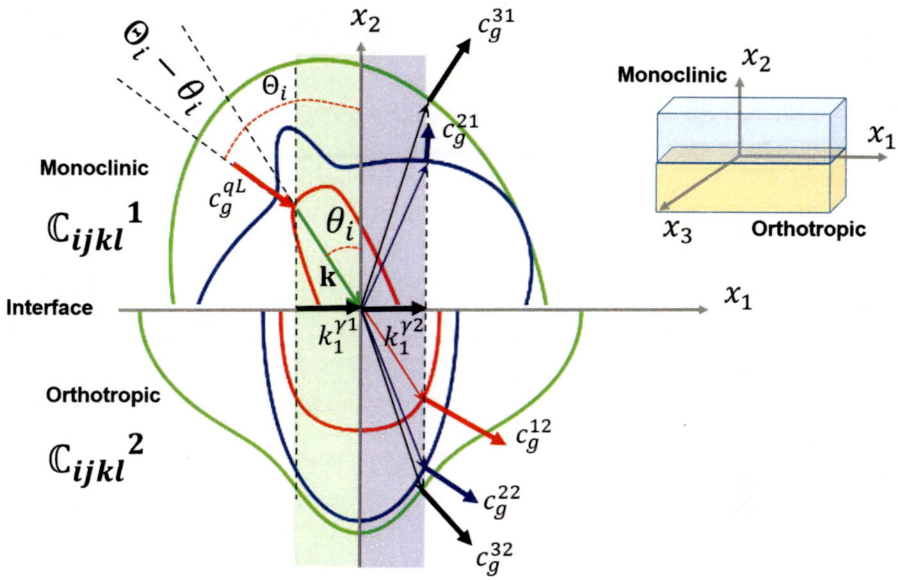

FIGURE 4.21 Wave interaction at the interface of two non isotropic material (here Monoclinic and Orthotropic), A qP wave incident from the monoclinic material at the interface is considered. Figure shows the slowness profiles of qL, qFS and qSS modes in monoclinic material on $x_1 - x_2$ plane placed above the interface, whereas it shows the relevant slowness profiles of qL, qFS and qSS modes in orthotropic material on $x_1 - x_2$ plane placed below the interface.

the material systems. In Fig. 4.21 slowness profiles of qL, qFS, and qSS modes in monoclinic material on $x_1 - x_2$ plane are placed above the interface, whereas slowness profiles of qL, qFS, and qSS modes in orthotropic material on $x_1 - x_2$ plane are placed below the interface.

With this setup let's assume a qL type mode was incident on the interface ($x_1 - x_3$ plane). A plane joining the incident wave and the x_2 axis is an incident plane. The plane of this page is aligned with the plane of the incident. Please note that the incident angle of qL wave is θ_i, i.e., the direction of the wave vector (**k**) but that is not the same as the energy flow direction or the direction of the incident ray. The angle of incident ray or the wave energy incidence marked with its group wave velocity c_g^{qL} is designated as Θ_i in Fig. 4.21. Wave incident angle and the respective ray incident angle are not the same. Next, as per the anisotropic Snell's law (refer Appendix) the reflected and refracted wave modes will be on the same plane ($x_1 - x_2$ plane) as we are on this page. However, to match the projection of the incident wave vector on the $x_1 - x_3$ plane (the **k**.**x**) which is $k_1^{\gamma1}$ in Fig. 4.21, all the slowness wave vectors of reflected wave modes in monoclinic above the interface and all the slowness wave vectors of refracted wave modes in orthotropic material below the interface have unique combination of alignments such that their projection of the wave vectors on the $x_1 - x_3$ plane is $k_1^{\gamma2}$ and is equal to $k_1^{\gamma1}$. This is shown in Fig. 4.21 along the x_1 axis. Here, the superscript γ is used to designate the quasi wave modes, qL, qFS, or qSS by 1, 2, and 3, respectively as they are similarly used in Appendix (please read Appendix in Section 4.5.3.2). Let's assume the projection of the incident wave vector on the $x_1 - x_3$ plane along x_1 axis is $k_1^{\gamma 1}$. Then it is necessary that the projections of all the reflected and refracted wave modes along x_1 axis should be equal to $k_1^{\gamma 1}$. So the best way to find the reflected and refracted wave vectors is to draw a vertical line at a $k_1^{\gamma 2} = k_1^{\gamma 1}$ distance from the origin as shown in Fig. 4.21. Next, the reflected and refracted wave vectors in respective materials can be found by joining the origin and the respective intersection points between the newly drawn vertical line and the slowness profiles. This understanding is universal and not limited to the material system selected here in this example. Fig. 4.21 also shows the direction of wave energy for each wave modes derived by drawing the normal to the respective slowness surfaces where they intersect with the slowness vector. The group velocity or wave energy vectors for reflected qFS and qSS modes in monoclinic material (material 1) are c_g^{21} and c_g^{31}, respectively shown in Fig. 4.21. The group velocity or wave energy vectors for refracted qL, qFS, and qSS modes in orthotropic material (material 2) are c_g^{12} c_g^{22} and c_g^{32}, respectively shown in Fig. 4.21. Please note that there are three wave modes in the refracted wave set in orthotropic material, but there are only two wave modes (qSS and qFS) in reflected wave mode set in monoclinic material. Here, in this example, the vertical line drawn at a distance $k_1^{\gamma 2} = k_1^{\gamma 1}$ did not intersect the qL slowness profile in the monoclinic material. This means that the reflected angle is greater than the critical angle of qL mode in monoclinic with this specific arrangement of incident wave. Such specific cases with some other cases on critical incidence are discussed at the end of this section separately. Now when the group velocity vector for each mode is found, the magnitude of the group velocity should also be found from the basic understanding presented in Fig. 4.13b. It is discussed that the projection of the group velocity of a mode along the direction of wave vector or slowness vector is the phase velocity of that mode. Here, the phase velocity of reflected or refracted wave modes are known from the slowness profiles, because they are already obtained from the Christoffel solution. Next, the direction of the group wave vectors (the steering angle) is found from the above geometrical analysis, which is discussed in detail in Appendix in Section 4.5.4. Hence, group wave velocity for each mode can also be found easily.

Similar concept is presented in Fig. 4.22 for the $x_2 - x_3$ plane, where $x_2 - x_3$ plane slowness profiles of qL, qFS, and qSS modes are placed above the interface for monoclinic material, and slowness profiles of qL, qFS, and qSS modes in transversely isotropic material on $x_2 - x_3$ plane are placed below the interface. With this setup let's assume a qFS type mode was incident on the interface ($x_2 - x_3$ plane). A plane joining the incident wave and the x_2 axis is the incident plane. An orthogonal plane to this page is aligned with the plane of the incident. So as per the anisotropic Snell's law (refer Appendix), the reflected and refracted wave modes will be on the same plane ($x_2 - x_3$ plane) orthogonal to this page passing through the reader. However to match the projection of the incident wave vector on the $x_2 - x_3$ plane (the **k.x**) along the negative x_3 axis, all the slowness wave vectors of reflected wave modes in monoclinic above the interface and the slowness wave vectors of refracted wave modes in orthotropic material below the interface will have a different but unique combination of alignments as shown in Fig. 4.22. Let's assume the projection of the incident wave vector on the $x_2 - x_3$ plane along

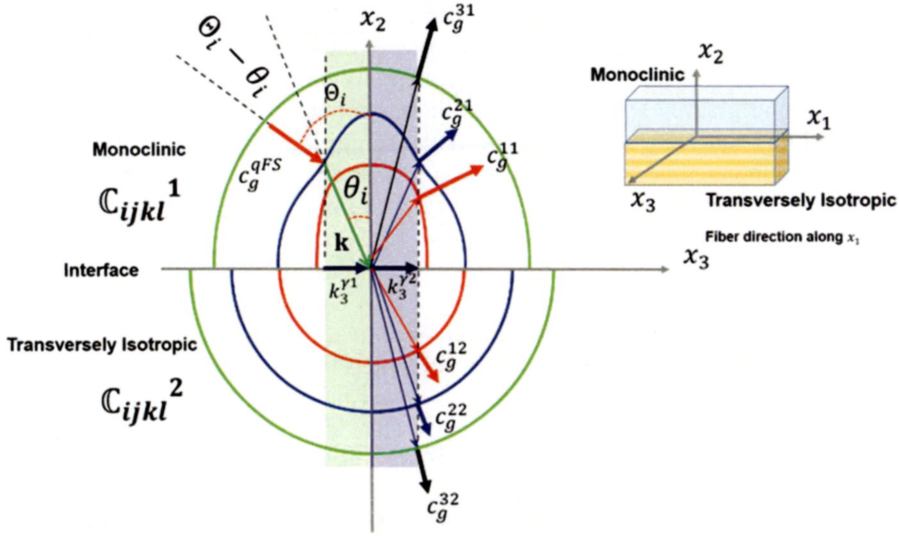

FIGURE 4.22 Wave interaction at the interface of two non isotropic material (here Monoclinic and Transversely Isotropic), A qFS wave incident from the monoclinic material at the interface is considered. Figure shows the slowness profiles of qL, qFS and qSS modes in monoclinic material on $x_2 - x_3$ plane placed above the interface, whereas it shows the relevant slowness profiles of qL, qFS and qSS modes in Transversely Isotropic material on $x_2 - x_3$ plane placed below the interface.

x_3 axis is $k_3^{\gamma 1}$. Then it is necessary that the projections of all the reflected and refracted wave modes along x_3 axis should be equal to $k_3^{\gamma 1}$. So the best way to find the reflected and refracted wave vectors is to draw a vertical line at a $k_3^{\gamma 2} = k_3^{\gamma 1}$ distance from the origin as shown in Fig. 4.22. Next, the reflected and refracted wave vectors in respective materials can be found by joining the origin and the respective intersection points between the newly drawn vertical line and the slowness profiles. This understanding is universal and not limited to the material system selected here in this example for the $x_2 - x_3$ plane. Fig. 4.22 also shows the direction of wave energy for each wave modes derived by drawing the normal to the respective slowness surfaces where they intersect with the slowness vector. The group velocity or wave energy vectors for reflected qL, qFS, and qSS modes in monoclinic material (material 1) are c_g^{11}, c_g^{21}, and c_g^{31}, respectively as shown in Fig. 4.22. The group velocity or wave energy vectors for refracted qL, qFS, and qSS modes in monoclinic material (material 2) are c_g^{12} c_g^{22} and c_g^{32}, respectively shown in Fig. 4.22. Please note that there are three wave modes in both reflected and refracted wave set in monoclinic and transversely isotropic. The three wave energy vectors in transversely isotropic material are perfectly aligned with their respective wave slowness direction. This is because the transversely isotropic material has the fiber direction along the x_1 axis and the $x_2 - x_3$ plane behaves like an isotropic medium. This understanding is universal and not limited to the material system selected here in this example.

Using similar setup reader can practice any different material combinations and different incident wave coming from any side of the interface.

A few more discussions are warrant here. Without describing the concept of critical and grazing angle, discussion on anisotropic wave behavior is not complete. Let's consider Fig. 4.23. It shows the same slowness profiles of three wave modes in monoclinic material. Based on this profile few points on the profile are marked. If an arbitrary incident wave vector (**k**) with incident angle θ_i and a specific local wave energy direction (i.e., the angle of the ray Θ_i) does not intersect with any slowness profile, then it means that the incident angle is higher than the critical angle for that wave mode. This case is already occurred in Fig. 4.21, where the qL reflected wave mode faced this similar issue. At critical angle the wave energy after refraction goes parallel to the interface. Hence, all the points on the slowness surface that have the surface tangent perpendicular to the interface or in other words the normal to the slowness surface is parallel to the interface, are responsible to demonstrate the critical angle phenomena. For example, in Fig. 4.23 six such points (qL^a, qL^b, qFS^a, qFS^b, qSS^a, qSS^b) are found that has the pointing direction of wave energy vector parallel to the interface. Here only 2D surfaces are drawn but similar concept is valid for 3D and the pointing wave vector will be parallel to the interface plane or $x_1 - x_3$ plane in Fig. 4.23 having both components

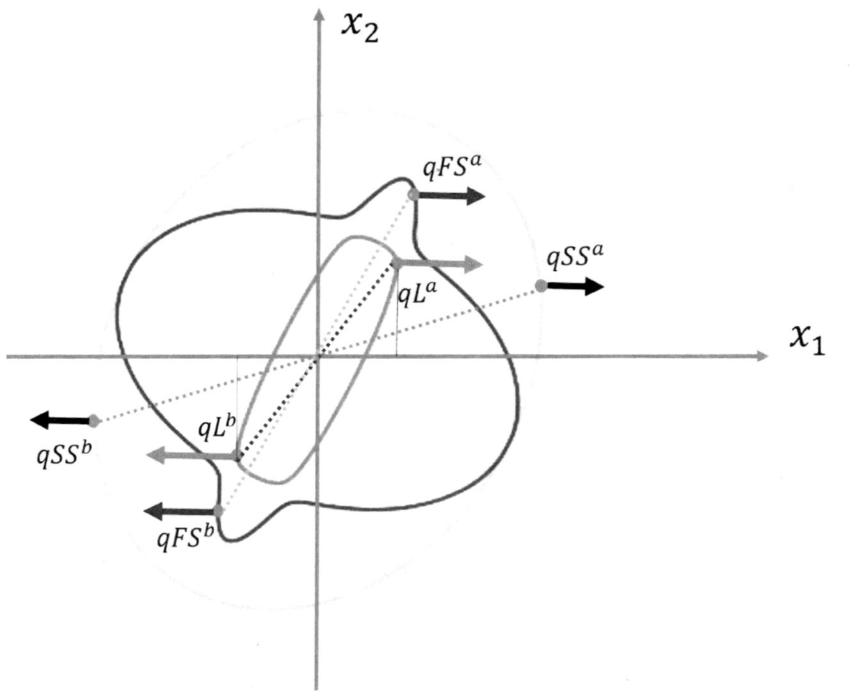

FIGURE 4.23 Schematic representation of the concept of critical angle and grazing angle in anisotropic material, here slowness profiles of three wave modes in monoclinic material are shown.

along x_1 and x_3 axis. As θ_i and Θ_i are not same when measured from the vertical axis (here x_2 axis), sometimes with specific slowness profile of a mode, it is possible that for certain θ_i, Θ_i becomes 90°, i.e., the wave energy incidence is along the interface. Those angles of incidence are called the grazing angle of incidence. These are same points identified earlier as qL^a, qL^b, $qFS^a, qFS^b, qSS^a, qSS^b$ where wave energy will incident parallel to the interface. Hence, whether these six points will serve as grazing points or the critical points, will depends on which side of the interface it belongs to. That means is the material it is used for, material 1 (incident material) or the material 2 (refractive material). Depending on the localized slowness profile, sometime it is possible that wave energy direction Θ_i of a mode is greater than 90° (i.e., away from the interface), when the θ_i is still less than 90° (the incident wave vector direction toward the interface). Such cases are not permissible. Hence, certain θ_i for certain wave modes are not allowable in anisotropic solids. Slowness profile gives very good understanding of such situations to avoid NDE inspection with such incident angles.

Exercise: Find the grazing angles, critical angles, and non-permissible zone of incident wave vectors from the wave slowness profiles of some materials shown in Fig. 4.24.

4.5 APPENDIX

4.5.1 Energy Flux & Group Velocity

Energy density of the propagating waves can be expressed as the strain energy density based on the discussions in Section 3.8.3 in Chapter 3. Further energy density (kinetic energy) can be

$$\varepsilon = \rho \dot{u}^2 \tag{A4.1}$$

where real part of the displacement u can be recollected from Eq. (4.12)

$$u = 2A\cos\left(kx - \omega t\right)\cos\left(\Delta kx - \Delta\omega t\right) \tag{A4.2}$$

Hence, Eq. (A4.1) can be modified to

$$\varepsilon = 4\rho\omega^2 A^2 \cos^2 0.5\left(\Delta kx - \Delta\omega t\right)\sin^2\left(kx - \omega t\right) + \ldots \tag{A4.3}$$

Taking the integration of the above expression over a period of time during when the modulation of the wave does not change the dominant part of the wave energy can be written as [7]

$$\varepsilon \cong 2\rho\omega^2 A^2 \cos^2 0.5\left(\Delta kx - \Delta\omega t\right) \tag{A4.4}$$

Which shows the velocity of the wave energy is $c_E = \Delta\omega/\Delta k \cong d\omega/dk$. We expressed the group velocity of the wave packet as $c_g = \frac{d\omega}{dk}$, and hence, the group velocity of the wave packet is equal to the velocity of the energy propagation, $c_g = c_E$.

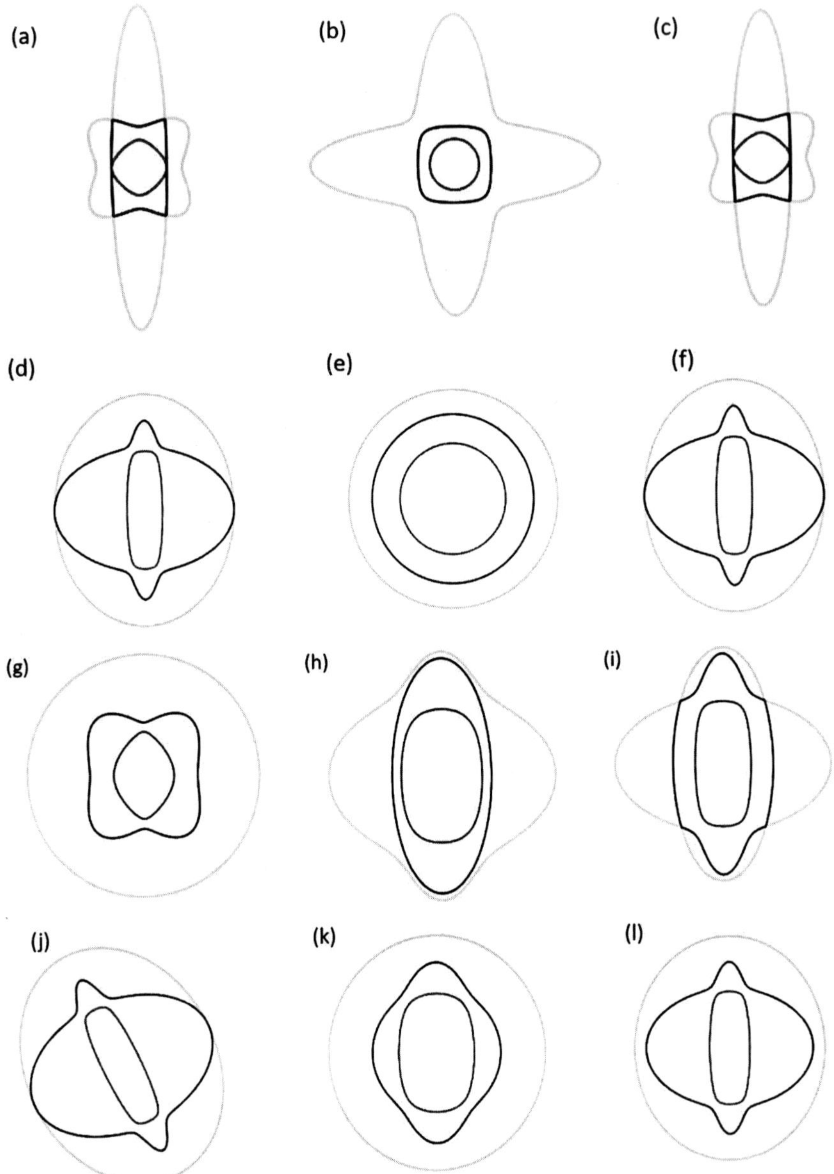

FIGURE 4.24 Exercise Problem to find grazing angles and critical angles and non permissible zone of incident wave vectors from the wave slowness profiles: In an anisotropic material on a) $x_1 - x_2$ plane b) $x_2 - x_3$ plane and c) $x_1 - x_3$ plane; In transversely isotropic material on d) $x_1 - x_2$ plane e) $x_2 - x_3$ plane and f) $x_1 - x_3$ plane. In orthotropic material on g) $x_1 - x_2$ plane h) $x_2 - x_3$ plane and i) $x_1 - x_3$ plane. In monoclinic material on j) $x_1 - x_2$ plane k) $x_2 - x_3$ plane and l) $x_1 - x_3$ plane. Color code: Red = quasi longitudinal, Green = quasi slow shear, Blue = quasi fast shear.

Using alternate definition velocity of energy transport is defined as the ratio of the time average of the power per cross sectional area and the time average of the total energy (addition of strain energy density and kinetic energy) per unit length of the media in which the wave propagates.

4.5.2 Integral Approach to Obtain Governing Elastodynamic Equation based on Classical Mechanics

The elastodynamic equation derived in Chapter 3 was from Newton's second law of motion and the approach is called the differential approach. Similar equation could be derived using integral approach of classical mechanics. The most prominent function that is used in classical mechanics is called the Lagrangian ($\mathcal{L}$) of a system or simply the Lagrangian density. Here, the system is a deformable body in which the analysis of wave propagation is sought. Lagrangian of a system is defined as a function of the generalized coordinates of the system and their respective temporal and spatial derivatives. For example, if the generalized coordinates of a system like displacement are designated as u_i in a multidimensional system, the Lagrangian can be defined as

$$\mathcal{L} = \mathcal{L}(u_i, \dot{u}_i, u_{i,j}) \tag{A4.5}$$

where $\dot{u}_i$ are the velocities (temporal derivatives) and $u_{i,j}$ are the displacement gradients (spatial derivates). The Lagrangian density of a deformable body can be defined as the difference of the kinetic energy density and the strain energy density.

When wave propagates in a material wave energy is transported from one point to the other. The action is a temporal and spatial action. That means that at two different time points, say at t_1 and t_2, the system is in different states and the values of the arguments of the Lagrangian would be different. To define the action which is a scalar function a spatial and temporal integral is necessary and can be written as

$$\mathfrak{L} = \int_{t_1}^{t_2}\int_{v} \mathcal{L}(u_i, \dot{u}_i, u_{i,j})\, dv\, dt = \int_{t_1}^{t_2}\int_{v} \mathcal{L}(u_i, \dot{u}_i, u_{i,j})\, dx_j\, dt \tag{A4.6}$$

According the classical mechanics theorem [15, 16] a system evolves from t_1 and t_2 such a way that the scalar function defined in Eq. (A4.6) is minimized. This principle is called principle of least action or the Hamilton's principle [15]. As the Lagrangian is a function of u_i, $\dot{u}_i$, and $u_{i,j}$, generalized coordinate system, which is related to the displacements for wave propagation, perturbation of this generalized coordinates will cause the change in the action. Thus, a small variation (δ) to these generalized coordinate systems (here displacements), while satisfying all the boundary conditions will cause a small variation (δ) in action $\mathfrak{L}$ as follows

$$\delta\mathfrak{L} = \mathfrak{L}\left(u_i + \delta u_i, \dot{u}_i + \delta\dot{u}_i, u_{i,j} + \delta u_{i,j}\right) - \mathfrak{L}(u_i, \dot{u}_i, u_{i,j}) \tag{A4.7}$$

According to the Hamilton's principle $\delta\mathfrak{L} \to 0$ to satisfy the principle of least action in classical mechanics. That means that the system could have evolved only in one way by minimizing the action and the action $\mathfrak{L}$ in minimum. Perturbation of the generalized coordinates will not perturb the action and will remain the same, making $\delta\mathfrak{L} \to 0$. Next by expanding the first term in Eq. (A4.7) using a Taylor's series the $\delta\mathfrak{L}$ can be written as

$$\delta\mathfrak{L} = \mathfrak{L}\left(u_i, \dot{u}_i, u_{i,j}\right) + \frac{\partial\mathfrak{L}}{\partial u_i}\delta u_i + \frac{\partial\mathfrak{L}}{\partial \dot{u}_i}\delta\dot{u}_i + \frac{\partial\mathfrak{L}}{\partial u_{i,j}}\delta u_{i,j} - \mathfrak{L}\left(u_i, \dot{u}_i, u_{i,j}\right)$$

$$or\ \delta\mathfrak{L} = \frac{\partial\mathfrak{L}}{\partial u_i}\delta u_i + \frac{\partial\mathfrak{L}}{\partial \dot{u}_i}\delta\dot{u}_i + \frac{\partial\mathfrak{L}}{\partial u_{i,j}}\delta u_{i,j} \tag{A4.8}$$

After substituting Eq. (A4.8) into the equation of action in Eq. (A4.6) the variation of action can be written as

$$\delta\mathfrak{L} = \int_{t_1}^{t_2}\int_{v}\left[\frac{\partial\mathcal{L}}{\partial u_i}\delta u_i + \frac{\partial\mathcal{L}}{\partial \dot{u}_i}\delta\dot{u}_i + \frac{\partial\mathcal{L}}{\partial u_{i,j}}\delta u_{i,j}\right]dx_j dt \tag{A4.9}$$

Now by interchanging the conventional derivatives and variational symbol one can write

$$\delta\mathfrak{L} = \int_{t_1}^{t_2}\int_{v}\left[\frac{\partial\mathcal{L}}{\partial u_i}\delta u_i + \frac{\partial\mathcal{L}}{\partial \dot{u}_i}\frac{d(\delta\dot{u}_i)}{dt} + \frac{\partial\mathcal{L}}{\partial u_{i,j}}\frac{d(\delta u_i)}{dx_j}\right]dx_j\, dt \tag{A4.10}$$

The second and third terms with respect to time and space, respectively in Eq. (A4.9) after performing the integral by parts can viz.

$$\int_{t_1}^{t_2}\frac{\partial\mathcal{L}}{\partial \dot{u}_i}\frac{d(\delta\dot{u}_i)}{dt}dt = \frac{\partial\mathcal{L}}{\partial u_i}\delta\dot{u}_i\Big|_{t_1}^{t_2} - \int_{t_1}^{t_2}\frac{d}{dt}\left(\frac{\partial\mathcal{L}}{\partial \dot{u}_i}\right)\delta u_i dt \tag{A4.11}$$

$$\int_{v1}^{v2}\frac{\partial\mathcal{L}}{\partial u_{i,j}}\frac{d(\delta u_i)}{dx_j}dx_j = \frac{\partial\mathcal{L}}{\partial u_{i,j}}\delta u_i\Big|_{v1}^{v2} - \int_{v1}^{v2}\frac{d}{dx_j}\left(\frac{\partial\mathcal{L}}{\partial u_{i,j}}\right)\delta u_i dx_j \tag{A4.12}$$

Substituting Eqs. (A4.11) and (4.12) into Eq. (A4.10), the variation of action modifies to

$$\delta\mathfrak{L} = \int_{t_1}^{t_2}\int_{v}\left[\frac{\partial\mathcal{L}}{\partial u_i} - \frac{d\mathcal{L}}{dt}\left(\frac{\partial\mathcal{L}}{\partial \dot{u}_i}\right) - \frac{d\mathcal{L}}{dx_j}\left(\frac{\partial\mathcal{L}}{\partial u_{i,j}}\right)\right]dx_j\,\delta u_i\, dt \tag{A4.13}$$

According to the Hamilton's principle, $\delta\mathfrak{L} \to 0$, and hence,

$$\frac{\partial\mathcal{L}}{\partial u_i} - \frac{d\mathcal{L}}{dt}\left(\frac{\partial\mathcal{L}}{\partial \dot{u}_i}\right) - \frac{d\mathcal{L}}{dx_j}\left(\frac{\partial\mathcal{L}}{\partial u_{i,j}}\right) = 0 \tag{A4.14}$$

Now defining the Lagrangian of the system in the light of the discussion in Section 3.8.3 in Chapter 3, the difference of the kinetic energy density and the strain energy density can be written as

$$\mathcal{L} = \mathcal{K} - \mathcal{E} \tag{A4.15}$$

The displacements in the body are already defined as u_i and hence, Eq. (A4.15) can be expanded to

$$\mathcal{L} = \frac{1}{2}\rho(\dot{u}_i)^2 - \frac{1}{2}\sigma_{ij}e_{ij} = \frac{1}{2}\rho(\dot{u}_i)^2 - \frac{1}{2}\mathbb{C}_{ijkl}e_{ij}e_{kl} \tag{A4.16}$$

Substituting Eqs. (A4.16) and (3.72) into Eq. (A4.14), ultimately the Hamilton principle gives

$$-\frac{1}{2}\frac{d}{dt}\left(\frac{\partial\rho(\dot{u}_i)^2}{\partial\dot{u}_i}\right) + \frac{d\mathcal{L}}{dx_j}\left(\frac{\partial\sigma_{ij}e_{ij}}{\partial u_{i,j}}\right) = 0 \tag{A4.17}$$

$$\rho\ddot{u}_i - \frac{d}{dx_j}\left(\frac{1}{2}\sigma_{ij}\frac{\partial(u_{i,j}+u_{j,i})}{\partial u_{i,j}}\right) = 0 \tag{A4.18}$$

Or the same equation written in Eq. (4.25) as

$$\sigma_{ij,j} - \rho\ddot{u}_i = 0 \tag{A4.19}$$

4.5.3 Understanding the Snell's Law in Isotropic and Anisotropic Media

Snell's law is the key to the analysis of wave propagation at the interfaces. Reflection and refraction of incident waves at the interfaces must satisfy an equal phase rule derived from phase-matching requirement described below. A metallic welded joint is an example of an interface where instead of boundary conditions, we define interface conditions. At an interface stresses and displacements are continuous. Out of six stresses at a point, three stresses must be continuous, which are one normal stress and two shear stresses at the interface.

4.5.3.1 Snell's Law at Isotropic Material Interface

Let's assume an interface between two isotropic materials of different material properties, i.e., different Lamé constants and densities ($\mathbb{L}_1$, m_1, ρ_1 and $\mathbb{L}_2$, m_2, ρ_2) as shown in Fig. A.4.1. If an incident wave ray from the material 1 hits the interface between the two materials, the wave will have two options, (a) reflected back to the material 1 and (b) refracted inside the material 2. However, both of this events will not happen arbitrarily. Interface conditions at this interface are $u_2{}^1 = u_2{}^2$; $u_1{}^1 = u_1{}^2$; $u_3{}^1 = u_3{}^2$; $\sigma_{22}{}^1 = \sigma_{22}{}^2$; $\sigma_{21}{}^1 = \sigma_{21}{}^2$; $\sigma_{23}{}^1 = \sigma_{23}{}^2$ where the superscripts are for the material no. It is inevitable that the wave at the interface between two materials cannot take different characteristics. It must have one

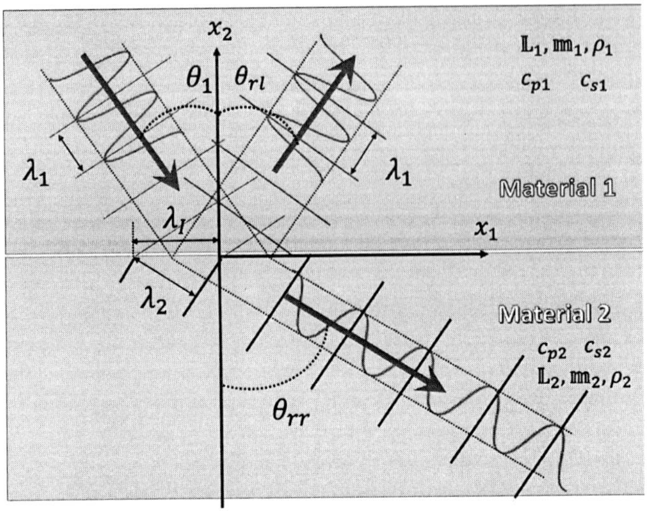

FIGURE A.4.1 Schematic representation of Snell's Law at the interface between two isotropic materials of different material properties.

wave vector with specific wave number along the interface in both side of the material, otherwise wave cannot propagate from the material 1 to the material 2. Wave number (k), frequency (ω), and phase velocity (c) of a media are related by an equation written in Eq. (4.7) or (4.63). Again wave numbers are related by wavelength as written in Eq. (4.5). Fig. A.4.1 shows how a propagating wave can gets reflected or refracted after an incident on an interface. Fig. A.4.1 also shows how a wave front can keep the same phase or the wave numbers along the interface while modulating the other wave parameters in respective materials beyond the interface. In material 1 and material 2 the wave lengths are λ_1 and λ_2 and let's assume the wavelength to enforce the phase-matching along the interface is λ_I. The incident angle of the incident wave with respect to the x_2 axis is say θ_i. The reflected and refracted wave angles with respect to x_2 axis are say θ_{rl} and θ_{rr}, respectively. Hence, to have equal wavelength along the interface λ_I, according to Fig. A4.1 we can write

$$\lambda_I = \frac{\lambda_1}{\sin\theta_i} = \frac{\lambda_1}{\sin\theta_{rl}} = \frac{\lambda_2}{\sin\theta_{rr}} \tag{A4.20}$$

Thus it is apparent that $\theta_i = \theta_{rl}$, and this fact is inevitably used in Section 4.3.3. Further applying Eqs. (4.5) and (4.63), we can write

$$k = k_1 \sin\theta_i = k_1 \sin\theta_{rl} = k_2 \sin\theta_{rr} \tag{A4.21}$$

and

$$c_I = \frac{c_1}{\sin\theta_i} = \frac{c_1}{\sin\theta_{rl}} = \frac{c_2}{\sin\theta_{rr}} \tag{A4.22}$$

where k is the wave number along the interface, c_I is the wave velocity along the interface, k_1 and k_2 are the wave numbers in material 1 and material 2, respectively, c_1 and c_2 are the phase wave velocities in material 1 and material 2, respectively. In this example in Fig. A4.1 it is visible that the phase and the wave numbers along the interface are unique, but in material 2 the wave numbers are smaller than the material 1, when the wavelength and wave velocity in material 2 are higher than the material 1. The above equations are valid for both P-waves and SV-waves.

4.5.3.2 Snell's Law at Anisotropic Material Interface

Let's assume an interface between two anisotropic materials of different constitutive matrices ($\mathbb{C}_{ijkl}{}^1$, $\mathbb{C}_{ijkl}{}^2$) and densities ρ_1 and ρ_2 as shown in Fig. A.4.2. If an incident ray coming from the material 1 hits the interface between the two materials, the wave will follow two process, (a) three reflected wave rays (qL, qFS, and qSS) will come back to the material 1 and (b) three refracted wave rays (qL, qFS, and qSS) will go inside the material 2. However, both events will not happen arbitrarily. These six wave rays may go into six different directions depending on their polarity (respective polarization vector $\mathbf{g}_L$, $\mathbf{g}_{FS}$, $\mathbf{g}_{SS}$ obtained from the Christoffel's solution) and the direction of propagation of the energy as discussed in Section 4.4.3. After the wave incidence, wave energy vectors, of the modes in two different materials may not lie on the same plane of incidence (Fig. A.4.2). Please note the word used here is, "wave energy vector" not the "wave vector." Wave energy vectors for all the modes qL, qFS, and qSS are shown in Fig. A.4.2. Please note that these directions are pointed by a normal on the slowness surface along a wave vector direction (i.e., the normal on the surface at the intersection of wave vector and slowness surface). It can be seen that they are not on the same plane and have their own independent directions. It can be explained further, drawing attention to Section 4.4. In that section, wave vector or the wave propagation direction is denoted by $\mathbf{k}$, where the direction cosine or the unit normal vector for the wave vector was $\hat{\mathbf{n}} = n_i\hat{\mathbf{e}}_i$. Whereas, the wave energy

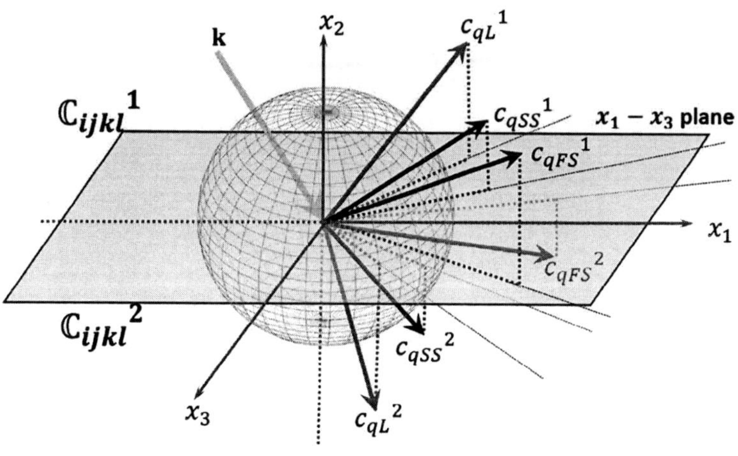

FIGURE A.4.2 Schematic representation of Snell's Law at the interface between two anisotropic materials of different constitutive matrices.

vector or the direction of energy propagation was denoted by $\hat{\mathbf{N}} = N_i\hat{\mathbf{e}}_i$. It is noted that $\hat{\mathbf{n}} = n_i\hat{\mathbf{e}}_i$ and $\hat{\mathbf{N}} = N_i\hat{\mathbf{e}}_i$ are not parallel to each other in anisotropic material. Let's focus on the interface for a moment. It is immaterial whether the wave vector $\mathbf{k}$ or the wave energy vector $\hat{\mathbf{N}}$ are nonparallel and take their own course in the two materials for all the modes, the displacements in three directions (u_m) in respective materials and the relevant interfacial stresses (σ_{ij}) at the interface must be continuous. Interface conditions at an anisotropic interface will not be simple as we discussed in Section 4.5.3.1 for an isotropic material. Let's assume a generalized displacement function in anisotropic solid as it was assumed in Eq. (4.94)

$$u_m = Ag_m e^{i(\mathbf{k}^{\gamma a}.\mathbf{x}-\omega t)} \tag{A4.23}$$

where γa parameter is introduced to specify the type of wave modes in the respective materials on either side of the interface. These wave modes include the incident wave, quasi longitudinal (qL), quasi fast shear (qFS) or quasi slow shear (qSS) wave modes in material 1, and similar qL, qFS, and qSS wave modes in material 2. Thus, there are seven wave rays to be considered at the anisotropic interface in Fig. A4.2. Let's define γa as 01, 11, 21, 31, 12, 22, and 32, where 0 is for the incident ray, the first index γ is to specify qL by 1, qFS by 2, and qSS by 3. The second index a is to specify the material type, material 1 or material 2. The generalized continuity condition at the interface are similar to the one written for isotropic interface $u_2{}^1 = u_2{}^2$; $u_1{}^1 = u_1{}^2$; $u_3{}^1 = u_3{}^2$; $\sigma_{22}{}^1 = \sigma_{22}{}^2$; $\sigma_{21}{}^1 = \sigma_{21}{}^2$; $\sigma_{23}{}^1 = \sigma_{23}{}^2$, however, at an anisotropic interface, displacements and stresses are contributed by all the modes in two materials designated by γ. Based on these specifications, displacement continuity condition for any displacement u_m along x_m direction at the interface can be written as follows

$$A^0 g_m{}^0 e^{i(\mathbf{k}^{01}.\mathbf{x})} + \sum_{\gamma=1}^{3} A^{\gamma 1} g_m{}^{\gamma 1} e^{i(\mathbf{k}^{\gamma 1}.\mathbf{x})} = \sum_{\gamma=1}^{3} A^{\gamma 2} g_m{}^{\gamma 2} e^{i(\mathbf{k}^{\gamma 2}.\mathbf{x})} \quad or$$

$$A^0 g_m{}^0 e^{i(k_j{}^{01}x_j)} = -\sum_{\gamma=1}^{3} A^{\gamma 1} g_m{}^{\gamma 1} e^{i(k_j{}^{\gamma 1}x_j)} + \sum_{\gamma=1}^{3} A^{\gamma 2} g_m{}^{\gamma 2} e^{i(k_j{}^{\gamma 2}x_j)} \tag{A4.24}$$

where the time harmonic part $e^{-i\omega t}$ is omitted from the equations, because the equation of continuity must be valid at all time. Thus, the above equation is time invariant. Substituting the above displacements in the stress equations in Eq. (4.92), similar continuity condition can be written for the stresses as follows

$$\begin{aligned} &\mathbb{C}_{ijml}{}^1 A^0 g_m{}^0 k_l{}^{01} e^{i(k_j{}^{01}x_j)} \\ &= -\sum_{\gamma=1}^{3} \mathbb{C}_{ijml}{}^1 A^{\gamma 1} g_m{}^{\gamma 1} k_l{}^{\gamma 1} e^{i(k_j{}^{\gamma 1}x_j)} + \sum_{\gamma=1}^{3} \mathbb{C}_{ijml}{}^2 A^{\gamma 2} g_m{}^{\gamma 2} k_l{}^{\gamma 2} e^{i(k_j{}^{\gamma 2}x_j)} \end{aligned} \tag{A4.25}$$

where left side of the equation gives the stresses σ_{ij} in material 1 due to the incident wave. The first term on the right side gives the stresses σ_{ij} in material 1 due to the three reflected wave modes. And the second term on the right side gives the stresses σ_{ij} in material 2 due to the three refracted wave modes. For Eqs. (A4.24) and (A4.25) to be valid, the exponential terms with the phase factors, are linearly independent and thus must be equal. It is necessary to satisfy the following equalities

$$k_j^{01}x_j = k_j^{11}x_j = k_j^{21}x_j = k_j^{31}x_j = k_j^{12}x_j = k_j^{22}x_j = k_j^{23}x_j \tag{A4.26}$$

Eq. (A4.26) is in fact the Snell's law at an anisotropic material interface. At the interface where the above equations (Eqs. (A4.24) and (A4.25)) are valid, is defined by the $x_1 - x_3$ plane with $x_2 = 0$. Hence, the index summation in Eq. (A4.26) is only for 1 and 3 directions can viz.

$$\mathbf{k}.\mathbf{x} = k_1^{01}x_1 + k_3^{01}x_3 \tag{A4.27.1}$$

$$\mathbf{k}.\mathbf{x} = k_1^{11}x_1 + k_3^{11}x_3 \tag{A4.27.2}$$

$$\mathbf{k}.\mathbf{x} = k_1^{21}x_1 + k_3^{21}x_3 \tag{A4.27.3}$$

$$\mathbf{k}.\mathbf{x} = k_1^{31}x_1 + k_3^{31}x_3 \tag{A4.27.4}$$

$$\mathbf{k}.\mathbf{x} = k_1^{12}x_1 + k_3^{12}x_3 \tag{A4.27.5}$$

$$\mathbf{k}.\mathbf{x} = k_1^{22}x_1 + k_3^{22}x_3 \qquad \text{A4.27.6)}$$

$$\mathbf{k}.\mathbf{x} = k_1^{23}x_1 + k_3^{23}x_3 \tag{A4.27.7}$$

In the above equation (Eq. (A4.27)) the term $\mathbf{k}.\mathbf{x}$ is introduced to show that the expressions in Snell's law in Eq. (A4.26) are in fact the dot product of two vectors, where $\mathbf{x}$ can be any arbitrary vector on the $x_1 - x_3$ plane passing through a point (x_1, x_3) as shown in Fig. A.4.3. Hence, it can be said that $\mathbf{x}.\hat{\mathbf{e}}_2 = 0$. Next, let's take any two arbitrary equations for any two arbitrary modes from Eq. (A4.27) to subtract to show

$$\mathbf{k}.\mathbf{x} = k_1^{11}x_1 + k_3^{11}x_3 \tag{A4.28.1}$$

$$\mathbf{k}.\mathbf{x} = k_1^{22}x_1 + k_3^{22}x_3 \tag{A4.28.2}$$

$$\cdots \quad \cdots \cdots \cdots \cdots \cdots \cdots \cdots \cdots$$

$$0 = \left(\mathbf{k}^{11} - \mathbf{k}^{22}\right).\mathbf{x} \tag{A4.28.3}$$

As the choice of the modes in Eq. (A4.28) was arbitrary it should be valid for any combination of the modes out of the seven wave vectors at the interface. Hence, it can be said that

$$\left(\mathbf{k}^{mp} - \mathbf{k}^{nq}\right).\mathbf{x} = 0 \tag{A4.29}$$

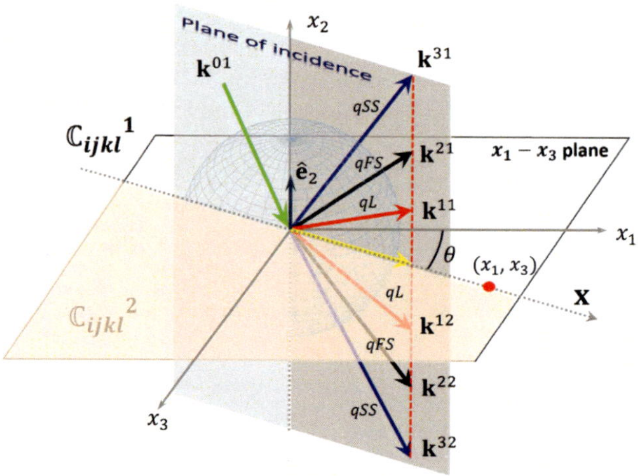

FIGURE A.4.3 Schematic representation of Snell's Law at the interface between two anisotropic materials of different constitutive matrices: Satisfying all rules on a plane of incident.

on the $x_1 - x_3$ plane, where m and n are two independent indices to replace γ with possible number 0, 1, 2, and 3 to designate incident or a specific wave modes and p and q are two independent indices to replace a with possible number 1 and 2 to designate material no 1 or 2. Comparing Eq. (A4.29) and $\mathbf{x}.\hat{\mathbf{e}}_2 = 0$ signifies that the vector $\left(\mathbf{k}^{mp} - \mathbf{k}^{nq}\right)$ must be parallel to the unit vector $\hat{\mathbf{e}}_2$ along x_2 axis. Hence, based on fundamentals of vector products discussed in Chapter 2 automatically infer that the cross product of any set of subtracted vector and unit vector $\hat{\mathbf{e}}_2$ along x_2 axis will be zero, i.e.

$$\left(\mathbf{k}^{mp} - \mathbf{k}^{nq}\right) \times \hat{\mathbf{e}}_2 = 0 \quad or \quad \mathbf{k}^{mp} \times \hat{\mathbf{e}}_2 = \mathbf{k}^{nq} \times \hat{\mathbf{e}}_2 = \mathbf{k}^{\gamma a} \times \hat{\mathbf{e}}_2 \tag{A4.30}$$

This means that the cross product of each wave vectors (incident or wave modes in respective materials) with the unit vector $\hat{\mathbf{e}}_2$ along x_2 axis must be equal. This also means that all these wave vectors $\mathbf{k}^{\gamma a}$ must lie on a same plane. This understanding is presented in Fig. A.4.3 where, the respective modal wave vectors (denoting the direction of wave propagation of respective wave modes $\mathbf{k}^{\gamma a}$ but not pointing to the direction of the wave energy propagation of the respective wave modes) are shown with an incident wave vector $\mathbf{k}^{01}$. The magnitude of the cross product of $\mathbf{k}^{\gamma a}$ and $\hat{\mathbf{e}}_2$ can be fundamentally written as

$$\left|\mathbf{k}^{\gamma a} \times \hat{\mathbf{e}}_2\right| = \left|\mathbf{k}^{\gamma a}\right| \sin\phi_{\gamma a} \tag{A4.31}$$

where $\phi_{\gamma a}$ is the respective angles between the wave vectors $\mathbf{k}^{\gamma a}$ and the unit vector $\hat{\mathbf{e}}_2$ along x_2 axis. The angles $\phi_{\gamma a}$ depend on the anisotropy of the material 1 and material 2. Ultimately the Snell's law for any two wave modes at an anisotropic interface takes the form

$$\frac{\left|\mathbf{k}^{mp}\right|}{\sin\phi_{nq}} = \frac{\left|\mathbf{k}^{nq}\right|}{\sin\phi_{mp}} \tag{A4.32}$$

where m and n are two independent indices to replace γ with possible number 0, 1, 2, and 3 to designate incident or a specific wave modes and p and q are two independent indices to replace a with possible number 1 and 2 to designate material no. 1 or 2. To discuss further, it can be seen that the incident wave ray and the unit vector $\hat{\mathbf{e}}_2$ along x_2 axis makes a unique plane which can be defined as an incident plane. Let's assume the plane identified as plane of incidence in Fig. A.4.3, makes an angle θ with x_1 axis. With this definition and requirement specified in Eqs. (A4.27) and (A4.32) it is apparent that the components of all the wave vectors along $\mathbf{x}$ in Fig. A.4.3 designated by a yellow vector are equal.

Drawing conclusion from this ongoing discussion Eqs. (A4.24) and (A4.25) can be further modified as follows to calculate the relation between the amplitude coefficients of each wave modes.

$$A^0 g_m{}^0 + \sum_{\gamma=1}^{3} A^{\gamma 1} g_m{}^{\gamma 1} = \sum_{\gamma=1}^{3} A^{\gamma 2} g_m{}^{\gamma 2} \tag{A4.33}$$

$$\mathbb{C}_{ijml}{}^1 A^0 g_m{}^0 k_l{}^{01} = -\sum_{\gamma=1}^{3} \mathbb{C}_{ijml}{}^1 A^{\gamma 1} g_m{}^{\gamma 1} k_l{}^{\gamma 1} + \sum_{\gamma=1}^{3} \mathbb{C}_{ijml}{}^2 A^{\gamma 2} g_m{}^{\gamma 2} k_l{}^{\gamma 2} \tag{A4.34}$$

Eq. (A4.33) has three equations from displacement continuity and Eq. (A4.34) also has three equations from stress continuity. There are six equations in total and seven amplitude of wave rays. Hence, normalizing all the equations with an incident amplitude of the incident wave A^0, the equations modified with reflection and refraction coefficients as follows

$$g_m{}^0 + \sum_{\gamma=1}^{3} \mathcal{R}^{\gamma 1} g_m{}^{\gamma 1} - \sum_{\gamma=1}^{3} \mathcal{T}^{\gamma 2} g_m{}^{\gamma 2} = 0 \tag{A4.35}$$

$$\mathbb{C}_{ijml}{}^1 g_m{}^0 k_l{}^{01} + \sum_{\gamma=1}^{3} \mathbb{C}_{ijml}{}^1 \mathcal{R}^{\gamma 1} g_m{}^{\gamma 1} k_l{}^{\gamma 1} - \sum_{\gamma=1}^{3} \mathbb{C}_{ijml}{}^2 \mathcal{T}^{\gamma 2} g_m{}^{\gamma 2} k_l{}^{\gamma 2} = 0 \tag{A4.36}$$

where $\mathcal{R}^{11}$, $\mathcal{R}^{21}$, $\mathcal{R}^{31}$ ($\mathcal{R}^{\gamma 1} = A^{\gamma 1}/A^0$) are three reflection coefficients in material 1 and $\mathcal{T}^{12}$, $\mathcal{T}^{22}$, $\mathcal{T}^{32}$ ($\mathcal{T}^{\gamma 2} = A^{\gamma 2}/A^0$) are three refraction coefficients in material 2. When the incident wave amplitude, constitutive matrix for material 1 and material 2 and polarization vectors of respective modes in respective materials from Christoffel solution are known, the above six linear algebraic equations and can be solved easily to find the reflection and refraction coefficients.

4.5.4 Slowness, Group Velocity and Steering Angle

Here in this section the calculation of steering angle is presented to find the group wave velocity direction from the phase velocity plot. As discussed before, one should not plot the phase velocity profile but instead plot the slowness profile. Slowness is the inverse of the phase velocity as discussed before. Once the phase velocity

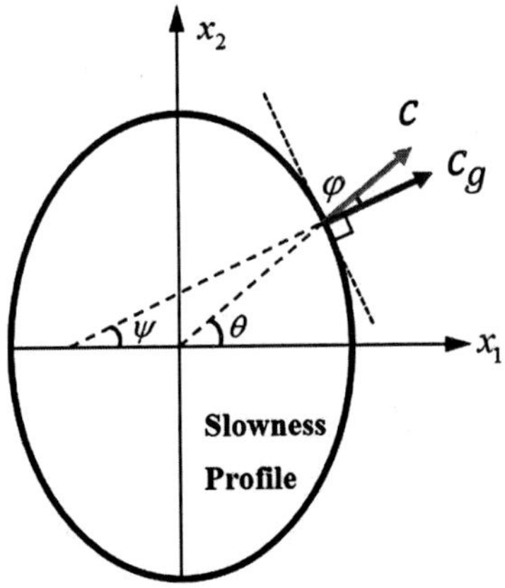

FIGURE A.4.4 The steering angle is defined as the angle between the two vectors $\hat{\boldsymbol{n}}$ and $\hat{N}$ on a dummy slowness profile.

is calculated along any wave vector direction or along the slowness direction, the phase velocity profile on any plane can be obtained. By taking the inverse of the phase velocities in respective slowness directions slowness profile can be calculated and plotted as shown in Figs. 4.15 through 4.24. Next, it is known that direction of slowness vector $\mathbf{k}$ along the unit vector $\hat{\mathbf{n}}$ and the group wave vector or the direction of the energy propagation $\hat{\mathbf{N}}$ may not be the same direction and are different at different spatial points as shown in Fig. 4.20. The steering angle is defined as the angle between the two vectors $\hat{\mathbf{n}}$ and $\hat{\mathbf{N}}$ as shown in Fig. A.4.4 on a dummy slowness profile. The steering angle φ is defined as the algebraic difference between the group wave velocity direction ψ and the phase wave velocity direction θ. Hence,

$$\varphi = \theta - \psi \tag{A4.37}$$

From previous discussions in Sections 4.4.2 and 4.4.3, it is apparent that the group wave velocity direction ψ and the phase wave velocity direction θ can be known from the geometric interpretation of the slowness profiles. Next taking tan of the angle φ one gets

$$\tan\varphi = \tan(\theta - \psi) = \frac{\tan\theta - \tan\psi}{1 + \tan\theta\tan\psi} \tag{A4.38}$$

After rearranging the steering angle can be found as

$$\varphi = \tan^{-1}\left(\frac{\tan\theta - \tan\psi}{1 + \tan\theta\tan\psi}\right) \tag{A4.39}$$

This formulation is valid only when the phase velocity and the group velocity are on the same plane. Let's assume that they are on the same plane. With this assumption, next using the steering angle, the magnitude of the group velocity can be found as follows. As the phase velocity is the projection of the group velocity, one can write $c = c_g \cos(\varphi)$. Hence, the group velocity $c_g = c\sec(\varphi)$.

4.6 SUMMARY

Wave propagation in elastic media is discussed in many books with greater detail in past few decades. Hence, in this chapter just an overview of the topic is presented for the reader to understand the wave propagation in different material types, e.g., fluid, isotropic solids, and anisotropic solid materials. This chapter helps understand the basic definition of the properties of the elastic waves in solid materials, wave interaction at the interfaces and Snell's laws for isotropic and anisotropic materials, necessary for understanding CNDE modeling.

REFERENCES

1. *Ultrasonic Nondestructive Evaluation: Engineering and Biological Material Characterization*. 2004, New York: CRC Press.
2. Computational and experimental methods in structures, in *Structural Health Monitoring for Advanced Compiste Structures*, F.M.H. Aliabadi, Editor, Vol. **8**. 2018, New Jersey: World Scientific.
3. Auld, B.A., *Acoustic Fields and Waves in Solids*, Vol. **I & II.** 1973, New York: John Wiley & Sons.
4. Gopalakrishnan, S., *Wave Propagation in Materials and Structures*. 2017, New York: CRC Press.
5. Kundu, T., Mechanics of elastic waves and ultrasonic nondestructive evaluation, in *Ultrasonic Nondestructive Evaluation*, T. Kundu, Editor. 2004, CRC Press: New York. p. 1–142.
6. Rokhlin, S.I., Chimenti, D.E., Nagy, P.B., *Physical Ultrasonics of Composites*. 2011, Oxford, UK: Oxford University Press.
7. Rose, J.L., *Ultrasonic Waves in Solid Media*. 1999, Cambridge: Cambridge University Press.
8. Haberman, R., *Elementary Applied Partial Differential Equations With Fourier Series and Boundary Value Problems*, 3rd ed. 1997, New Jersey: Prentice Hall.
9. Achenbach, J.D., *Wave Propagation in Elastic Solids*. 1999, New York: Elsevier.
10. Graff, K.F., *Wave Motion in Elastic Solids*. 1975, New York: Dover Publication.
11. Bhuiyan, M.Y., Giurgiutiu, V., Using the gauge condition to simplify the elastodynamic analysis of guided wave propagation. *INCAS Bulletin*, 2016. **8**(3): pp. 11–26.
12. Auld, B.A., *Acoustic Fields and Waves in Solids*. 1990, New York: Wiley.
13. Rokhlin, S.I.C., Chimenti, D., Nagy, P.B., *Physical Ultrasonics of Composites*. 2011, Oxford, UK: Oxford University Press.
14. Banerjee, S. and T. Kundu, Advanced application of distributed point source method – ultrasonic field modeling in solid media, in *DPSM for Modeling Engineering Problems*, T. Kundu and D. Placko, Editors. 2007, Hoboken, New Jersey: John & Willey Publication.
15. Castillo, G.F.T.d., *An Introduction to Hamiltonian Mechanics*. 2018, Switzerland: Springer.
16. Morin, D., *Introduction to Classical Mechanics*. 2007, Cambridge: Cambridge University Press.

5 Wave Propagation in Bounded Structures

5.1 BASIC UNDERSTANDING OF GUIDED WAVES AND ITS APPLICATION IN NDE

Guided waves are confined between two boundaries of a structure and guided inside the material by virtue of multiple total internal reflections from the boundaries. Fig. 5.1 shows fundamental differences between the bulk waves and guided waves that are used in ultrasonic nondestructive evaluation (NDE) based on their respective applications. Ultrasonic NDE in pulse-echo mode typically uses the nature of the bulk waves in solid materials. Ultrasonic NDE in pitch-patch mode typically uses the nature of guided waves in bounded solid materials. Due to several total internal reflections, the guided waves can travel longer distance away from the ultrasonic transducer or from a wave source inside or outside the material. This contrasts with the bulk wave propagation discussed in Chapter 4. Guided wave modes are frequently used for ultrasonic NDE of large structures in aerospace applications and for civil applications. Guided wave modes are also used for ultrasonic NDE of long pipelines. Fig. 5.1a shows the ultrasonic NDE of material in pulse-echo mode. Typically, the material is in air and a couplant/gel or water or any suitable fluid media is used in front of the transducer for efficient transmission of ultrasonic energy. Fig. 5.1b shows the typical application of ultrasonic NDE with guided wave inspection. In this mode, wedge-type fixtures are used. The materials that are used to make the wedges made by respective companies are typically a proprietary information. However, for research purpose, one could easily make such wedges in laboratory using poly methyl methacrylate (PMMA) or acrylic glass and quartz material. Of course, machining and shaping is a challenging task. However, there are several companies and vendors that make the ultrasonic wedges. In most of the cases, transducers are attached to the wedges for convenience. These wedges are sold with specific angle of incidence, e.g., 30°, 45°, 60°, 75°, etc. Based on a plate geometry, if these set angles are not appropriate (discussed later in this chapter) to generate guided wave modes, one needs flexibility in changing the angle on incidence and sensing for the guided wave modes. Fig. 5.1c provides such flexibility through under water inspection, where material is submerged under water. The transducer and received angles can be changed by fixtures with their respective rotational capability. In this mode, guided wave propagated through plate leaked into the fluid media and can be sensed by the ultrasonic receiver. Fig. 5.1d shows an additional case of recently developed NDE method, typically known as structural health monitoring (SHM) using surface mounted actuators and sensors. In this mode, actuators and sensors made of piezoelectric materials capable of electromechanical transduction are directly mounted on the surface of the structure by means of a glue of military/industry grade adhesives.

These piezoelectric actuators and sensors are not applicable for wide band of frequencies, their response is limited to a narrow band of frequency where ultrasonic inspection can be done effectively. Moreover, the actuators and plate geometry are tuned specifically to generate certain wave modes at certain range of frequencies. This means that for a given geometry of a plate, to generate (or we call resonate) a specific guided wave mode at a specific frequency, the geometrical and material property of an actuator must be designed. Or vice versa, with given geometry and material properties of a plate and given geometry and material properties of piezoelectric actuators, the specific range of applicable frequencies for ultrasonic

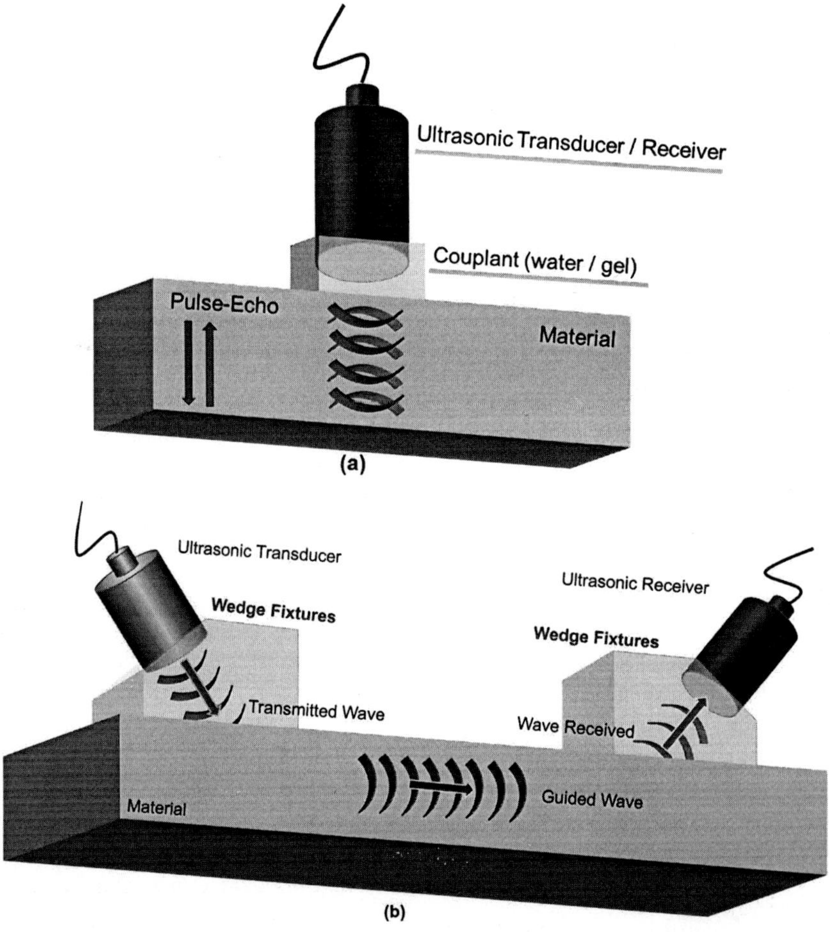

FIGURE 5.1 a) A typical pulse-echo NDE setup with couplant or ad-hoc material with low impedance mismatch, b) A Typical pitch-catch NDE experimental set up with wedge angle for angle beam incidence c) A typical NDE experimental setup for under water experiment (inside a water tank) to generate Guided waves in the structure but sense the Leaky waves in water, c) A Typical NDE/SHM experimental setup for generating Guided waves using embedded or surface mounted transducers or piezoelectric wafers.

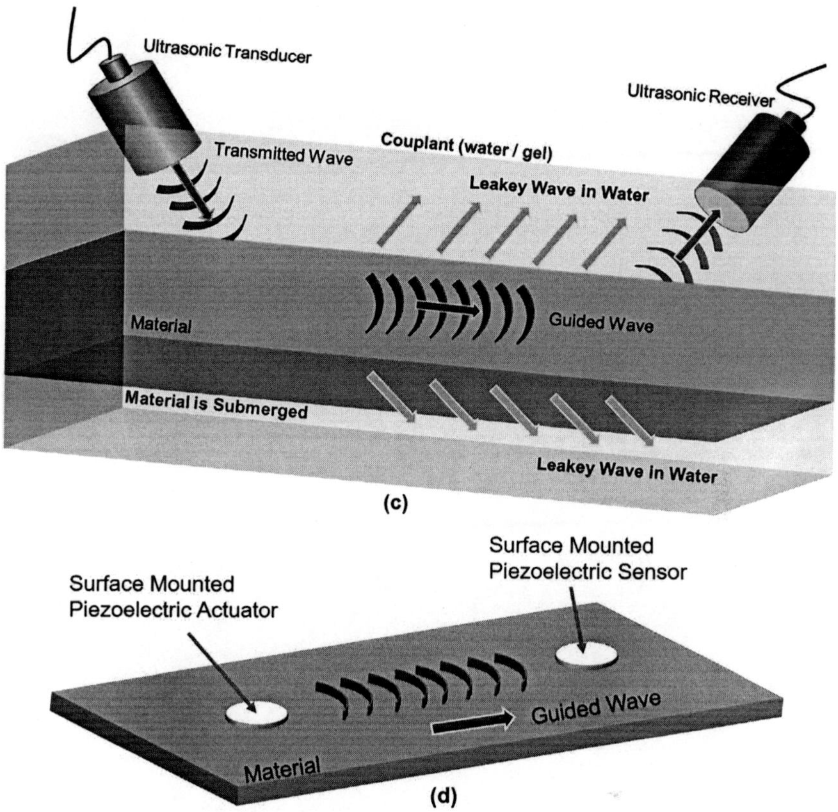

FIGURE 5.1 *(Continued)*

inspection must be found. This process is called guided wave tuning in SHM [1]. Hence, NDE of solid plates may be classified into four genres. (1) Material in air or in vacuum (mathematically, neglecting air) tested in pulse-echo mode, (2) Material in air or vacuum tested with wedged guided waves, (3) Material submerged in fluid or called under water scans where leaky waves are sensed, and (4) Material with surface mounted actuator and sensor for SHM. Nature of the guided wave modes in plates and cylindrical structures is different in nature and will be discussed in subsequent sections followed by mathematical solution of the guided wave modes. The above-mentioned categories are also valid for pipeline NDE. Additionally, for pipes sometimes they are buried under ground and boundary conditions may give additional 5th genre where, (5) Material is underneath an infinite solid layer with or without fluid on the other side (depending on the pipe if carrying liquid or not).

Guided wave energy is concentrated near the boundary, or between two parallel boundaries. By this definition, Rayleigh waves [2, 3] (Fig. 5.2a) composed of bulk P-wave and SV waves that propagate along the surface of the stress-free solid boundary with decaying amplitude, Love waves [4, 5] (Fig. 5.2b) composed of bulk SH-wave that propagate along the interface between two solid layers one being

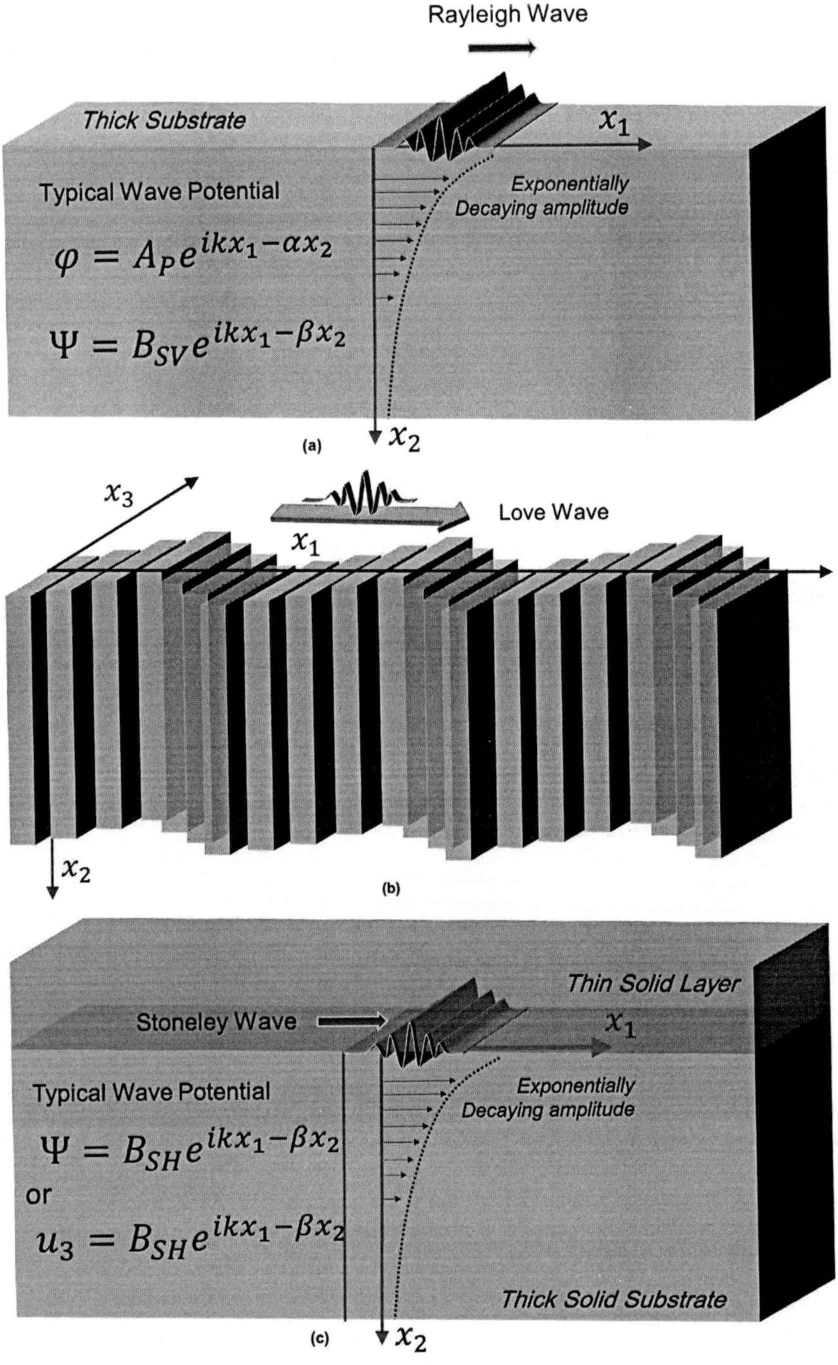

FIGURE 5.2 a) A schematic showing Rayleigh wave and its exponential decaying nature across the depth, the figure also shows the potentials to be used in Helmholtz decomposition b) A schematic of Love wave, c) A schematic of Stoneley wave, d) A schematic of Scholte wave.

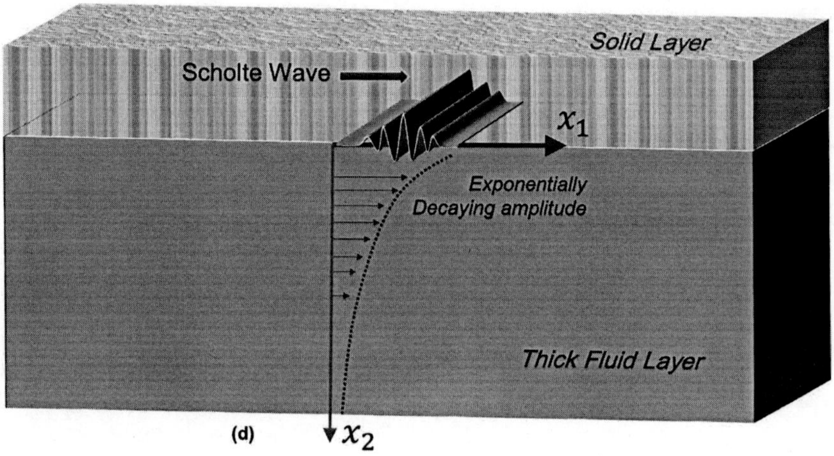

FIGURE 5.2 (*Continued*)

infinite, Stoneley waves [6, 7] (Fig. 5.2c) that propagate along the interface between two solid layers and Scholte waves [8, 7] (Fig. 5.2d) composed of bulk P-waves that propagate along the interface between infinite fluid and solid layers are all guided waves. All these linear guided waves need an interface to guide themselves while maintaining the maximum energy of the waves concentrated near the boundary and decaying amplitude with depth. These guided waves are used for underwater exploration, seismic analysis, and even near-surface inspection of materials and characterization at high frequencies. Surface acoustic waves (SAW) [2], which are nothing but Rayleigh waves at high frequencies (>5 MHz), are frequently used in electronic circuits and devices like filters, oscillators, and transformers. SAW waves utilize the electromechanical transduction of acoustic waves on piezoelectric substrate. Recently, highly nonlinear nondispersive solitary waves traveled through periodic array of phononic crystals are reported [9]. By definition, solitary waves also fall under the category of guided waves. Each of these guided wave modes has their valuable applications in engineering. But guided waves in plates in air or under water (bounded by two stress free or fluid loaded surfaces, respectively) have tremendous applications in ultrasonic NDE and SHM. Horace Lamb [10] in 1917 first explained the existence of such guided wave modes in plates, which was very complex. Unlike two modes in bulk media, plates bounded by two surfaces support infinite set of two modes. It was not used or not well understood till 1990s. Thus, to honor his name, such guided waves are called Lamb waves [4, 5]. Sometimes they are called Rayleigh-Lamb waves because both are governed by the interfaces. Similarly, guided ultrasonic waves in cylindrical structures [5] also have tremendous applications in ultrasonic NDE of long pipelines above ground or buried underground pipelines. In plates and cylindrical structures, guided wave propagates along the central axis of the structure with a necessary propagating wave number to satisfy the energy requirement for a wave mode to propagate. In these cases, thus it is necessary to understand when and how these requirements are satisfied and the wave modes get suitable environment to propagate in the waveguides successfully. A bounded structure that is capable of guiding the wave energy in a particular direction is called the waveguide. In the following sections, physics-based equations in a waveguide are solved and guiding characteristics with their necessary requirements are explained such that they can be suitably used for ultrasonic NDE.

5.2 GUIDED WAVES IN ISOTROPIC PLATES USING CLASSICAL APPROACH

From the previous chapter, it is clear that in isotropic material, P and SV waves are coupled and the SH wave is decoupled, which can support itself. Hence, guided waves in plates also have no exception. In isotropic plate, SH wave can guide itself. P wave and SV wave together can propagate in the plate to generate guided wave modes. In the following sections, SH waves and the union of P and SV waves are discussed mathematically.

5.2.1 Guided SH Wave Modes in Isotropic Plate

Following similar discussions and mathematical derivation presented in Section 4.3.3 in Chapter 4, on bulk wave interaction at isotropic interface, SH waves in plate like structure are discussed herein. Discussion on SH wave in a plate bounded by two stress-free surfaces is pictorially presented in Fig. 5.3. SH wave is decoupled from the P and SV waves. The SH wave potentials that interact in a bounded isotropic plate can be divided into two categories. (1) An upgoing wave, which propagates only to get totally reflected back to the plate from the top surface. (2) A down going wave, which propagates only to get totally reflected back to the plate from the bottom surface. These two waves interfere and create several possibilities of wave modes to propagate inside the wave along the central axis of the plate. These possibilities

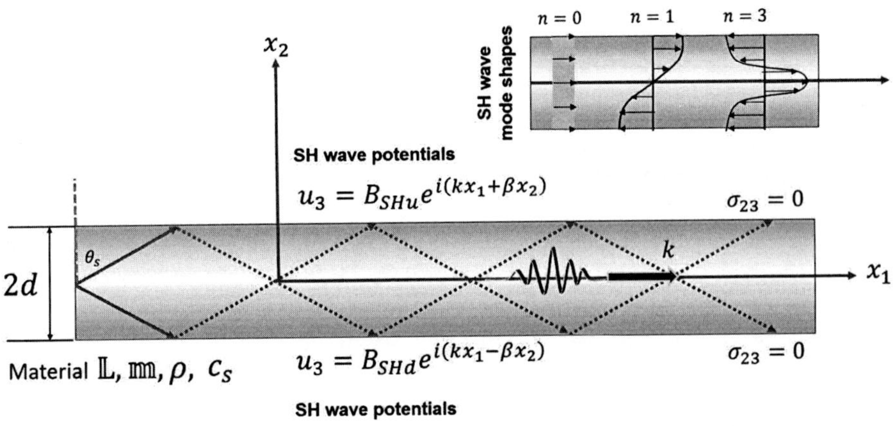

FIGURE 5.3 Schematic of SH Guided wave in a plate like structure bounded by two traction free boundary surfaces.

are frequency dependent. That means the wave modes are dispersive. At a specific frequency, there might be one or more than one SH guided wave modes propagating in the plate. A fundamental fact about a guided wave mode is that it must propagate along the x_1 -axis saturating the complete wave guide as shown in Fig. 5.3. Two potentials, one for upgoing wave and one for down going wave, are thus considered to solve guided SH wave in plate. In the following wave potentials, the time harmonic part $e^{-i\omega t}$ is omitted without the loss of generality.

$$\text{upgoing SH-wave} \qquad \Psi_{SHu}\left(x_j\right) = B_{SHu} e^{i\left(k_j^s x_j\right)} \tag{5.1}$$

$$\text{down going SH-wave} \qquad \Psi_{SHd}\left(x_j\right) = B_{SHd} e^{i\left(k_j^s x_j\right)} \tag{5.2}$$

The case with SH wave interaction presented in Fig. 5.3 is an antiplane problem but still it is a two-dimensional problem whereas, the particle displacements are orthogonal to the wave propagation direction. There will not be any out-of-plane wave vectors but there will be out-of-plane displacements. The wave vector $\mathbf{k}^s$ will have two components along x_1- and x_2-axes, respectively. Recalling the previous discussion on wave number vectors in Section 4.3.2.3, let us assume the magnitudes of the SH-wave wave number in the isotropic plate in Fig. 5.3 is k_s, where

$$k_s = \frac{\omega}{c_s} \quad ; \quad c_s = \sqrt{\frac{\mathrm{m}_1}{\rho_1}} \tag{5.3}$$

Next, the components of the wave number along x_1-axis are

<u>*Wave numbers along x_1-axis*</u>:

$$k_1^s = k_s \sin\theta_s \tag{5.4}$$

where θ_s is the angle of the wave vector with the x_2-axis. However, this angle may not be constant or fixed for SHM wave to propagate. There might be several options for this angle θ_s, and hence, it is not wise to keep θ_s in the formulation and definitely wise to keep it as a variable. From here onward, wave numbers along different axes will be identified separately without expressing them using the trigonometric function. However, their resultant must produce the wave number along the direction of the wave vector, irrespective of the angle of the propagating wave vector in the plate. Hence, let us assume a common wave number k along the x_1-axis and β along x_2-axis.

$$k = k_1^s \quad \text{and} \quad \beta = k_2^s = \sqrt{k_s^2 - k^2} \tag{5.5}$$

The forward going wave number along x_1-axis is same for both upgoing and down going waves. The components of the wave numbers along x_2-axis are also same but can be opposite in sign for upgoing and down going wave, respectively.

With the above notations, the wave vector components can be expressed as follows:

$$\text{upgoing } (+)\text{SH-wave} \quad : \quad \mathbf{k}^{s+} = k\hat{\mathbf{e}}_1 + \beta\hat{\mathbf{e}}_2 \tag{5.6}$$

$$\text{down going } (-) \text{ SH-wave} \quad : \quad \mathbf{k}^{s-} = k\hat{\mathbf{e}}_1 - \beta\hat{\mathbf{e}}_2 \tag{5.7}$$

where $|\mathbf{k}^s| = k_s$, and substituting the wave vectors written in Eqs. (5.6) and (5.7) into Eqs. (5.1) and (5.2) and performing the dot product between the respective wave vectors and **x** the wave potentials can be rewritten as

$$\text{upgoing SH-wave} \qquad \Psi_{SHu}(x_j) = B_{SHu}e^{i(kx_1+\beta x_2)} \tag{5.8}$$

$$\text{down going SH-wave} \qquad \Psi_{SHd}(x_j) = B_{SHd}e^{i(kx_1-\beta x_2)} \tag{5.9}$$

The problem depicted here is an antiplane two-dimensional problem. That means both u_1 and u_2 displacement fields are equal to zero. Only u_3 displacement field will survive. As there is no other displacement component present for SH-wave interaction with the interface, Helmholtz decomposition of the resultant displacement field is redundant and hence, the potential function assumed in Eqs. (5.8) and (5.9) can be directly assumed for the u_3 displacement. Thus, u_3 displacement potentials in the plate can be written as

$$u_3 = B_{SHu}e^{i(kx_1+\beta x_2)} + B_{SHd}e^{i(kx_1-\beta x_2)} \tag{5.10}$$

Recalling the stress equation (Eq. (4.28)) discussed in Section 4.3.1 and substituting the displacement field equations in Eq. (5.10), the stress components will be

<u>*In the plate, stress components* ($\sigma_{11} = \sigma_{22} = \sigma_{33} = \sigma_{12} = 0$) *but* σ_{13} *and* σ_{23} are</u>

$$\sigma_{13} = \mathrm{m}\left[\frac{\partial u_3}{\partial x_1}\right] \tag{5.11}$$

$$\sigma_{23} = \mathrm{m}\left[\frac{\partial u_3}{\partial x_2}\right] \tag{5.12}$$

Top and bottom surfaces ($x_2 = \pm d$) of the plate are stress free. Hence, it is not necessary that the displacements u_3 to be defined at the surfaces but it is necessary that the stresses σ_{13} and σ_{23} are equal to zero. However, in Eq. (5.10), there are two unknowns and with two stresses on two surfaces result four boundary conditions. As only two boundary conditions are necessary, stress along the particle displacement direction $\sigma_{23} = 0$ is imposed on top and bottom surfaces. The stress σ_{23} at the top and bottom surfaces can be written as

$$\sigma_{23}\big|_{x_2=d} = i\beta\mathrm{m}\left(B_{SHu}e^{i\beta d} - B_{SHd}e^{-i\beta d}\right)e^{ikx_1} = 0 \tag{5.13}$$

$$\sigma_{23}\big|_{x_2=-d} = i\beta\mathrm{m}\left(B_{SHu}e^{-i\beta d} - B_{SHd}e^{i\beta d}\right)e^{ikx_1} = 0 \tag{5.14}$$

Eqs. (5.13) and (5.14) result in an eigen value problem in the following matrix form

$$\begin{bmatrix} e^{i\beta d} & e^{-i\beta d} \\ e^{-i\beta d} & e^{i\beta d} \end{bmatrix} \begin{Bmatrix} B_{SHu} \\ B_{SHd} \end{Bmatrix} = \begin{Bmatrix} 0 \\ 0 \end{Bmatrix} \tag{5.15}$$

As $i\beta m \neq 0$ and $e^{ikx_1} \neq 0$.

Determinant of the matrix in Eq. (5.15) is set to zero to have a nontrivial solution and the determinant can be written as

$$e^{2i\beta d} - e^{-2i\beta d} = 2i\sin(2\beta d) = 0 \tag{5.16}$$

Hence, $\sin(2\beta d) = 0$ or $\sin(2\beta d) = \sin n\pi$, where $n = 0,1,2,3,\ldots.$

Further, the wave number β can be simplified to

$$\beta = \sqrt{k_s^2 - k^2} = \frac{n\pi}{2d} \quad ; \quad n = 0,1,2,3,\ldots. \tag{5.17}$$

In the above equation (Eq. (5.17)), d and k_s are constants and thus for different values of n, the wave number k along x_1-axis will be different. Hence, let us assume k_n is the nth solution of the wave numbers that can propagate in the plate. Recollecting Eq. (5.3), it is evident that the phase wave velocity along the x_1-axis should be different as follows with different values of k_n. Further, the equation for the permissible SH wave phase wave velocities in a plate can be written as

$$k_s^2 = \beta^2 + k_n{}^2; \; k_n = \frac{\omega}{c_n}; \; c_n = \frac{\omega}{k_n} = \frac{\omega}{\sqrt{k_s^2 - \left(\frac{n\pi}{2d}\right)^2}}; \; n = 0,1,2,3,\ldots. \tag{5.18}$$

Substituting different values of $n = 0,1,2,3,\ldots.$ different phase velocities of different wave modes can be achieved. But substituting $n = 0$, $c_n = \frac{\omega}{k_s} = c_s$, a constant wave velocity, and is not dependent of the frequency. Thus, it can be said that the first guided SH-wave mode is nondispersive. With further investigation, it can be seen that higher order modes are dispersive in nature but at their respective cut-off frequencies, phase wave velocities tend to merge to the shear phase wave velocity of the media. First three mode shapes of the SH wave modes are presented in Fig. 5.3. They are obtained from mathematical function of displacement along x_2-axis obtained from the displacement expression in Eq. (5.10) and $\frac{B_{SHu}}{B_{SHd}} = e^{+i\pi}$ from Eq. (5.15). As shown in Fig. 5.3, the guided SH wave modes across the thickness are alternatively symmetric and antisymmetric. A sample dispersion curve in frequency-wave number domain and in frequency-phase wave velocity domain is presented in Fig. 5.4a and b, respectively. To understand in more general sense about the guided SH wave phenomena, it is wise to obtain a nonbiased generalized solution where the frequency and wave number axes could be dimension less. In SI unit, angular frequency and wave numbers are *rad/sec.* and *1/m*, respectively. Hence, convenient dimensionless frequency and wave number parameters are $\Omega = 2fd/c_s$, $\overline{k} = 2kd$. A typical solution of the dispersion equation in Eq. (5.15) is presented herein for an aluminum plate with Young's

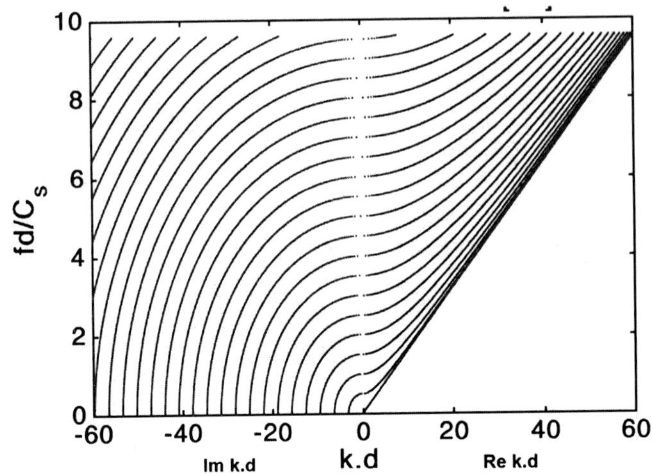

FIGURE 5.4 Typical Dispersion curve (normalized wave number vs. normalized frequency for the SH wave modes in a plate).

modulus ($E = 72.4$ GPa), Poisson's ratio ($\nu = 0.33$), and density $\rho = 2780$ kg/m^3. Wave velocities in aluminum plate are $c_p = 6212$ m/sec and $c_s = 3130$ m/sec. The dimensionless frequency axis (Ω) is divided into 0.1 interval and was considered up to 10. For example, applying $\Omega = 2fd/c_s$, $\Omega = 1$ means approximately $f \cong 626$ KHz in a 5 mm thick ($2d = 5$ mm) aluminum plate. The roots (k), which are the wave numbers at each normalized frequency, were calculated using a computer code plugging the material and geometrical properties. Fig. 5.4 shows the dispersion curve of the SH wave modes in the above-mentioned aluminum plate. However, considering appropriate half thickness (d) and material property parameter (c_s), actual dispersion or the frequency-wave number relationship can be obtained in any plate of different thickness.

5.2.2 Guided Rayleigh-Lamb Wave Modes in Isotropic Plate

In the previous section, guided SH wave formulation is presented. Following the similar approach is this section formulation for guided Lamb wave modes and their solution is presented. Modes generated from the union of P and SV waves guided in a plate like structure are called the Rayleigh-Lamb wave modes. Discussion on Lamb wave in a plate bounded by two stress-free surfaces is pictorially presented in Fig. 5.5. SH wave is decoupled from the P and SV waves; thus in this section, only the P and SV wave potentials are considered. The P and SV wave potentials that interact in a bounded isotropic plate can be divided into two categories. (1) Two upgoing waves, one for each P and SV waves, propagate only to get totally reflected back to the plate from the top surface. (2) Two down going waves, one for each P and SV waves, propagate only to get totally reflected back to the plate from the bottom surface. These four waves interfere and create several possibilities of wave modes to propagate inside the wave along the central axis of the plate.

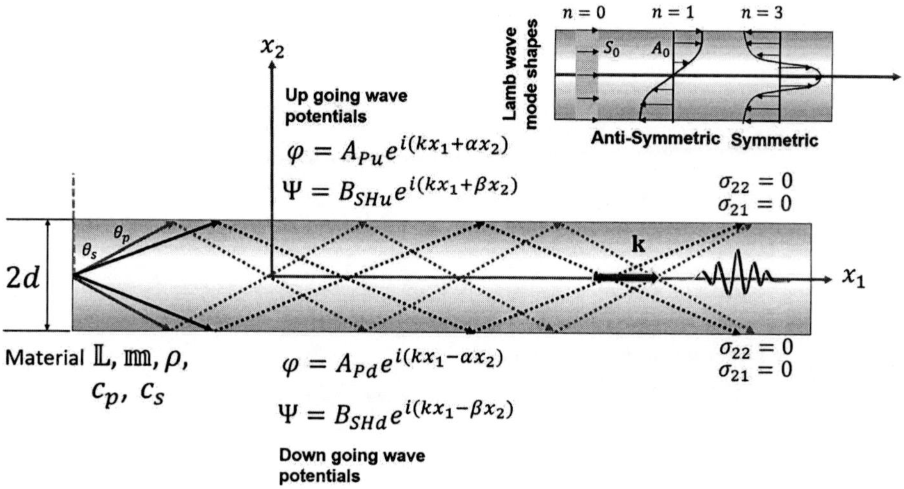

FIGURE 5.5 Schematic of Guided waves composed of P and SV waves in a plate like structure bounded by two traction free boundary surfaces, showing Symmetric and Antisymmetric wave mode shapes.

These possibilities are frequency dependent. That means the Rayleigh-Lamb wave modes are dispersive. At a specific frequency, there might be one or more than one Lamb wave modes propagating in the plate. A fundamental fact about a Lamb wave mode is also that it must propagate along the x_1-axis saturating the complete wave guide as shown in Fig. 5.5. Four potentials, two for upgoing wave and two for down going wave, are thus considered to solve Guided Lamb wave modes in plate. In the following wave potentials, the time harmonic part $e^{-i\omega t}$ is omitted without the loss of generality.

$$\text{upgoing P-wave} \qquad \varphi_{Pu}(x_j) = A_{Pu}e^{i(k_j^P x_j)} \tag{5.19.1}$$

$$\text{down going P-wave} \qquad \varphi_{Pd}(x_j) = A_{Pd}e^{i(k_j^P x_j)} \tag{5.19.2}$$

$$\text{upgoing SV-wave} \qquad \Psi_{SVu}(x_j) = B_{SVu}e^{i(k_j^s x_j)} \tag{5.19.3}$$

$$\text{down going SV-wave} \qquad \Psi_{SVd}(x_j) = B_{SVd}e^{i(k_j^s x_j)} \tag{5.19.4}$$

The case with Lamb wave interaction presented in Fig. 5.5 is an in-plane problem and is a two-dimensional problem. From discussions in Section 4.3.3, there are two wave vectors $\mathbf{k}^{\mathbf{P}}$ and $\mathbf{k}^{\mathbf{s}}$, and both will have two components along x_1- and x_2-axes, respectively. Recalling the previous discussion on wave number vectors in

Section 4.3.2.3, let us assume the magnitudes of the P-wave numbers and SV-wave numbers in isotropic plate shown in Fig. 5.5 are k_p and k_s, where

$$k_p = \frac{\omega}{c_p} \quad ; \quad c_p = \sqrt{\frac{(\mathbb{L}+2\mathrm{m})}{\rho}} \tag{5.20.1}$$

$$k_s = \frac{\omega}{c_s} \quad ; \quad c_s = \sqrt{\frac{\mathrm{m}}{\rho}} \tag{5.20.2}$$

Next, the components of the wave number along x_1-axis should be equal and let us assume they are equal to k.

Wave numbers along x_1-axis:

$$k = k_1^p = k_1^s = k_p \sin\theta_p = k_s \sin\theta_s \tag{5.21}$$

where θ_p and θ_s are the angle of the P-wave and SV-wave vectors, respectively, with the x_2-axis. However, this angle may not be constant or fixed for all wave modes that are capable of propagating in the plate. There might be several options/combinations for this angle θ_p and θ_s, and hence, it is not wise to keep θ_p and θ_s in the formulation and definitely wise to keep them as variables. From here onward, wave numbers along different axes will be identified separately without expressing them using the trigonometric functions. However, their resultant must produce the wave number along the direction of the wave vector, irrespective of the angle of the propagating wave vector in the plate. Hence, let us assume a common wave number k along the x_1-axis and β along x_2-axis for SV wave like it was done for SH waves in the previous section. Additionally, let us assume α as the wave number along the x_2-axis for the P wave

$$k = k_1^p \quad \text{and} \quad \alpha = k_2^p = \sqrt{k_p^2 - k^2} \tag{5.22.1}$$

$$k = k_1^s \quad \text{and} \quad \beta = k_2^s = \sqrt{k_s^2 - k^2} \tag{5.22.2}$$

The forward going wave number along x_1-axis is same for both upgoing and down going waves. The components of the wave numbers along x_2-axis are also same but can be opposite in sign for upgoing and down going wave, respectively.

With the above notations, the wave vector components can be expressed as follows:

$$\text{upgoing } (+) \text{ P-wave} \;:\; \mathbf{k}^{p+} = k\hat{\mathbf{e}}_1 + \alpha\hat{\mathbf{e}}_2 \tag{5.23.1}$$

$$\text{down going } (-) \text{ P-wave} \;:\; \mathbf{k}^{p-} = k\hat{\mathbf{e}}_1 - \alpha\hat{\mathbf{e}}_2 \tag{5.23.2}$$

$$\text{upgoing } (+) \text{ SH-wave} \;:\; \mathbf{k}^{s+} = k\hat{\mathbf{e}}_1 + \beta\hat{\mathbf{e}}_2 \tag{5.23.3}$$

$$\text{down going } (-) \text{ SH-wave} \;:\; \mathbf{k}^{s-} = k\hat{\mathbf{e}}_1 - \beta\hat{\mathbf{e}}_2 \tag{5.23.4}$$

where $|\mathbf{k}^{p\pm}| = k_p$ and $|\mathbf{k}^{s\pm}| = k_s$, and substituting the wave vectors written in Eq. (5.23.i) into Eq. (5.19.i) and performing the dot product between the respective wave vectors and **x,** the wave potentials can be rewritten as

$$\textit{upgoing P-wave} \qquad \varphi_{Pu}(x_j) = A_{Pu}e^{i(kx_1+\alpha x_2)} \tag{5.24.1}$$

$$\textit{down going P-wave} \qquad \varphi_{Pd}(x_j) = A_{Pd}e^{i(kx_1-\alpha x_2)} \tag{5.24.2}$$

$$\textit{upgoing SV-wave} \qquad \Psi_{SVu}(x_j) = B_{SVu}e^{i(kx_1+\beta x_2)} \tag{5.24.3}$$

$$\textit{down going SV-wave} \qquad \Psi_{SVd}(x_j) = B_{SVd}e^{i(kx_1-\beta x_2)} \tag{5.24.4}$$

Hence, total scalar and vector potential in the plate can be written as

$$\varphi = A_{Pu}e^{i(kx_1+\alpha x_2)} + A_{Pd}e^{i(kx_1-\alpha x_2)} \tag{5.25}$$

$$\Psi = B_{SVu}e^{i(kx_1+\beta x_2)} + B_{SVd}e^{i(kx_1-\beta x_2)} \tag{5.26}$$

The problem depicted here is an in-plane two-dimensional problem. That means both u_1 and u_2 displacement fields are not equal to zero but u_3 displacement field will be equal to zero. Hence, the Helmholtz decomposition of displacements will carry only two potentials in it. Following Eq. (4.38.2) and omitting the time harmonic part from the generalized plane wave potentials, the displacement wave fields in the plate can be written as

Displacements along x_1- and x_2-axes:

$$u_1 = \frac{\partial\varphi}{\partial x_1} + \frac{\partial\Psi}{\partial x_2} \tag{5.27.1}$$

$$u_2 = \frac{\partial\varphi}{\partial x_2} - \frac{\partial\Psi}{\partial x_1} \tag{5.27.2}$$

Recalling the stress equation Eq. (4.28) discussed in Section 4.3.1 and substituting the displacement field equations in Eq. (5.27.i) in the plate the stress components will be

Stress components σ_{11}, σ_{12}, and σ_{22} are

$$\sigma_{11} = \mathrm{m}\left[\frac{c_p^2}{c_s^2}\left(\frac{\partial^2\varphi}{\partial x_1^2} + \frac{\partial^2\varphi}{\partial x_2^2}\right) + 2\left(\frac{\partial^2\Psi}{\partial x_1 x_2} - \frac{\partial^2\varphi}{\partial x_2^2}\right)\right] = \mathrm{m}\left[\Gamma^2\ \nabla^2\varphi + 2\left(\Psi_{,12} - \varphi_{,22}\right)\right] \tag{5.28.1}$$

$$\begin{aligned}\sigma_{22} &= \mathrm{m}\left[\frac{c_p^2}{c_s^2}\left(\frac{\partial^2\varphi}{\partial x_1^2} + \frac{\partial^2\varphi}{\partial x_2^2}\right) - 2\left(\frac{\partial^2\varphi}{\partial x_1^2} + \frac{\partial^2\Psi}{\partial x_1 x_2}\right)\right] \\ &= \mathrm{m}\left[\Gamma^2\ \nabla^2\varphi - 2\left(\Psi_{,12} + \varphi_{,11}\right)\right]\end{aligned} \tag{5.28.2}$$

$$\sigma_{12} = \mathbb{m}\left[2\frac{\partial^2\varphi}{\partial x_1 x_2} + \left(\frac{\partial^2\Psi}{\partial x_2^2} - \frac{\partial^2\Psi}{\partial x_1^2}\right)\right] = \mathbb{m}\left[\ 2\varphi_{,12} + \left(\Psi_{,22} - \Psi_{,11}\right)\right] \quad (5.28.3)$$

where $c_p = \sqrt{\frac{(\mathbb{L}+2\mathbb{m})}{\rho}}$ and $c_s = \sqrt{\frac{\mathbb{m}}{\rho}}$ are the P-wave or longitudinal and S-wave or Shear wave velocities in the isotropic solid media; thus, $\frac{c_p^2}{c_s^2} = \Gamma^2 = \frac{\mathbb{L}+2\mathbb{m}}{\mathbb{m}}$; and $\nabla^2\varphi = \frac{\partial^2\varphi}{\partial x_1^2} + \frac{\partial^2\varphi}{\partial x_2^2}$ is the divergence of the gradient of the scalar field φ, which can be explained referring Sections 2.3.1 and 2.3.2 in Chapter 2. Next, here it is wise to remind the Helmholtz equations in Eqs. (4.49) and (4.50) and rewritten as follows, in order to simplify Eq. (5.28.i).

$$\nabla^2\varphi = -k_p^2\varphi \quad (5.29)$$

$$\nabla^2\Psi = -k_s^2\Psi \quad (5.30)$$

Although similar simplification was used before in Chapter 4, it was not discussed in detail. Treating the stress equations in the light of harmonic plane wave potentials and substituting the $\nabla^2\varphi$ by $-k_p^2\varphi$ in Eq. (5.28.i), we get

$$\sigma_{11} = \mathbb{m}\left[-\Gamma^2\ k_p^2\varphi + 2\left(\Psi_{,12} - \varphi_{,22}\right)\right] \quad (5.31.1)$$

$$\sigma_{22} = \mathbb{m}\left[-\Gamma^2\ k_p^2\varphi - 2\left(\Psi_{,12} + \varphi_{,11}\right)\right] \quad (5.31.2)$$

$$\sigma_{12} = \mathbb{m}\left[2\varphi_{,12} + \left(\Psi_{,22} - \Psi_{,11}\right)\right] \quad (5.31.3)$$

Further, $\frac{c_p^2}{c_s^2} = \Gamma^2 = \frac{k_s^2}{k_p^2}$; and hence, replacing $-\Gamma^2\ k_p^2 = -k_s^2$ Eq. (5.31.1), we write further

$$\sigma_{11} = \mathbb{m}\left[-\ k_s^2\varphi + 2\left(\Psi_{,12} - \varphi_{,22}\right)\right] \quad (5.32.1)$$

$$\sigma_{22} = \mathbb{m}\left[-\ k_s^2\varphi - 2\left(\Psi_{,12} + \varphi_{,11}\right)\right] \quad (5.32.2)$$

$$\sigma_{12} = \mathbb{m}\left[\ 2\varphi_{,12} + \left(\Psi_{,22} - \Psi_{,11}\right)\right] \quad (5.32.3)$$

Next, to satisfy the stress-free boundary conditions at the top and bottom surfaces, it is necessary to enforce $\sigma_{22} = 0$ and $\sigma_{12} = 0$ at $x_2 = \pm d$. Enforcing these stress-free boundary conditions at the top and bottom surfaces (i.e., substituting $x_2 = \pm d$ in Eqs. (5.25) and (5.26) and subsequent substitution in Eqs. (5.27.i), (5.28.2), and (5.28.3)), one can write the four following homogeneous equations

$$\sigma_{22}\big|_{x_2=+d} = -\mathbb{m}\left[\left(k_s^2 - 2k^2\right)\left(A_{Pu}e^{i\alpha d} + A_{Pd}e^{-i\alpha d}\right)\right] + 2\mathbb{m}k\beta\left(B_{SVu}e^{i\beta d} - B_{SVd}e^{-i\beta d}\right) = 0 \quad (5.29.1)$$

$$\sigma_{22}\big|_{x_2=-d} = -\mathrm{m}\left[\left(k_s^2 - 2k^2\right)\left(A_{Pu}e^{-i\alpha d} + A_{Pd}e^{i\alpha d}\right)\right] + 2\mathrm{m}k\beta\left(B_{SVu}e^{-i\beta d} - B_{SVd}e^{i\beta d}\right) = 0 \tag{5.29.2}$$

$$\sigma_{12}\big|_{x_2=+d} = -2\mathrm{m}k\alpha\left[\left(A_{Pu}e^{i\alpha d} - A_{Pd}e^{-i\alpha d}\right)\right] + \left(k^2 - \beta^2\right)\left(B_{SVu}e^{i\beta d} + B_{SVd}e^{-i\beta d}\right) = 0 \tag{5.29.3}$$

$$\sigma_{12}\big|_{x_2=-d} = -2\mathrm{m}k\alpha\left[\left(A_{Pu}e^{-i\alpha d} - A_{Pd}e^{i\alpha d}\right)\right] + \left(k^2 - \beta^2\right)\left(B_{SVu}e^{-i\beta d} + B_{SVd}e^{i\beta d}\right) = 0 \tag{5.29.4}$$

where e^{ikx_1} is intentionally omitted as it is common in all the above equations and $e^{ikx_1} \neq 0$. Additionally, using Eq. (5.22.2), $\left(k^2 - \beta^2\right) = k^2 - k_s^2 + k^2 = 2k^2 - k_s^2$.

Hence, the above equations can be written in a matrix form as follows:

$$\begin{bmatrix} \left(k^2-\beta^2\right)e^{i\alpha d} & \left(k^2-\beta^2\right)e^{-i\alpha d} & 2k\beta e^{i\beta d} & -2k\beta e^{-i\beta d} \\ \left(k^2-\beta^2\right)e^{-i\alpha d} & \left(k^2-\beta^2\right)e^{i\alpha d} & 2k\beta e^{-i\beta d} & -2k\beta e^{i\beta d} \\ -2k\alpha e^{i\alpha d} & 2k\alpha e^{-i\alpha d} & \left(k^2-\beta^2\right)e^{i\beta d} & \left(k^2-\beta^2\right)e^{-i\beta d} \\ -2k\alpha e^{-i\alpha d} & 2k\alpha e^{i\alpha d} & \left(k^2-\beta^2\right)e^{-i\beta d} & \left(k^2-\beta^2\right)e^{i\beta d} \end{bmatrix} \begin{Bmatrix} A_{Pu} \\ A_{Pd} \\ B_{SVu} \\ B_{SVd} \end{Bmatrix} = \begin{Bmatrix} 0 \\ 0 \\ 0 \\ 0 \end{Bmatrix} \tag{5.30}$$

For nontrivial solution of the above equations, A_{Pu}, A_{Pd}, B_{SVu}, B_{SVd} cannot be equal to zero but the determinant of the matrix is equal to zero.

$$\Delta = det\begin{vmatrix} \left(k^2-\beta^2\right)e^{i\alpha d} & \left(k^2-\beta^2\right)e^{-i\alpha d} & 2k\beta e^{i\beta d} & -2k\beta e^{-i\beta d} \\ \left(k^2-\beta^2\right)e^{-i\alpha d} & \left(k^2-\beta^2\right)e^{i\alpha d} & 2k\beta e^{-i\beta d} & -2k\beta e^{i\beta d} \\ -2k\alpha e^{i\alpha d} & 2k\alpha e^{-i\alpha d} & \left(k^2-\beta^2\right)e^{i\beta d} & \left(k^2-\beta^2\right)e^{-i\beta d} \\ -2k\alpha e^{-i\alpha d} & 2k\alpha e^{i\alpha d} & \left(k^2-\beta^2\right)e^{-i\beta d} & \left(k^2-\beta^2\right)e^{i\beta d} \end{vmatrix} = 0 \tag{5.31}$$

Using root searching method in MATLAB, the roots can be plotted in frequency-wave number domain. Eq. (5.31) has two variables, the angular frequency ω and the wave number k along x_1-axis of the Rayleigh-Lamb wave modes. Please note that the above matrix was obtained for only one monochromatic frequency by assuming plane harmonic wave solutions. However, as no specific frequency is mentioned in Eq. (5.31), the equation is valid for all possible frequencies. Hence, for root finding, it is wise to discretize the frequency axis into small segments and then find the roots for k from Eq. (5.31), which is called extended plane wave expansion approach [7]. It is clear from Eq. (5.31) that the solution depends just not only on frequencies but also on the material properties (c_p and c_s) and the geometrical properties such as half thickness of the plate (d). After substituting the material and geometrical properties in Eq. (5.31), for each angular frequency, one can find multiple wave numbers both in real and imaginary domain. Imaginary wave numbers are called evanescent wave modes. Referring plane wave potentials, the phase part are $e^{ik.x}$, where if the wave numbers are imaginary (ik), the term will result in an exponentially decaying function ($e^{i(ik)x} = e^{-kx}$) that will cause the wave not to propagate further in the media and dissipate quickly. However, for real wave numbers, this will not be the case and wave will propagate in the media. The wave number solution heavily depends on the material and geometrical properties of the structure. To understand in more general sense about the guided wave phenomena, it is wise to obtain a nonbiased generalized solution where the frequency and wave number axes could be dimensionless. In SI unit, angular frequency and wave numbers are *rad/sec.* and *1/m*, respectively. Hence, convenient dimensionless frequency and wave number parameters are

$$\Omega = \frac{2fd}{c_p} \quad ; \quad \bar{\mathrm{k}} = 2kd \tag{5.32}$$

A typical solution of the dispersion equation in Eq. (5.31) is presented herein for an aluminum plate with Young's modulus ($E = 72.4\ GPa$), Poisson's ratio ($\nu = 0.33$), and density $\rho = 2780\ kg/m^3$. The Lamé constants are written in terms of Young's modulus and Poisson's ratio as follows:

$$\mathbb{L} = \frac{E\nu}{(1+\nu)(1-2\nu)} \quad and \quad \mathrm{m} = \frac{E}{2(1+\nu)} \tag{5.33}$$

Wave velocities in aluminum plate are $c_p = 6212\ m/sec$ and $c_s = 3130\ m/sec$. The dimensionless frequency axis (Ω) is divided into 0.1 interval and was considered up to 5. For example, applying Eq. (5.32), $\Omega = 1$ means approximately $f \cong 1.24\ MHz$ in a $5mm$ thick ($2d = 5\ mm$) aluminum plate. The roots (k), which are the wave numbers at each normalized frequency, were calculated using a computer code substituting the material and geometrical properties. Fig. 5.6 shows the dispersion curve of the Rayleigh-Lamb wave modes in the above-mentioned aluminum plate. However, considering appropriate half thickness (d), and material property parameter (c_p), actual dispersion or the frequency-wave number relationship can be obtained in any plate of different material and thickness.

In Fig. 5.6, both symmetric and antisymmetric wave modes appear. However, if it is necessary to investigate only symmetric or only antisymmetric wave modes, it is possible to have separate equations for symmetric and antisymmetric Rayleigh-Lamb wave modes by assuming slightly different potentials omitting the imaginary part form the exponential term associated with the x_2-axis. Hence, Eqs. (5.25) and (5.26) will still have the $e^{i(kx_1)}$ terms but, along x_2-axis, potentials will be assumed to varying following hyperbolic functions, e.g., sinh or cosh. With new modified potentials, Eq. (5.29) will be modified subsequently. In the modified equations, keeping only the even terms in the equation of displacements (Eq. (5.27.1)) for u_1 but keeping only the odd terms in the u_2 (Eq. (5.27.1)) displacement will be responsible for manifesting the Symmetric wave modes as shown in Fig. 5.5 with n = 0 and n = 3.

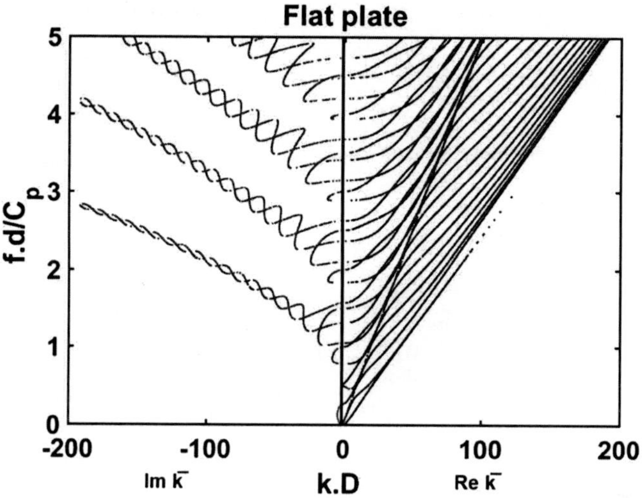

FIGURE 5.6 Typical Dispersion curve (normalized wave number vs. normalized frequency for the Rayleigh-Lamb wave modes in a plate).

This will only happen if A_{Pu} and B_{SVd} are zero. Vice versa, keeping only the odd terms in the equation of displacements (Eq. (5.27.1)) for u_1 but keeping only the even terms in the u_2 (Eq. (5.27.1)) displacement will be responsible for manifesting the antisymmetric wave modes as shown in Fig. 5.5 with n=1 and n=5 or odd numbers. This will only happen if A_{Pd} and B_{SVu} are zero. This discussion with detailed derivation can be found in Ref [7] and is omitted in this chapter. However, the symmetric and antisymmetric Rayleigh-Lamb equations and their respective symmetric and antisymmetric wave mode shapes are discussed using a CNDE problem in Chapter 11 in sections 11.4.2 and 11.4.4.

5.2.3 Generalized Guided Wave Modes in Isotropic Plate with Perturbed Geometry

5.2.3.1 Motivation

In the previous section, wave propagation in planar waveguide is discussed. However, there are many other geometries derived from the plate-like structure and are used in engineering applications. With much technological advancements, use of complicated structural geometries is getting more common for functional and architectural purposes in engineering. As the structural geometry of the plate is the most important characteristic for understanding the wave propagation in plate-like structures with varying boundary conditions, geometric perturbation could significantly alter this understating. Hence, to alleviate such situation, it is necessary to have a formulation for wave propagation in geometrically altered structures. At many instances in engineering applications, corrugated or periodic surface boundaries are used, for example, corrugated heat sink, reinforcement bars in concrete structures (Fig. 5.7), periodic guard rails etc. On the other hand, mimicking biological

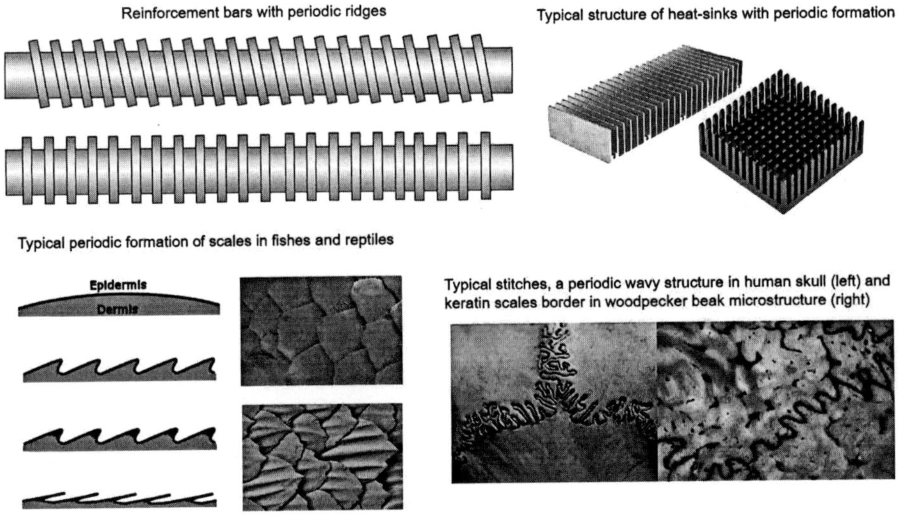

FIGURE 5.7 Example of different periodic structures.

creatures, periodic arrangement of keratin scales in reptiles (Fig. 5.7) have inspired new structures with lower drag and high speed that joined the new genre of structures and materials. NDE of such perturbed geometries requires understanding of wave propagation in periodically corrugated structures. It is neither possible nor cost-effective to have individual formulation for each problem. Hence, a generalized formulation for a periodic structure is presented herein. A sinusoidally corrugated periodic structure is considered for this analysis. Reason for approaching with a sinusoidally corrugated plate is that according to the understanding from Fourier transform, any arbitrary geometrical surface can be mathematically presented as a superposition of multiple sinusoidally corrugated surfaces with different wave lengths. Hence, solving the propagating wave modes in a sinusoidally corrugated plate can eventually help solve the propagating wave modes in any periodic media. Additionally, corrugated waveguides may have applications in impact mitigation and wave attenuation as discussed below.

For decades, researchers are trying to create innovative technologies to protect different industrial structures, cities, and even human organs from the harmful effects of acoustic and impact generated stress waves. Concept of impact mitigation, energy absorption, wave trapping, etc. is developed. Impact mitigation is one of the methods to reduce the detrimental effect of the shock waves on structural components and even on the human body [11]. There are several methods reported to improve shock mitigation and minimize the peak amplitude of shock waves since 1960s. For instance, sandwich structures have been utilized for blast mitigations with fluid impregnated foam [12], foam cores with aerogel, soda-lime glass beads, Glycerin, Tuff [13, 14] and multiple resonators [15], etc. Recently, manmade materials such as metamaterials and periodic structures contributed significantly in this direction. Acoustic metamaterials are recently used for mitigation of blast-waves using the property of effective negative mass density [16–18]. Utilizing the negative mass density features of metamaterials is an unique approach to handle the blasts generated shock waves.

Energy absorbers are normally used almost in all aerospace structures like in space shuttles, satellites, and rocket components. Structures can absorb energy in two ways [19]: (1) irreversible deformation, and (2) friction or damping. Energy-absorbing structures are made using energy absorbent materials such as foams [20], collapsible mechanisms [21, 22], bi-stable structures [23–25], periodic structures like honeycombs, and combinations of material foams and periodic structures [22]. Although human fleets are credible, our natural selection has made many energy absorbing mechanisms perfected over time in different parts of many creatures, like a bird.

Recently, researchers have studied the anatomy of the skulls and beak of woodpeckers. This is to understand how they withstand the strong shock wave after impacting the wood substrates at 1000g, without causing any harm to their skulls [26–28] or brain. The results showed that a woodpecker's skull has a significant capacity to absorb the shock and minimize the damage. Studies showed that woodpeckers' ability is due to the microstructure bits of their cancellous frontal skull bone and a unique architecture of a hyoid bone [29], which starts at the front bottom of the beak, then near the facial structure, it splits into two tapered arcs, which surrounds the woodpecker's skull but ends at the upper end of the facial structure at the top end of the beak. These structures are shown to be effective to absorb 75% of the stress amplitude [30] that propagates to the woodpecker's brain. Other studies also conducted at microscopic and nanoscopic scales on the woodpeckers' beaks. Additional investigation shows that at lower scale, the beak structure is composed of corrugated structural joints of keratin scale-like stitches (Fig. 5.7). These additional microstructures are also in the form of a corrugated mesh around their brain [27–29]. Similarly, a human's skull bones are also stitched to each other in a corrugated fashion (Fig. 5.7). These features must have a significant impact on blocking certain lower frequency stress waves otherwise they were not necessary. Hence, the design process of a bio-mechanism could utilize the nature of the architecture to mitigate the harmful effect from the high impact stress waves on structures and human organs. It may also prevent concussion in human brain. Additionally, if the physics of wave features in such corrugated media is properly understood, they could be utilized to design impact mitigation systems in vehicles such as cars or airplanes. Hence, understanding the wave propagation in corrugated media is valuable.

Corrugated waveguides are the topic of interest for decades. Researchers have studied the electromagnetic wave propagation in corrugated waveguides [31, 32] with very high corrugation and studied stopband in sinusoidally corrugated plates [33]. El- Bahrawy [34] studied elastic wave propagation in a sinusoidally corrugated plate for symmetric Rayleigh-Lamb modes. However, when the depth of corrugation approached zero, the passband and the stopband did not disappear, which is incompatible with experimental and theoretical results for plates with plane boundaries. Later, El- Bahrawy [35] derived the wave propagation equation in an elastic half-space with doubly corrugated surface. Nico et al. [36] studied the diffraction of homogeneous and inhomogeneous plane waves on a doubly corrugated liquid/solid interface. They showed that diffraction of sound theory on 1D corrugated surfaces can be extended to 2D corrugated surfaces. To overcome the limitation, Banerjee and Kundu [37] reformulated the problem by considering P-wave and S-wave potentials and studied the symmetric and antisymmetric Rayleigh-Lamb modes in the sinusoidally corrugated plates. They found the passbands of both symmetric and antisymmetric modes have cut-on resonance along the direction of wave propagation. Kundu et al. [38] carried out an experimental observation to quantify the stopband and passband in the corrugated waveguides and matched with the theoretical prediction in Reference [37]. The experimental results [38] confirmed that the elastic wave is able to propagate at the passband frequency and unable to propagate at the stopband. Later, Das et al. [39] studied the interactions between bounded ultrasonic beams and the corrugated plates using distributed point source method (DPSM). The results were also validated through their experiments. They reported that the strength of the reflected beams for the planar waveguides was higher than the corrugated plates. The backward scattering was weaker for the planar waveguides compared to the corrugated plates. Recently, the effect of the initial stress on shear waves in a periodic waveguide with small corrugation is reported by Hawwa [40]. In this research, the results showed that the corrugation surfaces with small wavelengths have a significant influence on wave phase speed compared to the corrugation surfaces with large wavelengths. The acoustic band gap in composite structures was introduced by Kushwaha [41]. Huang et al. [42] investigated the effect of plate symmetry on the defect states of shear horizontal waves propagation in the periodically corrugated piezoelectric plates with square corrugation. They reported that once the defect size is increased, the frequency of defect bands gets lower.

Although many results are available on corrugated waveguides, a generalized formulation for Rayleigh-Lamb and shear horizontal (SH) wave is still scarce. In most up-to-date formulation in References [37, 38], the stop bands disappeared when the corrugation depth was made to zero but it did not produce the well-known exact dispersion behavior in planar waveguides from the Rayleigh-Lamb equation. Hence, questions remain whether is it possible to have a generalized guided wave formulation that can produce results for both the planar waveguides ($\varepsilon = 0$) and corrugated plates ($\varepsilon > 0$). Additionally, stopband and passband of Rayleigh-Lamb wave modes in the corrugated waveguides are well studied in References [34, 35, 37] but the effect of corrugation height and periodicity are not fully understood. Therefore, a generalized formulation is necessary that should work for any depth of corrugation, any wavelength of corrugation, and for any mean thickness of the plate. More quantitative studies are also necessary to understand the effect of different types of corrugations on band gaps and on the wave modes that are reluctant to pass through the media.

In this chapter, a generalized mathematical formulation of Rayleigh-Lamb wave and SH wave propagation in corrugated media with and without corrugation is presented. Rayleigh-Lamb wave and SH wave propagation equations in the corrugated plates are analytically derived using scalar and vector potential functions. By solving the dispersion equations, the dispersion curves for the corrugated waveguides are obtained for the propagating and evanescent waves. Waveguides with different periodicities and different corrugation heights are also analyzed using the generalized formulation. To verify the formulation, first, the Rayleigh-Lamb wave and

SH wave dispersion band structures were compared using traditional well-known dispersion equation and newly formulated generalized dispersion equation plugging ($\varepsilon = 0$). Further, to validate the results, the dispersion curves from the analytical solution are compared to the finite element results. Next, the band gap for different corrugation heights and periodicities is studied. The governing equation for Rayleigh-Lamb wave and SH wave propagation and the boundary conditions in corrugated waveguides is discussed under theory. Verification of the results, the dispersion curve for different corrugations, and band gap study are explained in the results and discussion section.

5.2.3.2 Generalized Formulation

To study Rayleigh-Lamb wave propagation (union of P and SV waves) and shear horizontal (SH) wave propagation in a generalized geometry, a periodically corrugated elastic waveguide with isotropic homogenous material properties is considered. Fig. 5.8 depicts the schematic of the corrugated plate. As shown in Fig. 5.8, d is the half of the average thickness of the plate $2d$, the corrugation depth is ε, e is the corrugation coefficient to be multiplied with d to control the corrugation depth, and the wavelength of the periodic surface is D. Governing equation of motion is expressed in Cartesian coordinate system.

Equations to represent the top (+) and bottom (-) surfaces of the plate can be written as

$$x_2^+ = d + \varepsilon cos\left(\frac{2\pi x_1}{D}\right)$$
$$x_2^- = -d - \varepsilon cos\left(\frac{2\pi x_1}{D}\right) \tag{5.34}$$

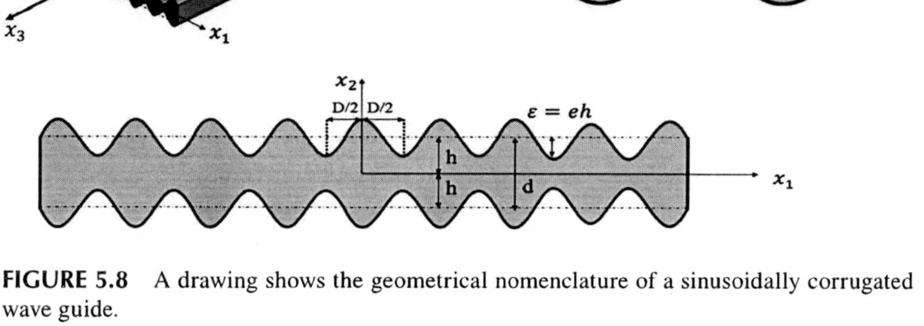

FIGURE 5.8 A drawing shows the geometrical nomenclature of a sinusoidally corrugated wave guide.

where $-\frac{D}{2} \le x_1 \le \frac{D}{2}$, $\varepsilon = ed$, the positive and negative superscripts mean the top surface and the bottom surface, respectively. Fundamental equation of motion or the Navier's equation in isotropic solids can be written as

$$(\mathbb{L} + 2\mathrm{m})u_{j,ji} - \mu\epsilon_{ijk}\epsilon_{kpq}u_{q,pj} + f_i = \rho\ddot{u}_i \tag{5.35}$$

where $\mathbb{L}$ and m are the Lamé constants, ϵ is a permutation symbol, and u_i is the displacement field. To derive the stress and displacement equations for Lamb wave propagation, both compressional wave (P-wave) and shear wave (SV-wave) are considered. Applying Helmholtz decomposition, $u_i = \phi_{,i} + \epsilon_{ijk}\psi_{k,j}$, where ϕ and ψ are two potential functions. Eq. (5.35) can be divided into two Helmholtz equations for a planar two-dimensional case where the out-of-plane displacement (u_3) is assumed to be zero.

$$\nabla^2\varphi + k_p^2\varphi = 0 \ and \ \nabla^2\psi + k_s^2\psi = 0 \tag{5.36}$$

To obtain Eq. (5.36), plane harmonic monochromatic waves are assumed to propagate in the structure in Fig. 5.8 with time harmonic part of the potentials as $e^{-i\omega t}$. Solution of Eq. (5.36) in periodically corrugated media can be solved using Bloch theorem [43]. Commonly, in the study of waves in phononic crystals [44], the Bloch theorem is directly applied to the displacement field $u_i(x_j) = A\ e^{i(\mathbf{k.x}+\mathbf{G.x})}$, where $\mathbf{G} = \frac{2\pi n}{D}$, D is the periodicity of the structure or one wave length of the periodic structure and n takes the values $0, \pm1, \pm2, \pm3, \ldots$ Here, the Bloch theorem is directly applied to the potential functions ϕ and ψ, which automatically satisfies the governing equation of motion.

All up-going and down-going P and SV waves can be written in terms of the scalar and vector potential functions as written in Eqs. (5.37) and (5.38), respectively. According to Fig. 5.8, periodicity of the structure is only along the x_1-axis and thus Bloch potentials are assumed along x_1-axis only to solve monochromatic waves. Time harmonic part of the potentials $e^{-i\omega t}$ is omitted in these equations.

$$\phi = \sum_{n=-\infty}^{n=+\infty} A_{Pun} e^{i\left(k x_1 + \frac{2\pi n}{D} x_1 + \alpha_n x_2\right)} + \sum_{n=-\infty}^{n=+\infty} A_{Pdn} e^{i\left(k x_1 + \frac{2\pi n}{D} x_1 - \alpha_n x_2\right)} \quad \text{for P} - \text{wave} \tag{5.37}$$

$$\psi = \sum_{n=-\infty}^{n=+\infty} B_{SVun} e^{i\left(k x_1 + \frac{2\pi n}{D} x_1 + \beta_n x_2\right)} + \sum_{n=-\infty}^{n=+\infty} B_{SVdn} e^{i\left(k x_1 + \frac{2\pi n}{D} x_1 - \beta_n x_2\right)} \quad \text{for SV} - \text{wave} \tag{5.38}$$

where k is the fundamental component of the wave number along x_1-axis, α_n is the n-th component of the P-wave wave number in x_2 direction (Eq. (5.37)), and β_n is the n-th wave number component in x_2 direction for SV waves (Eq. (5.38)). Amplitudes of the wave modes are designated with A and B for the longitudinal waves and the shear waves, respectively. Subscript P is for longitudinal wave, SV is for shear vertical wave in $x_1 - x_2$ plane, u is for up-going wave and d is for down-going wave, and n is used to distinguish the n-th wave mode.

$$\alpha_n = \sqrt{k_p^2 - \left(k + \frac{2\pi n}{D}\right)^2} = \sqrt{k_p^2 - k_n^2} \tag{5.39}$$

$$\beta_n = \sqrt{k_s^2 - \left(k + \frac{2\pi n}{D}\right)^2} = \sqrt{k_s^2 - k_n^2} \tag{5.40}$$

In Eqs. (5.37) and (5.38), $k_p = \omega/c_p$ is the P-wave number and $k_s = \omega/c_s$ is the SV-wave number. ω is the propagating angular frequency in *rad/sec*, the P-wave speed and the SV-wave speed in the medium are written in Eqs. (5.20.1) and (5.20.2), respectively, where mass density of elastic material is ρ. The displacement field can be defined using Stokes-Helmholtz decomposition for the in-plane wave propagation as follows, where the wave potentials are written in Eqs. (5.37) and (5.38).

$$\begin{aligned} u_1 &= \frac{\partial\phi}{\partial x_1} + \frac{\partial\psi}{\partial x_2} \\ u_2 &= \frac{\partial\phi}{\partial x_2} - \frac{\partial\psi}{\partial x_1} \end{aligned} \tag{5.41}$$

The strain-displacement relation and stress-strain relations in a linear elastic media are described in Chapter 3 but repeated herein as follows:

$$e_{ij} = \frac{1}{2}\left(\frac{\partial u_i}{\partial x_j} + \frac{\partial u_j}{\partial x_i}\right) \quad i, j = 1,2,3 \tag{5.42}$$

$$\sigma_{ij} = 2\mathrm{m}e_{ij} + \mathbb{L}e_{kk}\delta_{ij} \tag{5.43}$$

Utilizing the potentials in Eqs. (5.37) and (5.38) and by substituting Eqs. (5.41) and (5.42) into Eq. (5.43), the stresses along x_1 and x_2 directions can be written as follows:

$$\sigma_{11} = \sum_{n=-\infty}^{n=+\infty} e^{ik_n x_1} \begin{bmatrix} \left(-2\mathrm{m}k_n^2 - \mathbb{L}\left(k_n^2 + \alpha_n^2\right)\right) A_{Pun} e^{i\alpha_n x_2} + \left(-2\mathrm{m}k_n^2 - \mathbb{L}\left(k_n^2 + \alpha_n^2\right)\right) A_{pdn} e^{-i\alpha_n x_2} \\ + \left(-2\mathrm{m}k_n\beta_n\right) B_{SVun} e^{i\beta_n x_2} + \left(2\mathrm{m}k_n\beta_n\right) B_{SVdn} e^{-i\beta_n x_2} \end{bmatrix}$$

$$\sigma_{12} = \sum_{n=-\infty}^{n=+\infty} e^{ik_n x_1} \begin{bmatrix} \left(-2\mathrm{m}k_n\alpha_n\right) A_{pun} e^{i\alpha_n x_2} + \left(2\mathrm{m}k_n\alpha_n\right) A_{pdn} e^{-i\alpha_n x_2} \\ + \mathrm{m}\left(k_n^2 - \beta_n^2\right) B_{SVun} e^{i\beta_n x_2} + \mathrm{m}\left(k_n^2 - \beta_n^2\right) B_{SVdn} e^{-i\beta_n x_2} \end{bmatrix}$$

$$\sigma_{22} = \sum_{n=-\infty}^{n=+\infty} e^{ik_n x_1} \begin{bmatrix} \left(-2\mathrm{m}\alpha_n^2 - \mathbb{L}\left(k_n^2 + \alpha_n^2\right)\right) A_{Pun} e^{i\alpha_n x_2} + \left(-2\mathrm{m}\alpha_n^2 - \mathbb{L}\left(k_n^2 + \alpha_n^2\right)\right) A_{Pdn} e^{-i\alpha_n x_2} \\ + \left(2\mathrm{m}k_n\beta_n\right) B_{SVun} e^{i\beta_n x_2} + \left(-2\mathrm{m}k_n\beta_n\right) B_{SVdn} e^{-i\beta_n x_2} \end{bmatrix}$$

(5.44)

Similarly, to study the SH waves, it is assumed that the SH wave is propagating on $x_1 - x_2$ plane. But the particles have displacement along the x_3 direction only. In other words, $u_1 = u_2 = 0$ and u_3 is nonzero for the SH wave case. If these displacements are substituted in Eq. (5.35), the following wave equation is obtained for the harmonic SH waves.

$$\nabla^2 u_3 + k_s^2 u_3 = 0 \tag{5.45}$$

Solution of the $u_3$displacement field can be defined as

$$u_3 = \sum_{n=-\infty}^{n=+\infty} B_{SHun} e^{i\left(k x_1 + \frac{2\pi n}{D} x_1 + \beta_n x_2\right)} + \sum_{n=-\infty}^{n=+\infty} B_{SHdn} e^{i\left(k x_1 + \frac{2\pi n}{D} x_1 - \beta_n x_2\right)} \tag{5.46}$$

The displacement field for SH wave generates the following shear stresses that are relevant for applying the traction free boundary conditions on the top and the bottom surfaces.

$$\begin{aligned}\sigma_{13} &= \mathrm{m}\left(\frac{\partial u_3}{\partial x_1} + \frac{\partial u_1}{\partial x_3}\right) = \mathrm{m}\frac{\partial u_3}{\partial x_1} = \mathrm{m}\sum_{n=-\infty}^{n=+\infty}(ik_n)e^{ik_n x_1}\left[B_{SHun}e^{i\beta_n x_2} + B_{SHdn}e^{-i\beta_n x_2}\right] \\ \sigma_{23} &= \mathrm{m}\left(\frac{\partial u_3}{\partial x_2} + \frac{\partial u_2}{\partial x_3}\right) = \mathrm{m}\frac{\partial u_3}{\partial x_2} = \mathrm{m}\sum_{n=-\infty}^{n=+\infty}(i\beta_n)e^{ik_n x_1}\left[B_{SHun}e^{i\beta_n x_2} + B_{SHdn}e^{-i\beta_n x_2}\right]\end{aligned} \tag{5.47}$$

5.2.3.3 Boundary Conditions

The traction-free boundary conditions should be applied at the upper (x_2^+) and lower surfaces (x_2^-) of the corrugated waveguide. However, both the surfaces are perturbed and the direction cosines (normal to the surfaces) change along the x_1 -axis, continuously. Equations for the upper and the lower surfaces of the corrugated plate are given in Eq. (5.34). To derive the normal stress and shear stress on the upper and lower surfaces, the normal unit vector at the traction-free surfaces should be defined. Let us assume a point on the surface of the corrugated plate. At that point, the Cartesian coordinate system is like the one shown along the central plate line. However, at the same point, the surface is rotated due to the sinusoidally corrugated surface and there could be a new local transformed coordinate, which is rotated about the x_3-axis. The new axis that is normal to the surface is designated as x_1' and the orthogonal axis or the axis that is tangential to the surface is x_2'. The projections of these new x_1- and x_2-axes or the direction cosines of these axes with respect to x_1- and x_2-axes or the normal unit vector are defined by n_1 and n_2, respectively as shown in Fig. 5.8. Note that n_1 and n_2 are the projections of the normal unit vector along x_1 and x_2 directions, respectively. These equations for n_1 and n_2 can be derived from the equation of the surfaces written in Eq. (5.34) by taking their respective derivatives. n_1 and n_2 for the top and the bottom surfaces can be written as

$$\text{for the top surface } x_2^+ \quad \begin{aligned} n_1 &= \frac{(2\pi\varepsilon/D)\sin(2\pi x_1/D)}{\sqrt{((2\pi\varepsilon/D)\sin(2\pi x_1/D))^2+1}} \\ n_2 &= \frac{1}{\sqrt{((2\pi\varepsilon/D)\sin(2\pi x_1/D))^2+1}} \end{aligned} \tag{5.48}$$

$$\text{for the bottom surface } x_2^- \quad \begin{aligned} n_1 &= \frac{-(2\pi\varepsilon/D)\sin(2\pi x_1/D)}{\sqrt{((2\pi\varepsilon/D)\sin(2\pi x_1/D))^2+1}} \\ n_2 &= \frac{1}{\sqrt{((2\pi\varepsilon/D)\sin(2\pi x_1/D))^2+1}} \end{aligned} \tag{5.49}$$

Using Eqs. (5.48) and (5.49), a transformation matrix should be constructed to transform the local stresses into the transformed coordinate system of x_1', x_2', and x_3. Fig. 5.8 shows the transformed coordinate systems alongside the original coordinate system in a corrugated wave guide. The transformation matrix $[R]$ is constructed applying the understanding from Section 2.4.2 in Chapter 2.

$$[R]=\begin{bmatrix} n_1 & n_2 & 0 \\ -n_2 & n_1 & 0 \\ 0 & 0 & 1 \end{bmatrix} \tag{5.50}$$

The stresses are then transformed using Eq. (2.38) as follows:

$$\begin{bmatrix} \sigma_{11}' & \sigma_{12}' & \sigma_{13}' \\ \sigma_{21}' & \sigma_{22}' & \sigma_{23}' \\ \sigma_{31}' & \sigma_{32}' & \sigma_{33}' \end{bmatrix} = [R]\begin{bmatrix} \sigma_{11} & \sigma_{12} & \sigma_{13} \\ \sigma_{12} & \sigma_{22} & \sigma_{23} \\ \sigma_{13} & \sigma_{23} & \sigma_{33} \end{bmatrix}[R]^T =$$

$$\begin{bmatrix} \sigma_{11}n_1^2+\sigma_{22}n_2^2+2\sigma_{12}n_1n_2 & (\sigma_{22}-\sigma_{11})n_1n_2+\sigma_{12}\left(n_1^2-n_2^2\right) & \sigma_{13}n_1+\sigma_{23}n_2 \\ (\sigma_{22}-\sigma_{22})n_1n_2+\sigma_{12}\left(n_1^2-n_2^2\right) & -\sigma_{11}n_1^2+\sigma_{22}n_2^2-2\sigma_{12}n_1n_2 & -\sigma_{13}n_2+\sigma_{23}n_1 \\ \sigma_{13}n_1+\sigma_{23}n_2 & -\sigma_{13}n_2+\sigma_{23}n_1 & \sigma_{33} \end{bmatrix} \tag{5.51}$$

For Lamb wave and SH wave propagation, the general boundary conditions to obtain the traction-free boundary conditions at the upper and lower surfaces, appropriate stresses should be made to zero. For Lamb wave propagation, the stresses that are required to be zero are σ_{11}' and σ_{12}'. But, For SH wave propagation, only σ_{13}' stress needs to be equal to zero. The derivation is not complete yet. The transformed stress state that is written in Eq. (5.51) is not well defined to find the solution of the dispersion relation in a corrugated plate. Due to Bloch expansion of the wave function, here it is necessary to apply the orthogonality principle. Hence, the relevant stresses discussed above for Lamb wave and SH wave propagation are multiplied with the normalized eigen functions $\frac{e^{-ikx_1-\frac{2\pi n}{D}x_1}}{n_2^2}$ and $\frac{e^{-ikx_1-\frac{2\pi n}{D}x_1}}{n_2^2}$, respectively. Next, the transformed stresses need to be integrated along the x_1-axis over a period of corrugation of the periodic plate. By this procedure, the orthogonality conditions are imposed on the stresses and the traction-free boundary conditions can be applied on the transformed integrated stresses. For Lamb wave propagation, the orthogonality-imposed boundary conditions can be given by:

$$\begin{bmatrix} \overline{\sigma'}_{11} \\ \overline{\sigma'}_{12} \end{bmatrix} = \int_{-D/2}^{D/2} e^{-ik_n x_1} \begin{bmatrix} \sigma_{11}\left(\frac{n_1}{n_2}\right)^2 + \sigma_{22} - 2\sigma_{12}\left(\frac{n_1}{n_2}\right) \\ \sigma_{12}\left(\left(\frac{n_1}{n_2}\right)^2 - 1\right) + (\sigma_{11} - \sigma_{22})\left(\frac{n_1}{n_2}\right) \end{bmatrix} dx_1 = 0 \tag{5.52}$$

and for SH wave propagation, the orthogonality-imposed boundary conditions can be expressed by:

$$\overline{\sigma'}_{13} = \int_{-D/2}^{D/2} e^{-ik_n x_1} \left[\sigma_{13}\left(\frac{n_1}{n_2}\right) + \sigma_{23}\right] dx_1 = 0 \tag{5.53}$$

where $k_n = k + \frac{2\pi n}{D}$, $-\infty < n < +\infty$. One can do the following integrations and can find these integrations to be equal to zero for the first Bloch mode $n = 0$.

$$\int_{-D/2}^{D/2} e^{-ik_n x_1} \left(-2\sigma_{12}\left(\frac{n_1}{n_2}\right)\right) dx_1 = 0 \tag{5.54}$$

$$\int_{-D/2}^{D/2} e^{-ik_n x_1} \left(\sigma_{13}\left(\frac{n_1}{n_2}\right)\right) dx_1 = 0 \tag{5.55}$$

$$\int_{-D/2}^{D/2} e^{-ik_n x_1} \left((\sigma_{11} - \sigma_{22})\left(\frac{n_1}{n_2}\right)\right) dx_1 = 0 \tag{5.56}$$

Hence, the stress equations with the fundamental wave number k along x_1-axis using $n = 0$, with one period of corrugation at the upper and lower surfaces of the corrugated plate are expressed by Eqs. (5.57) and (5.58) for Lamb wave and SH wave propagations, respectively.

$$\begin{aligned} \overline{\sigma'}_{11} = \int_{-D/2}^{D/2} \Bigg[\frac{4\pi^2\varepsilon^2}{D^2} \sin^2\left(\frac{2\pi x_1}{D}\right) \Big[& \left(-2\mathbb{m}k^2 - \mathbb{L}\left(k^2 + \beta^2\right)\right) A_{Pu0} e^{i\alpha x_2} \\ & + \left(-2\mathbb{m}k^2 - \mathbb{L}\left(k^2 + \beta^2\right)\right) A_{Pd0} e^{-i\alpha x_2} \\ & + (-2\mathbb{m}k\beta) B_{SVu0} e^{i\beta x_2} + (2\mathbb{m}k\beta) B_{SVd0} e^{-i\beta x_2} \Big] \Bigg] dx_1 = 0 \end{aligned} \tag{5.57}$$

$$\overline{\sigma'}_{12} = \int_{-D/2}^{D/2}\Bigg[\left(\frac{4\pi^2\varepsilon^2}{D^2}\sin^2\left(\frac{2\pi x_1}{D}\right)-1\right)\Big[(-2\mathrm{m}k\alpha)A_{Pu0}e^{i\alpha x_2} + (2\mathrm{m}k\alpha)A_{Pd0}e^{-i\alpha x_2} + \mathrm{m}\left(k^2-\beta^2\right)B_{SVu0}e^{i\beta x_2} + \mathrm{m}\left(k^2-\beta^2\right)B_{SVd0}e^{-i\beta x_2}\Big]\Bigg]dx_1 = 0 \quad (5.58)$$

In the above equations, the wave number k along x_1-axis is the fundamental solution of the wave number in a corrugated waveguide. However, according to Bloch theorem, if k is a solution with $n = 0$, then all other wave numbers (k_n), following the rule $k_n = k + \frac{2\pi n}{D}$, are also the solutions of the system for all other values of n ($-\infty < n < +\infty$). Thus, inserting $n = 0$ in the Bloch solution for the fundamental modes, Eqs. (5.57) and (5.58) can be written for the top surface x_2^+ and the bottom surface x_2^- separately that will give four simultaneous equations, where the wave amplitudes (A_{Pu0}, A_{Pd0}, B_{SVu0}, B_{SVu0}) are unknowns. After substituting the surface profiles written in Eq. (5.34) and performing the integrals in Eqs. (5.57) and (5.58), four simultaneous linear equations can be found in a matrix form as written below in Eq. (5.56).

$$[\mathfrak{D}] = \begin{bmatrix} \mathfrak{D}_{11} & \mathfrak{D}_{12} & \mathfrak{D}_{13} & \mathfrak{D}_{14} \\ \mathfrak{D}_{21} & \mathfrak{D}_{22} & \mathfrak{D}_{23} & \mathfrak{D}_{24} \\ \mathfrak{D}_{31} & \mathfrak{D}_{32} & \mathfrak{D}_{33} & \mathfrak{D}_{34} \\ \mathfrak{D}_{41} & \mathfrak{D}_{42} & \mathfrak{D}_{43} & \mathfrak{D}_{44} \end{bmatrix}\begin{bmatrix} A_{Pu0} \\ A_{Pd0} \\ B_{SVu0} \\ B_{SVd0} \end{bmatrix} = 0 \quad (5.59)$$

Eq. (5.59) is an eigen value problem, where the generalized matrix coefficients with height of corrugation $\varepsilon = 0$ should result in the same dispersion equation written in Eq. (5.30). Hence, with $\varepsilon = 0$, Eq. (5.59) should result in the dispersion behavior of a plate. The generalized equations for the coefficients of the matrix in Eq. (5.59) with arbitrary corrugation depth ε can be expressed as

$$\mathfrak{D}_{11} = \mathrm{m}D\left[\left(-2k^2-\Gamma_r k_p^2\right)\frac{4\pi^2\varepsilon}{D^2}\frac{J_1(\varepsilon\alpha)}{\alpha} + \left(2k^2-k_s^2\right)J_0(\varepsilon\alpha)\right]e^{i\alpha d}$$

$$\mathfrak{D}_{12} = \mathrm{m}D\left[\left(-2k^2-\Gamma_r k_p^2\right)\frac{4\pi^2\varepsilon}{D^2}\frac{J_1(\varepsilon\alpha)}{\alpha} + \left(2k^2-k_s^2\right)J_0(\varepsilon\alpha)\right]e^{-i\alpha d}$$

$$\mathfrak{D}_{13} = \mathrm{m}D\left[-2k\beta\left(\frac{4\pi^2\varepsilon}{D^2}\frac{J_1(\varepsilon\beta)}{\beta} - J_0(\varepsilon\beta)\right)\right]e^{i\beta d}$$

$$\mathfrak{D}_{14} = \mathrm{m}D\left[2k\beta\left(\frac{4\pi^2\varepsilon}{D^2}\frac{J_1(\varepsilon\beta)}{\beta} - J_0(\varepsilon\beta)\right)\right]e^{-i\beta d}$$

$$\mathfrak{D}_{21} = \mathrm{m}D\left[\left(-2k^2-\Gamma_r k_p^2\right)\frac{4\pi^2\varepsilon}{D^2}\frac{J_1(\varepsilon\alpha)}{\alpha} + \left(2k^2-k_s^2\right)J_0(\varepsilon\alpha)\right]e^{-i\alpha d}$$

$$\mathfrak{D}_{22} = \mathrm{m}D\left[\left(-2k^2 - \Gamma_r k_p^2\right)\frac{4\pi^2\varepsilon}{D^2}\frac{J_1(\varepsilon\alpha)}{\alpha} + \left(2k^2 - k_s^2\right)J_0(\varepsilon\alpha)\right]e^{i\alpha d}$$

$$\mathfrak{D}_{23} = \mathrm{m}D\left[-2k\beta\left(\frac{4\pi^2\varepsilon}{D^2}\frac{J_1(\varepsilon\beta)}{\beta} - J_0(\varepsilon\beta)\right)\right]e^{-i\beta d}$$

$$\mathfrak{D}_{24} = \mathrm{m}D\left[2k\beta\left(\frac{4\pi^2\varepsilon}{D^2}\frac{J_1(\varepsilon\beta)}{\beta} - J_0(\varepsilon\beta)\right)\right]e^{i\beta d}$$

$$\mathfrak{D}_{31} = \mathrm{m}D\left[-2k\alpha\left(\frac{4\pi^2\varepsilon}{D^2}\frac{J_1(\varepsilon\alpha)}{\alpha} - J_0(\varepsilon\alpha)\right)\right]e^{i\alpha d}$$

$$\mathfrak{D}_{32} = \mathrm{m}D\left[2k\alpha\left(\frac{4\pi^2\varepsilon}{D^2}\frac{J_1(\varepsilon\alpha)}{\alpha} - J_0(\varepsilon\alpha)\right)\right]e^{-i\alpha d}$$

$$\mathfrak{D}_{33} = \mathrm{m}D\left[\left(2k^2 - k_s^2\right)\left(\frac{4\pi^2\varepsilon}{D^2}\frac{J_1(\varepsilon\beta)}{\beta} - J_0(\varepsilon\beta)\right)\right]e^{i\beta d}$$

$$\mathfrak{D}_{34} = \mathrm{m}D\left[\left(2k^2 - k_s^2\right)\left(\frac{4\pi^2\varepsilon}{D^2}\frac{J_1(\varepsilon\beta)}{\beta} - J_0(\varepsilon\beta)\right)\right]e^{-i\beta d}$$

$$\mathfrak{D}_{41} = \mathrm{m}D\left[-2k\alpha\left(\frac{4\pi^2\varepsilon}{D^2}\frac{J_1(\varepsilon\alpha)}{\alpha} - J_0(\varepsilon\alpha)\right)\right]e^{-i\alpha d}$$

$$\mathfrak{D}_{42} = \mathrm{m}D\left[2k\alpha\left(\frac{4\pi^2\varepsilon}{D^2}\frac{J_1(\varepsilon\alpha)}{\alpha} - J_0(\varepsilon\alpha)\right)\right]e^{i\alpha d}$$

$$\mathfrak{D}_{43} = \mathrm{m}D\left[\left(2k^2 - k_s^2\right)\left(\frac{4\pi^2\varepsilon}{D^2}\frac{J_1(\varepsilon\beta)}{\beta} - J_0(\varepsilon\beta)\right)\right]e^{-i\beta d}$$

$$\mathfrak{D}_{44} = \mathrm{m}D\left[\left(2k^2 - k_s^2\right)\left(\frac{4\pi^2\varepsilon}{D^2}\frac{J_1(\varepsilon\beta)}{\beta} - J_0(\varepsilon\beta)\right)\right]e^{i\beta d} \tag{5.60}$$

where $\Gamma_r = \mathbb{L}/\mathrm{m}$ is the Lame ratio.

Similarly, the SH-wave dispersion equation for the corrugated plate in a matrix form is

$$\mathcal{D}^{SH} = \begin{bmatrix} \mathcal{D}_{11}^{SH} & \mathcal{D}_{12}^{SH} \\ \mathcal{D}_{21}^{SH} & \mathcal{D}_{22}^{SH} \end{bmatrix}\begin{Bmatrix} B_{SHu0} \\ B_{SHd0} \end{Bmatrix} = 0 \tag{5.61}$$

where the generalized coefficients in the above matrix are expressed as

$$\begin{aligned}
\mathcal{D}_{11}^{SH} &= \mathrm{m}D\left[i\beta J_0(\varepsilon\alpha)e^{i\beta d}\right] \\
\mathcal{D}_{12}^{SH} &= \mathrm{m}D\left[i\beta J_0(\varepsilon\alpha)e^{-i\beta d}\right] \\
\mathcal{D}_{21}^{SH} &= \mathrm{m}D\left[i\beta J_0(\varepsilon\alpha)e^{-i\beta d}\right] \\
\mathcal{D}_{22}^{SH} &= \mathrm{m}D\left[i\beta J_0(\varepsilon\alpha)e^{i\beta d}\right]
\end{aligned} \tag{5.62}$$

Eqs. (5.60) and (5.62) are the eigen value problems. To obtain the dispersion solution (eigen values), the determinant of the matrix $[\mathfrak{D}]$ must be equal to zero. The Lamb wave dispersion curve (Fig. 5.6) obtained from Eq. (5.30) for a planar waveguide can be regenerated if the ε value set to zero in Eq. (5.60). First, it is verified if Eq. (5.59) could result in the same expression obtained from tradition approach for obtaining the dispersion equations for a planar waveguide. It was found that Eq. (5.59) exactly replicates Eq. (5.30) obtained for Rayleigh-Lamb wave dispersion equation available in wave propagation books, e.g. Reference [45]. Hence, Eq. (5.59) is a generalized dispersion equation for the Rayleigh-Lamb wave modes in corrugated or noncorrugated flat isotropic plates

Next, the dispersion relations in the frequency-wave number $(\omega - k)$ domain can be solved using any suitable root-finding algorithm by plugging appropriate structural and material properties of the wave guides. Extended solutions in both real and imaginary wave numbers domain are calculated. The procedure of root-finding for real and imaginary wave numbers is identical. This procedure is performed at a certain frequency for a range of real and imaginary wave numbers, separately. Using MATLAB code, each of the roots is calculated using Eqs. (5.59) and (5.61). The roots are found by following the bisection method. Here, the procedure is only explained for real wave numbers. At a certain frequency, the determinants are calculated for each successive real wave number. The calculated determinants could be in the form of real or imaginary numbers. If the calculated determinants are real, the expected roots are found between the wave numbers, which have negative multiplication of successive determinants. If the calculated determinants are imaginary, the roots are found between wave numbers, which result in negative values after multiplication of its real and imaginary parts. The procedure is performed for other frequencies until the entire frequency domain is swept. This procedure is identically followed for the imaginary wave numbers. All the calculated roots can be plotted on frequency-wave number domain plots.

5.2.3.4 Discussions on Generalized Rayleigh Lamb and SH Modes

In this section, Lamb wave propagation in planar and corrugated aluminum plates is studied analytically. First, the dispersion curves for Rayleigh-Lamb wave modes in a planar waveguide from the generalized approach presented in Section 5.2.3.2 with $\varepsilon = 0$ are graphically verified with the dispersion curve obtained from well-established Rayleigh-Lamb dispersion equation presented in Eq. (5.31) and Ref. [45]. In this study, aluminum material with Young's modulus ($E = 72.4$ GPa), Poisson's ratio ($\nu = 0.33$), and density $\rho = 2780\ kg/m^3$ are considered. To understand in a more general sense about the Guided Rayleigh-Lamb wave SH wave phenomena, it is wise to obtain a nonbiased generalized solution where the frequency and wave number axes could be dimensionless as discussed in Section 5.2.2. In SI unit, angular frequency and wave numbers are *rad/sec.* and *1/m*, respectively. Hence, convenient dimensionless frequency and wave number parameters are $\Omega = 2fd/c_p$, $\bar{k} = 2kd$. $\Omega = 2fd/c_s$ is considered for SH guided waves. Wave velocities in the aluminum plate are $c_p = 6212\ m/sec$ and $c_s = 3130\ m/sec$. The dimensionless frequency axis (Ω) was divided into 0.1 interval and was considered up to 5 for Rayleigh-Lamb wave and up to 10 for SH waves. For example, applying $\Omega = 2fd/c_s$, $\Omega = 1$ means approximately $f \cong 626\ KHz$ in a *5mm* thick ($2d = H = 5\ mm$) aluminum plate for SH wave. The roots (k), which are the wave numbers at each normalized frequency, were calculated using a computer code plugging the material and geometrical properties. Here, the pictorial verification of Rayleigh-Lamb and SH waves are illustrated in Fig. 5.9. Figs. 5.9 and 5.10 depict Rayleigh-Lamb and SH wave dispersion curves in a planar waveguide obtained from Eqs. (5.59) and (5.61) setting $\varepsilon=0$, respectively, and dispersion curves obtained from well-established dispersion equations in the literature. Please note that the normalized frequency (Ω) in Fig. 5.9 is normalized with longitudinal wave velocity (c_p) but the normalized frequency (Ω) in Fig 5.10 is obtained with shear wave velocity (c_s). The results showed that the dispersion curves obtained from the generalized formulation with $\varepsilon = 0$ exactly match with existing formulation for a planar waveguide. Additionally, dispersion curves are also plotted for a corrugated waveguide, with different corrugation heights and periodicities by solving analytically the obtained Eqs. (5.59) and (5.61). Dispersion curves in a planar waveguide are compared with the wave dispersion in the corrugated waveguide with corrugation of $\varepsilon = 0.5h$ and periodicity of $D = 3d$ in Fig. 5.11. A dimensionless variable $\bar{k}$ ($2kd$) is assumed to be either real or purely imaginary. As apparent in Fig. 5.11, the corrugated waveguide has several mode conversions in the real $\bar{k}$ domain. In corrugated waveguides, additional modes are appeared in the

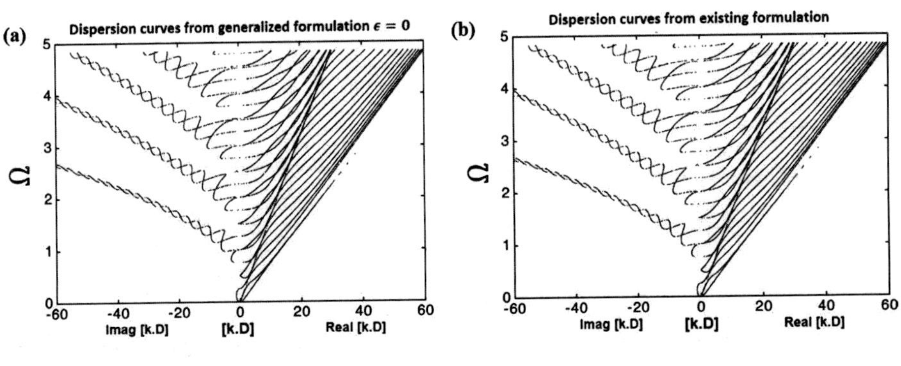

FIGURE 5.9 Typical Dispersion curve (normalized wave number vs. normalized frequency for the Rayleigh-Lamb wave modes in a corrugated wave guide a) resembles flat plate dispersion behavior with corrugation depth equal to zero plugged in to the current formulation, b) resembles flat plate dispersion behavior from existing formulation for Rayleigh-Lamb wave modes.

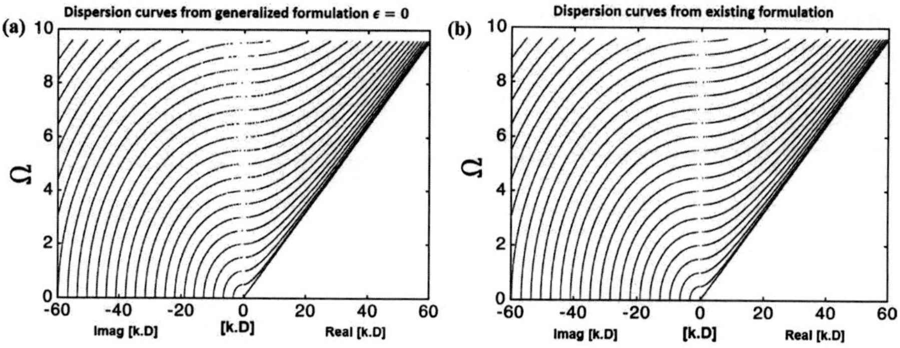

FIGURE 5.10 Typical Dispersion curve (normalized wave number vs. normalized frequency for the SH Guided wave modes in a corrugated wave guide a) resembles flat plate dispersion behavior with corrugation depth equal to zero plugged in to the current formulation, b) resembles flat plate dispersion behavior from existing formulation for SH Guided wave modes.

evanescent zone (imaginary $\bar{k}$) as a result of the corrugation, which are not present in a planar waveguide [46]. In this context, two red boxed areas in Fig. 5.11b are separately presented in Fig. 5.11c. This area is identified as the mode conversion zone for any corrugated waveguide. It was found that the mode conversion zone in frequency-wave number plots for corrugated waveguides decreases and becomes more collimated with the increasing depth of corrugation. A unique feature of this zone is that at a specific wave number for a specific wave mode, there are multiple frequency solutions, which may be up to three frequencies. This makes a specific mode to demonstrate wave propagation with both positive and negatively group velocities

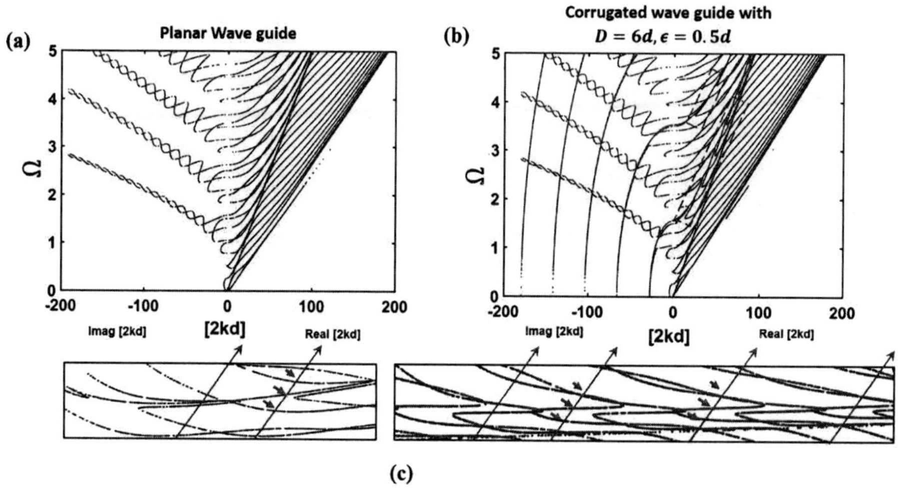

FIGURE 5.11 Dispersion curve for a) planar waveguide b) corrugated wave guide c) showing modal cross over and mode conversion in corrugated wave guides.

at a certain wave number. Such zones and such modes can be separately studied to understand the unique wave transport phenomena, which are not the objective of this chapter and are omitted herein.

Further, the effect of corrugation height on the Rayleigh-Lamb modes is investigated. As depicted in Fig. 5.12, the dispersion curve for $\epsilon = 0.1d$, $\epsilon = 0.2d$, $\epsilon = 0.3d$, and $\epsilon = 0.5d$ is plotted keeping the length of periodicity fixed ($D = 4d$). According to the results, additional evanescent modes are generated with increasing corrugation height. The number of additional modes is increased with increasing depth of corrugation. With increasing depth of corrugation mode, conversion zones are narrowed as seen in Fig. 5.12.

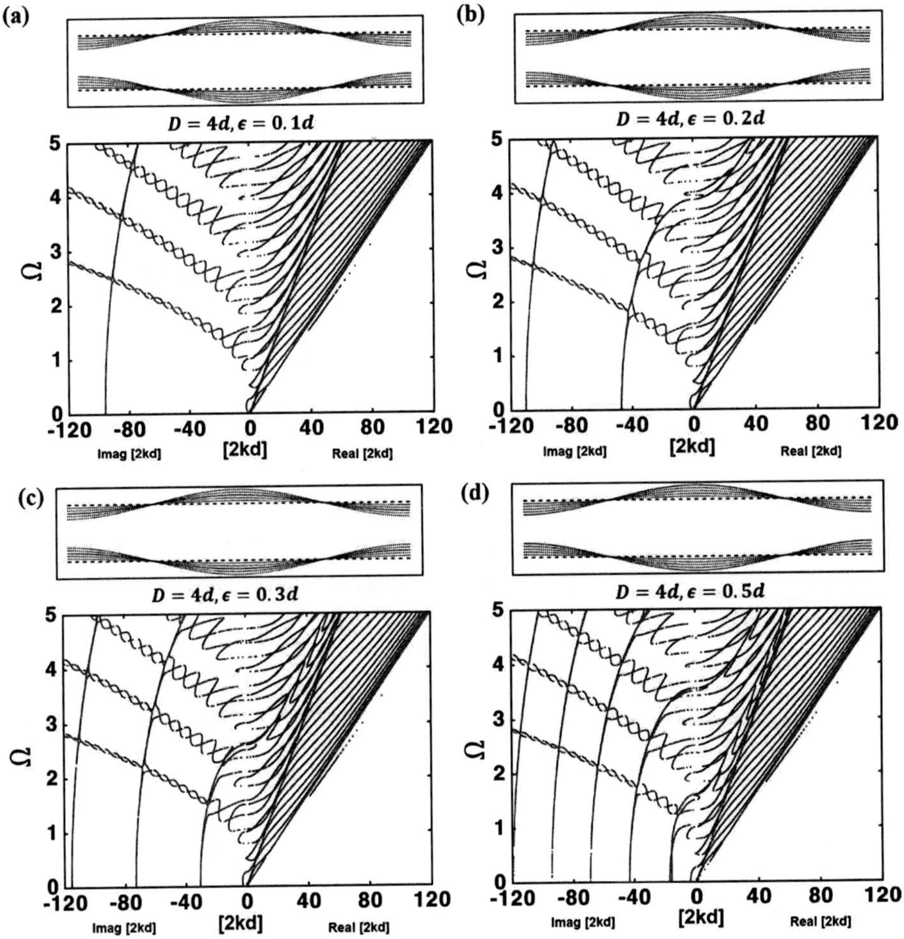

FIGURE 5.12 Effects of corrugation height on Lamb wave propagation in a corrugated wave guide with fixed period D = 4d of corrugation a) $\epsilon = 0.1d$, b) $\epsilon = 0.2d$, c) $\epsilon = 0.3d$, d) $\epsilon = 0.5d$.

Next, the effect of periodicity is studied on the dispersion curves keeping the corrugation height constant. Fig. 5.13 shows the Rayleigh-Lamb wave propagation in the corrugated waveguides with different periodicities ($D = 2d$, $D = 4d$, $D = 6d$, and $D = 8d$) while the corrugation height is $\epsilon = 0.5d$. In a corrugated waveguide with $D = 2d$ (Fig. 5.13a), the number of solution space are less compared to others in the frequency-wave number domain. Due to the increasing length of periodicity in corrugated waveguides, the possible range of wave number solutions is increased at certain frequencies.

To study the bandgap for impact mitigation problems, the dispersion curves in corrugate waveguides are required to be analyzed to find the band gaps. According to the Bloch theorem, all possible solutions of wave number vector (k) are obtained from the fundamental wave number with added factor of $2\pi n/D$, where $n = 0, \pm1, \pm2, \pm3$. Here, the solution is restricted within the first Brillouin zone [41] for the forgoing discussion of bandgaps. Propagation of elastic wave is considered along the ΓX direction [47] of

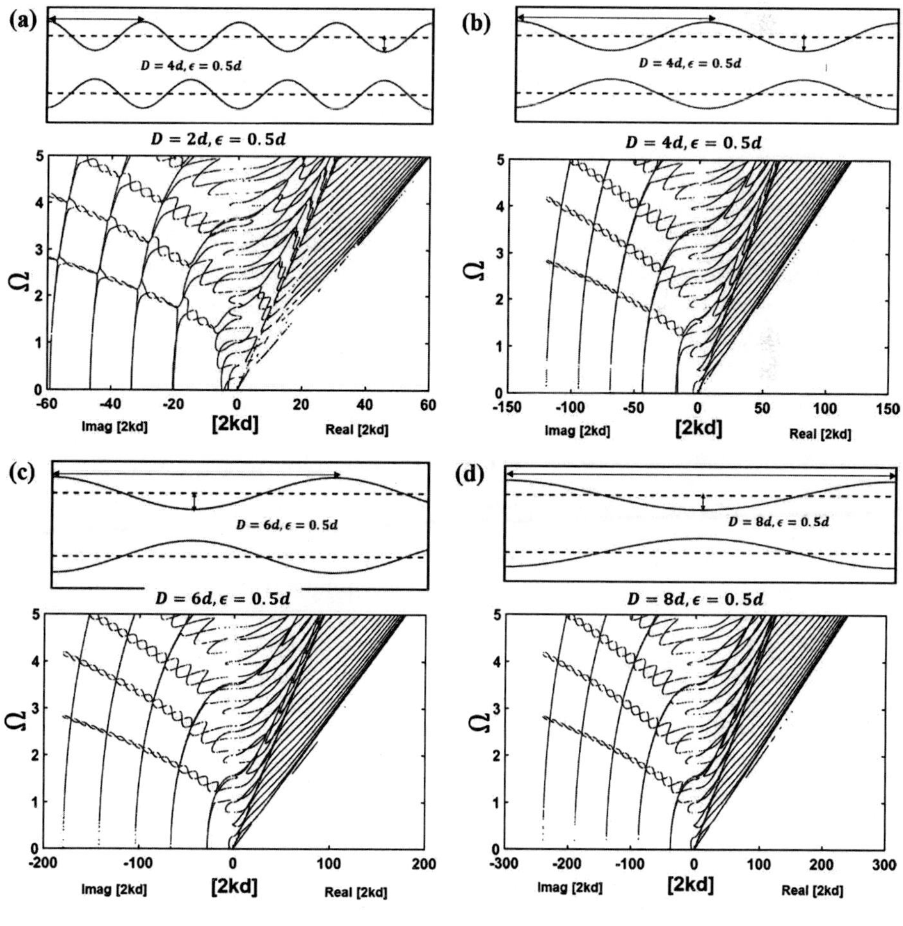

FIGURE 5.13 Effects of corrugation period on Lamb wave propagation in a corrugated wave guide with fixed corrugation depth $\epsilon = 0.5d$, a) $D = 2d$, b) $D = 4d$, c) $D = 6d$ d) $D = 8d$.

irreducible Brillouin zone for which $0 < Re(k) < \frac{\pi}{D}$ [48]. All the band gaps in corrugated waveguides are determined for $0 < Re(k.D = \bar{k}) < \pi$ within frequency range of 0 to 3 MHz. To understand the effect of the parameters (periodicity and corrugation height) on the band gaps, parametric study was carried out for different configurations of corrugated waveguides. The dispersion relations were solved for different periodicities ($D = 2d$, $D = 4d$, $D = 6d$, $D = 8d$, $D = 10d$, $D = 12d$) and different corrugation heights ($\epsilon = 0.1d$, $\epsilon = 0.15d$, $\epsilon = 0.2d$, $\epsilon = 0.25d$, $\epsilon = 0.3d$, $\epsilon = 0.35d$, $\epsilon = 0.4d$, $\epsilon = 0.45d$, $\epsilon = 0.5d$). Next, the bandgaps are determined in each corrugated waveguide. Corrugated waveguides with certain periodicity and corrugation height have multiple bandgaps. Each bandgap has its own lower and a higher frequency limits. To study the influence of the corrugation height and corrugation periodicity on the bandgaps, picking the upper and lower frequency limits of frequencies six bandgap plots are created where corrugation factor e changed from 0.1 to 0.5 and are presented in Fig. 5.14. Fig. 14 is obtained by connecting all the lower and the higher frequency limits of the first, second, third, and fourth band gap zones, respectively, in corrugated waveguides with a fixed periodicity but with different corrugation heights.

According to the results, there are four bandgaps in corrugated plates with $D = 2d$. However, the number of band gaps in waveguides with $D = 4d$ is decreased to three. Moreover, once the length of periodicity changed from $D = 6d$ to $D = 8d$, the bandgap zones are merged to each other. Thus, increasing the wavelength of the periodicity can affect the number of bandgaps in a waveguide. Here, the normalized wave number ($\bar{k} = 2kd$) is limited within 0 and π within the first Brillouin Zone. Once the length of the periodicity is increased, the wave number decreased to be in the range of 0 to π. Consequently, in a certain frequency, the number of existing wave number solutions become less in corrugated plates with a small length of periodicities compare to corrugated plates with a bigger length of periodicities. Therefore, the number of bandgaps is decreased with the increase of the length of the periodicity.

5.2.4 Exercise: Guided Waves in Isotropic Plate with Experimental NDE Situations

So far, in the above sections, guided waves in isotropic plate in air or vacuum (neglecting air) are discussed with detailed mathematical treatments of the governing differential equation. Plate in air is a typical case for a bulk-wave NDE problem, where predominantly handheld pulse-echo ultrasonic testing (HUT) are performed (Fig. 5.1a). A very thin layer of gel or fluid layer is used to transmit ultrasonic wave into the media. In case of guided waves shown in Fig. 5.1b, their depiction can utilize the solution from the above derivation. The cases for SHM shown in Fig. 5.1d can utilize the solution of the guided waves presented in the previous sections. However, cases for NDE shown in Fig 5.1c requires a bit more discussion. So far, it can be seen that boundary conditions played a key role in solving the dispersion behavior in the plates. NDE problems with material submerged under water will not take the stress-free boundary conditions that are applied in the previous section, whereas continuity conditions must be enforced at the fluid-solid interfaces. At the top and bottom surfaces of the plate,

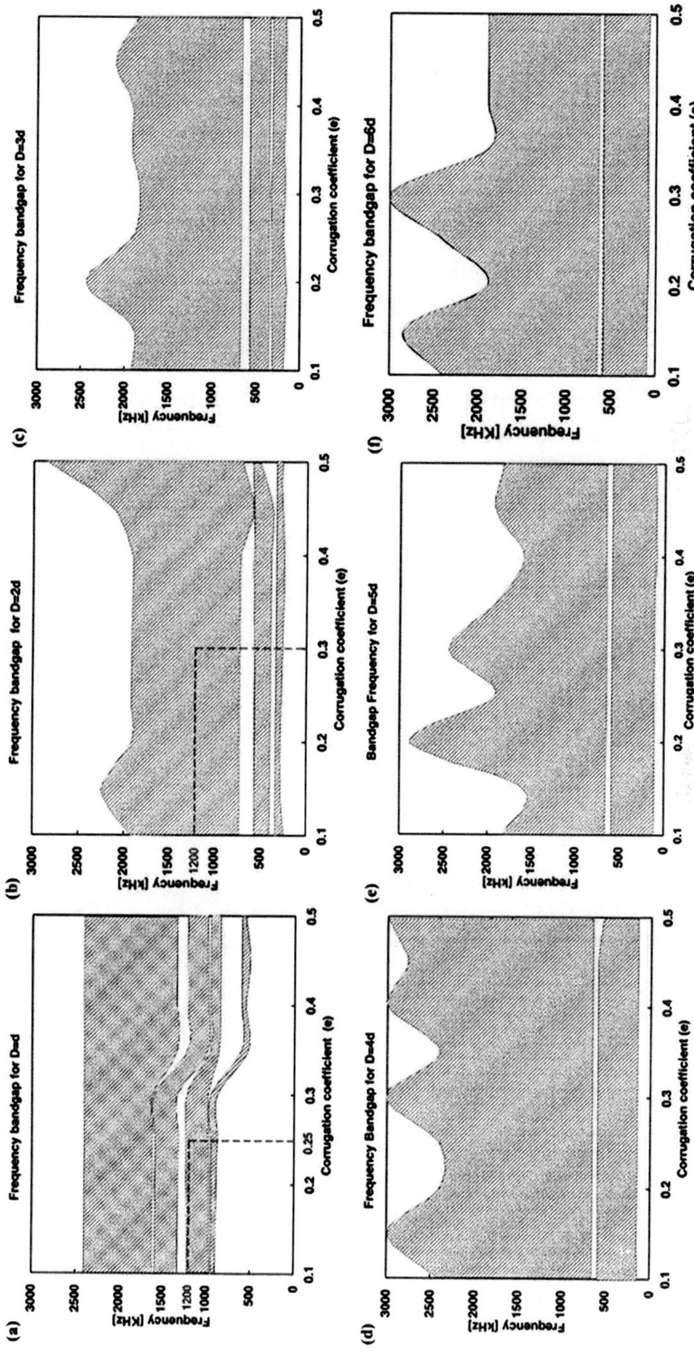

FIGURE 5.14 Frequency bandgaps for different configurations of corrugated plates.

fluid pressure must match with the normal stresses in the plate. Also, the normal displacements must match at any solid-fluid boundary. As fluid media cannot carry any shear stresses, no comment can be made on the shear stress continuity at the solid-fluid boundary. Hence, pressure and displacement functions in fluid should be brought to the calculation for the NDE case in Fig. 5.1c, like they were used before in Eqs. (4.20) through (4.24). Although the complete solution is not presented in this section, a preliminary approach to solve the problem is discussed with the necessary wave potentials. It is expected that reader could proceed with the solution based on the discussions presented in the previous section. Let us assume a plane P-wave (because fluid cannot take any shear stress) incidence through fluid on the fluid-sloid interface. After incidence, at the fluid-solid interface wave will break into two parts, P wave and SV wave in solid. As shown in Fig. 5.5, these two primary waves will totally internally reflect inside the plate and will guide inside the plate to form guided wave modes. While guiding, partially wave energy will leak into the fluid on either side of the plate as shown in Fig. 5.1c. Now, there are six wave potentials to be considered. Two fluid potentials on either side of the plate, two P-wave potentials (one for up going and one for down going) and two SV-wave potentials (one for up going and one for down going) should be considered. The potentials in the fluid (*a* for above and *b* for below the plate) can be considered as

$$\phi_{af} = A_{af} e^{i(k_i x_i)} \quad and \quad \mathbf{u}_{af} = \nabla \phi_{af} \tag{5.63}$$

$$\phi_{bf} = A_{bf} e^{i(k_i x_i)} \quad and \quad \mathbf{u}_{bf} = \nabla \phi_{bf} \tag{5.64}$$

where the subscript f signifies the fluid media. Similar potential functions that are assumed to solve guided waves in plates should be assumed here. The potentials for P waves and SV waves are

$$upgoing\ P-wave \qquad \varphi_{Pu}(x_j) = A_{Pu} e^{i\left(k_j^p x_j\right)} \tag{5.65.1}$$

$$down\ going\ P-wave \qquad \varphi_{Pd}(x_j) = A_{Pd} e^{i\left(k_j^p x_j\right)} \tag{5.65.2}$$

$$upgoing\ SV-wave \qquad \Psi_{SVu}(x_j) = B_{SVu} e^{i\left(k_j^s x_j\right)} \tag{5.65.3}$$

$$down\ going\ SV-wave \qquad \Psi_{SVd}(x_j) = B_{SVd} e^{i\left(k_j^s x_j\right)} \tag{5.65.4}$$

Hence, total scalar and vector potential in the plate can be written as

$$\varphi = A_{Pu} e^{i(kx_1 + \alpha x_2)} + A_{Pd} e^{i(kx_1 - \alpha x_2)} \tag{5.66}$$

$$\Psi = B_{SVu} e^{i(kx_1 + \beta x_2)} + B_{SVd} e^{i(kx_1 - \beta x_2)} \tag{5.67}$$

Using appropriate potential functions, continuity conditions at the solid fluid boundary can be enforced as follows:

Normal displacement along x_2:

$$u_2^{af} = u_2^{s} \; at \; top \; surface \; x_2 = +d \; ; \; u_2^{bf} = u_2^{s} \; at \; bottom \; surface \; x_2 = -d \quad (5.68)$$

Normal stress along x_2:

$$p^{af} = \sigma_{22}^{s} \; at \; top \; surface \; x_2 = +d \; ; \; p^{af} = \sigma_{22}^{s} \; at \; bottom \; surface \; x_2 = -d \quad (5.69)$$

Shear stress at the interfaces:

$$\sigma_{21}^{s} = 0 \; at \; top \; surface \; x_2 = +d \; \sigma_{21}^{s} = 0 \; at \; bottom \; surface \; x_2 = -d \quad (5.70)$$

where superscript s represents the solid material and f represents the fluid material.

Displacements and stresses in solids can be found using Eqs. (5.27.2), (5.28.2), and (5.28.3). From here onward, reader could derive the Rayleigh-Lamb equations for a guided wave in plates submerged in fluid following the steps described in Sections 5.2.2 and 5.2.3.3.

5.3 GUIDED WAVES PROPAGATION IN ANISOTROPIC PLATES

5.3.1 Analytical Approach for Single-Layered General Anisotropic Plate

Guided wave propagation in generalized anisotropic media is by far the most complex problem in wave propagation that researchers attempted to solve for decades. The very first reason for this complexity arises from the fact that the fundamental bulk wave modes (qL, qFS, qSS modes discussed in Section 4.4 in Chapter 4) in anisotropic media are all coupled. In isotropic plate, the SH wave is decoupled from the P and SV waves. However, this is not the case in an anisotropic plate media. This is the very reason that the Helmholtz decomposition described in Eq. (4.38.1) is not feasible in anisotropic media. The potentials ϕ and ψ cannot be uniquely separated. Hence, in 1961, Buchwald proposed three potential functions, say $\Theta_1(x_j)e^{-i\omega t}$, $\Theta_2(x_j)e^{-i\omega t}$ and $\Theta_3(x_j)e^{-i\omega t}$ [49], to describe the three displacement functions in anisotropic media. In isotropic media, the potential functions were directly assumed satisfying the equation of motion. But the Buchwald potentials are unknown in anisotropic media. They are assumed to describe the displacement functions. However, they need to be found out satisfying the fundamental equation of motion written in Eq. (4.91) without the forcing function F_i. The displacement function using the potential functions can be written as

$$\begin{aligned} u_1 &= \left(\frac{\partial \Theta_1}{\partial x_1}\right) e^{-i\omega t} \\ u_2 &= \left(\frac{\partial \Theta_2}{\partial x_2} + \frac{\partial \Theta_3}{\partial x_3}\right) e^{-i\omega t} \\ u_3 &= \left(\frac{\partial \Theta_2}{\partial x_3} - \frac{\partial \Theta_3}{\partial x_2}\right) e^{-i\omega t} \end{aligned} \quad (5.71)$$

Based on the displacement function, the stress equations in a transversely isotropic material can be written as

$$\begin{aligned}
\sigma_{11} &= \mathbb{C}_{11}\frac{\partial u_1}{\partial x_1}+\mathbb{C}_{12}\frac{\partial u_2}{\partial x_2}+\mathbb{C}_{13}\frac{\partial u_3}{\partial x_3}\\
\sigma_{22} &= \mathbb{C}_{12}\frac{\partial u_1}{\partial x_1}+\mathbb{C}_{22}\frac{\partial u_2}{\partial x_2}+\mathbb{C}_{23}\frac{\partial u_3}{\partial x_3}\\
\sigma_{33} &= \mathbb{C}_{13}\frac{\partial u_1}{\partial x_1}+\mathbb{C}_{23}\frac{\partial u_2}{\partial x_2}+\mathbb{C}_{33}\frac{\partial u_3}{\partial x_3}\\
\sigma_{23} &= \mathbb{C}_{44}\left(\frac{\partial u_2}{\partial x_3}+\frac{\partial u_3}{\partial x_2}\right)\\
\sigma_{31} &= \mathbb{C}_{55}\left(\frac{\partial u_3}{\partial x_1}+\frac{\partial u_1}{\partial x_3}\right)\\
\sigma_{12} &= \mathbb{C}_{66}\left(\frac{\partial u_1}{\partial x_2}+\frac{\partial u_2}{\partial x_1}\right)
\end{aligned} \tag{5.72}$$

Convention described in Section 3.9.3 in Chapter 3 and Eq. (3.83) was developed using x_3-axis as an axis of symmetry, whereas here x_1-axis is considered axis of symmetry. Hence, we follow the right convention here $\mathbb{C}_{44} = (\mathbb{C}_{22} - \mathbb{C}_{23})/2$ as it was considered in Eq. (4.102).

Substituting the displacement equations in Eqs. (5.71) and (5.72) and then substituting the stress equations in the governing equation of motion without body force or a forcing function (Eq. (4.25)), three comprehensive equations are derived

$$\begin{aligned}
&\frac{\partial}{\partial x_1}\Big[\mathbb{C}_{11}\Theta_{1,11}-\mathbb{C}_{12}\left(\Theta_{2,22}+\Theta_{3,32}+\Theta_{2,33}-\Theta_{3,23}\right)\\
&\quad+\mathbb{C}_{55}\left(\Theta_{1,22}+\Theta_{2,22}+\Theta_{3,32}+\Theta_{1,33}+\Theta_{2,33}-\Theta_{3,23}\right)+\rho\omega^2\Theta_1\Big]=0
\end{aligned} \tag{5.73.1}$$

$$\begin{aligned}
&\frac{\partial}{\partial x_2}\Big[\mathbb{C}_{12}\Theta_{1,11}+\mathbb{C}_{22}(\Theta_{2,22}+\Theta_{2,33})+\mathbb{C}_{55}\left(\Theta_{1,11}+\Theta_{2,11}\right)+\rho\omega^2\Theta_2\Big]\\
&\quad+\frac{\partial}{\partial x_3}\Big[\mathbb{C}_{55}\Theta_{3,11}+\mathbb{C}_{44}\left(\Theta_{3,22}+\Theta_{2,33}\right)+\rho\omega^2\Theta_3\Big]=0
\end{aligned} \tag{5.73.2}$$

$$\begin{aligned}
&-\frac{\partial}{\partial x_2}\Big[\mathbb{C}_{44}\left(\Theta_{3,33}+\Theta_{3,22}\right)+\mathbb{C}_{55}\left(\Theta_{3,11}\right)+\rho\omega^2\Theta_3\Big]\\
&\quad+\frac{\partial}{\partial x_3}\Big[\mathbb{C}_{12}\Theta_{1,11}+\mathbb{C}_{55}\left(\Theta_{1,11}+\Theta_{2,11}\right)+\rho\omega^2\Theta_2\Big]=0
\end{aligned} \tag{5.73.3}$$

After several steps of mathematical juggleries, one can get three independent and sufficient equations (each one obtained from each directional derivatives along the three directions x_1, x_2, and x_3) that will satisfy the equation of motion using the potentials $\Theta_1(x_j)e^{-i\omega t}$, $\Theta_2(x_j)e^{-i\omega t}$, and $\Theta_3(x_j)e^{-i\omega t}$. Hence, to solve the potentials, the three equations can be written as follows:

$$\Lambda_5\left[\frac{\partial^2\Theta_1}{\partial x_2^2}+\frac{\partial^2\Theta_1}{\partial x_3^2}\right]+\Lambda_1\frac{\partial^2\Theta_1}{\partial x_1^2}+\omega^2\Theta_1+\Lambda_3\left[\frac{\partial^2\Theta_2}{\partial x_2^2}+\frac{\partial^2\Theta_2}{\partial x_3^2}\right]=0$$

$$\Lambda_3\frac{\partial^2\Theta_1}{\partial x_1^2}+\Lambda_2\left[\frac{\partial^2\Theta_2}{\partial x_2^2}+\frac{\partial^2\Theta_2}{\partial x_3^2}\right]+\Lambda_5\left[\frac{\partial^2\Theta_2}{\partial x_1^2}\right]+\omega^2\Theta_2=0 \tag{5.74}$$

$$\Lambda_4\left[\frac{\partial^2\Theta_3}{\partial x_2^2}+\frac{\partial^2\Theta_3}{\partial x_3^2}\right]+\Lambda_5\left[\frac{\partial^2\Theta_3}{\partial x_1^2}\right]+\omega^2\Theta_3=0$$

Eq. (5.74) can be written in a matrix form

$$\begin{bmatrix}\left(\Lambda_5(D_2^2+D_3^2)+\Lambda_1D_1^2+\omega^2\right) & \Lambda_3(D_2^2+D_3^2) & 0\\ \Lambda_3D_1^2 & \left(\Lambda_2(D_2^2+D_3^2)+\Lambda_5D_1^2+\omega^2\right) & 0\\ 0 & 0 & \left(\Lambda_4(D_2^2+D_3^2)+\Lambda_5D_1^2+\omega^2\right)\end{bmatrix}$$

$$\times\begin{pmatrix}\Theta_1\\ \Theta_2\\ \Theta_3\end{pmatrix}=\begin{pmatrix}0\\ 0\\ 0\end{pmatrix} \tag{5.75}$$

where

$$\Lambda_1=\mathbb{C}_{11}/\rho$$
$$\Lambda_2=\mathbb{C}_{22}/\rho\ ;$$
$$\Lambda_3=(\mathbb{C}_{12}+\mathbb{C}_{55})/\rho$$
$$\Lambda_4=\mathbb{C}_{44}/\rho$$
$$\Lambda_5=\mathbb{C}_{55}/\rho$$
$$D_i^2=\frac{\partial^2}{\partial x_i^2}$$

Above three homogeneous equations for the potentials $\Theta_1(x_j)e^{-i\omega t}$, $\Theta_2(x_j)e^{-i\omega t}$, and $\Theta_3(x_j)e^{-i\omega t}$ are quite complex, which are very hard to solve. However, Mal et al. [50] proposed a solution of Eq. (5.75) in the following form. The following equations can also be found in Reference [51] with few typographical errors, but are presented in correct form herein.

$$\begin{Bmatrix}\Theta_1\\ \Theta_2\\ \Theta_3\end{Bmatrix}=\begin{bmatrix}K_{11} & K_{12} & 0\\ K_{21} & K_{22} & 0\\ 0 & 0 & 1\end{bmatrix}\begin{bmatrix}C_u^1e^{i\xi_1x_3}+C_d^1e^{-i\xi_1x_3}\\ C_u^2e^{i\xi_2x_3}+C_d^2e^{-i\xi_2x_3}\\ C_u^3e^{i\xi_3x_3}+C_d^3e^{-i\xi_3x_3}\end{bmatrix}e^{i(k_1x_1+k_2x_2)} \tag{5.76}$$

where

$$K_{11} = \Lambda_3 P_+$$
$$K_{12} = \Lambda_3 P_-$$
$$K_{21} = \omega^2 - \Lambda_2 k_1^2 - \Lambda_5 P_+$$
$$K_{22} = \omega^2 - \Lambda_2 k_1^2 - \Lambda_5 P_-$$
$$P_{+,-} = -\left(\frac{\beta}{2\alpha}\right) \mp \sqrt{\left[\left(\frac{\beta}{2\alpha}\right)^2 - \frac{\gamma}{\alpha}\right]}$$
$$\alpha = \Lambda_2 \Lambda_5$$
$$\beta = \left(\Lambda_1 \Lambda_2 + \Lambda_5^2 - \Lambda_3^2\right) k_1^2 - \omega^2 (\Lambda_2 + \Lambda_5)$$
$$\gamma = (\Lambda_1 k_1^2 - \omega^2)(\Lambda_5 k_1^2 - \omega^2)$$
$$\xi_1^2 = -k_2^2 + P_+$$
$$\xi_2^2 = -k_2^2 + P_-$$
$$\xi_3^2 = -k_2^2 + (\omega^2 - \Lambda_5 k_1^2)/\Lambda_4$$

Eq. (5.76) is the solution of the potential functions proposed by Buchwald [52]. Substituting the solution of the potential functions in Eqs. (5.71) and (5.72) will give the displacement and stress fields in the anisotropic plate, respectively.

Next, it is necessary to explain the results using a graphical presentation. Please refer Fig. 5.15, where a plate structure is shown with an incident wave $\mathbf{k}_i$. After incidence a part of the wave is reflected along the same plane of incidence following the similar understanding discussed in the Chapter 4 appendix. Rest of the energy is guided in the plate. It is assumed that the wave vector $\mathbf{k}$ is propagated along the

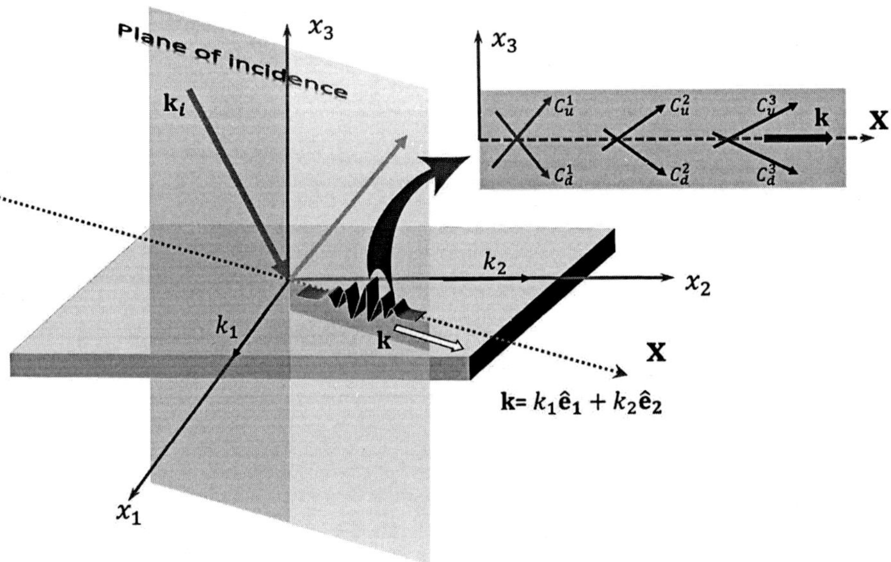

FIGURE 5.15 Guided wave propagation in anisotropic plate.

incident plane saturating the thickness of the plate as shown in Fig. 5.15. The wave vector has two components k_1 and k_2 along x_1- and x_2-axes. The term $e^{i(k_1x_1+k_2x_2)}$ in Eq. (5.76) is to represent the phase part of this guided wave along the **x** direction as shown in Fig. 5.15. However, in the guided wave plane, there are three potentials have both up (u) and down (d) going wave with amplitudes C_u^1, C_d^1, C_u^2, C_d^2, C_u^3, C_d^3, respectively. ξ_1, ξ_2, and ξ_3 are the wave number components of their wave potentials along the vertical x_3-axis.

5.3.2 Analytical Approach for Multilayered General Anisotropic Plate

From the above section, it is almost visible that the formulation with multiple layers will be complex and a daring task. These multiple layers could be made of isotropic layers, combination of isotropic and anisotropic layers of different degrees (refer Chapter 3), or all layers could be made of anisotropic layers with different degrees. In a typical NDE problem with fluid loaded media on either side of a multilayer structure, it is not necessary to know the displacement and stresses at each layer but the resultant sound amplitude and phase are required either on the top layer or on the bottom layer pertaining to the type of NDE problem depicted in Fig. 5.1. Originally, this method was proposed by Thomson and Haskell [53, 54] for solving wave propagation in multilayered geological systems.. This method is called the Transfer Matrix Method (TMM) and researchers used the same to solve ultrasonic NDE problems with multilayered isotropic and anisotropic media by scaling the frequency of interest many orders of magnitude compared to its original seismic frequency of interest. From the physics point of view, problem is similar and easily transferable. In this method, vector of displacements (u_1, u_2, and u_3) and relevant interfacial stresses (σ_{31}, σ_{32}, σ_{33}) at any layer can be written as a product of a propagator matrix and wave amplitude coefficients. Eliminating the wave amplitudes from the equation, displacement and stresses at the top of a layer are expressed by a modified propagator matrix and the displacement and stresses at the bottom of the layer. Such expressions can be written for all layers. By applying the interfacial continuity conditions, the propagator matrices can be combined. This can be continued from the top layer to the bottom layer, and size of the propagator matrix depends on the number of layers in the system. This will connect the displacement and stresses at the top layer with the bottom layer by the combined propagator matrix through transferring the properties. Thus, this method is called transfer matrix method.

Although this method seems quite intriguing and much smaller matrix to be handled for solution at the cost of having no option for knowing the displacement and stresses at the internal layers, TMM suffers from a computational flaw. Physics of such flaw comes from refraction of waves within a layer or multiple layers and internal reflection of the waves internally within a layer causing evanescent wave modes within those layers. These nonpropagating wave modes make the numerical computation harder and are discussed in many books and articles [50]. To solve this issue, Global Matrix Method (GMM) was proposed [55], and extended to anisotropic media by Mal [56]. In GMM, displacement and stress equations for each layer are assembled into a single big matrix called the global matrix. This method is unconditionally stable [51]. Alternatively, Stiffness

Matrix Method (SMM) [57] was also proposed to solve wave propagation in multilayered anisotropic media. Owing to the original concept of expressing two equations with displacements and stresses in a single equation for two consecutive layers SMM has similar approach. In SMM equations for displacements and stresses are separated in two different equations where two consecutive layers are together. This is achieved by rewriting the same equations in different forms. This gives the opportunity to form a stiffness matrix for each layer and the equations for propagator matrix with phase terms are written in much robust form. Detailed discussions on TMM, GMM, and SMM can be found in multiple books and articles [51, 53–58] and are omitted in this book. Only recently unified analytical method (UAM) [59] is proposed assuming straight crested guided wave mode in laminates and the mathematical derivation was made orthogonal axis invariant. In UAM approach, first guided-ultrasonic wave propagating in an arbitrary laminate media is considered, described by its particle displacement, which is the function of the angular frequency ω, wave number $|\mathbf{k}|$, and phase velocity $c = \omega / |\mathbf{k}|$; similarly, it is described in Chapter 4. In UAM method, the direction $\mathbf{x}$ in Fig. 5.15 is aligned along the direction x_1 (i.e., the incident plane is along $x_1 - x_3$ plane). Then only the straight crested guided wave modes were assumed to propagate along the x_1 direction. Further, it is assumed that the system will be under x_2 invariant condition. However, wave in multilayered anisotropic or isotropic media consists of superposition of waves propagating in multiple directions as described in Chapter 4 and in Section 5.3.1. Wave propagating in the $\mathbf{x}$ direction in Fig. 5.15 cannot be always under orthogonal to $\mathbf{x}$ invariant condition. The invariant condition in UAM only works for the infinite straight crested waves. In general, for traditional NDE applications, NDE of composite plates with multiple layers with arbitrary angle of incidence by finite dimension transducers at higher frequencies, assumption of straight crested waves is not always practical or feasible. UAM is selectively relevant for guided wave SHM and useful at lower frequencies, but bulk wave NDE such assumption is not valid. Although UAM [59] fundamentally formulates the Christoffel's equation for a plane-strain lamina, later for multilayer problem, it uses the TMM [53, 54] or GMM [51, 56] approach to construct the main matrix to solve for the eigenvalues and eigenvectors. Readers interested in detailed analytical method to solve wave dispersion in composite materials are recommended to further read and evaluate the TMM [53, 54, 58], GMM [50, 51, 56], SMM [57], and UAM [59] method described in respective references.

5.3.3 Semianalytical Approach for Single- and Multilayered Anisotropic Plates

If one refers books, research articles, and Section 5.3.1 on analytical methods to solve fundamental wave modes in multilayered anisotropic wave guide, it will immediately appear that the solution of guided wave modes in multilayered solid is a daunting task. Hence, many researchers including author Banerjee [60] have pursued a semianalytical approach to understand the guided wave dispersion in multilayered plate-like structure.

All multilayered situations could be handled by this semianalytical approach. By virtue of its name, it is self-explanatory that the semianalytical method uses partially analytical and partially numerical formulations. When the numerical approach is introduced, Finite Element Method (FEM) is most common to use and adopt for a problem. This method is no exception. Hence, finite element discretization is used in this method across the depth of the plate. Thus, this method is called Semi Analytical Finite Element (SAFE) method, or simply the SAFE. In SAFE approach, to model guided waves in composite structures assumes analytical guided wave propagation along the plate and performs a finite element discretization across the multilayer thickness of the plate. The SAFE formulation forms a stable eigenvalue problem for solving wave numbers and mode shapes in a given frequency range. In this book, numerical computation of wave field for NDE problems is discussed. Although there are many drawbacks discussed in Chapter 1, FEM is one of the prominent methods used in wave field modeling. Despite FEM being not discussed in this book explicitly, readers are recommended to refer any FEM text books for basic understanding of FEM, such as Reference [61].

Let us assume a multilayered anisotropic plate shown in Fig. 5.16. In Fig. 5.16, the conventional right-hand coordinate system used in Fig. 5.15 is flipped to define the thickness of the plate along the positive x_3-axis from the top surface of the plate. x_1-axis is the direction of guided wave propagation designated with a wave vector **k** at frequency ω. The cross section of the plate is on the $x_2 - x_3$ plane. Across depth, along x_3 -axis, multinodded (say n noded, where n could be any number between 3 and infinity) one-dimensional (1D) elements that are dedicated for each layer are used to discretize the cross section of the plate as shown in Fig. 5.16 (Fig. 5.16 shows

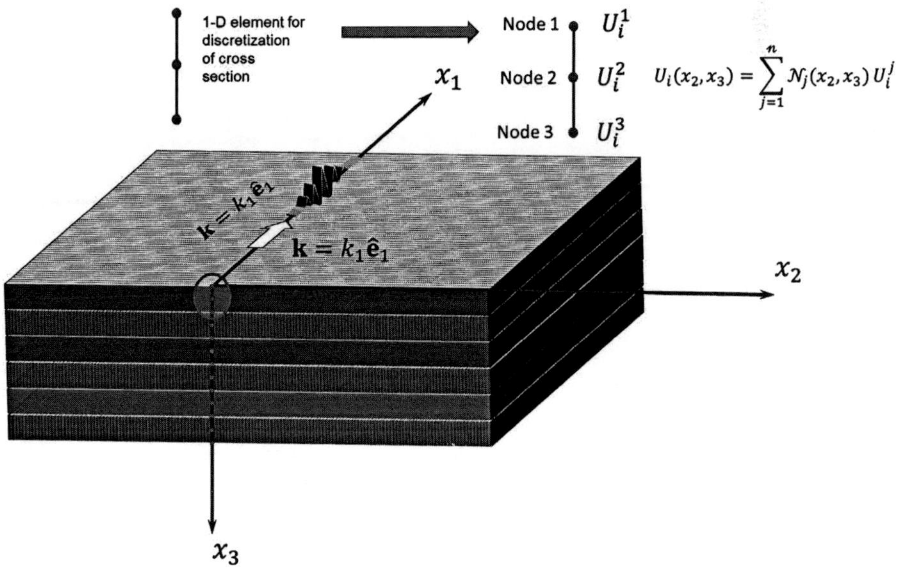

FIGURE 5.16 Semi-analytical modeling approach for simulating wave propagation in composite layered plate.

three nodded element for clarity). The discretization is also to describe the mode shapes of the guided wave modes in the plate. The harmonic displacement (u_i), stress (σ_{ij}), and strain (e_{ij}) field components at any point (x_k) of the waveguide in Cartesian coordinate system are expressed by

$$\mathbf{u} = u_i(x_k) = \begin{bmatrix} u_1 & u_2 & u_3 \end{bmatrix} \tag{5.77}$$

$$\sigma = \sigma_{ij}(x_k) = \begin{bmatrix} \sigma_{11} & \sigma_{22} & \sigma_{33} & \sigma_{23} & \sigma_{31} & \sigma_{12} \end{bmatrix} \tag{5.78}$$

$$\mathrm{e} = \mathrm{e}_{kl}(x_k) = \begin{bmatrix} \mathrm{e}_{11} & \mathrm{e}_{22} & \mathrm{e}_{33} & 2\mathrm{e}_{23} & 2\mathrm{e}_{31} & 2\mathrm{e}_{12} \end{bmatrix} \tag{5.79}$$

where standard index notation described in Chapter 2 is 3 and is valid. Please note that the factor 2 with each shear strain component is added to satisfy the constitutive relation described between Eqs. (3.78) and (3.81). Hence, recalling Eq. (3.75), the constitutive equation of the material at a point is given by

$$\sigma_{ij} = \mathbb{C}_{ijkl}\mathrm{e}_{kl} \tag{5.80}$$

where $\mathbb{C}_{ijkl}$ is the stiffness or constitutive matrix, described in Section 3.9. Recalling Eq. (3.72) the stress-displacement relation at any material point can be written as

$$\mathrm{e}_{ij} = \frac{1}{2}\left(u_{i,j} + u_{j,i}\right) \tag{5.81}$$

However, let us expand Eq. (5.81) like in Appendix in Chapter 3 and express each strain component in Eq. (5.79), with respect to the displacements in Eq. (5.77) at any arbitrary material point as follows:

$$\mathrm{e}_{11} = \frac{\partial u_1}{\partial x_1} \tag{5.82.1}$$

$$\mathrm{e}_{22} = \frac{\partial u_2}{\partial x_2} \tag{5.82.2}$$

$$\mathrm{e}_{33} = \frac{\partial u_3}{\partial x_3} \tag{5.82.3}$$

$$2\mathrm{e}_{23} = \left(\frac{\partial u_2}{\partial x_3} + \frac{\partial u_3}{\partial x_2}\right) \tag{5.82.4}$$

$$2\mathrm{e}_{31} = \left(\frac{\partial u_3}{\partial x_1} + \frac{\partial u_1}{\partial x_3}\right) \tag{5.82.5}$$

$$2\mathrm{e}_{12} = \left(\frac{\partial u_1}{\partial x_2} + \frac{\partial u_2}{\partial x_1}\right) \tag{5.82.6}$$

By scrutinizing Eqs. (5.82.1) and (5.82.6), it is realized that all of them have the derivative terms $\frac{\partial}{\partial x_1}, \frac{\partial}{\partial x_2}, \frac{\partial}{\partial x_3}$ with factors either 1 or 0. For example, in Eq. (5.82.1), $\frac{\partial}{\partial x_1}$ has factor 1, but Eq. (5.82.4) has factor 0 for $\frac{\partial}{\partial x_1}$. Similarly, in Eq. (5.82.6), $\frac{\partial}{\partial x_3}$ has factor 0, but both Eqs. (5.82.4) and (5.82.5) have factor 1 for $\frac{\partial}{\partial x_3}$. Hence, it is wise to construct three matrices as follows and to express the strain filed in a bit different form.

$$\begin{Bmatrix} e_{11} \\ e_{22} \\ e_{33} \\ 2e_{23} \\ 2e_{31} \\ 2e_{12} \end{Bmatrix} = \begin{bmatrix} \frac{\partial}{\partial x_1} & 0 & 0 \\ 0 & 0 & 0 \\ 0 & 0 & 0 \\ 0 & 0 & 0 \\ 0 & 0 & \frac{\partial}{\partial x_1} \\ 0 & \frac{\partial}{\partial x_1} & 0 \end{bmatrix} u_1 + \begin{bmatrix} 0 & 0 & 0 \\ 0 & \frac{\partial}{\partial x_2} & 0 \\ 0 & 0 & 0 \\ 0 & 0 & \frac{\partial}{\partial x_2} \\ 0 & 0 & 0 \\ \frac{\partial}{\partial x_2} & 0 & 0 \end{bmatrix} u_2 + \begin{bmatrix} 0 & 0 & 0 \\ 0 & 0 & 0 \\ 0 & 0 & \frac{\partial}{\partial x_3} \\ 0 & \frac{\partial}{\partial x_3} & 0 \\ \frac{\partial}{\partial x_3} & 0 & 0 \\ 0 & 0 & 0 \end{bmatrix} u_3 \quad (5.83)$$

Simplifying Eq. (5.83), one can write

$$\begin{Bmatrix} e_{11} \\ e_{22} \\ e_{33} \\ 2e_{23} \\ 2e_{31} \\ 2e_{12} \end{Bmatrix} = \begin{bmatrix} 1 & 0 & 0 \\ 0 & 0 & 0 \\ 0 & 0 & 0 \\ 0 & 0 & 0 \\ 0 & 0 & 1 \\ 0 & 1 & 0 \end{bmatrix} \frac{\partial u_1}{\partial x_1} + \begin{bmatrix} 0 & 0 & 0 \\ 0 & 1 & 0 \\ 0 & 0 & 0 \\ 0 & 0 & 1 \\ 0 & 0 & 0 \\ 1 & 0 & 0 \end{bmatrix} \frac{\partial u_2}{\partial x_2} + \begin{bmatrix} 0 & 0 & 0 \\ 0 & 0 & 0 \\ 0 & 0 & 1 \\ 0 & 1 & 0 \\ 1 & 0 & 0 \\ 0 & 0 & 0 \end{bmatrix} \frac{\partial u_3}{\partial x_3} \quad (5.84)$$

and further

$$\boldsymbol{\varepsilon} = \left[\mathbf{L}_1 \frac{\partial}{\partial x_1} \; \mathbf{L}_2 \frac{\partial}{\partial x_2} \; \mathbf{L}_3 \frac{\partial}{\partial x_3} \right] . \mathbf{u}^T \quad (5.85)$$

where T indicates transpose of the vector $\mathbf{u}$, and

$$\mathbf{L}_1 = \begin{bmatrix} 1 & 0 & 0 \\ 0 & 0 & 0 \\ 0 & 0 & 0 \\ 0 & 0 & 0 \\ 0 & 0 & 1 \\ 0 & 1 & 0 \end{bmatrix} \quad \mathbf{L}_2 = \begin{bmatrix} 0 & 0 & 0 \\ 0 & 1 & 0 \\ 0 & 0 & 0 \\ 0 & 0 & 1 \\ 0 & 0 & 0 \\ 1 & 0 & 0 \end{bmatrix} \quad \mathbf{L}_3 = \begin{bmatrix} 0 & 0 & 0 \\ 0 & 0 & 0 \\ 0 & 0 & 1 \\ 0 & 1 & 0 \\ 1 & 0 & 0 \\ 0 & 0 & 0 \end{bmatrix} \quad (5.86)$$

5.3.3.1 Hamilton's Principle and the Governing Equation

Before going into any further detail, readers are recommended to read the section on classical mechanics in Section 3.10.4 in Chapter 3. Wave propagation in a composite plate can be considered as a dynamic system that is evolving over time and must abide by the fundamental physics and principles of a classical system. According to the Hamilton's principle [62, 63], the system must satisfy the $\delta\mathfrak{L} \to 0$, which is the principle of least action in classical mechanics. That means that the system could have evolved only in one way my minimizing the action and the action $\mathfrak{L}$ in minimum. $\mathfrak{L}$ is called Lagrangian of a system, $\mathcal{L} = \mathcal{L}(u_i, \dot{u}_i, u_{i,j})$, as described in Eq. (A4.5). Lagrangian of a system in the light of Section 3.8.3 in Chapter 3 is the difference of the kinetic energy density and the strain energy density ($\mathcal{L} = \mathcal{K} - \mathcal{E}$) as described in Eq. (A4.15). Hence, the Hamilton's principle can be written as

$$\delta\mathfrak{L} = \int_{t_1}^{t_2}\int_{V} \delta(\mathcal{K} - \mathcal{E})\, dx_j\, dt \tag{5.87}$$

where $\mathcal{K}$ is the kinetic energy density and $\mathcal{E}$ is the strain energy density of the system as described in Eq. (A4.16) using index notation. In light of strain and stress vectors at a point described at the beginning of this section, the energy densities can be written in vector form as follows:

$$\mathcal{K} = \frac{1}{2}\dot{\mathbf{u}}\rho\dot{\mathbf{u}}^T \qquad \mathcal{E} = \frac{1}{2}\varepsilon\mathbb{C}\varepsilon^T \tag{5.88}$$

where ρ is the material density, and the dot represents a time derivative. After integrating by parts, Eq. (5.88) can be written as

$$\delta\mathfrak{L} = \int_{t_1}^{t_2}\left[\left[\int_{V} \delta\varepsilon\mathbb{C}(\varepsilon^T)\, dv\right] + \left[\int_{V} \delta\ddot{\mathbf{u}}\rho(\mathbf{u}^T)\, dv\right]\right] dt = 0 \tag{5.89}$$

Eq. (5.89) is the fundamental governing equation, which should be solved using SAFE approach. By now, readers are expected to see a pattern in solving wave propagation in solid and fluid, which should not change even if SAFE is used. That pattern is used in this chapter over and over again. The displacement filed is assumed to be harmonic along the propagation direction k. Similarly, here in this problem, the harmonic displacement function is assumed. However, here it is done in a bit different way for later convenience with discretization. As the thickness direction of the plate belongs to the $x_2 - x_3$ plane, and wave propagation is along the direction x_1, the phase term is specifically assigned for the x_1 direction only and kept a generic two dimensional function for the $x_2 - x_3$ plane as written below.

$$u_i\left(x_j, t\right) = u_i\left(x_1, x_2, x_3, t\right) = U_i(x_2, x_3)e^{i(k_1 x_1 - \omega t)} \tag{5.90}$$

The displacement field over the $x_2 - x_3$ plane, $U_i(x_2, x_3)$ a spatial function can be further expressed using discretized function along the x_3-axis with the values described at the nodal points (refer Fig. 5.16) only.

5.3.3.2 Discretization of Plate Thickness

Across thickness, i.e. along x_3-axis, each lamina of the plate is discretized [60] with multinodded element. In Fig. 5.16, a three nodded element in any arbitrary layer is shown. The $U_i(x_2,x_3)$ function (i.e., $U_1(x_2,x_3)$, $U_2(x_2,x_3)$, and $U_3(x_2,x_3)$ along x_1, x_2, and x_3, respectively) in Eq. (5.90) is described by its nodal values at the node points, e.g., U_i^j at the j-th node and respective discretization shape function $\mathcal{N}_j(x_2,x_3)$ for the j-th node. Hence, displacements at any point on the element can be expressed in terms of the shape functions $\mathcal{N}_j(x_2,x_3)$ and the nodal unknown displacements $U_i^j(x_2,x_3)$. Hence, displacement at any point on the element (e) $(u_i^{(e)}(x_1,x_2,x_3,t))$ is expressed by

$$\mathbf{u}^{(e)}(x_1,x_2,x_3,t)=\begin{Bmatrix} u_1^{(e)}(x_1,x_2,x_3,t) \\ u_2^{(e)}(x_1,x_2,x_3,t) \\ u_3^{(e)}(x_1,x_2,x_3,t) \end{Bmatrix}=\begin{bmatrix} \sum_{j=1}^{n}\mathcal{N}_j(x_2,x_3)U_1^j \\ \sum_{j=1}^{n}\mathcal{N}_j(x_2,x_3)U_2^j \\ \sum_{j=1}^{n}\mathcal{N}_j(x_2,x_3)U_3^j \end{bmatrix}^{(e)} e^{i(k_1x_1-\omega t)}$$
$$=\mathcal{N}(x_2,x_3)\mathbf{q}^{(e)}e^{i(k_1x_1-\omega t)} \tag{5.91}$$

where

$$\mathcal{N}(x_2,x_3)=\begin{bmatrix} \mathcal{N}_1 & 0 & 0 & \mathcal{N}_2 & 0 & 0 & \ddots & 0 & 0 & \mathcal{N}_n & 0 & 0 \\ 0 & \mathcal{N}_1 & 0 & 0 & \mathcal{N}_2 & 0 & 0 & \ddots & 0 & 0 & \mathcal{N}_n & 0 \\ 0 & 0 & \mathcal{N}_1 & 0 & 0 & \mathcal{N}_2 & 0 & 0 & \ddots & 0 & 0 & \mathcal{N}_n \end{bmatrix} \tag{5.92}$$

and

$$\mathbf{q}^{(e)}=\begin{bmatrix} U_1^1 & U_2^1 & U_3^1 & U_1^2 & U_2^2 & U_3^2 & \cdots\cdots\cdots & U_1^n & U_2^n & U_3^n \end{bmatrix} \tag{5.93}$$

5.3.3.3 Element Strain Equation

Substituting the newly found expression for $\mathbf{u}^{(e)}$ written in Eq. (5.91) in the strain equation expressed in Eq. (5.85), strain vector for each element can take the following mathematical shape.

$$\boldsymbol{\varepsilon}^{(e)}=\left[\mathbf{L_1}\frac{\partial}{\partial x_1}\; L_2\frac{\partial}{\partial x_2}\; L_3\frac{\partial}{\partial x_3}\right]\mathcal{N}(x_2,x_3)\mathbf{q}^{(e)}e^{i(k_1x_1-\omega t)} \tag{5.94}$$

Applying differential operator on the shape function matrix $\mathcal{N}(x_2,x_3)$, the strain vector for each element will be

$$\boldsymbol{\varepsilon}^{(e)} = \left[ik_1\mathbf{L}_1\mathcal{N} \quad \mathbf{L}_2\frac{\partial \mathcal{N}}{\partial x_2} \quad \mathbf{L}_3\frac{\partial \mathcal{N}}{\partial x_3} \right]\mathbf{q}^{(e)}e^{i(k_1x_1-\omega t)} \tag{5.95}$$

In some cases, the above matrix in Eq. (5.95) is written as

$$\boldsymbol{\varepsilon}^{(e)} = [ik_1\mathbf{B}_1 \quad \mathbf{B}_2]\mathbf{q}^{(e)}e^{i(k_1x_1-\omega t)} \tag{5.96}$$

where

$$\mathbf{B}_1 = \mathbf{L}_1\mathcal{N}\ ;\ \mathbf{B}_2 = \left[\mathbf{L}_2\frac{\partial \mathcal{N}}{\partial x_2} \quad \mathbf{L}_3\frac{\partial \mathcal{N}}{\partial x_3} \right]; \tag{5.97}$$

5.3.3.4 Governing Wave Equation

Let us assume along the thickness direction of the plate has total N_e numbers of layers. Hence, there are total N_e numbers of n nodded elements in the plate. Assuming $d\mathrm{v}^{(e)}$ being the elemental volume of the $e-th$ element and assembling all the elements in the light of Eq. (5.96), the Hamilton's principle or the governing equation of the system described in Eq. (5.89) can be further modified into

$$\delta\mathfrak{L} = \int_{t_1}^{t_2}\left\{ \bigcup_{e=1}^{N_e}\left[\left[\int_{\mathrm{v}^{(e)}} \delta\boldsymbol{\varepsilon}^{(e)}\mathbb{C}^{(e)}\left(\boldsymbol{\varepsilon}^{(e)T}\right)d\mathrm{v}^{(e)} \right] + \left[\int_{\mathrm{v}^{(e)}} \delta\ddot{\mathbf{u}}^{(e)}\rho^{(e)}\left(\mathbf{u}^{(e)T}\right)d\mathrm{v}^{(e)} \right]\right]\right\}dt = 0 \tag{5.98}$$

where $\mathbb{C}^{(e)}$ and $\rho^{(e)}$ are the constitutive matrix and the density of respective $e-th$ element. In this case, where multilayered plate is analyzed, material properties for each layer or each element (i.e., the $e-th$ element) must be separately specified. The symbol $\bigcup$ symbolizes that the elemental equations for each element are required to be assembled in a bigger matrix by matching the interfacial conditions. The assembly procedure is not discussed in this chapter and basic understanding of FEM [61] is assumed throughout this book. However, while presenting Spectral Element Method (SEM) in Chapter 10, the assembly procedure is briefly discussed.

Further substituting the strain vector in Eq. (5.96), the strain energy part of the governing equation takes the following form. Please note that being a nonzero term, the exponential phase and time harmonic part is removed without the loss of the generality from the equation intentionally.

$$\left[\int_{v^{(e)}} \delta\varepsilon^{(e)}\mathbb{C}^{(e)}\left(\varepsilon^{(e)T}\right)dv^{(e)}\right]$$

$$= \delta\mathbf{q}^{(e)}\int_{v^{(e)}}[ik_1\mathbf{B}_1\ \mathbf{B}_2]\mathbb{C}^{(e)}\begin{bmatrix} ik_1\mathbf{B}_1{}^T \\ \mathbf{B}_2{}^T \end{bmatrix}dv^{(e)}\mathbf{q}^{(e)T}\delta\mathbf{q}^{(e)}\int_{v^{(e)}}\left[k_1{}^2\mathbf{B}_1\mathbb{C}^{(e)}\mathbf{B}_1^T\right.$$
$$\left.+ik_1\mathbf{B}_2^T\mathbb{C}^{(e)}\mathbf{B}_1 - ik_1\mathbf{B}_1^T\mathbb{C}^{(e)}\mathbf{B}_2 + \mathbf{B}_2\mathbb{C}^{(e)}\mathbf{B}_2^T\right]dv^{(e)}\mathbf{q}^{(e)T} \tag{5.99}$$

Similarly, substituting the displacement vector in Eq. (5.91), the kinetic energy part of the governing equation takes the following form. Please note that like Eq. (5.99), being a nonzero term, the exponential phase and time harmonic part is removed from the equation intentionally, without the loss of the generality.

$$\left[\int_{v^{(e)}} \delta\ddot{\mathbf{u}}^{(e)}\rho^{(e)}\left(\mathbf{u}^{(e)T}\right)dv^{(e)}\right] = -\omega^2\delta\mathbf{q}^{(e)}\int_{v^{(e)}}\mathcal{N}\rho^{(e)}\mathcal{N}^T\,dv^{(e)}\mathbf{q}^{(e)T} \tag{5.100}$$

Eq. (5.98), the governing equation of the system, has two parts and those two parts are now expanded in Eqs. (5.99) and (5.100). The discretized governing equation while satisfying the Hamilton's principle can be further obtained by substituting Eqs. (5.99) and (5.100) in Eq. (5.98). Collecting the terms with integrals and rearranging the equation, governing equation will be

$$\int_{t_1}^{t_2}\left\{\bigcup_{e=1}^{N_e}\delta\mathbf{q}^{(e)}\left[k_1{}^2\mathrm{S}_1^{(e)} + ik_1\mathrm{S}_3^{(e)} + \mathrm{S}_2^{(e)} - \omega^2 m^{(e)}\right]\mathbf{q}^{(e)T}\right\}dt = 0 \tag{5.101}$$

where elemental expressions of the newly introduced terms are

$$\mathrm{S}_1^{(e)} = \int_{v^{(e)}}\left[\mathbf{B}_1\mathbb{C}^{(e)}\mathbf{B}_1^T\right]dv^{(e)}$$

$$\mathrm{S}_2^{(e)} = \int_{v^{(e)}}\left[\ \mathbf{B}_2\mathbb{C}^{(e)}\mathbf{B}_2^T\right]dv^{(e)}$$

$$\mathrm{S}_3^{(e)} = \int_{v^{(e)}}\left[\mathbf{B}_2^T\mathbb{C}^{(e)}\mathbf{B}_1 - \mathbf{B}_1^T\mathbb{C}^{(e)}\mathbf{B}_2\right]dv^{(e)}$$

$$m^{(e)} = \int_{v^{(e)}}\mathcal{N}\rho^{(e)}\mathcal{N}^T dv^{(e)}$$

Next applying the assembling procedure and joining the nodes of two adjacent elements, the above equation will become

$$\int_{t_1}^{t_2} \left\{ \delta \mathbf{U} \left[k_1{}^2 \mathbf{K}_1 + ik_1 \mathbf{K}_3 + \mathbf{K}_2 - \omega^2 \mathbf{M} \right] \mathbf{U}^T \right\} dt = 0 \tag{5.102}$$

where **U** is the global displacement vector of unknown nodal displacements across the depth of the plate. Like $\mathbf{q}^{(e)}$ written in Eq. (5.93) consists of the nodal displacements from each n nodded element, the **U** consists of nodal displacements for all the nodes in assembled system, where nodes at the element interface are shared and has only one unknown value assigned to each displacement (u_1, u_2, u_3). With this understanding, the new terms introduced in Eq. (5.102) cane be expressed by

$$\mathbf{K}_1 = \bigcup_{e=1}^{N_e} \mathrm{S}_1^{(e)} \quad \mathbf{K}_2 = \bigcup_{e=1}^{N_e} \mathrm{S}_2^{(e)}, \quad \mathbf{K}_3 = \bigcup_{e=1}^{N_e} \mathrm{S}_3^{(e)}, \quad \mathbf{M} = \bigcup_{e=1}^{N_e} m^{(e)} \tag{5.103}$$

From Eq. (5.102), it is easily realized that to satisfy the Hamilton's principle, the integral should be equal to zero, and moreover, the integrand must be equal to zero. This should be valid for any arbitrary virtual displacement $\delta\mathbf{U}$. Hence, the governing equation of the system with wave propagating along x_1-axis follows the following homogeneous equation

$$\left[k_1{}^2 \mathbf{K}_1 + ik_1 \mathbf{K}_3 + \mathbf{K}_2 - \omega^2 \mathbf{M} \right] \mathbf{U} = 0 \tag{5.104}$$

Further, Eq. (5.104) can be written as follows [64]:

$$\left(\begin{bmatrix} 0 & \mathbf{K}_2 - \omega^2 M \\ \mathbf{K}_2 - \omega^2 \mathbf{M} & i\mathbf{K}_3 \end{bmatrix} - k_1 \begin{bmatrix} \mathbf{K}_2 - \omega^2 \mathbf{M} & 0 \\ 0 & -\mathbf{K}_1 \end{bmatrix} \right) \begin{bmatrix} \mathbf{U} \\ k_1 \mathbf{U} \end{bmatrix} = 0 \tag{5.105}$$

5.3.3.5 Eigen Value Problem: Wave Dispersion Solution and Phase Velocity

From the above expression in Eq. (5.105), it can be immediately realized that the governing equation has now become an eigenvalue problem. Here, the eigen values are also solved using a root-finding approach at a given frequency. First, lower and upper bounds of the frequency axis are selected and the axis is finely discretized to input different frequency values to the system to find the roots. If the dimension of vector **U** is assumed to be D ($D = nN_e - (L-1)$, where L is the number of layers in the plate), then at each frequency ω, $2D$ eigenvalues and $2D$ eigenvectors are obtained. The eigen values are obtained as pair of wave numbers that are both positive and negative ($\pm k_1 = \pm(k_1^r + ik_1^i)$, where k_1^r is the real part of the wave number and k_1^i is the imaginary part of the wave number), representing wave propagating along positive or negative x_1-axis as shown in Fig. 5.16. The wave numbers could be only real, pure, imaginary, and complex conjugates. Real wave numbers are the ones that represent the real propagating wave modes with no decaying term associated.

Complex conjugates always have both real and imaginary parts, which signifies propagating wave modes with decaying amplitude. This could be explained as follows: In the exponential term in Eq. (5.91), ik_1x_1 part contributes to the phase part of the propagating wave mode. When an eigen value for the wave number is found to be $(k_1 = (k_1^r + ik_1^i)$, ik_1x_1 will result in $ik_1x_1 = (ik_1^r x_1 - k_1^i x_1)$, where the imaginary oscillatory part is contributed by the k_1^r real part of the wave number and the exponentially decaying part is contributed by the $-k_1^i$ imaginary part of the wave number. Hence, complex conjugate wave numbers are representative of damped propagative waves decaying in the $\pm x_1$ directions. With similar explanation, if the eigen wave numbers are found to be pure imaginary, then they will only contribute to the decaying part of the wave mode and will not propagate; such wave modes are called evanescent wave modes. Based on our previous discussion in Chapter 4 and beginning of Chapter 5, the phase velocity of the propagating wave modes can be found as

$$c = \frac{\omega}{k_1^r} \tag{5.106}$$

5.3.3.6 Dispersion Behavior

Here in this section, few sample dispersion plots are presented for multilayered plates. Fig. 5.17 shows the typical dispersion curves in a (a) eight-layered transversely isotropic composite with all 0° layers, (b) $(0/90/0/90)_s$ eight-layered composite plate, (c) $(45/\text{-}45/0/90)_s$ composite laminates, and (d) $(0/0/90/90)_s$ composite laminates with the material properties for transversely isotropic material presented in Eq. (4.102). In order to substitute right material properties for 0° and 90°, the axis of symmetry was changed from x_1-axis to x_2-axis, which became their dominant fiber direction, respectively. All the dispersion curves presented are for the propagation direction along the x_1-axis as shown in Fig. 5.16. Please note that with the change in the direction of propagation of the wave (**k** vector) the dispersion behavior changes. The dispersion curves in Fig. 5.17 are obtained using LAMSS-Composites software developed by Laboratory for Active Materials and Smart Structures (LAMSS) laboratory at the University of South Carolina, Columbia, SC, USA, developed under the project funded by NASA Langley Research Center, contract no. NNL15AA16C.

5.3.3.7 Group Velocity of Propagating Wave Modes

As discussed before in Chapter 4, the group velocity of the propagating wave modes can be found by taking derivatives of the frequency-wave number dispersion relations. Conventionally, group velocity is found based on the differences calculated in the wave number values at two adjacent points on the same mode but at two different frequencies. For example, from a dispersion solution, let us take two adjacent points, p and q, from a same mode. The difference in frequency and wave numbers is, say $\omega_q - \omega_p$ and $k_1{}^q - k_1{}^p$. Then the group velocity is $c_g = \omega_q - \omega_p / k_1{}^q - k_1{}^p$. Thus, it is apparent that the accuracy of the group velocity calculation heavily depends on the frequency discretization or the frequency resolution of the dispersion solution. Hence, to solve the eigen values from Eq. (105), the frequency axis must be finely discretized. An alternative method to find the group velocity was proposed by

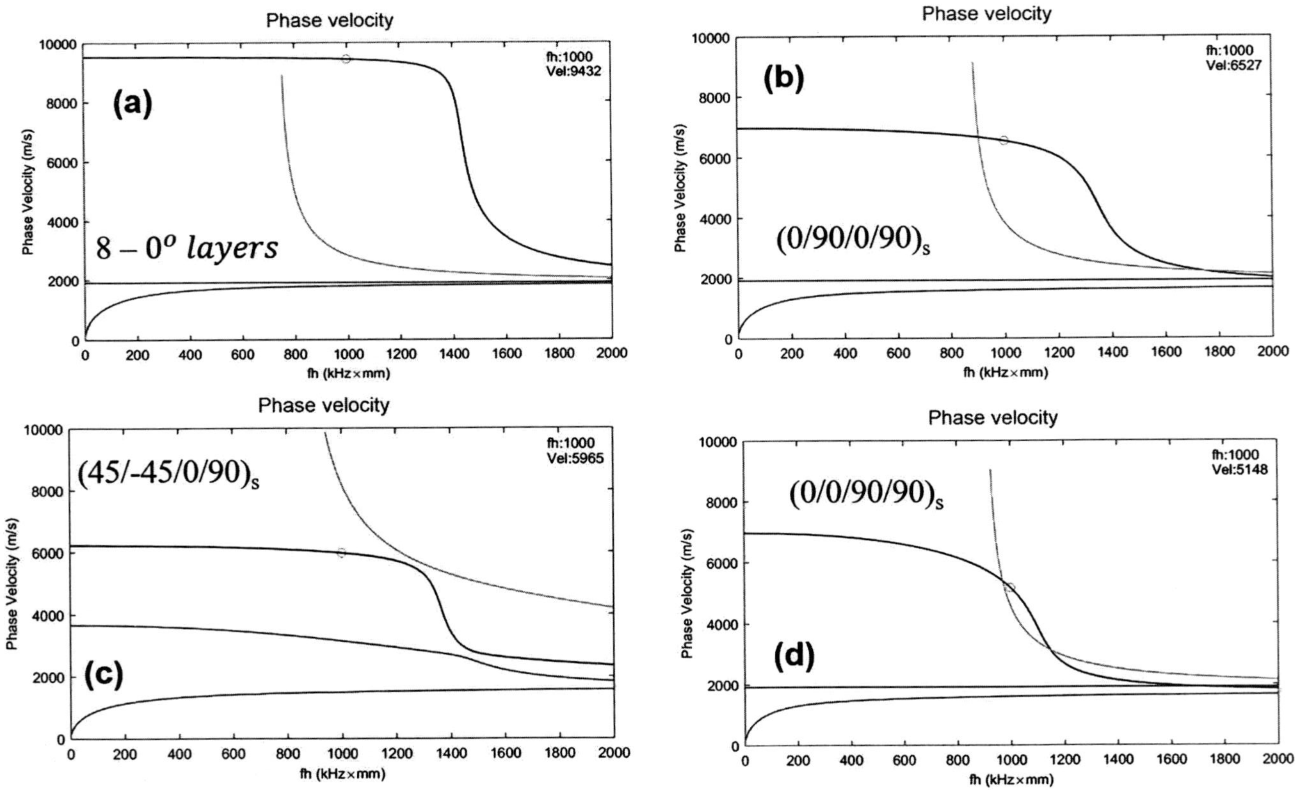

FIGURE 5.17 Typical dispersion curves in a) 8 layered transversely isotropic composite with all 0° layers, b) $(0/90/0/90)_s$ 8 layered composite plate, c) $(45/-45/0/90)_s$ composite laminates, d) $(0/0/90/90)_s$ composite laminates with the material properties for transversely isotropic material presented in Eq. (4.102).

Svante Finnveden. [65]. By virtue of this new method, without any contribution from adjacent points, the group velocity can be directly calculated at each (k_1, ω) point anywhere in the frequency-wave number solution space. The approach starts from the governing equation written in Eq. (5.104). Derivative of Eq. (5.104) with respect to the wave number along the propagation direction can be written as

$$\frac{\partial}{\partial x_1}\left(\left[k_1^2\mathbf{K}_1 + ik_1\mathbf{K}_3 + \mathbf{K}_2 - \omega^2\mathbf{M}\right]\mathbf{U}\right) = 0 \tag{5.107}$$

Upon simplification and rearrangement, the governing equation is

$$\left[2k_1\mathbf{K}_1 + i\mathbf{K}_3 - 2\omega\frac{\partial\omega}{\partial x_1}\mathbf{M}\right]\mathbf{U} + \left[k_1^2\mathbf{K}_1 + ik_1\mathbf{K}_3 + \mathbf{K}_2 - \omega^2\mathbf{M}\right]\frac{\partial\mathbf{U}}{\partial x_1} = 0 \tag{5.108}$$

Next, multiplying Eq. (5.108) by the transpose of the left eigenvectors $\mathbf{U}_L^T$, one can write

$$\mathbf{U}_L^T\left[2k_1\mathbf{K}_1 + i\mathbf{K}_3 - 2\omega\frac{\partial\omega}{\partial x_1}\mathbf{M}\right]\mathbf{U}_R + U_L^T\left[k_1^2\mathbf{K}_1 + ik_1\mathbf{K}_3 + \mathbf{K}_2 - \omega^2\mathbf{M}\right]\frac{\partial\mathbf{U}}{\partial x_1} = 0 \tag{5.109}$$

From Eq. (5.104), $k_1^2\mathbf{K}_1 + ik_1\mathbf{K}_3 + \mathbf{K}_2 - \omega^2\mathbf{M} = 0$, left with only one choice to satisfy the following equation

$$\mathbf{U}_L^T\left[2k_1\mathbf{K}_1 + i\mathbf{K}_3 - 2\omega\frac{\partial\omega}{\partial x_1}\mathbf{M}\right]\mathbf{U}_R = 0 \tag{5.110}$$

In Eq. (5.110), the derivative term $\frac{\partial\omega}{\partial x_1}$ is a scalar quantity. Hence, the group velocity can be found as

$$c_g = \frac{\partial\omega}{\partial x_1} = \frac{\mathbf{U}_L^T(2k_1\mathbf{K}_1 + i\mathbf{K}_3)\mathbf{U}_R}{2\omega U_L^T\mathbf{M}U_R} \tag{5.111}$$

Previously in Chapter 4, how the group velocity direction can be found from the known phase velocity profile (in fact from slowness profile) is discussed in detail. Readers are recommended to follow the discussion on finding the steering angle direction in Section 4.5.4 in Chapter 4.

5.4 GUIDED WAVE PROPAGATION IN CYLINDRICAL RODS AND PIPES

Here, in this section, basic mathematical derivation for the guided wave mode solution in cylindrical structures rod or pipes is presented briefly. Detailed discussion of wave propagation in cylindrical structures can be found in Reference [66]. As a rod or a pipe cylindrical in nature, a cylindrical coordinate system should be used. A typical pipe structure is shown in Fig. 5.18 and respective coordinate systems are depicted considering coordinate transformation rules presented in Fig. A.3.1

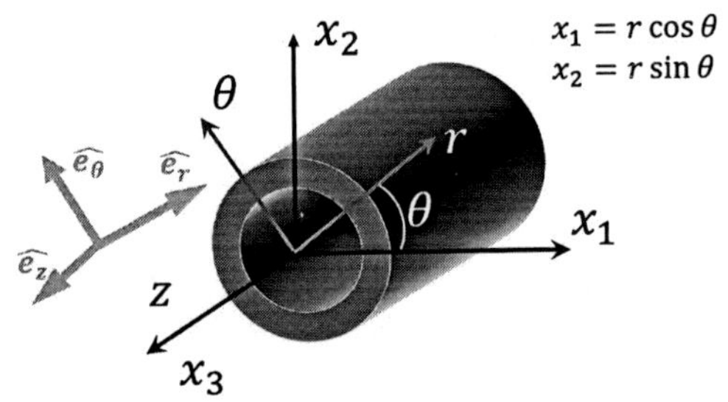

FIGURE 5.18 Pipe structure for analysis of wave propagation in cylindrical coordinate system.

in Chapter 3. The new coordinate system is represented by r, θ, and z. Using transformed displacement and stresses as presented in Section 3.10.2 in cylindrical coordinate system, the governing differential equation of motion in any cylindrical structure can be written as [4, 5]

$$\begin{aligned}
\nabla^2 u_r - \frac{u_\theta}{r^2} - \frac{2}{r^2}\frac{\partial u_\theta}{\partial \theta} + \frac{1}{1-2\upsilon}\frac{\partial \Delta}{\partial r} &= \frac{1}{c_T^2}\frac{\partial^2 u_r}{\partial t^2} \\
\nabla^2 u_\theta - \frac{u_\theta}{r^2} + \frac{2}{r^2}\frac{\partial u_r}{\partial \theta} + \frac{1}{1-2\upsilon}\frac{\partial \Delta}{\partial \theta} &= \frac{1}{c_T^2}\frac{\partial^2 u_\theta}{\partial t^2} \\
\nabla^2 u_z + \frac{1}{1-2\upsilon}\frac{\partial \Delta}{\partial z} &= \frac{1}{c_T^2}\frac{\partial^2 u_z}{\partial t^2}
\end{aligned} \tag{5.112}$$

where c_T is the wave velocity and υ is the Poisson's ratio

$$\begin{aligned}
\nabla^2 &= \frac{\partial}{\partial r^2} + \frac{1}{r}\frac{\partial}{\partial r} + \frac{1}{r^2}\frac{\partial^2}{\partial \theta^2} + \frac{\partial^2}{\partial z^2} \\
\Delta &= \frac{\partial u_r}{\partial r} + \frac{1}{r}\left(\frac{\partial u_\theta}{\partial r} + u_r\right) + \frac{\partial u_z}{\partial z}
\end{aligned} \tag{5.113}$$

where Δ is called dilatation [4, 5].

Like it is presented for plate structures, according to the Stokes-Helmholtz decomposition, the displacement vector can be defined in terms of scalar and a vector potential function ϕ and $\psi(\psi_r, \psi_\theta, \psi_z)$. After substituting the potentials in Eq. (4.38.1) in cylindrical coordinate system, one can get the following displacement equations as also written in Reference [66]

$$\begin{aligned}
u_r &= \frac{\partial \phi}{\partial r} + \frac{1}{r}\frac{\partial \psi_z}{\partial \theta} - \frac{\partial \psi_\theta}{\partial z} \\
u_\theta &= \frac{1}{r}\frac{\partial \phi}{\partial \theta} + \frac{\partial \psi_r}{\partial z} - \frac{\partial \psi_z}{\partial r} \\
u_z &= \frac{\partial \phi}{\partial z} + \frac{1}{r}\frac{\partial (r\psi_\theta)}{\partial r} - \frac{1}{r}\frac{\partial \psi_r}{\partial \theta}
\end{aligned} \tag{5.114}$$

Next, the displacement functions written in Eq. (5.114) are substituted into Eq. (5.112). The displacement equations of motion are identically satisfied if the potential functions satisfy the following equations

$$\begin{aligned}
\nabla^2\phi &= \frac{1}{c_L^2}\frac{\partial^2\phi}{\partial t^2} \\
\nabla^2\psi_r - \frac{\psi_r}{r^2} - \frac{2}{r^2}\frac{\partial\psi_\theta}{\partial\theta} &= \frac{1}{c_T^2}\frac{\partial^2\psi_r}{\partial t^2} \\
\nabla^2\psi_\theta - \frac{\psi_\theta}{r^2} + \frac{2}{r^2}\frac{\partial\psi_r}{\partial\theta} &= \frac{1}{c_T^2}\frac{\partial^2\psi_\theta}{\partial t^2} \\
\nabla^2\psi_z &= \frac{1}{c_T^2}\frac{\partial^2\psi_z}{\partial t^2}
\end{aligned} \tag{5.115}$$

where c_L and c_T are the longitudinal and shear wave velocities. For guided waves in cylindrical structure, the wave propagation in the cylinder must be solved for wave propagating along the z direction as shown in Fig. 5.18. Dispersion equations for axially propagating wave in cylinders are quite complex. Using the time harmonic exponential term for the wave propagating along z direction as $e^{i(k_z z-\omega t)}$, through intuitive separation of variables, the possible potential function that satisfies Eq. (5.115) can be written as [4]

$$\begin{aligned}
\phi &= \Phi(r)\cos(n\theta+\theta_0)e^{i(k_z z-\omega t)} \\
\psi_r &= \Psi_r(r)\sin(n\theta+\theta_0)e^{i(k_z z-\omega t)} \\
\psi_\theta &= \Psi_\theta(r)\cos(n\theta+\theta_0)e^{i(k_z z-\omega t)} \\
\psi_z &= \Psi_z(r)\sin(n\theta+\theta_0)e^{i(k_z z-\omega t)}
\end{aligned} \tag{5.116}$$

where an arbitrary constant θ_o is introduced and n could be 0, 1, 2, 3, ... This expression is written because the potential functions must be periodic in the circumferential direction for wave propagating along the z direction with wave number k_3.This requirement comes from the nature of the θ dependence for longitudinal, torsional, and flexural modes. This dependence regulates the potential function and helps to discard one of the periodic functions ($\cos(n\theta+\theta_0)$ or $\sin(n\theta+\theta_0)$) both of which appear in the solution [5]. Nevertheless $\Phi(r)$, $\Psi_r(r)$, $\Psi_\theta(r)$, and $\Psi_z(r)$ are the radial functions, which are unknown to be solved. There are four radial functions associated with the components of displacements; however, there are only three equations available. The property of gauge invariance discussed in Chapter 4 can now be used to eliminate the constant without the loss of generality, and we can set

$$\Psi_\theta(r) = -\Psi_r(r) \tag{5.117}$$

By substituting Eq. (5.116) into Eq. (5.115) in conjugation with Eq. (5.117), the following ordinary differential equations are obtained:

$$\frac{d^2\Phi}{dr^2}+\frac{1}{r}\frac{d\Phi}{dr}+\left(\eta^2-\frac{n^2}{r^2}\right)\Phi=0$$
$$\frac{d^2\Psi_z}{dr^2}+\frac{1}{r}\frac{d\Psi_z}{dr}+\left(\beta^2-\frac{n^2}{r^2}\right)\Psi_z=0 \quad (5.118)$$
$$\frac{d^2\Psi_r}{dr^2}+\frac{1}{r}\frac{d\Psi_r}{dr}+\left(\beta^2-\frac{(n+1)^2}{r^2}\right)\Psi_r=0$$

where

$\eta=\sqrt{\frac{\omega^2}{c_L^2}-k_z^2}$, $\beta=\sqrt{\frac{\omega^2}{c_T^2}-k_z^2}$ are the wave numbers for the L wave and T waves, respectively.

$c_z=\omega/k_z$ is the phase velocity along the z direction. The set of equations in Eq. (5.118) are the form of Bessel equations [5] of order n for Φ and Ψ_z but of order $m=n+1$ for Ψ_r. Hence, their solutions are the Bessel functions

$$\Phi(r)=A_1J_n(\eta r)+A_2Y_n(\eta r)$$
$$\Psi_z(r)=C_1J_{mn}(\beta r)+C_2Y_n(\beta r) \quad (5.119)$$
$$\Psi_r(r)=B_1J_{n+1}(\beta r)+B_2Y_{n+1}(\beta r)$$

where there are six arbitrary constants (A_1, A_2, B_1, B_2, C_1, C_2) to be solved using boundary conditions. Substituting Eq. (5.119) into Eq. (5.116), the potential functions can be written as

$$\phi=\left[A_1J_n(\eta r)+A_2Y_n(\eta r)\right]\cos(n\theta+\theta_0)e^{i(k_zz-\omega t)}$$
$$\psi_r=\left[B_1J_{n+1}(\beta r)+B_2Y_{n+1}(\beta r)\right]\sin(n\theta+\theta_0)e^{i(k_zz-\omega t)}$$
$$\psi_\theta=-\left[B_1J_{n+1}(\beta r)+B_2Y_{n+1}(\beta r)\right]\cos(n\theta+\theta_0)e^{i(k_zz-\omega t)} \quad (5.120)$$
$$\psi_z=\left[C_1J_n(\beta r)+C_2Y_n(\beta r)\right]\sin(n\theta+\theta_0)e^{i(k_zz-\omega t)}$$

Please note that $\Psi_\theta(r)=-\Psi_r(r)$.

Let us consider a solid isotropic cylinder of a circular cross-section of radius a. At the boundary $r=a$, the surface stresses, i.e., σ_{rr}, $\sigma_{r\theta}$, and σ_{rz} are zero. Again, when the cylinder is not hollow, the wave motion must be finite at $r=0$. Hence, to avoid singularity, the coefficients of the Bessel function of the second kind in the general solution must be zero. Thus, $Y_n(\eta r)$, $Y_n(\beta r)$, and $Y_{n+1}(\beta r)$ could be discarded from the solution in Eq. (5.119).

$$\begin{aligned}
\phi &= A_1 J_n(\eta r)\cos(n\theta+\theta_0)e^{i(k_z z-\omega t)} \\
\psi_r &= B_1 J_{n+1}(\beta r)\sin(n\theta+\theta_0)e^{i(k_z z-\omega t)} \\
\psi_\theta &= -B_1 J_{n+1}(\beta r)\cos(n\theta+\theta_0)e^{i(k_z z-\omega t)} \\
\psi_z &= C_1 J_n(\beta r)\sin(n\theta+\theta_0)e^{i(k_z z-\omega t)}
\end{aligned} \tag{5.121}$$

As the material is isotropic and homogeneous, the stress expressions in cylindrical coordinate system to apply the boundary condition can be written as

$$\sigma_{rr} = \mathbb{L}\nabla^2\phi + 2\mathrm{m}\left(\frac{\partial^2\phi}{\partial r^2} - \frac{\partial^2\psi_\theta}{\partial r\,\partial z} - \frac{1}{r^2}\frac{\partial\psi_z}{\partial\theta} + \frac{1}{r}\frac{\partial^2\psi_z}{\partial r\,\partial\theta}\right)$$

$$\sigma_{r\theta} = \mathrm{m}\left(\frac{2}{r}\frac{\partial^2\phi}{\partial r\,\partial\theta} - \frac{2}{r^2}\frac{\partial\phi}{\partial\theta} - \frac{1}{r}\frac{\partial\psi_r}{\partial z} + \frac{\partial^2\psi_r}{\partial r\,\partial z} + \frac{1}{r}\frac{\partial^2\psi_\theta}{\partial\theta\,\partial z} + \frac{1}{r^2}\frac{\partial^2\psi_z}{\partial\theta^2} + \frac{1}{r}\frac{\partial\psi_z}{\partial r} - \frac{\partial^2\psi_z}{\partial r^2}\right)$$

$$\sigma_{rz} = \mathrm{m}\left(-\frac{1}{r^2}\psi_\theta - \frac{\partial^2\psi_\theta}{\partial z^2} + \frac{1}{r^2}\frac{\partial\psi_r}{\partial\theta} + \frac{1}{r}\frac{\partial^2\psi_z}{\partial\theta\,\partial z} + \frac{1}{r}\frac{\partial\psi_\theta}{\partial r} + 2\frac{\partial^2\phi}{\partial z\,\partial r} - \frac{1}{r}\frac{\partial^2\psi_r}{\partial r\,\partial\theta} + \frac{\partial^2\psi_\theta}{\partial r^2}\right)$$

$$\sigma_{zz} = \mathbb{L}\nabla^2\phi + 2\mathrm{m}\left(\frac{\partial^2\phi}{\partial z^2} + \frac{1}{r}\frac{\partial^2\psi_r}{\partial\theta\,\partial z} + \frac{1}{r}\frac{\partial\psi_\theta}{\partial z} + \frac{\partial^2\psi_\theta}{\partial r\,\partial z}\right)$$

$$\sigma_{\theta z} = \mathrm{m}\left(\frac{\partial^2\psi_r}{\partial z^2} + \frac{1}{r^2}\frac{\partial\psi_\theta}{\partial\theta} + \frac{2}{r}\frac{\partial^2\phi}{\partial z\,\partial\theta} - \frac{1}{r^2}\frac{\partial^2\psi_r}{\partial\theta^2} + \frac{\partial^2\psi_z}{\partial r\,\partial z} + \frac{1}{r}\frac{\partial^2\psi_\theta}{\partial r\,\partial\theta}\right)$$

$$\sigma_{\theta\theta} = \mathbb{L}\nabla^2\phi + 2\mathrm{m}\left(\frac{1}{r}\frac{\partial\phi}{\partial r} + \frac{1}{r^2}\frac{\partial^2\phi}{\partial\theta^2} + \frac{1}{r}\frac{\partial^2\psi_r}{\partial\theta\,\partial z} - \frac{1}{r}\frac{\partial\psi_\theta}{\partial z} + \frac{1}{r^2}\frac{\partial\psi_z}{\partial\theta} - \frac{1}{r}\frac{\partial^2\psi_z}{\partial r\,\partial\theta}\right) \tag{5.122}$$

To apply the boundary conditions, exact expressions for these stresses are necessary. Substituting the potential functions (Eq. (5.121)) in the stress equations written in Eq. (5.122) can be expressed as follows.

The stress expressions are written below in the order as it is expressed in Eq. (5.122)

$$\sigma_{rr} = \cos(n\theta)e^{ik_z z}\left[\begin{aligned}
&A_1\left[\left(-\mathbb{L}\left(\eta^2+k_z^2\right) - 2\mathrm{m}\eta^2 - 2\mathrm{m}\frac{n}{r^2} + 2\mathrm{m}\frac{n^2}{r^2}\right)J_n(\eta r)\right. \\
&\left.+\frac{2\mathrm{m}p}{r}J_{n+1}(\eta r)\right] + \mathrm{m}B_1\left[(2ik_z\beta)J_n(\beta r) + \left(\frac{-2ik_z(n+1)}{r}\right)J_{n+1}(\beta r)\right] \\
&+\mathrm{m}C_1\left[\left(\frac{2n(n-1)}{r^2}\right)J_n(\beta r) - \frac{2n\beta}{r}J_{n+1}(\beta r)\right]
\end{aligned}\right] \tag{5.123}$$

$$\sigma_{r\theta} = \mathrm{m}\sin(n\theta)e^{ik_z z}\begin{bmatrix} A_1\left[\left(\dfrac{2n(-n+1)}{r^2}\right)J_n(\eta r)+\dfrac{2n\eta}{r}J_{n+1}(\eta r)\right] \\ +B_1\left[ik_z\beta J_n(\beta r)-\dfrac{2ik_z}{r}J_{n+1}(\beta r)\right] \\ +C_1\left[\left(\beta^2-\dfrac{2n(n-1)}{r^2}\right)J_n(\beta r)-\dfrac{2\beta}{r}J_{n+1}(\beta r)\right] \end{bmatrix} \tag{5.124}$$

$$\sigma_{rz} = \mathrm{m}\cos(n\theta)e^{ik_z z}\begin{bmatrix} A_1\left[\dfrac{2ik_z n}{r}J_n(\eta r)-2ik_z\eta J_{n+1}(\eta r)\right] \\ +B_1\left[-\dfrac{\beta n}{r}J_n(\beta r)+\left(\beta^2+k_z^2\right)J_{n+1}(\beta r)\right] \\ +C_1\left[\dfrac{in\,k_z}{r}J_n(\beta r)\right] \end{bmatrix} \tag{5.125}$$

$$\sigma_{zz} = \mathrm{m}\cos(n\theta)e^{ik_z z}\begin{bmatrix} A_1\left[-\dfrac{\mathbb{L}}{\mu}k_p^2-2k^2\right]J_n(\eta r) \\ +B_1\left[-2ikqJ_n(\beta r)+4i\dfrac{n}{r}J_{n+1}(\beta r)\right] \end{bmatrix} \tag{5.126}$$

$$\sigma_{\theta z} = \mathrm{m}\sin(n\theta)e^{ik_z z}\begin{bmatrix} A_1\left[-2ik\dfrac{n}{r}J_n(\eta r)\right] \\ +B_1\left[\beta\dfrac{n}{r}J_n(\beta r)-k^2J_{n+1}(\beta r)\right] \\ +C_1\left[ik\dfrac{n}{r}J_n(\beta r)-ikq\,J_{n+1}(\beta r)\right] \end{bmatrix} \tag{5.127}$$

$$\sigma_{\theta\theta} = \cos(n\theta)e^{ik_z z}\begin{bmatrix} A_1\left[\left(-\mathbb{L}\left(\eta^2+k_z^2\right)+2\mathrm{m}\dfrac{n}{r^2}\right)J_n(\eta r)-\dfrac{2\mathrm{m}\eta}{r}J_{n+1}(\eta r)\right] \\ +\mathrm{m}B_1\left[\left(\dfrac{2ik_z(n+1)}{r}\right)J_{n+1}(\beta r)\right] \\ +\mathrm{m}C_1\left[\left(\dfrac{2n(1-\mathrm{n})}{r^2}\right)J_n(\beta r)+\dfrac{2n\beta}{r}J_{n+1}(\beta r)\right] \end{bmatrix} \tag{5.128}$$

Next enforcing the boundary conditions on $\sigma_{rr} = 0$, $\sigma_{r\theta} = 0$, and $\sigma_{rz} = 0$, at $r = a$, will result in three equations with three unknown coefficients and an eigen value problem as follows:

$$\begin{bmatrix} D_{11} & D_{12} & D_{13} \\ D_{21} & D_{22} & D_{23} \\ D_{31} & D_{32} & D_{33} \end{bmatrix} \begin{Bmatrix} A_1 \\ B_1 \\ C_1 \end{Bmatrix} = \begin{Bmatrix} 0 \\ 0 \\ 0 \end{Bmatrix} \tag{5.129}$$

where

$$D_{11} = \cos(n\theta)e^{i(k_z z)}\left[\left(-\mathbb{L}\left(\eta^2 + k_z^2\right) - 2\mathrm{m}\eta^2 - 2\mathrm{m}\frac{n}{a^2} + 2\mathrm{m}\frac{n^2}{a^2}\right)J_n(\eta a) + \frac{2\mathrm{m}p}{a}J_{n+1}(\eta a)\right] \tag{5.130}$$

$$D_{12} = \mathrm{m}\cos(n\theta)e^{i(k_z z)}\left[(2ik_z\beta)J_n(\beta a) + \left(\frac{-2ik_z(n+1)}{a}\right)J_{n+1}(\beta a)\right] \tag{5.131}$$

$$D_{13} = \mathrm{m}\cos(n\theta)e^{i(k_z z)}\left[\left(\frac{2n(n-1)}{a^2}\right)J_n(\beta a) - \frac{2n\beta}{a}J_{n+1}(\beta a)\right] \tag{5.132}$$

$$D_{21} = \mathrm{m}\sin(n\theta)e^{ik_z z}\left[\left(\frac{2n(-n+1)}{a^2}\right)J_n(\eta a) + \frac{2n\eta}{a}J_{n+1}(\eta a)\right] \tag{5.133}$$

$$D_{22} = \mathrm{m}\sin(n\theta)e^{ik_z z}\left[ik_z\beta J_n(\beta a) - \frac{2ik_z}{a}J_{n+1}(\beta a)\right] \tag{5.134}$$

$$D_{23} = \mathrm{m}\sin(n\theta)e^{ik_z z}\left[\left(\beta^2 - \frac{2n(n-1)}{a^2}\right)J_n(\beta a) - \frac{2\beta}{a}J_{n+1}(\beta a)\right] \tag{5.135}$$

$$D_{31} = \mathrm{m}\cos(n\theta)e^{ik_z z}\left[\frac{2ik_z n}{a}J_n(\eta a) - 2ik_z\eta J_{n+1}(\eta a)\right] \tag{5.136}$$

$$D_{32} = \mathrm{m}\cos(n\theta)e^{ik_z z}\left[-\frac{\beta n}{a}J_n(\beta a) + \left(\beta^2 + k_z^2\right)J_{n+1}(\beta a)\right] \tag{5.137}$$

$$D_{33} = \mathrm{m}\cos(n\theta)e^{ik_z z}\left[\frac{ink_z}{a}J_n(\beta a)\right] \tag{5.138}$$

For a nontrivial solution, determinant of the matrix in Eq. (5.128) should be zero, which will be

$$det\begin{vmatrix} D_{11} & D_{12} & D_{13} \\ D_{21} & D_{22} & D_{23} \\ D_{31} & D_{32} & D_{33} \end{vmatrix} = 0 \tag{5.139}$$

Eq. (5.139) is a general dispersion equation for all possible modes propagating in a cylinder, which is like the one obtained in Eq. (5.31) for a plate-like structure. For different orders of Bessel function, a set of solution of wave numbers for each frequency will be obtained from Eq. (5.139), which will be pertaining to the torsional, longitudinal, and flexural wave modes. These three modes can be separately understood and analyzed from the above dispersion equation as follows:

5.4.1 Torsional Wave Modes in Cylindrical Wave Guides

Torsional wave modes are assumed in a system when only u_θ displacement function will exist in the cylindrical wave guide. For only u_θ to survive according to Eq. (5.114), $\phi = 0$, $\psi_r = 0$, and $\psi_\theta = 0$ are the necessary conditions. Hence, using only the surviving potential ψ_z with zeroth-order Bessel function $\psi_z = C_1 J_0(\beta r)e^{i(k_z z-\omega t)}$, the u_θ displacement function can be written as

$$u_\theta = -\frac{\partial \psi_z}{\partial r} = -C_1\beta J_1(\beta r)e^{i(k_z z-\omega t)} = C\,J_1(\beta r)e^{i(k_z z-\omega t)} \tag{5.140}$$

where $C = -C_1\beta$.

For this mode of wave, to survive only one boundary condition on $\sigma_{r\theta} = 0$ is necessary. Referring Eq. (5.122) in the light of surviving ψ_z potential, the $\sigma_{r\theta}$ takes the form

$$\sigma_{r\theta} = \mathrm{m}\left(\frac{1}{r}\frac{\partial \psi_z}{\partial r} - \frac{\partial^2 \psi_z}{\partial r^2}\right) = r\frac{\partial}{\partial r}\left(\frac{u_\theta}{r}\right) = 0 \ at \ r = a \tag{5.141}$$

Thus, one can get the dispersion equation as

$$\beta a\,J_0(\beta a) = 2J_1(\beta a) \tag{5.142}$$

For a given m-th root of βa as $\beta_m a$, the frequency-wave number relation can be written as

$$(\beta_m a)^2 = \left(\frac{\omega a}{c_T}\right)^2 - (k_z a)^2 \tag{5.143}$$

The dispersion curves obtained from the above spectral equation are very similar to the SH waves in plate-like structures as presented in Fig. 5.4.

5.4.2 Exercise: Longitudinal and Flexural Wave Modes in Cylindrical Structures

5.4.2.1 Longitudinal Wave

Unlike torsional mode where u_θ is the only displacement that survives in the formulation, in case of longitudinal wave modes, $u_\theta = 0$ and variation of any potential function along θ direction is zero, i.e., $\partial/\partial\theta = 0$. Referring Eq. (5.114), it is evident that to have u_r and u_z survive, while $u_\theta = \partial/\partial\theta = 0$, $\phi \neq 0$ and $\psi_\theta \neq 0$. Based on these conditions and following steps provided in the previous sections, please derive the dispersion equation for longitudinal wave modes in cylindrical structure.

5.4.2.2 Flexural Wave

The dispersion equation presented in Eq. (5.139) if solved for the zeroth order Bessel function, i.e., $n = 0$, then both torsional and longitudinal wave modes could be obtained and can be investigated separately, like they are discussed above. For $n = 0$ in cylindrical wave guide one ($u_\theta = 0$) or two ($u_r = 0$ and $u_z = 0$), displacement components were found to be zero depending on if they are longitudinal or torsional wave modes, respectively. For the case of flexural wave modes, all displacement components must survive. Such situation can only be obtained for $n = 1$ or higher. Thus, Eq. (5.139) already includes the flexural wave modes and first few of them can be found when the order of the Bessel function is 1. Higher order will result more flexural wave modes. Details on these modes and plots for the wave dispersion can be found in a few classical books [4, 5] and are omitted in this book.

5.5 SUMMARY

This chapter provides an overview of guided waves in plate and cylindrical structures. Guided wave modes in isotropic and anisotropic layered structures are discussed with few graphics to visualize the wave phenomena. A generalized guided wave formulation in corrugated periodic isotropic media is presented, which in fact results in the same classical dispersion equation when the corrugation depth is made to zero. To find the dispersion behavior in layer anisotropic material transfer matrix method, global matrix method, stiffness matrix method, and semianalytical finite element methods are briefly reviewed.

REFERENCES

1. Giurgiutiu, V., *Structural Health Monitoring with Piezoelectric Wafer Active Sensors, 2nd ed.* 2014, Amsterdam: Elsevier Academic Press.
2. Auld, B.A., *Acoustic Fields and Waves in Solids.* Vol. I & II. 1973, New York: John Wiley & Sons.
3. Rose, J.L., *Ultrasonic Waves in Solid Media.* 1999, Cambridge: Cambridge University Press.

4. Achenbach, J.D., *Wave Propagation in Elastic Solids*. 1999, New York: Elsevier.
5. Graff, K.F., *Wave Motion in Elastic Solids*. 1975, New York: Dover Publication.
6. Stoneley, R., Elastic waves at the surface of separation of two solids. *Proceedings of the Royal Society of London Series A*, 1924. **106**: pp. 416–428.
7. Kundu, T., Mechanics of elastic waves and ultrasonic nondestructive evaluation, in *Ultrasonic Nondestructive Evaluation*, T. Kundu, Editor. 2004, New York: CRC Press. pp. 1–142.
8. Scholte, J.G., The range and existence of Rayleigh and Stonely waves. *Monthly Notices of the Royal Astronomical Society Geophysical Supplements*, 1947. **5**: pp. 120–126.
9. Daraio, C, Nesterenko, V.F., Herbold, E.B., Jin, S., Tunability of solitary wave properties in one-dimensional strongly nonlinear phononic crystals. *Physical Review. E, Statistical, Nonlinear, and Soft Matter Physics*, 2006. **73**(2 Pt 2): p. 026610.
10. Horace, L., On waves in an elastic plate. *Proceedings of the Royal Society A*, 1917. **93**(648).
11. Sahler, C.S. and B.D. Greenwald, Traumatic brain injury in sports: a review. *Rehabilitation Research and Practice*, 2012.
12. Dawson, M., Composite plates with a layer of fluid-filled, reticulated foam for blast protection of infrastructure. *International Journal of Impact Engineering*, 2009. **36**(10–11): pp. 1288–1295.
13. Schimizze, B., et al., An experimental and numerical study of blast induced shock wave mitigation in sandwich structures. *Applied Acoustics*, 2013. **74**(1): pp. 1–9.
14. Christou, G.A., et al., Shock attenuation of PMMA sandwich panels filled with soda-lime glass beads: a fluid-structure interaction continuum model simulation. *International Journal of Impact Engineering*, 2012. **47**: pp. 48–59.
15. Chen, H., et al., Wave propagation and absorption of sandwich beams containing interior dissipative multi-resonators. *Ultrasonics*, 2017. **76**: pp. 99–108.
16. Tan, K.T., H. Huang, and C. Sun, Blast-wave impact mitigation using negative effective mass density concept of elastic metamaterials. *International Journal of Impact Engineering*, 2014. **64**: pp. 20–29.
17. Kim, E., Y.H.N. Kim, and J. Yang, Nonlinear stress wave propagation in 3D woodpile elastic metamaterials. *International Journal of Solids and Structures*, 2015. **58**: pp. 128–135.
18. Kim, E., et al., Impact and blast mitigation using locally resonant woodpile metamaterials. *International Journal of Impact Engineering*, 2017. **101**: pp. 24–31.
19. Li, M., et al., Crashworthiness design optimisation of metal honeycomb energy absorber used in lunar lander. *International Journal of Crashworthiness*, 2011. **16**(4): pp. 411–419.
20. Reglero, J., et al., Design and testing of an energy absorber prototype based on aluminum foams. *Materials & Design*, 2010. **31**(7): pp. 3568–3573.
21. Alghamdi, A., Collapsible impact energy absorbers: an overview. *Thin-Walled Structures*, 2001. **39**(2): pp. 189–213.
22. Schaedler, T.A., et al., Designing metallic microlattices for energy absorber applications. *Advanced Engineering Materials*, 2014. **16**(3): pp. 276–283.
23. Whitman, Z. and V. La Saponara, Bistable structures for energy absorption, I: Metallic structures. *Journal of Mechanics of Materials and Structures*, 2007. **2**(2): pp. 347–358.
24. Whitman, Z. and V. La Saponara, Bistable structures for energy absorption, II: Composite structures under tension. *Journal of Mechanics of Materials and Structures*, 2007. **2**(2): pp. 359–375.
25. Winkelmann, C., S.S. Kim, and V. La Saponara, Design and development of hybrid composite bistable structures for energy absorption under quasi-static tensile loading. *Composite Structures*, 2010. **93**(1): pp. 171–178.

26. Gibson, L., Woodpecker pecking: how woodpeckers avoid brain injury. *Journal of Zoology*, 2006. **270**(3): pp. 462–465.
27. Yoon, S.-H. and S. Park, A mechanical analysis of woodpecker drumming and its application to shock-absorbing systems. *Bioinspiration & Biomimetics*, 2011. **6**(1): p. 016003.
28. Wang, L., H. Zhang, and Y. Fan, Comparative study of the mechanical properties, micro-structure, and composition of the cranial and beak bones of the great spotted woodpecker and the lark bird. *Science China Life Sciences*, 2011. **54**(11): pp. 1036–1041.
29. Oda, J., J. Sakamoto, and K. Sakano, Mechanical evaluation of the skeletal structure and tissue of the woodpecker and its shock absorbing system. *JSME International Journal Series A Solid Mechanics and Material Engineering*, 2006. **49**(3): pp. 390–396.
30. San Ha, N., G. Lu, and X. Xiang, Energy absorption of a bio-inspired honeycomb sandwich panel. *Journal of Materials Science*, 2019. **54**(8): pp. 6286–6300.
31. Bostrom, A., Passbands and stopbands for an electromagnetic waveguide with a periodically varying cross section. *IEEE Transactions on Microwave Theory and Techniques*, 1983. **31**(9): pp. 752–756.
32. Boström, A., Propagating, damped, and leaky surface waves on the corrugated traction-free boundary of an elastic half-space. *The Journal of the Acoustical Society of America*, 1989. **85**(4): pp. 1549–1555.
33. Sandström, S.E., Stopbands in a corrugated parallel plate waveguide. *The Journal of the Acoustical Society of America*, 1986. **79**(5): pp. 1293–1298.
34. El-Bahrawy, A., Stopbands and passbands for symmetric Rayleigh-Lamb modes in a plate with corrugated surfaces. *Journal of Sound and Vibration*, 1994. **170**(2): pp. 145–160.
35. El-Bahrawy, A., Point force excitation of surface waves along the doubly corrugated traction-free boundary of an elastic half-space. *The Journal of the Acoustical Society of America*, 1994. **96**(5): pp. 3167–3176.
36. Declercq, N.F., et al., Diffraction of homogeneous and inhomogeneous plane waves on a doubly corrugated liquid/solid interface. *Ultrasonics*, 2005. **43**(8): pp. 605–618.
37. Banerjee, S. and T. Kundu, Elastic wave propagation in sinusoidally corrugated waveguides. *Journal of the Acoustical Society of America*, 2006. **119**(4): pp. 2006–2017.
38. Kundu, T., Banerjee, S., Kumar, V.J., An experimental investigation of guided wave propagation in corrugated plates showing stop bands and pass bands. *Journal of Acoustical Society of America*, 2006. **120**(3): pp. 1217–1226.
39. Das, S., et al., DPSM modeling for studying interaction between bounded ultrasonic beams and corrugated plates with experimental verification. *IEEE Transactions on Ultrasonics, Ferroelectrics, and Frequency Control*, 2007. **54**(9): pp. 1860–1872.
40. Hawwa, M.A., Shear waves in an initially stressed elastic plate with periodic corrugations. *Arabian Journal for Science and Engineering*, 2017. **42**(5): pp. 1831–1840.
41. Kushwaha, M. and P. Halevi, Band-gap engineering in periodic elastic composites. *Applied Physics Letters*, 1994. **64**(9): pp. 1085–1087.
42. Huang, Y., C. Zhang, and W. Chen, Elastic wave band structures and defect states in a periodically corrugated piezoelectric plate. *Journal of Applied Mechanics*, 2014. **81**(8): p. 081005.
43. Collet, M., et al., Floquet–Bloch decomposition for the computation of dispersion of two-dimensional periodic, damped mechanical systems. *International Journal of Solids and Structures*, 2011. **48**(20): pp. 2837–2848.
44. Indaleeb, M.M., et al., Deaf band based engineered Dirac cone in a periodic acoustic metamaterial: a numerical and experimental study. *Physical Review B*, 2019. **99**(2): p. 024311.
45. Kundu, T., *Ultrasonic Nondestructive Evaluation: Engineering and Biological Material Characterization*. 2003, New York: CRC press.

46. Graff, K.F., *Wave Motion in Elastic Solids*. 2012, New York: Dover Publications Inc.
47. Brillouin, L., *Wave Propagation in Periodic Structures*, 2nd ed. 1953, New York: Dover Publication.
48. Deymier, P.A., *Acoustic Metamaterials and Phononic Crystals*. Vol. **173**. 2013, Berlin: Springer Science & Business Media.
49. Banerjee, S. and T. Kundu, Elastic wave propagation in symmetrically periodic sinusoidal waveguide. *NDE for Health Monitoring and Diagnostics*. Vol. **5394**. 2004, San Diego, CA, USA: SPIE.
50. Mal, A.K., Yin, C. C., Bar-Cohen, Y., Ultrasonic nondestructive evaluation of cracked composite laminates. *Composites Engineering*, 1991. **1**: pp. 85–101.
51. Kundu, T., Guided waves for plate inspection, in *Ultrasonic Nondestructive Evaluation*, T. Kundu, Editor. 2004, New York: CRC Press. pp. 223–310.
52. Buchwald, V.T., Rayleigh waves in transversely isotropic media. *The Quarterly Journal of Mechanics and Applied Mathematics*, 1961. **14**: pp. 293–317.
53. Haskell, N.A., The dispersion of surface waves on multilayered media. *Bulletin of the Seismological Society of America*, 1953. **43**: pp. 17–34.
54. Thomson, W.T., Transmission of elastic waves through a stratified solid medium .*Journal of Applied Physics*, 1950. **21**: pp. 89–93.
55. Knopoff, L., A matrix method for elastic wave problems. *Bulletin of the Seismological Society of America*, 1964. **54**: pp. 431–438.
56. Mal, A.K., Wave propagation in layered composite laminates under periodic surface loads. *Wave Motion*, 1988. **10**: pp. 257–266.
57. Rokhlin, S.I., Wang, L., Stable recursive algorithm for elastic wave propagation in layered anisotropic media: stiffness matrix method. *Journal of Acoustic Society of America*, 2002. **112**(822–834).
58. Wang, L., Rokhlin, S.I., Stable reformulation of transfer matrix method for wave propagation in layered anisotropic media. *Ultrasonics*, 2001. **39**: pp. 413–424.
59. Barazanchy, D., *Non-destructive evaluation of composites: predictive ultrasonic guided-waves modeling, non-destructive material characterization, and the application to aerospace structures, in Mechanical Engineering*. (Doctoral dissertation). 2017, University of South Carolina.
60. Banerjee, S., Banerji, P., Berning, F., Eberle, K., Lamb wave propagation and scattering in layered composite plates, in *Smart Nondestructive Evaluation and Health Monitoring of Structural and Biological Systems II*, T. Kundu, Editor. 2003, San Diego, CA, USA: SPIE.
61. Krishnamoorthy, C.S., *Finite Element Analysis: Theory and Programming*. 2000, New Delhi, India: Tata McGraw Hill.
62. Castillo, G.F.T.d., *An Introduction to Hamiltonian Mechanics*. 2018, Berlin: Springer.
63. Morin, D., *Introduction to Classical Mechanics*. 2007, Cambridge: Cambridge University Press.
64. Bartoli, I., Marzani, A., Lanza di Scalea F., Viola, E., Modeling wave propagation in damped waveguides of arbitrary cross-section. *Journal of Sound and Vibration*, 2006. **295**: pp. 685–707.
65. Finnveden, S., Evaluation of modal density and group velocity by a finite element method. *Journal of Sound and Vibration*, 2004. **273**(1): pp. 51–75.
66. Qu, J., Jacobs, L., Cylindrical waveguides and their applications in ultrasonic evaluation in *Ultrasonic Nondestructive Evaluation*, T. Kundu, Editor. 2004, New York: CRC Press. pp. 311–362.

6 Overview of Basic Numerical Methods and Parallel Computing

6.1 UNDERSTANDING ERROR

Understanding error is fundamental to any numerical analysis and important for CNDE. Let us talk about how to justify a numerical number when it is estimated from an experiment or from a numerical model. It is necessary to measure any physical parameter using a device or a meter or it can be estimated from a model. Whenever a number is employed in a computation, e.g., CNDE simulations, it is necessary to assure that the parameters in simulation should be used with certain level of confidence. For example, let us consider a material property matrix that is frequently used in CNDE simulations. Experimental inspection of the material indicates that the material has Elastic modulus ($\mathbb{E}$) between 245 and 253 GPa, possibly a normal distribution within that bound. We can say that the material has mean elastic modulus equal to 249 GPa. Question is, is it correct? Or how much error one could incur if the simulations are run with these numbers, and the simulations are to be validated with experiment on a material, which does not have the mean material property? We can have confidence on the result only if multiple experiments are conducted on different batch of materials, and the error bounds are properly defined. Error bounds can also be normally distributed. In a numerical model, error arises from multiple avenues. The use of approximations in representing exact condition through approximate physics and mathematical operations causes most of the fundamental errors. If the physics implemented for a problem is correct and is implemented in a code through nearly good conditioned mathematical model, such errors can be mitigated significantly. Here, in this book, many such physics-based models are discussed, which can simulate the CNDE problem using correct physics. Moreover, after employing correct physics and better numerical solver discussed further in this chapter, one can still have errors from the models. Each model may have its convergence criterion and may have respective stability criterion that must be satisfied. After satisfying all criteria, one may still have errors in matching the results with experiments. Another approach to verify the numerical results is to compare them with the analytical solutions. However, analytical solution of most CNDE problems is rather complicated and not possible to solve. Thus, one must rely on only numerical simulations. For some problems, some mathematical parameters are not known and for those parameters one may need optimization. Optimization is synonymous to the minimization of the error [1, 2]. Generally, for numerical model, error could be defined as follows:

$$U_a - \tilde{U}_n = Error = E \tag{6.1}$$

where U_a = Analytical Solution or experimental observation and $\tilde{U}_n$ = Numerical estimation of the observables. Objective will be to minimize this error to have a valid numerical solution. Hence, it is necessary to know and identify the sources of errors. Once the primary sources of error from the physics and the respective numerical technique are mitigated, one should consider the following other sources of errors: (1) *Truncation error*: These errors occur when approximations are used to represent exact mathematical operations and quantities, (2) *Round-off Error*: These errors occur when numbers have limited significant figures. (3) Inherent errors from the measurement are that used to represent exact numbers for verification and validation of a simulation. Total error is thus

$$E_m + E_t + E_r = E \tag{6.2}$$

where E_m is the error from experimental measurements, E_t is the truncation error, *and* E_r is rounding off error. As discussed earlier, these errors in CNDE problems are not absolute number but it should be a distribution with appropriate mean and standard deviation. When error is defined in terms of percentage, which is called the relative error must also have a distribution. If U_a is the exact value (not known) of a parameter and $\tilde{U}_n$ is the known approximate value of a parameter, the relative error [1] could be defined as

$$\left|\frac{U_a - \tilde{U}_n}{\tilde{U}_n}\right| \le \epsilon_a; \left|\frac{\tilde{U}_n - U_a}{\tilde{U}_n}\right| \le \epsilon_a; \left|\frac{U_a - \tilde{U}_n}{U_a}\right| \le \epsilon_a \text{ or } \left|\frac{\tilde{U}_n - U_a}{U_a}\right| \le \epsilon_a \tag{6.3}$$

Knowing distribution of exact solution and approximate solution, distribution of relative error could also be found and should be used in any CNDE modeling technique.

6.2 ERROR PROPAGATION: TAYLOR SERIES

The study of error propagation [1] shows how errors associated with numbers can propagate through mathematical functions. For example, if two independent variables that have errors are multiplied, error in the calculated new product will not be same as the individual two variables. It is necessary to estimate the error in the product. This answer could be given by Taylor series expansion.

6.2.1 Taylor Series Expansion

Say, for example, a function $f(x)$ is a function of an independent variable, x. Δx is a small deviation of x (Fig. 6.1), where the value of the function is to be found, which means that the value of the function at $x + \Delta x$ $\left(f(x + \Delta x)\right)$ is unknown. However, the exact value of the function $f(x)$ at x is known. A Taylor Series expansion [1, 2]

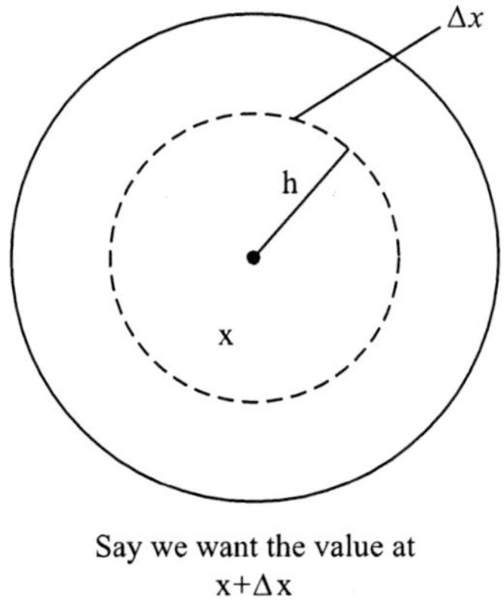

FIGURE 6.1 Visual justification for using Taylor series.

could be written to express the unknown function value at $x+\Delta x$, $f(x+\Delta x)$ using the function values at x, if the function $f(x)$ is differentiable in the neighborhood of x. If the function $f(x)$ and its first $n+1$ derivatives are continuous in an interval Δx, then the value of the function at $x+\Delta x$ is given by

$$f(x+\Delta x)=f(x)+\frac{\Delta x f'(x)}{1!}+\Delta x^2\frac{f''(x)}{2!}+\Delta x^3\frac{f'''(x)}{3!}+\cdots+\Delta x^n\frac{f^n(x)}{n!}+\cdots \quad (6.4)$$

where $f'(x)$, $f''(x)$, $f'''(x)$, and $f^n(x)$ are the first, second, third, and n-th derivatives of the function $f(x)$. A good way to understand the Taylor series is to explain it term by term. For example, the *first term* in the series is

$$f(x+\Delta x)\approx f(x) \quad (6.5)$$

It is the *zeroth order approximation*. It indicates that the value of x at the new point is the same as the old point. If the function changes within the interval, then this approximation bound to have tremendous error. However, additional terms of the Taylor series could provide better estimate of the function $f(x)$ at $x+\Delta x$. The *first order term* added to the zeroth order may get a better estimate but still not accurate. With *first order approximation* the function value at $x+\Delta x$ is estimated as

$$f(x+\Delta x)\approx f(x)+\frac{\Delta x f'(x)}{1!} \quad (6.6)$$

Although the Taylor series with the *first-order approximation* can predict a change, a second-order term could be further added to visualize if some curvature is present in the function. The expression with *second-order approximation* will look like

$$f(x+\Delta x)\approx f(x)+\frac{\Delta x f'(x)}{1!}+\Delta x^2\frac{f''(x)}{2!} \quad (6.7)$$

Similarly, *fourth-order approximation* means that one should retain up to fourth order derivative of the function $f(x)$

$$f(x+\Delta x) \approx f(x)+\frac{\Delta x f'(x)}{1!}+\Delta x^2\frac{f''(x)}{2!}+\Delta x^3\frac{f'''(x)}{3!}+\Delta x^4\frac{f''''(x)}{4!} \tag{6.8}$$

The Taylor series is can be expressed in a general form as

$$f(x+\Delta x) = f(x)+\frac{\Delta x f'(x)}{1!}+\Delta x^2\frac{f''(x)}{2!}+\Delta x^3\frac{f'''(x)}{3!}+\cdots+R_n \tag{6.9}$$

Here, if $\Delta x = x_{i+1} - x_i$, $x_{i+1} = x + \Delta x$ and $x_i = x$ and R_n is the remainder term, R_n can be defined as

$$R_n = \frac{f^{n+1}(x_i)\Delta x^{n+1}}{(n+1)!} \tag{6.10}$$

In other words, we could refine the variables slightly differently but achieve similar results. Let, $f(x)$ be a continuous function, i.e., n^{th} order derivative exists. Thus, f^n is continuous around x.

Let us assume that $\tilde{x}$ is an approximate solution of x, i.e., $f(\tilde{x}) = approximate$ (Fig. 6.2a)

Hence, $|f(x) - f(\tilde{x})| = \Delta\tilde{f}(x)$, and $f(x)$ is not known as shown in Fig. 6.2b.

So, if $\Delta x = x - \tilde{x}$, the Taylor series expansion will look like

$$f(x) = f(\tilde{x}) + (x-\tilde{x})\frac{f'(\tilde{x})}{1!}+(x-\tilde{x})^2\frac{f''(\tilde{x})}{2!}+\cdots \tag{6.11}$$

$$\epsilon = f(x) - f(\tilde{x}) \cong (x-\tilde{x})\frac{(f'(\tilde{x}))}{1!}+(x-\tilde{x})^2\frac{(f''(\tilde{x}))}{2!}+\cdots \tag{6.12}$$

The error could be estimated as $\epsilon = f(x) - f(\tilde{x})$. Convergence of $f(\tilde{x})$ to $f(x)$ will be evaluated with the error ϵ. Higher the order of approximation included

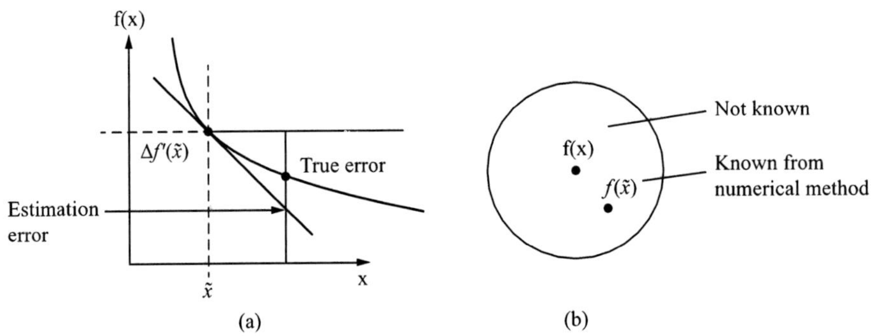

FIGURE 6.2 Projected value from first order Taylor series shows magnified error compared to true error from projected from $\tilde{x}$ to x.

in Eq. (6.12), lower the error will be and more converged results will be obtained. Similar definition of Taylor series expansion is valid for multivariate functions. Let, $f(x.y)$ is the function of 2 independent variables. Then the Taylor series expansion of the function to find its value at (x_{i+1}, y_{i+1}) or $(x+\Delta x, y+\Delta y)$ can be written as

$$f(x_{i+1}, y_{i+1}) \approx f(x_i, y_i) + (x_{i+1} - x_i)\frac{\partial f}{\partial x}\bigg|_{x_i, y_i} + (y_{i+1} - y_i)\frac{\partial f}{\partial y}\bigg|_{x_i, y_i} + \cdots + R_n \tag{6.13}$$

where $|f(x_{i+1}, y_{i+1}) - f(x_i, y_i)| = \Delta\tilde{f}(x_i, y_i) = \Delta\tilde{x}\frac{\partial f}{\partial x} + \Delta\tilde{y}\frac{\partial f}{\partial y}$, $\Delta\tilde{x} = x_{i+1} - x_i$ and $\Delta\tilde{y} = y_{i+1} - y_i$ are with first-order approximation. Similarly, for n dimensional function could be expanded with first-order approximation as follows:

$$f(x_1, x_2, x_3, \ldots, x_n) = f(x_j) = f(x_j - \Delta x_j) + \sum_{j=1}^{n} \Delta x_j \frac{\partial f}{\partial x_j} \tag{6.14}$$

6.2.2 Stability Condition

The stability condition measures the sensitivity of an approximate solution with respect to the input values [1]. The computation is said to be numerically unstable if the uncertainty of the input values is grossly magnified by the numerical method. These ideas can be studied using a first-order Taylor series:

$$f(x) \cong f(\tilde{x}) + f'(\tilde{x})(x - \tilde{x}) \tag{6.15}$$

The above relationship could be used to estimate the relative error of $f(x)$ as follows:

$$\epsilon_a = \delta\tilde{f} = \left|\frac{\tilde{f} - f}{\tilde{f}}\right| = \left|\frac{f'(\tilde{x})(x - \tilde{x})}{f(\tilde{x})}\right| \tag{6.16}$$

On the contrary, the relative error of x is given by

$$\delta\tilde{x} = \frac{x - \tilde{x}}{\tilde{x}} \tag{6.17}$$

Using Eqs. (6.16) and (6.17), a condition number can be formulated, which can be defined as the ratio of the relative errors with the function (dependent variable) and variables (independent variable). Condition number can be written as

$$C_n = \frac{\delta\tilde{f}}{\delta\tilde{x}} = \frac{\textit{Relative error in the output}}{\textit{Relative error in the input}} \tag{6.18}$$

where, $\delta\tilde{f} = \frac{f'(\tilde{x})(x-\tilde{x})}{f(\tilde{x})}$; $\delta\tilde{x} = \frac{x-\tilde{x}}{\tilde{x}}$, and thus

$$C_n = \frac{\delta\tilde{f}}{\delta\tilde{x}} = \frac{\tilde{x}f'(x - \tilde{x})}{f(\tilde{x})} \tag{6.19}$$

The condition number provides a measure of the extent to which an uncertainty in x is magnified in to $f(x)$. There are three possibilities, which bring the following statements:

$C_n = 1$ Relative errors are identical (*Not good- No judgement*)
$C_n > 1$ Relative error is amplified (*Not good, but judgment is possible*)
$C_n < 1$ Relative error is attenuated (*good & judgement is possible*)

Condition number could be used to estimate the permissible error in the input value for a given permissible error in a function.

6.2.3 Summary from Error Propagation

- True error:

$$\in = x_t - x_a = u_a - \tilde{u}_n$$

- True percentage relative error:

$$\in_a = \left| \frac{u_a - \tilde{u}_n}{u_a} \right| 100\%$$

- Approximate % relative Error:

$$\epsilon_a = \frac{\%approximation - previous\ approximation}{\%approximation} \times 100\%$$

- Stopping criteria for a computation:
  Let, $\in_s$ be the desired percentage of relative error called *tolerance.*
  Terminate computation when $\in_a < \in_s$
- Taylor series:

$$f(x_{i+1}) = f(x_i) + f'(x_i)h + \frac{f''(x_i)}{2!}h^2 + \cdots\cdots + \frac{f^n(x_i)}{n!}h^n + O\left(h^{n+1}\right)$$

Note:

$$f'(x_i) = \frac{f(x_{i+1}) - f(x_i)}{h} + O\left(h^2\right) \quad \text{here}, h = x_{i+1} - x_i$$

Reminder:

$$R_n = \frac{f^{n+1}(x_i)h^{n+1}}{(n+1)!}$$

Or

$$R_n = O\left(h^{n+1}\right)$$

For "n" independent variables $x_1, x_2, x_3 \ldots\ldots x_n$ with respective errors $\Delta\tilde{x}_1, \Delta\tilde{x}_2, \Delta\tilde{x}_3, \Delta\tilde{x}_4 \ldots\ldots\ldots \Delta\tilde{x}_n$, the error in the function f can be written as

$$\Delta f = \left|\frac{\partial f}{\partial x_1}\right| \Delta\tilde{x}_1 + \left|\frac{\partial f}{\partial x_2}\right| \Delta\tilde{x}_2 + \left|\frac{\partial f}{\partial x_3}\right| \Delta\tilde{x}_3 + \left|\frac{\partial f}{\partial x_4}\right| \Delta\tilde{x}_4 + \cdots + \left|\frac{\partial f}{\partial x_n}\right| \Delta\tilde{x}_n$$

$$R_n = \frac{f^{n+1}(x_{i+1})}{(n+1)!} \Delta\tilde{x}_n^{\,n+1}$$

where $\Delta\tilde{x}_n = (x_{i+1} - x_i)_n$ for the n-th variable.

6.3 FINITE DIFFERENCE METHOD (FDM)

Finite Difference Method (FDM) [3, 4] is used to find the value of the unknown function at different points along the axis of independent variable. When the governing differential equation describing the unknown dependent variable with respect to the independent variable is known, FDM can discretize the equation to solve for the unknown variable. Let us assume a function $f(x)$ and its derivatives $\left(f'(x), f''(x), f'''(x), f''''(x)\ldots f^n(x)\right)$ that are continuous over a domain x (independent variable) as shown in Fig. 6.3. The x-axis is discretized into several points (x_{i-p} to x_{i+p}), where $p = 5$ in Fig. 6.3. Each interval between two neighboring points in Fig. 6.3 is Δx. The point of interest x_i is the point where the unknown primary function $f(x_i)$ and its derivatives, $f'(x_i)$, $f''(x_i)$, $f'''(x_i)$, etc. are to be found or expressed using

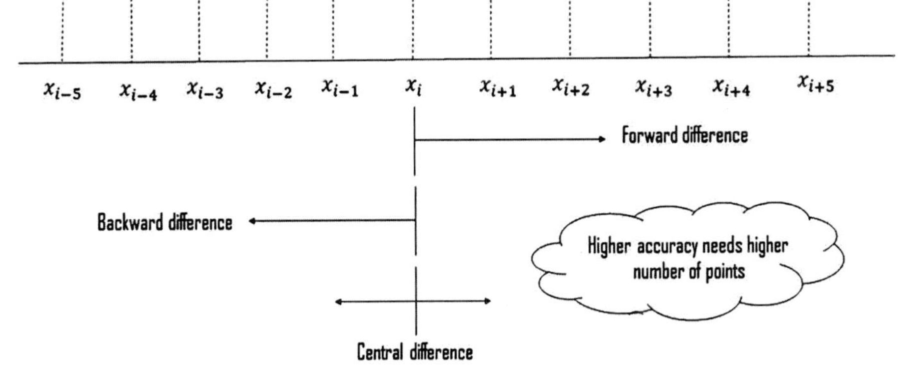

FIGURE 6.3 Schematic presentation of finite difference discretization for forward difference, backward difference, and central difference.

TABLE 6.1
Basic Definition of the Three Finite Difference Methods (ref. Fig. 6.3)

Forward Difference	$f'(x_i)$, $f''(x_i)$, $f'''(x_i)$, $f''''(x_i)$ are expressed by the points in the forward direction only, i.e., x_{i+1}, x_{i+2}, x_{i+3}, x_{i+4}, x_{i+5}, etc.
Backward Difference	$f'(x_i)$, $f''(x_i)$, $f'''(x_i)$, $f''''(x_i)$ are expressed by the points in the backward direction only, i.e., x_{i-1}, x_{i-2}, x_{i-3}, x_{i-4}, x_{i-5}, etc.
Central Difference	$f'(x_i)$, $f''(x_i)$, $f'''(x_i)$, $f''''(x_i)$ are expressed by the points in both directions, i.e., x_{i-5}, x_{i-4}, x_{i-3}, x_{i-2}, $x_{i-1}x_{i+1}$, x_{i+2}, x_{i+3}, x_{i+4}, x_{i+5}, etc.

the function values at the neighboring points between x_{i-p} to x_{i+p}. The finite difference method is classified [2] into three types,

1. Forward Difference (FD)
2. Backward Difference (BD)
3. Central Difference (CD)

The three difference methods are enumerated in Table 6.1.

Referring the Taylor series expansion in Eq. (6.9) with first order approximation, the value of the primary unknown function at x_{i+1}, i.e., $f(x_{i+1})$ can be expressed as

$$f(x_{i+1}) = f(x_i) + \Delta x\, f'(x_i) + \frac{\Delta x^2}{2!} f''(x_i) + \cdots \tag{6.20}$$

Using the above equation, the first derivative of the primary variable $f(x)$ at x_i, i.e., $f'(x_i)$ can be expressed as

$$f'(x_i) = \frac{f(x_{i+1}) - f(x_i)}{\Delta x} - \frac{\Delta x}{2} f''(x_i) + O(\Delta x) \tag{6.21}$$

Retaining only the lower order terms in the above equation that is subjected to higher error and less accurate can be expressed as simply

$$f'(x_i) = \frac{f(x_{i+1}) - f(x_i)}{\Delta x} + O(\Delta x) \tag{6.22}$$

This says that to express the derivative of a function at a point, only one additional value of the function at a neighboring point is necessary. The neighboring point is in the forward direction, and according to Table 6.1, the formula in Eq. (6.22) is a forward difference formula. Similarly, using the forward difference expressions, the following Taylor series expansion can be written for the primary unknown function at x_{i+2}, i.e., $f(x_{i+2})$ as follows:

$$f(x_{i+2}) = f(x_i) + \frac{2\Delta x}{1!} f'(x_i) + \frac{4\Delta x^2}{2!} f''(x_i) + \cdots\ldots\ldots \tag{6.23}$$

Multiplying 2 with Eq. (6.20) and then subtracting from Eq. (6.23), the following expressions can be written in a sequence:

$$\begin{aligned} f(x_{i+2}) &= f(x_i) + \frac{2\Delta x}{1!} f'(x_i) + \frac{4\Delta x^2}{2!} f''(x_i) + \cdots \ldots\ldots \\ 2f(x_{i+1}) &= 2f(x_i) + 2\Delta x f'(x_i) + \Delta x^2 f''(x_i) + \cdots \ldots\ldots \\ \hline f(x_{i+2}) - 2f(x_{i+1}) + f(x_i) &= \Delta x^2 f''(x_i) \\ f''(x_i) &= \frac{f(x_{i+2}) - 2f(x_{i+1}) + f(x_i)}{\Delta x^2} + O(\Delta x) \end{aligned} \tag{6.24}$$

This says that to express the second derivative of a function at a point, two additional values of the function at two neighboring points are necessary. Similarly, using Forward difference formula with first-order approximation, the third derivative will need three points in the forward direction. If order of approximation is increased by one order, one additional point would be necessary, i.e., a second-order derivative using second-order FD approximation will require three additional points. Although derivations for all different approximations are not shown in the chapter but can be found elsewhere [1], the FD, BD, and CD formulas with second-order approximations are written below. It can be noticed that in case of CD, to express the first derivative of the function with second-order approximation, only two neighboring points are necessary, which is one point less than the FD or BD formulas. Hence, with fourth-order approximation of CD, we need four points for first derivative, five points for second derivative, six points for third derivative, and so on. Table 6.2 shows the number of points necessary for respective difference formulas, where O is the order of approximation and D is the order of derivatives.

6.3.1 FD Formula with $O(\Delta x^2)$

$$f'(x_i) = \frac{-f(x_{i+2}) + 4f(x_{i+1}) - 3f(x_i)}{2\Delta x} \tag{6.25.1}$$

TABLE 6.2
Number of Points Required in Finite Difference Formula

Difference	Number of Points Required
FD	$O + D$
BD	$O + D$
CD	$O + D - 1$

$$f''(x_i)=\frac{-f(x_{i+3})+4f(x_{i+2})-5f(x_{i+1})+2f(x_i)}{\Delta x^2} \quad (6.25.2)$$

$$f'''(x_i)=\frac{-3f(x_{i+4})+14f(x_{i+3})-24f(x_{i+2})+18f(x_{i+1})-5f(x_i)}{2\Delta x^3} \quad (6.25.3)$$

$$f''''(x_i)=\frac{-2f(x_{i+5})+11f(x_{i+4})-24f(x_{i+3})+26f(x_{i+2})-14f(x_{i+1})+3f(x_i)}{\Delta x^4} \quad (6.25.4)$$

6.3.2 BD Formula with O(Δx^2)

$$f'(x_i)=\frac{3f(x_i)-4f(x_{i-1})+f(x_{i-2})}{2\Delta x} \quad (6.26.1)$$

$$f''(x_i)=\frac{2f(x_i)-5f(x_{i-1})+4f(x_{i-2})-f(x_{i-3})}{\Delta x^2} \quad (6.26.2)$$

$$f'''(x_i)=\frac{5f(x_i)-18f(x_{i-1})+24f(x_{i-2})-14f(x_{i-3})+3f(x_{i-4})}{2\Delta x^3} \quad (6.26.3)$$

$$f''''(x_i)=\frac{-3f(x_i)-14f(x_{i-1})+26f(x_{i-2})-24f(x_{i-3})+11f(x_{i-4})-2f(x_{i-5})}{\Delta x^4} \quad (6.26.4)$$

6.3.3 CD Formula with O(Δx^2)

$$f'(x_i)=\frac{f(x_{i+1})-f(x_{i-1})}{2\Delta x} \quad (6.27.1)$$

$$f''(x_i)=\frac{f(x_{i+1})-2f(x_i)+f(x_{i-1})}{\Delta x^2} \quad (6.27.2)$$

$$f'''(x_i)=\frac{f(x_{i+2})-2f(x_{i+1})+2f(x_{i-1})-f(x_{i-2})}{2\Delta x^3} \quad (6.27.3)$$

$$f''''(x_i)=\frac{f(x_{i+2})-4f(x_{i+1})+6f(x_i)-4f(x_{i-1})+f(x_{i-2})}{\Delta x^4} \quad (6.27.4)$$

6.3.4 CD Formula with O(Δx^4)

$$f'(x_i)=\frac{f(x_{i-2})-8f(x_{i-1})+8f(x_{i+1})-f(x_{i+2})}{12\Delta x} \quad (6.28.1)$$

$$f''(x_i)=\frac{-f(x_{i-2})+16f(x_{i-1})-30f(x_i)+16f(x_{i+1})-f(x_{i+2})}{12\Delta x^2} \quad (6.28.2)$$

$$f'''(x_i)=\frac{f(x_{i-3})-8f(x_{i-2})+13f(x_{i-1})-13f(x_{i+1})+8f(x_{i+2})-f(x_{i+3})}{8\Delta x^3} \quad (6.28.3)$$

$$f''''(x_i)=\frac{f(x_{i-3})+12f(x_{i-2})-39f(x_{i-1})+56f(x_i)-39f(x_{i+1})+12f(x_{i+2})-f(x_{i+3})}{6\Delta x^4} \quad (6.28.4)$$

6.4 TIME INTEGRATION: EXPLICIT FDM SOLUTION OF DIFFERENTIAL EQUATIONS

Here, in this section, for more generalized discussion, the nomenclature of the primary variable is changed from f to x. It is assumed that any arbitrary primary variable x is a function of time t and the function can be written as $x(t)$. Derivative of the function $x(t)$ with respect to time t is written with dots as shown below:

$$\frac{dx}{dt} = \dot{x}(t)\,;\ \frac{d^2x}{dt^2} = \ddot{x}(t),\ \frac{d^3x}{dt^3} = \dddot{x}(t) \tag{6.29}$$

Time derivative cases are frequently occurring in CNDE problems. Some of them are discussed in chapters 7, 8, 9, 10, and 11. In almost all chapters, one way of other solutions of dynamic system is presented using numerical technique, where the fundamental governing differential equation of an unknown variable $x(t)$ is known or derived using physics as written below:

$$\mathbb{L}x(t) = f(t) \tag{6.30}$$

where $\mathbb{L}$ is the differential operator, but the values of the primary variable $x(t)$ should be obtained at different points along the axis of time t. The unknown primary variable $x(t)$ should be found when the system is subjected to an external loading function $f(t)$ boundary condition. Let us assume a function $x(t)$ and its derivatives ($\dot{x}(t)$, $\ddot{x}(t)$, $\dddot{x}(t)$, etc.) that are continuous over a domain t as shown in Fig. 6.4. The t-axis is discretized into several points (t_{i-p} to t_{i+p}), where $t = 2$ in Fig. 6.4.

A typical dynamic system will have the following differential operator $\mathbb{L}$ as follows:

$$\mathbb{L} = M\frac{d^2}{dt^2} + C\frac{d}{dt} + K \tag{6.31}$$

and the dynamic equation could be written as

$$M\ddot{x} + C\dot{x} + Kx = f(t) \tag{6.32}$$

where M is the mass of the system, C is the damping coefficient of the system, and K is the stiffness of the system. For a single degree-of-freedom system, the above parameters are single value parameters but for a multidegrees-of-freedom system, they will be matrices. To solve Eq. (6.32), i.e., the value of the function $x(t)$ at different time steps, one should utilize the FDM formulas discussed in the previous section. Eq. (6.32) is valid at every time step. In explicit method, the response quantities are expressed in terms of previously determined values of the function, first derivative of the function and the second derivative of the function. Drawing relevance to CNDE, in most cases, displacement function is the primary unknown in the dynamic equation. Hence, in explicit method, the

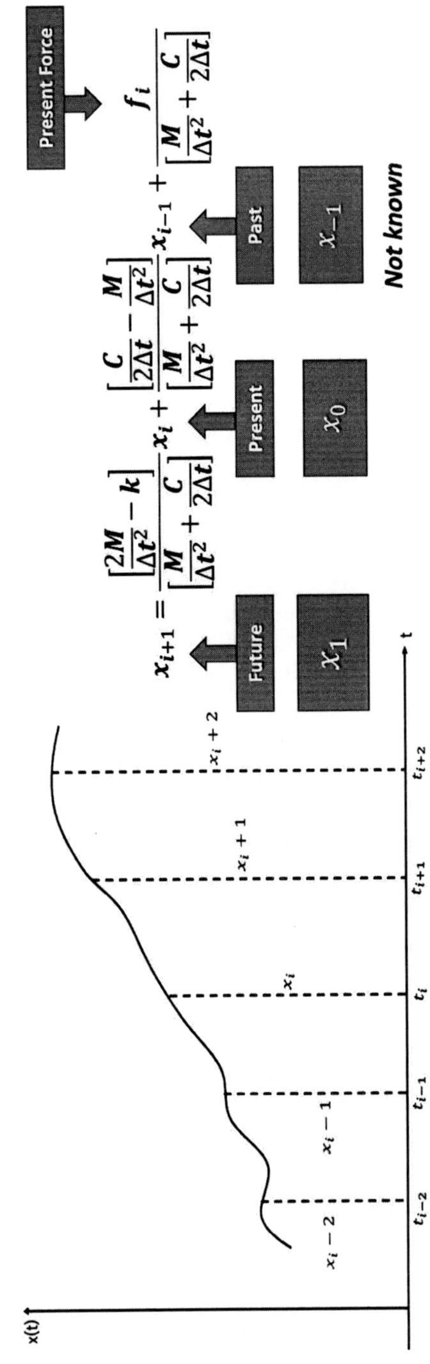

FIGURE 6.4 Central difference methods is not self starting.

primary variable displacement at an arbitrary time point can be found using previously determined values of displacement, velocity, and acceleration. Eq. (6.32) could be explicitly expressed at t_i. If the value of the primary variable at t_i is x_i, then one can write [1, 3, 5]

$$M\ddot{x}_i + C\dot{x}_i + Kx_i = f_i(t) \tag{6.33}$$

Similarly, if the value of the primary variable at any t_j is x_j, Taylor series expansion at x_{i+1} and x_{i-1} can be written as

$$x_{i+1} = x_i + \Delta t\dot{x}_i + \frac{\Delta t^2}{2!}\ddot{x}_i + \frac{\Delta t^3}{3!}\dddot{x}_i \tag{6.34}$$

$$x_{i-1} = x_i - \Delta t\dot{x}_i - \frac{\Delta t^2}{2!}\ddot{x}_i - \frac{\Delta t^3}{3!}\dddot{x}_i \tag{6.35}$$

Deriving the CD formula from the above two expressions and referring Eqs. (6.27.1) and (6.27.2), the first and second derivatives of $x(t_i)$ could be written as

$$\left.\begin{aligned} \dot{x}_i &= \frac{1}{2\Delta t}(x_{i+1} - x_{i-1}) \\ \ddot{x}_i &= \frac{1}{\Delta t^2}(x_{i-1} - 2x_i + x_{i+1}) \end{aligned}\right\} \tag{6.36}$$

Substituting Eq. (6.36) in Eq. (6.33),

$$M\left\{\frac{x_{i-1} - 2x_i + x_{i+1}}{\Delta t^2}\right\} + C\left\{\frac{x_{i+1} - x_{i-1}}{2\Delta t}\right\} + Kx_i = f_i \tag{6.37}$$

where

$$x_i = x(t_i) \qquad f_i = f(t_i)$$

In the above equation, the future point x_{i+1} is unknown; however, the present point x_i and the past point x_{i-1} are known. Hence, expressing the future unknown point in terms of the known past and present points can be written as

$$x_{i+1}\left[\frac{M}{\Delta t^2} + \frac{C}{2\Delta t}\right] + x_i\left[-\frac{2M}{\Delta t^2} + k\right] + x_{i-1}\left[\frac{M}{\Delta t^2} - \frac{C}{2\Delta t}\right] = f_i \tag{6.38.1}$$

or

$$x_{i+1}M^e = f_i - x_i\left[K - \frac{2M}{\Delta t^2}\right] - x_{i-1}\left[\frac{M}{\Delta t^2} - \frac{C}{2\Delta t}\right] = F^e \tag{6.38.2}$$

where new terms M^e and F^e are introduced, which are effective mass and effective force at the location of the degree of freedom. Alternatively, the above equation can also be written as

$$x_{i+1} = \frac{\left[\frac{2M}{\Delta t^2} - k\right]}{\left[\frac{M}{\Delta t^2} + \frac{C}{2\Delta t}\right]} x_i + \frac{\left[\frac{C}{2\Delta t} - \frac{M}{\Delta t^2}\right]}{\left[\frac{M}{\Delta t^2} + \frac{C}{2\Delta t}\right]} x_{i-1} + \frac{f_i}{\left[\frac{M}{\Delta t^2} + \frac{C}{2\Delta t}\right]} \tag{6.39}$$

Fig. 6.4 shows the visual impression of future, present, and past data points in Eq. (6.39). Solution of a differential equation requires initial conditions for the primary variable and the first derivative of the primary variable at the beginning of the time ($t = 0$) to start the algorithm. Let us assume the initial conditions are

$$x(t=0) = x_0 \qquad \dot{x}(t=0) = \dot{x}_0$$

If one is intended to find the value of the function at the first future point, i.e., x_1 to start the algorithm, the algorithm will not start in its own, because according to Eq. (6.39), the value of the function at x_{-1} is necessary like it is shown in Fig. 6.4. It is apparent that the CD algorithm is not self-starting [1] and thus following mathematical treatment is necessary. First, expressing Eq. (6.33) at the present point that is at x_0 at the time of initial condition, the dynamic equation viz.

$$M\ddot{x}_0 + C\dot{x}_0 + Kx_0 = f_0(t) \tag{6.40}$$

or

$$M\ddot{x}_0 = f(t=0) - C\dot{x}_0 - Kx_0$$

or

$$\ddot{x}_0 = \frac{f_0 - C\dot{x}_0 - kx_0}{M} \tag{6.41}$$

Using Taylor series expansion (Eq. (6.11)) of the function about present point x_0 and expressing the past point x_{-1} will result in

$$x_{-1} = x_0 - \Delta t\dot{x}_0 + \frac{\Delta t^2}{2!}\ddot{x}_0 \tag{6.42}$$

Next, substituting Eq. (6.41) in Eq. (6.42) x_{-1} can be expressed as

$$x_{-1} = x_0 - \Delta t\dot{x}_0 + \frac{\Delta t^2}{2!}\left(\frac{f_0 - C\dot{x}_0 - Kx_0}{M}\right) \tag{6.43}$$

In Eq. (6.43), the past point x_{-1} is completely expressed using the values of the function and its derivatives at the present point. Substituting Eq. (6.43) in Eq. (6.39) at the beginning of the time, the value of the function at the future point x_1 can be written as

$$x_1 = \left[\frac{\left[\frac{2M}{\Delta t^2} - k\right]}{\left[\frac{M}{\Delta t^2} + \frac{C}{2\Delta t}\right]} + \frac{\left[\frac{C}{2\Delta t} - \frac{M}{\Delta t^2}\right]}{\left[\frac{M}{\Delta t^2} + \frac{C}{2\Delta t}\right]} \left(1 - \frac{\Delta t^2 K}{2M}\right) \right] x_0 - \left[\frac{\left[\frac{C}{2\Delta t} - \frac{M}{\Delta t^2}\right]}{\left[\frac{M}{\Delta t^2} + \frac{C}{2\Delta t}\right]} \left(\Delta t + \frac{\Delta t^2 C}{2M}\right) \right] \dot{x}_0$$

$$+ f_0 \left[\frac{1}{\left[\frac{M}{\Delta t^2} + \frac{C}{2\Delta t}\right]} + \frac{\Delta t^2}{2M} \frac{\left[\frac{C}{2\Delta t} - \frac{M}{\Delta t^2}\right]}{\left[\frac{M}{\Delta t^2} + \frac{C}{2\Delta t}\right]} \right] \tag{6.44}$$

Hence, now the algorithm can be started and the subsequent values of the function in all time points can be calculated using the explicit CD-based FDM.

6.5 TIME INTEGRATION: EXPLICIT SOLUTION OF MULTIDEGREES-OF-FREEDOM SYSTEM

In the above section, equations are written for a single degree-of-freedom system, where mass is assumed to be lumped at the location of the system where the motion of the particle is unknown. Similar solution can be made for multiple degrees-of-freedom system. In multidegrees-of-freedom system, at each degree of freedom, the mass is assumed to be lumped at the location of the particle in the system where the motions of the particles are unknown. The governing dynamic equation for n-degrees of freedom system, which is a set of linear equations of second-order ordinary differential equations, can be written as

$$\begin{bmatrix} m_1 & 0 & 0 & 0 \\ 0 & m_2 & 0 & 0 \\ 0 & 0 & \ddots & 0 \\ 0 & 0 & 0 & m_n \end{bmatrix} \begin{Bmatrix} \ddot{x}_1 \\ \ddot{x}_2 \\ \ddot{x}_3 \\ \vdots \\ \ddot{x}_n \end{Bmatrix} + \begin{bmatrix} c_{11} & c_{12} & \cdots & c_{1n} \\ c_{21} & c_{22} & \cdots & c_{2n} \\ \vdots & \cdots & \ddots & \vdots \\ c_{n1} & \cdots & \cdots & c_{nn} \end{bmatrix} \begin{Bmatrix} \ddot{x}_1 \\ \ddot{x}_2 \\ \ddot{x}_3 \\ \vdots \\ \dot{x}_n \end{Bmatrix} + \begin{bmatrix} k_{11} & k_{12} & \cdots & k_{1n} \\ k_{21} & k_{22} & \cdots & k_{2n} \\ \vdots & \cdots & \ddots & \vdots \\ k_{n1} & \cdots & \cdots & k_{nn} \end{bmatrix} \begin{Bmatrix} \ddot{x}_1 \\ \ddot{x}_2 \\ \vdots \\ \vdots \\ x_n \end{Bmatrix} = \begin{Bmatrix} f_1 \\ f_2 \\ \vdots \\ \vdots \\ f_n \end{Bmatrix}$$

or

$$\mathbf{M}\underline{\ddot{x}} + \mathbf{C}\underline{\dot{x}} + \mathbf{K}\underline{x} = \underline{\mathbf{f}} \tag{6.45}$$

where **M, C,** and **K** are the mass matrix, damping matrix, and stiffness matrix, respectively. $\underline{\mathbf{f}}$ is a vector of forces acting at the location of the particle along the degrees of freedom. $\underline{\ddot{x}}$, $\underline{\dot{x}}$, and $\underline{x}$ are the vectors of accelerations (second derivative of primary unknown variable), velocities (first derivative of primary unknown variable), displacements (primary unknown variables) at the degrees of freedom, respectively. The explicit CD method is also applicable for multidegrees-of-freedom system and Eqs. (6.33) through (6.44) are valid when the mass (M), damping coefficient (C), and stiffness (K) are replaced by the mass matrix (**M**), damping matrix (**C**), and stiffness matrix (**K**), respectively. Next, the necessary mathematical steps or the algorithm is described using Eqs. (6.33) through (6.44).

6.5.1 Explicit Solution Algorithm for Multidegrees-of-Freedom System [3]

Step 1: Form **K, M,** and **C** matrices as written in Eq. (6.45)

Step 2: Initialize $\underline{\ddot{x}}$, $\underline{\dot{x}}$, and $\underline{x}$ at $t = 0$, i.e., $\underline{\dot{x}_0}$, $\underline{x_0}$ and calculate $\underline{\ddot{x}_0}$ using Eq. (6.41) replacing the matrices formed in Step 1.

$$\underline{\ddot{x}_0} = \frac{\underline{\mathbf{f}_0} - \mathbf{C}\underline{\dot{x}_0} - \mathbf{k}\underline{x_0}}{\mathbf{M}} \tag{6.46}$$

Step 3: Select time step = Δt

Step 4: Calculate $a_0 = \frac{1}{\Delta t^2}$; $a_1 = \frac{1}{2\Delta t}$; $a_2 = 2a_0$; $a_3 = \frac{1}{a_2} = \frac{1}{2a_0} = \frac{\Delta t^2}{2}$

Step 5: Using Eq. (6.43), calculate $\underline{x_{-1}}$ where $\underline{x_{-1}} = \underline{x_0} - \Delta t\underline{\dot{x}_0} + a_3\underline{\ddot{x}_0}$. This makes the algorithm start.

Step 6: Next, form an effective mass matrix as follows: $\mathbf{M}^e = \frac{\mathbf{M}}{\Delta t^2} + \frac{\mathbf{C}}{2\Delta t}$

$$\mathbf{M}^e = a_0[\mathbf{M}] + a_1[\mathbf{C}] \tag{6.47}$$

Step 7: Triangularize the effective mass matrix: $\mathbf{M}^e = [\mathbf{L}][\mathbf{D}][\mathbf{L}]^T$

Step 8: Using Eq. (6.38.2) for each time step, calculate effective force vector at time t

$$\mathbf{F}^e = \underline{\mathbf{f}} - ([\mathbf{K}] - a_2[\mathbf{M}])\underline{x_i} - (a_0[\mathbf{M}] - a_1[\mathbf{C}])x_{i-1} \tag{6.48}$$

Step 9: Solve for displacement at time $+\Delta t$, i.e., at x_{i+1}

$$\mathbf{M}^e\ \underline{x_{i+1}} = \mathbf{F}^e \ or \ \underline{x_{i+1}} = \mathbf{F}^e / \mathbf{M}^e \tag{6.49}$$

Step 10: Using the following equations, next calculate the velocity vector and the acceleration vector, i.e., $\underline{\dot{x}}$ & $\underline{\ddot{x}}$ at time t

$$\underline{\dot{x}_i} = a_1(-\underline{x_{i-1}} - \underline{x_{i+1}}) \tag{6.50.1}$$

$$\underline{\ddot{x}_i} = a_0(\underline{x_{i-1}} - 2\underline{x_i} + \underline{x_{i+1}}) \tag{6.50.2}$$

An important consideration for using CD method is that the timestep Δt should be less than Δt_{cr} and the critical value is expressed as $2/\omega_{max}$.

$$\Delta t \le \Delta t_{cr} = \frac{2}{\omega_{max}} \tag{6.51}$$

If $\Delta t > \Delta t_{cr}$, the algorithm is unstable.

ω_{max} is the maximum natural frequency of the system. Using the stiffness matrix **K** and the mass matrix **M,** an eigen value analysis of the homogeneous system of equations results in the following equation [1, 5] in the matrix form.

$$\left[\mathbf{K}-\omega^2\mathbf{M}\right]=\mathbf{0} \tag{6.52}$$

Eigen solution for the n degrees of freedom system will result in a vector of natural frequencies containing n numbers of natural frequencies. The maximum frequency selected from that vector is considered ω_{max} and can be substituted in Eq. (6.51) for selecting Δt for explicit CD algorithm. This is the most basic but very useful algorithm to solve a dynamic system.

6.5.2 Runge-Kutta (RK4) Algorithm for Multidegrees-of-Freedom System

RK4 method, which is the short form of fourth-order Runge-Kutta Method [1], is a versatile method to solve a set of first-order ordinary differential equations (Eq. (6.53)) simultaneously. The first-order method to solve first-order ordinary differential equation is called the Euler's Method, the second-order method is called the Heun's method, the following third- and fourth-order methods are Runge-Kutta methods, RK3 and RK4, respectively. In the first-order ordinary differential equation, the derivative of the unknown function ($\underline{x}$) with respect to the independent variable (t) is equal to a nonhomogeneous function ($\underline{\mathbf{g}}$) of the independent variable as written in Eq. (6.53).

$$\frac{d\underline{x}}{dt}=\dot{\underline{x}}=\underline{\mathbf{g}}\left(\underline{x},t\right) \tag{6.53}$$

The function on the right side of the equation can be a nonlinear nonhomogeneous function and to the numerical method, linear or nonlinear functions are equivalent because the value of the function in discretized domain is used in the algorithm. Making it relevant to the dynamic system, the unknown variables could be the displacements or velocities and the independent variable is time. The future value of the primary variable at $t+\Delta t$ or at $i+1$-th step can be calculated using the following generic equation:

$$\underline{x}_{i+1}=\underline{x}_i+\Delta t\ \underline{\mathbf{S}} \tag{6.54}$$

where Eq. (6.54) is derived from Eq. (6.6). The factor $\underline{\mathbf{S}}$ is called the slope function, which changes based on the order of the solution method, e.g., Euler, Heun's, or RK4. Hence, RK4 is a FD explicit method. Euler's and Heun's methods are not discussed in this book because they are subjected to more errors and are not suitable for CNDE. $\underline{\mathbf{S}}$ is uniquely calculated for RK4 method, which will be described later.

Utilizing the flexibility of RK4 method, first it is convenient to translate Eq. (6.45) into two sets of equations of first-order ordinary differential equations.

Let us assume the first-order derivative of the unknown function vector $\underline{x}$ as $\underline{\dot{x}} = \underline{v}$. Hence, two sets of the dynamic equations could be written as follows:

$$\mathbf{M}\underline{\ddot{x}} + \mathbf{C}\underline{\dot{x}} + \mathbf{K}\underline{x} = \mathbf{f} \tag{6.55.1}$$

$$\underline{\dot{x}} - \underline{v} = \underline{\mathbf{0}} \tag{6.55.2}$$

where both the above sets have n equations each. Substituting, $\underline{\dot{x}}$ in Eq. (6.54.2) into Eq. (6.54.1), and substituting $\underline{\ddot{x}} = \underline{\dot{v}}$, the above two equations can further viz.

$$\begin{bmatrix} \mathbf{M}_{n\times n} & \mathbf{0}_{n\times n} \\ \mathbf{0}_{n\times n} & \mathbf{I}_{n\times n} \end{bmatrix} \begin{Bmatrix} \underline{\dot{v}}_{nx1} \\ \underline{\dot{x}}_{nx1} \end{Bmatrix} + \begin{bmatrix} \mathbf{C}_{n\times n} & \mathbf{K}_{n\times n} \\ -\mathbf{I}_{n\times n} & \mathbf{0}_{n\times n} \end{bmatrix} \begin{Bmatrix} \underline{v} \\ \underline{x} \end{Bmatrix} = \begin{Bmatrix} \underline{\mathbf{f}} \\ \underline{\mathbf{0}} \end{Bmatrix} \tag{6.56}$$

With careful observation, it can be seen that Eq. (6.56) is a $2n$ set of first-order ordinary differential equations. Hence, introducing a new variable $\underline{q} = \{\ \underline{v}\ \underline{x}\ \}^T$, Eq. (6.56) can be rewritten as

$$\mathbf{R}\underline{\dot{q}} = -\mathbf{K}^{rk}\underline{q} + \mathbf{f}^{rk} \tag{6.57.1}$$

or

$$\underline{\dot{q}} = \left[\mathbf{f}^{rk}\mathbf{R}^{-1}\right] - \left[\mathbf{K}^{rk}\mathbf{R}^{-1}\right]\underline{q} \tag{6.57.2}$$

or

$$\underline{\dot{q}} = \mathbf{F} \tag{6.57.3}$$

where

$$\mathbf{F}\left(\underline{q},t\right) = \left[\mathbf{f}^{rk}\mathbf{R}^{-1}\right] - \left[\mathbf{K}^{rk}\mathbf{R}^{-1}\right]\underline{q} \tag{6.57.4}$$

To start the RK4 method like other methods, it requires initial condition. Here, the initial conditions are required for $\underline{q}$ vector, which consists of velocity and displacements along all the degrees of freedom. As before in the previous section, let us assume the initial conditions are

$$\underline{x}(t=0) = \underline{x_0}; \quad \underline{\dot{x}}(t=0) = \underline{\dot{x}_0} = \underline{v}_0; \quad \underline{q}_0 = \{\underline{v}_0 \quad \underline{x}_0\}^T \tag{6.58}$$

To predict the future value at $t + \Delta t$ step or at $i + 1$-th step, i.e., $\underline{q}_{i+1} = \underline{q}_i + \Delta t\ \underline{\mathbf{S}}$, RK4 method needs four different slopes to construct the slope function $\underline{\mathbf{S}}$ as described in Fig. 6.5. The slope function in RK4 is constructed as follows:

$$\underline{\mathbf{S}} = \frac{1}{6}\left(k_1 + 2k_2 + 2k_3 + k_4\right) \tag{6.59}$$

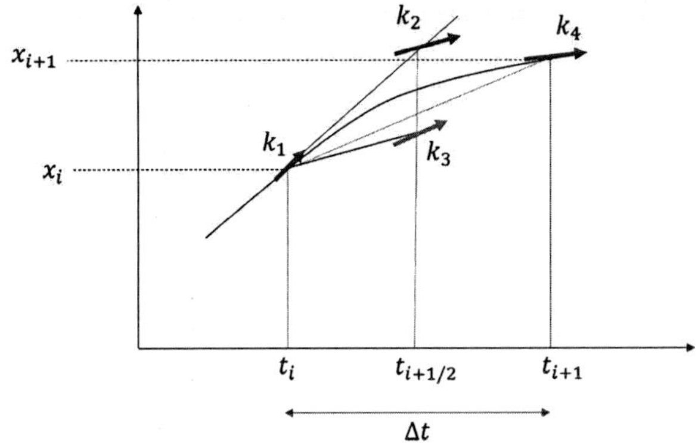

FIGURE 6.5 Schematics showing the slopes used in Runge-Kutta method.

where

$$k_1 = \mathbf{F}\left(\underline{q}_i, t_i\right) \tag{6.60.1}$$

$$k_2 = \mathbf{F}\left(\underline{q}_i + \Delta t \frac{k_1}{2}, t_{i+1/2}\right) \tag{6.60.2}$$

$$k_3 = \mathbf{F}\left(\underline{q}_i + \Delta t \frac{k_2}{2}, t_{i+1/2}\right) \tag{6.60.3}$$

$$k_4 = \mathbf{F}\left(\underline{q}_i + \Delta t \; k_3, t_{i+1}\right) \tag{6.60.4}$$

As shown in Fig. 6.5, first the slope at t_i is calculated, which is k_1 and using the initial slope, a new slope is calculated at $t_{i+1/2}$, which is k_2. Using k_2 slope from t_i, function is again predicted at $t_{i+1/2}$ and a new slope is calculated, which is k_3. Using the k_3 slope, function is predicted from t_i at t_{i+1} and the slope is calculated which is k_4. Given proper weight to each slope as written in Eq. (6.57), the final slope function $\underline{\mathbf{S}}$ is obtained to calculate $\underline{q}_{i+1} = \underline{q}_i + \Delta t \ \underline{\mathbf{S}}$, at t_{i+1}, when the values at the previous point at t_i are known, which is $\underline{q}_i$. To start the algorithm first value at time t_1 which $\underline{q}_1$ is calculated using the equation $\underline{q}_1 = \underline{q}_0 + \Delta t \ \underline{\mathbf{S}}$ with the initial condition written in Eq. (6.58).

6.6 TIME INTEGRATION: IMPLICIT FDM SOLUTION OF DIFFERENTIAL EQUATIONS

Dynamic equation discussed above is valid at every time point. Hence, revisiting Eq. (6.33) for multidegrees-of-freedom system and expressing the dynamic equation at the future time step at t_{i+1}, one can write the implicit form of dynamic equation as follows:

$$\mathbf{M}\ddot{\underline{x}}_{i+1} + \mathbf{C}\dot{\underline{x}}_{i+1} + \mathbf{K}\underline{x}_{i+1} = \underline{\mathbf{f}}_{i+1} \tag{6.61}$$

In implicit method, BD formulas are used and the response quantities are expressed at the future time step where more than one past point is used including the present. First, the Taylor series expansion is written about the future point x_{i+1} and expressed at present and past two points as follows:

$$\underline{x}_i = \underline{x}_{i+1} - \Delta t \dot{\underline{x}}_{i+1} + \frac{\Delta t^2}{2} \ddot{\underline{x}}_{i+1} - \frac{\Delta t^3}{6} \dddot{\underline{x}}_{i+1} + O\left(\Delta t^3\right) \quad (6.62.1)$$

$$\underline{x}_{i-1} = \underline{x}_{i+1} - 2\Delta t \dot{\underline{x}}_{i+1} + \frac{(2\Delta t)^2}{2} \ddot{\underline{x}}_{i+1} - \frac{(2\Delta t)^3}{6} \dddot{\underline{x}}_{i+1} + O\left(\Delta t^3\right) \quad (6.62.2)$$

$$\underline{x}_{i-2} = \underline{x}_{i+1} - 3\Delta t \dot{\underline{x}}_{i+1} + \frac{(3\Delta t)^2}{2} \ddot{\underline{x}}_{i+1} - \frac{(3\Delta t)^3}{6} \dddot{\underline{x}}_{i+1} + O\left(\Delta t^3\right) \quad (6.62.3)$$

Using the above three equations and eliminating $\dddot{\underline{x}}_{i+1}$, the first and second derivatives (i.e., velocity and acceleration vectors) can be written as

$$\dot{\underline{x}}_{i+1} = \frac{1}{6\Delta t}\left(11\underline{x}_{i+1} - 18\underline{x}_i + 9\underline{x}_{i-1} - 2\underline{x}_{i-2}\right) \quad (6.63.1)$$

$$\ddot{\underline{x}}_{i+1} = \frac{1}{\Delta t^2}\left(2\underline{x}_{i+1} - 5\underline{x}_i + 4\underline{x}_{i-1} - \underline{x}_{i-2}\right) \quad (6.63.2)$$

Next, Eqs. (6.63.1) and (6.63.2) can be substituted in Eq. (6.61) and will be:

$$\mathbf{M}\left[\frac{1}{\Delta t^2}\left(2\underline{x}_{i+1} - 5\underline{x}_i + 4\underline{x}_{i-1} - \underline{x}_{i-2}\right)\right] + \mathbf{C}\left[\frac{1}{6\Delta t}\left(11\underline{x}_{i+1} - 18\underline{x}_i + 9\underline{x}_{i-1} - 2\underline{x}_{i-2}\right)\right] + \mathbf{K}\left[\underline{x}_{i+1}\right] = \underline{\mathbf{f}}_{i+1} \quad (6.64)$$

Like in Fig. 6.4, also using implicit method the primary unknown value at the future point $\underline{x}_{i+1}$ can be expressed using the value at the present point $\underline{x}_i$ and the past two values $\underline{x}_{i-1}$ and $\underline{x}_{i-2}$ are as follows:

$$\underline{x}_{i+1}\left[\frac{2}{\Delta t^2}[\mathbf{M}] + \frac{11}{6\Delta t}[\mathbf{C}] + [\mathbf{K}]\right] = \underline{\mathbf{f}}_{i+1} + \left[\frac{5}{\Delta t^2}[\mathbf{M}] + \frac{3}{\Delta t}[\mathbf{C}]\right]\underline{x}_i - \left[\frac{4}{\Delta t^2}[\mathbf{M}] + \frac{3}{2\Delta t}[\mathbf{C}]\right]\underline{x}_{i-1} + \left[\frac{1}{\Delta t^2}[\mathbf{M}] + \frac{1}{3\Delta t}[\mathbf{C}]\right]\underline{x}_{i-2} \quad (6.65)$$

This method is called the Houbolt Method [1] and requires the initial conditions and special treatment to start the algorithm.

6.6.1 Implicit Solution Algorithm (Houbolt Method) [3, 4]

Step 1: Form $\boldsymbol{K}$, $\boldsymbol{M}$, and $\boldsymbol{C}$ matrices as written in Eq. (6.45)
Step 2: Initialize $\ddot{\underline{x}}$, $\dot{\underline{x}}$, and $\underline{x}$ at $t = 0$, i.e., $\dot{\underline{x}}_0$, $\underline{x}_0$ and calculate $\ddot{\underline{x}}_0$ using Eq. (6.41) replacing the matrices formed in Step 1 like in Eq. (6.46).

Step 3: Select step Δt & calculate integration constants

$$a_0 = \frac{2}{\Delta t^2}; a_1 = \frac{11}{6\Delta t}; a_2 = \frac{5}{\Delta t^2};$$
$$a_3 = \frac{3}{\Delta t}; a_4 = -2a_0; a_5 = -\frac{a_3}{2};$$
$$a_6 = \frac{a_0}{2}; a_7 = \frac{a_3}{9}$$

Step 4: Calculate the effective stiffness matrix

$$[\mathbb{K}] = [\mathbf{K}] + a_0[\mathbf{M}] + a_1[\mathbf{C}] \tag{6.66}$$

Step 6: Triangularize $[\mathbb{K}] \doteq [\mathbf{L}][\mathbf{D}][\mathbf{L}]^T$
Step 7: For each time step, calculate the effective force vector from Eq. (6.65)

$$\mathbb{F}_{i+1} = \underline{\mathbf{f}}_{i+1} + [\mathbf{M}](a_2\underline{x}_i + a_4\underline{x}_{i-1} + a_6\underline{x}_{i-2}) + [\mathbf{C}](a_3\underline{x}_i + a_5\underline{x}_{i-1} + a_7\underline{x}_{i-2}) \tag{6.67}$$

Step 8: Solve

$$[\mathbb{K}]\underline{x}_{i+1} = \mathbb{F}_{i+1} \tag{6.68.1}$$

$$\underline{x}_{i+1} = [\mathbb{K}]^{-1}\{\mathbb{F}_{i+1}\} \tag{6.68.2}$$

Step 9: Calculate $\underline{\dot{x}}$ & $\underline{\ddot{x}}$ @ $t + \Delta t$ using Eqs. (6.63.1) and (6.63.2), respectively

$$\underline{\dot{x}}_{i+1} = a_2\underline{x}_{i+1} - a_3\underline{x}_i - a_5\underline{x}_{i-1} - a_7\underline{x}_{i-2} \tag{6.69.1}$$

$$\underline{\ddot{x}}_{i+1} = a_0\underline{x}_{i+1} - a_2\underline{x}_i - a_4\underline{x}_{i-1} - a_6\underline{x}_{i-2} \tag{6.69.2}$$

Next, repeat the steps from Step 2.

6.6.2 Implicit Newmark β Method

As Newmark β method is an implicit method, the governing dynamic equation is expressed at the future time step as written in Eq. (6.61), $\mathbf{M}\underline{\ddot{x}}_{i+1} + \mathbf{C}\underline{\dot{x}}_{i+1} + \mathbf{K}\underline{x}_{i+1} = \underline{\mathbf{f}}_{i+1}$. However, in Newmark β method, a slightly different approach is followed. In this method, both the primary unknown function and its first derivatives are expressed at the future point ($\underline{x}_{i+1}$) about the present point (x_i) using the Taylor series expansion as follows:

$$x_{i+1} = x_i + \Delta t\dot{x}_i + \frac{\Delta t^2}{2}\ddot{x}_i + \frac{\Delta t^3}{3!}\dddot{x}_i + O(\Delta t^3) \tag{6.70.1}$$

$$\dot{x}_{i+1} = \dot{x}_i + \Delta t\ddot{x}_i + \frac{\Delta t^2}{2}\dddot{x}_i + O(\Delta t^2) \tag{6.70.2}$$

The above equations are further expressed by introducing two new parameters β and γ as follows:

$$\underline{x}_{i+1} = \underline{x}_i + \Delta t\dot{\underline{x}}_i + \frac{\Delta t^2}{2}\ddot{\underline{x}}_i + \beta\Delta t^3\dddot{\underline{x}}_i \qquad (6.71.1)$$

$$\dot{\underline{x}}_{i+1} = \dot{\underline{x}}_i + \Delta t\ddot{\underline{x}}_i + \gamma\Delta t^2\dddot{\underline{x}}_i \qquad (6.71.2)$$

where there exists $0 \le \gamma \le 1$, and $0 \le \beta \le 1$ must be valid for the above equations to be true. In Eqs. (6.71.1) and (6.71.2), $\dddot{x}_i$ is unknown, which was eliminated in Houbolt method discussed in the previous section by introducing two additional past values at $\underline{x}_{i-1}$ and $\underline{x}_{i-2}$. However, in Newmark's method, the gradient of acceleration or the derivative of acceleration $\dddot{x}_i$ is expressed by assuming how the acceleration is changing between two consecutive time points t_i and t_{i+1}. For a linear system, $\dddot{x}_i$ can be expressed as

$$\dddot{x}_i = \frac{\ddot{x}_{i+1} - \ddot{x}_i}{\Delta t} \qquad (6.72)$$

Substituting Eq. (6.72) in Eqs. (6.71.1) and (6.71.2), the displacement and velocity vectors can be expressed as

$$\underline{x}_{i+1} = \underline{x}_i + \Delta t\dot{\underline{x}}_i + \left(\frac{1}{2} - \beta\right)\Delta t^2\ddot{\underline{x}}_i + \beta\Delta t^2\ddot{\underline{x}}_{i+1} \qquad (6.73.1)$$

$$\dot{\underline{x}}_{i+1} = \dot{\underline{x}}_i + (1-\gamma)\Delta t\ddot{\underline{x}}_i + \gamma\Delta t\ddot{\underline{x}}_{i+1} \qquad (6.73.2)$$

where there exists $0 \le \gamma \le 1$, and $0 \le \beta \le 1/2$ must be valid for the above equations to be true. γ and β are parameters are introduced in the equation such that they can be chosen depending on the desired accuracy and stability of the algorithm discussed later in this section. There are many options to choose how the acceleration is changing between two time points t_i and t_{i+1}, but the two most basic options are to assume (a) an average acceleration and (b) the acceleration in changing linearly between the time step $t_{i+1} - t_i = \Delta t$. Table 6.3 shows the selection of the values of γ and β for linear acceleration and average acceleration methods.

For zero damping, Newmark's method is conditionally stable if

$$\gamma \ge \frac{1}{2},\ \beta \le \frac{1}{2}\ \textit{and}\ \Delta t \le \frac{1}{\omega_{max}\sqrt{\gamma/2 - \beta}} \qquad (6.74)$$

TABLE 6.3
Parameters for Newmark β Methods

Acceleration	β	γ
Average acceleration	1/4	1/2
Linear acceleration	1/2	1/6

where ω_{max} is the maximum natural frequency in the structural system. Newmark's method is unconditionally stable if

$$2\beta \geq \gamma \geq \frac{1}{2} \tag{6.75}$$

However, if γ is greater than $\frac{1}{2}$, errors are introduced. These errors are associated with "numerical damping" and "period elongation" [chapra]. For a large multidegrees-of-freedom system, the time should be chosen more carefully as discussed in Section 6.5.1. The limit for time step in Eq. (6.74) can be written in a more generic form as

$$\frac{\Delta t}{T_{min}} \leq \frac{1}{2\pi\sqrt{\gamma/2 - \beta}} \tag{6.76}$$

Simulation models for any large real structure or for any CNDE problem normally contain a large number of periods, which are smaller than the integration time step; therefore, it is essential that one selects a numerical integration method for CNDE problem that is unconditional for all the time steps. Satisfying all the above conditions, the future value of the function ($\underline{x}_{i+1}$) can be ultimately solved using the following equation, which is similar to Eq. (6.49) as in Eq. (6.77), except the effective matrices are written elaborately in Eqs. (6.78) and (6.79), respectively

$$\mathbf{M}^e\ \underline{x}_{i+1} = \mathbf{F}^e{}_{i+1}\ or\ \underline{x}_{i+1} = \mathbf{F}^e{}_{i+1} / \mathbf{M}^e \tag{6.77}$$

where

$$\mathbf{M}^e = \frac{1}{\beta\Delta t^2}[\mathbf{M}] + \frac{\gamma}{\beta\Delta t}[\mathbf{C}] + [\mathbf{K}] \tag{6.78}$$

$$\begin{aligned}\mathbf{F}^e{}_{i+1} = \underline{\mathbf{f}}_{i+1} &+ \left[\left(\frac{1}{2\beta} - 1\right)[\mathbf{M}] + \Delta t\left(\frac{\gamma}{2\beta} - 1\right)[\mathbf{C}]\right]\ddot{x}_i \\ &+ \left[\frac{1}{\beta\Delta t}[\mathbf{M}] + \left(\frac{\gamma}{\beta} - 1\right)[\mathbf{C}]\right]\dot{x}_i + \left[\frac{1}{\beta\Delta t^2}[\mathbf{M}] + \frac{\gamma}{\beta\Delta t}[\mathbf{C}]\right]\underline{x}_i\end{aligned} \tag{6.79}$$

Further, after solving the value of the primary variable at the future time step, velocity and accelerations at the future time step are solved as follows:

$$\underline{\dot{x}}_{i+1} = \frac{1}{\beta\Delta t}(\underline{x}_{i+1} - \underline{x}_i) - \left(\frac{\gamma}{\beta} - 1\right)\underline{\dot{x}}_i - \Delta t\left(\frac{\gamma}{2\beta} - 1\right)\underline{\ddot{x}}_i \tag{6.80}$$

$$\underline{\ddot{x}}_{i+1} = \frac{1}{\beta\Delta t^2}(\underline{x}_{i+1} - \underline{x}_i) - \frac{1}{\beta\Delta t}\underline{\dot{x}}_i - \left(\frac{1}{2\beta} - 1\right)\underline{\ddot{x}}_i \tag{6.81}$$

Further in the next time steps, the above values are used as present values and the algorithm is proceeded to solve the system entirely over a desired duration of time.

The Newmark method is adopted to solve CNDE problem using spectral element method as described in Chapter 10.

6.6.3 Implicit Wilson θ Method

Wilson θ method is an extension of the linear acceleration method that is presented under Newmark β method in Section 6.6.2. Wilson θ method is an implicit method to find the unknown function and its derivatives at any arbitrary time step $\theta\Delta t$ where $\theta < 1$. When $\theta = 1$, with linear acceleration assumption, it will result the Newmark β method. In Wilson θ method, the second-order derivative or acceleration of the primary unknown variable function at any arbitrary time step $t + \tau$ is assumed as

$$\ddot{\underline{x}}_{t+\tau} = \ddot{\underline{x}}_t + \frac{\tau}{\theta\Delta t}\left(\ddot{\underline{x}}_{t+\theta\Delta t} - \ddot{\underline{x}}_t\right) \tag{6.82}$$

where $\ddot{\underline{x}}_{t+\theta\Delta t}$ is the acceleration later $\theta\Delta t$ (with $\theta < 1$) from the known time point t. Now integrating Eq. (6.82), applying initial conditions and taking τ as any arbitrary time step, the velocity and displacement functions at $t + \tau$ time point can be written as

$$\int \ddot{\underline{x}}_{t+\tau} = \int \ddot{\underline{x}}_t + \int \frac{\tau}{\theta\Delta t}\left(\ddot{\underline{x}}_{t+\theta\Delta t} - \ddot{\underline{x}}_t\right) \tag{6.83.1}$$

$$\dot{\underline{x}}_{t+\tau} = \dot{\underline{x}}_t + \ddot{\underline{x}}_t\tau + \frac{\tau^2}{2\theta\Delta t}\left(\ddot{\underline{x}}_{t+\theta\Delta t} - \ddot{\underline{x}}_t\right) \tag{6.83.2}$$

$$\int \dot{\underline{x}}_{t+\tau} = \int \left(\dot{\underline{x}}_t + \ddot{\underline{x}}_t\tau\right) + \int \frac{\tau^2}{2\theta\Delta t}\left(\ddot{\underline{x}}_{t+\theta\Delta t} - \ddot{\underline{x}}_t\right) \tag{6.84.1}$$

$$\underline{x}_{t+\tau} = \underline{x}_t + \dot{\underline{x}}_t\tau + \frac{1}{2}\ddot{\underline{x}}_t\tau^2 + \frac{\tau^3}{6\theta\Delta t}\left(\ddot{\underline{x}}_{t+\theta\Delta t} - \ddot{\underline{x}}_t\right) \tag{6.84.2}$$

Considering $\tau = \theta\Delta t$ a θ fraction of original time step Δt using in the previous algorithm and replaced in Eqs. (6.83.2) and (6.84.2), one gets

$$\dot{\underline{x}}_{t+\theta\Delta t} = \dot{\underline{x}}_t + \frac{\theta\Delta t}{2}\left(\ddot{\underline{x}}_{t+\theta\Delta t} + \ddot{\underline{x}}_t\right) \tag{6.85.1}$$

$$\underline{x}_{t+\theta\Delta t} = \underline{x}_t + \dot{\underline{x}}_t\theta\Delta t + \frac{(\theta\Delta t)^2}{6}\left(\ddot{\underline{x}}_{t+\theta\Delta t} + 2\ddot{\underline{x}}_t\right) \tag{6.85.2}$$

As Wilson θ method is an implicit method, the governing dynamic equation is expressed at the future arbitrary time step as follows:

$$\mathbf{M}\ddot{\underline{x}}_{t+\theta\Delta t} + \mathbf{C}\dot{\underline{x}}_{t+\theta\Delta t} + \mathbf{K}\underline{x}_{t+\theta\Delta t} = \underline{\mathbf{f}}_{t+\theta\Delta t} \tag{6.86}$$

Substituting Eqs. (6.85.1) and (6.85.2) in Eq. (6.86), like before in other method presented in the previous sections, the primary variable at the future time step at $t + \theta\Delta t$ can be found using the following equation:

$$\mathbf{M}^e\ \underline{x}_{t+\theta\Delta t} = \mathbf{F}^e{}_{t+\theta\Delta t}\ \textit{or}\ \underline{x}_{t+\theta\Delta t} = \mathbf{F}^e{}_{t+\theta\Delta t} / \mathbf{M}^e \tag{6.87}$$

where

$$\mathbf{M}^e = \frac{6}{\theta^2\Delta t^2}[\mathbf{M}] + \frac{3}{\theta\Delta t}[\mathbf{C}] + [\mathbf{K}] \tag{6.88}$$

$$\mathbf{F}^e{}_{t+\theta\Delta t} = \underline{\mathbf{f}}_t + \theta\left(\underline{\mathbf{f}}_{t+\theta\Delta t} - \underline{\mathbf{f}}_t\right) + \left[2[\mathbf{M}] + \frac{\theta\Delta t}{2}[\mathbf{C}]\right]\ddot{x}_t + \left[\frac{6}{\theta\Delta t}[\mathbf{M}] + 2[\mathbf{C}]\right]\dot{x}_t + \left[\frac{6}{(\theta\Delta t)^2}[\mathbf{M}] + \frac{3}{\theta\Delta t}[\mathbf{C}]\right]\underline{x}_t \tag{6.89}$$

Solving Eq. (6.87), displacement function at any arbitrary time step $\underline{\dot{x}}_{t+\theta\Delta t}$ is obtained. Then the algorithm proceeds like other algorithms discussed for other methods in previous sections.

6.7 VELOCITY VERLET INTEGRATION SCHEME

Verlet method or simply the velocity verlet integration scheme is frequently used in molecular dynamics simulation to solve the second-order Newton's equation, which involves force and acceleration. In Perielastodynamic simulation method for CNDE, similar set of equations are obtained as discussed in Chapter 11. Hence, solution of perielastodynamic approach requires solving equations similar to the molecular dynamic formulation. The equation of motion (or Newton's equation of motion) for a conservative multidegrees-of-freedom system with second-order ordinary differential equations can be written as

$$\mathbf{M}\underline{\ddot{x}} = \underline{\mathbf{f}}(\underline{x}) = -\nabla V(\underline{x}) \tag{6.90}$$

where $\mathbf{M}$ is the mass matrix, $\underline{\mathbf{f}}(\underline{x})$ is the force vector composed of forces acting on each degrees of freedom, $\underline{x}$ is the vector or ensemble of the displacements of all points in the system. $V(\underline{x})$ is the scalar potential function. Denoting $\underline{\ddot{x}} = \underline{a}$, which is acceleration, $\underline{\dot{x}} = \underline{v}$, which is velocity, acceleration at any time t_{i+1} can be written as

$$\underline{a}_{i+1} = \underline{\mathbf{f}}(\underline{x}_{i+1})/\mathbf{M} \tag{6.91}$$

Hence, similarly from the above equation, the acceleration at the beginning of the time at $t = 0$ is derived as follows:

$$\underline{a}_0 = \underline{\mathbf{f}}(\underline{x}_0)/\mathbf{M} \tag{6.92}$$

Using the acceleration obtained at the previous time step, the initial conditions for displacements and velocities $\underline{x}_0$ and $\underline{v}_0$, the velocity at the half time step can be written as

$$\underline{v}_{0+1/2} = \underline{v}_0 + \frac{\Delta t}{2}\underline{a}_0 \tag{6.93}$$

At any later time step, a generalized equation replicating the above situation can be written as

$$\underline{v}_{j+1/2} = \underline{v}_j + \frac{\Delta t}{2}\underline{a}_i \tag{6.94}$$

Using the velocity calculated above, the displacements at the full-time step can be calculated as

$$\underline{x}_{i+1} = \underline{x}_{j+1} + \Delta t \ \underline{v}_{j+1/2} \tag{6.95}$$

Using the displacement value calculated at the end of the full-time step and substituting the values, new force vector can be obtained. Next, the acceleration at the end of the full-time step can be obtained using Eq. (6.91) $\underline{a}_{j+1} = \underline{\mathbf{f}}\left(\underline{x}_{i+1}\right) / \mathbf{M}$

Using the acceleration obtained $\underline{a}_{j+1}$, the velocity at the end of the full-time step is calculated as follows:

$$\underline{v}_{i+1} = \underline{v}_{j+1/2} + \frac{\Delta t}{2}\underline{a}_{j+1} \tag{6.96}$$

Once the velocity and accelerations, $\underline{v}_{j+1}$ and $\underline{a}_{j+1}$, are found at the future time step at t_{i+1}, the algorithm can proceed to solve for the displacement, velocities, and accelerations at the other time steps.

6.8 OVERVIEW OF PARALLEL COMPUTING FOR CNDE

Many CNDE methods including few discussed in this book in the following chapters are computationally demanding. It has been realized in past few years that it is impossible to tackle or simulate a reasonable size NDE problem that is conducted in the laboratories without exploiting the computational power in a parallel manner. Hence, CNDE needs parallel computing. Parallel computing for CNDE is conceptualized in a manner where it processes a part(s) of the CNDE algorithms in parallel that are independent from each other and can rejoin afterward to understand the physics in detail. Necessary parallel computing platforms and respective architectures are discussed under each method in respective chapters, but here in this section, a basic overview of parallel computing is presented.

6.8.1 What is Parallel Computing

Parallel computing is essentially a type of computation in which multiple calculations are performed at the same time. This operation is performed on the basis that a big sequential or serial problem can be split up into smaller parallel problems such that the problem can be solved simultaneously in a more efficient manner [6]. Development of computers in past few decades has helped advance the technological frontier, scientific breakthroughs in many ways and helped humans to simulate the unknown natural phenomena, which were not in the grasp of humans even a hundred years back by leaping the computing capabilities and enhancing the computational bounds [7]. Even more so, the development of parallel computing has exponentially increased the speed with which the information can be processed. This made possible to tackle and solve previously unsolvable complex scientific research problems in various field such as science, engineering, and information technology. Parallel computing has been able to turn a new leaf with the appearance of hardware with multicore designs [8]. The use of parallel hardware has been global as of current time. All the newly developed laptops, desktops, and even servers use multicore processors. And these new platforms require development of software in a new manner; one that can fully exercise the benefits of multiple cores.

6.8.2 Historical Background of Parallel Computing

The idea of development of parallel hardware and software comes from shortcomings of the conventional serial hardware and software that runs a single job at a time but needs faster computation and better overall performance. The shift in computing architectures from traditional serial machines to a model with multiple not-so-fast processors put together started in early 1980s. Although designing parallel computers was difficult, it was put into motion due to very high predicted gains. Over time, researchers found ways to utilize parallel machines for various scientific applications. The upsurge of parallel computing was a conceptual parting from the expensive-to-build supercomputer since it was able to accomplish the foundation of better computing power by making use of hundreds of thousands—of microprocessors, all running calculations concurrently [9]. These small processors working together with memory and an interconnect were the building blocks of a new computing paradigm, leading to a development and advancement of diverse parallel computing architectures within the next decade.

The systems with multiple cores and processors were no longer just the domain of supercomputing but rather ubiquitous [7]. The new laptops and even mobile phones contain more than one processing core. The mainstream adoption of parallel computing is a result of the cost of components dropping due to Moore's law that the number of transistors on a microchip doubles every two years, though the cost of computers halved [10]. Due to this, the cost of hardware such as CPU has gone down and various multicore chips have been released, such as the 64-core Tilera TILE64 and the Cell BE. These chips were a natural evolution of the multisocket platforms, i.e., machines that could host several CPUs each on a separate chip, of the mid-to-late 1990s [11]. The manufacturers such as Dell and Apple have produced even faster machines for the home market that easily outperform the supercomputers of old that once took a room to house. Devices that contain multiple cores allow us to explore parallel-based programming on a single machine.

After the development of multiple core CPU hardware was the arrival of GPGPU (General Purpose Graphical Processing Unit) computing, i.e., the concept of using Graphical Processing Unit (GPU) for General Purpose computing [12]. Although a single GPU core cannot be compared with a contemporary CPU core, the massively parallel architectures with hundreds or thousands of cores connected with high-bandwidth and high-speed RAM of the GPUs are able to overcome this disadvantage. As a result, the computational speed of GPU is several magnitudes faster. In addition, GPGPU also has a definite advantage in terms of energy consumption. In other words, you can get more computation done with less energy consumption. This is especially critical in the server and cloud infrastructure domain, where the energy consumed by a CPU over its operational lifetime can be much higher than its actual price. The development of GPGPU technology is a revolutionary step. It enables the solving of problems that were previously not possible with contemporary single or even multicore CPU technology. Although these multicore architectures have significantly improved the computation performance, it requires an explicit redesign of algorithms such that the traditional serial program can be transformed into a parallel program.

6.8.3 Serial vs. Parallel Computing for CNDE

Before the development of multiple core processors, most programs were written for sequential operation, which utilized the then standard, single-core systems. Even after the progress in the hardware, many researchers in various fields still are mostly writing sequential programs since they are unaware of the presence and prospects of multiple cores and parallel computing. Even though many programs can obtain acceptable performance on a single core, the researchers need to be made aware of the enormous runtime improvements that can be obtained with implementation of parallel computing. The researchers unknowingly try to utilize the multiple cores by running multiple instances of the same program. In addition to that, simply running serial programs in a computer with multiple core processors will not improve the performance of program either. The reason for that is very simple: the serial programs are not designed to utilize multiple processors and hence are ignorant of their existence. Therefore, the effectiveness of such a program on a system with multiple processors will be the same as its performance on a single processor of the multiprocessor system. However, that is not what is required for CNDE. Rather, for CNDE, it is necessary to have a faster execution and runtime for the program that one is running in an efficient and timely manner. To achieve this goal, it is required to transform the serial programs into parallel programs and utilize multiple cores in the current computers to the fullest extent [13]. This can be done in two ways: (1) develop software and libraries that can automatically convert serial programs into parallel programs or (2) manually rewrite the serial program so that it can be executed in parallel fashion.

6.8.4 Methods for Parallel Programs

It is known that for faster computation, parallel computation is necessary. One can do so by writing parallel program that supports parallel computing. There are many theoretical ideas as to how we proceed with writing a parallel program. The most fundamental and philosophical concept is that to break a bigger problem into cluster of smaller problems, solve one at a time or parallelly. Hence, a concept that is used to accomplish the task is by apportioning the at-hand work into smaller works to be completed among the cores. There are two extensively used approaches: (1) task-parallelism and (2) data-parallelism [7].

6.8.4.1 Task-Parallelism

In task-parallelism, various tasks are partitioned and carried out in solving the problem among the cores, i.e., one has multiple tasks that need to be completed. This form of parallelism covers the simultaneous execution of computer programs across multiple processors on same or multiple machines. It is the execution on multiple cores of many different functions across the same or different datasets. It focuses on executing different operations or tasks in parallel to fully utilize the available computing resources in form of processors and memory. One way to do so would be creating threads for doing parallel processing where each thread is responsible for performing a different operation. For example, if one has a large data set, it is required to know the minimum, maximum, and the average value from the data set. One can have different processors, each look at the same data set and compute three different answers. Hence, in task parallelism, tasks are divided into several tasks to be processed.

6.8.4.2 Data-Parallelism

In data-parallelism, data used are repartitioned in solving the problem among the cores, and each core carries out similar operations on its part of the data. This form of parallelism focuses on distribution of data sets across the multiple computation programs. It is the simultaneous execution on multiple cores of the same function across the elements of a dataset. One way to do so is to divide the input data into subsets and pass it to the threads performing same task on different CPUs. For example, one can have a lot of data that is required to be processed quickly. Data may be in the form of a lot of pixel data from an image or a lot of payroll cheques to be updated. Taking that data and dividing it up among multiple processors is a method of getting data parallelism.

6.8.4.3 Simple Example of Parallelization

For a clear distinction between the two parallelisms, here is an example showcasing both in a single operation. Suppose that a class with one hundred students had a midterm exam consisting of four questions. The class has four graders: G1, G2, G3, and G4. In order to grade the exam, four graders can use the following two approaches: (1) Each of them can grade all one hundred students on four different questions, i.e., G1 grades question 1, G2 grades question 2, and so on. (2) They can divide the one hundred students' exam papers into four subsets of twenty-five exams each, and each of them can grade all the papers in one of the subsets, i.e., G1 grades the papers in the first subset, G2 grades the papers in the second subset, and so on.

In both approaches, the "cores" are the graders. The first approach can be considered as the task-parallelism. There are four tasks to be carried out: grading the first question, grading the second question, and so on. So, the graders will be "executing different tasks in parallel." On the other hand, the second approach can be considered as data-parallelism. The "data" are the students' exam papers, which are divided among the cores, and each core grades all four questions on each paper.

6.8.5 Understanding the Patterns in Parallel Program Structure

To be able to select the appropriate parallelization approach, it is of great importance that we understand the patterns of the program structure. We can distinguish the parallel program structure patterns into two major categories [14]:

Globally Parallel, Locally Sequential (GPLS): GPLS implies that the program can perform multiple tasks in parallel; however, each of the tasks runs sequentially. Some of the distinguished patterns that fall into this category include are Single program, multiple data and Multiple program, multiple data.

Globally Sequential, Locally Parallel (GSLP): GSLP implies that the program runs as a sequential program; however, some of the tasks can run in parallel when required. Some of the distinguished patterns that fall into this category include are Fork/join and Loop parallelism.

6.8.6 Types of Parallel Hardware

It is essential to understand the architectural characteristics of parallel machines. In 1966, Michael Flynn introduced a taxonomy of computer architectures, which distinguishes between the number of instruction streams and the number of data streams a system can handle, i.e., how many data and instructions they can execute simultaneously. For both cases, it can either be single or multiple, which means that their combination can produce four possible outcomes:

6.8.6.1 Single Instruction, Single Data (SISD)

A simple system that executes one instruction at a time, operating on a single data item. A von Neumann system has a single instruction stream and a single data stream, so it is classified as a single instruction, single data, or SISD, system. Nowadays, most of the CPUs have multiple core processors configuration and each of those cores can be considered a SISD machine.

6.8.6.2 Single Instruction, Multiple Data (SIMD)

A system that executes a single instruction at a time, but the instruction can be applied on multiple data items. This type of system often executes its instructions in lockstep. The first instruction is applied to all the data items simultaneously; only then the consequent instructions are applied. This type of parallel system is usually employed in data parallel programs, programs in which the data are divided among the processors and each data item is subjected to the same set of instructions. Vector processors were the first systems that followed this concept. Similarly, graphics processing units are also classified as SIMD systems; Streaming Multiprocessor (SM for Nvidia) or the SIMD unit (for AMD).

6.8.6.3 Multiple Instructions, Single Data (MISD)

A system that executes multiple instructions on single data item, i.e., performing different operations on the same data. Systems built using the MISD model are not useful in most of the application. However, it is useful when fault tolerance is required in a system.

6.8.6.4 Multiple Instructions, Multiple Data (MIMD)

A system that executes multiple independent instructions and each of those instructions can have its own multiple data. It is considered the most versatile system. MIMD system, practically, is an assemblage of autonomous processors that perform execution independently. Multicore machines, including GPUs, follow this concept. GPUs are made from a collection of SM/SIMD units, whereby each can execute its own program. So, although each unit is considered SIMD system, collectively they conduct as a MIMD system.

Parallel computers can be roughly classified according to the level at which the hardware supports parallelism, with multicore and multiprocessor computers having multiple processing elements within a single machine, while clusters, massively parallel processors (MPPs), and grids use multiple computers to work on the same task. Specialized parallel computer architectures are sometimes used alongside traditional processors, for accelerating specific tasks.

6.8.7 Type of Parallel Software

Parallel processing software is an application that manages the execution of tasks in a program on a parallel computing architecture by dispensing huge application calls between multiple CPU and GPU cores within an underlying architecture reducing runtime. Specific algorithms are built for efficient task processing. It is used to solve large and complex back-end computations and programs. Parallel processing manages division and distribution of task between processors. Its primary purpose is to utilize processors to ensure that throughput, application availability, and scalability provide optimal end-user processing through the usage of multiple core processors.

6.8.7.1 Parallel Programming Languages

Various parallel programming languages, libraries, Application Programming Interfaces (APIs), and parallel programming models are developed for parallel computing. These programming languages are based on the underlying memory architecture—shared memory and distributed memory. In a shared-memory system, the cores can share access to the computer's memory, i.e., each core can read and write each memory location. In a distributed memory system, conversely, each core has its own, private memory, and the cores must communicate explicitly by doing something like sending messages across a network. Shared memory programming languages communicate by manipulating shared memory variables. Distributed memory uses message passing. POSIX Threads (defined by IEEE Std 1003.1c-1995) and OpenMP are two of the most widely used shared memory APIs, whereas Message Passing Interface (MPI) is the most widely used message-passing system API.

6.8.7.2 Automatic Parallelization

Automatic parallelization or simply called auto-parallelization refers to converting serial program into parallel program by employing multiple processors simultaneously in a shared-memory multiprocessor (SMP) system [15]. The objective of automatic parallelization, as the name implies, is to automate the process of transforming serial program to parallel such that the programmers do not have to go through the hectic and error-prone manual parallelization process [16]. Though the quality of automatic parallelization has improved in the past several decades, fully automatic parallelization of sequential programs by compilers is still far from being a standard norm. The auto-parallelization mostly focuses on loops since most of the execution time of a program takes place inside of loop.

Mainstream parallel programming languages remain either explicitly parallel or (at best) partially implicit, in which a programmer gives the compiler directives for parallelization. A few fully implicit parallel programming languages exist—SISAL [17], Parallel Haskell, System C (for FPGAs), Mitrion-C, VHDL, and Verilog etc.

6.8.8 CPU vs GPU Parallel Computing

With the advancement in the field of parallel computing, most prominently the serial programs are being parallelized using two different approaches: CPU and GPU. A CPU primarily decides what to do to a data item depending on the tasks that are already completed. Parallel programming for CPUs is about differentiating instructions that can take place simultaneously from those that take place in sequence and interpreting them accordingly. A GPU primarily decides what to do to a data item based on its location among other data items. Parallel programming for GPUs is about subdividing the input data using a coordinate system that one invents to distinguish between data items that need to be processed with different instructions. A CPU can do parallel computing using its cores. Each core is strong with significant processing power. So, a CPU core can execute a big task few times due to hardware limit implemented for a core and the core count. If one compares CPU with a GPU, GPU will have hundreds of cores with limited processing power. However, all weak GPU cores execute a single instruction at a time, depending on the calculations they need to do; the GPU architecture is suitable to finish the specific job much faster. The important part is that the GPU is a system, not just a processor (singular). The GPU system is organized for graphics problems with a massively parallel architecture consisting of thousands of smaller efficient cores designed for handling multiple tasks simultaneously. A CPU, on the other hand, is organized to address sequential processing and has a lot of cache and coherency and isolation between the two 12 CPU cores (72 in a new Intel announcement). The CPU parallel computing is done by writing parallel programs using Message-Passing Interface (MPI), POSIX threads or Pthreads, and OpenMP—three of the most widely used application programming interfaces (APIs) for parallel programming [7]. MPI and Pthreads are libraries of type definitions, functions, and macros that can be used in various compilers. Pthreads and OpenMP were designed for programming shared-memory systems. They provide mechanisms for accessing shared-memory locations. MPI, on the other hand, was designed for programming distributed-memory systems. It provides mechanisms for sending messages. OpenMP is a relatively high-level extension if used in C/C++. It can parallelize a loop with a single directive. On the other hand, Pthreads provides some coordination constructs that are unavailable in OpenMP. OpenMP allows us to parallelize many programs with relative ease, while Pthreads provides us with some constructs that make other programs easier to parallelize.

Similarly, GPU parallel computing has been advancing with leaps and bounds since the first early attempts. Current tools cover a wide range of capabilities as far as problem decomposition and expressing parallelism are concerned. On one side of the spectrum, we have tools that require explicit problem decomposition, such as Compute Unified Device Architecture (CUDA) and OpenCL, and on the other extreme, we have tools like OpenACC that let the compiler take care of all the data migration and thread spawning necessary to complete a task on a GPU [11].

6.8.8.1 CPU Parallel Computing using OpenMP

OpenMP, which stands for Open Multi-Processing, is parallel computing API that can transform the serial program into parallel without the need to rewrite it completely, taking advantage of multicore hardware [18]. If one has an existing unparalleled computer code in C/C++, it is obvious that rewriting the whole program will encounter numerous difficulties in terms of both cost and correctness. Hence, utilizing OpenMP is the solution to this problem. OpenMP is proficient in allowing the gradual conversion of sequential programs to parallel ones. In OpenMP, the compiler is in charge of processing all the parallelizing specifics such as spawning, initiating, and terminating threads. The OpenMP API consists of compiler directives, library routines, and environmental variables. The execution outline of OpenMP supports the Globally Sequential, Locally Parallel (GSLP) structure. Essentially, the compiler transforms the computation intensive portion of a sequential program into parallel. For example, implementation of a CNDE modeling problem using DPSM, discussed in Chapter 7, used C/C++ compiler, which comes with OpenMP support. The instructions to the compiler come in the form of #pragma preprocessor directives. Pragma directives allow a programmer to access compiler-specific preprocessor extensions [19]. After the implementation of OpenMP, with DPSM, a speedup of less than 5 times was achieved. However, higher speedup was necessary and other parallel coding opportunities were explored. Thus, next approach was to proceed to implement GPU parallel computing using CUDA.

6.8.8.2 GPU Parallel Computing using CUDA

GPU parallel computing is a newly developed innovative platform. It is utilized in many computation intensive fields such as applied computation, engineering design and analysis, simulation, machine learning, and vision and imaging systems. CUDA is currently the best open platform for exploiting into the capability of GPU parallel computing. CUDA is a hardware/software platform for parallel computing introduced by Nvidia in late 2006 using GPU for general purpose computing. The CUDA hardware entails graphics cards equipped with one or more CUDA-enabled graphics processing units (GPUs) exclusively developed by NVIDIA [20]. The NVIDIA CUDA Toolkit software offers a development environment with APIs and libraries, which are supported in C/C++ compilers. It can be implemented in Windows, Mac OS X, and Linux operating systems. The basis for parallel computing with CUDA is memory transfers between the host and device. In CUDA, CPU is called host and its memory is called host memory and

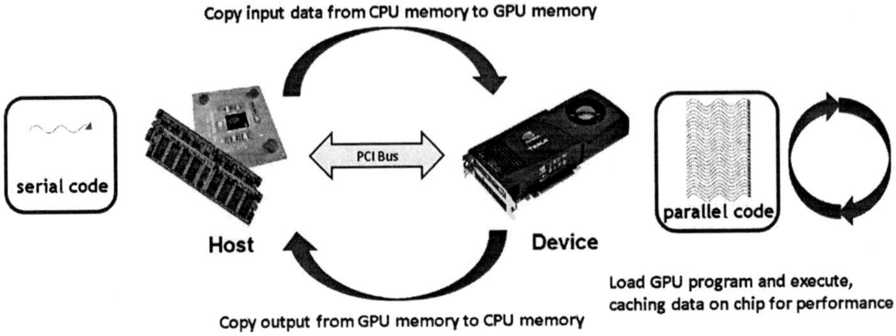

FIGURE 6.6 Parallel computing architecture.

GPU is the device and its memory is called device memory. The execution of a program or code starts from the host, where the users have all the required input data. The input data are then copied from host memory to device memory. Then the CUDA program is executed in parallel in the device and the user can compute the output results. Finally, the output is copied back to host memory. This is how the CUDA program is executed and one can see that it follows the Globally Sequential Locally Parallel pattern. The framework for the CUDA implementation adopted for several CNDE methods discussed in following chapters is shown in Fig. 6.6. Next, a simple example shows how CUDA code is written and implemented in a computer code. It showcases a CUDA program that executes a simple multiplication of two numbers, a and b.

```
CUDA Code:
        __global__void multiply(int *p1, int *p2, int *p3) {
                *m = *p1 + *p2;
        }
Main Code:
        Int main(void) {
                int p1, p2, m;        //host copies of p1, p2, m
                int *d_p1, *d_p2, *d_m;    //device copies of
                p1, p2, m
                int size = sizeof(int);

                // Allocate space for device copies of p1, p2, m
                cudaMalloc((void **)&d_p1,size);
                cudaMalloc((void **)&d_p2,size);
                cudaMalloc((void **)&d_m,size);
                //Setup input values
                a=2;
                b=7;

                // Copy inputs to device
                cudaMemcpy(d_p1, &a, size, cudaMemcpyHostToDevice);
                cudaMemcpy(d_p2, &b, size, cudaMemcpyHostToDevice);
```

```
        // Launch multiply() CUDA function on GPU
        Add<<<1,1>>>(d_p1,d_p2,d_m);

        // Copy result back to host
        cudaMemcpy(&m, d_m, size, cudaMemcpyDeviceToHost);
        // Cleanup
        cudaFree(d_p1); cudaFree(d_p2); cudaFree(d_m);
        return 0;
}
```

In the above code, there is a main code that is executed in CPU or host and CUDA code that is executed in GPU or device in parallel.

REFERENCES

1. Chapra, S.C., Canale, R. P., *Numerical Methods for Engineers*, 7th ed. 2010, New York: McGraw-Hill Education.
2. Lindfield, G., Penny, J., *Numerical Methods using Matlab*, 4th ed. 2018, New York: Academic Press.
3. Dukkipati, R.V., *MATLAB for Engineers*. 2009, New Delhi: New Age Science.
4. LeVeque, R.J., *Finite Difference Methods for Ordinary and Partial Differential Equations: Steady-State and Time-Dependent Problems*. 2007, Philadelphia, PA: Society for Industrial and Applied Mathematics (SIAM).
5. Chopra, A.K., *Dynamics of Structures*, 5th ed. International Series, Civil Engineering and Engineering Mechanics. 2016, India: Prentice-Hall.
6. Gottlieb, A. and G. Almasi, *Highly Parallel Computing*, 1989, Redwood City, CA: Benjamin/Cummings.
7. Pacheco, P., *An Introduction to Parallel Programming*, 2011, Burlington, VT: Elsevier.
8. Asanovic, K., et al., *The Landscape of Parallel Computing Research: A View from Berkeley*. 2006, Technical Report UCB/EECS-2006-183, EECS Department, University of California at Berkeley.
9. Ringman, N., *Using Graphical Processors to Implement Radio Base Station Control Plane Functions*, 2019.
10. Schaller, R.R., Moore's law: past, present and future. *IEEE Spectrum*, 1997. **34**(6): pp. 52–59.
11. Barlas, G., *Multicore and GPU Programming: An Integrated Approach*. 2014, Amsterdam: Elsevier.
12. Nickolls, J. and D. Kirk, *Graphics and computing GPUs*, in *Computer Organization and Design: The Hardware/Software Interface*, D.A. Patterson and J.L. Hennessy, Editors, 4th ed. 2009, San Francisco, CA: Morgan Kaufmann, pp. A2–A77.
13. Barney, B., Introduction to parallel computing. *Lawrence Livermore National Laboratory*, 2010. **6**(13): p. 10.
14. Berekovic, M., et al., *Architecture of Computing Systems—ARCS 2018*. 2009.
15. Shen, J.P. and M.H. Lipasti, *Modern Processor Design: Fundamentals of Superscalar Processors*. 2013, Long Grove, Illinois: Waveland Press.
16. Yehezkael, R.B., et al. *Experiments in separating computational algorithm from program distribution and communication*, in *International Workshop on Applied Parallel Computing*. 2000, Berlin, Heidelberg: Springer.
17. Cann, D., Retire fortran? a debate rekindled. *Communications of the ACM*, 1992. **35**(8): pp. 81–90.

18. Costa, J.J., et al. *Running OpenMP applications efficiently on an everything-shared SDSM*, in *18th International Parallel and Distributed Processing Symposium, 2004. Proceedings*. 2004, New York: IEEE.
19. Eigenmann, R. and B.R. de Supinski. *Openmp in a new era of parallelism*, in *4th International Workshop, IWOMP*. 2008, Berlin, Heidelberg: Springer.
20. Zibula, A., *General Purpose Computation on Graphics Processing Units (GPGPU) using CUDA*. Winter Term 2009, 2012. **1010**.

7 Distributed Point Source Method for CNDE

7.1 BASIC PHILOSOPHY OF DISTRIBUTED POINT SOURCE METHOD (DPSM)

7.1.1 DPSM and Other Methods

DPSM is very useful to simulate real time ultrasonic nondestructive evaluation (NDE) and structural health monitoring (SHM) problems. In NDE/SHM, an active source of energy or actuation of ultrasonic wave is fundamental to a problem. In traditional NDE this ultrasonic energy is emitted in water or in couplant. Then the energy carrying fluid interacts with the structure to be inspected and transmit the energy inside the solid material. Hence, DPSM should be capable of simulating bounded ultrasonic beams discussed in Chapter 1, emitting active energy from the ultrasonic transducers. Also, DPSM should be capable to simulating the fluid-structure interaction of waves at the boundaries and at the interfaces. In fact, DPSM is versatile in fulfilling the above requirements. DPSM is a semi analytical technique where numerical computation is performed utilizing analytical Green's functions. At the very heart of any numerical method, the fundamental process requires discretization of the problem domain where the governing differential equations are valid. From the discussion presented in Chapter 3, it is clear that the governing wave propagation equation in Eq. (3.49) was obtained from an integral equation and is valid at any point located anywhere in the problem domain. DPSM is also not an exception. However, at this point it is wise to present the unique differences of DPSM compared to the most popular and traditional numerical methods. One of those popular methods is Finite Element Method (FEM). Basic understanding of readers on FEM is assumed and is not explicitly discussed in this book. In FEM, the whole problem domain is discretized with multiple nodal points and respective shape functions along their connecting lines. Mathematically the primary unknowns at any points inside the element (i.e., anywhere in the space between the nodal points) are defined by respective shape functions multiplied with the nodal values of the primary unknowns. The line joining the nodal points ensures the connectivity of the space in between, where required accuracy and the order of continuity are ensured. In wave propagation problems, primary unknowns are the particle displacements or the particle velocities. So, the displacements and velocities are the unknowns at the nodal points. After solving the system of equation, primary variable anywhere in the domain can be found using the shape functions and the secondary variables like stresses are found at the Gauss points through numerical integration.

In DPSM however, it is not necessary to discretize the whole domain. Only the wave sources and wave interfaces are discretized. Hence, it can be said that all the boundaries and the interface associated with a problem are required to be discretized. But DPSM is not similar to the conventional Boundary Element Method (BEM) [1, 2] either. In BEM, the nodal points are distributed along the boundary, but in DPSM the point sources are distributed around the boundary to avoid singularity.

7.1.2 Characteristics of DPSM Sources, Active and Passive

DSPM uses two types of sources: "Active" and "Passive." In DPSM, "Active" sources are used to discretize the wave actuators, transducers, and any other form of energy sources that sends energy into the material. Most of these sources are responsible of generating bounded ultrasonic beams using traditional NDE setup discussed in Chapter 1 and Chapter 4. "Passive" sources, on the other hand, are used to discretize the problem boundaries and interfaces which may act as passive wave sources during the wave energy interactions. Hence, the main different between "Active" and "Passive" sources can be specified as follows: "Active" sources are distributed at the interfaces or boundaries where initial conditions (in time domain simulation) and boundary conditions (in frequency domain simulations) can be specified. These are the active wave sources. It is not necessary for the transducers to be the only active sources. Internal sources inside the materials like, crack edge as acoustic emission sources must also utilize the "Active" DPSM sources. On the contrary, "Passive" sources are distributed at the interfaces or boundaries where matching interface conditions or homogeneous boundary conditions can be specified. The "Active" and "Passive" sources and their actions are discussed below in the following paragraph but, the "Active" and "Passive" sources will not be in quotes.

First let us visualize a point source and its characteristics, used in DPSM. Fig. 7.1a shows a typical point source in fluid media with radiated wave field and potential control lines. Spherical potential control lines are the equi-phase lines that are like the wave fronts. Irrespective of being an active or passive source, all point sources are enclosed by a boundary layer or a boundary surface. The spherical boundary layer enclosing a point source should be in contact with an interface. Hence, an actual point source is always away from the material interface radiating or monochromatically vibrating continuously. Next in DPSM it is necessary to define the front and back of the point sources, because only the frontal contribution of the radiation fields is used in the calculation. Leeward side contributions of the point sources are proved [3–5] to be negligible and are neglected from the analysis. Now it can be immediately realized that there are infinite possibilities of ray directions to consider front or back of the point sources. Front and Leeward of the sources are defined as follows and are shown in Fig. 7.1b. When a point source is in contact with a material interface, the vector joining the center of the point source and the contact point between the boundary layer and the interface is considered propagating wave vector. A surface perpendicular to this vector is the separator between the frontal and leeward side of the point sources. The shaded area in Fig. 7.1b is the only area that is assumed to be seen by the point sources. Any material points behind the shaded area are assumed to be invisible by the point sources and contribution to those points by the point sources is ignored in wave field computation.

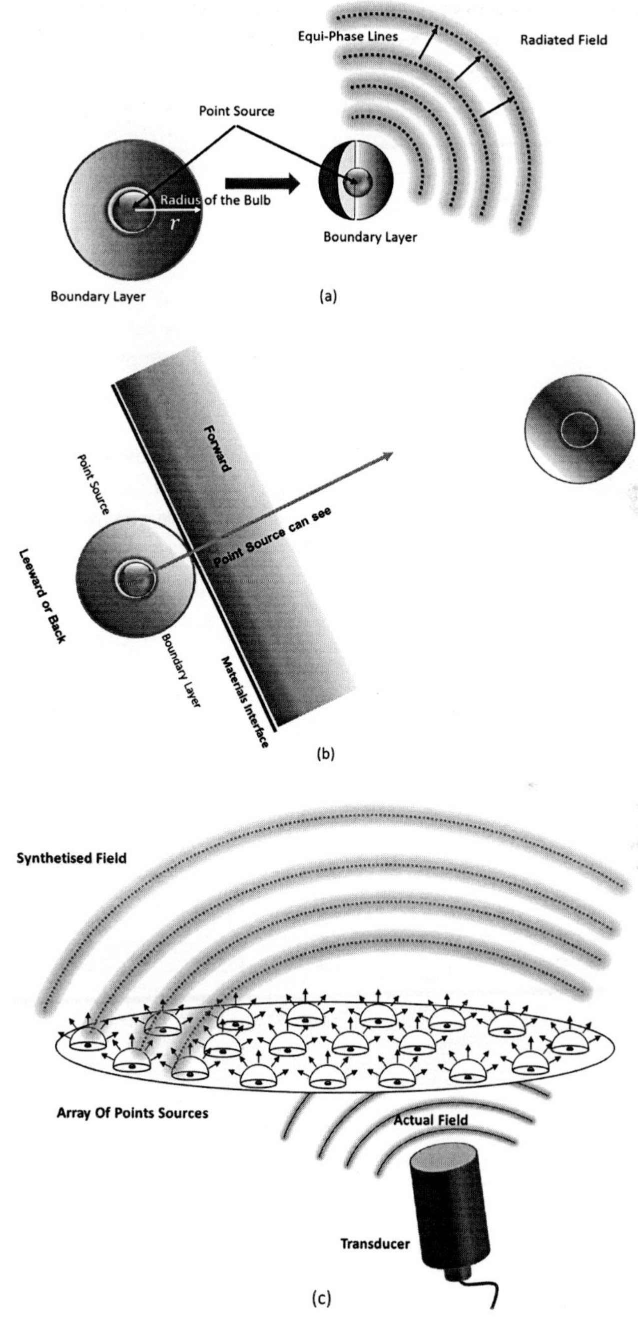

FIGURE 7.1 a) Spherical wave front generated by a point source and its boundary layer b) Forward and Leeward side of a point source, also shows the base point where the source is located and second is a sphere, which is the boundary layer enclosing the source, c) Schematic of an infinite plane, generating plane waves, simulating wave generated by a transducer.

Next, based on the above understanding let us assume an NDE problem setup at a certain instant when there is no probing energy into the material. At that instant none of the sources are ultrasonically illuminated. However, at an instant when the ultrasonic pulse is generated by the transducer placed in a fluid, the active sources at the transducer end will be illuminated. This active illumination will cause the passive sources at the material interface to be ultrasonically illuminated almost instantly. If the material is removed the passive sources will disappear. A field produced by an actual transducer is simulated by these active sources and presented by a synthesized field contributed by the superposition of all the individual field produced by all the point sources as shown in Fig. 7.1c. This is similar for the passive sources. These point sources active or passive are like radiating or locally vibrating mini hemispheres. These point sources can only perceive or see the problem domain lies in front of them. Front or leeward is defined by a vector pointing toward an interface connecting the location of the point source and the interface. This concept is shown in Fig. 7.2. If there is no material in front of the transducer, there is no need for the passive sources except the active sources just behind the transducer surface. However, if a scatterer is placed in front of the transducer, one set of passive sources just behind the scatterer are illuminated to scatter the ultrasonic field. If there is a material interface along the ultrasonic path (Fig. 7.3a), then the interface will illuminate two set of sources. One set of sources will emit the reflected ultrasonic filed placed behind the interface (Fig. 7.3b) and another set of sources that will transmit the ultrasonic field (transmitted field) beyond the interface placed in front of the interface (Fig. 7.3c). Hence, an interface has two layers of imaginary point sources. These point sources can only emit the ultrasonic field in front of them toward the domain that they can see. This scenario is explained in Fig. 7.3.

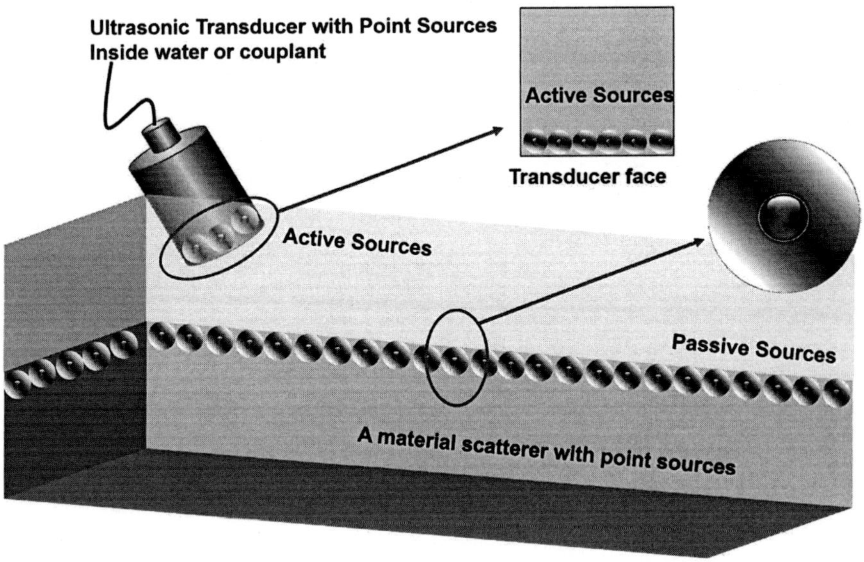

FIGURE 7.2 A material scatterer submerged into water while inspected by a transducer with distributed point sources.

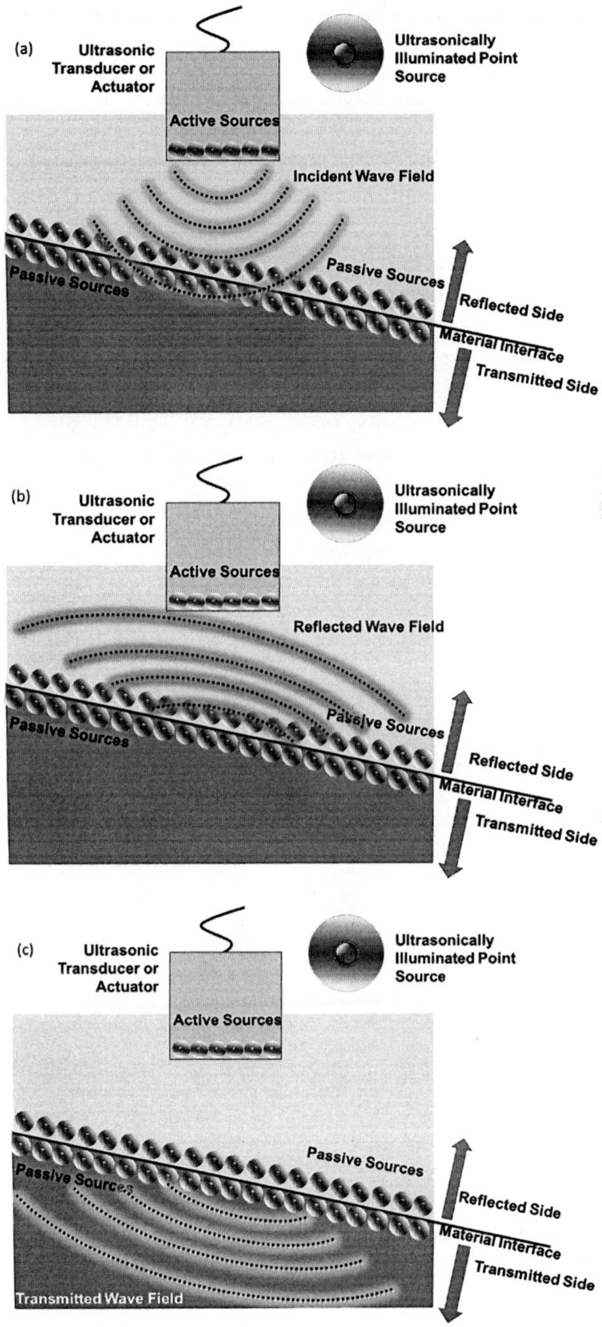

FIGURE 7.3 A material submerged into water while inspected by a transducer with distributed point sources a) showing incident wavefield by the illuminated active point sources b) showing reflected wavefield by the illuminated passive point sources below the interface, c) showing transmitted wavefield by the illuminated passive point sources above the interface.

7.1.3 Synthesis of Ultrasonic Field by Multiple Point Sources

Philosophically, this section is already discussed in Section 7.1.2; however, it is timely to discuss the concept more mathematically for practical implementation. It is discussed earlier that DPSM requires Green's function for the computation of the wave field. First, it is necessary to know what a Green's function is. In most general sense, Green's function is nothing but an influence function. Mathematically, it can be discussed with much rigor but first it is necessary to conceptually know the Green's function. In most general sense, a Green's function of a primary variable of a system with defined initial and/or boundary conditions is described by a response of the system at a certain point at a certain distance from the location of the applied load where an unit load is assumed to be applied to the system. Hence, with superposition principle of a linear system, we can say that the total response at that point in the system is the integrated response obtained by multiplying the Green's function with the actual load applied to the system where the load was specified by the boundary conditions. Mathematically, Green's function ($G(\mathbf{x},\mathbf{s})$) at $\mathbf{x}$ is the solution or the response to an impulse function of an inhomogeneous linear differential operator ($L = L(\mathbf{x})$) acting at $\mathbf{s}$, which is defined on a problem domain with specified boundary and initial conditions. Following equation summarizes the above statement

$$LG(\mathbf{x},\mathbf{s}) = \delta(\mathbf{s}-\mathbf{x}) \tag{7.1}$$

where δ is the Dirac delta function and by definition $\int_{-\infty}^{\infty}\delta(\mathbf{r})d\mathbf{r} = 1$; $\mathbf{r} = \mathbf{s}-\mathbf{x}$ and for any arbitrary loading function $f(\mathbf{s})$ at $\mathbf{s}$ on the system, we can write $\int_{-\infty}^{\infty} f(\mathbf{s})\delta(\mathbf{s}-\mathbf{x})\,ds = f(\mathbf{x})$. Exploiting this property and Eq. (7.1), the above integral can be further written as

$$\int_{-\infty}^{\infty} LG(\mathbf{x},\mathbf{s})f(\mathbf{s})ds = f(\mathbf{x}) \quad or \quad L\int_{-\infty}^{\infty} G(\mathbf{x},\mathbf{s})f(\mathbf{s})ds = f(\mathbf{x}) \quad or \quad Lu(\mathbf{x}) = f(\mathbf{x}) \tag{7.2}$$

Hence

$$u(\mathbf{x}) = \int_{-\infty}^{\infty} G(\mathbf{x},\mathbf{s})f(\mathbf{s})ds \tag{7.3}$$

where $u(\mathbf{x})$ is the primary unknown for the system $L = L(\mathbf{x})$ is the governing differential operator that describe the physics of operation of the system and $f(\mathbf{x})$ is the inhomogeneous part of the equation specify the external influence, or the external loading on the system, when other initial and boundary conditions are specified, and ds is the elemental space (area or volume) of the system. Readers those who are aware of Green's function may find the above representation of the equations is reversed herein, however, it was done with a purpose. The final equation of the unknown primary variable in Eq. (7.3) is most valuable for us. In CNDE problems, the ultrasonic displacement field generated in solids and pressure field generated in fluid are the primary unknown. Once the displacement field is calculated, strain, and stress field in solids can be easily calculated further on. Similarly, in fluid once the pressure field is calculated, displacement and velocity fields can be calculated. Hence, $u(\mathbf{x})$ is the displacement or pressure field in CNDE problems in respective material type. The differential operator $L = L(\mathbf{x})$ is the governing differential operator that describes the wave propagation in respective media. For example, in fluid media, the operator is written as $(\nabla^2 + k_f^2)$ as it was derived in Eq. (4.22) without inhomogeneous term. Such operators in solid media will be derived, discussed and solved in this chapter. The $f(\mathbf{x})$ is the loading function and in CNDE, it may signify the actuation of a transducer (emitting active sources). Similarly, in DPSM, activation of the passive sources can also signify the function $f(\mathbf{x})$. So, to find a synthesized field (displacement or pressure) in respective media at a point $\mathbf{x}$, the integral in Eq. (7.3) describes the utility of a Green's function which can be multiplied with the corresponding loading function and can be integrated or summer over. Let us rework on Eq. (7.3) in numerical perspective. Instead of having continuous force function $f(\mathbf{s})$ on elemental space $d\mathbf{s}$, multiple sources with localized force at their respective core point can be assumed with specific source strengths representative of their respective forces. In this light, Eq. (7.3) can be revised as follows:

$$u(\mathbf{x}) = \int_{-\infty}^{\infty} G(\mathbf{x},\mathbf{s}) f(\mathbf{s}) d\mathbf{s} = \sum_{m=1}^{N} A_m G(\mathbf{x},\mathbf{s}_m) \tag{7.4}$$

where unknown A_m is the source strength of the $m-th$ point source located at point $\mathbf{s}_m$, provided $G(\mathbf{x},\mathbf{s}_m)$, the Green's function at $\mathbf{x}$ due to a source located at $\mathbf{s}_m$ is known. It is assumed that numerically N numbers of such sources are sufficient to produce the same wave field that could be produced originally by an actual source, within an acceptable error bound. If the error bound is specified, the maximum number of point sources that should be used in the solution could be predicted via convergence analysis. The radius of the point sources (see Fig. 7.1b) or the diameter of the source bulb is governed by the number of point sources required to simulate a virtual ultrasonic field. As a thumb rule, higher the number of sources on a defined space, smaller the diameter of the source bulb should be used. After using specific number of point sources, converged $u(\mathbf{x})$ will be the DPSM computed numerical synthesized field produced by a real ultrasonic source. This is further pictographically explained in a set of multiple source and target points as shown in Fig. 7.4.

In general, three most common ultrasonic wave fields are simulated in CNDE. Their respective synthesized fields are either spherical, cylindrical or plane waves. Spherical waves are generated by a point source in an infinite medium as shown in Fig. 7.1a, the cylindrical waves are generated by a line source, while an infinite plane generates plane waves, as shown in Fig. 7.1c. As discussed in this section, the wave field $u(\mathbf{x})$ due to a finite plane source emitting bounded ultrasonic beam can be assumed to be the summation of wave fields generated by N number of converged point sources distributed over the finite source area as shown in Figs. 7.2 and 7.3. As shown in these figures, the source bulbs are assumed to be located near the front face of a transducer. This assumption can be further explained in a different manner as follows: It is assumed that a harmonic point source radiates or vibrates while

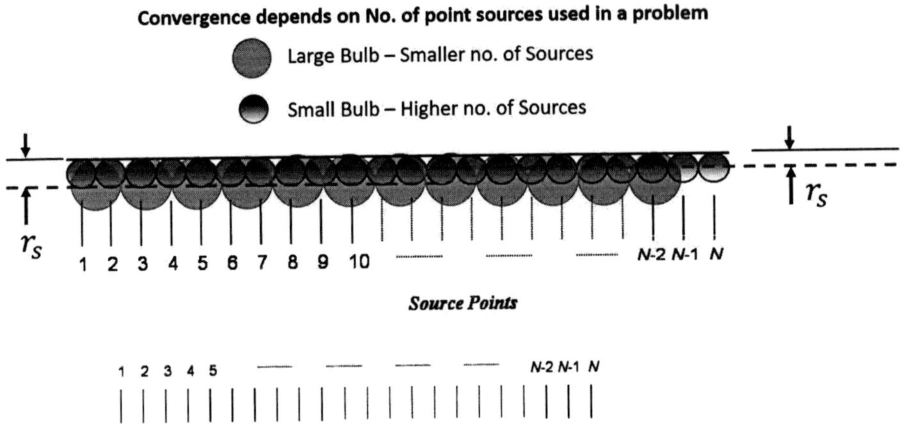

FIGURE 7.4 A schematic showing the use of higher vs. lower number of point sources and its effect on source bulb.

they expand and contract alternately giving a harmonic function. This expansion and contraction can be represented by two stages of a point source. One is the base point where the source is located and second is a sphere, which is the boundary layer enclosing the source as shown in Fig. 7.1b, respectively. The point represents the contracted position and the sphere with the boundary bulb represents the expanded position. When many of these point sources are placed side-by-side on a plane surface like they are placed in front and behind an interface in Fig. 7.3, then the contracted and expanded positions of the point sources may continuously overlap each other. The combined effect of many point sources, placed side by side, represent a line source when the contracted and represents an inflated boundary tube touching the interface boundary when expanded. The combined effect of this expansion causes the interface or the boundary to vibrate the particles in the direction normal to the interface. Orthogonal components of the vibration at a point on the interface, and the vibration of the neighboring source points, cancel each other. However, components along the wave vector (shown in Fig. 7.1b) do not vanish along the edge of the surface. If this edge effect does not have a significant contribution on the total motion of the surface, then the normal vibration of a finite plane or an interface can be approximately modeled by replacing the finite surface by a judicially selected number of point sources distributed over the interface or on the vibrating surface.

7.2 MODELING ULTRASONIC TRANSDUCER IN A FLUID

In most common NDE problems, ultrasonic transducer is placed under water to scan a test article. First for a common CNDE problem, it is wise to see, how virtually an ultrasonic bounded beam can be generated at a specific angular frequency (ω) in fluid media using a virtual ultrasonic transducer. An ultrasonic transducer is described in Chapters 1 and 3, where it is mentioned that a piezoelectric crystal vibrates under electromechanical transduction and the transducer face generates motion in fluid in front of the transducer causing a certain velocity of the surface. This velocity could be the boundary condition to the surface of the transducer. Although the vibration is a time domain phenomenon, it contains many frequencies based on the understanding of Fourier transform. Hence, practically, boundary conditions to the ultrasonic transducer surface can be provided at each frequency. Solution of the ultrasonic fields is thus also obtained at different frequencies. Here onwards in this chapter we will restrict our discussion of frequency domain ultrasonic field at a specific frequency and the process could be repeated for multiple other frequencies. Using DPSM an ultrasonic transducer in fluid can be activated by the distributed point sources and the synthesized field in front of the transducer could be produced by superposing the field produced by each point sources as written in Eq. (7.4) at a certain angular frequency (ω). In fluid, the wave field is a pressure wave field and Eq. (7.4) can be more generally written for pressure as follows:

$$\mathrm{p}(\mathbf{x}) = \int_{-\infty}^{\infty} G(\mathbf{x},\mathbf{s},\omega) f(\mathbf{s}) ds = \sum_{m=1}^{N} A_m G(\mathbf{x},\mathbf{s}_m,\omega) \tag{7.5}$$

where $p(\mathbf{x})$ is the pressure field at any point $\mathbf{x}$ in front of the transducer and $G(\mathbf{x},\mathbf{s}_m,\omega)$ is the elastodynamic pressure Green's function obtained from the governing elastodynamic equation in fluid. Please note that the Green's function $G(\mathbf{x},\mathbf{s}_m,\omega)$ is monochromatic and changes with frequency. Hence, $G(\mathbf{x},\mathbf{s}_m,\omega)$ is a frequency domain Green's function. Here, it is clear that a mathematical expression for the Green's function in fluid is required in a canonical form. In the following subsection, Green's function in a Newtonian fluid media is discussed.

7.2.1 Elastodynamic Green's Function in Fluid

Fundamental elastodynamic equation in fluid is derived in Chapter 4 and is written in Eq. (4.18), and is reiterated here, in slightly modified form

$$\nabla^2 \mathrm{p} - \frac{1}{c_f^2}\frac{\partial^2 \mathrm{p}}{\partial t^2} = f(\mathbf{x},t) \quad \text{or} \quad \left(\nabla^2 - \frac{1}{c_f^2}\frac{\partial^2}{\partial t^2}\right)\mathrm{p} = f(\mathbf{x},t) \quad or \quad L\mathrm{p}(\mathbf{x},t) = f(\mathbf{x},t) \tag{7.6}$$

Referring Eqs. (7.1) and (7.2), equation for elastodynamic Green's function can be written as

$$LG(\mathbf{x},\mathbf{0},t) = \delta(t)\delta(\mathbf{x}-\mathbf{0}) \tag{7.7}$$

where the point source with impulse force is assumed to be located at $\mathbf{s} = \mathbf{0}$. It is needless to mention that a point source in fluid will act as an axisymmetric source and will generate concentric spherical wave fronts.

7.2.1.1 Reciprocal and Causal Green's Function from Green's Formula

To solve Eq. (7.7), for the elastodynamic operator or wave operator L, it is necessary to initiate our discussion from Green's formula [6]. For time-dependent problem,

L has both space and time variables. The Green's formula for the elastodynamic equation after integral by parts can be written as

$$\int_0^t \iiint_0^{\mathbf{x}} [uL(v) - vL(u)] d\mathrm{v}\, dt = \iiint \left(u\frac{\partial v}{\partial t} - v\frac{\partial u}{\partial t} \right)\Bigg|_0^t d\mathrm{v} - c_f^2 \int_0^t \left(\oiint (u\nabla v - v\nabla u).\hat{\mathbf{n}}\, ds \right) dt \tag{7.8}$$

where u and v are two possible solutions of the operator L in Eq. (7.6), $\iiint$ represents three-dimensional volume integral and $\oiint$ indicates surface integral over the problem boundary. If both the solutions satisfy the homogeneous boundary conditions, then $\left(\oiint (u\nabla v - v\nabla u).\hat{\mathbf{n}}\, ds\right) = 0$, the first part of the equation in Eq. (7.8), may not be equal to zero. Although not rigorously derived in this chapter, the above formula can help prove [6] that the Green's function, a solution of Eq. (7.7) is reciprocal. That means that the response at $\mathbf{x}$ at time t due to a source located at $\mathbf{0}$ after initiating at $t = 0$, will be the same response at $\mathbf{x} = \mathbf{0}$ at time t due to a source located at $\mathbf{x}$ initiated at time $t = 0$. Green's reciprocity formula can be written as follows:

$$G(\mathbf{x}, \mathbf{0}, t, 0) = G(\mathbf{0}, \mathbf{x}, t, 0) \tag{7.9}$$

Additionally, the Green's function is source varying Green's function satisfying causality principle $G(\mathbf{x}, \mathbf{0}, t, 0) = 0$, for $t < 0$.

7.2.1.2 Generalized Equation for Green's Function

Next, let us assume two possible solutions, u and v, that satisfy the nonhomogeneous wave equation, Eq. (7.6), utilizing Eq. (7.9) as follows:

$$u = u(\mathbf{x}, t)\ ;\ v = G(\mathbf{x}, \mathbf{0}, t, 0) = G(\mathbf{0}, \mathbf{x}, t, 0) \tag{7.10}$$

Satisfying initial conditions for $u(\mathbf{x}, 0)$ and $\partial u(\mathbf{x}, 0)/\partial t$ and $v = G(\mathbf{x}, \mathbf{0}, t, 0)$ is the causal source varying Green's function satisfying the reciprocity theorem. $v = G(\mathbf{x}, \mathbf{0}, t, 0)$ satisfies the homogeneous boundary conditions (at $t = t$, $v = 0$; $\partial v/\partial t = 0$) but $u = u(\mathbf{x}, t)$ may not satisfy the same. Based on this light, Eq. (7.8) can be modified to

$$\int_0^t \iiint_0^{\mathbf{x}} [uL(G) - GL(u)] d\mathrm{v}\, dt$$
$$= \iiint \left(u\frac{\partial G}{\partial t} - G\frac{\partial u}{\partial t} \right)\Bigg|_0^t d\mathrm{v} - c_f^2 \int_0^t \left(\oiint (u\nabla G - G\nabla u).\hat{\mathbf{n}}\, ds \right) dt \tag{7.11}$$

Substituting, Eqs. (7.6) and (7.7), we get

$$\int_0^t \iiint_0^{\mathbf{x}} [u\delta(t)\delta(\mathbf{x} - \mathbf{0}) - Gf(\mathbf{x}, t)] d\mathrm{v}\, dt$$
$$= \iiint \left(u\frac{\partial G}{\partial t} - G\frac{\partial u}{\partial t} \right)\Bigg|_0^t d\mathrm{v} - c_f^2 \int_0^t \left(\oiint (u\nabla G - G\nabla u).\hat{\mathbf{n}}\, ds \right) dt \tag{7.12}$$

Utilizing the property of Dirac delta function $\int_{-\infty}^{\infty} f(\mathbf{s})\delta(\mathbf{s}-\mathbf{x})d\mathbf{s} = f(\mathbf{x})$, the above equation further modifies to

$$u(0,t)=\int_0^t\int\int\int_0^{\mathbf{x}}[Gf(\mathbf{x},t)]d\mathrm{v}\,dt$$
$$+\int\int\int\left(u(\mathbf{x},t)\frac{\partial G(\mathbf{x},t)}{\partial t}-G(\mathbf{x},t)\frac{\partial u(\mathbf{x},t)}{\partial t}-u(\mathbf{x},0)\frac{\partial G(\mathbf{x},0)}{\partial t}+G(\mathbf{x},0)\frac{\partial u(\mathbf{x},0)}{\partial t}\right)d\mathrm{v}$$
$$-c_f^2\int_0^t\left(\oiint(u\nabla G-G\nabla u).\hat{\mathbf{n}}\;ds\right)dt \tag{7.13}$$

The boundary conditions at $t=t,\ v=0;\ \partial v/\partial t=0$ yeilds

$$u(0,t)=\int_0^t\int\int\int_0^{\mathbf{x}}[Gf(\mathbf{x},t)]d\mathrm{v}\,dt+\int\int\int\left(G(\mathbf{x},0)\frac{\partial u(\mathbf{x},0)}{\partial t}-u(\mathbf{x},0)\frac{\partial G(\mathbf{x},0)}{\partial t}\right)d\mathrm{v}$$
$$-c_f^2\int_0^t\left(\oiint(u\nabla G-G\nabla u).\hat{\mathbf{n}}\;ds\right)dt \tag{7.14}$$

It can be shown that role of $\mathbf{x}$ and $\mathbf{0}$ can be interchanged using reciprocity property and we get

$$u(\mathbf{x},t)=\int_0^t\int\int\int_0^{\mathbf{x}}[Gf(0,0)]d\mathrm{v}\,dt$$
$$+\int\int\int\left(G(\mathbf{x},\mathbf{0},t,0)\frac{\partial u(0,0)}{\partial t}-u(0,0)\frac{\partial G(\mathbf{x},0,t,0)}{\partial t}\right)d\mathrm{v}$$
$$-c_f^2\int_0^t\left(\oiint(u(0,0)\nabla G-G\nabla u(0,0)).\hat{\mathbf{n}}\;ds\right)dt \tag{7.15}$$

Next, the same above equation can be written for a homogeneous equation (i.e., $f(\mathbf{x},t)=0$) with homogeneous boundary conditions can be written as

$$u(\mathbf{x},t)=\int\int\int\left(G(\mathbf{x},\mathbf{0},t,0)\frac{\partial u(\mathbf{0},0)}{\partial t}-u(0,0)\frac{\partial G(\mathbf{x},\mathbf{0},t,0)}{\partial t}\right)d\mathrm{v} \tag{7.16}$$

Comparing, Eq. (7.16) with Eq. (7.3) it is evident that the Green's function is also the influence function for the initial condition for derivative $\frac{\partial u}{\partial t}$ and $-\frac{\partial G}{\partial t}$ is the influence function for the initial condition u. Referring to Eq. (7.7), if we solve the wave equation with initial conditions $u=0$ and $\partial u/\partial t=\delta(\mathbf{x}-\mathbf{0})$, the solution will be the Green's function in most general sense. Hence, Eq. (7.7) can be further written as

$$LG(\mathbf{x},\mathbf{0},t)=\left(\nabla^2-\frac{1}{c_f^2}\frac{\partial^2}{\partial t^2}\right)G=0 \tag{7.17}$$

Subjected to homogeneous boundary conditions and an initial condition at $t=0$ as follows: $G=0$ and $\frac{\partial G}{\partial t}=\delta(\mathbf{x}-\mathbf{0})$. Hence, the problem has reduced to solve the wave equation in Eq. (7.17) subjected to the initial conditions. Utilizing a general solution for a wave equation in a uniform homogeneous isotropic infinite media in Eq. (4.8) in Chapter 4 for one dimension, solution of Eq. (7.17) can be written as a superposition of two general solutions with wave propagating from two opposite directions. Solution of G will be

$$G=f_1\left(x-c_ft\right)+f_2(x+c_ft) \tag{7.18}$$

where $f_1\left(x-c_ft\right)$ and $f_2(x+c_ft)$ are two arbitrary wave functions moving to the right and left, respectively, with velocity c_f.

7.2.1.3 Solution of Green's Function with Spherical Wave Front, Huygens' Principle

In the above subsection, generalized wave equation and generalized function for the Green's functions are obtained. It is needless to mention that a point source in fluid will act as an axisymmetric source and will generate concentric spherical wave fronts. On concentric spherical wave fronts the phase of the particles are equal and they are not θ or ϕ dependent as introduced in Section 3.10.3. In this case, the wave amplitude and phases are the only function of the radial distance r. Thus, in this subsection, Eq. (7.17) is transformed to spherical coordinate system with respect the radial distance (r) only. Using Eq. (A.3.20) the transformed equation can be written as

$$\left(\nabla^2-\frac{1}{c_f^2}\frac{\partial^2}{\partial t^2}\right)G=\left[\frac{1}{r^2}\left[\frac{\partial}{\partial r}\left(r^2\frac{\partial}{\partial r}\right)\right]-\frac{1}{c_f^2}\frac{\partial^2}{\partial t^2}\right]G=0 \tag{7.19}$$

Applying a very well-known transformation $G(r,t)=\mathbb{G}(r,t)/r$ [6], the above equation simplifies to one-dimensional wave equation

$$\left[\frac{\partial}{\partial r^2}-\frac{1}{c_f^2}\frac{\partial^2}{\partial t^2}\right]\mathbb{G}(r,t)=0 \tag{7.20}$$

which must give the same solution as written in Eq. (7.18). The solution of radial Green's function can be

$$G(r,t)=\frac{f_1\left(r-c_ft\right)+f_2(r+c_ft)}{r} \tag{7.21}$$

where $f_1(r-c_f t)$ is spherical expanding and $f_2(r+c_f t)$ is spherically contracting at velocity c_f. Considering a source radiating outward in DPSM, the spherically contracting wave function is redundant and thus Eq. (7.21) modifies to

$$G(r,t)=\frac{f_1(r-c_f t)}{r} \tag{7.22}$$

Please recollect Eq. (7.17) with initial conditions $G=0$ and $\frac{\partial G}{\partial t}=\delta(\mathbf{x}-0)$, which says Eq. (7.22) must satisfy $\frac{\partial G}{\partial t}=\delta(\mathbf{r})$. Taking time derivative of Eq. (7.22) and applying the initial condition we get

$$\left.\frac{\partial G}{\partial t}\right|_{t=0}=-\frac{c_f}{r}\left[\frac{df_1(r)}{dr}\right]_{t=0}=\delta(r) \tag{7.23}$$

Utilizing the property of Dirac delta function $\int_{-\infty}^{\infty}\delta(\mathbf{r})d\mathbf{r}=1$, integrating both side of Eq. (7.23) over a spherical domain of arbitrary radius r, and performing integration by parts we get

$$1=-c_f\int_0^r \frac{1}{r}\frac{df_1(r)}{dr}4\pi r^2 dr=4\pi c_f\int_0^r f_1(r)dr \tag{7.24}$$

where $f_1(r)=0$ for $r>0$ is used. Just because integral of $f_1(r)$ gives unity, in a reverse notion we can say $f_1(r)$ is itself a form of Dirac delta function and can be written as

$$f_1(r)=\frac{1}{4\pi c_f}\delta(r) \tag{7.25}$$

Consequently, Eq. (7.22) modifies to

$$G(r,t)=\frac{\delta(r-c_f t)}{4\pi c_f r} \tag{7.26}$$

The time domain Green's function $G(r,t)$ can be transformed to frequency domain Green's function by taking Fourier transform of Eq. (7.26) and can be written as

$$G(r,\omega)=\mathcal{F}\left[G(r,t)\right]=\int_{-\infty}^{\infty}\frac{\delta(r-c_f t)}{4\pi c_f r}e^{-i\omega t}dt=\frac{e^{i\frac{\omega}{c_f}r}}{4\pi r}=\frac{e^{ik_f r}}{4\pi r} \tag{7.27}$$

It is shown that a concentrated source at any arbitrary point $\mathbf{x}=\mathbf{x}_0$ at any time t_0 in an infinite media, influence another arbitrary point $\mathbf{x}$ at time t only if $|\mathbf{x}-\mathbf{x}_0|=c_f(t-t_0)$. A point source in DPSM emits a wave moving in all possible directions with velocity c_f and at any arbitrary time laps $(t-t_0)$ and later the effect of the point source is located on a spherical distance $c_f(t-t_0)$ away. This is called the well known as

Huygens' principle [7–9]. Although it is explained in time domain similar concept is valid for the frequency domain Green's function except it affects almost instantly. Due to an impulse with central frequency ω at $\mathbf{x} = \mathbf{x_0}$ by a point source influence any arbitrary point $\mathbf{x}$ almost instantly at that frequency ω and the influence is governed by the Green's function which is a function of distance between $\mathbf{x}$ and $\mathbf{x_0}$. In all the above derivations, distance between two points making such source-target pairs are assumed to be r and $\mathbf{x_0}$ is considered conveniently located at the origin, i.e., $\mathbf{x_0} = \mathbf{0}$.

7.2.2 DPSM in Lieu of Surface Integral Technique

Previous subsections discussed how the Green's function in fluid media can be calculated. Eq. (7.27) gives the expression for radially propagating wave in fluid media at a distance r due to a point source located at the origin. If several of these point sources are distributed over a transducer face, as discussed in Section 7.1.3, then the pressure field at target point $\mathbf{x}$ in the fluid, due to the point sources located at point $\mathbf{x_0}$ distributed over the transducer surface, can be given by integrating Eq.(7.27) over the transducer surface with elemental surface area ds,

$$\mathrm{p}(\boldsymbol{x}) = \int_{s} V \frac{\exp(ik_f r)}{4\pi r} ds(\mathbf{x_0}) \tag{7.28}$$

where V is directly proportional to the amplitude of the velocity of the point source and $r = |\mathbf{x} - \mathbf{x_0}|$. Employing the similar concept depicted in Eq. (7.4), the integral form in Eq. (7.28) can be written in the following form with superposition rule

$$\mathrm{p}(\boldsymbol{x}) = \sum_{m=1}^{N} \left(\frac{V}{4\pi} \Delta S_m \right) \frac{\exp(ik_f r_m)}{r_m} = \sum_{m=1}^{N} A_m \frac{\exp(ik_f r_m)}{r_m} \tag{7.29}$$

On the contrary, employing Rayleigh-Sommerfeld theory [10], the pressure field can be written as

$$\mathrm{p}(\boldsymbol{x}) = -\frac{i\omega\rho}{2\pi} \int_{s} v_3(\mathbf{x_0}) \frac{\exp(ik_f r)}{r} ds(\mathbf{x_0}) \tag{7.30}$$

where $v_3(\boldsymbol{x_0})$ is the particle velocity component normal to the transducer surface when the transducer is pointed toward the x_3 direction as shown in Fig. 7.5. ρ is the density of the fluid. Particle velocity on the transducer surface could be different or constant played by the function $v_3(\mathbf{x_0})$. For constant velocity of the transducer surface ($v_3(\mathbf{x_0}) = V_0$) Eq. (7.30) simplifies to

$$\mathrm{p}(\boldsymbol{x}) = -\frac{i\omega\rho V_0}{2\pi} \int_{s} \frac{\exp(ik_f r)}{r} ds(\mathbf{x_0}) \tag{7.31}$$

A comparison between Eqs. (7.28) and (7.31) gives

$$V = -2i\omega\rho V_0 \tag{7.32}$$

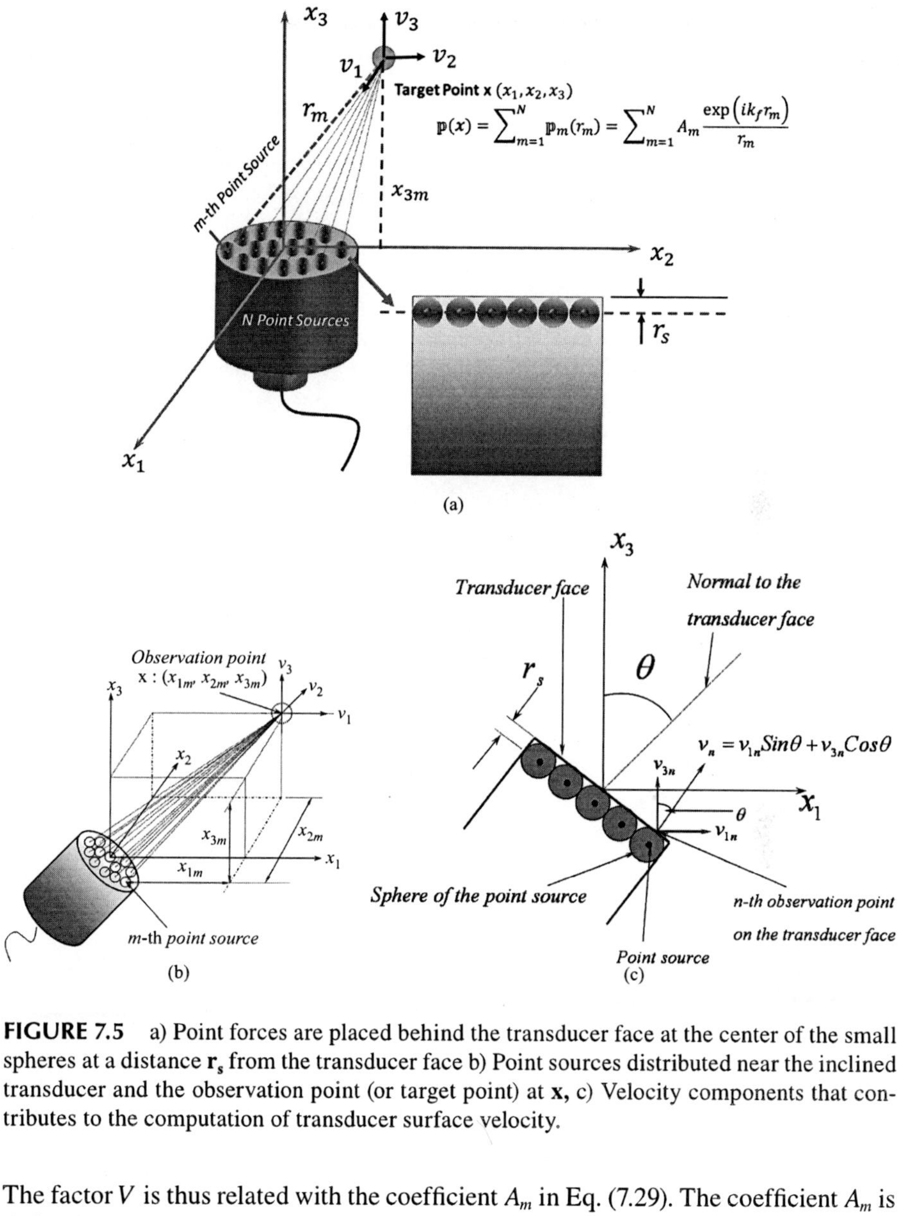

FIGURE 7.5 a) Point forces are placed behind the transducer face at the center of the small spheres at a distance $\mathbf{r_s}$ from the transducer face b) Point sources distributed near the inclined transducer and the observation point (or target point) at **x**, c) Velocity components that contributes to the computation of transducer surface velocity.

The factor V is thus related with the coefficient A_m in Eq. (7.29). The coefficient A_m is defined as the source strength of the m-th distributed point source behind the transducer surface. Pressure at a target point **x** due to a discrete point source with source strength A_m vibrating at a an angular frequency ω located at $\mathbf{x}_0^m$ point in fluid media with wave velocity c_f results a radial distance r_m will simply be the multiplication of Green's function and the source strength as follows:

$$\mathbb{p}_m(r_m) = A_m \frac{\exp(ik_f r_m)}{r_m} \quad ; \quad k_f = \frac{\omega}{c_f} \tag{7.33}$$

7.2.2.1 Computing Pressure and Velocity Field: Mathematical Expressions

As discussed before, if N numbers of point sources are distributed over the transducer surface, as shown in Fig. 7.5a, then the total pressure at point $\mathbf{x}$ is given by

$$\mathrm{p}(x) = \sum_{\mathrm{m}=1}^{N} \mathrm{p}_m(r_m) = \sum_{m=1}^{N} A_m \frac{\exp(ik_f r_m)}{r_m} \tag{7.34}$$

where r_m is the distance of the m-th point source from the target point $\mathbf{x} = (x_1, x_2, x_3)$.

In a Newtonian fluid, directional gradient of the pressure field at a point $\mathbf{x}$ is directly proportional to the temporal gradient of the velocity of the particles $\mathbf{x}$ and can be expressed by

$$-\frac{\partial \mathrm{p}}{\partial n} = \rho \frac{\partial v_n}{\partial t} \tag{7.35}$$

Let us assume that the particle velocity at a point is harmonic ($v_n = v_n e^{\pm i\omega t}$) if the point source excited in the fluid is harmonic. Thus, the time derivative of spatial velocity will be $\frac{\partial v_n}{\partial t} = \pm i\omega v_n e^{\pm i\omega t}$. Hence, v_n can be expressed by

$$v_n = \mp \frac{1}{i\omega\rho} \frac{\partial \mathrm{p}}{\partial n} \tag{7.36}$$

Velocity in the radial direction, at a distance r from the m-th point source, is given by

$$\begin{aligned} v_m(r) &= \frac{A_m}{i\omega\rho} \frac{\partial}{\partial r}\left(\frac{\exp(ik_f r)}{r}\right) = \frac{A_m}{i\omega\rho}\left(\frac{ik_f \exp(ik_f r)}{r} - \frac{\exp(ik_f r)}{r^2}\right) \\ &= \frac{A_m}{i\omega\rho} \frac{\exp(ik_f r)}{r}\left(ik_f - \frac{1}{r}\right) \end{aligned} \tag{7.37}$$

The i-th component of the velocity is

$$v_{im}(r) = \frac{A_m}{i\omega\rho} \frac{\partial}{\partial x_i}\left(\frac{\exp(ik_f r)}{r}\right) = \frac{A_m}{i\omega\rho} \frac{x_i \exp(ik_f r)}{r^2}\left(ik_f - \frac{1}{r}\right) \tag{7.38}$$

Hence, according to Fig. 7.5 when contributions of all the N sources are summed, the total velocity in x_3 direction at point $\mathbf{x}$ can be

$$v_3(x) = \sum_{m=1}^{N} v_{3m}(r_m) = \sum_{m=1}^{N} \frac{A_m}{i\omega\rho} \frac{x_{3m} \exp(ik_f r_m)}{r_m^2}\left(ik_f - \frac{1}{r_m}\right) \tag{7.39}$$

where x_{3m} is the component of r_m projected on the x_3-axis, measured from the m-th source as shown in Fig. 7.5a. Let us assume that the transducer velocity is specified as a boundary condition in a CNDE problem by V_0 at point **x**, then it is obvious that the expression in Eq. (7.39) will be equal to V_0 and can be written as

$$\sum_{m=1}^{N} \frac{A_m}{i\omega\rho} \frac{x_{3m} \exp(ik_f r_m)}{r_m^2} \left(ik_f - \frac{1}{r_m} \right) = V_0 \tag{7.40}$$

Next, in this problem it is intended to model the ultrasonic field in front of the transducer when the velocity of the transducer face is prescribed. There are N point sources assumed just below the transducer face with N unknown source strengths $(A_1, A_2, A_3, \ldots .A_N)$. There are N points on the transducer surface, where the boundary condition in Eq. (7.40) can be applied and it is possible to obtain a system of N linear equations to solve for N unknowns. The point sources as discussed before are not placed at the transducer surface but slightly below the face where the source bulb and its boundary layer barely touch the transducer surface as shown in Fig. 7.5, sources are placed r_s distance away from the face. This arrangement is also shown in Fig. 7.1b and can be explained by Eq. (7.40). To have a real normal velocity on the transducer face, the r_m distance cannot be zero for any sources, which will make the equation singular. Hence, the arrangement in Fig. 7.5 will safely create N number of simultaneous linear equations to solve for the unknown source strengths. For nonviscous perfect fluids boundary condition prescribed for v_3 component of the velocity is sufficient, but for viscous fluid, other two velocity components have to be prescribed. For ultrasonic testing using an ultrasonic transducer, it is known that other two velocity components v_1 and v_2 in Fig. 7.5 must be equal to zero. However, prescribing additional two velocity components at N points on the transducer face generates $3N$ equations. Whereas, only N sources are defined with N unknown source strengths. To circumvent this problem, a triangular source arrangement, called triplet source can be defined at the same r_s distance away from the transducer face. This will generate $3N$ equations and $3N$ unknowns. The other two velocity components can be written as

$$v_1 = \sum_{m=1}^{N} \frac{A_m}{i\omega\rho} \frac{x_{1m} \exp(ik_f r_m)}{r_m^2} \left(ik_f - \frac{1}{r_m} \right) = 0 \tag{7.41}$$

$$v_2 = \sum_{m=1}^{N} \frac{A_m}{i\omega\rho} \frac{x_{2m} \exp(ik_f r_m)}{r_m^2} \left(ik_f - \frac{1}{r_m} \right) = 0 \tag{7.42}$$

Next, if the transducer is rotated in Cartesian coordinate system as shown in Fig. 7.5b then the velocity component prescribed normal to the transducer face will be the component of all the three velocities (v_1, v_2, and v_3) along the direction of the normal to the transducer face. If the normal to the transducer face is defined as $\hat{\mathbf{n}} = n_1\hat{\mathbf{e}}_1 + n_2\hat{\mathbf{e}}_2 + n_3\hat{\mathbf{e}}_3$. Then the velocity on the transducer face can be written simply as a dot product discussed in Chapter 2 and can be. $V_0 = v_1n_1 + v_2n_2 + v_3n_3$, where n_i is the direction cosine of the i-th direction. Fig. 7.5c shows a case where the transducer is rotated by θ angle from x_3 -axis about x_2 -axis pointing away from the reader.

In such specific case, the prescribed velocity V_0 on the transducer face will expressed as $V_0 = v_1 \sin\theta + v_2 \cos\theta$.

7.2.2.2 Computing Pressure and Velocity Field: Matrix Formulation

Matrix formulation is fundamental to DPSM. A brief discussion and clarifications on the vector and matrix notations are discussed here in this subsection and will be used throughout this chapter. Without the loss of generality, similar vector and matrix notations will be used later in this chapter, without explicitly explaining their meaning, and will be assumed that reader have read this subsection. Following the original notations in Reference [11] but with slight modification in this book, velocity vectors in fluid is defined as $\mathbf{V}_F$. Similarly, the matrix for velocity Green's function in fluid is expressed as $\mathbf{M}_{FF}$, pressure Green's function matrix in fluid is expressed by $\mathbf{Q}_{FF}$, and the source strength vector is defined as $\mathbf{A}_F$. The $\mathbf{M}_{FF}$ and $\mathbf{Q}_{FF}$ matrixes are square matrices with equal number of target points and source points. They are good for applying boundary conditions and solving the unknown source strengths. However, to compute ultrasonic field (pressure or velocity) at numerous other target points when the source strengths are known the velocity Green's function and pressure Green's function matrices will not be square but will be governed by the number of target points distributed over a domain where the wave field is intended to compute. In such cases, the velocity Green's function and pressure Green's function matrices are written as $\mathbf{M}_{TF}$ and $\mathbf{Q}_{TF}$, rectangular matrices with nonequal target and source points.

To compute the ultrasonic field (pressure or velocity) in front of the transducer, it is necessary to solve the system of linear equations that are generated by Eq. (7.40) through Eq. (7.42) and find the unknown source strengths. The following formulation is presented for viscous fluid with triplet sources when all three velocity components are matched at the transducer face and fluid interface.

Eq. (7.40) has N equations with 3N unknowns. If they are arranged in a matrix form, a $Nx3N$ matrix will be generated. Adding Eqs. (7.41) and (7.42) will add $2N$ rows to the set of the equations and will generate a $3N \times 3N$ matrix. These equations can be combined into the following matrix equation

$$\mathbf{M}_{FF}\mathbf{A}_F = \mathbf{V}_F \tag{7.43}$$

$\mathbf{V}_F$ is a vector of velocity components in fluid with the $3N$ rows at N number of surface points $\mathbf{x}$ on the transducer face. At each point there are three velocities. $\mathbf{A}_F$ is $(3N \times 1)$ vector containing the source strengths of $3N$ number of triplet point sources. $\mathbf{M}_{FF}$ is a $(3N \times 3N)$ matrix that establish a relation between two vectors $\mathbf{V}_F$ and $\mathbf{A}_F$. Here $\mathbf{V}_F$, $\mathbf{A}_F$, and $\mathbf{M}_{FF}$ are expanded as follows:

$$\{\mathbf{V}_F\}^T = \begin{bmatrix} v_1^1 & v_2^1 & v_3^1 & v_1^2 & v_2^2 & v_3^2 & \ldots\ldots & v_1^N & v_2^N & v_3^N \end{bmatrix} \tag{7.44}$$

Elements of $\mathbf{V}_F$ vector are denoted by v_j^n where the subscript j and subscript n can take values 1, 2 or 3 indicating the direction of the velocity component and take values 1 through N corresponding to the point on the transducer face at which the velocity component is defined. Based on the previous discussion and Eq. (7.40) through (Eq. 7.42) the $\mathbf{V}_F$ vector modifies to

$$\{\mathbf{V}_F\}^T = \begin{bmatrix} 0 & 0 & V_0 & 0 & 0 & V_0 & \dots\dots & 0 & 0 & V_0 \end{bmatrix} \tag{7.45}$$

Similarly, vector $\mathbf{A}_F$ of the source strengths can be expressed by

$$\{\mathbf{A}_F\}^T = \begin{bmatrix} A_1 & A_2 & A_3 & A_4 & A_5 & A_6 & \dots\dots & A_{3N-2} & A_{3N-1} & A_{3N} \end{bmatrix} \tag{7.46}$$

The velocity Green's function matrix $\mathbf{M}_{FF}$ is obtained from Eqs. (7.40), (7.41), and (7.42), taken from [12]

$$\mathbf{M}_{FF} = \begin{bmatrix}
g(x_{11}^1, r_1^1) & g(x_{12}^1, r_2^1) & g(x_{13}^1, r_3^1) & g(x_{14}^1, r_4^1) & \cdots\cdots & g(x_{1(3N-1)}^1, r_{3N-1}^1) & g(x_{1(3N)}^1, r_{3N}^1) \\
g(x_{21}^1, r_1^1) & g(x_{22}^1, r_2^1) & g(x_{23}^1, r_3^1) & g(x_{24}^1, r_4^1) & \cdots\cdots & g(x_{2(3N-1)}^1, r_{3N-1}^1) & g(x_{2(3N)}^1, r_{3N}^1) \\
g(x_{31}^1, r_1^1) & g(x_{32}^1, r_2^1) & g(x_{33}^1, r_3^1) & g(x_{34}^1, r_4^1) & \cdots\cdots & g(x_{3(3N-1)}^1, r_{3N-1}^1) & g(x_{3(3N)}^1, r_{3N}^1) \\
g(x_{11}^2, r_1^2) & g(x_{12}^2, r_2^2) & g(x_{13}^2, r_3^2) & g(x_{14}^2, r_4^2) & \cdots\cdots & g(x_{1(3N-1)}^2, r_{3N-1}^2) & g(x_{1(3N)}^2, r_{3N}^2) \\
g(x_{21}^2, r_1^2) & g(x_{22}^2, r_2^2) & g(x_{23}^2, r_3^2) & g(x_{24}^2, r_4^2) & \cdots\cdots & g(x_{2(3N-1)}^2, r_{3N-1}^2) & g(x_{2(3N)}^2, r_{3N}^2) \\
g(x_{31}^2, r_1^2) & g(x_{32}^2, r_2^2) & g(x_{33}^2, r_3^2) & g(x_{34}^2, r_4^2) & \cdots\cdots & g(x_{3(3N-1)}^2, r_{3N-1}^2) & g(x_{3(3N)}^2, r_{3N}^2) \\
\cdots & \cdots & \cdots & \cdots & \cdots\cdots & \cdots & \cdots \\
g(x_{31}^N, r_1^N) & g(x_{32}^N, r_2^N) & g(x_{33}^N, r_3^N) & g(x_{34}^N, r_4^N) & \cdots\cdots & g(x_{3(3N-1)}^N, r_{3N-1}^N) & g(x_{3(3N)}^N, r_{3N}^N)
\end{bmatrix}_{3N\times 3N} \tag{7.47}$$

where

$$g\left(x_{jm}^n, r_m^n\right) = \frac{x_{jm}^n \exp(ik_f r_m^n)}{i\omega\rho\left(r_m^n\right)^2}\left(ik_f - \frac{1}{r_m^n}\right) \tag{7.48}$$

First subscript j of x can take values 1, 2, or 3. x_{jm}^n defines the component of r_m^n along j-th direction to find the Green's function at n-th target point due to the m-th source point. The subscript m takes values from 1 to $3N$ depending on which point source is considered and the superscript n can take values through 1 to N to designate a target point on the transducer surface where the velocity components are defined in $\mathbf{V}_F$ vector. Source strengths can be solved by taking inverse of the velocity Green's function matrix $\mathbf{M}_{FF}$ and multiplied with the velocity boundary condition vector $\mathbf{V}_F$ as follows:

$$\mathbf{A}_F = [\mathbf{M}_{FF}]^{-1}\,\mathbf{V}_F \tag{7.49}$$

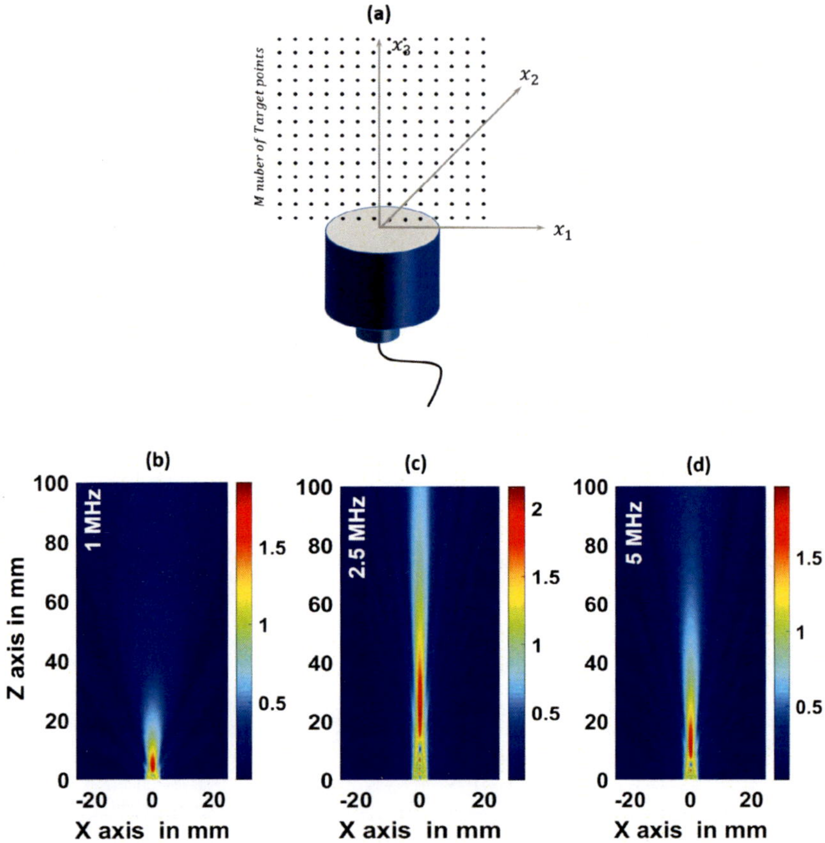

FIGURE 7.6 a) Discretization of a domain to plot pressure field in front of a transducer b) Pressure field with 1 MHz transducer, c) 2.5 MHz Transducer and d) 5 MHz Transducer.

After knowing the source strengths, employing Eq. (7.34) the pressure field in front of the transducer can be obtained. However, to find the pressure field in three dimension a three-dimensional grid of target points in front of the transducer should be articulated as shown in Fig. 7.6a on a $x_1 - x_3$ plane. Pressure and velocity fields in front of the transducer can be found as

$$\mathbf{P}_T = \mathbf{Q}_{TF}\mathbf{A}_F \tag{7.50}$$

$$\mathbf{V}_T = \mathbf{M}_{TF}\mathbf{A}_F \tag{7.51}$$

where $\mathbf{P}_T$ is a $(M \times 1)$ vector containing pressure values at M number of target points and $\mathbf{V}_T$ is a $(3M \times 1)$ vector containing three velocity components at every target point. Expressions inside a velocity Green's function matrix $\mathbf{M}_{TF}$ is similar to the expression written in Eq. (7.48), however, number of target points will M and source points will be 3N. On the contrary the pressure Green's function matrix $\mathbf{Q}_{TF}$ can be written as

$$\mathbf{Q}_{TF} = \begin{bmatrix} \frac{\exp(ik_f r_1^1)}{r_1^1} & \frac{\exp(ik_f r_2^1)}{r_2^1} & \frac{\exp(ik_f r_3^1)}{r_3^1} & \cdots & \cdots & \frac{\exp(ik_f r_{3N}^1)}{r_{3N}^1} \\ \frac{\exp(ik_f r_1^2)}{r_1^2} & \frac{\exp(ik_f r_2^2)}{r_2^2} & \frac{\exp(ik_f r_3^2)}{r_3^2} & \cdots & \cdots & \frac{\exp(ik_f r_{3N}^2)}{r_{3N}^2} \\ \frac{\exp(ik_f r_1^3)}{r_1^3} & \frac{\exp(ik_f r_2^3)}{r_2^3} & \frac{\exp(ik_f r_3^3)}{r_3^3} & \cdots & \cdots & \frac{\exp(ik_f r_{3N}^3)}{r_{3N}^3} \\ \cdots & \cdots & \cdots & \cdots & \cdots & \cdots \\ \cdots & \cdots & \cdots & \cdots & \cdots & \cdots \\ \frac{\exp(ik_f r_1^{N_T})}{r_1^{N_T}} & \frac{\exp(ik_f r_2^{N_T})}{r_2^{N_T}} & \frac{\exp(ik_f r_3^{N_T})}{r_3^{N_T}} & \cdots & \cdots & \frac{\exp(ik_f r_{3N}^{N_T})}{r_{3N}^{N_T}} \end{bmatrix}_{M\times 3N} \tag{7.52}$$

where the mathematical function inside the matrix is the pressure Green's function obtained from Eq. (7.34) and r_m^n defines the magnitude of the distance between the n-th target points due to the m-th source point.

7.2.2.3 Case Study: Modeling Pressure Field in Front of a Transducer

A 1 MHz and 2.5 MHz with 5 mm diameter and 5 MHz transducer with diameter 5 mm are modeled using DPSM. The pressure fields are plotted in front of the transducers over an area of 50 mm × 100 mm on the $x_1 - x_3$ or synonymously $x - z$ plane. Fig. 7.6b–d shows all the pressure fields in GPa, respectively, for 1 MHz, 2.5 MHz, 5 MHz.

7.3 MODELING ULTRASONIC WAVE FIELD IN ISOTROPIC SOLIDS

In the previous section Distributed Point Source Method is presented for fluid media. Similar approach with DPSM is generalized in this section to model ultrasonic filed in both solid and fluid simultaneously. This section is derived from the chapter on advanced application of DPSM in for isotropic solid media presented in a book in Reference [12]. Hence, readers are recommended to refer Reference [13] written by one of the author. In this section, only isotropic solid media is considered. Please note that fundamentally the nature of the point sources in fluids and solids are different in characteristics. As discussed in Chapter 1 and 5, an actual ultrasonic NDE problem may have both fluid and solid media to be considered in CNDE. For example, during the NDE of solid materials, often the solid specimen is immersed in a fluid medium for inspection because the fluid serves as a good coupling medium for ultrasonic waves as discussed in Chapter 1. Ultrasonic waves travel from the transducer to the solid specimen through a coupling fluid. Thus, in real life applications the acoustic bounded beams often strike the fluid-solid interface; therefore, the interaction between the bounded ultrasonic beams and fluid-solid interfaces need to be properly modeled.

The problem of fluid-solid interface is much more complex than a fluid-fluid interface discussed in Reference [14]. Realizing no direct connection with CNDE problems, such discussions on DPSM modeling of multi layered fluids is omitted in this book. However, wave fields in multilayered fluid structure might be of interest for biomedical applications. Pressure and three velocity components are the only four parameters of interest for computation of ultrasonic field in a fluid media as shown in Section 7.2. However, six stress components (σ_{ij}) and three displacement components (u_i) a total nine parameters becomes equally important in a solid media. Like pressure and velocity Green's functions were required in fluid, in solid, displacement and stress Green's functions will be required. The fluid-solid interface modeling has been a challenging problem in different fields of science and engineering. In many commercial finite element codes, the fluid is considered as a special case of solid that has a very small value of shear modulus. DPSM however does not require such simplifying assumptions. In this section, DPSM for isotropic solid is first discussed with mathematical derivation of necessary displacement and stress Green's function. Then step by step, a number of practical CNDE problems are discussed. The ultrasonic field modeling in the following problem geometries are presented 1) a fluid-solid interface between a solid half-space and a fluid half-space is considered, simulating the NDE of bulk material, 2) CNDE of a solid plate, with and without crack immersed in a fluid, 3) simulation of NDE of a multilayered solid plate immersed in a fluid, and 4) a corrugated waveguide immersed in a fluid are discussed. As briefly pointed out that the DPSM would need displacement and stress Green's functions in isotropic solids, it is timely to discuss the computation of solid Green's function mathematically. Calculation of elastodynamic Green's functions for a solid medium is presented in the following subsection.

7.3.1 Elastodynamic Displacement and Stress Green's Functions in Isotropic Solids

7.3.1.1 Elemental Point Source in Solid

The basic theory of DPSM is described in Section 7.1.2. The nature of active and passive elemental point sources in fluid media is illustrated in Fig. 7.1a. The active point sources in a fluid medium are those sources that are adjacent to the ultrasonic transmitter. This concept becomes more complicated when point sources in a solid medium is considered. In real ultrasonic applications, such as the ultrasonic NDE of materials and structures, the solid specimens generally remain in contact with a fluid medium (coupling liquid). The active medium (wave generator) is often not in the direct contact with the solid specimen during ultrasonic inspection. The point sources distributed at the solid boundary of the specimen are called passive point sources because the solid boundary does not generate any wave, it simply reflects and transmits the incident waves to satisfy the appropriate continuity conditions at the boundaries and interfaces.

The point sources distributed at the solid boundary should have three different force magnitudes in three different directions. This is because at any

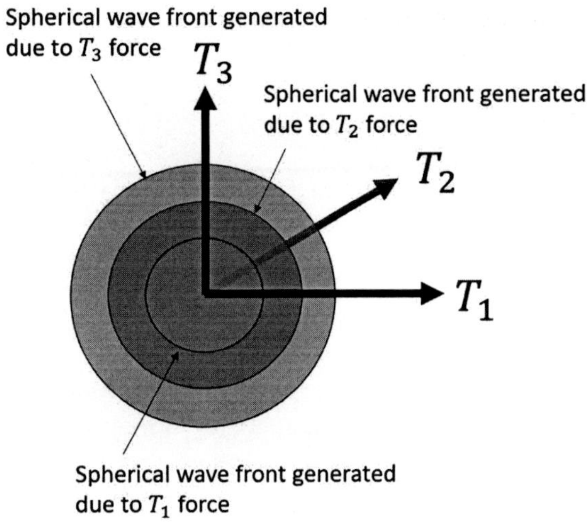

FIGURE 7.7 A point source with three different force of actuation in solid material.

interface or boundary, there are three interface, or three boundary conditions must be satisfied. In contrast, the point sources in fluid media use only one value of the source strength to compute the pressure field at any point. Force magnitudes (in later sections introduced as **T**) of the point sources in the solid medium are equivalent to the source strengths of the point sources in the fluid medium. Every point source in the solid medium has three force values or three source strengths in three mutually perpendicular directions. The schematic of a point source in the solid medium is presented in Fig. 7.7. Note that although in this figure for clarity three different spherical wave fronts are shown for the three-point forces acting in three mutually perpendicular directions these three wave fronts can coincide but having different strength distributions along the wave fronts. Different wave fronts for propagating P-waves and S-waves are not shown separately.

7.3.1.2 Navier's Equation of Motion with Body Force

This discussion requires understanding of continuum mechanics and wave equations, which are discussed in Chapter 3 and Chapter 4 in detail. The governing differential equation of motion which is also known as the equilibrium equation in a solid medium derived in Eq. (3.49) can be written as

$$\sigma_{ij,j} + F_i = \rho \ddot{u}_i \tag{7.53}$$

where σ_{ij} is the stress and u_i is the displacement at a point in the solid. $F_i = \rho f_i$ represents the body force per unit volume. $i, j = 1,2,3$. For homogeneous solid media the density (ρ) remains constant; however, stress (σ_{ij}) and displacement (u_i) are in general functions of both space (x_j) and time (t). Thus the body force (F_i) can be expressed as $F_i(x_j, t)$.

The constitutive law for any linear material iterated after Eq. (3.75) can be written as

$$\sigma_{ij} = \mathbb{C}_{ijkl}e_{kl} \tag{7.54}$$

where e_{kl} is the strain in solid, $k,l = 1,2,3$. $\mathbb{C}_{ijkl}$ is the matrix of elastic constants for any linear elastic isotropic or anisotropic material. For isotropic materials the stress-strain relation takes the following form reiterated herein after Eq. (3.86) in Chapter 3

$$\sigma_{ij} = \mathbb{L}\delta_{ij}e_{kk} + 2\mathrm{m}e_{ij} \tag{7.55}$$

where $\mathbb{L}, \mathrm{m}$ are the two Lamé constants and δ_{ij} is the Kronecker Delta function. After substituting the expression of strain $e_{ij} = \frac{1}{2}(u_{i,j} + u_{j,i})$ into Eq. (7.55) and then that into Eq. (7.53), the Navier's equation of motion can be obtained as derived in Chapter 3 and written in Eqs. (4.36) and (4.37).

The Navier's equation can be simplified further by Helmholtz decomposition discussed in Section 4.3.2.1 reiterated herein with additional modification.

$$\mathbf{u} = \nabla(\nabla.\mathbf{\Phi}) - \nabla\times(\nabla\times\mathbf{\Psi}) = \nabla\phi - \nabla\times\psi \tag{7.56}$$

where ϕ is a scalar potential while $\mathbf{\Phi}$, $\mathbf{\Psi}$, and ψ are vector potentials. Longitudinal-wave speed or the P-wave speed in an isotropic homogeneous solid is $c_p = \sqrt{\frac{(\mathbb{L}+2\mathrm{m})}{\rho}}$ and the transverse-wave speed or the S-wave speed is $c_s = \sqrt{\frac{\mathrm{m}}{\rho}}$. Dividing both sides of the Navier's equation by ρ, Eq. (4.36) with body force term intact modifies to

$$c_p^2(\nabla(\nabla.\mathbf{u})) - c_s^2(\nabla\times(\nabla\times\mathbf{u})) + \frac{\boldsymbol{F}}{\rho} = \ddot{\mathbf{u}} \tag{7.57}$$

Substituting Eq. (7.56) in Eq. (7.57),

$$\nabla\nabla.(c_p^2\nabla^2\mathbf{\Phi} - \ddot{\mathbf{\Phi}}) - \nabla\times\nabla\times(c_s^2\nabla^2\mathbf{\Psi} - \ddot{\mathbf{\Psi}}) + \frac{\boldsymbol{F}}{\rho} = 0 \tag{7.58}$$

Here, the vector potentials $\mathbf{\Phi}$ and $\mathbf{\Psi}$ are both functions of space and time. In deriving Eq. (7.58) following vector identities are used, $\nabla\cdot\nabla\times\mathbf{A} = 0, \nabla\times\nabla\mathbf{A} = 0$ for any vector $\mathbf{A}$ and $\nabla\left(\nabla\cdot\nabla\left(\nabla\cdot\mathbf{\Phi}\right)\right) = \nabla\left(\nabla^2\nabla\cdot\mathbf{\Phi}\right) = \nabla\nabla\cdot\nabla^2\mathbf{\Phi}$ and $\nabla\times\nabla\times\nabla\times\nabla\times\mathbf{\Psi} = \nabla\times\nabla\times\left(\nabla\left(\nabla\cdot\mathbf{\Psi}\right) - \nabla^2\mathbf{\Psi}\right) = -\nabla\times\nabla\times\left(\nabla^2\mathbf{\Psi}\right)$

7.3.1.3 Point Source Excitation in a Solid

Objective of this section is to obtain displacement and stress Green's function in a solid due to a point source excitation in solid. When a point source is acting inside a solid medium the body force term in Eq. (7.58) becomes a concentrated force with impulsive time dependence as discussed in Section 7.2. This force can be represented by the Dirac delta function in space. Decoupling the independent variables – time (t) and space ($\mathbf{x}$) – the body force can be written as

$$\boldsymbol{F}(\boldsymbol{x},t) = \mathbf{T}.f(t)\delta(\boldsymbol{x}) \quad or \quad F_i = T_i f(t)\delta(x_j) \tag{7.59}$$

where $\boldsymbol{T}$ is the force vector without the time and space dependency. Here, it is wise to bring a well-known solution of Poisson's equation. The general solution of the Poisson's equation $[\nabla^2\eta = q(\boldsymbol{x})]$ for a unknown primary variable $\eta(\mathbf{x})$ is given by Auld in his book in Ref. [7].

$$\eta = -\frac{1}{4\pi}\int_V \frac{q(\mathbf{y})}{|\mathbf{x}-\mathbf{y}|}dv \tag{7.60}$$

If $q(\boldsymbol{x})$ in Poisson's equation is replaced by Dirac delta function $\delta(\boldsymbol{x})$, then using a property of Dirac delta function $\int_v \delta(\boldsymbol{x})g(x)dv = g(0)$ one can write

$$\eta = -\frac{1}{4\pi|\boldsymbol{x}|} \tag{7.61}$$

Therefore, for $q(\boldsymbol{x}) = \delta(\boldsymbol{x})$, the Poisson's equation can be written as $\nabla^2\left(-\frac{1}{4\pi r}\right) = \delta(\boldsymbol{x})$, (where $r = |\boldsymbol{x}|$ considering the source is at the origin).

Now, substituting the expression of $\delta(\boldsymbol{x})$ in Eq. (7.59) and using the vector identity $\nabla^2\mathbf{A} = \nabla\nabla\cdot\mathbf{A} - \nabla\times\nabla\times\mathbf{A}$, the body force will be

$$\boldsymbol{F} = -\mathbf{T}\nabla^2\left(\frac{f(t)}{4\pi r}\right) = -\mathbf{T}\left[\nabla\left(\nabla.\left(\frac{f(t)}{4\pi r}\right)\right) - \nabla\times\left(\nabla\times\left(\frac{f(t)}{4\pi r}\right)\right)\right] \tag{7.62}$$

If the vector potentials are expressed in terms of two scalar potentials in the form $\mathbf{\Phi} = \mathbf{T}\phi$, $\mathbf{\Psi} = \mathbf{T}\psi$ and Eq. (7.62) is substituted into Eq. (7.58) the Navier's equation of motion takes the following form

$$\nabla\nabla.\mathbf{T}\left(c_p{}^2\nabla^2\phi - \ddot{\phi} - \frac{f(t)}{4\pi r\rho}\right) - \nabla\times\nabla\times\mathbf{T}\left(c_s{}^2\nabla^2\psi - \ddot{\psi} - \frac{f(t)}{4\pi r\rho}\right) = 0 \tag{7.63}$$

For time harmonic $\left(e^{-i\omega t}\right)$ excitation at the point source, the time dependence and space dependence of all the variables can be separated, as follows $\phi(x_j,t) = \phi(x_j)e^{-i\omega t}$, $\psi(x_j,t) = \psi(x_j)e^{-i\omega t}$, and $u_i(x_j,t) = U_i(x_j)e^{-i\omega t}$. Substituting these expressions in Eq. (7.63) one gets

$$\nabla\nabla.\mathbf{T}\left(c_p{}^2\nabla^2\varphi + \omega^2\varphi - \frac{1}{4\pi r\rho}\right)e^{-i\omega t} - \nabla\times\nabla\times\mathbf{T}\left(c_s{}^2\nabla^2\psi + \omega^2\psi - \frac{1}{4\pi r\rho}\right)e^{-i\omega t} = 0 \tag{7.64}$$

Hence, further one can write

$$\nabla^2\phi + \frac{\omega^2}{c_p{}^2}\phi = \frac{1}{4\pi\rho c_p{}^2 r} \tag{7.65}$$

$$\nabla^2\psi + \frac{\omega^2}{c_s{}^2}\psi = \frac{1}{4\pi\rho c_s{}^2 r} \tag{7.66}$$

Particular solutions of Eqs. (7.65) and (7.66) are given in Ref. [15]

$$\varphi = \frac{1-e^{ik_p r}}{4\pi\rho\omega^2 r} \tag{7.67}$$

$$\textit{and}\ \psi = \frac{1-e^{ik_s r}}{4\pi\rho\omega^2 r} \tag{7.68}$$

respectively, where $k_p = \frac{\omega}{c_p}$ and $k_s = \frac{\omega}{c_s}$.

7.3.1.4 Formulation of Displacement Green's Function

Substituting Eqs. (7.67) and (7.68) in Eq. (7.56) the time harmonic displacement function is expresses as

$$\boldsymbol{u} = \boldsymbol{U}e^{-i\omega t} = \left[\nabla\left(\nabla.\mathbf{T}\left(\frac{1-e^{1k_p r}}{4\pi\rho\omega^2 r}\right)\right) - \nabla\times\left(\nabla\times\mathbf{T}\left(\frac{1-e^{ik_s r}}{4\pi\rho\omega^2 r}\right)\right)\right]e^{-i\omega t} \tag{7.69}$$

Again, applying the vector identity ($\nabla^2\boldsymbol{A} = \nabla\nabla\cdot\boldsymbol{A} - \nabla\times\nabla\times\boldsymbol{A}$) in the above equation, the displacement equation takes the following form

$$\begin{aligned}\boldsymbol{u} = \boldsymbol{U}e^{-i\omega t} &= \nabla\left(\nabla.\mathbf{T}\left(\frac{1-e^{1k_p r}}{4\pi\rho\omega^2 r}\right)\right)e^{-i\omega t} + \nabla^2\boldsymbol{T}\left(\frac{1-e^{ik_s r}}{4\pi\rho\omega^2 r}\right)e^{-i\omega t}\\ &\quad - \nabla\left(\nabla.\boldsymbol{T}\left(\frac{1-e^{ik_s r}}{4\pi\rho\omega^2 r}\right)\right)e^{-i\omega t}\end{aligned} \tag{7.70}$$

Applying Laplace operator on $\boldsymbol{T}\left(\frac{1-e^{ik_s r}}{4\pi\rho\omega^2 r}\right)$ and using Eq. (7.66)

$$\boldsymbol{u} = \boldsymbol{U}e^{-i\omega t} = -\nabla\nabla.\left(\mathbf{T}\frac{e^{ik_p r} - e^{ik_s r}}{4\pi\rho\omega^2 r}\right)e^{-i\omega t} + \frac{\mathbf{T}e^{ik_s r}}{4\pi\rho\omega^2 r}k_s^2 e^{-i\omega t} = \underline{\mathbf{G}}(\mathbf{x};\mathbf{0}).\mathbf{T}.e^{-i\omega t} \tag{7.71}$$

In index notation Eq. (7.71) will be.

$$U_i = \frac{1}{4\pi\rho\omega^2}\left(k_s^2\frac{e^{ik_s r}}{r}T_i - \frac{\partial^2}{\partial x_i\,\partial x_j}\left(T_j\frac{e^{ik_p r} - e^{ik_s r}}{r}\right)\right)\text{or}$$

$$U_i = \frac{1}{4\pi\rho\omega^2}\left(k_s^2\frac{e^{ik_s r}}{r}\delta_{ij} - \frac{\partial^2}{\partial x_i\,\partial x_j}\left(\frac{e^{ik_p r} - e^{ik_s r}}{r}\right)\right)T_j = G_{ij}(x_k;0)T_j \tag{7.72}$$

Above expression is written assuming a point source is at the origin. However, if point source is located at **y** and the response at **x** is to be determined then one can write

$$u_i = U_i e^{-i\omega t} = G_{ij}(\mathbf{x};\mathbf{y},\omega)T_j e^{-i\omega t} \tag{7.73}$$

Here, $G_{ij}(\mathbf{x};\mathbf{y},\omega)$ is called the space-dependent frequency domain Green's function of displacement in isotropic homogeneous solid. Substituting $r=|\mathbf{x}-\mathbf{y}|$, the displacement Green's function can be written as [15]

$$G_{ij}(\mathbf{x};\mathbf{y},\omega)=\frac{1}{4\pi\rho\omega^2}\left[\frac{e^{ik_pr}}{r}\left(k_p^{\,2}R_iR_j+\left(3R_iR_j-\delta_{ij}\right)\left(\frac{ik_p}{r}-\frac{1}{r^2}\right)\right)\right.$$
$$\left.+\frac{e^{ik_sr}}{r}\left(k_s^{\,2}\left(\delta_{ij}-R_iR_j\right)-\left(3R_iR_j-\delta_{ij}\right)\left(\frac{ik_s}{r}-\frac{1}{r^2}\right)\right)\right] \tag{7.74}$$

where $R_i=\frac{x_i-y_i}{r}$

In matrix form

$$\mathbf{U}=\underline{G}(\mathbf{x};\mathbf{y},\omega)\mathbf{T} \tag{7.75}$$

If a point source is excited at $\mathbf{y}$ with a forcing direction along the $j-$th direction having a time harmonic frequency ω, then the displacement at $\mathbf{x}$ in the $i-$th direction will be represented by $G_{ij}\left(\mathbf{x};\mathbf{y},\omega\right)$.

7.3.1.5 Formulation of Stress Green's Function

For isotropic homogeneous solids the expression for stresses are given in Eqs. (3.86) and (7.55). Substituting the expression for displacement $u_i=U_ie^{-i\omega t}=G_{ij}(\mathbf{x};\mathbf{y},\,\omega)T_je^{-i\omega t}$ in the strain-displacement relation, $\mathrm{e}_{ij}=\frac{1}{2}(u_{i,j}+u_{j,i})$ we get

$$\mathrm{e}_{ij}=\frac{1}{2}\left(G_{ik,j}+G_{jk,i}\right)T_ke^{-i\omega t} \tag{7.76}$$

Substituting the expression for strains in Eq. (7.54), the stress Green's function at $\mathbf{x}$ due to a concentrated time harmonic force at $\mathbf{y}$ can be obtained. The harmonic time dependence $\left(e^{-i\omega t}\right)$ is implied and is not shown for convenience hereafter. For a general anisotropic material, the stress Green's function can be written as follows, which is discussed later in this chapter

$$S_{ij}(x;y)=\frac{1}{2}\mathbb{C}_{ijkl}\mathrm{e}_{kl}=\frac{1}{2}\mathbb{C}_{ijkl}\left(G_{kq,l}+G_{lq,k}\right)T_q \tag{7.77}$$

A special case of the above equation for isotropic homogeneous linearly elastic material, the expression for the stress Green's function at $\mathbf{x}$ due to a concentrated harmonic force at $\mathbf{y}$ can be written as

$$S_{ij}(\mathbf{x};\mathbf{y})=\mathrm{m}\left(G_{ik,j}+G_{jk,i}\right)T_k+\mathbb{L}\delta_{ij}G_{kq,k}T_q \tag{7.78}$$

Or

$$S_{ij}(\mathbf{x};\mathbf{y})=\left(\mathrm{m}\left(G_{ik,j}+G_{jk,i}\right)\delta_{kq}+\mathbb{L}\delta_{ij}G_{kq,k}\right)T_q \tag{7.79}$$

By rigorous differentiation, the expressions of all stress components represented in Eqs. (7.78) and (7.79) are obtained and are presented in Section 7.3.1.6.

7.3.1.6 Detailed Expressions for Displacement and Stress Green's Functions

The displacement Green's function presented in Eq. (7.74) can be written in the following form explicitly.

Let

$$ep = \frac{e^{ik_p r}}{r} \quad ; \quad es = \frac{e^{ik_s r}}{r} \text{ and } rp = \left(\frac{ik_p}{r} - \frac{1}{r^2}\right) ; \; rs = \left(\frac{ik_s}{r} - \frac{1}{r^2}\right) \tag{7.80}$$

Substituting the above expressions of Eq. (7.80) into Eq. (7.74), we get

$$G_{ij} = Gp_{ij} + Gs_{ij} \tag{7.81}$$

where

$$\begin{aligned} Gp_{11} &= \frac{1}{4\pi\rho\omega^2}\left[ep\left(k_p{}^2 R_1{}^2 + \left(3R_1{}^2 - 1\right)rp\right)\right] \; ; \\ Gs_{11} &= \frac{1}{4\pi\rho\omega^2}\left[es\left(k_s{}^2\left(1 - R_1{}^2\right) - \left(3R_1{}^2 - 1\right)rs\right)\right] \end{aligned} \tag{7.82}$$

$$\begin{aligned} Gp_{12} &= \frac{1}{4\pi\rho\omega^2}\left[ep\left(k_p{}^2 R_1 R_2 + (3R_1 R_2)rp\right)\right] \; ; \\ Gs_{12} &= \frac{1}{4\pi\rho\omega^2}\left[es\left(k_s{}^2(-R_1 R_2) - (3R_1 R_2)rs\right)\right] \end{aligned} \tag{7.83}$$

$$\begin{aligned} Gp_{13} &= \frac{1}{4\pi\rho\omega^2}\left[ep\left(k_p{}^2 R_1 R_3 + (3R_1 R_3)rp\right)\right] \; ; \\ Gs_{13} &= \frac{1}{4\pi\rho\omega^2}\left[es\left(k_s{}^2(-R_1 R_3) - (3R_1 R_3)rs\right)\right] \end{aligned} \tag{7.84}$$

$$\begin{aligned} Gp_{22} &= \frac{1}{4\pi\rho\omega^2}\left[ep\left(k_p{}^2 R_2{}^2 + \left(3R_2{}^2 - 1\right)rp\right)\right] \; ; \\ Gs_{22} &= \frac{1}{4\pi\rho\omega^2}\left[es\left(k_s{}^2\left(1 - R_2{}^2\right) - \left(3R_2{}^2 - 1\right)rs\right)\right] \end{aligned} \tag{7.85}$$

$$\begin{aligned} Gp_{23} &= \frac{1}{4\pi\rho\omega^2}\left[ep\left(k_p{}^2 R_2 R_3 + (3R_2 R_3)rp\right)\right] \; ; \\ Gs_{23} &= \frac{1}{4\pi\rho\omega^2}\left[es\left(k_s{}^2(-R_2 R_3) - (3R_2 R_3)rs\right)\right] \end{aligned} \tag{7.86}$$

$$Gp_{33} = \frac{1}{4\pi\rho\omega^2}\left[ep\left(k_p{}^2R_3{}^2 + \left(3R_3{}^2 - 1\right)rp\right)\right] \;;$$
$$Gs_{33} = \frac{1}{4\pi\rho\omega^2}\left[es\left(k_s{}^2\left(1 - R_3{}^2\right) - \left(3R_3{}^2 - 1\right)rs\right)\right] \tag{7.87}$$

$$Gp_{ij} = Gp_{ji} \quad ; \quad Gs_{ij} = Gs_{ji} \tag{7.88}$$

for any values of i and j. Using the above expressions in Eq. (7.81) for the Green's function the total displacement at any point $\mathbf{x}$ in the solid due to a point force acting at $\mathbf{y}$ can be written as

$$\begin{aligned} u_1 &= G_{11}T_1 + G_{12}T_2 + G_{13}T_3 \\ u_2 &= G_{21}T_1 + G_{22}T_2 + G_{23}T_3 \\ u_3 &= G_{31}T_1 + G_{32}T_2 + G_{33}T_3 \end{aligned} \tag{7.89}$$

where at $\mathbf{y}$, T_1, T_2, and T_3 are the magnitudes of the point forces acting along x_1, x_2, and x_3 directions, respectively. Similarly, u_1, u_2, and u_3 are the displacements at $\mathbf{x}$ along x_1, x_2, and x_3 directions, respectively. Hence, Eq. (7.89) can be expanded as follows:

$$\boldsymbol{u} = \begin{bmatrix} u_1 & u_2 & u_3 \end{bmatrix}^T \quad and \quad \boldsymbol{T} = \begin{bmatrix} T_1 & T_2 & T_3 \end{bmatrix}^T \tag{7.90}$$

$$\underline{\mathbf{G}}(\boldsymbol{x};\boldsymbol{y}) = \begin{bmatrix} \underline{\mathbf{G}_1}(\boldsymbol{x};\boldsymbol{y}) \\ \underline{\mathbf{G}_2}(\boldsymbol{x};\boldsymbol{y}) \\ \underline{\mathbf{G}_3}(\boldsymbol{x};\boldsymbol{y}) \end{bmatrix} \; and$$

$$\underline{\mathbf{G}_1}(\boldsymbol{x};\boldsymbol{y}) = [\,G_{11} \quad G_{12} \quad G_{13}\,]$$

$$\underline{\mathbf{G}_2}(\boldsymbol{x};\boldsymbol{y}) = [\,G_{21} \quad G_{22} \quad G_{23}\,]$$

$$\underline{\mathbf{G}_3}(\boldsymbol{x};\boldsymbol{y}) = [\,G_{31} \quad G_{32} \quad G_{33}\,]$$

Thus,

$$\begin{Bmatrix} u_1 \\ u_2 \\ u_3 \end{Bmatrix} = \begin{bmatrix} G_{11} & G_{12} & G_{13} \\ G_{21} & G_{22} & G_{23} \\ G_{31} & G_{32} & G_{33} \end{bmatrix} \begin{Bmatrix} T_1 \\ T_2 \\ T_3 \end{Bmatrix} \tag{7.91}$$

7.3.1.7 Differentiation of Displacement Green's Function with respect to x_1, x_2, x_3.

Let di denote the differentiation of a variable with respect to x_i.

Thus, $rpd1 = \frac{\partial r_p}{\partial x_i} = \frac{\partial}{\partial x_i}\left(\frac{ik_p}{r} - \frac{1}{r^2}\right)$. Therefore,

$$rpd1 = \left(\frac{2R_1}{r^3} - \frac{ik_p R_1}{r^2}\right) \;;\; rsd1 = \left(\frac{2R_1}{r^3} - \frac{ik_s R_1}{r^2}\right) \tag{7.92}$$

$$rpd2 = \left(\frac{2R_2}{r^3} - \frac{ik_p R_2}{r^2}\right) \;;\; rsd2 = \left(\frac{2R_2}{r^3} - \frac{ik_s R_2}{r^2}\right) \tag{7.93}$$

$$rpd3 = \left(\frac{2R_3}{r^3} - \frac{ik_p R_3}{r^2}\right) \;;\; rsd3 = \left(\frac{2R_3}{r^3} - \frac{ik_s R_3}{r^2}\right) \tag{7.94}$$

where R_i can be found after Eq. (7.74).

After introducing the symbols $eo_{ij} = R_i R_j$ and $et_{ij} = 3eo_{ij} = 3R_i R_j$ and keeping in mind that the repeated indices in the following expressions do not imply summation one can write

$$eo_{ii}di = \frac{\partial}{\partial x_i}\left(R_i^2\right) = -\frac{2R_i^3}{r} + \frac{2R_i}{r} \;;\; eo_{ij}di = -\frac{2R_i^2 R_j}{r} + \frac{R_j}{r} = eo_{ji}di \tag{7.95}$$

$$eo_{ij}dk = -\frac{2R_i R_j R_k}{r} \;;\; eo_{ii}dj = -\frac{2R_i R_i R_j}{r} \tag{7.96}$$

$$et_{ii}di = -\frac{6R_i^3}{r} + \frac{6R_i}{r} \;;\; et_{ij}di = -\frac{6R_i^2 R_j}{r} + \frac{3R_j}{r} = et_{ji}di \tag{7.97}$$

$$et_{ij}dk = -\frac{6R_i R_j R_k}{r} \;;\; et_{ii}dj = -\frac{6R_i R_i R_j}{r} \tag{7.98}$$

Substituting the above expressions in the differentiation of displacement Green's functions we get

$$\begin{aligned} Gp_{ii}di = {} & \frac{1}{4\pi\rho\omega^2}\left[ep\left(k_p^2 eo_{ii}di + (-1 + et_{ii})rpdi + rp.et_{ii}di\right)\right] \\ & + Gp_{ii}\left(ik_p R_i ep - \frac{R_i ep}{r}\right) \end{aligned} \tag{7.99}$$

$$\begin{aligned} Gs_{ii}di = {} & \frac{1}{4\pi\rho\omega^2}\left[es\left(-k_s^2 eo_{ii}di - (-1 + et_{ii})rsdi + rs.et_{ii}di\right)\right] \\ & + Gs_{ii}\left(ik_s R_i es - \frac{R_i es}{r}\right) \end{aligned} \tag{7.100}$$

$$Gp_{ii}dj = \frac{1}{4\pi\rho\omega^2}\left[ep\left(k_p{}^2 eo_{ii}dj + (-1+et_{ii})rpdj + rp.et_{ii}dj\right)\right] + Gp_{ii}\left(ik_pR_jep - \frac{R_jep}{r}\right) \tag{7.101}$$

$$Gs_{ii}dj = \frac{1}{4\pi\rho\omega^2}\left[es\left(-k_s{}^2 eo_{ii}dj - (-1+et_{ii})rsdj + rs.et_{ii}dj\right)\right] + Gs_{ii}\left(ik_sR_jes - \frac{R_jes}{r}\right) \tag{7.102}$$

$$Gp_{ij}di = \frac{1}{4\pi\rho\omega^2}\left[ep\left(k_p{}^2 eo_{ij}di + (et_{ij})rpdi + rp.et_{ij}di\right)\right] + Gp_{ij}\left(ik_pR_iep - \frac{R_iep}{r}\right) \tag{7.103}$$

$$Gs_{ij}di = \frac{1}{4\pi\rho\omega^2}\left[es\left(-k_s{}^2 eo_{ij}di - (et_{ij})rsdi + rs.et_{ij}di\right)\right] + Gs_{ij}\left(ik_sR_ies - \frac{R_ies}{r}\right) \tag{7.104}$$

$$Gp_{ji}di = \frac{1}{4\pi\rho\omega^2}\left[ep\left(k_p{}^2 eo_{ji}di + (et_{ji})rpdi + rp.et_{ji}di\right)\right] + Gp_{ji}\left(ik_pR_iep - \frac{R_iep}{r}\right) \tag{7.105}$$

$$Gs_{ji}di = \frac{1}{4\pi\rho\omega^2}\left[es\left(-k_s{}^2 eo_{ji}di - (et_{ji})rsdi + rs.et_{ji}di\right)\right] + Gs_{ji}\left(ik_sR_ies - \frac{R_ies}{r}\right) \tag{7.106}$$

$$Gp_{ij}dk = \frac{1}{4\pi\rho\omega^2}\left[ep\left(k_p{}^2 eo_{ij}dk + (et_{ij})rpdk + rp.et_{ij}dk\right)\right] + Gp_{ij}\left(ik_pR_kep - \frac{R_kep}{r}\right) \tag{7.107}$$

$$Gs_{ij}dk = \frac{1}{4\pi\rho\omega^2}\left[es\left(-k_s{}^2 eo_{ij}dk - (et_{ij})rsdk + rs.et_{ij}dk\right)\right] + Gs_{ij}\left(ik_sR_kes - \frac{R_kes}{r}\right) \tag{7.108}$$

Therefore

$$G_{ii}di = Gp_{ii}di + Gs_{ii}di \tag{7.109}$$

$$G_{ii}dj = Gp_{ii}dj + Gs_{ii}dj \tag{7.110}$$

$$G_{ij}di = Gp_{ij}di + Gs_{ij}di \tag{7.111}$$

$$G_{ji}di = Gp_{ji}di + Gs_{ji}di \tag{7.112}$$

$$G_{ij}dk = Gp_{ij}dk + Gs_{ij}dk \tag{7.113}$$

As noted earlier, from Eqs. (7.99) to (7.108) repeated indices do not mean summation and *di* represents differentiation with respect to x_i. Thus, $Gp_{ij}di = \frac{\partial}{\partial x_i}\left(Gp_{ij}\right)$, where, *i*, $j = 1$, 2 or 3 but not added.

Stress components (σ_{ij}) at point **x** due to a concentrated unit force acting in the x_m direction at point **y** can be written as σ_{ij}^{m}, where

$$\sigma_{33}^{1} = (2\mathrm{m} + \mathbb{L})(G_{31}d3) + \mathbb{L}(G_{11}d1 + G_{21}d2) \tag{7.114.1}$$

$$\sigma_{33}^{2} = (2\mathrm{m} + \mathbb{L})(G_{32}d3) + \mathbb{L}(G_{12}d1 + G_{22}d2) \tag{7.114.2}$$

$$\sigma_{33}^{3} = (2\mathrm{m} + \mathbb{L})(G_{33}d3) + \mathbb{L}(G_{13}d1 + G_{23}d2) \tag{7.114.3}$$

$$\sigma_{11}^{1} = (2\mathrm{m} + \mathbb{L})(G_{11}d1) + \mathbb{L}(G_{21}d2 + G_{31}d3) \tag{7.115.1}$$

$$\sigma_{11}^{2} = (2\mathrm{m} + \mathbb{L})(G_{12}d1) + \mathbb{L}(G_{22}d2 + G_{32}d3) \tag{7.115.2}$$

$$\sigma_{11}^{3} = (2\mathrm{m} + \mathbb{L})(G_{13}d1) + \mathbb{L}(G_{23}d2 + G_{33}d3) \tag{7.115.3}$$

$$\sigma_{31}^{1} = \mathrm{m}(G_{31}d1 + G_{11}d3) \tag{7.116.1}$$

$$\sigma_{31}^{2} = \mathrm{m}(G_{32}d1 + G_{12}d3) \tag{7.116.2}$$

$$\sigma_{31}^{3} = \mathrm{m}(G_{33}d1 + G_{13}d3) \tag{7.116.3}$$

$$\sigma_{32}^{1} = \mathrm{m}(G_{31}d2 + G_{21}d3) \tag{7.117.1}$$

$$\sigma_{32}^{2} = \mathrm{m}(G_{32}d2 + G_{22}d3) \tag{7.117.2}$$

$$\sigma_{32}^{3} = \mathrm{m}(G_{33}d2 + G_{23}d3) \tag{7.117.3}$$

Thus, the expressions of stress components at any point **x** due to a concentrated force acting at another point **y** in the solid in any direction are obtained. These expressions are useful to calculate the total ultrasonic field in solid for CNDE. For a group of concentrated forces acting at the boundary or at the interface (Fig. 7.8), as modeled

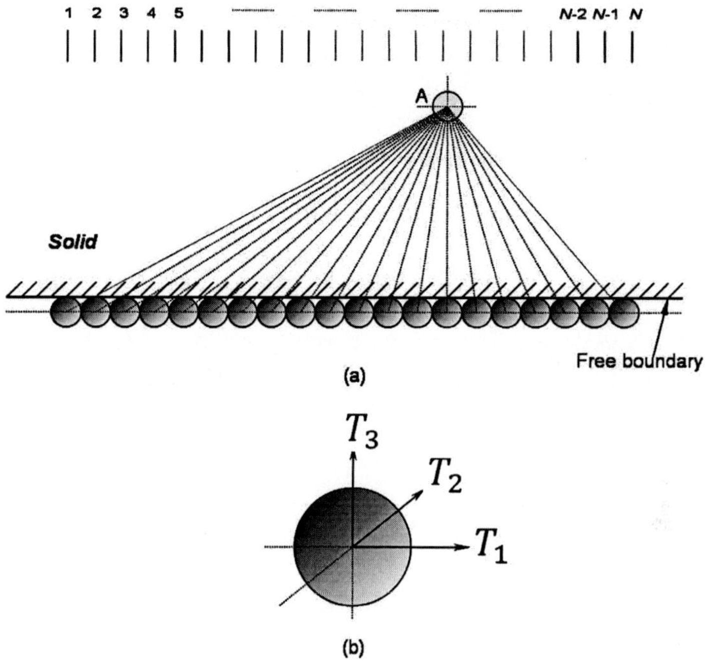

FIGURE 7.8 (a) Point sources distributed along the boundary of a solid medium. Total field at point A is computed by adding the contributions of all point sources. (b) A point source for the solid modeling consists of three-point forces acting in three mutually perpendicular directions.

by the Distributed Point Source Method, the total field is obtained by simple superposition as follows:

$$S_{ij} = \sum_{m=1}^{N} S_{ij}^{m}(x; y_m) = \sum_{m=1}^{M}\left[\left(\sigma_{ij}^{1}\right)^{m} T_1^{m} + \left(\sigma_{ij}^{2}\right)^{m} T_2^{m} + \left(\sigma_{ij}^{3}\right)^{m} T_3^{m}\right] = \sum_{m=1}^{M} \boldsymbol{s}_{ij}^{m}\boldsymbol{T}^{m} \quad (7.118)$$

where

$$\boldsymbol{s}_{ij}^{m} = \left[\begin{array}{ccc}\left(\sigma_{ij}^{1}\right)^{m} & \left(\sigma_{ij}^{2}\right)^{m} & \left(\sigma_{ij}^{3}\right)^{m}\end{array}\right] \quad and \quad \boldsymbol{T}^{m} = \left[\begin{array}{ccc}T_1^{m} & T_2^{m} & T_3^{m}\end{array}\right]^{T} \quad (7.119)$$

M is the total number of point sources and m corresponds to the m-th point source that has three force components T_1^m, T_2^m, and T_3^m.

7.3.2 Computation of Displacements and Stresses in the Solid for Multiple Point Sources

7.3.2.1 Displacement and Stresses at a Single Point

Let us consider a solid boundary in a two-dimensional space where multiple point sources are distributed along the boundary. When the point sources at the solid boundary are excited, the response at any point inside the solid medium can be computed by simply superimposing the contributions of all point sources. The solid half space and N number of point sources at the free boundary are shown in Fig. 7.8. Next objective is to compute the total response at A due to these point sources at the free boundary. Displacement value at any point **x** due to a point source at **y** is given in Eq. (7.74). Therefore, the total displacement at point A due to N point sources distributed along the solid boundary can be written as

$$u_1 = \sum_{m=1}^{N} (G_{11}^{m} T_1^{m} + G_{12}^{m} T_2^{m} + G_{13}^{m} T_3^{m}) = \sum_{m=1}^{M} \underline{G_1}^{m} T^{m} \tag{7.120}$$

$$u_2 = \sum_{m=1}^{N} \left(G_{21}^{m} T_1^{m} + G_{22}^{m} T_2^{m} + G_{23}^{m} T_3^{m}\right) = \sum_{m=1}^{M} \underline{\boldsymbol{G}_2}^{m} \boldsymbol{T}^{m} \tag{7.121}$$

$$u_3 = \sum_{m=1}^{N} \left(G_{31}^{m} T_1^{m} + G_{32}^{m} T_2^{m} + G_{33}^{m} T_3^{m}\right) = \sum_{m=1}^{M} \underline{\boldsymbol{G}_3}^{m} \boldsymbol{T}^{m} \tag{7.122}$$

Similarly, the normal stress and shear stress components at A can be written utilizing Eq. (7.118) as:

$$S_{11} = \sum_{m=1}^{N} \left[\left(\sigma_{11}^{1}\right)^{m} T_1^{m} + \left(\sigma_{11}^{2}\right)^{m} T_2^{m} + \left(\sigma_{11}^{3}\right)^{m} T_3^{m}\right] = \sum_{m=1}^{M} \underline{s_{11}}^{m} \boldsymbol{T}^{m} \tag{7.123}$$

$$S_{22} = \sum_{m=1}^{N} \left[\left(\sigma_{22}^{1}\right)^{m} T_1^{m} + \left(\sigma_{22}^{2}\right)^{m} T_2^{m} + \left(\sigma_{22}^{3}\right)^{m} T_3^{m}\right] = \sum_{m=1}^{M} \underline{s_{22}}^{m} \boldsymbol{T}^{m} \tag{7.124}$$

$$S_{33} = \sum_{m=1}^{N} \left[\left(\sigma_{33}^{1}\right)^{m} T_1^{m} + \left(\sigma_{33}^{2}\right)^{m} T_2^{m} + \left(\sigma_{33}^{3}\right)^{m} T_3^{m}\right] = \sum_{m=1}^{M} \underline{s_{33}}^{m} \boldsymbol{T}^{m} \tag{7.125}$$

$$S_{32} = \sum_{m=1}^{M} \left[\left(\sigma_{32}^{1}\right)^{m} T_1^{m} + \left(\sigma_{32}^{2}\right)^{m} T_2^{m} + \left(\sigma_{32}^{3}\right)^{m} T_3^{m}\right] = \sum_{m=1}^{M} \underline{s_{32}}^{m} \boldsymbol{T}^{m} \tag{7.126}$$

$$S_{31} = \sum_{m=1}^{M} \left[\left(\sigma_{31}^{1}\right)^{m} T_1^{m} + \left(\sigma_{31}^{2}\right)^{m} T_2^{m} + \left(\sigma_{31}^{3}\right)^{m} T_3^{m}\right] = \sum_{m=1}^{M} \underline{s_{31}}^{m} \boldsymbol{T}^{m} \tag{7.127}$$

$$S_{12} = \sum_{m=1}^{M} \left[\left(\sigma_{12}^{1}\right)^{m} T_1^{m} + \left(\sigma_{12}^{2}\right)^{m} T_2^{m} + \left(\sigma_{12}^{3}\right)^{m} T_3^{m}\right] = \sum_{m=1}^{M} \underline{s_{12}}^{m} \boldsymbol{T}^{m} \tag{7.128}$$

where σ_{ij}^{p} is defined in Eq. (7.1114) through (7.117) where, subscripts i, j, and superscript p can take values *1,2* or *3*.

7.3.2.2 Displacement and Stresses at a Multiple Points: Matrix Formulation

Computation of displacements and stresses at a single point in the solid are presented in the previous section. However, for CNDE it is of interest to find ultrasonic field over a problem domain. Therefore, it is necessary to compute the field at multiple points. For a group of observation points (or target points *T*) let us compute the displacement and stress fields inside the solid. Let us assume that there are *M* number of target points (*T*). The displacement and stress expressions at *M* target points due to *N* source points (*S*) are given in Eq. (7.129.1) through (7.130.6) in matrix forms. In these equations, the subscript *T* stands for the target points inside the solid medium and *S* stands for the set of source points at the boundary of the solid medium. In the following equations, **ui** are the displacements at *M* target points along the x_i direction and **sij** are the stresses at *M* target points (**i** and **j** can take values **1**,**2** or **3**). The *M* target points, and *N* source pointes are shown in Fig. 7.9.

$$\mathbf{u1}_T = \mathbf{DS1}_{TS}.\mathbf{A}_S \tag{7.129.1}$$

$$\mathbf{u2}_T = \mathbf{DS2}_{TS}.\mathbf{A}_S \tag{7.129.2}$$

$$\mathbf{u3}_T = \mathbf{DS3}_{TS}.\mathbf{A}_S \tag{7.129.3}$$

$$\mathbf{s33}_T = \mathbf{S33}_{TS}.\mathbf{A}_S \tag{7.130.1}$$

$$\mathbf{s11}_T = \mathbf{S11}_{TS}.\mathbf{A}_S \tag{7.130.2}$$

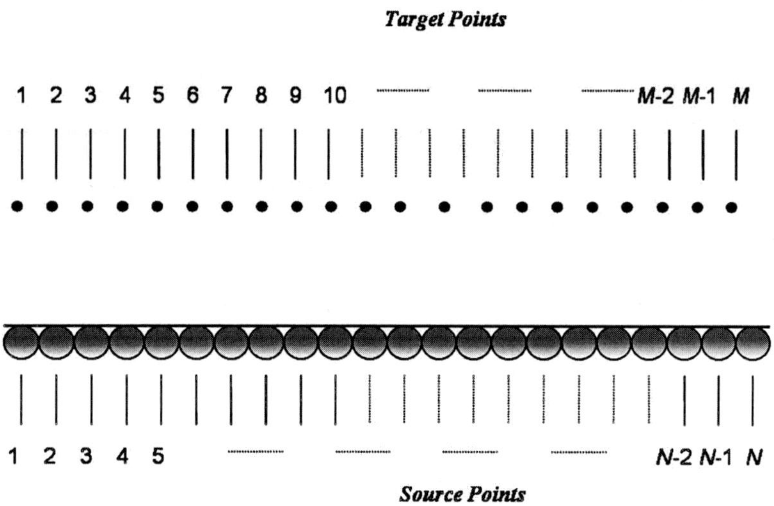

FIGURE 7.9 A schematic diagram of *N* source points along the boundary (or interface) and *M* target points inside a solid.

$$\mathbf{s22}_T = \mathbf{S22}_{TS}.\mathbf{A}_S \tag{7.130.3}$$

$$\mathbf{s31}_T = \mathbf{S31}_{TS}.\mathbf{A}_S \tag{7.130.4}$$

$$\mathbf{s32}_T = \mathbf{S32}_{TS}.\mathbf{A}_S \tag{7.130.5}$$

$$\mathbf{s12}_T = \mathbf{S12}_{TS}.\mathbf{A}_S \tag{7.130.6}$$

where

$$\{A_S\}^T = \begin{bmatrix} (T^1)^T & (T^2)^T & (T^3)^T & (T^4)^T & \dots\dots & (T^{M-2})^T & (T^{M-1})^T & (T^M)^T \end{bmatrix}^T_{(3Mx1)} \tag{7.131}$$

$(T^i)^T$ is a (1x3) vector, which is the transpose of $\mathbf{T}^i$, defined in Eqs. (7.90) and (7.119), where the superscript i indicates the i-th point source; for the m-th point source $i=m$. Displacement and stress matrices from Eqs. (7.129.1) to (7.130.6) have the form as shown in Eqs. (7.132.1) and (7.132.2), respectively, where only $\mathbf{S33}_{TS}$ and $\mathbf{DS3}_{TS}$ matrices are written in detail for illustration purposes. The 1x3 vector s_{ij}^m used in Eq. (7.132.2) is defined in Eq. (7.119). The additional subscript n of $s_{ij}^{m}{}_{n}$ indicates the n-th target point. Note that n varies from 1 to M while the superscript m varies from 1 to N. Similarly, the element $G_i^m{}_n$ of Eq. (7.131.1) is defined in Eq. (7.91). Superscript m stands for the m-th point source and the second subscript n corresponds to the n-th target point.

$$\mathbf{S33}_{TS} = \begin{bmatrix} s_{33}{}^1{}_1 & s_{33}{}^2{}_1 & s_{33}{}^3{}_1 & s_{33}{}^4{}_1 & s_{33}{}^5{}_1 & \dots\dots & s_{33}{}^{N-2}{}_1 & s_{33}{}^{N-1}{}_1 & s_{33}{}^N{}_1 \\ s_{33}{}^1{}_2 & s_{33}{}^2{}_2 & s_{33}{}^3{}_2 & s_{33}{}^4{}_2 & s_{33}{}^5{}_2 & \dots\dots & s_{33}{}^{N-2}{}_2 & s_{33}{}^{N-1}{}_2 & s_{33}{}^N{}_2 \\ s_{33}{}^1{}_3 & s_{33}{}^2{}_3 & s_{33}{}^3{}_3 & s_{33}{}^4{}_3 & s_{33}{}^5{}_3 & \dots\dots & s_{33}{}^{N-2}{}_3 & s_{33}{}^{N-1}{}_3 & s_{33}{}^N{}_3 \\ \dots & \dots & \dots & \dots & \dots & \dots\dots & \dots & \dots & \dots \\ \dots & \dots & \dots & \dots & \dots & \dots\dots & \dots & \dots & \dots \\ s_{33}{}^1{}_{M-2} & s_{33}{}^2{}_{M-2} & s_{33}{}^3{}_{M-2} & s_{33}{}^4{}_{M-2} & s_{33}{}^5{}_{M-2} & \dots\dots & s_{33}{}^{N-2}{}_{M-2} & s_{33}{}^{N-1}{}_{M-2} & s_{33}{}^N{}_{M-2} \\ s_{33}{}^1{}_{M-1} & s_{33}{}^2{}_{M-1} & s_{33}{}^3{}_{M-1} & s_{33}{}^4{}_{M-1} & s_{33}{}^5{}_{M-1} & \dots\dots & s_{33}{}^{N-2}{}_{M-1} & s_{33}{}^{N-1}{}_{M-1} & s_{33}{}^N{}_{M-1} \\ s_{33}{}^1{}_M & s_{33}{}^2{}_M & s_{33}{}^3{}_M & s_{33}{}^4{}_M & s_{33}{}^5{}_M & \dots\dots & s_{33}{}^{N-2}{}_M & s_{33}{}^{N-1}{}_M & s_{33}{}^N{}_M \end{bmatrix}_{(Mx3N)} \tag{7.132.1}$$

and

$$\mathbf{DS3}_{TS} = \begin{bmatrix} \mathbf{G}_3{}^1{}_1 & \mathbf{G}_3{}^2{}_1 & \mathbf{G}_3{}^3{}_1 & \dots & \mathbf{G}_3{}^{N-2}{}_1 & \mathbf{G}_3{}^{N-1}{}_1 & \mathbf{G}_3{}^N{}_1 \\ \mathbf{G}_3{}^1{}_2 & \mathbf{G}_3{}^2{}_2 & \mathbf{G}_3{}^3{}_2 & \dots & \mathbf{G}_3{}^{N-2}{}_2 & \mathbf{G}_3{}^{N-1}{}_2 & \mathbf{G}_3{}^N{}_2 \\ \mathbf{G}_3{}^1{}_3 & \mathbf{G}_3{}^2{}_3 & \mathbf{G}_3{}^3{}_3 & \dots & \mathbf{G}_3{}^{N-2}{}_3 & \mathbf{G}_3{}^{N-1}{}_3 & \mathbf{G}_3{}^N{}_3 \\ \dots & \dots & \dots & \dots & \dots & \dots & \dots \\ \mathbf{G}_3{}^1{}_{M-1} & \mathbf{G}_3{}^2{}_{M-1} & \mathbf{G}_3{}^3{}_{M-1} & \dots & \mathbf{G}_3{}^{N-2}{}_{M-1} & \mathbf{G}_3{}^{N-1}{}_{M-1} & \mathbf{G}_3{}^N{}_{M-1} \\ \mathbf{G}_3{}^1{}_M & \mathbf{G}_3{}^2{}_M & \mathbf{G}_3{}^3{}_M & \dots & \mathbf{G}_3{}^{N-2}{}_M & \mathbf{G}_3{}^{N-1}{}_M & \mathbf{G}_3{}^N{}_M \end{bmatrix}_{(Mx3N)} \tag{7.132.2}$$

$s_{33}{}^m{}_n$ and $\boldsymbol{G}_3{}^m{}_n$ for the n-th target point and m-th point source can be written as

$$s_{33}{}^m{}_n = \left[\left(\sigma_{33}^1\right)^m \; \left(\sigma_{33}^2\right)^m \; \left(\sigma_{33}^3\right)^m \right]_n \quad and \quad \mathbf{G}_3{}^m{}_n = \left[G_{31}{}^m \; G_{32}{}^m \; G_{33}{}^m \right]_n \tag{7.133}$$

Similarly, other stress and displacement matrices can be written following the similar pattern in Eq. (7.129.1) through (7.130.6).

7.3.2.3 Matrix Representation of Fluid Displacements

In Section 7.2.2, velocity field in front of a transducer is calculated however, displacement field was not discussed as there was no need for solid-fluid interface modeling. However, CNDE of real solid structure would need the interface conditions to be enforced at the fluid-solid interface. In solids, the displacements and stresses are obtained using the formulations presented in Section 7.3.2.2. However, to match these displacements and stresses at the fluid-solid interface one would need to define the displacements and pressure in fluid. Pressure in fluid is already discussed in Section 7.2.2.2. Hence, displacement in fluid media which was skipped before is presented in this section. The definition of r_n^m in the following equations is identical to that used in Section 7.2. It is the distance between the m-th point source and the n-th target point. In the same manner, the matrix expression for displacements at T set of target points due to S set of source points in the fluid can be written as:

$$\mathbf{U1}_T = \mathbf{DF1}_{TS}\,\mathbf{A}_S \tag{7.134.1}$$

$$\mathbf{U2}_T = \mathbf{DF2}_{TS}\,\mathbf{A}_S \tag{7.134.2}$$

$$\mathbf{U3}_T = \mathbf{DF3}_{TS}\,\mathbf{A}_S \tag{7.134.3}$$

where

$$\mathbf{DF}i_{TS} = \begin{bmatrix} g(R_{i1}^1,r_1^1) & g(R_{i1}^2,r_1^2) & g(R_{i1}^3,r_1^3) & \dots & g(R_{i1}^{N-1},r_1^{N-1}) & g(R_{i1}^N,r_1^N) \\ g(R_{i2}^1,r_2^1) & g(R_{i2}^2,r_2^2) & g(R_{i2}^3,r_2^3) & \dots & g(R_{i2}^{N-1},r_2^{N-1}) & g(R_{i2}^N,r_2^N) \\ g(R_{i3}^1,r_3^1) & g(R_{i3}^2,r_3^2) & g(R_{i3}^3,r_3^3) & \dots & g(R_{i3}^{N-1},r_3^{N-1}) & g(R_{i3}^N,r_3^N) \\ g(R_{i4}^1,r_4^1) & g(R_{i4}^2,r_4^2) & g(R_{i4}^3,r_4^3) & \dots & g(R_{i4}^{N-1},r_4^{N-1}) & g(R_{i4}^N,r_4^N) \\ \dots & \dots & \dots & \dots & \dots & \dots \\ g(R_{iM}^1,r_M^1) & g(R_{iM}^2,r_M^2) & g(R_{iM}^3,r_M^3) & \dots & g(R_{iM}^{N-1},r_M^{N-1}) & g(R_{iM}^N,r_M^N) \end{bmatrix}_{(MxN)} \tag{7.135}$$

where

$$g(R_{in}^m, r_n^m) = \frac{1}{\rho\omega^2}\left[\frac{1}{r_n^m} ik_f R_{in}^m e^{ik_f r_n^m} - \frac{e^{ik_f r_n^m}}{\left(r_n^m\right)^2} R_{in}^m \right] \tag{7.136}$$

$R_{in}^m = \frac{x_{in}^m - y_{in}^m}{r_n^m}$ and i takes values 1, 2, and 3.

Here, in Eq. (7.135), size of the matrix must be noticed. The size of the matrix is M x N, not M x $3N$ like it was used in solids. Point sources that contribute ultrasonic field in fluid has only one source strength unlike three required in solids in three orthogonal directions.

7.4 CNDE CASE STUDIES FOR ISOTROPIC SOLIDS USING DPSM

7.4.1 Computational Wave Field Modeling at Fluid-Solid Interface [4]

7.4.1.1 NDE Problem Statement

Let us consider a plane interface between a solid half-space and a fluid half-space analogous to a thick metallic structure to be inspected in ultrasonic water tank. A schematic diagram of the system for CNDE analysis is shown in Fig. 7.10, where the fluid media is drawn below the solid material. Only a few point sources are shown along the interface in the diagram to keep it simple; however, in the actual model the point sources are distributed over the entire interface. To simulate a real NDE problem, a circular transducer is assumed to be immersed in the fluid. A number of small point sources are distributed below the transducer face and on both sides of the interface. These sources should produce the total ultrasonic field in fluid and solid media. $\mathbf{A}_I$ is the source strength vector of the point sources that are placed above the solid-fluid interface that are responsible for generating the reflected ultrasonic field in the fluid. Similarly, $\mathbf{A}_I$* is the source strength vector of the sources that are distributed below the solid-fluid interface and model the transmitted field in the solid. The point sources that have been distributed below the transducer face have source strength vector $\mathbf{A}_S$. For CNDE it is customary to compute the field in both solid and fluid media. In Fig. 7.10, two points C and D are shown for the illustration purpose.

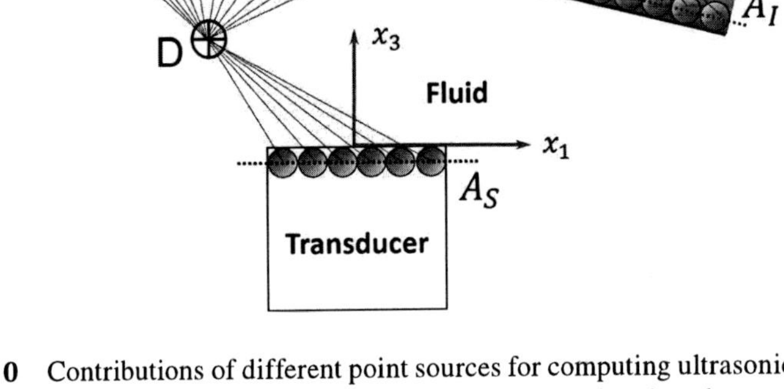

FIGURE 7.10 Contributions of different point sources for computing ultrasonic fields in the fluid (point D) and the solid (point C) are shown by lines connecting the relevant point sources to the point of interest (C or D).

The ultrasonic field at point C is the superposition of contributions by all the point sources distributed below the interface. Similarly, the ultrasonic field at point D will be the superposition of contributions by all the point sources distributed above the solid-fluid interface and below the transducer face.

7.4.1.2 Matrix formulation

Velocity and pressure fields in the fluid medium and stress and displacement fields in the solid medium can be expressed in the matrix form when we take a group of target points instead of single points C or D in Fig. 7.10. Referring to Eq. (7.43) we can write

$$\mathbf{M}_{SS}\mathbf{A}_S = \mathbf{V}_S \tag{7.137}$$

where $\mathbf{V}_S$ is the $(N \times 1)$ vector of the velocity components at the transducer surface corresponding to N number of source points distributed slightly behind the transducer face, $\mathbf{A}_S$ is the $(N \times 1)$ vector containing the source strengths of the transducer sources. The elements of $\mathbf{M}_{SS}$ are given in Eq. (7.48) and the matrix is defined when all target points are distributed on the transducer surface. Similarly, the velocity at any set of target points due to the interface sources (see Fig. 7.10) can be written as

$$\mathbf{M}_{TI}\mathbf{A}_I = \mathbf{V}_T \tag{7.138}$$

The interface has N source points distributed on each side of the interface. Hence, $\mathbf{A}_I$ has $(N\text{x}1)$ elements. Following Eq. (7.50) and in addition to Eq. (7.50), pressure at the same set of target points in the fluid medium due to the interface sources can be written as

$$\mathbf{Q}_{TI}\mathbf{A}_I = \mathbf{P}_T{}^I \tag{7.139}$$

Therefore, at a common set of target points in fluid, which are visible by both the transducer active sources and interface passive sources, the total pressure field generated by the transducer and interface sources can be written as

$$\mathbf{P}_T + \mathbf{P}_T{}^I = \mathbf{Q}_{TS}\mathbf{A}_S + \mathbf{Q}_{TI}\mathbf{A}_I = \mathbf{P}_T \tag{7.140}$$

Similarly, at any set of target points, the displacement along the x_3 direction in fluid can be written as

$$\mathbf{DF3}_{TS}\mathbf{A}_S + \mathbf{DF3}_{TI}\mathbf{A}_I = \mathbf{U3}_T \tag{7.141}$$

It should be reminded herein that each point source that is considered to calculate the transmitted field in solid, has three different point forces in three different directions as unknowns.

Using Eq. (7.130.1), the normal stress in x_3 direction at the interface can be written as

$$\mathbf{s33}_{II^*} = \mathbf{S33}_{II^*}\mathbf{A}_I{}^* \tag{7.142}$$

Similarly, the shear stresses, using Eqs. (7.130.4) and (7.130.5), we can write

$$\mathbf{s31}_{11^*} = \mathbf{S31}_{11^*}\mathbf{A}_1^* \tag{7.143}$$

$$\mathbf{s32}_{11^*} = \mathbf{S32}_{11^*}\mathbf{A}_1^* \tag{7.144}$$

where $\mathbf{A}_1^*$ is the source strength vector of dimension ($3N$ x 1) contains the source strengths of the sources distributed below the fluid-solid interface.

7.4.1.3 Boundary Conditions

Across the fluid-solid interface the normal displacement (u_3 component) should be continuous. Also, at the interface negative of the normal stress ($-\sigma_{33}$) in solid and the pressure in the fluid should be continuous. Shear stresses at the fluid-solid interface should vanish. If the normal velocity of the transducer face is assumed to be V_0 the boundary conditions on the transducer surface can be written as

$$\mathbf{M}_{SS}\mathbf{A}_S + \mathbf{M}_{S1}\mathbf{A}_1 = \mathbf{V}_{S0} \tag{7.145}$$

Similarly, the interface conditions at fluid-solid interface can be written as

$$\mathbf{Q}_{1S}\mathbf{A}_S + \mathbf{Q}_{11}\mathbf{A}_1 = -\mathbf{S33}_{11^*}\mathbf{A}_1^* \tag{7.146}$$

$$\mathbf{DF3}_{1S}\mathbf{A}_S + \mathbf{DF3}_{11}\mathbf{A}_1 = \mathbf{DS3}_{11^*}\mathbf{A}_1^* \tag{7.147}$$

$$\mathbf{S31}_{11^*}\mathbf{A}_1^* = 0 \tag{7.148}$$

$$\mathbf{S32}_{11^*}\mathbf{A}_1^* = 0 \tag{7.149}$$

Eqs. (7.146) through (7.149) can be written in matrix form

$$\begin{bmatrix} M_{SS} & M_{S1} & 0 \\ Q_{1S} & Q_{11} & S33_{11^*} \\ DF3_{1S} & DF3_{11} & -DS3_{11^*} \\ 0 & 0 & S31_{11^*} \\ 0 & 0 & S32_{11^*} \end{bmatrix}_{(M+4N)x(M+4N)} \begin{Bmatrix} A_S \\ A_1 \\ A_1^* \end{Bmatrix}_{(M+4N)} = \begin{Bmatrix} V_{S0} \\ 0 \\ 0 \end{Bmatrix}_{M+4N} \tag{7.150}$$

$$or\ [\mathbf{MAT}]\{\Lambda\} = \{\mathbf{V}\} \tag{7.151}$$

7.4.1.4 Solution

The source strength vector $\{\Lambda\}$ of the total system can be obtained from Eq. (7.151),

$$\{\Lambda\} = [\mathbf{MAT}]^{-1}\{\mathbf{V}\} \tag{7.152}$$

After the source strengths are obtained, the ultrasonic, pressure, velocity, stress, and displacement fields can be calculated. For example, the pressure field in the fluid can be written as:

$$\mathbf{P}_{(F)} = \mathbf{Q}_{(F)S}\mathbf{A}_S + \mathbf{Q}_{(F)1}\mathbf{A}_1 \quad (7.153)$$

where F is a set of target points inside the fluid medium. Similarly, in the solid, the stress, and displacement fields can be calculated as

$$\begin{aligned} \mathbf{s33}_{(S)} &= \mathbf{S33}_{(S)1^*}\mathbf{A}_1^* \\ \mathbf{s11}_{(S)} &= \mathbf{S11}_{(S)1^*}\mathbf{A}_1^* \\ &\cdots \\ \mathbf{u3}_{(S)} &= \mathbf{DS3}_{(S)1^*}\mathbf{A}_1^* \\ \mathbf{u1}_{(S)} &= \mathbf{DS1}_{(S)1^*}\mathbf{A}_1^* \end{aligned} \quad (7.154)$$

where S is a set of target points inside the solid media.

7.4.1.5 Numerical Results Near Fluid Solid Interface

A computer code is written to implement the above equations and, in this section, numerical results for the above CNDE problem are presented for a thick aluminum solid slab immersed in water. For convenience in subsequent discussions x_1, x_2, x_3, and x, y, z has been used interchangeably. Therefore, please keep in mind that $x_1 = x$, $x_2 = y$, and $x_3 = z$. In CNDE using DPSM, while placing the point sources, one needs to satisfy the requirement of the maximum spacing allowed between neighboring point sources as discussed in the Appendix. This distance depends on the wavelength of the ultrasonic signal. The wave speed and other material properties of aluminum and water are given in Table 7.1.

TABLE 7.1
Material Properties and Critical Angle Calculation

Wave speed in water (c_f)	1.48 km/sec
Density of water (ρ_f)	1 gm/c.c.
P-wave speed in aluminum (c_p)	6.5 km/sec
S-wave speed in aluminum (c_s)	3.13 km/sec
Density of aluminum (ρ_s)	2.7 gm/c.c.
First Lamé constant ($\mathbb{L}$)	$\rho_s(c_p^2 - 2c_s^2) = 61.17$ GPa
Second Lamé constant ($\mathbb{m}$)	$\rho_s c_s^2 = 26.45$ GPa
Poisson´s ratio (ν)	$\frac{\mathbb{L}}{2(\mathbb{L}+\mathbb{m})} = 0.349$
Rayleigh-wave speed in aluminum (c_r)	$\frac{0.862+1.14\nu}{1+\nu}c_s = 2.923$ km/sec
Rayleigh critical angle (θ_c)	$Sin^{-1}\left(\frac{c_f}{c_r}\right) = 30.4196°$

Ultrasonic waves are generated by an ultrasonic transducer that is placed in water at different inclinations relative to the interface. The diameter of the transducer was 4 mm.

In the following figures, the gray scale bars are provided to give an idea about the magnitudes of the ultrasonic fields in the contour plots. Note that the grey scale bars are not identical in all figures. In Fig. 7.11, the normal stresses σ_{11} and σ_{33}

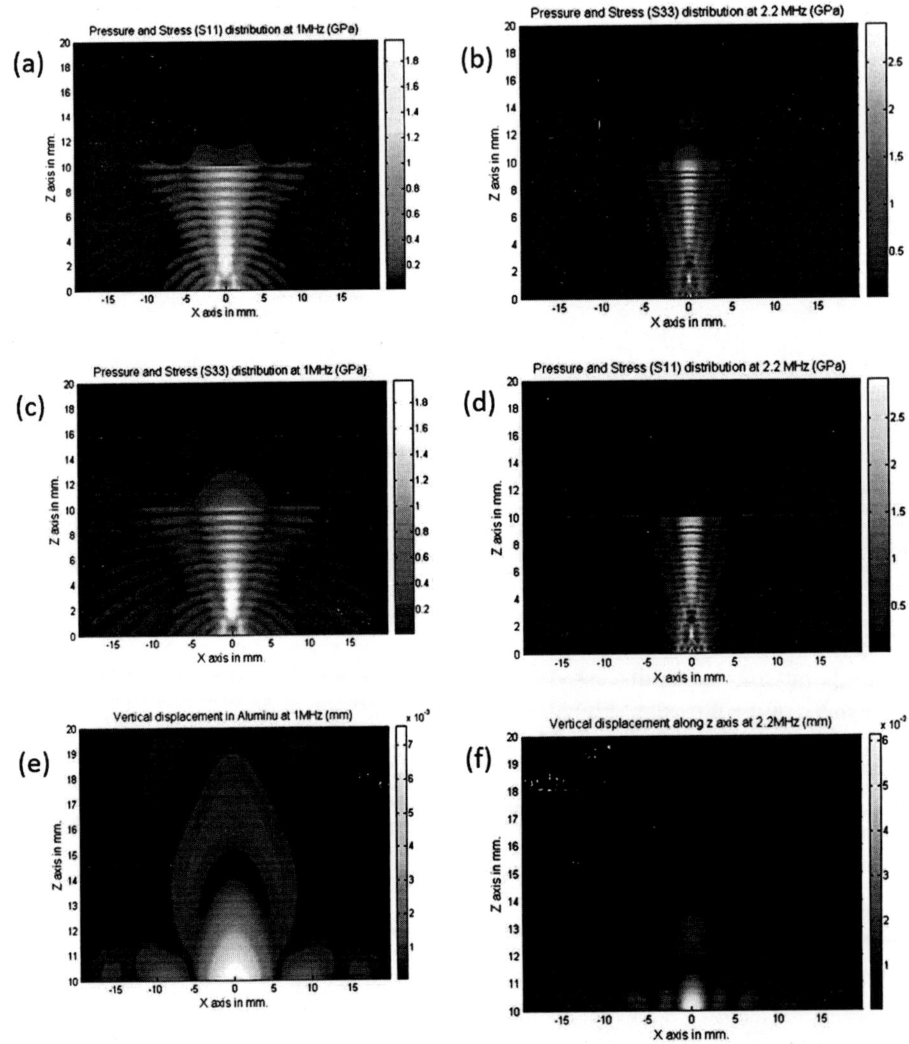

FIGURE 7.11 (Place the figures a to f in two columns as shown above) Ultrasonic field for normal incidence of the wave on the interface (a) The normal stress along x axis (**s**11) in aluminum and pressure in water at 1 MHz. (b) The normal stress along x axis (**s**11) in aluminum and pressure in water at 2.2 MHz. (c) The normal stress along z axis (**s**33) in aluminum and pressure in water at 1 MHz. (d) The normal stress along z axis (**s**33) in aluminum and pressure in water at 2.2 MHz. (e) The vertical displacement along z axis (**u**3) in aluminum at 1 MHz. (f) The vertical displacement along z axis (**u**3) in aluminum at 2.2 MHz.

in aluminum are plotted in addition to the pressure field in water for normal incidence of the ultrasonic wave at the interface. Ultrasonic field in aluminum and water are shown in Fig. 7.11a–f. The left column of Fig. 7.11a, c, and e is generated with 1 MHz ultrasonic transducer and the right column (Fig. 7.11b, d, and f) is for 2.2 MHz transducer. Top and middle rows show horizontal (σ_{11}) and vertical (σ_{33}) stresses, respectively. These two stress components are different in aluminum but same in water. The vertical displacement (u_3) in aluminum is plotted in the bottom row. Higher number of peaks and dips are visible for the 2.2 MHz signal in comparison to that for the 1 MHz signal. It can be seen that the beam is more collimated at 2.2 MHz as expected.

Leaky guided waves in the fluid are also modeled for critical inclination of a transducer which is frequently used in NDE problems to generate the Rayleigh waves in the test articles. Formula for the critical angle calculation is shown in Table 7.1. For the critical angle (30.42°) of incidence, the ultrasonic fields in aluminum and water are shown in Fig. 7.12a–f. For convenience, the pressure in water and stresses in aluminum are plotted in the same figures. The top row shows the horizontal normal stress (σ_{11}) and middle row shows the vertical normal stress (σ_{33}) components. The vertical displacement component in the solid are plotted in the bottom row. As before, the left column (Fig. 7.12a, c, and e) is generated with a 1 MHz transducer and the right column (Fig. 7.12b, d, and f) is generated with a 2.2 MHz transducer. It is well known that as the frequency of the signal increases its wavelength decreases and the depth of guided wave penetration decreases since it is proportional to the wavelength. Comparing Fig. 7.12a and b, it can be seen that the thickness of the disturbance (penetration depth in solid) due to the Rayleigh wave propagating along the interface is higher with 1 MHz signal compared to 2.2 MHz signal. The leaky wave in water is clearly visible in Fig. 7.12a–d. The vertical displacement fields in the solid are plotted in Fig. 7.12e and f for 1 MHz and 2.2 MHz signals, respectively. It can be seen that the vertical displacements in Fig. 7.12e and f are more confined near the interface compared to Fig. 7.12e and f because the guided wave mostly excites the particles near the interface. It is interesting to note that even the null region predicted by {Bertoni, 1973 #10} in the reflected beam profile can be seen in the computed results (see Fig. 7.12d).

Fig. 7.13a–f shows the ultrasonic fields in aluminum and water, for 45.42° angle of incidence, which is 15° greater than the Rayleigh angle. The left column (Fig. 7.13a, c, and e) are generated with 1 MHz transducer and the right column (Fig. 7.13b, d, and f) are generated with 2.2 MHz transducer. It is well known that when the angle of incidence of is greater than the critical angle, most of the incident energy is reflected by the interface. Therefore, very small amount of energy is transmitted inside the solid material. This phenomenon is clearly visible in Fig. 7.13a–f. The pressure distribution in water and normal stresses σ_{11} and σ_{33} in aluminum is shown in Fig. 7.13a and c, respectively, for 1 MHz signal frequency. Similarly, Fig. 7.13b and d shows those for 2.2 MHz signal frequency. In Fig. 7.13a–d a weak guided wave is observed although the incident angle is not a critical angle. The guided wave is observed because the incident beam is not perfectly collimated. Therefore, a small amount of energy still strikes the interface at a critical angle. Similar wave fields are calculated with an incident angle 15.42°, which is 15° less than the critical angle. Wave fields

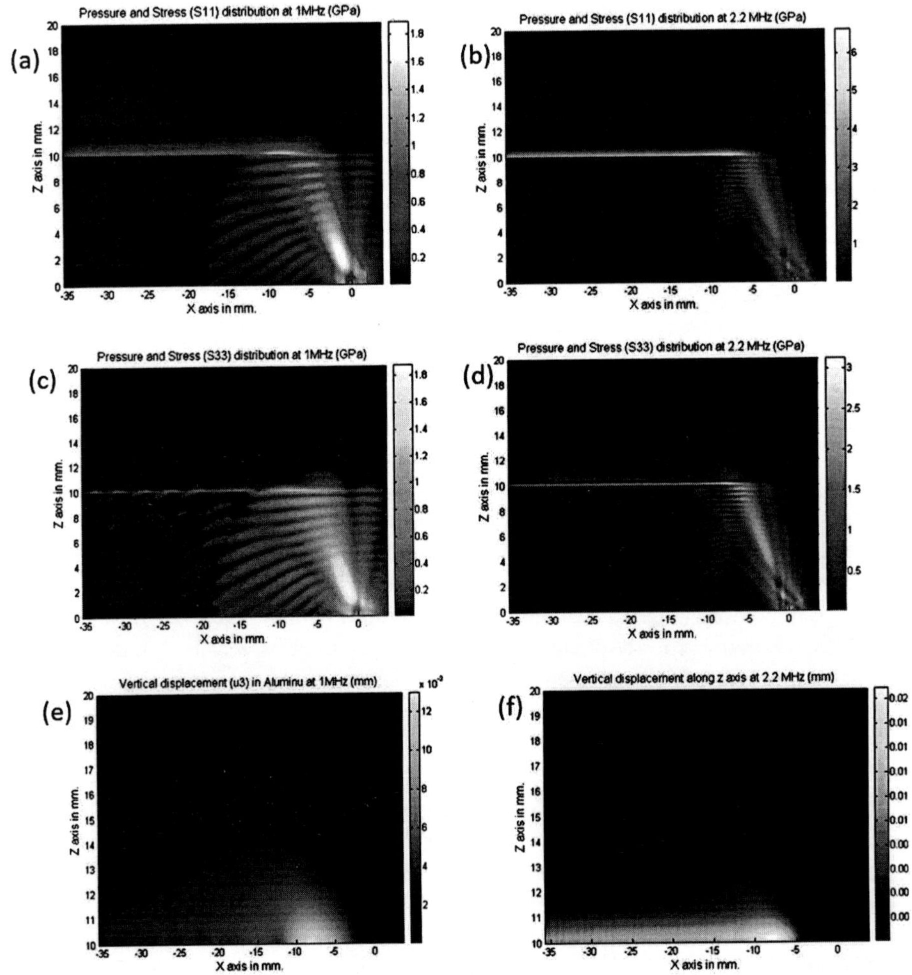

FIGURE 7.12 Ultrasonic field for critical angle (30.42°) of incidence of the wave on the interface (a) The normal stress along *x* axis (*s*11) in aluminum and pressure in water at 1 MHz. (b) The normal stress along *x* axis (*s*11) in aluminum and pressure in water at 2.2 MHz. (c) The normal stress along *z* axis (*s*33) in aluminum and pressure in water at 1 MHz. (d) The normal stress along *z* axis (*s*33) in aluminum and pressure in water at 2.2 MHz. (e) The vertical displacement along *z* axis (***u***3) in aluminum at 1 MHz. (f) The vertical displacement along *z* axis (***u***3) in aluminum at 2.2 MHz.

are arranged in same manner in Fig. 7.14a–f like they were arranged in Fig. 7.13. It can be seen that when the ultrasonic beam hits the interface at an angle larger than a critical angle, the transmitted field propagate inside aluminum with a higher transmission angle than the angle of incidence in water. Hence, together, Figs. 7.11–7.14 establish the accuracy of the DPSM method that were able to replicate the most well-known physical phenomena through CNDE.

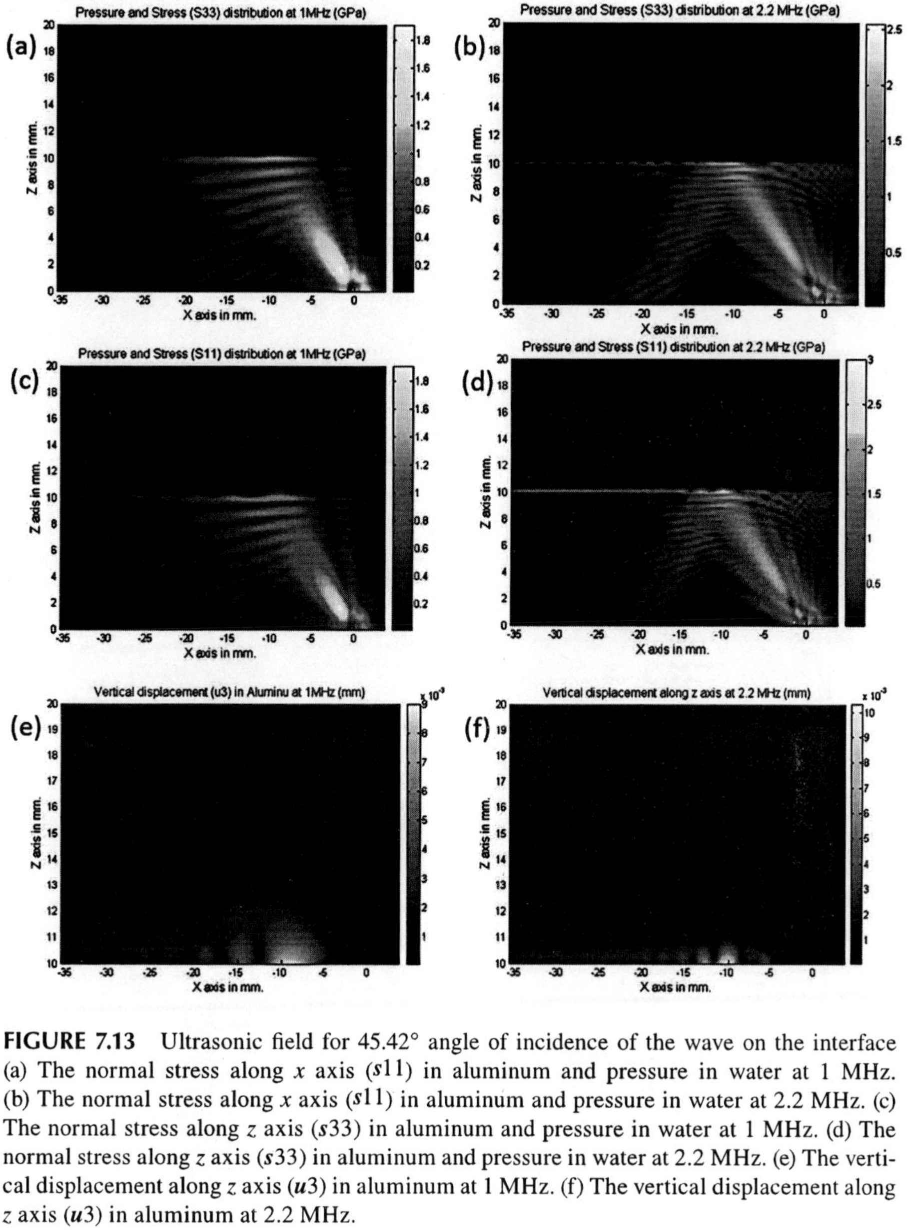

FIGURE 7.13 Ultrasonic field for 45.42° angle of incidence of the wave on the interface (a) The normal stress along x axis ($s11$) in aluminum and pressure in water at 1 MHz. (b) The normal stress along x axis ($s11$) in aluminum and pressure in water at 2.2 MHz. (c) The normal stress along z axis ($s33$) in aluminum and pressure in water at 1 MHz. (d) The normal stress along z axis ($s33$) in aluminum and pressure in water at 2.2 MHz. (e) The vertical displacement along z axis ($u3$) in aluminum at 1 MHz. (f) The vertical displacement along z axis ($u3$) in aluminum at 2.2 MHz.

7.4.2 Computational Wave Field Modeling in a Solid Plate Immersed in Fluid [3]

7.4.2.1 NDE Problem Statement

A most common NDE problem is considered herein where ultrasonic field is generated by an ultrasonic transducer of finite dimension in the vicinity of a solid plate

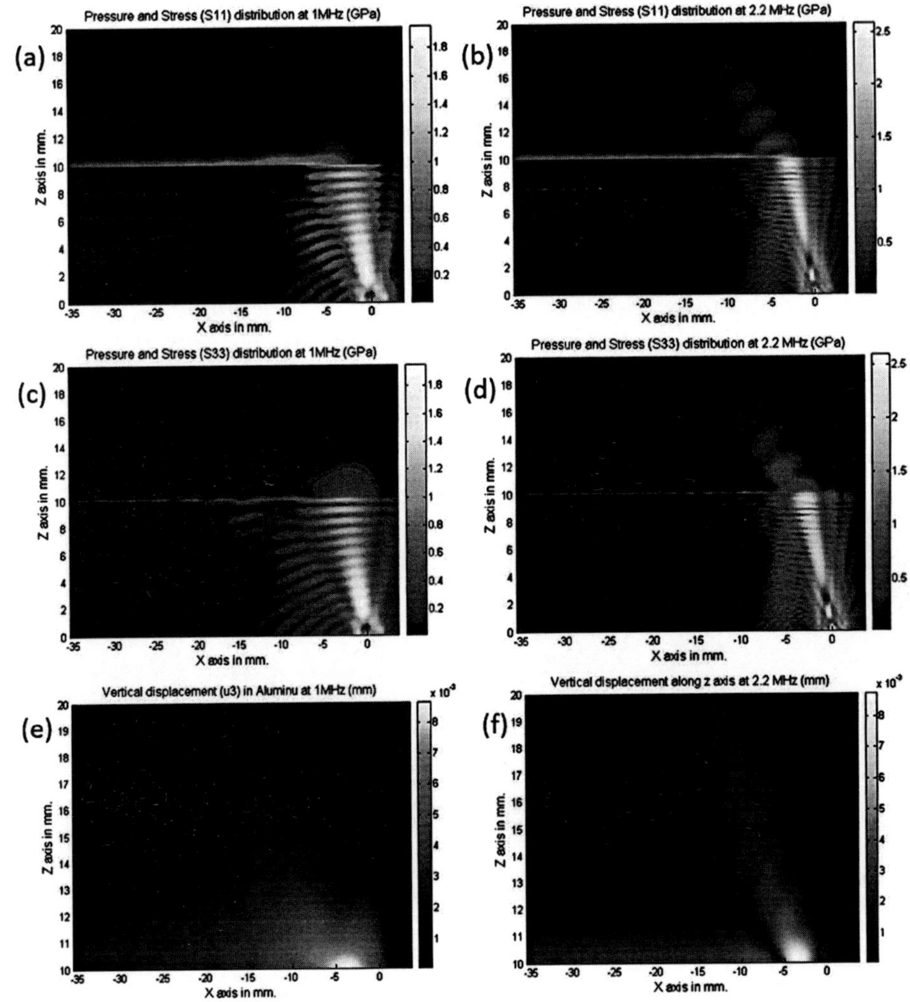

FIGURE 7.14 Ultrasonic field for 15.42° angle of incidence of the wave on the interface (a) The normal stress along x axis ($s11$) in aluminum and pressure in water at 1 MHz. (b) The normal stress along x axis ($s11$) in aluminum and pressure in water at 2.2 MHz. (c) The normal stress along z axis ($s33$) in aluminum and pressure in water at 1 MHz. (d) The normal stress along z axis ($s33$) in aluminum and pressure in water at 2.2 MHz. (e) The vertical displacement along z axis ($u3$) in aluminum at 1 MHz. (f) The vertical displacement along z axis ($u3$) in aluminum at 2.2 MHz.

when both the plate and the transducers are immersed in a fluid. Using critical angle of inclination of the transducers, guided waves are generated in the plate (refer Fig. 5.1c). This situation is considered in this section by placing two transducers on either side of the plate, which will be solved using DPSM based CNDE. The ultrasonic fields in the solid plate are calculated for critical angles corresponding to the symmetric and antisymmetric guided wave modes in the plate.

7.4.2.2 Matrix Formulation and Boundary Conditions

Two circular transducers are placed in the fluid on two sides of a plate as shown in Fig. 7.15. Behind the transducer faces and on both sides of the interfaces, a number of point sources are distributed. These sources when superimposed, should produce the total ultrasonic field in fluid and solid media. The lower interface is denoted by "Interface 1." $\mathbf{A}_1$ is the source strength vector of the point sources that are placed above the Interface 1 and generate part of the ultrasonic field in the fluid below the plate. $\mathbf{A}_1^*$ is the source strength vector of the sources that are distributed below the Interface 1, responsible for the transmitted field in the solid plate. Similarly, $\mathbf{A}_2$ and $\mathbf{A}_2^*$ are the source strength vectors of the point sources that are distributed above and below Interface 2, respectively. Transducer faces have source strength vectors $\mathbf{A}_S$ and $\mathbf{A}_R$. In Fig. 7.15, three points (C, D, and E) are shown to analyze the effect of the point sources. The ultrasonic field at point C is the summation of the contributions of all point sources with source strengths $\mathbf{A}_1^*$ and $\mathbf{A}_2$ distributed below and above the Interfaces 1 and 2, respectively. Ultrasonic field at point D is the summation of contributions of the point sources with source strengths $\mathbf{A}_1$ and $\mathbf{A}_S$ distributed above the solid-fluid interface 1 and behind the front face of transducer S, respectively. Similarly, the Ultrasonic field at point E is the summation of contributions of the point sources with source strengths $\mathbf{A}_2^*$ and $\mathbf{A}_R$ distributed below the solid-fluid interface 2 and behind the front face of transducer R, respectively. Here, in this section, matrix formulation presented in the previous sections is used using similar notations and reader's familiarity with the nations is assumed.

Across the fluid-solid interfaces the displacement normal to the interface should be continuous. Also, across the interfaces, the normal stress (**s33**) in solid and fluid media should be continuous and the shear stresses at the interfaces must vanish.

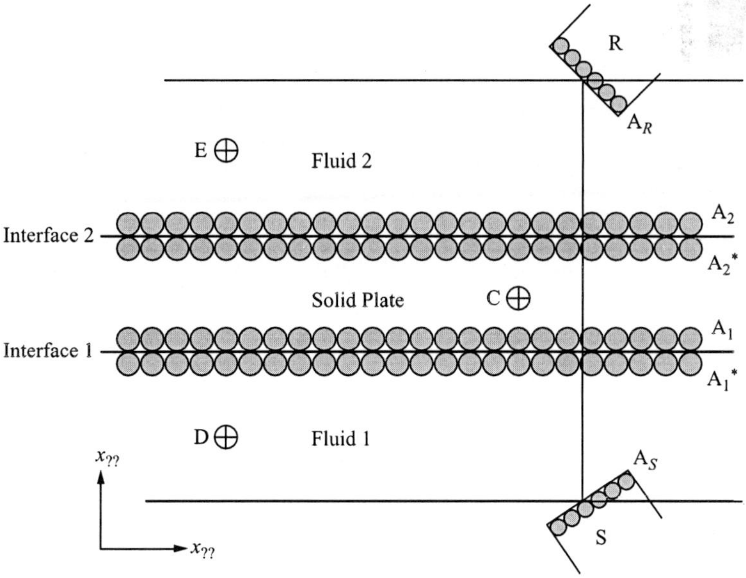

FIGURE 7.15 Distribution of the point sources near the inclined transducers and the fluid-solid interfaces of the plate.

If the normal velocities of the two transducer faces are assumed to be $\mathbf{V}_{S0}$ and $\mathbf{V}_{R0}$, then the velocity of the surface of the transducer (designated as *S*) can be expressed as

$$\mathbf{M}_{SS}\mathbf{A}_S + \mathbf{M}_{S1}\mathbf{A}_1 = \mathbf{V}_{S0} \tag{7.155}$$

and on the surface of the transducer designated as *R*.

$$\mathbf{M}_{R2^*}\mathbf{A}_2^* + \mathbf{M}_{RR}\mathbf{A}_R = \mathbf{V}_{R0} \tag{7.156}$$

At the interfaces, from the continuity of the normal stress the following equations are obtained:

$$\mathbf{Q}_{1S}\mathbf{A}_S + \mathbf{Q}_{11}\mathbf{A}_1 = -\mathbf{S33}_{11^*}\mathbf{A}_1^* - \mathbf{S33}_{12}\mathbf{A}_2 \tag{7.157}$$

$$\mathbf{Q}_{22^*}\mathbf{A}_2^* + \mathbf{Q}_{2R}\mathbf{A}_R = -\mathbf{S33}_{21^*}\mathbf{A}_1^* - \mathbf{S33}_{22}\mathbf{A}_2 \tag{7.158}$$

Continuity of the normal displacement across the two interfaces gives

$$\mathbf{DF3}_{1S}\mathbf{A}_S + \mathbf{DF3}_{11}\mathbf{A}_1 = \mathbf{DS3}_{11^*}\mathbf{A}_1^* + \mathbf{DS3}_{12}\mathbf{A}_2 \tag{7.159}$$

$$\mathbf{DF3}_{22^*}\mathbf{A}_2^* + \mathbf{DF3}_{2R}\mathbf{A}_R = \mathbf{DS3}_{21^*}\mathbf{A}_1^* + \mathbf{DS3}_{22}\mathbf{A}_2 \tag{7.160}$$

and from the vanishing shear stresses at the fluid-solid interface, one can get

$$\mathbf{S31}_{11^*}\mathbf{A}_1^* + \mathbf{S31}_{12}\mathbf{A}_2 = 0 \tag{7.161}$$

$$\mathbf{S32}_{11^*}\mathbf{A}_1^* + \mathbf{S32}_{12}\mathbf{A}_2 = 0 \tag{7.162}$$

Please note that the matrices used in the above equations are all respective Green's function matrices that are discussed in the previous sections. Next, Eqs. (7.155) through (7.162) can be written in the matrix form

$$\begin{bmatrix} \mathbf{M}_{SS} & \mathbf{M}_{S1} & \mathbf{0} & \mathbf{0} & \mathbf{0} & \mathbf{0} \\ \mathbf{Q}_{1S} & \mathbf{Q}_{11} & \mathbf{S33}_{11^*} & \mathbf{S33}_{12} & \mathbf{0} & \mathbf{0} \\ \mathbf{DF3}_{1S} & \mathbf{DF3}_{11} & -\mathbf{DS3}_{11^*} & -\mathbf{DS3}_{12} & \mathbf{0} & \mathbf{0} \\ \mathbf{0} & \mathbf{0} & \mathbf{S31}_{11^*} & \mathbf{S31}_{12} & \mathbf{0} & \mathbf{0} \\ \mathbf{0} & \mathbf{0} & \mathbf{S32}_{11^*} & \mathbf{S32}_{12} & \mathbf{0} & \mathbf{0} \\ \mathbf{0} & \mathbf{0} & \mathbf{S32}_{21^*} & \mathbf{S32}_{22} & \mathbf{0} & \mathbf{0} \\ \mathbf{0} & \mathbf{0} & \mathbf{S31}_{21^*} & \mathbf{S31}_{22} & \mathbf{0} & \mathbf{0} \\ \mathbf{0} & \mathbf{0} & \mathbf{S33}_{21^*} & \mathbf{S33}_{22} & \mathbf{Q}_{22^*} & \mathbf{Q}_{2R} \\ \mathbf{0} & \mathbf{0} & -\mathbf{DS3}_{21^*} & -\mathbf{DS3}_{22} & \mathbf{DF3}_{22^*} & \mathbf{DF3}_{2R} \\ \mathbf{0} & \mathbf{0} & \mathbf{0} & \mathbf{0} & \mathbf{M}_{R2^*} & \mathbf{M}_{RR} \end{bmatrix}_{(2M+8N)x(2M+8N)} \begin{Bmatrix} \mathbf{A}_S \\ \mathbf{A}_1 \\ \mathbf{A}_1^* \\ \mathbf{A}_1 \\ \mathbf{A}_2^* \\ \mathbf{A}_R \end{Bmatrix}_{(2M+8N)} = \begin{Bmatrix} \mathbf{V}_{S0} \\ \mathbf{0} \\ \mathbf{0} \\ \mathbf{0} \\ \mathbf{0} \\ \mathbf{0} \\ \mathbf{0} \\ \mathbf{0} \\ \mathbf{0} \\ \mathbf{V}_{R0} \end{Bmatrix}_{(2M+8N)} \tag{7.163}$$

$$or \quad [\mathbf{MAT}]\{\mathbf{\Lambda}\} = \{\mathbf{V}\} \tag{7.164}$$

7.4.2.3 Solution

The source strength vector $\{\Lambda\}$ of the total system can be calculated from Eq. (7.164)

$$\{\Lambda\} = [\mathbf{MAT}]^{-1}\{\mathbf{V}\} \tag{7.165}$$

After calculating the source strengths, the pressure, velocity, stress, and displacement values at any point in the NDE system can be calculated. For example, the pressure field in the fluid is

$$\underline{\mathbf{PR}}_{(F1)} = \mathbf{Q}_{(F1)S}\mathbf{A}_S + \mathbf{Q}_{(F1)1}\mathbf{A}_1 \tag{7.166}$$

$$\underline{\mathbf{PR}}_{(F2)} = \mathbf{Q}_{(F2)R}\mathbf{A}_R + \mathbf{Q}_{(F2)2^*}\mathbf{A}_2^* \tag{7.167}$$

where $F1$ is a set of target points inside the fluid medium below the plate and $F2$ is a set of target points inside the fluid medium above the plate. Similarly, in the solid plate, the stress, and displacement fields can be obtained from the following set of equations.

$$\begin{aligned}
\mathbf{s33}_{(S)} &= \mathbf{S33}_{(S)1^*}\mathbf{A}_1^* + \mathbf{S33}_{(S)2}\mathbf{A}_2 \\
\mathbf{s11}_{(S)} &= \mathbf{S11}_{(S)1^*}\mathbf{A}_1^* + \mathbf{S11}_{(S)2}\mathbf{A}_2 \\
&\cdots \\
\mathbf{u3}_{(S)} &= \mathbf{DS3}_{(S)1^*}\mathbf{A}_1^* + \mathbf{DS3}_{(S)2}\mathbf{A}_2 \\
\mathbf{u1}_{(S)} &= \mathbf{DS1}_{(S)1^*}\mathbf{A}_1^* + \mathbf{DS1}_{(S)2}\mathbf{A}_2
\end{aligned} \tag{7.168}$$

where S is a set of target points inside the solid plate.

7.4.2.4 Numerical Results: Ultrasonic Fields in Solid Plate

A computer codes is developed to model the ultrasonic field in the solid plate by using the formulation presented above. Rayleigh-Lamb equations discussed in Chapter 5 guides the Lamb wave modes to propagate in a solid plate at certain frequencies with a certain phase velocity. Their relationship is defined as dispersion behavior and is discussed in detail in Chapter 5. All wave modes inside the plate are dispersive and are the solutions of Eq. (5.31) depicted in Chapter 5. As a most popular and useful NDE problem, in this CNDE simulation, fundamental symmetric and antisymmetric modes in plates are modeled.

DPSM modeling is used to model the ultrasonic field in an aluminum plate immersed in water and the dispersion plot for such a plate is shown in Fig. 7.16. The dispersion curves in Fig. 7.16 are obtained solving Eq. (5.31) with material properties listed in Table 7.1. 4 mm diameter transducers are used to generate the ultrasonic signals. To generate guided wave modes transducer orientations are different for different frequencies. To obtain the maximum pressure of the striking beam at the plate surface position the plate specimens are placed at different distances from the transducers exciting different frequencies. The results (stress and displacement

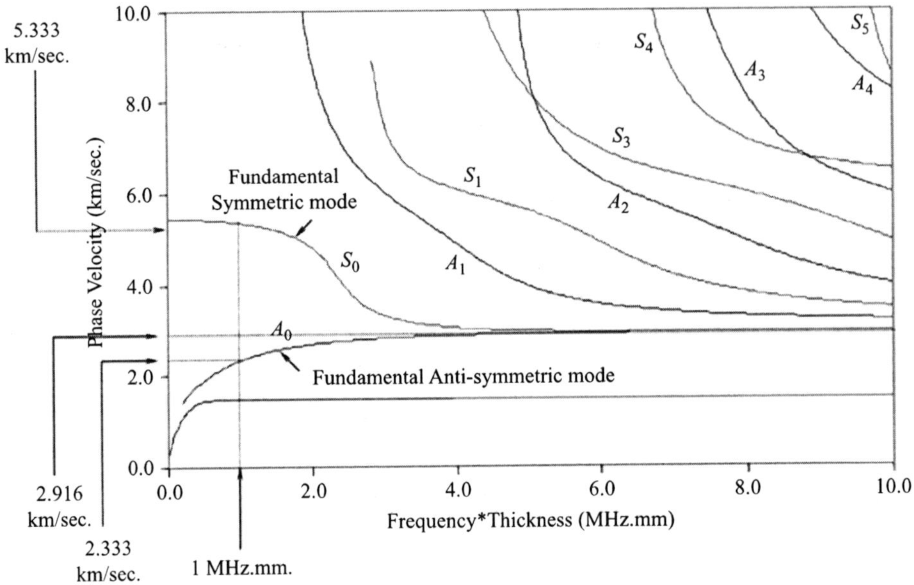

FIGURE 7.16 Dispersion curve for an Aluminum plate with material properties in Table 7.1.

fields inside the plate) are presented in Figs. 7.17–7.20. The properties (the density and wave speeds) of the aluminum plate are given in Table 7.1. 50 mm length of the plate is simulated that is immersed in water. Unit along the abscissa of Fig. 7.16 is the product of frequency and thickness in MHz-mm. Along the ordinate the phase velocity is in km/sec. It can be seen from this figure that for 10 mm thick plate and 1 MHz transducer (or for 1 mm thick plate and 10 MHz transducer) Rayleigh waves are produced for an inclination angle of $sin^{-1}\left(\frac{c_f}{c_r}\right) = sin^{-1}\left(\frac{1.48}{2.916}\right) = 30.50°$.

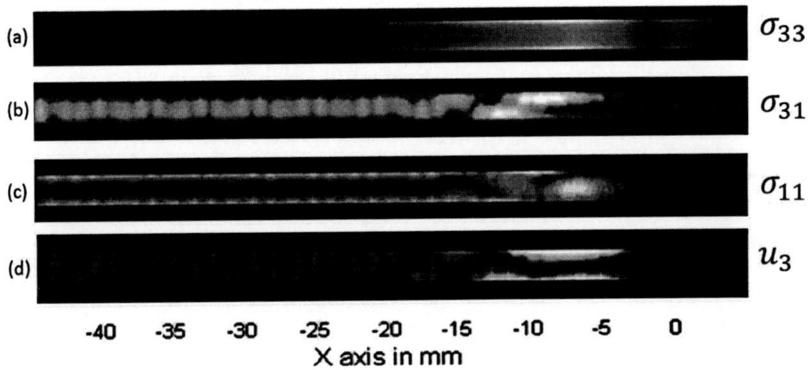

FIGURE 7.17 Ultrasonic fields inside a 1 mm thick aluminum plate for A_0 mode of excitation, generated by 1 MHz transducers inclined at 39.43° angle as shown in Fig. 4.10. Four plots show the variations of (a) the vertical normal stress s_{33}, (b) the shear stress s_{31}, (c) the horizontal normal stress s_{11} and (d) the vertical displacement u_3 component.

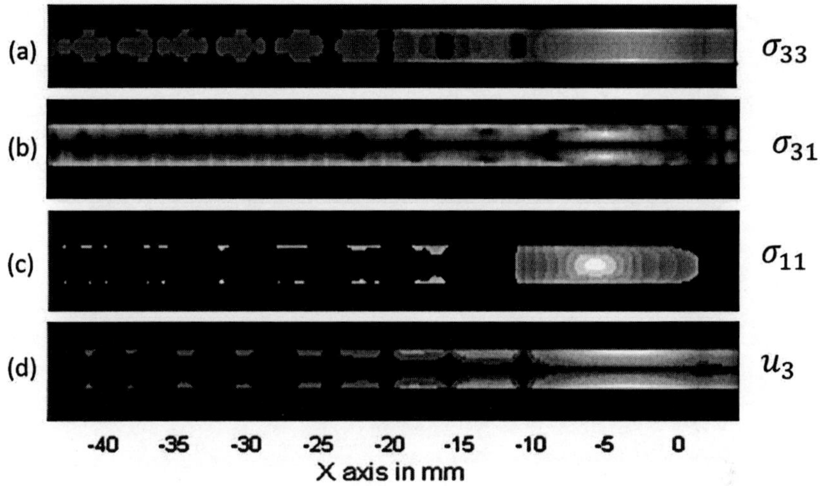

FIGURE 7.18 Ultrasonic fields inside a 1 mm thick aluminum plate for S_0 mode of excitation, generated by 1 MHz transducers inclined at 16.11° angle as shown in Fig. 4.10. Four plots show the variations of (a) the vertical normal stress s_{33}, (b) the shear stress s_{31}, (c) the horizontal normal stress s_{11} and (d) the vertical displacement u_3 component.

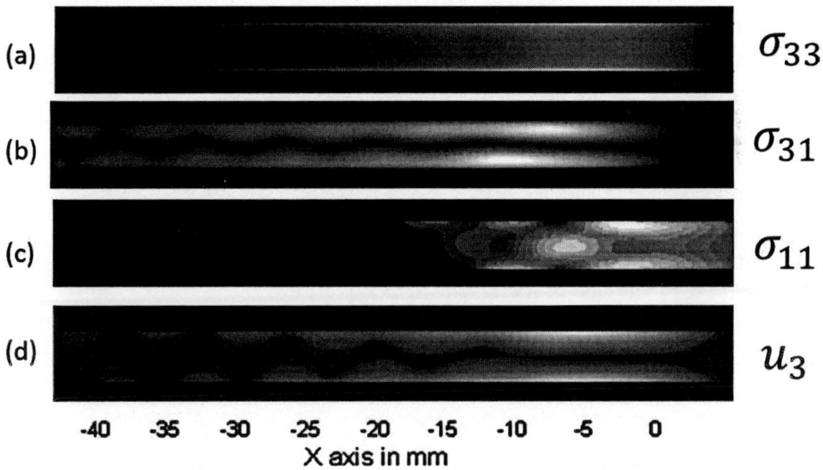

FIGURE 7.19 Ultrasonic fields inside a 2 mm thick aluminum plate for A_0 mode of excitation, generated by 0.5 MHz transducers inclined at 39.43° angle as shown in Fig. 4.10. Four plots show the variations of (a) the vertical normal stress s_{33}, (b) the shear stress s_{31}, (c) the horizontal normal stress s_{11} and (d) the vertical displacement u_3 component.

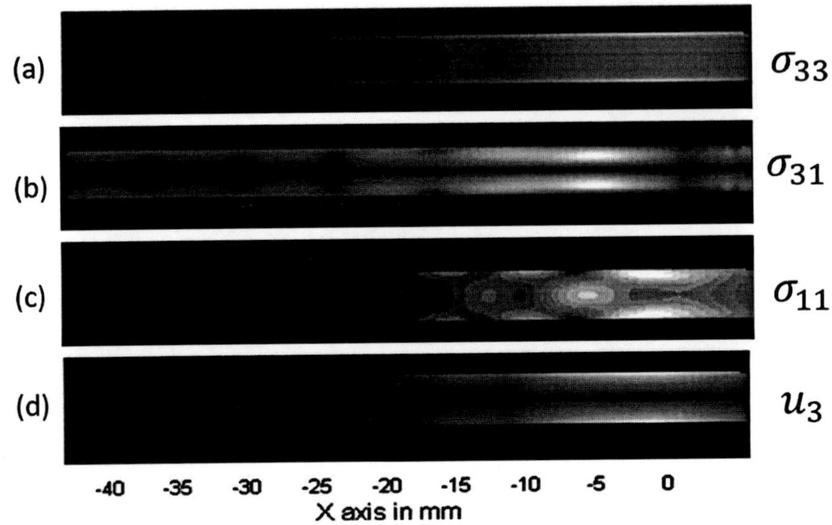

FIGURE 7.20 Ultrasonic fields inside a 2 mm thick aluminum plate for S_0 mode of excitation, generated by 0.5 MHz transducers inclined at 16.11° angle as shown in Fig. 4.10. Four plots show the variations of (a) the vertical normal stress s_{33}, (b) the shear stress s_{31}, (c) the horizontal normal stress s_{11} and (d) the vertical displacement u_3 component.

A 1 mm thick aluminum plate is considered, and two 1 MHz transducers are used for generating guided waves in a plate. Fig. 7.16 shows that the symmetric mode in a 1 mm thick aluminum plate propagates with a velocity of 4.733 km/sec at 1 MHz frequency. Therefore, to generate this symmetric mode in the plate, transducers must be inclined at a critical angle of $sin^{-1}\left(\frac{c_f}{c_l}\right) = sin^{-1}\left(\frac{1.48}{5.333}\right) = 16.11°$. The phase velocity of the antisymmetric mode is 2.333 km/sec. Therefore, to generate the antisymmetric mode the transducers must be inclined at a critical angle of $sin^{-1}\left(\frac{c_f}{c_a}\right) = sin^{-1}\left(\frac{1.48}{2.333}\right) = 39.4343°$. Figs. 7.17 and 7.18 show the ultrasonic fields in the 1 mm thick aluminum plate for 39.43° and 16.11° angles of incidence, respectively.

The antisymmetric mode generated in the plate is shown in Fig. 7.17 and the symmetric mode is shown in Fig. 7.18. With the same critical angles of inclination and 0.5 MHz transducers antisymmetric and symmetric modes in a 2 mm thick aluminum plate are then generated. The ultrasonic fields for these two modes are presented in Figs. 7.19 and 7.20. Fig. 7.19 clearly shows the antisymmetric mode in the plate and Fig. 7.20 shows the symmetric mode. To view the detailed wave field with magnitude of stresses in GPa readers are recommended to refer the reference [13].

7.4.3 Computational Wave Field Modeling in a Solid Plate with Inclusion or Crack [16]

7.4.3.1 Problem Geometry

A plate containing an internal anomaly and immersed in a fluid medium is considered here as shown in Fig. 7.21. One can see from this figure that a cavity or an

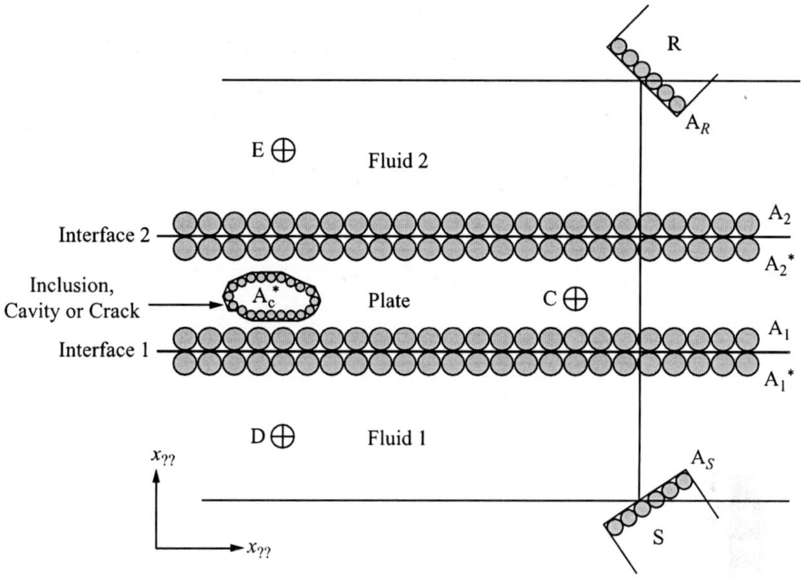

FIGURE 7.21 Geometry of the isotropic plate with an anomaly (crack, cavity or inclusion) immersed in water. Small circles show the distributed point sources used in the DPSM formulation.

inclusion is entrapped inside the solid plate. A cavity may take stress free boundary conditions but an inclusion at their boundary must take continuity conditions enforced on displacements and normal stress components. Along the fluid-solid interfaces two layers of point sources are distributed shown by the small circles in the diagram. Along the inclusion boundary another two sets of point sources should be distributed. However, if the internal anomaly is a crack or cavity, instead of an inclusion then only one set of point sources is necessary inside the anomaly boundary, as shown in Fig. 7.21.

Two circular piston transducers are immersed in the fluid, and symmetrically placed on two sides of the plate specimen. Contributions of the source strengths are very similar to that of the3 problem discussed in Section 7.4.2 hence, detailed discussion on their contributions are omitted. Only the contributions of $\mathbf{A}_c$ and $\mathbf{A}_c^*$ that are the source strength vectors representing the sources around the anomaly boundary are discussed. Transducer faces have source strength vectors $\mathbf{A}_S$ and $\mathbf{A}_R$. In Fig. 7.21, three points (C, D, and E) are considered for the illustration purpose. The ultrasonic field at point C inside the solid plate is the summation of the contributions of the point source $\mathbf{A}_1^*$, $\mathbf{A}_2$, and $\mathbf{A}_c^*$. Ultrasonic field at point D inside Fluid 1 is the summation of contributions of the point sources $\mathbf{A}_1$ and $\mathbf{A}_S$. Similarly, the Ultrasonic field at point E inside Fluid 2 is the summation of contributions of the point sources $\mathbf{A}_2^*$ and $\mathbf{A}_R$. It is possible to ignore the effect of the point sources that are located in the shadow region by introducing a fictitious interface in the plate going through the anomaly and thus transforming the homogeneous plate into a plate made of two layers of identical material properties. Such discussions are omitted here and can be found in reference [5] under the discussion of controlled source radiation. However, the following results are obtained without introducing such fictitious interfaces.

7.4.3.2 Matrix Formulation: Boundary and Continuity Conditions

If the normal velocities of the two transducer-faces are known, then the particle velocities on the surface of the transducer designated as *S*, can be computed and equated to the known value

$$\mathbf{M}_{SS}\mathbf{A}_S + \mathbf{M}_{SI}\mathbf{A}_I = \mathbf{V}_{S0} \tag{7.169}$$

and similarly, the computed particle velocities on the surface of the transducer designated as *R* can be equated to the known velocity value for that transducer

$$\mathbf{M}_{R2^*}\mathbf{A}_2^* + \mathbf{M}_{RR}\mathbf{A}_R = \mathbf{V}_{R0} \tag{7.170}$$

Other continuity conditions are achieved at the fluid-solid interfaces, from the continuity of the normal stresses

$$\mathbf{Q}_{1S}\mathbf{A}_S + \mathbf{Q}_{11}\mathbf{A}_1 = -\mathbf{S33}_{11^*}\mathbf{A}_1^* - \mathbf{S33}_{12}\mathbf{A}_2 - \mathbf{S33}_{1c^*}\mathbf{A}_c^* \tag{7.171}$$

$$\mathbf{Q}_{22^*}\mathbf{A}_2^* + \mathbf{Q}_{2R}\mathbf{A}_R = -\mathbf{S33}_{21^*}\mathbf{A}_1^* - \mathbf{S33}_{22}\mathbf{A}_2 - \mathbf{S33}_{2c^*}\mathbf{A}_c^* \tag{7.172}$$

and the continuity of the normal displacement gives

$$\mathbf{DF3}_{1S}\mathbf{A}_S + \mathbf{DF3}_{11}\mathbf{A}_1 = \mathbf{DS3}_{11^*}\mathbf{A}_1^* + \mathbf{DS3}_{12}\mathbf{A}_2 + \mathbf{DS3}_{1c^*}\mathbf{A}_c^* \tag{7.173}$$

$$\mathbf{DF3}_{22^*}\mathbf{A}_2^* + \mathbf{DF3}_{2R}\mathbf{A}_R = \mathbf{DS3}_{21^*}\mathbf{A}_1^* + \mathbf{DS3}_{22}\mathbf{A}_2 + \mathbf{DS3}_{2c^*}\mathbf{A}_c^* \tag{7.174}$$

From the vanishing shear stress condition at the fluid-solid interface, one can write

$$\mathbf{S31}_{11^*}\mathbf{A}_1^* + \mathbf{S31}_{12}\mathbf{A}_2 + \mathbf{S31}_{1c^*}\mathbf{A}_c^* = 0 \tag{7.175}$$

$$\mathbf{S32}_{11^*}\mathbf{A}_1^* + \mathbf{S32}_{12}\mathbf{A}_2 + \mathbf{S32}_{1c^*}\mathbf{A}_c^* = 0 \tag{7.176}$$

The vanishing normal and shear stresses at a horizontal crack or cavity boundary gives

$$\mathbf{S33}_{CI1^*}\mathbf{A}_1^* + \mathbf{S33}_{CI2}\mathbf{A}_2 + \mathbf{S33}_{CIc^*}\mathbf{A}_c^* = 0 \tag{7.177}$$

$$\mathbf{S31}_{CI1^*}\mathbf{A}_1^* + \mathbf{S31}_{CI2}\mathbf{A}_2 + \mathbf{S31}_{CIc^*}\mathbf{A}_c^* = 0 \tag{7.178}$$

$$\mathbf{S32}_{CI1^*}\mathbf{A}_1^* + \mathbf{S32}_{CI2}\mathbf{A}_2 + \mathbf{S32}_{CIc^*}\mathbf{A}_c^* = 0 \tag{7.179}$$

where *CI* indicates the target points on the crack surface.

For any general shape of inclusions and cavities the σ_{33}, σ_{31}, and σ_{32} stresses at every point on the inclusion or cavity boundary needs to be transformed to the corresponding normal stress and parallel shear stresses at that point of the boundary. Eq. (7.169) through (7.179) will still be good for such situations but the matrices used in the equation will have different values due to the transformation operation. The boundary and continuity equations can be written in the matrix form similar to Eq. (7.163) except the vector elements corresponding to the $\boldsymbol{A}_c^*$ source strength vector appear in the equation.

$$\begin{bmatrix} \mathbf{M}_{SS} & \mathbf{M}_{S1} & \mathbf{0} & \mathbf{0} & \mathbf{0} & \mathbf{0} & \mathbf{0} \\ \mathbf{Q}_{1S} & \mathbf{Q}_{11} & \mathbf{S33}_{11^*} & \mathbf{S33}_{12} & \mathbf{0} & \mathbf{0} & \mathbf{S33}_{1c^*} \\ \mathbf{DF3}_{1S} & \mathbf{DF3}_{11} & -\mathbf{DS3}_{11^*} & -\mathbf{DS3}_{12} & \mathbf{0} & \mathbf{0} & -\mathbf{DS3}_{1c^*} \\ \mathbf{0} & \mathbf{0} & \mathbf{S31}_{11^*} & \mathbf{S31}_{12} & \mathbf{0} & \mathbf{0} & \mathbf{S31}_{1c^*} \\ \mathbf{0} & \mathbf{0} & \mathbf{S32}_{11^*} & \mathbf{S32}_{12} & \mathbf{0} & \mathbf{0} & \mathbf{S32}_{1c^*} \\ \mathbf{0} & \mathbf{0} & \mathbf{S32}_{21^*} & \mathbf{S32}_{22} & \mathbf{0} & \mathbf{0} & \mathbf{S32}_{2c^*} \\ \mathbf{0} & \mathbf{0} & \mathbf{S31}_{21^*} & \mathbf{S31}_{22} & \mathbf{0} & \mathbf{0} & \mathbf{S31}_{2c^*} \\ \mathbf{0} & \mathbf{0} & \mathbf{S33}_{21^*} & \mathbf{S33}_{22} & \mathbf{Q}_{22^*} & \mathbf{Q}_{2R} & \mathbf{S33}_{2c^*} \\ \mathbf{0} & \mathbf{0} & -\mathbf{DS3}_{21^*} & -\mathbf{DS3}_{22} & \mathbf{DF3}_{22^*} & \mathbf{DF3}_{2R} & -\mathbf{DS3}_{2c^*} \\ \mathbf{0} & \mathbf{0} & \mathbf{0} & \mathbf{0} & \mathbf{M}_{R2^*} & \mathbf{M}_{RR} & \mathbf{0} \\ \mathbf{0} & \mathbf{0} & \mathbf{S33}_{CI1^*} & \mathbf{S33}_{CI2} & \mathbf{0} & \mathbf{0} & \mathbf{S33}_{CIc^*} \\ \mathbf{0} & \mathbf{0} & \mathbf{S31}_{CI1^*} & \mathbf{S31}_{CI2} & \mathbf{0} & \mathbf{0} & \mathbf{S31}_{CIc^*} \\ \mathbf{0} & \mathbf{0} & \mathbf{S32}_{CI1^*} & \mathbf{S32}_{CI2} & \mathbf{0} & \mathbf{0} & \mathbf{S32}_{CIc^*} \end{bmatrix}_{(2N+8M+C) x (2N+8M+C)} \begin{Bmatrix} \mathbf{A}_S \\ \mathbf{A}_1 \\ \mathbf{A}_1^* \\ \mathbf{A}_1 \\ \mathbf{A}_2^* \\ \mathbf{A}_R \\ \mathbf{A}_c^* \end{Bmatrix}_{(2N+8M+C)} = \begin{Bmatrix} \mathbf{V}_{S0} \\ \mathbf{0} \\ \mathbf{0} \\ \mathbf{0} \\ \mathbf{0} \\ \mathbf{0} \\ \mathbf{0} \\ \mathbf{0} \\ \mathbf{0} \\ \mathbf{V}_{R0} \\ \mathbf{0} \end{Bmatrix}_{(2N+8M+C)} \tag{7.180}$$

7.4.3.3 Solution

The vector of source strengths $\{\mathbf{\Lambda}\}$ of the total system can be obtained by solving the system of linear equations

$$\{\mathbf{\Lambda}\} = [\mathbf{MAT}]^{-1}\{\mathbf{V}\} \tag{7.181}$$

After calculating the source strengths, the pressure, velocity, stress and displacement values at any point can be obtained from the equations like it is previously discussed in Sections 7.4.1 and 7.4.2.

7.4.3.4 Numerical Results: Ultrasonic Fields in Solid Plate with Horizontal Crack

The numerical results are generated for an isotropic aluminum plate containing a thin cavity or crack when the plate is placed under water. The material properties of this aluminum plate are shown in Table 7.1. Fig. 7.22a, b, c, and d shows σ_{11}, σ_{13}, σ_{33}, and u_3 distributions, respectively, in the defective aluminum plate for the transducers inclined at the Rayleigh critical angle, 30.50°. It is verified that the boundary conditions at the cavity boundary are perfectly satisfied. More details about this problem can be found in {Banerjee, 2007 #10}.

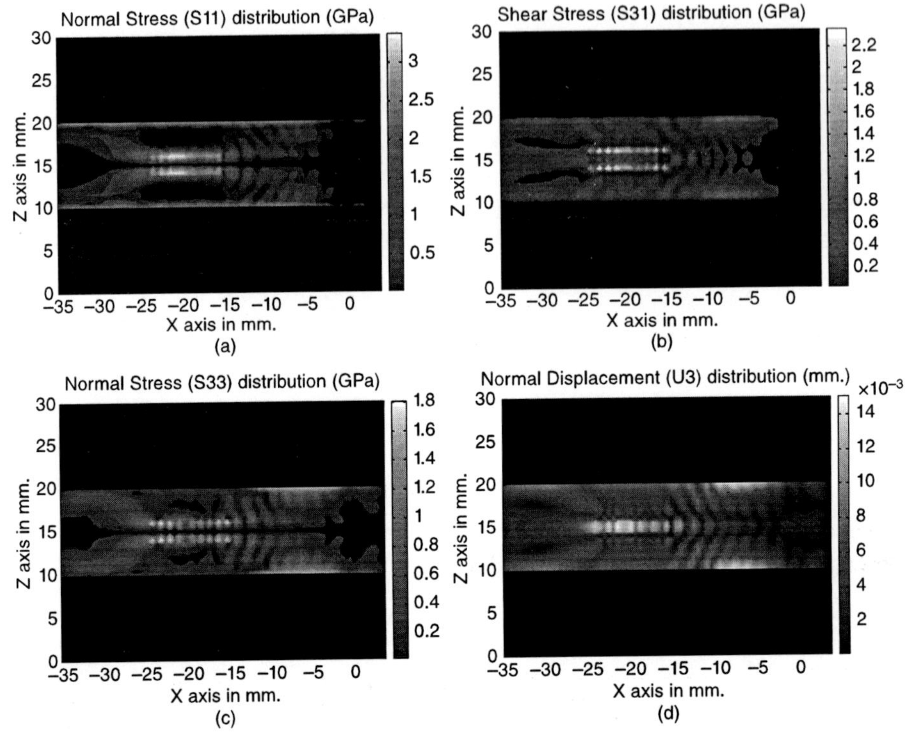

FIGURE 7.22 Ultrasonic fields developed in a 10 mm thick aluminum plate with a horizontal cavity/crack for the Rayleigh angle of incidence: (a) s_{11} distribution, (b) s_{13} distribution, (c) s_{33} distribution, (d) u_3 distribution.

7.5 MODELING ULTRASONIC FIELD IN ANISOTROPIC SOLIDS (E.G., COMPOSITES)

Studies of the elastic waves propagation in anisotropic media have been of great interest to researchers in the field of NDE. This interest is further intensified by the recent expansion in the use of anisotropic materials in the form of composite materials, super alloys, and crystal structures in the wide variety of applications in aerospace, mechanical, chemical, and biomedical industries. NDE of composites and anisotropic materials are frequently occurring which requires additional understanding of wave propagation in anisotropic materials, which are discussed in Chapter 1, 4, and 5. Here in this section CNDE of anisotropic material using DPSM will be discussed. As it is introduced that DPSM needs elastodynamic Green's function for its formulation, anisotropic displacement and stress Green's functions are essential to further our discussion. Development of the mathematical models for the waves in anisotropic media has different approach than modeling wave propagation in isotropic media. Their differences were first identified with the widely used bulk wave techniques. Later, Green's function solution, used in the formulation of integral representations and boundary integral equations for scattering problems and understanding of the elastodynamic response of an unbounded homogeneous linearly elastic solid, was used in modeling the wave in anisotropic media [17]. Various techniques have been used in the derivation of Green's function. It has been derived using Cagniard-de Hoop method by Kraut [18] for transversely isotropic half-space and by Burridge [19] for an anisotropic half-space. Simplified method to Cagniard-de Hoop has been used by Willis [20, 21] and Payton [22] for several anisotropic half-space problems. The Green's function for unbounded, anisotropic elastic medium has been developed by Yeatts [23]. Moreover, the method has been further developed by Wang and Achenbach [24, 25] to develop 2-D and 3-D Green's functions for anisotropic solids. Summarizing different techniques following subsection presents the detailed mathematical treatment to calculate displacement and stress elastodynamic Green's function is anisotropic solids.

7.5.1 Elastodynamic Displacement and Stress Green's Function in General Anisotropic Solids

Based on discussions in Chapter 4 and Chapter 5 on wave propagation in anisotropic media, clear understanding of the phase information associated with the wave modes in bulk anisotropic media are assumed. However, before we advance to perform CNDE of anisotropic medium using DPSM, with the very requirement of DPSM, the elastodynamic Green's function in the anisotropic media [23, 26–30] are calculated. In the following paragraphs, it is shown that the fundamental solutions of Christoffel's equation presented in Chapter 4 are very crucial in the calculation of the Green's function. Now to develop mathematical formulation of Green's function, let us revisit Eq. (4.93). In contrast to the transformation of Eq. (4.93), into homogeneous Christoffel's equation (Eq. (4.96)), an impulse force in introduced as a body force replacing F_i with the help of Dirac Delta function in the media due to a point source, which by definition contributes to the development of the solution of the elastodynamic Green's function as discussed earlier in this chapter. Based on this development, the Cauchy-Navier Equation is developed as follows:

$$\left[\mathbb{C}_{ijml}\frac{\partial^2}{\partial x_j\,\partial x_l}-\rho\delta_{im}\frac{\partial^2}{\partial t^2}\right]G_{mp}\left(x_n,t\right)=-\delta_{ip}\delta(x_n)\delta(t)F_p \tag{7.182}$$

where $G_{mp}\left(x_n,t\right)$ is the time domain displacement Green's function along *m-th* direction the at x_n point, due to a point source actuation along the *p-th* direction with force amplitude F_p.

Next, transforming the spatio-temporal elastodynamic equation using Fourier transformation ($\mathcal{F}$) the displacement Green's function in Eq. (7.182) is transformed into an elastodynamic equation to find the displacement Green's function at a point x_n in frequency-wave number domain. The frequency-wavenumber domain equation can be written as

$$\left[\mathbb{C}_{ijml}k_jk_l-\rho\omega^2\delta_{im}\right]\tilde{G}_{mp}\left(k_n,\omega\right)=-\delta_{ip}\frac{1}{\left(2\pi\right)^3}f_p\left(\omega\right) \tag{7.183}$$

It is important to note that $\mathcal{F}\left[\delta(x_n)\right]=1/(2\pi)^3$ is used in the above Fourier transformation. Also, if it is considered that an operator exist in $\mathcal{L}^2$ Hilbert space, the above equation can be written as

$$\mathbf{L}_{im}\tilde{G}_{mp}(k_n,\omega)=\delta_{ip}\frac{1}{(2\pi)^3}f_p(\omega) \tag{7.184}$$

where

$$\tilde{G}_{mp}(k_n,\omega)=\frac{1}{(2\pi)^4}\int\int\int_{-\infty}^{\infty}\int_{-\infty}^{\infty}G_{mp}(x_n,t)e^{i(k_l x_l-\omega t)}\,dx^3\,dt \tag{7.185}$$

$$\mathbf{L}_{im}(k_n,\omega)=\left[\mathbb{C}_{ijml}k_jk_l-\rho\omega^2\delta_{ij}\right] \tag{7.186}$$

Please note that the $\mathbf{L}_{im}(k_n,\omega)$ will be called the Christoffel operator here, which was earlier found in Eq. (4.96) in slightly different form. Next it is possible to further transform the Green's function in frequency-wavenumber domain (Eq. (7.185)) into frequency-space domain by the applying inverse Fourier transform with respect to the wavenumber k_n. Thus, the frequency-space domain displacement Green's function with Christoffel operator can be written as

$$g_{mp}(x_n,\omega)=\frac{1}{(2\pi)^3}\int\int\int_{-\infty}^{\infty}\left[\mathbf{L}_{mp}(k_n,\omega)^{-1}\right]e^{-ik_lx_l}\,dk^3 \tag{7.187}$$

It can be seen that, the above equation is very thought-provoking, since it indicates that the displacement Green's function at any point x_n in space will be the integral of all possible wave numbers that can propagate in an anisotropic medias. It means that the total displacement Green's function at any point x_n is the superposition of the influences of all the waves vectors propagating in all possible directions. Hence, it is essential to compute the wave propagation in all possible directions due to a point source which is already performed in Chapter 4 (Section 4.4.2) to understand the ultrasonic wave propagation in three-dimensional bulk anisotropic media. In a nutshell, solutions or the extraction of the eigen values from the Christoffel operator $\mathbf{L}_{im}(k_n,\omega)$ in Eq. (7.187) has already been perfromed in Section 4.4.2.

7.5.2 Exact Mathematical Expression for the Green's Function

In Section 7.5.1, using Fourier transformation, it was possible to gain a physical and mathematical meaning of the displacement Green's function in Eq. (7.187). However, exploring Eq. (7.187), it is realized that the equation is still not in the form, which can be effortlessly implemented into a computer code for CNDE applications. Hence, the next step would be to solve the elastodynamic equations for Green's function equation using alternate approach and develop a more convenient form such that one computes the Green's function in frequency domain. To achieve that, in this section, few techniques and theorems are utilized to find the mathematical expression of the Green's function which can be used in the numerical computation and modeling of the wave field. To develop this new more tractable form of Green's function two different approaches: Radon transform approach implementing residue evaluation and Fourier transform approach implementing Cauchy integral identity are used. Both these approaches are described in Sections 7.5.2.1 and 7.5.2.2, respectively. Meaning of the essential terms used in the following calculation of the Green's function are explained along with their respective derivation.

7.5.2.1 Radon Transform Approach: Solution of Elastodynamic Green's Function

A monochromatic harmonic impulse force or a Dirac delta function is introduced as a body force simulating a point source in Eq. 7.182 which is further modified to

$$\left[\mathbb{C}_{ijml}\frac{\partial^2}{\partial x_j\,\partial x_l}-\rho\delta_{im}\frac{\partial^2}{\partial t^2}\right]G_{mp}\left(x_n,t\right)=-\delta_{ip}\delta(x_n)e^{-i\omega t} \tag{7.188}$$

where $G_{mp}\left(x_n,t\right)$ is time domain displacement Green's function representing displacement along *m-th* direction at a point ($\mathbf{y}$) away from a point source located at $\mathbf{y_0}$, due to an unit force amplitude acting along the *p-th* direction and $\mathbf{x}=\left|\mathbf{y}-\mathbf{y_0}\right|$. After transforming Eq. 7.188 into frequency domain, the elastodynamic equation will be

$$\left[\mathbb{C}_{ijml}\frac{\partial^2}{\partial x_j\,\partial x_l}+\rho\omega^2\delta_{im}\right]g_{mp}\left(x_n,\omega\right)=-\delta_{ip}\delta(x_n) \tag{7.189}$$

The above equation is a second order partial differential equation. Therefore, it is required to find a method that can transform this complex partial differential equation into a set of ordinary differential equations. This is required because ordinary differential equations are more easier to track and solve. One such method is Radon's transform [31] method. Radon's transform is an elegant technique that transforms any n-dimensional function into an n-dimensional sinogram, where the sinogram is defined on the spaces of infinite n-1-dimensional planes in the n-dimensional space, whole value at a particular n-1-dimensional plane is equal to the n-1-dimensional surface integral of the n-dimensional function over that plane. To clarify the above statement in a simpler form a two-dimensional example would be suitable. For example, please imagine a 2D plane on which a two-dimensional function $f\left(x_1,x_2\right)$ is defined. On this 2D plane we could imagine infinite numbers of straight lines (which is n-1-z dimensional) at an angle θ with the x_1-axis, on which, projection of the function $f\left(x_1,x_2\right)$ could be taken along another infinite number of lines that are perpendicular to the earlier lines but are equally spaced and their perpendicular distance from the origin along the angle θ is h. This will create a space of straight lines on this 2D plane as explained in Fig. 7.23. Let us assume just one straight line on this plane with angle θ_r marked as *sl* in Fig. 7.23. Another set of discrete lines orthogonal to the line *sl* are imagined marked with h_j. Where h_j is orthogonal distance of the *j-th* straight line from the origin. Imagine a bunch of straight rays are incident on the object $f\left(x_1,x_2\right)$ from a source and a photo

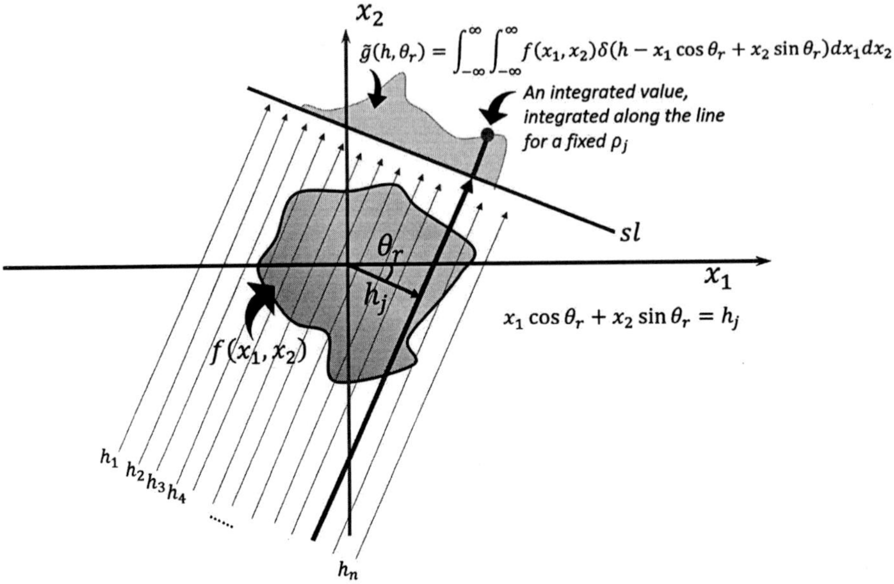

FIGURE 7.23 A Schematic showing the process of Radon transform in 2D.

frame is held on the other side of the object, then the image of the object formed on the frame is the summation of the intensities along each ray path. To explain this further an arbitrary bold line is shown in Fig. 7.23, along which the function is integrated to get a value shown in the figure with a red dot. If such integration is performed along each line h_j where, j is an index takes values 1 to infinity, then the intensity plot along sl for angle θ_r, $\tilde{g}(h,\theta_r)$ could be achieved. The $\tilde{g}(h,\theta_r)$ function can be written as

$$\tilde{g}(h,\theta_r) = \int_{-\infty}^{\infty}\int_{-\infty}^{\infty} f(x_1,x_2)\delta\left(h-(x_1\cos\theta_r + x_2\sin\theta_r)\right)dx_1dx_2 \tag{7.190}$$

where the Dirac delta function is used to initiate the integration along a line if the line is at a distance h, else the integration will be equal to zero. Using this process, $\tilde{g}(h,\theta)$ function can be obtained for all possible θ_rs on the plane. Hence, a two-dimensional function called sinogram $\tilde{g}(h,\theta)$ is achieved after transforming a two-dimensional function $f(x_1,x_2)$ into ρ and θ space. Similar process can also be performed in a three-dimensional space. In case of 3D, the 1D lines in 2D will become 2D plane. Hence, in 3D, a sinogram can be more conveniently expressed in terms of h and $\hat{\mu}$, where $\hat{\mu}$ is the unit normal direction, consists of direction cosines to specify the orientation of the planes. Hence, in 3D assuming x_1, x_2 and x_3 being the coordinates of a point ($\mathbf{x}$) in an Euclidean space $\Re^3$, Radon transform [24, 25, 28, 32] of any function $f(\mathbf{x})$, which is defined and absolutely integrable over all space, can be written as

$$\tilde{g}\left(h,\hat{\mu}\right) = R\{f(\mathbf{x})\} = \int f(\mathbf{x})\delta(h-\hat{\mu}.\mathbf{x})dv \tag{7.191}$$

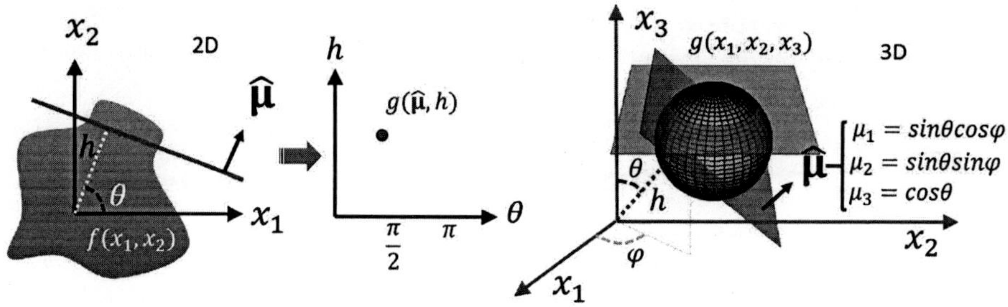

$$\delta(h - \hat{\mu}.\mathbf{x}) = \delta(h - x_1 sin\theta cos\varphi + x_2 sin\theta sin\varphi + x_3 cos\theta)$$

FIGURE 7.24 A schematic diagram to visualize the Radon transform in 2D and 3D. The figure shows that the integral is performed over a circular domain with different radius in 2D and over a sphere with different radius in 3D.

where dv is the volume element ($dv = dx_1 dx_2 dx_3$) and Eq. (7.191) is a surface integral over the plane $h = \mu_k x_k$. The physical meaning of the integral could be briefly described in 2D and 3D using Fig. 7.24. In case of 2D, as discussed above, infinite numbers of lines can be defined as the function of distance h from the origin with their local orientation of the lines designated by their unit normal $\hat{\mu}$ such that line integral along that line can be represented by a point in the Radon space as shown in Fig. 7.24 and also in Fig. 7.23. In 3D, let us assume a spherical function $f(x_1, x_2, x_3)$ or object that is to be transformed in Radon domain. As shown in Fig. 7.24 two arbitrary planes are shown that are designated by $\hat{\mu}$ at a distance h. In case of 3D, the infinite numbers of planes can be defined as the function of distance h from the origin with their local orientation of the planes designated by their unit normal $\hat{\mu}$ such that 3D image can be represented by a 3D stack of 2D plane images in the Radon space. The Dirac δ function in Eq. (7.191) signifies that the values exist only on the plane $\hat{\mu}, h$ and everywhere else the integral ceases to exist, i.e., equal to zero.

Radon transform is frequently used in the field of computed tomography (CT) and 3D image reconstruction from a cluster of 2D projection images taken on different planes. The transformation of 3D function into multiple functions projected on 2D planes is made possible by Radon transform by parameterizing the orientation of the planes ($\hat{\mu}$). Contrary to that, the transformation of images on the 2D planes to a 3D image is made possible by inverse Radon transform. For example, in X-Ray scan and MRI scans [33], source and detectors are kept opposite side of the patient's body part and by rotating the source-target setup, i.e., changing the θ, sinogram is achieved. Then through inverse Radon transform, the patient's body part is visualized. However, in this case for anisotropic elastodynamic Green's function, first a forward Radon transform is applied. Radon transform gives a set of coupled ordinary differential equations. Here, the 2D planes in 3D are parameterized by $\hat{\mu}$ as shown in Fig. 7.24. Eq. (7.189) was originally solved by Wang and Achenbach using Radon Transform [24, 25, 28, 32].

Radon transform has many elegant mathematical features and here we mention a few of them that was adopted in the process of transforming the partial differential equation in Eq. (7.189) into the coupled ordinary differential equations. The inverse Radon transform of the above equation [28] is a two-step process and can be written as

$$\bar{g}\left(h,\hat{\mu}\right)=\frac{\partial^2}{\partial\rho^2}\tilde{g}(h,\hat{\mu}) \tag{7.192}$$

$$g\left(x_n\right)=\boldsymbol{R}^*\left(\bar{g}\right)=-\frac{1}{8\pi^2}\int_{|\mu|=1}^{sphere}\bar{g}\left(\hat{\mu}.\,\boldsymbol{x},\,\hat{\mu}\right)d\Omega\left(\mu\right)\quad at\quad \mu.\,\boldsymbol{x}=h \tag{7.193}$$

Eq. (7.193) is an integral over a unit sphere with parameter θ and φ as described in Section 3.10.3 in Chapter 3 and in Fig. A.3.2. Discretization along θ and φ contributes to the convergence of the results and it is discussed briefly in the result section. High discretization (small values of θ and φ) contribute to heavy computation and low discretization (higher values of θ and φ) compromised the convergence. Hence, there is an optimized value of θ and φ that should be used for computation, which is not discussed earlier [34]. Another identity of Radon's transform [28] written in Eq. (7.192) was operated on Eq. (7.189) which is an equation for solving frequency domain Green's function ($g_{mp}(x_n,\omega)$), to get coupled ordinary differential equations in Radon domain $\tilde{g}_{mp}\left(h,\hat{\mu}\right)$.

$$R\left\{\left(\frac{\partial^2 g}{\partial x_i\,\partial x_j}\right)\right\}=\mu_i\mu_j\frac{\partial\tilde{g}\left(h,\mu\right)}{\partial h^2} \tag{7.194}$$

$$\Gamma_{im}(\mu_n)\frac{d^2\tilde{g}_{mp}}{dh^2}+\omega^2\delta_{im}\tilde{g}_{mp}=-\delta_{ip}\frac{1}{\rho}\delta(h) \tag{7.195}$$

where $\Gamma_{im}(\mu_n)=\frac{C_{ijml}\mu_j\mu_l}{\rho}$.

The coupled ordinary differential equations are then further decoupled using the eigen solution obtained from the Christoffel's equation Eq. (4.96) presented in Section 4.4 in Chapter 4 through Eq. (4.97). Please note that $\mu_j\mu_l$ in Eq. (7.195) are synonymous to n_jn_l in Eq. (4.96), where they are the product of the direction cosines of a wave vector. There are three eigen values and three respective eigen vectors for each wave vector direction. Where z is the index to take any eigen values in the system. γ_z is the z-th eigen value and $\left(P_{im}\right)^{(z)}$ is the projection matrix of the z-th eigen mode. The correspondence between the eigen modes with the wave velocities are c_{qL}, c_{qFS} and c_{qSS} and their respective eigen vectors $\mathbf{g_L}$, $\mathbf{g_{FS}}$, $\mathbf{g_{SS}}$ (described in Section 4.4 in Chapter 4) can be explicitly expressed in terms of γ_z and $\left(P_{im}\right)^{(z)}$ as follows:

$$\gamma_1=\left(c_{qL}\right)^2\ ;\ \gamma_2=\left(c_{qFS}\right)^2\ ;\ \gamma_3=\left(c_{qSS}\right)^2 \tag{7.196}$$

$$\left(P_{im}\right)^{(1)}=\varphi_{Li}\varphi_{Lm}\ ;\ \left(P_{im}\right)^{(2)}=\varphi_{FSi}\varphi_{FSm}\ ;\ \left(P_{im}\right)^{(3)}=\varphi_{SSi}\varphi_{SSm} \tag{7.197}$$

Please note that the notations used in Eqs. (7.196) and (7.197) are comparable to the notations used in Section 4.4 in Chapter 4 and in Eq. (4.97). Only change in notation was the components of the eigen vectors. In Eq. (4.97), the components of the eigen vectors are described as follows:

$$\mathbf{g_L} = g_{L1}\hat{\mathbf{e}}_1 + g_{L2}\hat{\mathbf{e}}_2 + g_{L3}\hat{\mathbf{e}}_3 \tag{4.97.1}$$

$$\mathbf{g_{FS}} = g_{FS1}\hat{\mathbf{e}}_1 + g_{FS2}\hat{\mathbf{e}}_2 + g_{FS3}\hat{\mathbf{e}}_3 \tag{4.97.2}$$

$$\mathbf{g_{SS}} = g_{SS1}\hat{\mathbf{e}}_1 + g_{SS2}\hat{\mathbf{e}}_2 + g_{SS3}\hat{\mathbf{e}}_3 \tag{4.97.3}$$

However, here in this chapter not to confuse with the Green's function notation in Frequency domain we described the eigen vectors as follows:

$$\mathbf{g_L} = \varphi_{L1}\hat{\mathbf{e}}_1 + \varphi_{L2}\hat{\mathbf{e}}_2 + \varphi_{L3}\hat{\mathbf{e}}_3 \tag{7.198.1}$$

$$\mathbf{g_{FS}} = \varphi_{FS1}\hat{\mathbf{e}}_1 + \varphi_{FS2}\hat{\mathbf{e}}_2 + \varphi_{FS3}\hat{\mathbf{e}}_3 \tag{7.198.2}$$

$$\mathbf{g_{SS}} = \varphi_{SS1}\hat{\mathbf{e}}_1 + \varphi_{SS2}\hat{\mathbf{e}}_2 + \varphi_{SS3}\hat{\mathbf{e}}_3 \tag{7.198.3}$$

The eigen values and projection matrix in Eqs. (7.196) and (7.197) are used to decouple Eq. (7.195). To understand the γ_z and $(P_{im})^{(z)}$ better, the equation for eigen solution is rewritten with notations in Eq. (7.195) which is like Eq. (4.96) already described in Chapter 4.

$$\Gamma_{im}\left(\hat{\mu}\right)\varphi_{mn} = \rho\gamma_n\varphi_{in}\ (n = 1,2,3) \tag{7.199}$$

where $\varphi_{jm}\varphi_{jn} = \varphi_{mj}\varphi_{nj} = \delta_{mn}$

Next, by defining a new variable $\hat{g}^*_{np}$ in transformed eigen coordinate system one could write Eq. (7.195) by pre and post multiplying the equation with eigen vectors as follows:

$$\left\{\varphi_{im}\Gamma_{im}\left(\hat{\mu}\right)\varphi_{mn}\frac{\partial^2}{\partial h^2} + \rho\omega^2\varphi_{im}\delta_{im}\varphi_{jn}\right\}\hat{g}^*_{np} = -\varphi_{im}\delta_{ip}\delta(h) \tag{7.200}$$

Further using the following identities, Eq. (7.200) becomes decoupled equations as written in Eq. (7.202)

$$\varphi_{im}\Gamma_{ij}\left(\hat{\mu}\right)\varphi_{jn} = \varphi_{im}\gamma_n\varphi_{in} = \gamma_n\varphi_{im}\varphi_{in} = \gamma_n\delta_{mn}\ ;\ \varphi_{im}\delta_{ij}\varphi_{jn} = \varphi_{im}\varphi_{in} = \delta_{mn} \tag{7.201}$$

$$\left\{\gamma_n\frac{\partial^2}{\partial h^2} + \omega^2\right\}\hat{g}^*_{np} = -\varphi_{pm}\frac{\delta(h)}{\rho} \tag{7.202}$$

Solution of Eq. (7.202) can be written as

$$\hat{g}^*_{np} = \frac{i\varphi_{pm}}{2\rho\gamma_n k_n} e^{ik_n|h|} \tag{7.203}$$

where $c_n = \sqrt{\gamma_n}$; $k_n = \frac{\omega}{c_n} = \frac{\omega}{\sqrt{\gamma_n}}$

And the solution of Eq. (7.195) can be written using the above relations as

$$\tilde{g}_{mp} = \sum_{z=1}^{3} \frac{i\left(P_{mp}\right)^{(z)}}{2\rho\ \gamma_z k_z} e^{ik_z|h|} \tag{7.204}$$

Further, applying the two-step process of inverse Radon transform described in Eqs. (7.192) and (7.193), first we get

$$\bar{g}_{mp}(h,\mu) = \frac{\partial^2}{\partial h^2}\tilde{g}_{mp} = \sum_{z=1}^{3} \frac{-\left(P_{mp}\right)^{(z)}}{2\rho\gamma_z}\left[2\delta(h) + i\frac{\omega}{\sqrt{\gamma_n}} e^{ik_z|h|}\right] \tag{7.205}$$

Or

$$\bar{g}_{mp}(h,\mu) = \bar{g}^D_{mp}(h,\mu) + \bar{g}^S_{mp}(h,\mu) \tag{7.206}$$

where

$$\bar{g}^D_{mp}(h,\mu) = \sum_{z=1}^{3} \frac{-\left(P_{mp}\right)^{(z)}}{2\rho\gamma_z}\, i\frac{\omega}{\sqrt{\gamma_n}} e^{ik_z|h|} \tag{7.207}$$

$$\bar{g}^S_{pk}(h,\mu) = \sum_{z=1}^{3} \frac{-\left(P_{mp}\right)^{(z)}}{2\rho\gamma_z}\delta\ (h) \tag{7.208}$$

Then using the second step of inverse Radon transform (Eq. (7.193)), expression of frequency domain Green's function is obtained as follows:

$$g_{mp}(x_n,\omega) = \left[g_{mp}(x_n,\omega)\right]^D + \left[g_{mp}(x_n,\omega)\right]^S \tag{7.209}$$

where

$$\left[g_{mp}(x_n,\omega)\right]^D = \frac{i\omega}{4(2\pi)^2\rho}\sum_{z=1}^{3}\int\int_{\theta=0;\ \phi=0}^{\theta=\pi;\ \phi=2\pi}\left[\left(1/\sqrt{\gamma_z}\right)^3\left(P_{mp}\right)^{(z)}\exp\left(i\left(k_i x_i\right)^{(z)}\right)ds(\mathbf{S})\right]$$

$$\left[g_{mp}(x_n,\omega)\right]^S = \frac{1}{8\pi^2}\int_{|n|=1}\Gamma^{-1}_{mp}(\mathbf{S})\delta(\mathbf{S}\ .\ \boldsymbol{x})ds(\mathbf{S}) = \frac{1}{8\pi^2|\mathbf{x}|}\int_{|d|=1}\Gamma^{-1}_{mp}(\boldsymbol{d})dv(\boldsymbol{d})$$

$\boldsymbol{d}$ is the normal vector to the direction of source-target combination along $\mathbf{x}$ [32].

$k_z|h| = k_z|\hat{\mu}.\mathbf{x}| = (k_i x_i)^{(z)}$ represents the dot product between the wave propagation direction $\mathbf{k}$ and the direction of source-target combination $\mathbf{x}$ for the z-th wave mode obtained from eigen solution of Eq. (4.96) or (7.199). This representation is conveniently present for programming purposes. Please note that the modal wave number $k_z = \omega / \sqrt{\gamma_z}$. It can be seen that the inverse Radon transform of Eq. (7.205) resulted two parts of the Green's function, a dynamic (D) regular part and a static singular (S) part [32].

To further simplify the static part of the solution, the residue evaluation method by Wang and Achenbach [35] is applied to get computationally more efficient equation as follows:

$$[g_{mp}]^s = -\frac{Im}{2\pi r \sum_{n=1}^{3} \left[\frac{A_{mp}(p+\alpha q)}{\partial_\alpha D(p+\alpha q)} \right]_{\alpha=\alpha_n}} \tag{7.210}$$

where Im = imaginary parts of the residue; $A_{mp} = adj[\Gamma_{mp}]$; $D = det[\Gamma_{mp}]$; $(p+\alpha q)$ is the plane perpendicular to the wave propagation direction in the sphere; are the three roots satisfying $D(p+\alpha_n q) = 0$ with Im >0 and (n=1,2,3).

After the computation of displacement Green's function, the stress Green's tensor was calculated. Rewriting the displacement function as $g_{mp} = u_m^p$ which says the displacement of a target particle in *m-th* direction due to a force given at the source point in the *p-th* direction. The strains (like in Eq. (4.92)) at the *m-th* target point can be written as

$$\mathbf{e}^p{}_{mj} = \frac{1}{2}\left(u_{m,j}^p + u_{j.m}^p\right) \tag{7.211}$$

where

$$u_{m,j}^p(x_n,\omega) = \left[u_{m,j}^p(x_n,\omega)\right]^D + \left[u_{m,j}^p(x_n,\omega)\right]^S \tag{7.212}$$

$$\left[u_{m,j}^p\right]^D = \left[-\frac{(\omega)^2}{4(2\pi)^2\rho} \sum_{z=1}^{3} \int \int_{\theta=0;\,\phi=0}^{\theta=\pi;\,\phi=2\pi} \left[(s_z)^4 (P_{mp})^{(z)} k_j \exp\left(i(k_i x_i)^{(z)}\right) sin\theta d\theta d\phi \right] \right] \tag{7.213}$$

$$[u_{m,j}^p]^S = \frac{Imx_j}{2\pi r^3} \sum_{n=1}^{3} \left[\frac{A_{mp}(p+\alpha q)}{\partial_\alpha D(p+\alpha q)} \right]_{\alpha=\alpha_n} \tag{7.214}$$

Here, in above equations, the previously described slowness in Chapters 4 and 5 is introduced. Please note that the γ_z are the eigen values and are the square of the wave velocities of the respective wave modes. Hence, inverse of the square root of the eigen values will be the respective slowness as described in the above equations, slowness $s_z = 1/\sqrt{\gamma_z}$. A_{mp} and D are the cofactor and determinant of Γ_{mp} matrix (Eq. (7.195)).

Like it is written in Eq. (4.92), the stress Green's tensor in anisotropic medium can be written as

$$\sigma^{P}{}_{ik}(\mathbf{x}) = \mathbb{C}_{ikmj}\, \mathbf{e}^{P}{}_{mj}(\mathbf{x}) \tag{7.215}$$

where $\sigma^{P}{}_{ik}$ is the stress tensor at the target point ($\mathbf{x}$) due to the source point with force along the direction p.

7.5.2.2 Fourier Transform Approach: Solution of Elastodynamic Green's Function

Let us considered an operator that exist in the $\mathcal{L}^2$ Hilbert space when the Fourier transform of Eq. (7.183) in frequency-wavenumber domain can be written after refs. [27, 36] as

$$\mathbf{L}_{im}\tilde{G}_{mp}(k_n,\omega) = \delta_{ip}\frac{1}{(2\pi)^3} f_p(\omega) \quad ; \quad \mathbf{L}_{im}(k_n,\omega) = \left[\mathbb{C}_{ijml}k_jk_l - \rho\omega^2\delta_{ij}\right] \tag{7.216}$$

where

$$f(\omega) = -\frac{1}{2\pi i\omega} + \frac{\delta(\omega)}{2} \tag{7.217}$$

$$\tilde{G}_{mp}(k_n,\omega) = \frac{1}{(2\pi)^4}\int\int_{-\infty}^{\infty}\int_{-\infty}^{\infty} G_{mp}(x_n,t)e^{i(k_lx_l-\omega t)}\, d\Omega\, dt \tag{7.218}$$

Frequency-space domain Green's function applying the inverse transform is

$$g_{mp}(x_n,\omega) = \frac{1}{(2\pi)^3}\int\int\int_{-\infty}^{\infty}\left[\mathbf{L}_{mp}(k_n,\omega)^{-1}\right]e^{-ik_lx_l}\, dk^3 \tag{7.219}$$

According to the spectral resolution theorem, we can express the inverse Christoffel's operator $\mathbf{L}_{mp}(k_n,\omega)$ [23, 29, 30] as

$$\mathbf{L}_{mp}(k_n,\omega)^{-1} = \sum_{z=1}^{3}\frac{\varphi_{mi}{}^{(z)}\varphi_{ip}{}^{(z)}}{\rho\gamma_z|\mathbf{k}|^2 - \rho\omega^2} = \sum_{z=1}^{3}\frac{\gamma_z^{-2}\left(P_{mp}\right)^{(z)}}{\rho\left(|\mathbf{k}|^2 - \omega^2/\sqrt{\gamma_z}^{\,2}\right)} \tag{7.220}$$

where z is the index to take any eigen values in the system. γ_z is the z-th eigen value and $(P_{im})^{(z)}$ is the projection matrix of the z-th eigen mode described explicitly in Eqs. (7.196) and (7.197). Substituting Eq. (7.220) into Eq. (7.219), we get an alternative expression for the displacement Green's function in spherical coordinate system as

$$g_{mp}(x_n,\omega) = \frac{1}{(2\pi)^3 \rho} \sum_{z=1}^{3} \iint \int_{|\mathbf{r}|=1}^{sphere} \left[\left(1/\sqrt{\gamma_z}\right)^2 \left(P_{mp}\right)^{(z)} sin\theta d\theta d\phi \right]$$

$$\times \int_{-\infty}^{\infty} \left[\frac{\exp(-i|\mathbf{k}|\,|\mathbf{x}|\,cos\theta)}{\left(|\mathbf{k}|^2 - \left(\omega/\sqrt{\gamma_z}\right)^2\right)} \right] |\mathbf{k}|^2\, dk \tag{7.221}$$

Next for convenient operation [36], k_3-axis is aligned along the direction of **x** and the integral is transformed into polar coordinate system.

where $\mathbf{k} = |\mathbf{k}|(sin\theta cos\varphi,\ sin\theta sin\varphi,\ cos\theta)$ and $dk^3 \to |\mathbf{k}|^2\, sin\theta d\theta d\phi dk$.

Earlier in Radon transform method similar spherical integral was achieved in Eq. (7.209) which was judicially separated and carried over to the calculation of strains and further in calculation of stresses. However, in Eq. (7.221) the latter part of the integral has a pole at $k|^2 = a^2$, where $a = \omega/\sqrt{\gamma_z}$ which can be further simplified by the application of Cauchy's integral identity [30].

$$\int_{-\infty}^{\infty} \frac{\exp(iku)}{\left(k^2 - a^2\right)} |\mathbf{k}|^2\, dk = \pi i a e^{ia|u|} + 2\pi\delta(u) \tag{7.222}$$

Substituting Cauchy's integral identity (Eq. (7.222)) into Eq. (7.221), followed by mathematical simplifications, one can get the final expression for the displacement Green's functions in frequency domain. Here, the surface integral in the first part of the equations was performed over a hemisphere pointing toward the k_3-axis along the direction of **x** and the line integral in the second part is performed along the perimeter of that hemisphere. Hence, it can be said that the integral only exists when the dot product between the wave vector **k** and the source-target line along **x** is greater than zero, i.e., $\mathbf{k.x} \geq 0$ or $cos\beta \geq 0$ when β is the angle made by the wave vector with the direction **x**.

$$g_{mp}(x_n,\omega) = \frac{i\omega}{8\pi^2 \rho} \sum_{z=1}^{3} \int \int_{\theta=0;\ \phi=0}^{\theta=\pi/2;\ \phi=2\pi} \left[\left(1/\sqrt{\gamma_z}\right)^3 \left(P_{mp}\right)^{(z)} \exp\left(i\left(s_i x_i\right)^{(z)}\right) sin\theta d\theta d\phi \right]$$

$$+ \frac{1}{2(2\pi)^2 \rho|\mathbf{x}|} \int_0^{2\pi} \left(1/\sqrt{\gamma_z}\right)^2 \left(P_{mp}\right)^{(z)} d\phi \tag{7.223}$$

After the computation of the displacement Green's function, the stress Green's function is then calculated. The displacement function is rewritten as $g_{mp} = u_m^p$, like it was previously expressed in Radon transform method. Like Eq. (7.211), strains at the target points can be written as

$$\mathbf{e}^p{}_{mj} = \frac{1}{2}\left(u^p_{m,j} + u^p_{j.m}\right) \tag{7.224}$$

where

$$u^p_{m,j}(x_n,\omega) = -\frac{(\omega)^2}{2(2\pi)^2\rho}\sum_{z=1}^{3}\int\limits_{\theta=0;\ \phi=0}^{\theta=\pi;\ \phi=2\pi}\left[(s_z)^4\left(P_{mp}\right)^{(z)}k_j\exp\left(i\left(s_i x_i\right)^{(z)}\right)\sin\theta d\theta d\phi\right]$$
$$-\frac{x_j}{2(2\pi)^2\rho|\mathbf{x}|^3}\int\limits_{0}^{2\pi}(s_z)^2\left(P_{mp}\right)^{(z)}d\phi \tag{7.225}$$

Expression for the derivatives of the displacement Green's function was not available in the literature presented herein. Using Eq. (7.224) the stress Green's tensor in the anisotropic medium can be written as

$$\sigma^p{}_{ik}(\mathbf{x}) = \mathbb{C}_{ikmj}e^p{}_{mj}(\mathbf{x}) \tag{7.226}$$

where $\sigma^p{}_{ik}$ is the stress tensor at $\boldsymbol{y}$ point due to the source at $\boldsymbol{y}_0$ with force acting along the p-th direction. Please note that the distance vector in all the above equations and in Eq. (7.226) is $\mathbf{x} = \boldsymbol{y} - \boldsymbol{y}_0$. The Constitutive matrix $\mathbb{C}_{ikmj}$ for different material type are discussed in Chapter 2 where necessary number of material constants are discussed and Chapter 4 where the slowness profiles in different material types are plotted.

In the final expression (Eq. (7.223)) above, we can see that the Green's function is calculated by integrating over the sphere and summing all three wave modes. It incorporates the influence of all the wave propagation vectors in all possible directions. For the visual understating of the expression, we investigate a schematic (Fig. 7.25) in detail. It is required to calculate the wave field in the intended direction of wave propagation (blue arrow) at any target point (black dot with blue ring) due to any source point (red sphere). For that purpose, the influence of the wave field at the target, due to all three wave modes in all possible wave direction (green arrow) is considered. Also, only the forward propagating waves, i.e., the grid points on the

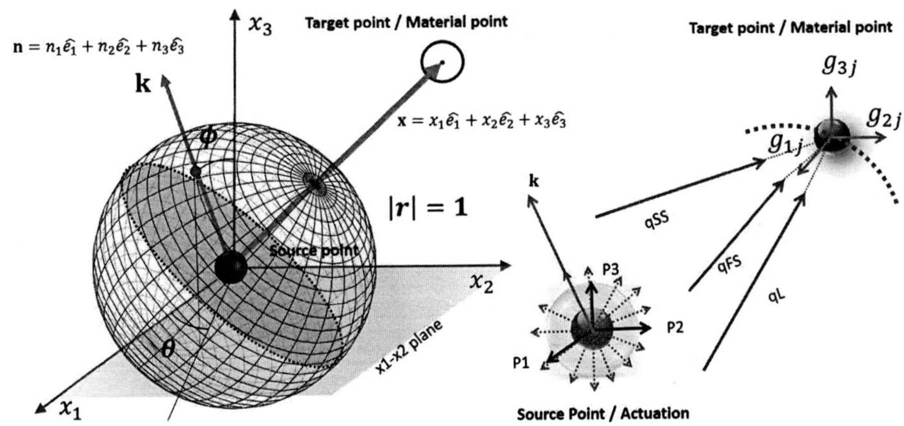

FIGURE 7.25 A schematic diagram to visualize the integral in Eq. (7.133).

sphere above the midplane (gray plane) are considered which is like the existence of Heaviside step function. Furthermore, the normal to the gray plane can be visualized as a radon plane $h = \mu_k x_k$ in Fig. 7.25 where μ_k is identical to n_k. The second part of the integral in Eq. (7.223) pertains to the perimeter of the circle on the midplane (gray plane) perpendicular to the blue arrow.

After the computation of displacement Green's function, one can proceed to compute the stress Green's function. Here, the nomenclature of displacement Green's function is bit changed to avoid the confusion in the derivation and the simplicity of understanding of the equation; It is written $g_{mp} = u_m^p$. Before one can calculate the stress Green's Function, we need to calculate the strain. Hence, finally both the displacement and stress Green's Functions required for the modeling wave fields in anisotropic media are available for use.

7.5.2.3 Comparison of Green's Function: Fourier vs. Radon Transform

The Green's functions were calculated in a line configuration using the two different approaches mentioned above to verify the identicality of the solution. A MATLAB code was written to implement the solution. In Fig. 7.26, a configuration of source target combinations is assumed for calculation of the displacement Green's function along a straight line inside the material. The configuration is composed of 51 target points with a spacing of 0.25 mm and point source with an excitation frequency of 1MHz. The target points and point source is separated by 10 mm. The Green's functions are calculated for transversely isotropic and monoclinic material that were considered previously in Chapter 4.

From the comparison of displacement Green's functions in Figs. 7.27 and 7.28 calculated using Radon and Fourier transform approach, it is proved that both methods

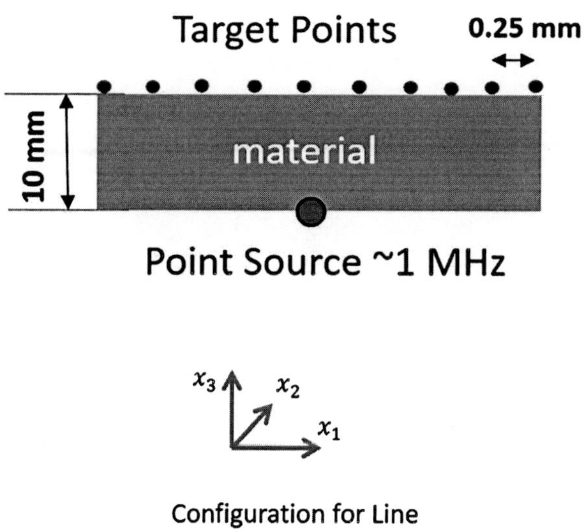

FIGURE 7.26 A schematic showing the line configuration of problem to calculate Green's function.

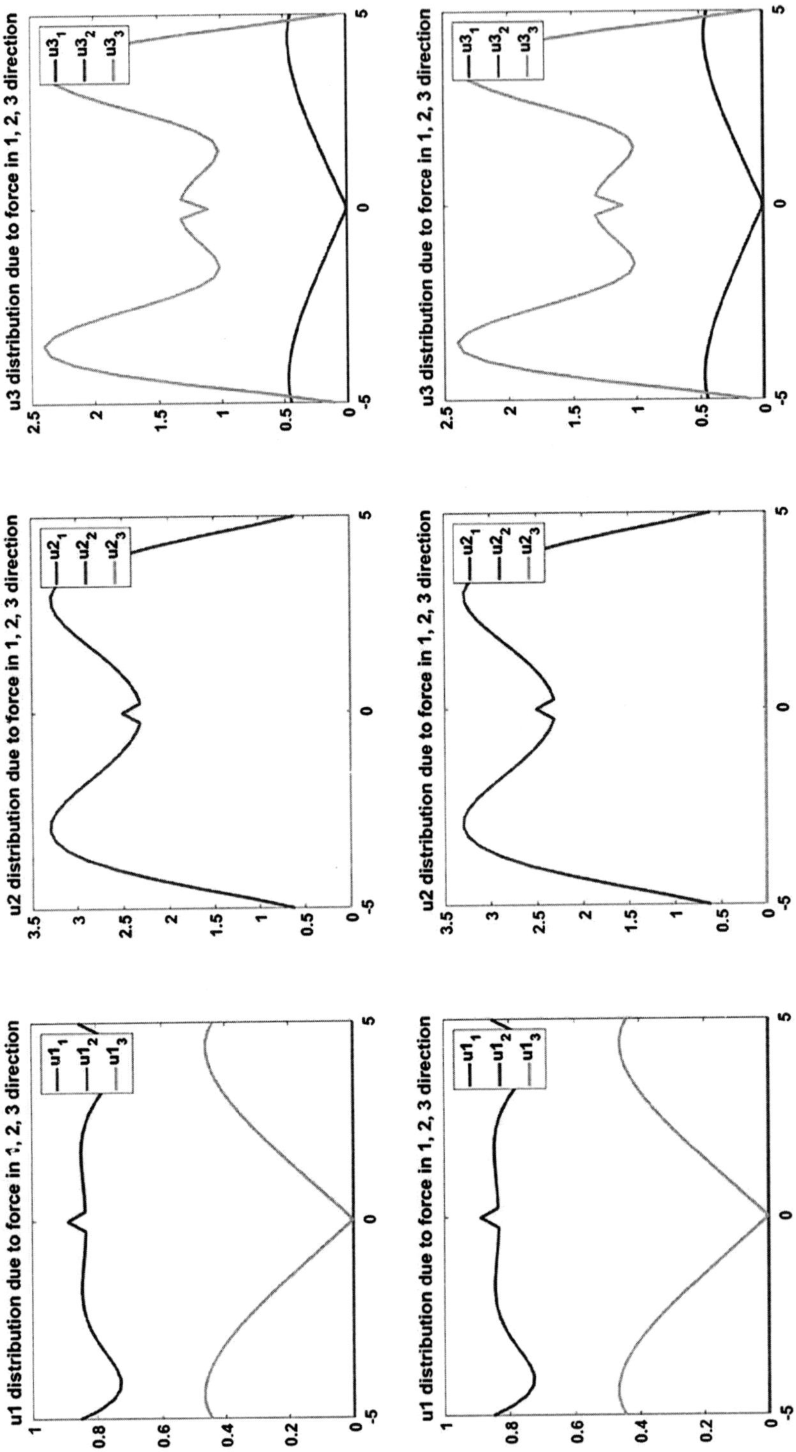

FIGURE 7.27 Displacement Green's function calculated using Radon Transform (top row) and Fourier Transform (bottom row) for transversely isotropic material.

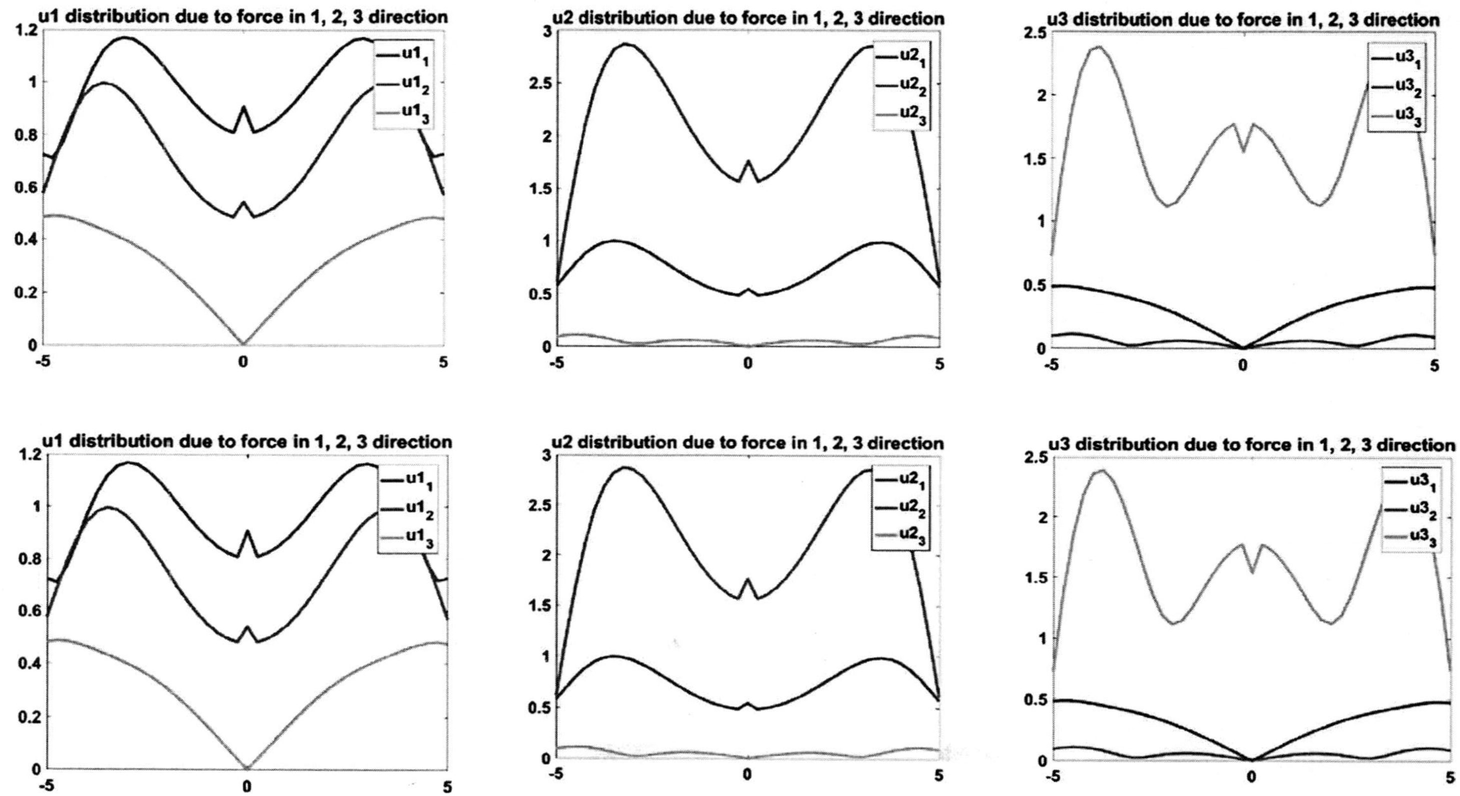

FIGURE 7.28 Displacement Green's function calculated using Radon Transform (top row) and Fourier Transform (bottom row) for monoclinic material.

give similar solutions or the solutions from the two techniques are identical. Same graphs were obtained for stress Green's functions and they were too identical. Hence, they can be used interchangeably.

7.5.2.4 Relation between Radon Transform and Fourier Transform

An obvious question may arise, if the displacement and stress Green's functions are equivalent obtained from Radon transform method and Fourier Transform method, then they must be related to each other mathematically. Is Radon transform related to Fourier transform? The answer is yes, they are related to each other. This can be explained through Fig. 7.23 in case of 2D more easily. However, similar is valid for 3D. Let us consider the projection of the function $f(x_1,x_2)$ on the line sl for a fixed θ_r. The sinogram function is written in Eq. (7.190), is rewritten here with modified form applying the direction cosines through the unit vector $\hat{\mu}$ introduced in the later part of Section 7.5.2.1 and in Fig. 7.24.

$$\tilde{g}(h,\theta_r)=\int\limits_{-\infty}^{\infty}\int\limits_{-\infty}^{\infty} f(x_1,x_2)\delta(h-\mu_i x_i)\,dx_1dx_2 \tag{7.227}$$

Next, an innovative approach is introduced. Let us take the one-dimensional Fourier transformation of the projection function with respect to h. The expression of the Fourier transformation ($\mathfrak{F}$) can be written as

$$\mathcal{G}(k,\theta_r)=\mathfrak{F}(\tilde{g}(h,\theta_r))=\int\limits_{-\infty}^{\infty}\tilde{g}(h,\theta_r)e^{-2\pi ihk}dh \tag{7.228}$$

The above transformation is very similar to the transformation from space to wave number domain transformation. Now, let us substitute Eq. (7.227) in Eq. (7.228) and it may be

$$\mathcal{G}(k,\theta_r)=\mathfrak{F}(\tilde{g}(h,\theta_r))=\int\limits_{-\infty}^{\infty}\left[\int\limits_{-\infty}^{\infty}\int\limits_{-\infty}^{\infty} f(x_1,x_2)\delta(h-\mu_i x_i)\,dx_1dx_2\right]e^{-2\pi ihk}dh \tag{7.229}$$

Interchanging the integration, we get

$$\mathcal{G}(k,\theta_r)=\mathfrak{F}(\tilde{g}(h,\theta_r))=\left[\int\limits_{-\infty}^{\infty}\int\limits_{-\infty}^{\infty} f(x_1,x_2)\int\limits_{-\infty}^{\infty}\delta(h-\mu_i x_i)e^{-2\pi ihk}dh\right]dx_1dx_2 \tag{7.230}$$

And further

$$\mathcal{G}(k,\theta_r)=\mathfrak{F}(\tilde{g}(h,\theta_r))=\left[\int\limits_{-\infty}^{\infty}\int\limits_{-\infty}^{\infty} f(x_1,x_2)e^{-2\pi ik\mu_i x_i}\right]dx_1dx_2 \tag{7.231}$$

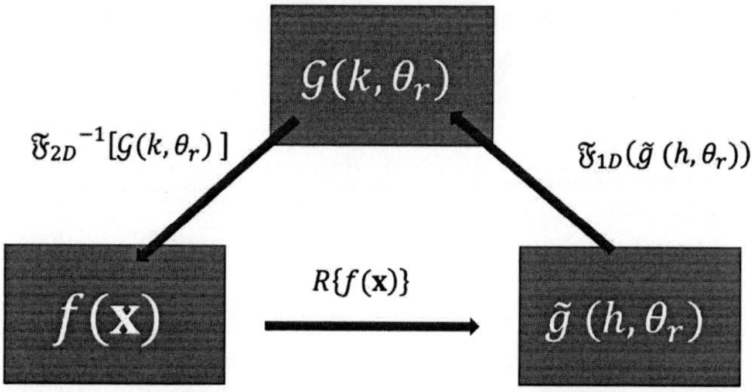

FIGURE 7.29 Interconnectivity of Radon Transform with Fourier Transform.

After carefully investigating Eq. (7.231), it can be seen that the right side of the equation is a 2D Fourier transform of the function $f(x_1,x_2)$. Hence, it can be summarized that one-dimensional Fourier transform of the projection of a function on a line at a fixed angle θ_r, i.e., the Radon transform of the function at a fixed angle θ_r is equal to the values on the line going through the two-dimensional Fourier transform of the same function at the same angle θ_r. This could be further explained by a self-explanatory Fig. 7.29. Mathematically, one can write

$$\mathfrak{F}_{1D}\left(R\{f(\mathbf{x})\}\right)=\mathfrak{F}_{2D}\left(f(\mathbf{x})\right) \tag{7.232}$$

$$f(\mathbf{x})\rightarrow R\{f(\mathbf{x})\}\rightarrow \mathfrak{F}_{1D}\left(R\{f(\mathbf{x})\}\right)\rightarrow \mathfrak{F}_{2D}^{-1}\left[\mathfrak{F}_{1D}\left(R\{f(\mathbf{x})\}\right)\right] \tag{7.233}$$

where $\mathfrak{F}_{1D}$ is one-dimensional Fourier transform, $\mathfrak{F}_{2D}$ is two-dimensional Fourier transform, and $\mathfrak{F}_{2D}^{-1}$ is two-dimensional inverse Fourier transform.

7.6 CNDE CASE STUDIES FOR ANISOTROPIC SOLIDS USING DPSM

For numerical computation of ultrasonic wave field in anisotropic media, we institute the analytical model developed for the Green's function in Section 7.3 into a numerical technique called Distributed point source method (DPSM) discussed earlier in this chapter. In DPSM, the Green's function is essential to calculate the displacement and stress profile and simulate the wave propagation behavior. Fundamentals of DPSM are discussed in Section 7.1 and the mathematical processes of using the Green's functions are discussed in subsequent sections in detail through examples. Briefly speaking, DPSM, as the name suggests, is the numerical technique based on distributing the source and target points in the actuator such as transducer and boundary and interfaces of the problem geometry. The ultrasonic field generated by the source such as transducer is the summation of the field generated due to all the point sources distributed at the transducer. Similarly, at the interface, there are transmitted and reflected field which is represented by placing the two layers of point sources on either side of the interface. Then, the source strengths of the point sources distributed over the transducer and the interface is calculated by satisfying the boundary and the interface continuity conditions as required. Finally, displacement and stress profile in an anisotropic medium and the pressure profile in a fluid medium can be calculated with the help of the source strengths.

The computational modeling can be performed in both frequency and time domain. Frequency domain CNDE is performed to visualize the NDE situation and wave filed (including wave interaction, propagation, scattering and attenuation) in the material as a snap shot at a given frequency. Similarly, the NDE problem could be solved at different frequencies. The time-domain modeling provides an actual ultrasonic signal with wave amplitudes of the signal varying with respect to time, which we actually obtain from an NDE experiment. However, in many cases it is necessary to know the frequency content in a signal rather than the amplitudes of the individual samples. Although DPSM can perform time domain analysis, most of the results in this book from DPSM are presented in frequency domain. How to calculate time domain signal from DPSM is also discussed toward the end of this chapter. However, other methods are more suitable for time domain analysis, e.g., EFIT, LISA, SEM, Peri-elastodynamic, etc. which are discussed in later chapters.

7.6.1 Numerical Computation of Wave Field in Anisotropic Half-space

A schematic diagram shown in Fig. 7.30 illustrates an NDE setup that is used for numerical simulation using DPSM. Here again, please note that x_1-, x_2-, and x_3- axes notations in Cartesian coordinate system introduced in Chapters 2 and 3 are synonymously used here with x, y and z or X, Y and Z coordinate notations, respectively. A circular transducer and an anisotropic half-space are immersed in fluid (water) is considered.

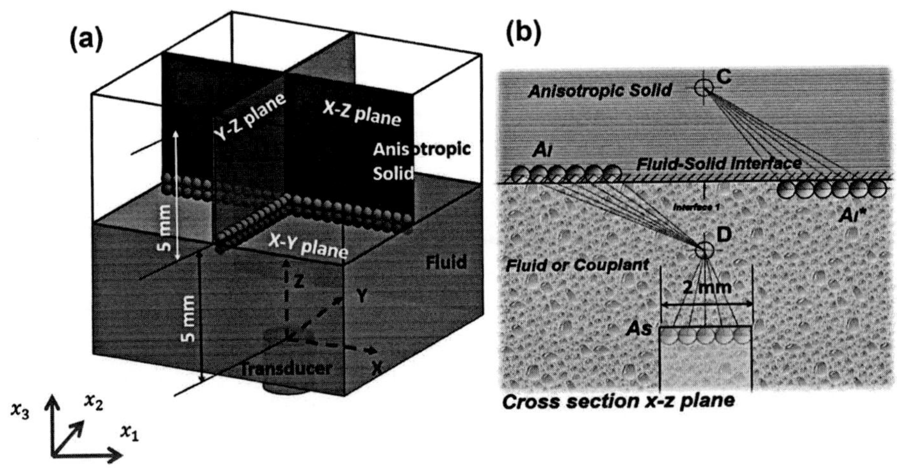

FIGURE 7.30 a) Schematics (not to scale) of the wave field computation problem in anisotropic solid half space, point sources distributed over the x-y plane, however, only two orthogonal line of point sources are shown b) cross-section view of the NDE problem along x-z plane.

The point sources are distributed below the transducer and on both sides of the interface between the two media. The contribution of different point sources is as shown by the lines connecting the relevant point sources to the points of interest (C, D). The total ultrasonic field at any arbitrary point C in the anisotropic solid is produced by the superposition of the strength of all the point sources denoted by $\mathbf{A}_{I*}$ whereas the total ultrasonic field at any arbitrary point D in the fluid is produced by the superposition of the strength of all the point sources denoted by $\mathbf{A}_s$ and $\mathbf{A}_I$. In this section, for anisotropic modeling, *N* number of target points and *M* number of source points are considered. This process is described before in Section 7.4.1 with *M* number of target points and *N* number of source points. This change is done intentionally, to avoid stereotype designation of the number of points, such that during formulation reader can track the actual number of point sources and target points, without losing over the nomenclature. Here in this problem, a 1 MHz transducer with the diameter of 2 mm is submerged into the water and 100 point sources were distributed concentrically just below the transmitting face. The distance between the transducer and interface is taken as 5 mm. For the actuation of the transducer, unit velocity is prescribed on the transducer face like it was done in Section 7.4.1. Following the similar approach but implementing anisotropic Green's function the matrix equation that is obtained from the boundary and interface conditions is written as

$$\begin{bmatrix} \mathbf{M}_{SS} & \mathbf{M}_{SI} & \mathbf{0} \\ \mathbf{Q}_{IS} & \mathbf{Q}_{II} & \mathbf{S33}_{II*} \\ \mathbf{DF3}_{IS} & \mathbf{DF3}_{I1} & -\mathbf{DS3}_{II*} \\ \mathbf{0} & \mathbf{0} & \mathbf{S31}_{II*} \\ \mathbf{0} & \mathbf{0} & \mathbf{S32}_{II*} \end{bmatrix} \begin{Bmatrix} \mathbf{A}_S \\ \mathbf{A}_I \\ \mathbf{A}_I * \end{Bmatrix} = \begin{Bmatrix} \mathbf{V}_{S0} \\ \mathbf{0} \\ \mathbf{0} \end{Bmatrix} \tag{7.234}$$

In Eq. (7.234), **M** and **Q** matrices are like the matrix defined in Eqs. (7.47) and (7.52), respectively. **DS3** matrix and **S3i** matrices are expressed as follows:

$$\mathbf{DS3} = \begin{bmatrix} g_3^{1_1} & g_3^{2_1} & \ldots\ldots & g_3^{M_1} \\ g_3^{1_2} & g_3^{2_2} & \ldots\ldots & g_3^{M_2} \\ \ldots & \ldots & \ldots\ldots & \ldots \\ \ldots & \ldots & \ldots\ldots & \ldots \\ g_3^{1_N} & g_3^{2_N} & \ldots\ldots & g_3^{M_N} \end{bmatrix}_{NX3M} \tag{7.235}$$

where $g_3^{m_n} = [g_{31}^{\ m} \quad g_{32}^{\ m} \quad g_{33}^{\ m}]_n$, and g_{3p} is expressed in Eq. (7.223) substituting, $m = 3$.

$$\mathbf{S3i} = \begin{bmatrix} s_{3i}^{1_1} & s_{3i}^{2_1} & \ldots\ldots & s_{3i}^{M_1} \\ s_{3i}^{1_2} & s_{3i}^{2_2} & \ldots\ldots & s_{3i}^{M_2} \\ \ldots & \ldots & \ldots\ldots & \ldots \\ \ldots & \ldots & \ldots\ldots & \ldots \\ s_{3i}^{1_N} & s_{3i}^{2_N} & \ldots\ldots & s_{3i}^{M_N} \end{bmatrix}_{NX3M} \tag{7.236}$$

where $s_{3i}^{m_n} = [\sigma_{3i}^{m} \;\; \sigma_{3i}^{m} \;\; \sigma_{3i}^{m}]_n$, i take values 1, 2, and 3 and, σ_{ki}^{p} substituting $p=m$, $k=3$, and, $i=1,2,3$, expressed in Eq. (7.226). Hence, comparing Eqs. (7.150) and (7.236), it is easily identifiable that the only changes were with the elements in the matrix. By substituting appropriate expression for the Green's function in the matrix the unknown source strengths vectors can be calculated as discussed in Eq. (7.152). After obtaining the source strengths, one can compute the wave field in anisotropic half-space media using similar steps discussed in Eq. (7.140) through Eq. (7.144), except appropriate Green's function matrices are replaced by the expressions written in Eqs. (7.235) and (7.236). Next, computed wave fields in three different anisotropic materials, (a) Transversely isotropic, (b) Fully orthotropic, and (c) Monoclinic material, are presented here.

7.6.1.1 Verification of Boundary Condition and Convergence

First the wave field in transversely isotropic half-space is calculated. The source strengths are appropriately computed by solving the governing matrix in Eq. (7.234). While doing so, it is necessary to confirm if number of source strengths used in the simulation are appropriate for convergence and the boundary conditions are accurately satisfied. Hence, both the pressure and normal stresses (in GPa) at the fluid-solid interface is checked if the boundary conditions are matched (Fig. 7.31). The verification was performed by applying the convergence criteria ($r \le \lambda/2\pi$) where distance between two adjacent point sources is $2r$ [13] and λ is the wavelength of the ultrasonic wave in the low speed media. Referring to Eqs. (4.5) and (4.7) in Chapter 4. It can be said that the wavelength (λ) is directly proportional to the phase wave velocity (c) at a fixed angular frequency (ω). Here in this CNDE simulation, an ultrasonic excitation of 1 MHz from the virtual transducer was used. Wave speed in water is lower that the phase wave velocity of quasi slow shear (*qSS*) wave mode, which is lowest in transversely isotropic solid. Hence, smallest wavelength (λ) was obtained using wave speed in water, which is 1.48 km/sec mentioned in Table 7.1. With 1 MHz excitation frequency and

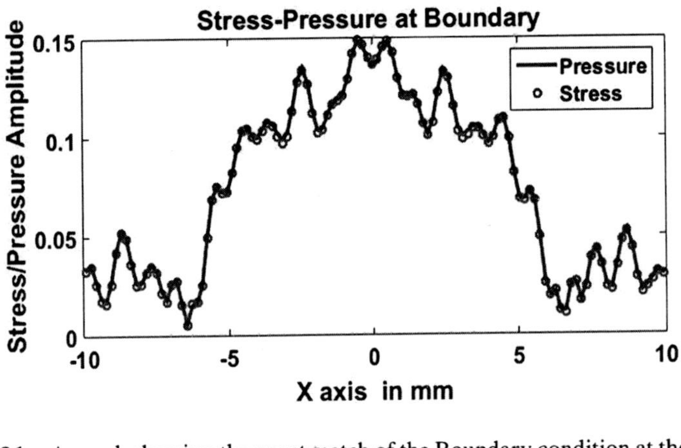

FIGURE 7.31 A graph showing the exact match of the Boundary condition at the solid-fluid interface.

1.48 km/sec. wave speed, the minimum wave length generated in this problem was found to be 0.235 mm. Hence, the $\lambda/2\pi$ was found to be 0.037 mm. As per the convergence criteria, $r \leq \lambda/2\pi$ presented before by Banerjee and Kundu [13] distance between two point sources were kept 0.074 mm. Thus 270-point sources were distributed over a 20 mm length of solid interface simulated in this problem. Based on this convergence criteria, it can be seen that the boundary conditions are exactly satisfied (Fig. 7.31) at any arbitrarily spatial points distributed along the interface.

7.6.1.2 Computed Wave Field in Anisotropic Solids

After verification, the wave fields on $x - z$ plane (alternatively $x_1 - x_3$ plane) in all three material types mentioned above were calculated. In this section, like before, $x-$, $y-$, and z-axes are synonymous to x_1-, x_2-, and x_3-axes. In all the following figures, the lower half of the plots represent the pressure field inside water and the upper half represents the stress fields inside solids. First stresses (σ_{11} and σ_{33}) in transversely isotropic material are shown in Fig. 7.32a and b, respectively. A transversely isotropic composite material with fiber direction along the x-axis is considered, when the $y - z$ plane is the plane of isotropy. It is known that the wave energy tends to bifurcate along the fiber direction in transversely isotropic media due to symmetry along the x direction when the wave actuation or the **k** wave vector is along the z direction. In Fig. 7.32b, two high intensity σ_{33} stress zones are identified on the $x - z$ plane. These zones are like a Torus shape with increasing diameter along the increasing z-axis inside the solid material. It shows that the actual wave energy propagation inside transversely isotropic material is not along the actuation direction, i.e., along the direction of $\boldsymbol{k}$ vector but at angle the group velocity direction. Next the wave fields in orthotropic and monoclinic materials on $x - z$ and $y - z$ planes were computed, such that the difference in wave fields are apparent in oppose to isotropic material which exhibits same wave field on both $x - z$ and $y - z$ planes. The stress fields in orthotropic material on $x - z$ plane (Fig. 7.33a and b) and $y - z$ planes (Fig. 7.34a and b) are presented herein. Similarly, the stress fields in monoclinic material on $x - z$ plane (Fig. 7.35a and b) and $y - z$ planes (Fig. 7.36a and b) are computed using appropriate material properties listed in Chapter 4 in Section 4.4.2. Comparing the stress fields

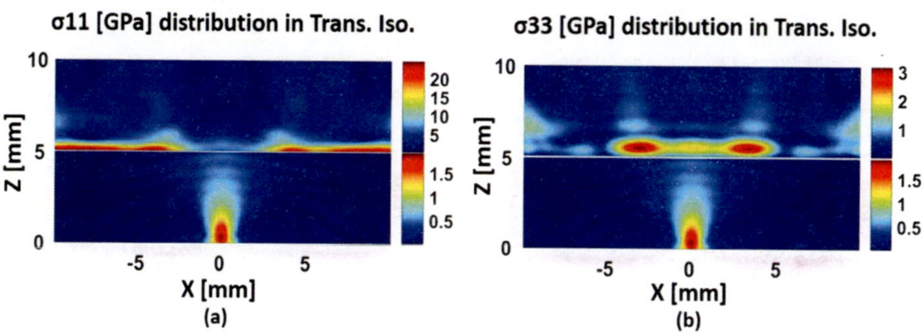

FIGURE 7.32 a) Stress11 (σ11) distribution in transversely isotropic half space b) Stress33 (σ33) distribution in transversely isotropic half space in GPa.

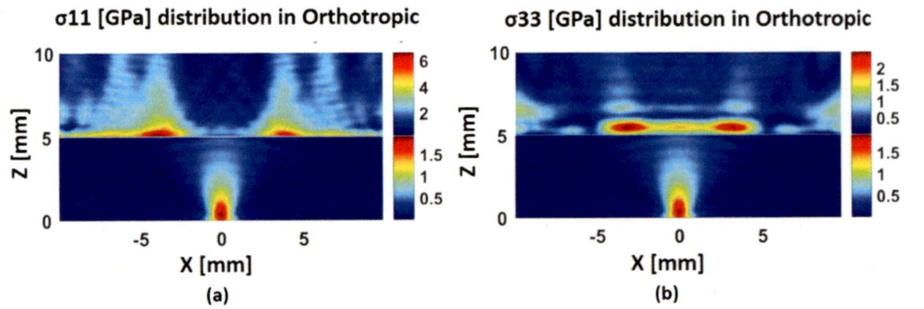

FIGURE 7.33 Wave field plot on x-z plane a) Stress11 (σ11) distribution in fully orthotropic half space b) Stress33 (σ33) distribution in fully orthotropic half space in GPa.

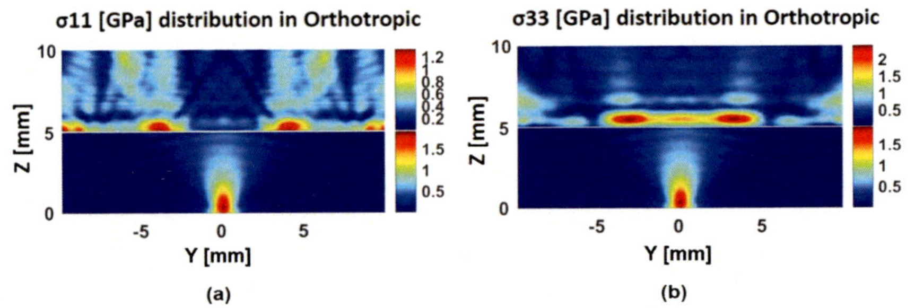

FIGURE 7.34 Wave field plot on y-z plane a) Stress11 (σ11) distribution in fully orthotropic half space b) Stress33 (σ33) distribution in fully orthotropic half space in GPa.

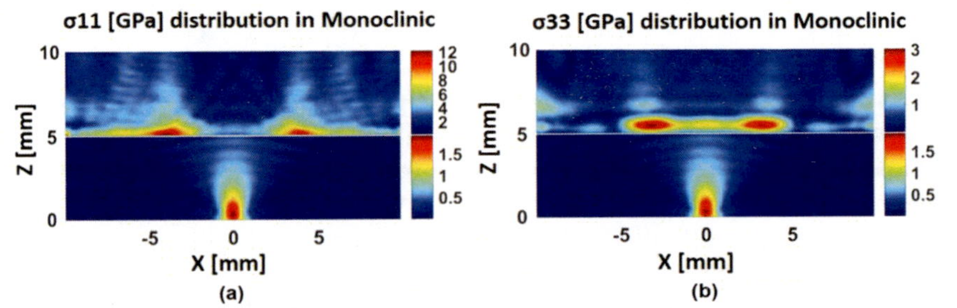

FIGURE 7.35 Wave field plot on x-z plane a) Stress11 (σ11) distribution in monoclinic half space b) Stress33 (σ33) distribution in monoclinic half space in GPa.

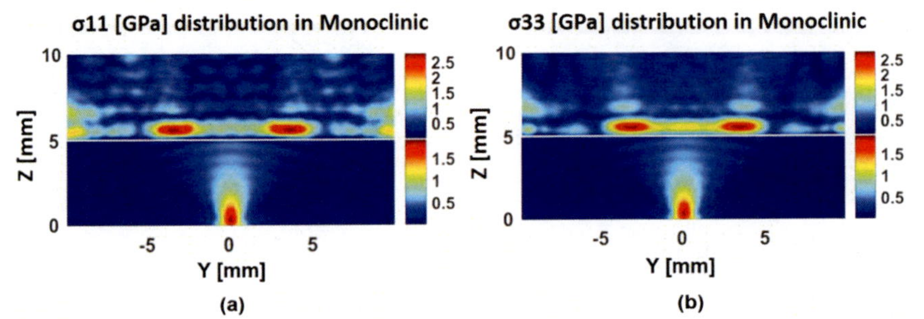

FIGURE 7.36 Wave field plot on y-z plane a) Stress11 (σ11) distribution in monoclinic half space b) Stress33 (σ33) distribution in monoclinic half space in GPa.

in transversely isotropic, fully orthotropic and monoclinic materials on $x-z$ and $y-z$ planes, the qualitative and quantitative differences are evident. The differences could be explained by the modal slowness behavior showed by the wave mode on the $x-z$ plane for all the three materials plotted in Figs. 4.15–4.19 in Chapter 4. The maximum wave energy on $x-z$ plane is contributed mostly by the quasi longitudinal (qL) wave mode and minimally influenced by the other quasi shear modes, hence to the qualitative similarities of the wave fields between the orthotropic and monoclinic material over the $x-z$ plane is evident. Similarly, the qualitative differences of the wave fields between the orthotropic and monoclinic material over the $y-z$ plane are evident from the modal slowness behavior shown by the wave mode on the $y-z$ plane for all the three materials discussed in Chapter 4. Hence, DPSM was able to simulate the wave filed in different anisotropic materials for frequency domain visualization of bulk CNDE experiments.

7.6.2 Numerical Computation of Wave Field in Anisotropic Plate

A CNDE problem shown in Fig. 7.37 is modeled using DPSM. A 3-mm thick composite plate immersed in water is considered to investigate using traditional pulse-echo & through transmission NDE mode. Thus, the plate is bounded by two solid–fluid interfaces. Two ~1 MHz transducers with 2 mm diameter are placed symmetrically on either side of the plate in water at 5 mm distance from the fluid-solid interface. To visualize the frequency domain wave phenomena inside different anisotropic media, ultrasonic wave fields were simulated in transversely isotropic, fully orthotropic and monoclinic materials using their respective Green's functions like it was performed for anisotropic half-space presented in the earlier section. Following standard DPSM approach [11, 13, 34], the point sources are distributed behind the transducer faces (Fig. 7.37) and distributed on either side of the interface between the solid and fluid media. The total ultrasonic field in fluid and solid media are the superposition of contributions of all the relevant point sources. $\mathbf{A}_1$ is the source strength vector of the point sources that are placed above the first solid–fluid interface. $\mathbf{A}_1$ generates the ultrasonic field in fluid below the plate simulating the reflection from the plate. The interface is called the "Interface 1". $\mathbf{A}_{1*}$ is the source strength vector of the sources that are distributed below the first solid–fluid interface and model the transmitted field in the solid plate. Similarly, $\mathbf{A}_2$ and $\mathbf{A}_{2*}$ are the source strength vectors of the point sources that have been distributed above and below the second fluid–solid interface, respectively. This interface is called the "Interface 2". Transducer faces have source strength vectors $\mathbf{A}_S$ and $\mathbf{A}_R$, respectively. The total ultrasonic wave field at any arbitrary point C in the fluid 1 is computed by superposing the contributions from all the point sources denoted by $\mathbf{A}_s$ and $\mathbf{A}_1$. The total ultrasonic wave field at any arbitrary point D inside an anisotropic solid is computed by superposing the contributions from each point sources with Green's function multiplied with their respective source strength denoted by $\mathbf{A}_{1*}$ and $\mathbf{A}_2$ (Fig. 7.37b). Finally, the total ultrasonic wave field at any arbitrary point E in the fluid 2 is computed by superposing the contributions from all the point sources denoted by $\mathbf{A}_{2*}$ and $\mathbf{A}_R$. The source

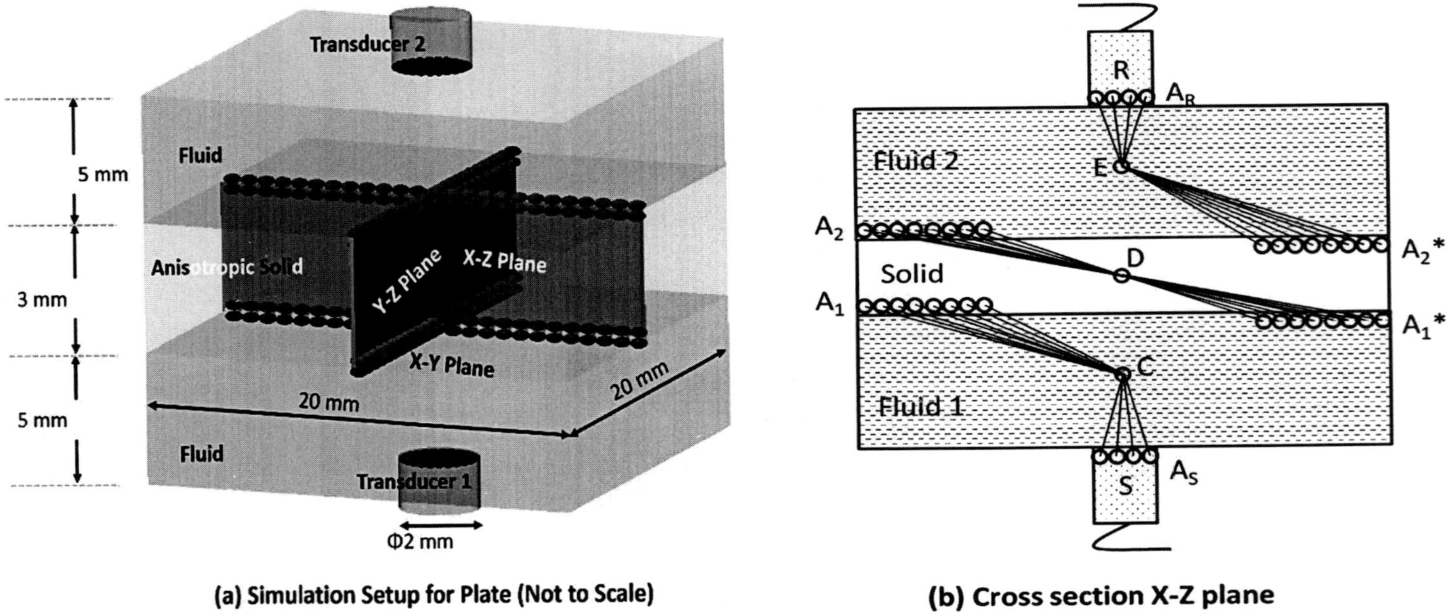

FIGURE 7.37 a) Schematics (not to scale) of the wave field computation problem in anisotropic plate, point sources distributed over the x-y plane, however, only two orthogonal line of point sources are shown b) cross-section view of the NDE problem along x-z plane.

strength depends on the 3D force vectors to actuate the point sources. The displacements or the stresses at any target point are computed by superposing the effect of each point source, that is by multiplying the source strengths with distance and material-dependent Green's function for that point. Hence the source strengths $\mathbf{A}_s$, $\mathbf{A}_1$, $\mathbf{A}_{1*}$, $\mathbf{A}_2$, $\mathbf{A}_{2*}$, and $\mathbf{A}_R$ needs to be calculated which are unknown. The source strengths can be calculated by solving a set of linear equations obtained by satisfying the boundary conditions discussed below.

The bottom and the top transducers are simulated with a prescribed velocity on the circular face equal to V_{S0} and V_{R0}, respectively. If pulse-echo or through transmission are to be simulated, only one transducer can be prescribed with initial velocity. In case of pulse-echo NDE mode, wave field is calculated at the same face of the transducer which is actuated. Whereas, in case of through transmission NDE mode, the wave field is computed on the other transducer face which was not actuated but placed inside the fluid to act as a sensor. As discussed before in Section 7.4, at the solid-fluid interfaces 1 and 2, the normal displacement and the normal stresses along z direction should be continuous, but the shear stresses must vanish. The prescribed normal velocity V_{S0} and V_{R0} on the transducer faces can be obtained by superposing the contribution of the velocity fields created by $\mathbf{A}_s$ and $\mathbf{A}_I$ and $\mathbf{A}_{2*}$, and $\mathbf{A}_R$ sources respectively. By calculating the fluid domain Green's functions [11, 34] written in Eq. (7.47) at the target points, the velocity equation will be

$$\mathbf{VG}_{SS}\mathbf{A}_S + \mathbf{VG}_{S1}\mathbf{A}_2 = \mathbf{V}_{S0} \tag{7.237}$$

$$\mathbf{VG}_{R2*}\mathbf{A}_{2*} + \mathbf{VG}_{RR}\mathbf{A}_R = \mathbf{V}_{R0} \tag{7.238}$$

The velocity Green's function matrices in fluid **VG** are explicitly written as

$$\mathbf{VG} = \begin{pmatrix} v(x_{t1}^1, r_1^1) & \cdots & v(x_{t1}^N, r_1^N) \\ \vdots & \ddots & \vdots \\ v(x_{tM}^1, r_M^1) & \cdots & v(x_{tM}^N, r_M^N) \end{pmatrix} \tag{7.239}$$

where $\left(x_{tn}^m, r_n^m\right) = \frac{x_{tn}^m \exp(ik_f r_n^m)}{i\omega\rho\left(r_n^m\right)^2}\left(ik_f - \frac{1}{r_n^m}\right)$; $k_f = \omega/c_f$, c_f is the wave velocity in fluid and $r = |\mathbf{y} - \mathbf{y}_0|$.

On the other hand, the normal displacement at the fluid-solid interface should be continuous. The following displacement equations using displacement Green's function matrices composed of displacement Green's function in fluid and anisotropic solid in the z direction can be written as

$$\mathbf{UfZ}_{1S}\mathbf{A}_S + \mathbf{UfZ}_{11}\mathbf{A}_1 = \mathbf{UAZ}_{11}\mathbf{A}_{1^*} + \mathbf{UAZ}_{12}\mathbf{A}_2 \tag{7.240}$$

$$\mathbf{UfZ}_{22*}\mathbf{A}_{2*} + \mathbf{UfZ}_{2R}\mathbf{A}_R = \mathbf{UAZ}_{21}\mathbf{A}_{1^*} + \mathbf{UAZ}_{22}\mathbf{A}_2 \tag{7.241}$$

where $\mathbf{UfZ}_{IS}$ is the displacement Green's function matrix for the target points at any interface I due to any point sources S located inside the fluid. Similarly, $\mathbf{UAZ}_{IS}$ is the displacement Green's function matrix at any interface I due to any point sources S located inside the anisotropic solid. The respective Green's function matrices are explicitly written in Eqs. (7.242) and (7.243), respectively

$$\mathbf{Ufi} = \begin{pmatrix} uf(R_{i1}^{1}, r_1^{1}) & \cdots & uf(R_{i1}^{N}, r_1^{N}) \\ \vdots & \ddots & \vdots \\ uf(R_{iM}^{1}, r_M^{1}) & \cdots & uf(R_{iM}^{N}, r_M^{N}) \end{pmatrix} \tag{7.242}$$

where $uf\left(R_{in}^{m}, r_n^{m}\right) = \frac{1}{\rho\omega^2}\left[\frac{1}{r} ik_f R_{in}^{m} \exp\left(ik_f r_n^{m}\right) - \frac{\exp(ik_f r_n^{m})}{\left(r_n^{m}\right)^2} R_{in}^{m}\right]$; $R_{in}^{m} = \frac{y_{in}^{m} - y_{0in}^{m}}{r_n^{m}}$; $i = 1, 2$ & 3

$$\mathbf{UAi}^{p} = \begin{pmatrix} u_i^{p}(R_{i1}^{1}) & \cdots & u_i^{p}(R_{m1}^{N}) \\ \vdots & \ddots & \vdots \\ u_i^{p}(R_{mM}^{1}) & \cdots & u_i^{p}(R_{iM}^{N}) \end{pmatrix} \tag{7.243}$$

where $u_i^{p}\left(R_{im}^{n}\right) = g_{ip}$; $i = 1, 2$ & 3 is written in Eq. (7.223) and $R_{in}^{m} = \frac{y_{in}^{m} - y_{0in}^{m}}{r_n^{m}}$, where r_n^{m} is synonymous to x_n for the m-th point source located at the origin. However, in these cases, locations of the sources (y_{0in}^{m}) and the target points (y_{in}^{m}) are changing.

The compressive normal stress in the anisotropic solid and the fluid pressure at the fluid-solid interfaces should be continuous, and thus equations for stress and pressure can be written as

$$\mathbf{P}_{1S}\mathbf{A}_S + \mathbf{P}_{11}\mathbf{A}_1 = -\sum \mathbf{33}_{11}\mathbf{A}_{1^*} - \sum \mathbf{33}_{12}\mathbf{A}_2 \tag{7.244}$$

$$\mathbf{P}_{22^*}\mathbf{A}_{2^*} + \mathbf{P}_{2R}\mathbf{A}_R = -\sum \mathbf{33}_{21^*}\mathbf{A}_{1^*} - \sum \mathbf{33}_{22}\mathbf{A}_2 \tag{7.245}$$

Additionally, the shear stresses at the fluid-solid interface should vanish because shear stress does exist in fluid. Thus

$$\sum \mathbf{31}_{11^*}\mathbf{A}_{1^*} + \sum \mathbf{31}_{12}\mathbf{A}_2 = 0 \tag{7.246}$$

$$\sum \mathbf{32}_{11^*}\mathbf{A}_{1^*} + \sum \mathbf{32}_{12}\mathbf{A}_2 = 0 \tag{7.247}$$

$\mathbf{P}$, is the pressure Green's function matrix in fluid, $\sum\mathbf{33}$, $\sum\mathbf{31}$ and $\sum\mathbf{32}$ are the stress Green's function matrix in the anisotropic plate for σ_{33}, σ_{31}, and σ_{32} stresses obtained from Eq. (7.226), respectively. Rearranging the equations from Eqs. (7.237) to (7.247), we get the following matrix in Eq. (7.248). After solving the linear algebra problem, the source strengths $\mathbf{A}_{l^*}$, $\mathbf{A}_s$, and $\mathbf{A}_l$ were obtained to calculate the wave field inside the fluid and anisotropic solid.

$$\begin{bmatrix} \mathbf{VG}_{SS} & \mathbf{VG}_{S1} & \mathbf{Zeros} & \mathbf{Zeros} & \mathbf{Zeros} & \mathbf{Zeros} \\ \mathbf{P}_{1S} & \mathbf{P}_{11} & \Sigma 33_{11^*} & \Sigma 33_{12} & \mathbf{Zeros} & \mathbf{Zeros} \\ \mathbf{UfZ}_{IS} & \mathbf{UfZ}_{II} & -\mathbf{UAZ}_{11^*} & -\mathbf{UAZ}_{12} & \mathbf{Zeros} & \mathbf{Zeros} \\ \mathbf{Zeros} & \mathbf{Zeros} & \Sigma 31_{11^*} & \Sigma 31_{12} & \mathbf{Zeros} & \mathbf{Zeros} \\ \mathbf{Zeros} & \mathbf{Zeros} & \Sigma 32_{11^*} & \Sigma 32_{12} & \mathbf{Zeros} & \mathbf{Zeros} \\ \mathbf{Zeros} & \mathbf{Zeros} & \Sigma 32_{21^*} & \Sigma 32_{22} & \mathbf{Zeros} & \mathbf{Zeros} \\ \mathbf{Zeros} & \mathbf{Zeros} & \Sigma 31_{21^*} & \Sigma 31_{22} & \mathbf{Zeros} & \mathbf{Zeros} \\ \mathbf{Zeros} & \mathbf{Zeros} & \Sigma 33_{21^*} & \Sigma 33_{22} & \mathbf{P}_{22^*} & \mathbf{P}_{2R} \\ \mathbf{Zeros} & \mathbf{Zeros} & -\mathbf{UAZ}_{21^*} & -\mathbf{UAZ}_{22} & \mathbf{UfZ}_{22^*} & \mathbf{UfZ}_{2R} \\ \mathbf{Zeros} & \mathbf{Zeros} & \mathbf{Zeros} & \mathbf{Zeros} & \mathbf{VG}_{R2^*} & \mathbf{VG}_{RR} \end{bmatrix} \begin{Bmatrix} \mathbf{A}_S \\ \mathbf{A}_1 \\ \mathbf{A}_{1^*} \\ \mathbf{A}_2 \\ \mathbf{A}_{2^*} \\ \mathbf{A}_R \end{Bmatrix} = \begin{Bmatrix} \mathbf{V}_{SO} \\ \mathbf{Zeros} \\ \mathbf{Zeros} \\ \mathbf{Zeros} \\ \mathbf{Zeros} \\ \mathbf{Zeros} \\ \mathbf{Zeros} \\ \mathbf{Zeros} \\ \mathbf{Zeros} \\ \mathbf{V}_{SO} \end{Bmatrix}$$

$$[\mathbf{MAT}]\{\Lambda\} = \{\mathbf{V}\} \tag{7.248}$$

where **Zeros** are the matrices with appropriate size to make the matrix a square matrix for linear algebra operation. After obtaining the source strengths $\left(\{\Lambda\} = [\mathbf{MAT}]^{-1}\{\mathbf{V}\}\right)$, wave fields inside fluid and solid half space were computed. The pressure field inside the fluid was calculated using the following equation, where F is a set of target points in fluid.

$$\mathbf{PR}_F = \mathbf{P}_{FS}\mathbf{A}_S + \mathbf{P}_{FI}\mathbf{A}_I \tag{7.249}$$

The displacement and stress fields (together called wave fields) inside the anisotropic half space were calculated using the following equations where m is the set of target points distributed inside the solid where the wave fields are intended to be computed. The stress and displacement wave fields from Eqs. (7.250) and (7.251) are presented in Section 7.6.2.1.

$$\mathbf{U1}_m = \mathbf{UAX}_{ml^*}\mathbf{A}_{l^*}\ ;\ \ \mathbf{U2}_m = \mathbf{UAY}_{ml^*}\mathbf{A}_{l^*}\ ;\ \ \mathbf{U3}_m = \mathbf{UAZ}_{ml^*}\mathbf{A}_{l^*} \tag{7.250}$$

$$\mathbf{S33}_m = \sum \mathbf{33}_{ml^*}\mathbf{A}_{l^*};\ \ \mathbf{S31}_m = \sum \mathbf{31}_{ml^*}\mathbf{A}_{l^*};\ \ \mathbf{S32}_m = \sum \mathbf{32}_{ml^*}\mathbf{A}_{l^*} \tag{7.251}$$

7.6.2.1 Computed Wave Field in Anisotropic Plate

In this section, $x-, y-$, and z-axes are synonymous to x_1-, x_2-, and x_3-axes. The stress and displacement wave fields on the $x-z$ plane were calculated for normal

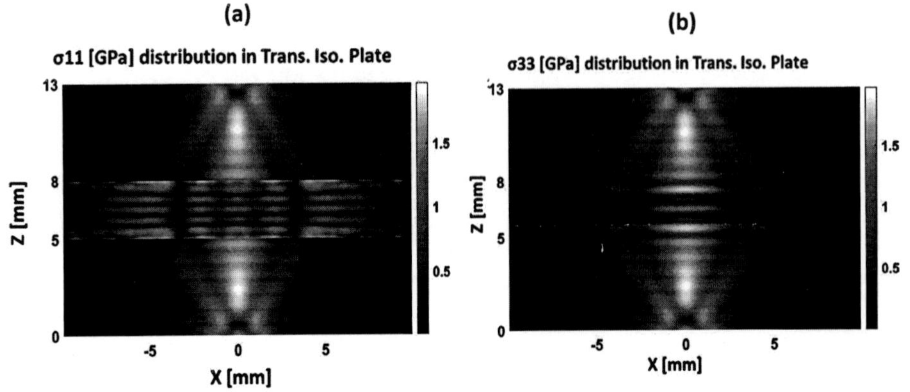

FIGURE 7.38 a) Stress11 (σ_{11}) distribution in transversely isotropic plate b) Stress33 (σ_{33}) distribution in transversely isotropic plate in GPa.

and oblique incidence (30 degree) of wave. In stress field plots, the lower and upper section (5 mm each) of the plots represent the pressure fields inside the water and the midsection (3 mm) represents the stress field inside the solid plate. A transversely isotropic composite material with fiber direction along the x-axis is considered, when the $y-z$ plane is the plane of isotropy. The wave fields are first calculated for normal incidence. The stresses (σ_{11} and σ_{33}) and pressure fields in transversely isotropic material, which are shown in Fig. 7.38a and b, respectively, were computed. Similarly, the displacement fields (u_1 and u_3) are shown in Fig. 7.39a and b. It is known that the wave energy tends to bifurcate along the fiber direction in transversely isotropic media due to symmetry along the x direction when the wave

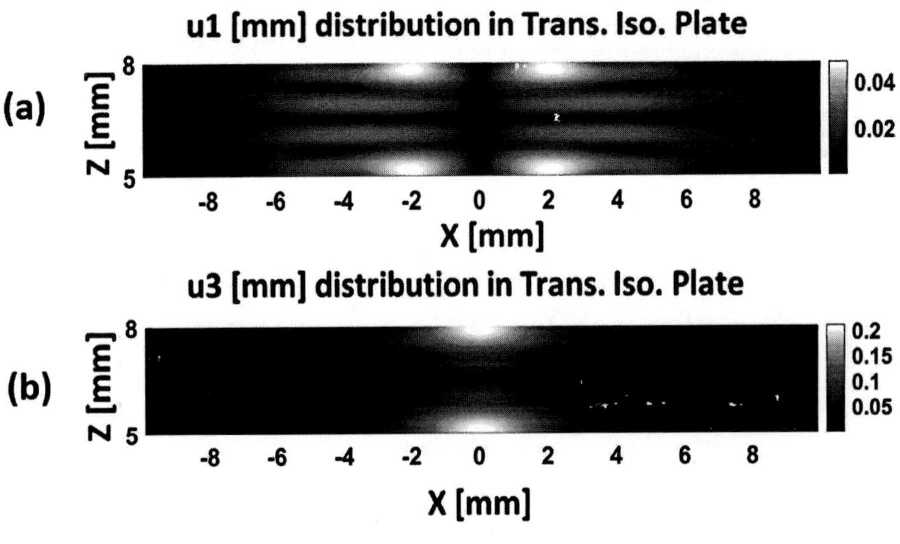

FIGURE 7.39 a) Displacement (u_1) distribution in transversely isotropic plate b) Displacement (u_3) distribution in transversely isotropic plate in GPa.

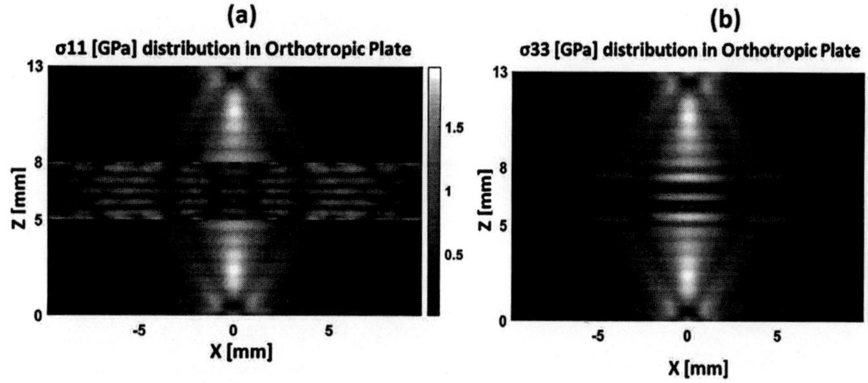

FIGURE 7.40 Wave field plot on x-z plane a) Stress11 (σ_{11}) distribution in fully orthotropic plate b) Stress33 (σ_{33}) distribution in fully orthotropic plate.

actuation direction or the **k** vector is along the z direction. Next the wave fields in orthotropic and monoclinic materials on $x-z$ planes are computed. The stress (Fig. 7.40a and b) and displacement (Fig. 7.41a and b) fields in orthotropic material on $x-z$ plane are presented. Similarly, the stress (Fig. 7.42a and b) and displacement (Fig. 7.43a and b) fields in monoclinic material on x-z plane are computed below. In the same way, the wave fields are next shown for the oblique incidence (30 degree) of wave. The stresses (σ_{11} and σ_{33}) and pressure fields in transversely isotropic material, which are shown in Fig. 7.44a and b, respectively, were computed. Similarly, the displacement fields (u_1 and u_3) are shown in Fig. 7.45a and b.

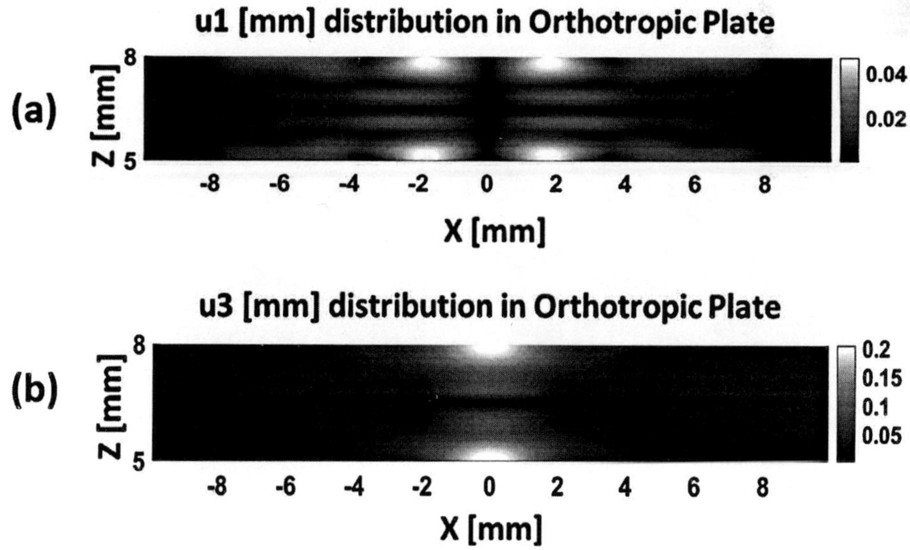

FIGURE 7.41 a) Displacement (u_1) distribution in fully orthotropic plate b) Displacement (u_3) distribution in fully orthotropic plate in Gpa.

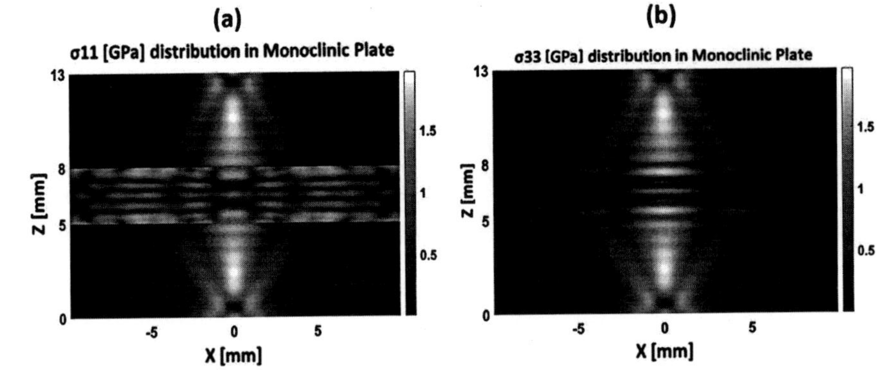

FIGURE 7.42 Wave field plot on x-z plane a) Stress11 (σ_{11}) distribution in monoclinic plate b) Stress33 (σ_{33}) distribution in monoclinic plate.

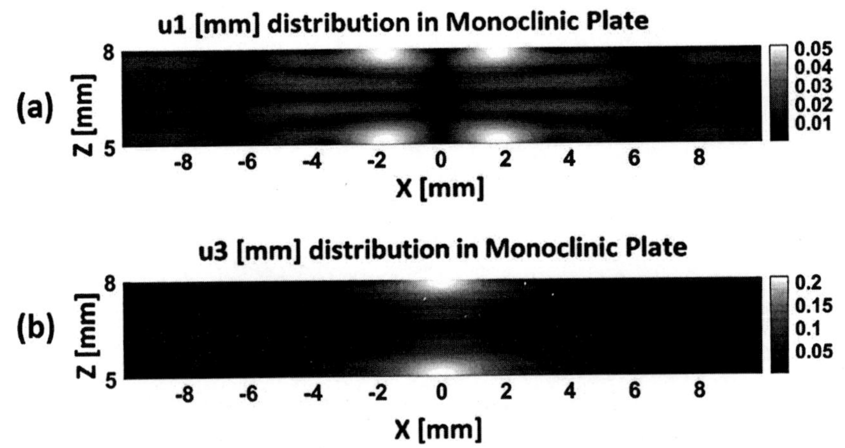

FIGURE 7.43 a) Displacement (u_1) distribution in monoclinic plate b) Displacement (u_3) distribution in fully monoclinic plate in GPa.

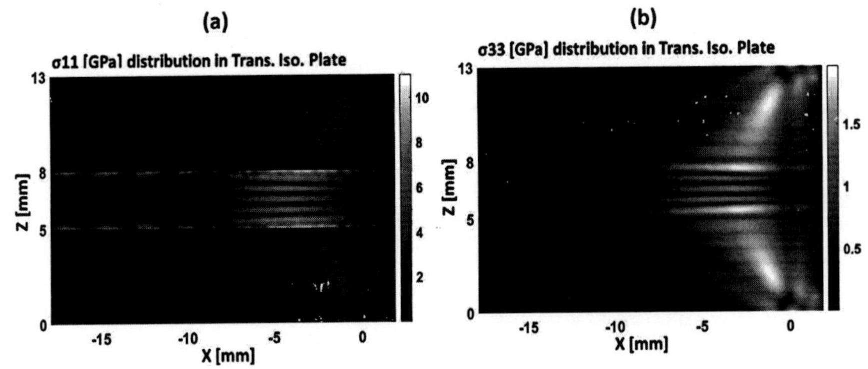

FIGURE 7.44 a) Stress11 (σ_{11}) distribution in transversely isotropic plate b) Stress33 (σ_{33}) distribution in transversely isotropic plate in GPa.

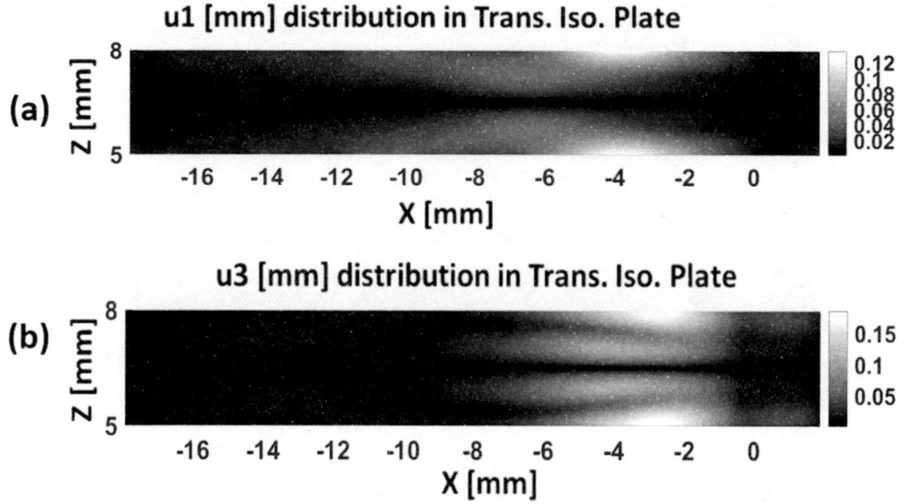

FIGURE 7.45 a) Displacement (u_1) distribution in transversely isotropic plate b) Displacement (u_3) distribution in transversely isotropic plate in GPa.

From the observation of stress and displacement fields, it can be seen that the wave tends to propagate along the length of the plate and guide. With appropriate critical angle of incidence it is possible to generate the Guided wave modes as it is shown in Section 7.4.3.4. It is also observed that the wave energy decays as the wave travels further away from the ultrasonic incident beam. Partial leaky Lamb waves are visible in pressure field for water which is not fully developed as 30 degree may not be the critical angle of incidence. Next the wave fields in orthotropic and monoclinic materials on $x - z$ planes were computed. The stress (Fig. 7.46a and b) and displacement (Fig. 7.47a and b) fields in orthotropic material are presented on the $x - z$ plane. Similarly, the stress (Fig. 7.48a and b) and displacement (Fig. 7.49a and b)

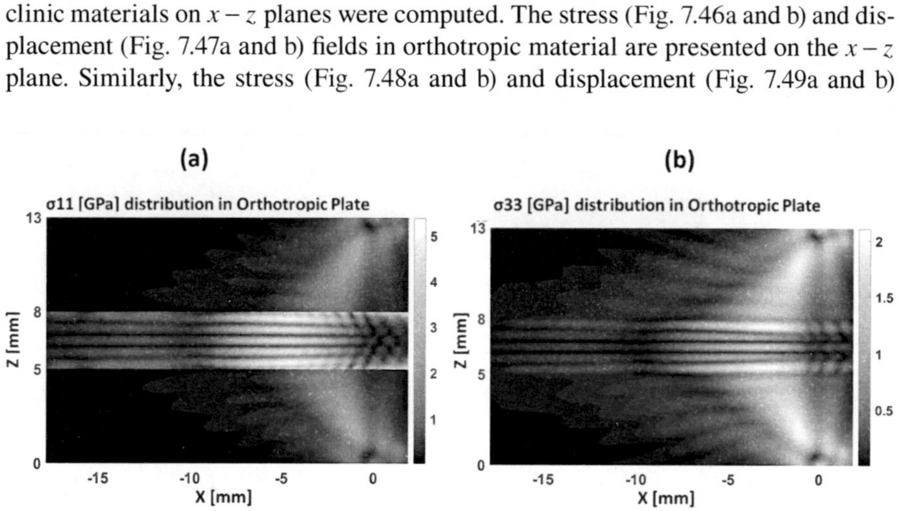

FIGURE 7.46 Wave field plot on x-z plane a) Stress11 (σ_{11}) distribution in fully orthotropic plate b) Stress33 (σ_{33}) distribution in fully orthotropic plate.

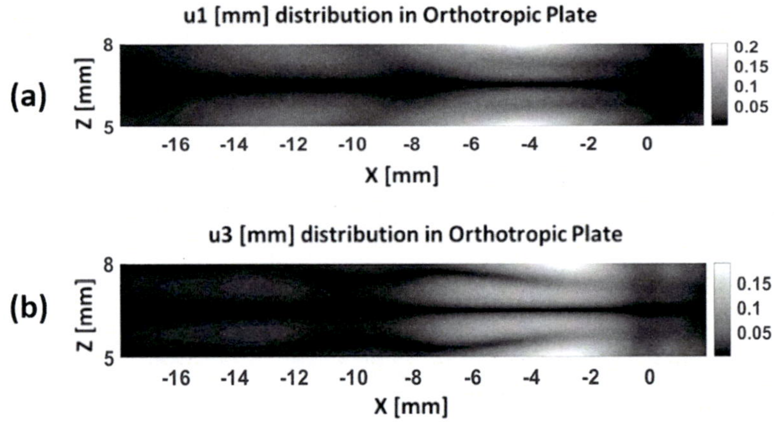

FIGURE 7.47 a) Displacement (u_1) distribution in fully orthotropic plate b) Displacement (u_3) distribution in fully orthotropic plate in GPa.

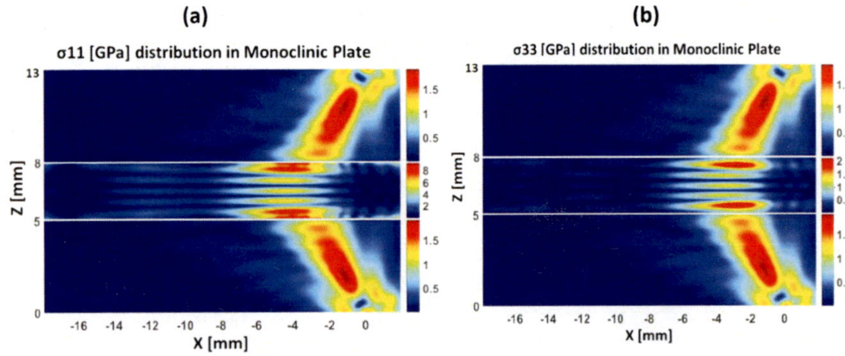

FIGURE 7.48 Wave field plot on x-z plane a) Stress11 (σ_{11}) distribution in monoclinic plate b) Stress33 (σ_{33}) distribution in monoclinic plate.

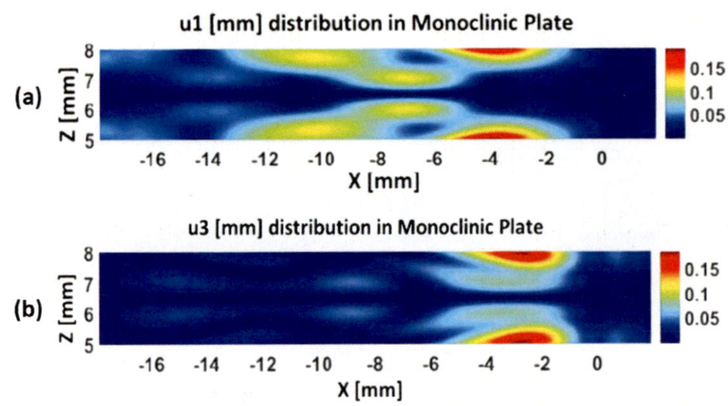

FIGURE 7.49 a) Displacement (u_1) distribution in monoclinic plate b) Displacement (u_3) distribution in fully monoclinic plate in GPa.

fields in monoclinic material are computed on the $x-z$ plane. Interestingly, it can be seen that the thirty degree angle of incidence is very close to the critical angle of incidence to develop symmetric guided wave modes in monoclinic material and it is evident through Figs. 7.48 and 7.49.

7.7 ENHANCING THE COMPUTATIONAL EFFICIENCY OF DPSM FOR ANISOTROPIC SOLIDS

7.7.1 Symmetry Informed Sequential Mapping of Anisotropic Green's function (SISMAG)

After implementing the Green's function expressions in Eqs. (7.223) or (7.209) and (7.226), it was realized that the calculation of Green's function due to multiple point sources is computationally demanding which makes DPSM slow and inefficient. This inefficiency comes precisely from the integration of the Green's function equation that must be performed over fine discretization of the spherical surface with θ and ϕ. This section is based on a publication by Shrestha and Banerjee [37] and modified in this present context. It is virtually impossible to compute the full wave field in the anisotropic media in a time efficient manner, even for a plate problem presented above in Section 7.6, was challenging. Calculated computation time for such a problem was ~180 hours in a standard desktop computer with 8 GB RAM, Intel i7 @2.8GHz processor. Therefore, to increase the time efficiency, a method called Symmetry Informed Sequential Mapping of Anisotropic Green's function (SISMAG) was implemented to create the matrix in Eq. (7.248). The technique is based on following two processes that increased the computational efficiency by 90 times. This is not even implementing the parallel computing. Implementation of parallel computing could enhance the capability further discussed later in this chapter.

7.7.1.1 SISMAG Step 1

The first proposition is based on a simple yet obvious argument that the Green's function at a target point due to a unit load acting on the source point is always same if the direction cosine of the line joining the source-target combination is constant. Based on this argument, the step 1 can be implemented in a computer code to automatically process the Green's function. Let us assume a set of n number of source points and m number of target points in a 2D system as shown in Figs. 7.37 and 7.50. To implement the sequential mapping of Green's function, the number of target points is extended from m to $(2m-1)$ as shown in Fig. 7.50a and b. According to the SISMAG process [37] the Green's functions at $(2m-1)$ extended target points, due to a point source at the central location is first calculated only once as shown in Fig. 7.50c. When the point source is located at the center of the pool of n source points then relevant or necessary target points are located at the middle of the pool of m target points as shown in Fig. 7.50b and c. However, if the central point source is moved all the way to the end of the pool of n source points, like it is shown in Fig. 7.50d, the necessary number of target points are equivalently located inside the domain, but rest of the target points are

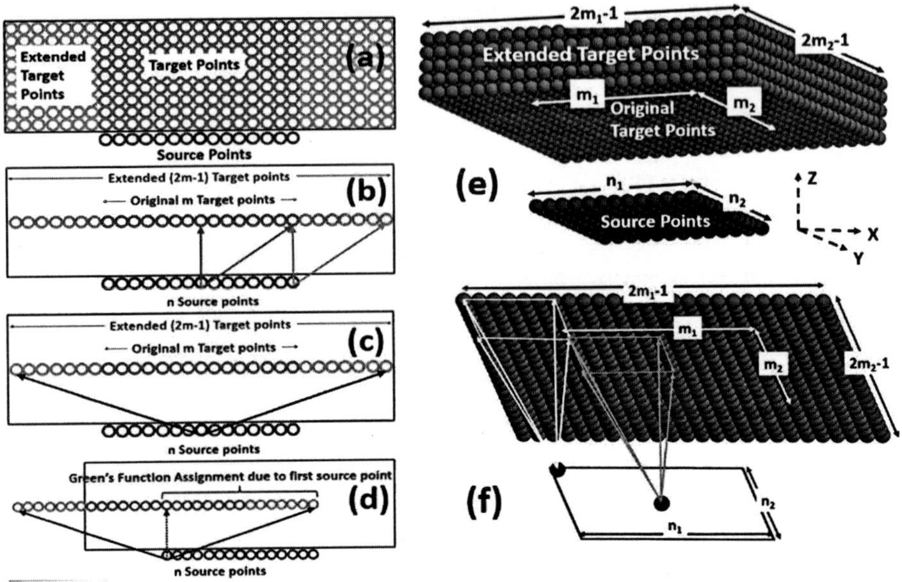

FIGURE 7.50 A schematics showing the sequential mapping of Green's function, a) source and target point combinations b) extended target points in 2D c) Calculated Green's function on extended point sources d) Assignment of Green's function for a first source e) source-target combinations in 3D, f) sequential mapping of 3D Green's function.

outside the domain and can be discarded. Similarly, in a step by step process the point sources can be moved inside the domain and respective usable target points, where the Green's functions are already calculated can be equivalently used in the calculation to compute the matrix. Hence, Green's function at m original target points is sequentially mapped and appropriately assigned by mimicking the central point source located at the location of the point sources below the actuation plane. For illustration, the allocation of $(2m-1)$ Green's function to the original m target point due to the first source point is shown in Fig. 7.50d. In case of first or end source points, the Green's function for m-th to $(2m-1)$-th point are allocated to the original target points as illustrated in Fig. 7.50d. Due to this technique, the process with 'conventional DPSM,' which originally demanded $(m*n)$ computations of Green's function is reduced to only $(2m-1)$ computations and hence reduced the computation time by the factor of $((2m-1)/m*n)$. Similarly, the mapping technique is implemented in 3D. In case of a 3D system, the number of target points is extended from $m1*m2$ to $(2m1-1)*(2m2-1)$ as shown in Fig. 7.50e and f. It can be seen that the Green's function at the target points inside the internal main block due to the central source point is same as the Green's function at the target points extended beyond the original domain of interest due to a corner source point as shown in Fig. 7.50f. Now the Green's functions at the $(2m1-1)*(2m2-1)$ extended target points are calculated due to a central point source,. Then $(2m1-1)*(2m2-1)$ Green's function is sequentially mapped at the original set of target points due to all $n1*n2$ number of point sources in a similar manner described in a 2D system. For 3D system, the computation time is reduced by the factor of $((2m1-1)*(2m2-1))/((m1*m2)*(n1*n2))$.

7.7.1.2 SISMAG Step 2

The second proposition to expedite the computation is with the symmetry of the Green's function at the bottom and top the fluid-solid interfaces due to respective point sources as shown in Fig. 7.51. It is true only if the material property between the interfaces is constant and the material is homogeneous. Hence, with the implementation of code, based on this argument, the Green's function needs to be calculated only at one interface, and it can be sequentially and symmetrically mapped on the other interfaces. To clarify the condition required for the implementation of this proposition, it is shown in Fig. 7.51b–e that it can be only implemented if the material is identical. Regarding a composite lamina, the symmetry can only be executed for lamina with identical fiber direction, i.e., 90° to 90°, 60° to 60°, 45° to 45° and so on. With this implementation, for each symmetric point source, an additional computational speedup is possible. This proposition holds true for any anisotropic material in general.

The relationship for the SISMAG implementation after employing the symmetry is tabulated in Table 7.2.

Where $\left(\mathbf{UAZ}_{ij}\right)_m^n$ is the displacement Green's function in solid at interface "i" due to point sources at interface "j" in the direction "m" due to the force in the direction

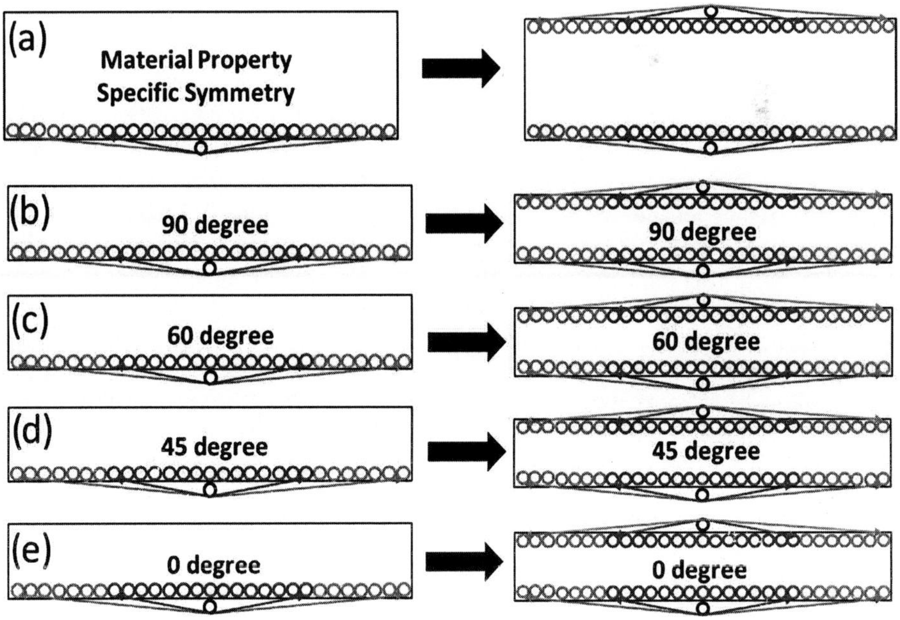

FIGURE 7.51 a) A schematics showing the SISMAG implementation in 2D b) SISMAG implementation in 90-degree c) SISMAG implementation in 60 degree d) SISMAG implementation in 45-degree e) SISMAG implementation in 0 degree.

TABLE 7.2
The Relation between Wave Fields at Interface 1 due to 1* Layer of Sources and Wave Field at Interface 1 due to Layer 2 Sources Following the Nomenclature in Fig. 7.27b

$(\mathbf{UAZ}_{11^*})_1^{\,1} = (\mathbf{UAZ}_{12})_1^{\,1}$	$(\mathbf{UAZ}_{11^*})_2^{\,1} = (\mathbf{UAZ}_{12})_2^{\,1}$	$(\mathbf{UAZ}_{11^*})_3^{\,1} = (\mathbf{UAZ}_{12})_3^{\,1}$
$(\mathbf{UAZ}_{11^*})_1^{\,2} = (\mathbf{UAZ}_{12})_1^{\,2}$	$(\mathbf{UAZ}_{11^*})_2^{\,2} = (\mathbf{UAZ}_{12})_2^{\,2}$	$(\mathbf{UAZ}_{11^*})_3^{\,2} = (\mathbf{UAZ}_{12})_3^{\,2}$
$(\mathbf{UAZ}_{11^*})_1^{\,3} = (\mathbf{UAZ}_{12})_1^{\,3}$	$(\mathbf{UAZ}_{11^*})_2^{\,3} = (\mathbf{UAZ}_{12})_2^{\,3}$	$(\mathbf{UAZ}_{11^*})_3^{\,3} = (\mathbf{UAZ}_{12})_3^{\,3}$
$(\Sigma\mathbf{11}_{11^*})^1 = (\Sigma\mathbf{11}_{12})^1$	$(\Sigma\mathbf{22}_{11^*})^1 = (\Sigma\mathbf{22}_{12})^1$	$(\Sigma\mathbf{33}_{11^*})^1 = (\Sigma\mathbf{33}_{12})^1$
$(\Sigma\mathbf{11}_{11^*})^2 = (\Sigma\mathbf{11}_{12})^2$	$(\Sigma\mathbf{22}_{11^*})^2 = (\Sigma\mathbf{22}_{12})^2$	$(\Sigma\mathbf{33}_{11^*})^2 = (\Sigma\mathbf{33}_{12})^2$
$(\Sigma\mathbf{11}_{11^*})^3 = (\Sigma\mathbf{11}_{12})^3$	$(\Sigma\mathbf{22}_{11^*})^3 = (\Sigma\mathbf{22}_{12})^3$	$(\Sigma\mathbf{33}_{11^*})^3 = (\Sigma\mathbf{33}_{12})^3$
$(\Sigma\mathbf{12}_{11^*})^1 = (\Sigma\mathbf{12}_{12})^1$	$(\Sigma\mathbf{31}_{11^*})^1 = (\Sigma\mathbf{31}_{12})^1$	$(\Sigma\mathbf{32}_{11^*})^1 = (\Sigma\mathbf{32}_{12})^1$
$(\Sigma\mathbf{12}_{11^*})^2 = (\Sigma\mathbf{12}_{12})^2$	$(\Sigma\mathbf{31}_{11^*})^2 = (\Sigma\mathbf{31}_{12})^2$	$(\Sigma\mathbf{32}_{11^*})^2 = (\Sigma\mathbf{32}_{12})^2$
$(\Sigma\mathbf{12}_{11^*})^3 = (\Sigma\mathbf{12}_{12})^3$	$(\Sigma\mathbf{31}_{11^*})^3 = (\Sigma\mathbf{31}_{12})^3$	$(\Sigma\mathbf{32}_{11^*})^3 = (\Sigma\mathbf{32}_{12})^3$

"n" and $\left(\Sigma \boldsymbol{RS}_{ij}\right)^n$ is the stress Green's function, σ_{rs} in solid at interface "i" due to point sources at interface "j" due to the force in the direction "n." In a similar manner, the relation for 21* and 22 can be formulated.

7.7.1.3 SISMAG Step 3

Although the symmetries mentioned above holds true for any generic anisotropic material, there exists additional symmetry for transversely isotropic and orthotropic materials as illustrated in Fig. 7.52. It is known that the transversely isotropic and orthotropic materials have three planes of symmetry, i.e., symmetry about x–, y–, and z-axes. Hence an additional quarter symmetry can be implemented for Green's functions at a set of target points on a plane due to any source points as shown in Fig. 7.52b. With this implementation, there is an additional computational speedup along with the previous cumulative speedups.

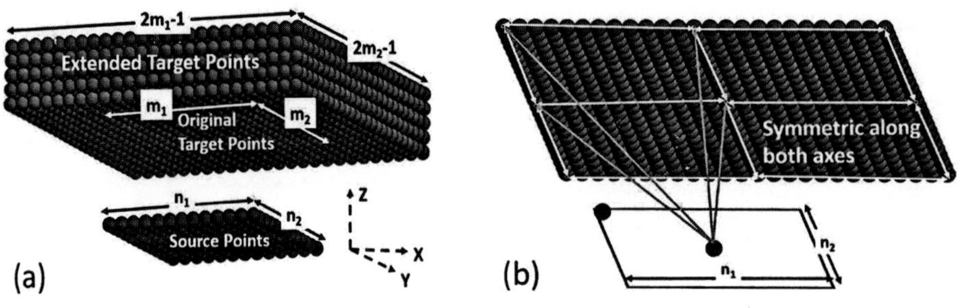

FIGURE 7.52 a) Source and target point distribution in 3D b) A 3D schematics illustrating additional symmetry in transversely isotropic and orthotropic materials.

7.8 COMPUTATION OF WAVE FIELDS IN MULTILAYERED ANISOTROPIC SOLIDS

A CNDE problem for a multilayered anisotropic plate is considered herein as shown in Fig. 7.53. Please note that x_1-, x_2-, and x_3- axes notations in Cartesian coordinate system introduced in Chapters 2 and 3 are synonymously used here with x, y, and z or X, Y, and Z coordinate notations, respectively. The scenario is simulated using DPSM. A 3-mm thick multilayered $(90/0)_2$ 4-ply composite plate immersed in water is considered to investigate using traditional pulse-echo and through transmission NDE mode. The plate is bounded by two solid–fluid interfaces. Two ~1 MHz transducers with 2 mm diameter are placed symmetrically on either side of the plate in water at 5 mm distance away from the fluid-solid interface. This problem is computationally demanding problem. Without SISMAG implementation it is almost impossible to solve. Hence, for this problem SISMAG is implemented as discussed in Section 7.7. To visualize the frequency domain wave phenomena inside anisotropic media, ultrasonic wave fields were simulated in the multilayered structure where each layer is considered transversely isotropic. Hence, based on the previous discussions, respective transversely isotropic Green's functions are used for the simulation. Following similar steps of DPSM described before in this chapter, $\mathbf{A}_l$ is the source strength vector of the point sources that are placed above the bottom of the fluid-solid interface, is responsible to generate ultrasonic field in the fluid below the plate due to the presence of the plate. The interface is called "Interface 1". $\mathbf{A}_{l*}$ is the source strength vector of the sources that are distributed below the bottom fluid-solid interface and models the transmitted field inside the solid plate. $\mathbf{A}_2$ and $\mathbf{A}_{2*}$ are the source strength vectors of the point sources that are distributed above

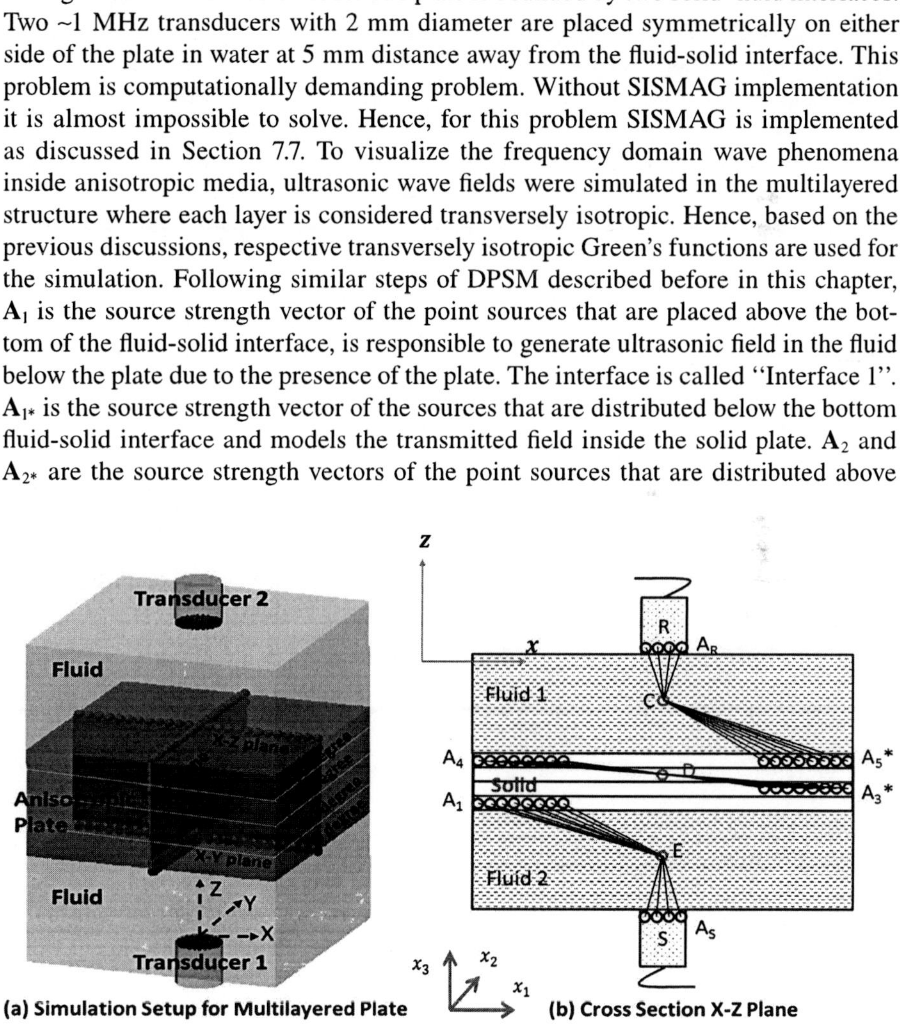

FIGURE 7.53 a) Schematics (not to scale) of the wave field computation problem in multilayered anisotropic plate, point sources distributed over the x-y plane, however, only two orthogonal line of point sources are shown b) cross-section view of the NDE problem along x-z plane.

and below the first solid–solid interface or second interface from the bottom, respectively. This interface is called "Interface 2". Similarly, $\mathbf{A}_3$ and $\mathbf{A}_{3*}$ are the source strength vectors of the point sources that are distributed above and below the second solid–solid interface or the third interface from the bottom, respectively. This interface is called "Interface 3". Also, $\mathbf{A}_4$ and $\mathbf{A}_{4*}$ are the source strength vectors of the point sources that have been distributed above and below the third solid–solid interface or fourth interface from the bottom, respectively. This interface is called "Interface 4". Finally, $\mathbf{A}_5$ and $\mathbf{A}_{5*}$ are the source strength vectors of the point sources that are distributed above and below the top fluid–solid interface, respectively. This interface is called "Interface 5". Transducer faces have source strength vectors $\mathbf{A}_S$ and $\mathbf{A}_R$. This process may continue for multilayered plates with more than four layers depicted in this problem in hand.

The total ultrasonic wave field at any arbitrary point C in the fluid 1 is computed by superposing the contributions from all the point sources denoted by $\mathbf{A}_s$ and $\mathbf{A}_1$. The total ultrasonic wave fields contributed by various set of point sources and their contributions at any arbitrary point D inside an anisotropic solid depends on where the arbitrary point is located. For illustration purpose, the total ultrasonic wave field at an arbitrary point D, which is in between the interface 3 and 4 are computed by superposing the contributions from each point sources with Green's function multiplied with their respective source strength denoted by $\mathbf{A}_{3*}$ and $\mathbf{A}_3$ (Fig. 7.53b). Similarly, the ultrasonic field can be computed inside solid at any points in between other interfaces as well. Finally, the total ultrasonic wave field at any arbitrary point E in the fluid 2 is computed by superposing the contributions from all the point sources denoted by $\mathbf{A}_{5*}$ and $\mathbf{A}_R$.

After distribution of point sources, the source strengths of the point sources distributed over the transducer and the interface are calculated by satisfying the boundary and the interface continuity conditions. The source strength depends on 3D force vectors for the actuation of point sources. The displacements or stresses at any target point are computed with the help of source strengths and Green's functions computed for that point. Hence, the source strengths $\mathbf{A}_s$, $\mathbf{A}_1$, $\mathbf{A}_{1*}$, $\mathbf{A}_2$, $\mathbf{A}_{2*}$, $\mathbf{A}_3$, $\mathbf{A}_{3*}$, $\mathbf{A}_4$, $\mathbf{A}_{4*}$, $\mathbf{A}_5$, $\mathbf{A}_{5*}$, and $\mathbf{A}_R$ needs to be calculated which are unknowns. The source strengths can be calculated by solving a set of linear equations obtained enforcing the boundary conditions which are discussed below.

Velocity boundary condition at the transducer faces can be written similar to Eqs. (7.237) and (7.238). On the other hand, the normal displacement at the fluid-solid interface should be continuous. The following displacement equations using displacement Green's function matrices composed of displacement Green's function in fluid and anisotropic solid in the z direction can be written as

$$\mathbf{UfZ}_{1S}\mathbf{A}_S + \mathbf{UfZ}_{11}\mathbf{A}_1 = \mathbf{UAZ}_{11}\mathbf{A}_{1^*} + \mathbf{UAZ}_{12}\mathbf{A}_2 \tag{7.252}$$

$$\mathbf{UfZ}_{55*}\mathbf{A}_{5*} + \mathbf{UfZ}_{5R}\mathbf{A}_R = \mathbf{UAZ}_{54*}\mathbf{A}_{4^*} + \mathbf{UAZ}_{55}\mathbf{A}_5 \tag{7.253}$$

where $\mathbf{UfZ}_{IS}$ is the displacement Green's function matrix for the target points at any interface I due to any point sources S spanning inside the fluid. In a similar manner, the displacement along all three directions at the solid-solid interfaces should also be continuous which can be written as

$$\mathbf{UAD}_{ij}\mathbf{A}_j + \mathbf{UAD}_{ii}\mathbf{A}_i = \mathbf{UAD}_{ik}\mathbf{A}_k + \mathbf{UAD}_{il}\mathbf{A}_l \tag{7.254}$$

where $\mathbf{UAD}_{IS}$ is the displacement Green's function matrix along "D" direction at any interface I due to any point sources S spanning inside the anisotropic solid. $\mathbf{D}$ = X,Y,Z representing 3 directions; i = 2, 3, and 4; $j = (i\ \text{-}1)^*$; $k = (i)^*$; $l = i + 1$. The respective Green's function matrices are explicitly written in Eqs. (7.242) and (7.243)

The compressive normal stress in the anisotropic solid and the fluid pressure at the fluid-solid interfaces should be continuous and thus equations for stresses and pressure can be written as

$$\mathbf{P}_{1S}\mathbf{A}_S + \mathbf{P}_{11}\mathbf{A}_1 = -\sum \mathbf{33}_{11}\mathbf{A}_{1^*} - \sum \mathbf{33}_{12}\mathbf{A}_2 \tag{7.255}$$

$$\mathbf{P}_{55^*}\mathbf{A}_{5^*} + \mathbf{P}_{5R}\mathbf{A}_R = -\sum \mathbf{33}_{54^*}\mathbf{A}_{4^*} - \sum \mathbf{33}_{55}\mathbf{A}_5 \tag{7.256}$$

Similarly, the compressive normal stresses at anisotropic solid-solid interfaces should be continuous and thus equations for stresses can be written as

$$\sum \mathbf{33}_{ij}\mathbf{A}_j + \sum \mathbf{33}_{ii}\mathbf{A}_i = \sum \mathbf{33}_{ik}\mathbf{A}_k + \sum \mathbf{33}_{il}\mathbf{A}_l \tag{7.257}$$

where i = 2, 3, and 4; $j = (i\ \text{-}1)^*$; $k = (i)^*$; $l = i + 1$.

Additionally, the shear stresses at the fluid-solid interfaces should vanish because fluid cannot take any shear stresses. Thus

$$\sum \mathbf{IJ}_{11^*}\mathbf{A}_{1^*} + \sum \mathbf{IJ}_{12}\mathbf{A}_2 = 0 \tag{7.258}$$

$$\sum \mathbf{IJ}_{54^*}\mathbf{A}_{4^*} + \sum \mathbf{IJ}_{55}\mathbf{A}_5 = 0 \tag{7.259}$$

where $\mathbf{IJ}$ = 31 and 32 represent the shear components.

However, the shear stresses at the solid-solid interfaces should be continuous. Thus

$$\sum \mathbf{IJ}_{ij}\mathbf{A}_j + \sum \mathbf{IJ}_{ii}\mathbf{A}_i = \sum \mathbf{IJ}_{ik}\mathbf{A}_k + \sum \mathbf{IJ}_{il}\mathbf{A}_l \tag{7.260}$$

where $\mathbf{IJ}$ = 31 AND 32 representing the shear components; i = 2, 3, and 4; $j = (i\ \text{-}1)^*$; $k = (i)^*$; $l = i + 1$.

$\mathbf{P}$ is the pressure Green's function matrix in fluid, $\Sigma\mathbf{33}$, $\Sigma\mathbf{31}$, and $\Sigma\mathbf{32}$ are the stress Green's function matrix in the anisotropic plate for σ_{33}, σ_{31}, and σ_{32} stresses obtained from Eq. (7.226), respectively. Rearranging the equations the following matrix can be obtained. After solving the linear algebra problem, the source strengths $\mathbf{A}_{l*}$, $\mathbf{A}_s$, and $\mathbf{A}_l$ can be obtained to calculate the wave field inside the fluid and solid.

$$
\begin{bmatrix}
VG_{SS} & VG_{S1} & Zeros & Zeros & Zeros & Zeros & Zeros & Zeros & Zeros & Zeros & Zeros & Zeros \\
P_{1S} & P_{11} & \Sigma33_{11*} & \Sigma33_{12} & Zeros & Zeros & Zeros & Zeros & Zeros & Zeros & Zeros & Zeros \\
UFZ_{1S} & UFZ_{11} & -UAZ_{11*} & -UAZ_{12} & Zeros & Zeros & Zeros & Zeros & Zeros & Zeros & Zeros & Zeros \\
Zeros & Zeros & \Sigma31_{11*} & \Sigma31_{12} & Zeros & Zeros & Zeros & Zeros & Zeros & Zeros & Zeros & Zeros \\
Zeros & Zeros & \Sigma32_{11*} & \Sigma32_{12} & Zeros & Zeros & Zeros & Zeros & Zeros & Zeros & Zeros & Zeros \\
Zeros & Zeros & \Sigma31_{21*} & \Sigma31_{22} & -\Sigma31_{22*} & -\Sigma31_{23} & Zeros & Zeros & Zeros & Zeros & Zeros & Zeros \\
Zeros & Zeros & \Sigma32_{21*} & \Sigma32_{22} & -\Sigma32_{22*} & -\Sigma32_{23} & Zeros & Zeros & Zeros & Zeros & Zeros & Zeros \\
Zeros & Zeros & \Sigma33_{21*} & \Sigma33_{22} & -\Sigma33_{22*} & -\Sigma33_{23} & Zeros & Zeros & Zeros & Zeros & Zeros & Zeros \\
Zeros & Zeros & UAX_{21*} & UAX_{22} & -UAX_{22*} & -UAX_{23} & Zeros & Zeros & Zeros & Zeros & Zeros & Zeros \\
Zeros & Zeros & UAX_{21*} & UAX_{22} & -UAX_{22*} & -UAX_{23} & Zeros & Zeros & Zeros & Zeros & Zeros & Zeros \\
Zeros & Zeros & UAX_{21*} & UAX_{22} & -UAX_{22*} & -UAX_{23} & Zeros & Zeros & Zeros & Zeros & Zeros & Zeros \\
Zeros & Zeros & Zeros & Zeros & \Sigma31_{32*} & \Sigma31_{33} & -\Sigma31_{33*} & -\Sigma31_{34} & Zeros & Zeros & Zeros & Zeros \\
Zeros & Zeros & Zeros & Zeros & \Sigma32_{32*} & \Sigma32_{33} & -\Sigma32_{33*} & -\Sigma32_{34} & Zeros & Zeros & Zeros & Zeros \\
Zeros & Zeros & Zeros & Zeros & \Sigma33_{32*} & \Sigma33_{33} & -\Sigma33_{33*} & -\Sigma33_{34} & Zeros & Zeros & Zeros & Zeros \\
Zeros & Zeros & Zeros & Zeros & UAX_{32*} & UAX_{33} & -UAX_{33*} & -UAX_{34} & Zeros & Zeros & Zeros & Zeros \\
Zeros & Zeros & Zeros & Zeros & UAY_{32*} & UAY_{33} & -UAY_{33*} & -UAY_{34} & Zeros & Zeros & Zeros & Zeros \\
Zeros & Zeros & Zeros & Zeros & UAZ_{32*} & UAZ_{33} & -UAZ_{33*} & -UAZ_{34} & Zeros & Zeros & Zeros & Zeros \\
Zeros & Zeros & Zeros & Zeros & Zeros & Zeros & \Sigma31_{43*} & \Sigma31_{44} & -\Sigma31_{44*} & -\Sigma31_{45} & Zeros & Zeros \\
Zeros & Zeros & Zeros & Zeros & Zeros & Zeros & \Sigma32_{43*} & \Sigma32_{44} & -\Sigma32_{44*} & -\Sigma32_{45} & Zeros & Zeros \\
Zeros & Zeros & Zeros & Zeros & Zeros & Zeros & \Sigma33_{43*} & \Sigma33_{44} & -\Sigma33_{44*} & -\Sigma33_{45} & Zeros & Zeros \\
Zeros & Zeros & Zeros & Zeros & Zeros & Zeros & UAX_{43*} & UAX_{44} & -UAX_{44*} & -UAX_{44*} & Zeros & Zeros \\
Zeros & Zeros & Zeros & Zeros & Zeros & Zeros & UAY_{43*} & UAY_{44} & -UAY_{44*} & -UAY_{44*} & Zeros & Zeros \\
Zeros & Zeros & Zeros & Zeros & Zeros & Zeros & UAZ_{43*} & UAZ_{44} & -UAZ_{44*} & -UAZ_{44*} & Zeros & Zeros \\
Zeros & Zeros & Zeros & Zeros & Zeros & Zeros & Zeros & Zeros & \Sigma31_{54*} & \Sigma31_{55} & Zeros & Zeros \\
Zeros & Zeros & Zeros & Zeros & Zeros & Zeros & Zeros & Zeros & \Sigma32_{54*} & \Sigma32_{55} & Zeros & Zeros \\
Zeros & Zeros & Zeros & Zeros & Zeros & Zeros & Zeros & Zeros & -\Sigma33_{54*} & -\Sigma33_{55} & P_{55*} & P_{5R} \\
Zeros & Zeros & Zeros & Zeros & Zeros & Zeros & Zeros & Zeros & -UAZ_{54*} & -UAZ_{55} & UFZ_{55*} & UFZ_{5R} \\
Zeros & Zeros & Zeros & Zeros & Zeros & Zeros & Zeros & Zeros & Zeros & Zeros & VG_{R5*} & VG_{RR}
\end{bmatrix}
\times
\begin{Bmatrix}
A_S \\ A_1 \\ A_{1*} \\ A_2 \\ A_{2*} \\ A_3 \\ A_{3*} \\ A_4 \\ A_{4*} \\ A_5 \\ A_{5*} \\ A_R
\end{Bmatrix}
=
\begin{Bmatrix}
V_{S0} \\ Zeros \\ Zeros \\ Zeros \\ Zeros \\ Zeros \\ Zeros \\ Zeros \\ Zeros \\ Zeros \\ Zeros \\ V_{R0}
\end{Bmatrix}
\tag{7.261}
$$

where *Zeros* indicate the matrix of 0s as required in the equation.

After obtaining the source strengths, wave field inside fluid and solid media are computed like the process described in this chapter in Sections 7.4 and 7.5.

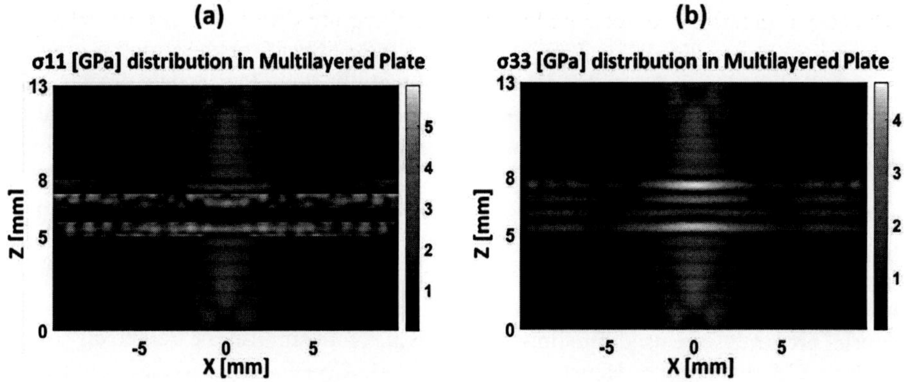

FIGURE 7.54 a) Stress11 (σ11) distribution in Pristine Multilayered plate b) Stress33 (σ33) distribution in Multilayered plate in GPa.

7.8.1 Wave Field Modeling in Pristine 4-ply Composite Plate

The stress and displacement wave fields are calculated in a healthy or pristine 4-ply plate, i.e., (90/0/90/0) plate. In the stress field plot, the lower and upper section (5 mm each) of the plots represent the pressure fields inside the water and the midsection (3 mm) represents the stress field inside the solid plate. The wave fields are shown for normal incidence from both side of the plate for illustration purpose. The stresses (σ_{11} and σ_{33}) and pressure fields in 4-ply plate, which are shown in Fig. 7.54a and b, respectively, were computed. Similarly, the displacement fields (u_1 and u_3) are shown in Fig. 7.55a and b.

7.8.2 Wave Field Modeling in Degraded 4-ply Composite Plate

7.8.2.1 Material Degradation

Damage and Degradation are two different terms associated with materials. Let us define that a material is said to be damaged when there are visible or measurable discontinuities or inclusions in the material at macro scale (say for example, size > 1 mm). Discontinuities and inclusions may occur in the materials from material usage or from manufacturing defects and they may span across different scales, starting from submicron to micron to millimeter to several millimeters. Usually macro scale defects or discontinuities are traditionally called damage or defects in NDE terminologies. These damages in respective material types are in the form of cracks,

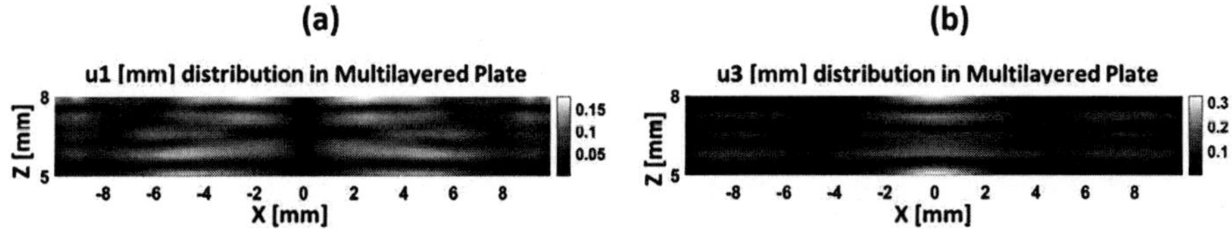

FIGURE 7.55 a) Displacement (u1) distribution in Pristine Multilayered plate b) Displacement (u3) distribution in Multilayered plate in GPa.

voids, dents, delamination etc. At lower scales there are different types of defects, like micron scale voids, dislocations, grain boundaries, fiber breakage, matrix cracking, etc., in their respective material systems. Generally, these lower scale defects are not considered as damages in NDE and are ignored. It is assumed that such small-scale defects do not affect the ultrasonic signals. Primary basis for this assumption is that, if the defects are several orders smaller than the wave length of the ultrasonic wave used in the NDE inspection, then their effects on the signals would be negligible. Which might be true, however, under CNDE this assumption could be tested for accuracy. Or a new frame work can be developed where, ensemble effect of such lower scale defects can be included in the models. In CNDE, always the macro scale material properties are used for simulations. Inclusion of small-scale defects in wave simulation in challenging. Hence, a new multiscale CNDE frame work is required where material properties of bulk material can be analyzes at the smaller scales, where small scale defects are included in a representative volume elements (RVE) and representative ensemble material properties can be calculated. Unlike material modeling or failure modeling where only few representative material properties are evaluated, in CNDE to simulate wave propagation in three dimensions, a complete constitutive matrix with each material coefficients are required to be evaluated under the influence of the defects. A, such framework in presented in Fig. 7.56. How such frame work can link the Digital NDE pipeline with progressive failure model is also demonstrated. Ensemble material properties with defects are called degraded material properties. Such degraded material properties can be used for CNDE simulation. A brief method to find degraded material properties is presented in the Appendix of this chapter. Using the method presented in Appendix, 0° and 90° alternate plies are considered to be degraded in the following simulation.

7.8.2.2 Wave Field in 4 ply Composite Plate with 0° and 90° Degraded Plies

A 5% void content is considered in the layers and following degraded percentages were used to degrade or reduce the respective material coefficients that are originally used in Chapter 4 for transversely isotropic material in Section 4.4.2.4.

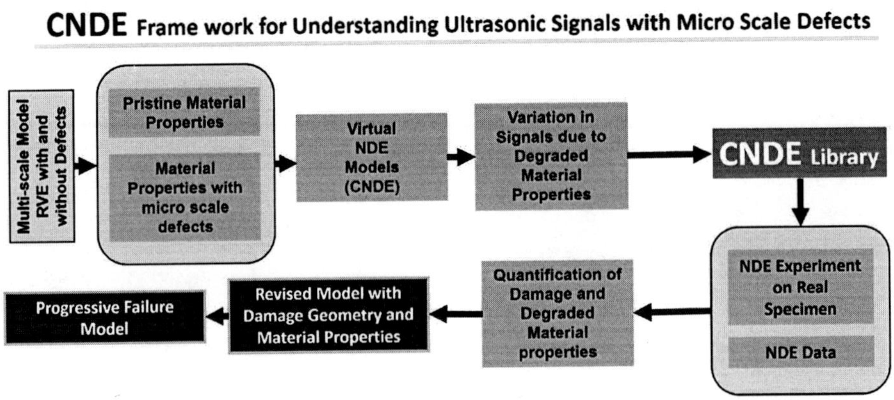

FIGURE 7.56 A CNDE framework for incorporating microscale defects in model.

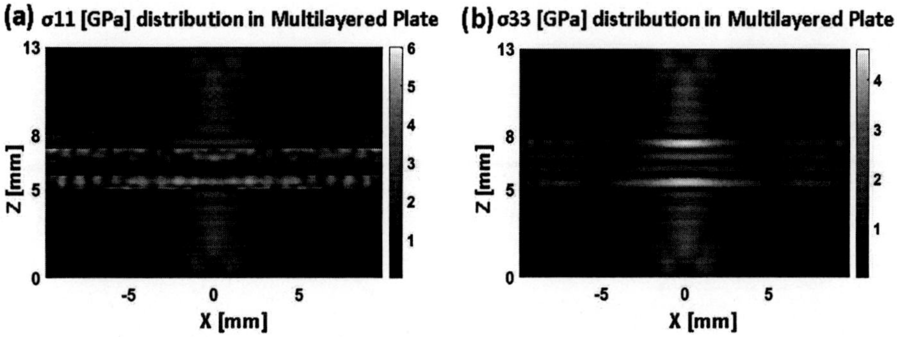

FIGURE 7.57 a) Stress11 (σ11) distribution in 0 and 90 degree degraded Multilayered plate b) Stress33 (σ33) distribution in Multilayered plate in GPa.

$$C^{5\%void} = \begin{bmatrix} -8.8\% & -4.2\% & -10.1\% & 0 & 0 & 0 \\ -4.2\% & -8.2\% & -10.1\% & 0 & 0 & 0 \\ -10.1\% & -10.1\% & -11.9\% & 0 & 0 & 0 \\ 0 & 0 & 0 & -8.5\% & 0 & 0 \\ 0 & 0 & 0 & 0 & -8.3\% & 0 \\ 0 & 0 & 0 & 0 & 0 & -8.8\% \end{bmatrix} \tag{7.262}$$

After calculating wave field in pristine composite material, the stress and displacement wave fields are calculated in a 4-ply plate, i.e., (90/0-degraded/90-degraded/0) plate where two adjacent 0° and 90° layers are considered degraded. In the stress field plot, the lower and upper section (5 mm each) of the plots represent the pressure fields inside the water and the midsection (3 mm) represents the stress field inside the solid plate. The wave fields are shown for normal incidence case. The stresses (σ_{11} and σ_{33}) and pressure fields are presented in composite material with transversely isotropic layers in Fig. 7.57a and b. Similarly, the displacement fields (u_1 and u_3) are shown in Fig. 7.58a and b. Comparing the wave fields in Figs. 7.54 and 7.57, not much difference could be identified except very small change in magnitude. Hence, to understand if there is any difference, more simulations are conducted (not shown here) with only 0° ply, only 90° ply, and both 0° and 90° ply degraded. Wave fields are plotted along a central line shown in Fig. 7.59a. The plot is compared between pristine, 0° degraded, 90° degraded, and 0° and 90° degraded plate as shown in Fig. 7.59b.

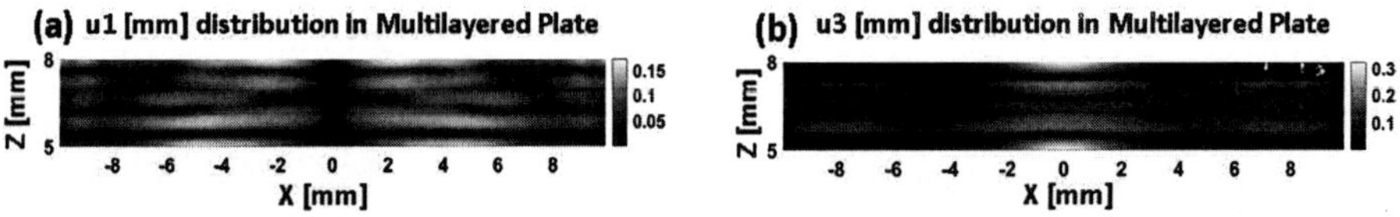

FIGURE 7.58 a) Displacement (u1) distribution in 0 and 90 degree degraded Multilayered plate b) Displacement (u3) distribution in Multilayered plate in GPa.

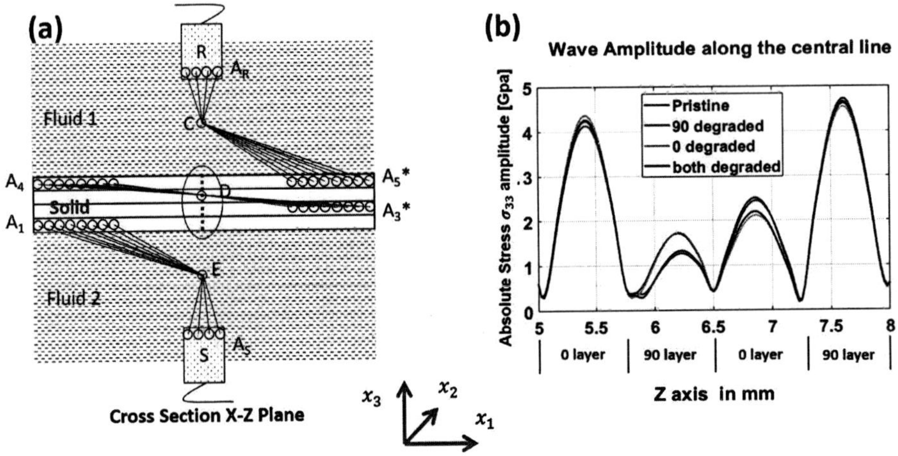

FIGURE 7.59 a) cross-section view of the NDE problem along x-z plane showcasing the central line where the wave amplitude is plotted. b) Line Plot of wave amplitude along the central line.

7.9 COMPUTATION OF WAVE FIELDS IN THE PRESENCE OF DELAMINATION IN COMPOSITE

In the previous section, material degradation as a damage scenario is simulated using DPSM to demonstrate that even small material perturbation could affect the wave field. However, if such changes can really affect the ultrasonic signal is still to be determined and a topic of ongoing research. Previous section presents a framework for DPSM where understanding the presence of different damage scenarios is possible. In this section, most critical type of damage scenario, i.e., delamination is considered and how to incorporate them in DPSM simulation is demonstrated followed by computed wave fields in two-layer delaminated composite. If a composite plate with multiple layers is considered, delamination is the defect/damage that is caused by the separation of the adjoining layers. The delamination mostly occurs due to consequence of repeated impacts, notches, manufacturing defects, stress concentrations due to structural defects or failure of the adhesive or matrix between the lamina. Please note that in the following discussion, again x_1, x_2, and x_3 axes notations in Cartesian coordinate system introduced in Chapters 2 and 3 are synonymously used here with x, y, and z or X, Y, and Z coordinate notations, respectively.

7.9.1 DELAMINATION IN DPSM

Delamination occurs between two layers. When implementing the DPSM technique, it is necessary to consider the effect of both delaminated and nondelaminated portions of the layers to calculate the total ultrasonic wave field. To schematically demonstrate this consideration, a 2-layered plate with the delamination between the layers is considered. For demonstration purpose, Fig. 7.60 shows a 2-layered plate with an exaggerated delamination between the layers. It has three interfaces 1, 2, and 3.

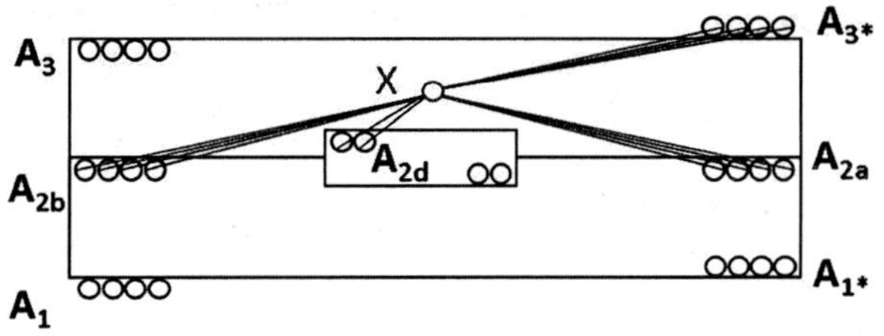

FIGURE 7.60 Schematics of 2-layered plate with delamination showcasing the point source distribution and ultrasonic field calculation.

Following the very nature of DPSM method shown in Fig. 7.8, the point sources below the interfaces, i.e., A_1, A_{2b}, A_{2d}, A_{2a}, and A_3 are responsible for the transmitted wave field and contribute to the displacements and stresses at X. Similarly point sources above the interfaces, i.e., A_{1*} and A_{3*} are responsible for the transmitted wave field and contribute to the displacements and stresses at X. The point sources at the interface with delamination are separated into several smaller components such as A_{2b}, A_{2d}, and A_{2a} where "b," "d," and "a" represents point sources before, during, and after delamination respectively. Therefore, the total ultrasonic field at any point "X" is the summation of ultrasonic field generated from nondelaminated ("b," "a") and delaminated ("d") sections of the interfaces with delamination and other required interfaces. Hence, in the interface with delamination, instead of distributing the point source as a single component, it is distributed into three components, i.e., before delamination, during delamination, and after delamination. Hence, the total ultrasonic field at point "X" is calculated as follows:

$$\mathbf{T}_X = \mathbf{T}_{X2b}\mathbf{A}_{2b} + \mathbf{T}_{X2d}\mathbf{A}_{2d} + \mathbf{T}_{X2a}\mathbf{A}_{2a} + \mathbf{T}_{X3*}\mathbf{A}_{3*} \tag{7.263}$$

It can be interpreted as the total ultrasonic field at "X" which is the summation of the ultrasonic field contributed by "2b," "2d," "2a," and "3*" where "2b," "2d," and "2a" generates the transmitted signal and "3*" cause the reflected signal.

7.9.2 Incorporation of Delamination Formulation in DPSM for CNDE

A CNDE problem for a 2-layered composite plate is shown in Fig. 7.61. The problem is modeled using DPSM. A 3-mm thick (90/0) 2-ply composite plate immersed in water with delamination between the ply is considered to investigate using traditional pulse-echo or through transmission NDE mode. Thus, the plate is bounded by two solid–fluid interfaces. Two ~1 MHz transducers with 2 mm diameter is placed symmetrically on either side of the plate in water at 5 mm distance from the fluid-solid interface. The point sources are distributed behind the transducer faces (Fig. 7.61) and distributed on either side of the interface between the solid and fluid media by implementing DPSM approach discussed throughout this chapter. The total ultrasonic field in fluid and solid media is the superposition of contribution of the point sources. $\mathbf{A}_1$ is the source strength vector of the point sources that are placed above the bottom solid–fluid interface and generate the ultrasonic reflected field in

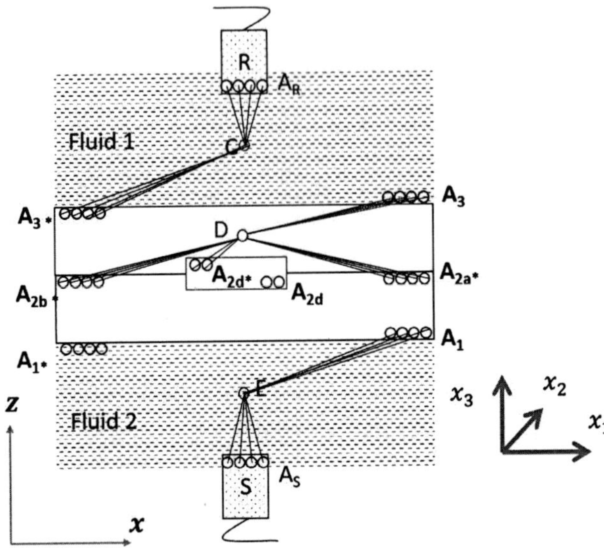

FIGURE 7.61 CNDE wavefield computation problem: A 2D Schematics (not to scale) of a (0/90) configuration 2-layered plate, point sources distributed above and below the interfaces.

the fluid below the plate. The interface is called "Interface 1". $\mathbf{A}_{1*}$ is the source strength vector of the sources that are distributed below the bottom solid–fluid interface and model the transmitted field in the solid plate like the problems discussed previously. However here, $\mathbf{A}_{2b}$, $\mathbf{A}_{2d}$, $\mathbf{A}_{2a}$, and $\mathbf{A}_{2b*}$, $\mathbf{A}_{2d*}$, $\mathbf{A}_{2a*}$ are the source strength vectors of the point sources that are distributed above and below the first solid–solid interface or second interface from the bottom, respectively. The point sources are distributed in a different manner to accommodate the presence of delamination in the interface. This interface is called "Interface 2". $\mathbf{A}_3$ and $\mathbf{A}_{3*}$ are the source strength vectors of the point sources that are distributed above and below the second solid–solid interface or third interface from the bottom, respectively. This interface is called "Interface 3". Transducer faces have source strength vectors $\mathbf{A}_S$ and $\mathbf{A}_R$ like their distribution are discussed in some previous sections. The total ultrasonic wave field at any arbitrary point E in the fluid 1 is computed by superposing the contributions from all the point sources denoted by $\mathbf{A}_s$ and $\mathbf{A}_1$. The total ultrasonic wave fields at any arbitrary point D inside an anisotropic solid depends on where the arbitrary point is located. For illustration purpose, the total ultrasonic wave field at the arbitrary point D which lies in between interface 2 and 3 are computed by superposing the contributions from each point sources with Green's function multiplied with their respective source strength denoted by $\mathbf{A}_{2b*}$, $\mathbf{A}_{2d*}$, $\mathbf{A}_{2a*}$, and $\mathbf{A}_3$ (Fig. 7.61). Similarly, the ultrasonic field can be computed for the solid points in between other interfaces as well considering relevant point source vectors that can see the point. Visibility of a point by point sources is discussed and depicted in Fig. 7.3. Finally, the total ultrasonic wave field at any arbitrary point C in the fluid 2 is computed by superposing the contributions from all the point sources denoted by $\mathbf{A}_{3*}$ and $\mathbf{A}_R$. The source strength depends on 3D force vectors for the actuation of the point sources. The displacement or stress field at any target point is computed with the help of source strengths and Green's function at that point. Hence, the source strengths $\mathbf{A}_s$, $\mathbf{A}_1$, $\mathbf{A}_{1*}$, $\mathbf{A}_{2b}$, $\mathbf{A}_{2d}$, $\mathbf{A}_{2a}$, $\mathbf{A}_{2b*}$, $\mathbf{A}_{2d*}$, $\mathbf{A}_{2a*}$, $\mathbf{A}_3$, $\mathbf{A}_{3*}$, and $\mathbf{A}_R$ needs to be calculated which are unknown. The source strengths can be calculated by satisfying the boundary conditions which are discussed below.

The bottom transducer is simulated with a prescribed velocity on the circular face equal to V_{S0} and top transducer with V_{R0}. At the solid-fluid interfaces 1 and 3, the normal displacement and the normal stresses along z direction should be continuous, however the shear stresses must vanish. The prescribed normal velocities V_{S0} and V_{R0} on the transducer faces can be obtained by superposing the contribution of the velocity fields created by $\mathbf{A}_s$, $\mathbf{A}_I$ and $\mathbf{A}_{3*}$, $\mathbf{A}_R$ sources, respectively. By calculating the fluid domain Green's functions on the target points, the velocity equation will be

$$\mathbf{VG}_{SS}\mathbf{A}_S + \mathbf{VG}_{S1}\mathbf{A}_2 = \mathbf{V}_{S0} \tag{7.264}$$

$$\mathbf{VG}_{R3*}\mathbf{A}_{3*} + \mathbf{VG}_{RR}\mathbf{A}_R = \mathbf{V}_{R0} \tag{7.265}$$

The velocity Green's function matrices in fluid **VG** are explicitly written in Eq. (7.239). On the other hand, the normal displacement at the fluid-solid interface should be continuous. The following displacement equations using displacement Green's function matrices composed of displacement Green's function in fluid and anisotropic solid in the z direction can be written as

$$\mathbf{UfZ}_{1S}\mathbf{A}_S + \mathbf{UfZ}_{11}\mathbf{A}_1 = \mathbf{UAZ}_{11}\mathbf{A}_{1^*} + \mathbf{UAZ}_{12}\mathbf{A}_2 \tag{7.266}$$

$$\mathbf{UfZ}_{33*}\mathbf{A}_{3*} + \mathbf{UfZ}_{3R}\mathbf{A}_R = \mathbf{UAZ}_{32*}\mathbf{A}_{2^*} + \mathbf{UAZ}_{33}\mathbf{A}_3 \tag{7.267}$$

In a similar manner, the displacement along all three directions at the solid-solid interfaces should also be continuous in all three portions, i.e., before, during, and after delamination which can be written as

$$\mathbf{UAD}_{2b1*}\mathbf{A}_{1*} + \mathbf{UAD}_{2b2b}\mathbf{A}_{2b} + \mathbf{UAD}_{2b2a}\mathbf{A}_{2a} = \mathbf{UAD}_{2b2b*}\mathbf{A}_{2b^*} + \mathbf{UAD}_{2b2a*}\mathbf{A}_{2b^*} + \mathbf{UAD}_{2b3}\mathbf{A}_3 \tag{7.268}$$

$$\mathbf{UAD}_{2a1*}\mathbf{A}_{1*} + \mathbf{UAD}_{2a2b}\mathbf{A}_{2b} + \mathbf{UAD}_{2a2a}\mathbf{A}_{2a} = \mathbf{UAD}_{2a2b*}\mathbf{A}_{2b^*} + \mathbf{UAD}_{2a2a*}\mathbf{A}_{2b^*} + \mathbf{UAD}_{2a3}\mathbf{A}_3 \tag{7.269}$$

where $\mathbf{UfZ}_{IS}$ is the displacement Green's function matrix for the target points at any interface I due to any point sources S spanning inside the fluid. Similarly, $\mathbf{UAD}_{IS}$ is the displacement Green's function matrix along "D," i.e., "X," "Y," or "Z" direction at any interface I due to the any point source S spanning inside the anisotropic solid. The respective Green's function matrices are explicitly written in Eqs. (7.242) and (7.243), respectively. The compressive normal stress in the anisotropic solid and the fluid pressure at the fluid-solid interfaces should be continuous and thus equations for stress and pressure can be written as

$$\mathbf{P}_{1S}\mathbf{A}_S + \mathbf{P}_{11}\mathbf{A}_1 = -\sum \mathbf{33}_{11}\mathbf{A}_{1^*} - \sum \mathbf{33}_{12b}\mathbf{A}_{2b} - \sum \mathbf{33}_{12d}\mathbf{A}_{2d} - \sum \mathbf{33}_{12a}\mathbf{A}_{2a} \tag{7.270}$$

$$\mathbf{P}_{33^*}\mathbf{A}_{3^*} + \mathbf{P}_{3R}\mathbf{A}_R = -\sum \mathbf{33}_{32b^*}\mathbf{A}_{2b^*} - \sum \mathbf{33}_{32d^*}\mathbf{A}_{2d^*} - \sum \mathbf{33}_{32a^*}\mathbf{A}_{2a^*} - \sum \mathbf{33}_{33}\mathbf{A}_3 \tag{7.271}$$

Similarly, the compressive normal stresses in the anisotropic solid at the solid-solid interfaces should be continuous and thus equations for stresses can be written as

$$\sum \mathbf{33}_{2b1^*}\mathbf{A}_{1^*} + \sum \mathbf{33}_{2b2b}\mathbf{A}_{2b} + \sum \mathbf{33}_{2b2a}\mathbf{A}_{2a} = \sum \mathbf{33}_{2b2b^*}\mathbf{A}_{2b^*} + \sum \mathbf{33}_{2b2a^*}\mathbf{A}_{2b^*} + \sum \mathbf{33}_{2b3}\mathbf{A}_3 \tag{7.272}$$

$$\sum \mathbf{33}_{2d1^*}\mathbf{A}_{1^*} + \sum \mathbf{33}_{2d2b}\mathbf{A}_{2b} + \sum \mathbf{33}_{2d2d}\mathbf{A}_{2d} + \sum \mathbf{33}_{2d2a}\mathbf{A}_{2a} = 0 \tag{7.273}$$

$$\sum \mathbf{33}_{2a1^*}\mathbf{A}_{1^*} + \sum \mathbf{33}_{2a2b}\mathbf{A}_{2b} + \sum \mathbf{33}_{2a2a}\mathbf{A}_{2a} = \sum \mathbf{33}_{2a2b^*}\mathbf{A}_{2b^*} + \sum \mathbf{33}_{2a2a^*}\mathbf{A}_{2b^*} + \sum \mathbf{33}_{2a3}\mathbf{A}_3 \tag{7.274}$$

$$\sum \mathbf{33}_{2d^*2b^*}\mathbf{A}_{2b^*} + \sum \mathbf{33}_{2d^*2d^*}\mathbf{A}_{2d^*} + \sum \mathbf{33}_{2d^*2a^*}\mathbf{A}_{2a^*} + \sum \mathbf{33}_{2d^*3}\mathbf{A}_3 = 0 \tag{7.275}$$

Additionally, the shear stresses at the fluid-solid interface should vanish because fluid cannot take any shear stresses. Thus

$$\sum \mathbf{IJ}_{11^*}\mathbf{A}_{1^*} + \sum \mathbf{IJ}_{12b}\mathbf{A}_{2b} + \sum \mathbf{IJ}_{12d}\mathbf{A}_{2d} + \sum \mathbf{IJ}_{12a}\mathbf{A}_{2a} = 0 \tag{7.276}$$

$$\sum \mathbf{IJ}_{32b}\mathbf{A}_{2b^*} + \sum \mathbf{IJ}_{32d}\mathbf{A}_{2d^*} + \sum \mathbf{IJ}_{32a}\mathbf{A}_{2a^*} + \sum \mathbf{IJ}_{33}\mathbf{A}_3 = 0 \tag{7.277}$$

where IJ = 31 and 32 representing the shear components.

However, the shear stresses at the solid-solid interfaces should be continuous. Thus

$$\sum \mathbf{IJ}_{2b1^*}\mathbf{A}_{1^*} + \sum \mathbf{IJ}_{2b2b}\mathbf{A}_{2b} + \sum \mathbf{IJ}_{2b2a}\mathbf{A}_{2a} = \sum \mathbf{IJ}_{2b2b^*}\mathbf{A}_{2b^*} + \sum \mathbf{IJ}_{2b2a^*}\mathbf{A}_{2b^*} + \sum \mathbf{IJ}_{2b3}\mathbf{A}_3 \tag{7.278}$$

$$\sum \mathbf{IJ}_{2d1^*}\mathbf{A}_{1^*} + \sum \mathbf{IJ}_{2d2b}\mathbf{A}_{2b} + \sum \mathbf{IJ}_{2d2d}\mathbf{A}_{2d} + \sum \mathbf{IJ}_{2d2a}\mathbf{A}_{2a} = 0 \tag{7.279}$$

$$\sum \mathbf{IJ}_{2a1^*}\mathbf{A}_{1^*} + \sum \mathbf{IJ}_{2a2b}\mathbf{A}_{2b} + \sum \mathbf{IJ}_{2a2a}\mathbf{A}_{2a} = \sum \mathbf{IJ}_{2a2b^*}\mathbf{A}_{2b^*} + \sum \mathbf{IJ}_{2a2a^*}\mathbf{A}_{2b^*} + \sum \mathbf{IJ}_{2a3}\mathbf{A}_3 \tag{7.280}$$

$$\sum \mathbf{IJ}_{2d^*2b^*}\mathbf{A}_{2b^*} + \sum \mathbf{IJ}_{2d^*2d^*}\mathbf{A}_{2d^*} + \sum \mathbf{IJ}_{2d^*2a^*}\mathbf{A}_{2a^*} + \sum \mathbf{IJ}_{2d^*3}\mathbf{A}_3 = 0 \tag{7.281}$$

where IJ = 31 and 32 representing the shear components.

P is the pressure Green's function matrix in fluid, $\boldsymbol{\Sigma 33}$, $\boldsymbol{\Sigma 31}$, and $\boldsymbol{\Sigma 32}$ are the stress Green's function matrix in the anisotropic plate for σ_{33}, σ_{31}, and σ_{32} stresses like they are described in Sections 7.5.2.1 and 7.6.2. Rearranging the equations from Eqs. (7.264) to (7.281), we get the matrix depicted in Eq. (7.282). After solving the linear algebra problem, the source strengths are obtained to calculate the wave field inside the fluid and solid.

$$
\begin{bmatrix}
VG_{SS} & VG_{S1} & Zeros & Zeros & Zeros & Zeros & Zeros & Zeros & Zeros & Zeros & Zeros & Zeros \\
P_{1S} & P_{11} & \Sigma 33_{11*} & \Sigma 33_{12} & Zeros & Zeros & Zeros & Zeros & Zeros & Zeros & Zeros & Zeros \\
UFZ_{1S} & UFZ_{11} & -UAZ_{11*} & -UAZ_{12} & Zeros & Zeros & Zeros & Zeros & Zeros & Zeros & Zeros & Zeros \\
Zeros & Zeros & \Sigma 31_{11*} & \Sigma 31_{12} & Zeros & Zeros & Zeros & Zeros & Zeros & Zeros & Zeros & Zeros \\
Zeros & Zeros & \Sigma 32_{11*} & \Sigma 32_{12} & Zeros & Zeros & Zeros & Zeros & Zeros & Zeros & Zeros & Zeros \\
Zeros & Zeros & \Sigma 31_{21*} & \Sigma 31_{22} & -\Sigma 31_{22*} & -\Sigma 31_{23} & Zeros & Zeros & Zeros & Zeros & Zeros & Zeros \\
Zeros & Zeros & \Sigma 32_{21*} & \Sigma 32_{22} & -\Sigma 32_{22*} & -\Sigma 32_{23} & Zeros & Zeros & Zeros & Zeros & Zeros & Zeros \\
Zeros & Zeros & \Sigma 33_{21*} & \Sigma 33_{22} & -\Sigma 33_{22*} & -\Sigma 33_{23} & Zeros & Zeros & Zeros & Zeros & Zeros & Zeros \\
Zeros & Zeros & UAX_{21*} & UAX_{22} & -UAX_{22*} & -UAX_{23} & Zeros & Zeros & Zeros & Zeros & Zeros & Zeros \\
Zeros & Zeros & UAX_{21*} & UAX_{22} & -UAX_{22*} & -UAX_{23} & Zeros & Zeros & Zeros & Zeros & Zeros & Zeros \\
Zeros & Zeros & UAX_{21*} & UAX_{22} & -UAX_{22*} & -UAX_{23} & Zeros & Zeros & Zeros & Zeros & Zeros & Zeros \\
Zeros & Zeros & Zeros & Zeros & \Sigma 31_{32*} & \Sigma 31_{33} & -\Sigma 31_{33*} & -\Sigma 31_{34} & Zeros & Zeros & Zeros & Zeros \\
Zeros & Zeros & Zeros & Zeros & \Sigma 32_{32*} & \Sigma 32_{33} & -\Sigma 32_{33*} & -\Sigma 32_{34} & Zeros & Zeros & Zeros & Zeros \\
Zeros & Zeros & Zeros & Zeros & \Sigma 33_{32*} & \Sigma 33_{33} & -\Sigma 33_{33*} & -\Sigma 33_{34} & Zeros & Zeros & Zeros & Zeros \\
Zeros & Zeros & Zeros & Zeros & UAX_{32*} & UAX_{33} & -UAX_{33*} & -UAX_{34} & Zeros & Zeros & Zeros & Zeros \\
Zeros & Zeros & Zeros & Zeros & UAY_{32*} & UAY_{33} & -UAY_{33*} & -UAY_{34} & Zeros & Zeros & Zeros & Zeros \\
Zeros & Zeros & Zeros & Zeros & UAZ_{32*} & UAZ_{33} & -UAZ_{33*} & -UAZ_{34} & Zeros & Zeros & Zeros & Zeros \\
Zeros & Zeros & Zeros & Zeros & Zeros & Zeros & \Sigma 31_{43*} & \Sigma 31_{44} & -\Sigma 31_{44*} & -\Sigma 31_{45} & Zeros & Zeros \\
Zeros & Zeros & Zeros & Zeros & Zeros & Zeros & \Sigma 32_{43*} & \Sigma 32_{44} & -\Sigma 32_{44*} & -\Sigma 32_{45} & Zeros & Zeros \\
Zeros & Zeros & Zeros & Zeros & Zeros & Zeros & \Sigma 33_{43*} & \Sigma 33_{44} & -\Sigma 33_{44*} & -\Sigma 33_{45} & Zeros & Zeros \\
Zeros & Zeros & Zeros & Zeros & Zeros & Zeros & UAX_{43*} & UAX_{44} & -UAX_{44*} & -UAX_{44*} & Zeros & Zeros \\
Zeros & Zeros & Zeros & Zeros & Zeros & Zeros & UAY_{43*} & UAY_{44} & -UAY_{44*} & -UAY_{44*} & Zeros & Zeros \\
Zeros & Zeros & Zeros & Zeros & Zeros & Zeros & UAZ_{43*} & UAZ_{44} & -UAZ_{44*} & -UAZ_{44*} & Zeros & Zeros \\
Zeros & Zeros & Zeros & Zeros & Zeros & Zeros & Zeros & Zeros & \Sigma 31_{54*} & \Sigma 31_{55} & Zeros & Zeros \\
Zeros & Zeros & Zeros & Zeros & Zeros & Zeros & Zeros & Zeros & \Sigma 32_{54*} & \Sigma 32_{55} & Zeros & Zeros \\
Zeros & Zeros & Zeros & Zeros & Zeros & Zeros & Zeros & Zeros & -\Sigma 33_{54*} & -\Sigma 33_{55} & P_{55*} & P_{5R} \\
Zeros & Zeros & Zeros & Zeros & Zeros & Zeros & Zeros & Zeros & -UAZ_{54*} & -UAZ_{55} & UFZ_{55*} & UFZ_{5R} \\
Zeros & Zeros & Zeros & Zeros & Zeros & Zeros & Zeros & Zeros & Zeros & Zeros & VG_{R5*} & VG_{RR}
\end{bmatrix}
\times
\begin{Bmatrix}
A_S \\ A_1 \\ A_{1*} \\ A_2 \\ A_{2*} \\ A_3 \\ A_{3*} \\ A_4 \\ A_{4*} \\ A_5 \\ A_{5*} \\ A_R
\end{Bmatrix}
=
\begin{Bmatrix}
V_{S0} \\ Zeros \\ Zeros \\ Zeros \\ Zeros \\ Zeros \\ Zeros \\ Zeros \\ Zeros \\ Zeros \\ Zeros \\ V_{R0}
\end{Bmatrix}
\quad (7.282)
$$

After solving Eq. (7.282), the pressure field inside the fluid can be calculated using the following equation, where F is a set of target points in fluid as shown in Fig. 7.6a.

$$\mathbf{PR}_F = \mathbf{P}_{FS}\mathbf{A}_S + \mathbf{P}_{FI}\mathbf{A}_I \tag{7.283}$$

Similarly, the displacement and stress fields (together called wave fields) inside the anisotropic plate with delamination can be calculated using appropriate source strengths as described in Section 7.6.2 in addition to the contribution by the source strengths from the delaminated area.

7.9.3 Wave Field Modeling of (0/0) 2-ply Plate with Delamination

The stress and displacement wave fields are first calculated for pristine (0/0) 2-ply plate for comparison. The stresses (σ_{11} and σ_{33}) and pressure fields, which are shown in Fig. 7.62a and b, respectively, are computed. Similarly, the displacement fields (u_1 and u_3) are shown in Fig. 7.63a and b. The stress and displacement wave fields are then calculated for (0/0) 2-ply plate with a delamination. The delamination is 5 mm long and 0.5 mm thick located in between the two layers. The stresses (σ_{11} and σ_{33}) and pressure fields, which are shown in Fig. 7.64a and b, respectively, are computed using the formulation presented in Section 7.9.2. Similarly, the displacement fields (u_1 and u_3) are shown in Fig. 7.65a and b. From the above figures in the wave field, it can be seen that the wave behavior changes when the delamination is present. Stress concentration at the tips of the delamination is visible clearly from the stress plots in Fig. 7.64a and b. It can also be seen that the wave is reflected due to the presence of the delamination near delaminated surface compared to the pristine state. However, from the current figures, it is hard to obtain information regarding the change in magnitude as well as the phase of the wave field due to the interference of delamination in the NDE ultrasonic signals. To investigate further, the wave field (σ_{11} stress) is plotted along a central line of the (0/0) 2-ply plate configuration shown in Fig. 7.66a. The plot is compared between pristine (0/0), delaminated (0/0) plate as shown in Fig. 7.66b. By observing the line plot of the wave amplitude along the central line as shown in Fig. 7.66b, it can be clearly seen that the wave amplitude

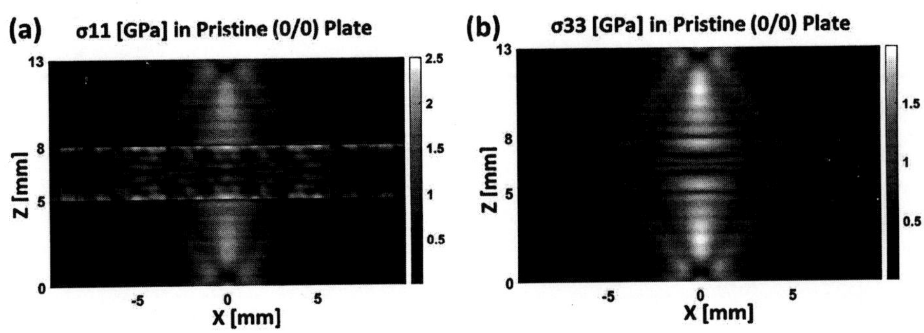

FIGURE 7.62 a) Stress11 (σ11) distribution b) Stress33 (σ33) distribution in pristine (0/0) 2-ply plate in GPa.

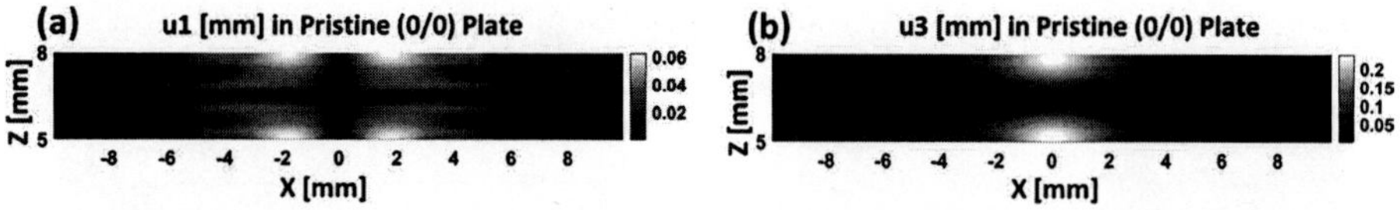

FIGURE 7.63 a) Displacement (u1) distribution b) Displacement (u3) distribution in pristine (0/0) 2-ply plate in mm.

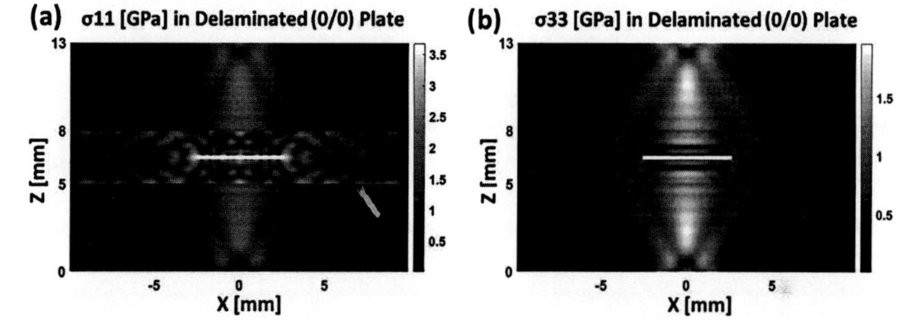

FIGURE 7.64 a) Stress11 (σ11) distribution b) Stress33 (σ33) distribution in (0/0) 2-ply plate with delamination in GPa.

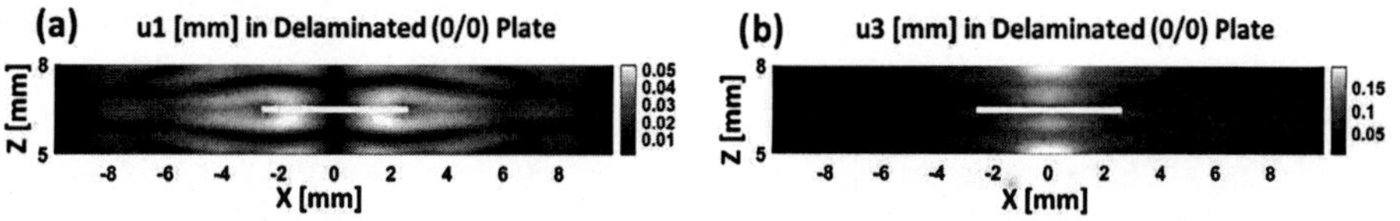

FIGURE 7.65 a) Displacement (u1) distribution b) Displacement (u3) distribution in (0/0) 2-ply plate with delamination in mm.

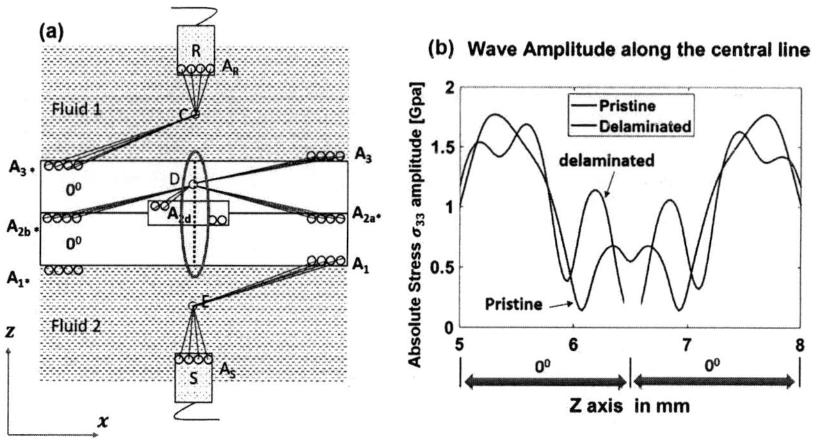

FIGURE 7.66 a) cross-section view of the CNDE problem along x-z plane showcasing the central line where the wave amplitude is plotted. b) Line Plot of wave amplitude (σ_{11} stress) along the central line.

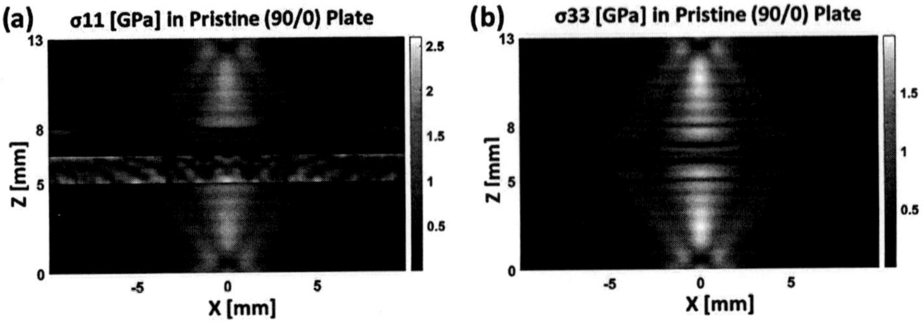

FIGURE 7.67 a) Stress11 (σ11) distribution b) Stress33 (σ33) distribution in pristine (90/0) 2-ply plate in GPa.

changes due to the presence of delamination. In addition, change in phase is also observed when comparing the stress field from pristine and delaminated states.

7.9.4 Wave Field Modeling of (90/0) 2-ply Plate with Delamination

First the stress and displacement wave fields were calculated for pristine (90/0) 2-ply plate. The stresses (σ_{11} and σ_{33}) and pressure fields are computed shown in Fig. 7.67a and b, respectively. Similarly, the displacement fields (u_1 and u_3) are shown in Fig. 7.68a and b. The stress and displacement wave fields were calculated for (90/0) 2-ply plate with delamination. The delamination is 5 mm long and 0.5 mm thick located in between the two layers. The stresses (σ_{11} and σ_{33}) and pressure fields are computed shown in Fig. 7.69a and b, respectively. Similarly, the displacement fields (u_1 and u_3) are shown in Fig. 7.70a and b. Next the wave field (σ_{11} stress) is plotted along a central line for (90/0) 2-ply plate configuration shown in Fig. 7.71a. The plot is compared between pristine (90/0), delaminated (90/0) plate as shown in Fig. 7.71b. By observing the line plot of the wave amplitude along the central line as shown in Fig. 7.71b, it can be clearly seen that the wave amplitude changes due to the presence of delamination. In addition, change in phase is also observed when comparing the stress field from pristine and delaminated states. The stress amplitude after interacting with delamination is higher in the 0° layer compared to 90° layer. Therefore, it can be concluded that the presence of delamination in a laminated structure can affect the wave field significantly. DPSM is able to corroborate the statement with the help of newly incorporated model to simulate the effect of the delamination in on wave propagation.

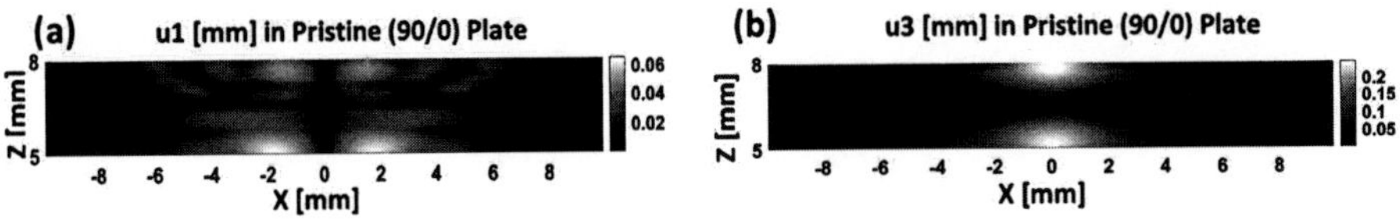

FIGURE 7.68 a) Displacement (u1) distribution b) Displacement (u3) distribution in pristine (90/0) 2-ply plate in mm.

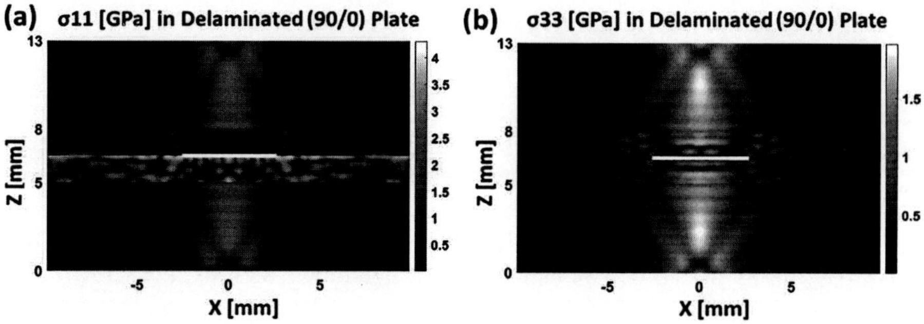

FIGURE 7.69 a) Stress11 (σ11) distribution b) Stress33 (σ33) distribution in (90/0) 2-ply plate with delamination in GPa.

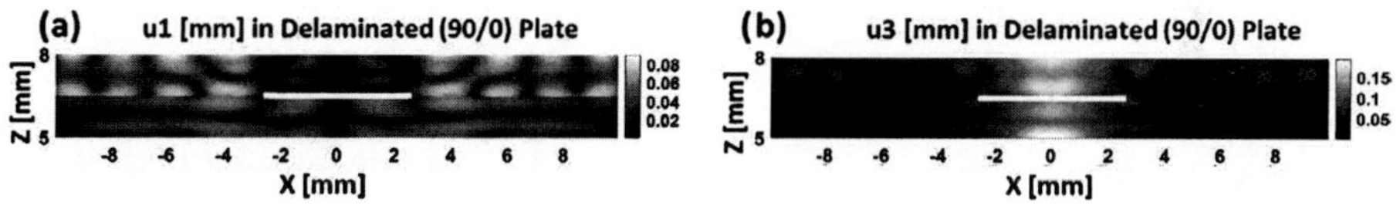

FIGURE 7.70 a) Displacement (u1) distribution b) Displacement (u3) distribution in (90/0) 2-ply plate with delamination in mm.

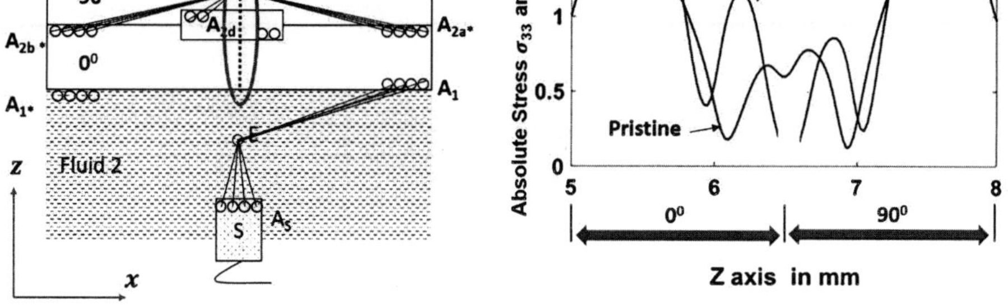

FIGURE 7.71 a) cross-section view of the NDE problem along x-z plane showcasing the central line where the wave amplitude is plotted. b) Line Plot of wave amplitude (σ_{11} stress) along the central line.

7.10 IMPLEMENTATION OF DPSM IN COMPUTER CODE FOR AUTOMATION

After successfully developing the code for modeling the wave field in the anisotropic media, it is realized that it is necessary to make the code generalized and versatile such that one could incorporate any number of layers and any number of delamination at any interface with single input. However, this is a daunting task to make the code fully generalized and automated. Even after generalization of the code it is expected that personnel with expert knowledge on the wave propagation and NDE techniques will be able to conduct CNDE simulation with the help of the DPSM code. Hence, this book and this chapter is written for the guidelines to provide that knowledge to the user as quickly as possible. In short, DPSM could provide an automated and computationally fast and efficient one-stop wave simulation platform for any anisotropic media with damage scenario such as material degradation, delamination, etc. There are four components essential for the proposal to come to fruition: Automation of the computational algorithm (discussed below), Parallelization of computationally taxing algorithms (discussed in Chapter 6), Development of Computation Speedup Components (discussed in Section 7.7), and Experimental Validation of the Computational technique.

7.10.1 Automation for Pristine and Degraded N-layered Media

DPSM requires the Green's functions for its implementation as discussed in the previous sections. In addition, the boundary and interface conditions need to be satisfied at each interface. After satisfying the boundary and interface conditions a matrix composed of Green's functions is formed. The size of the Green's function matrix increases with the increasing layers in the plate. Hence, it is obvious that additional interface conditions need to be added with every additional layer in the plate. It is necessary that the Green's function matrix is updated automatically and smartly for any N- layer plate such that the user doesn't have to go through each layer one by one. Such versatile automation can be made possible by implementation of following important components.

a. *Digitization of layer stacking sequence*
b. *Calculation of Christoffel Solution based on n unique layers*
c. *Calculation of Solid Green's function based on n unique layers*
d. *Automated Green's function matrix based on digitized stacking sequence and understanding of DPSM.*

7.10.1.1 Digitization of Layer Stacking Sequence

A new parameter, stacking sequence is introduced. From this parameter, user will be able to gain new information: such as how many unique layers are there? After this, user will be able to digitize the stacking sequence which is essential for the automation process.

Example: Assume the user had a 8-layered plate of (0/90/45/60) s such that
Stacking sequence = [0 90 45 60 60 45 90 0]
Unique layers = 4

Now to digitize the sequence with numeric number like 0, 1, 2, and so on, Digitized stacking sequence can be written as [0 1 2 3 3 2 1 0] for the case of unique layers presented above. Here, 0 degree layers is indexed with 0, 90 degree layer indexed with 1, 45 degree layer is indexed with 2 and the 60 degree layer is indexed with 3, as it is used in the C++ code. Different indexing is also possible but user needs to keep track of the indexing throughout the DPSM code and has to remain consistent. 90So, digitized stacking sequence for [0 90 45 60 60 45 90 0] stacking sequence is [0 1 2 3 3 2 1 0].

7.10.1.2 Calculation of Christoffel Solution based on n Unique Layers

Christoffel parameters are required to be calculated according to the digitization of the unique layers. Christoffel solutions are discussed in Section 7.5. User can solve the parameters such as CV(i) and FI (i), that are modal wave velocity vector and respective eigen vector matrix, respectively, for each material type (unique layer) where i = 0, 1, 2, 3

7.10.1.3 Calculation of Solid Components based on n Unique Layers

Similarly, the solid Green's function matrices are calculated according to the digitization of the unique layers. Matrices such as **DSm**{i} matrix and **Smn**{i} should be populated for each unique layer, where m, n = 1, 2, 3 and i = 0, 1, 2, 3

7.10.1.4 Automated DPSM Matrix based on Digitized Stacking Sequence

To automate the bigger Green's function matrix called DPSM matrix, user first need to find the total number of equations in the matrix satisfying the boundary and interface conditions. It has been found that number of equations per transducer (neqT) = 1, number of equations per fluid-solid interface (neqFS) = 4, and number of equations per solid-solid interface(neqSS) = 6.

For any specific CNDE setup, user first must know the number of transducer(nT), number of fluid-solid interface(nFS), and number of total layers in the plate. User can calculate the number of solid-solid interface(nSS) which is equal to no. of total layers – 1.

Next, user can calculate the total number of equations using the following expression:

$$\text{neqTotal} = \text{neqT} * \text{nT} + \text{neqFS} * \text{nFS} + \text{neqSS} * \text{nSS} \tag{7.284}$$

Further, user can find the size of the Green's function matrix or the DPSM matrix (matSize). By knowing the number of point sources in each transducers (nTr), and knowing the total number of point sources (nTo) in each layer the matSize can be calculated as follows:

$$\text{matSize} = (\text{neqT} * \text{nT}) * \text{nTr} + (\text{neqFS} * \text{nFS} + \text{neqSS} * \text{nSS}) * \text{nTo} \tag{7.285}$$

and the DPSM matrix is a matrix of size (matSize x matSize).

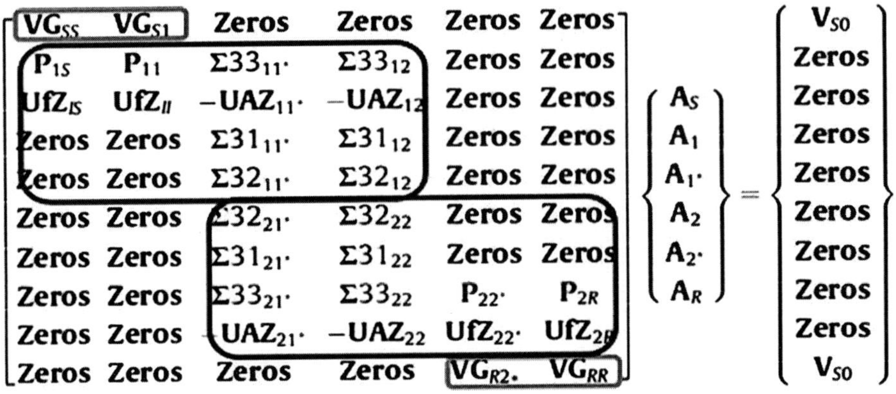

FIGURE 7.72 Green's Function Matrix for the CNDE problem in Fig. 7.37.

Now, to automate the process of populating the DPSM matrix, it is necessary that user should visualize and understand how the complete matrix is formed and how the equations are arranged. For example, the Green's function or the DPSM matrix for 1 layer and 4 layer plate shown in Figs. 7.72 and 7.73, respectively. By looking at two DPSM matrices, user can immediately realize that there is a pattern and such pattern can be easily automated. The first and the last equations are contributed by the transducer equations. The next four equations from the top and the bottom are the equations contributed by the fluid solid interface. The six set of equations in between are from the solid-solid interface in the setup. In 1-layer plate, there is

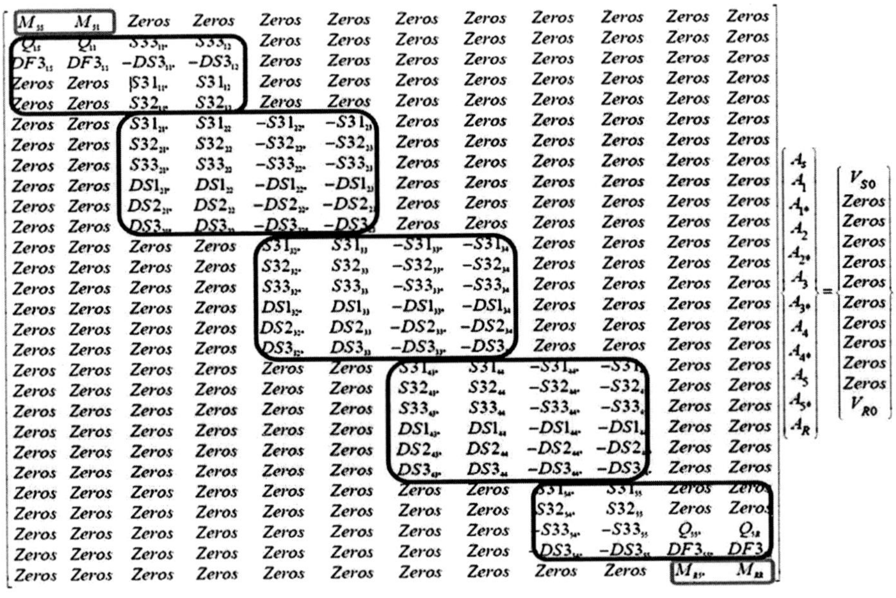

FIGURE 7.73 Green's Function Matrix for the NDE problem in Fig. 7.53.

no solid-solid interface, and hence the six set of equations does not exist. However, for 4-layer plate, there are 3 solid-solid interfaces hence there are three set of six equations representing the 3 solid-solid interfaces. With the understanding of these crucial components, the full automation of Green's function or the DPSM matrix for any N layered plate can be implemented.

7.10.2 Automation for N-layered Plate with Delamination

It is obvious that the DPSM formulation needs to be modified with the presence of delamination as described in Section 7.9. Next for a generalized DPSM code capable of incorporating delamination at any interface automatically, a mush sophisticated automation is necessary as described in this section. It is known that to implement DPSM user will need to a) calculate the Green's function, b) satisfy the boundary and interface conditions, c) populate the Green's function matrix and form the DPSM matrix, d) calculate the source strengths, and calculate the ultrasonic wave field for the specific CNDE problem. It can be realized that the components of green's function matrix increase with the occurrence of the delamination. Hence, the interface conditions need to be adjusted for each interface depending on whether a delamination is present or not. Given that the adjustment process of the interface conditions for a N-layered plate with delamination is a tedious and inefficient process if done manually, there is a need to automate the process. Such versatile automation can be made possible by implementation of the similar important components as a continuation of the other processes discussed in the previous section.

e. *Delamination sequence*
f. *Calculation of Solid Components in a delaminated layer*
g. *Automated DPSM matrix based on Delamination sequence.*

7.10.2.1 Delamination Sequence

A new parameter, called "delamination sequence" should be introduced. From this parameter, user will be able to gain new information: such as, which interfaces has the delamination? After this, user will digitize the delamination sequence which is essential for the automating the implementation of delamination in the code.

Example: Assume that the user has an 8-layered plate of $(0/90)_4$ such that it has nine interfaces. It has delamination on interfaces 2, 5, 6, 7 as shown in Fig. 7.74. If the delamination sequence is digitized such that no delamination = 0 and delamination = 1, the delamination sequence can be written as follows: Delamination sequence = [0 1 0 0 1 1 1 0 0] in a binary format. With the help of delamination sequence, user can calculate the required number of divisions of different category of point sources along the interface originally shown in Fig. 7.60. Number of segments of point sources at each interface will be one or three, depending on whether the interface has the delamination or not. For example, for interface 1, user can calculate the source strengths components such as A_I representing the whole interface. However, for interface 2, user should calculate the source strengths components such as A_{Ib}, A_{Id}, and A_{Ia} representing before delamination,

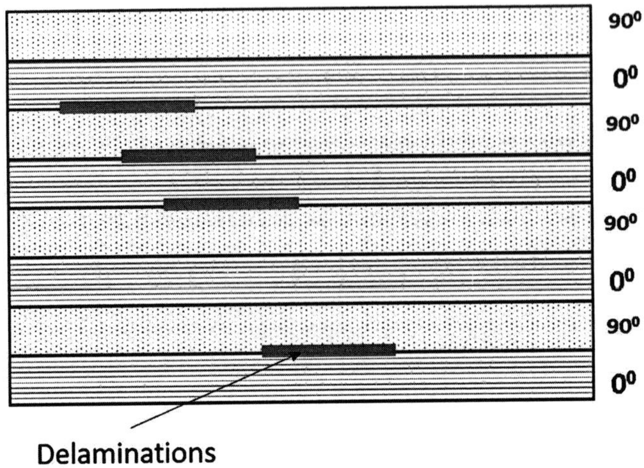

FIGURE 7.74 $(0/90)_4$ configuration of a 8-layered composite plate with delamination on the interfaces 2,5,6,7.

during delamination, and after delamination portions of interface, respectively, which is previously explained in Fig. 7.60.

7.10.2.2 Calculation of Solid Components based on n Unique Layers

Now, the required solid components such as displacements (**DS**) and stresses (**S**) are calculated according to the delamination sequence. User will have parameters such as **DSmn** and **Smn** for interfaces with no delamination like they are described in Section 7.10.1.3. And, user will have parameters such as **DSmnb**, **DSmnd**, and **DSmna** for displacement and **Smnb**, **Smnd**, **Smna** for interface with delamination, where m, n = 1, 2, 3.

7.10.2.3 Automated Population of DPSM Matrix based on Delamination Sequence

To automate the filling of DPSM matrix with the Green's function matrices, a user first need to know the interfaces that has delamination. User may get that information from delamination sequence. In addition, he/she should also calculate the total number of equations in the matrix and the number of components at each interface. For the number of equations in the matrix, it has been found that: number of equations per transducer (neqT) = 1, number of equations per fluid-solid interface (neqFS) = 4, number of equations per solid-solid interface without delamination (neqSS) = 6, and number of equations per solid-solid interface with delamination (neqSSD) = 18. For any specific CNDE setup, user must collect the number of transducer (nT), number of fluid-solid interface (nFS), number of solid-solid interface without delamination (nSS), and number of solid-solid interface with delamination (nSSD). Next the user can calculate the total no of equations

$$(neqTotal) = neqT * nT + neqFS * nFS + neqSS * nSS + neqSSD * nSSD \quad (7.286)$$

TABLE 7.3
Index to Control the Number of Segments at any Interface due the Presence of Delamination

	No-Delamination = 0; Delamination = 1							
Case	1	2	3	4	5	6	7	8
i-1	0	0	1	0	1	1	0	1
i	0	1	0	0	0	1	1	1
i+1	0	0	0	1	1	1	1	1

Next, the user can calculate the size the big Green's function matrix or the DPSM matrix, matSize. Also, user must know the maximum frequency of operation in the targeted CNDE model, and hence must know the number of point sources to be used in each transducer (nTr) and the total number of point sources (nTo) at the interface based on the convergence criteria presented in this chapter earlier such that he/she can calculate the matSize as follows:

$$matSize = (neqT * nT) * nTr + (neqFS * nFS + neqSS * nSS + neqSSD * nSSD) * nTo \quad (7.287)$$

The DPSM matrix is a matrix of size (matSize x matSize).

Next to find the number of segments at each interface, the user needs to look at the interface condition. To satisfy an interface condition at any interface, "i" the user need components from the interfaces "i-1," "i," and "i+1." Depending on whether there is a delamination in any of these interfaces, there can be differing numbers of components in the equations. Such definitions are presented in Table 7.3. Given that the equation depends on three interfaces, there are eight different possibilities which are tabulated below:

Case 1: This is the case with no delamination. Interface condition should be developed with the help of Fig. 7.75a.

For the interface, "I," i.e., 2, the interface condition can be written as follows:

$$\mathbf{UF}_{21*}\mathbf{A}_{1*} + \mathbf{UF}_{22}\mathbf{A}_2 = \mathbf{UF}_{22*}\mathbf{A}_{2^*} + \mathbf{UF}_{23}\mathbf{A}_3 \quad (7.288)$$

where $\mathbf{UF}_{st}$ is the ultrasonic field at source "s" due to target "t" and $\mathbf{A}_s$ is the source strength of the point sources "s." As it can be seen that the interface "i" has no delamination and one equation is sufficient to represent the whole interface.

Case 2: This is the case with delamination at the middle interface

For the interface, "i," i.e., 2, the interface condition can be written as follows:

$$\mathbf{UF}_{2b1*}\mathbf{A}_{1*} + \mathbf{UF}_{2b2b}\mathbf{A}_{2b} + \mathbf{UF}_{2b2a}\mathbf{A}_{2a} = \mathbf{UF}_{2b2b*}\mathbf{A}_{2b^*} + \mathbf{UF}_{2b2a*}\mathbf{A}_{2b^*} + \mathbf{UF}_{2b3}\mathbf{A}_3 \quad (7.289.1)$$

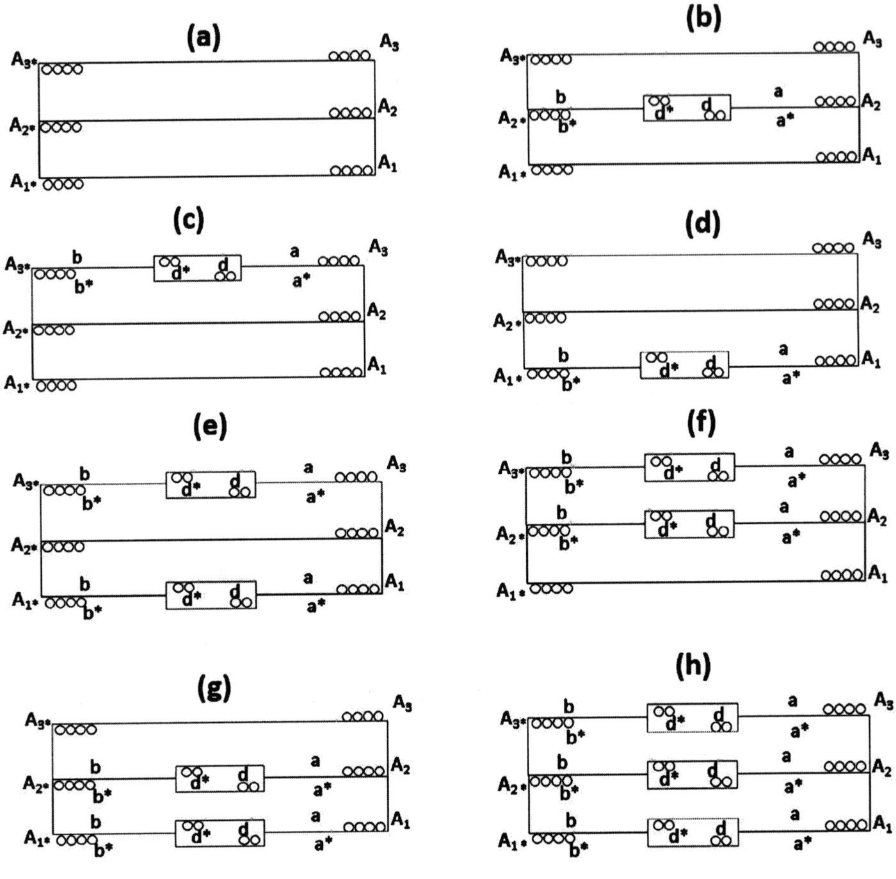

FIGURE 7.75 a) 2-layered plate showcasing Case 1, b) 2-layered plate showcasing Case 2, c) 2-layered plate showcasing Case 3, d) 2-layered plate showcasing Case 4, e) 2-layered plate showcasing Case 5, f) 2-layered plate showcasing Case 6, g) 2-layered plate showcasing Case 7, h) 2-layered plate showcasing Case 8.

$$\mathbf{UF}_{2d1*}\mathbf{A}_{1*} + \mathbf{UF}_{2d2b}\mathbf{A}_{2b} + \mathbf{UF}_{2d2d}\mathbf{A}_{2d} + \mathbf{UF}_{2d2a}\mathbf{A}_{2a} = 0 \tag{7.289.2}$$

$$\mathbf{UF}_{2a1*}\mathbf{A}_{1*} + \mathbf{UF}_{2a2b}\mathbf{A}_{2b} + \mathbf{UF}_{2a2a}\mathbf{A}_{2a} = \mathbf{UF}_{2a2b*}\mathbf{A}_{2_{b}^{*}} + \mathbf{UF}_{2a2a*}\mathbf{A}_{2_{b}^{*}} + \mathbf{UF}_{2a3}\mathbf{A}_{3} \tag{7.289.3}$$

As it can be seen that, the interface "i" has the delamination the component is separated into "b," "d," and "a" segments. Similarly, the interface conditions are also satisfied separately for these three segments.

Case 3: This is the case with delamination at the top interface

For the interface, "i," i.e., 2, the interface condition can be written as follows:

$$\mathbf{UF}_{21*}\mathbf{A}_{1*} + \mathbf{UF}_{22}\mathbf{A}_{2} = \mathbf{UF}_{22*}\mathbf{A}_{2^{*}} + \mathbf{UF}_{23b}\mathbf{A}_{3b} + \mathbf{UF}_{23b}\mathbf{A}_{3d} + \mathbf{UF}_{23a}\mathbf{A}_{3a} \tag{7.290}$$

Case 4: This is the case with delamination is at the bottom of the interface

For the interface, "i," i.e., 2, the interface condition can be written as follows:

$$\mathbf{UF}_{21b^*}\mathbf{A}_{1b^*} + \mathbf{UF}_{21d^*}\mathbf{A}_{1d^*} + \mathbf{UF}_{21a^*}\mathbf{A}_{1a^*} + \mathbf{UF}_{22}\mathbf{A}_2 = \mathbf{UF}_{22^*}\mathbf{A}_{2^*} + \mathbf{UF}_{23}\mathbf{A}_3 \quad (7.291)$$

Case 5: This is the case with delamination at the top and at the bottom interfaces

For the interface, "i," i.e., 2, the interface condition can be written as follows:

$$\mathbf{UF}_{21b^*}\mathbf{A}_{1b^*} + \mathbf{UF}_{21d^*}\mathbf{A}_{1d^*} + \mathbf{UF}_{21a^*}\mathbf{A}_{1a^*} + \mathbf{UF}_{22}\mathbf{A}_2 = \mathbf{UF}_{22^*}\mathbf{A}_{2^*} + \mathbf{UF}_{23b}\mathbf{A}_{3b} + \mathbf{UF}_{23b}\mathbf{A}_{3d} + \mathbf{UF}_{23a}\mathbf{A}_{3a} \quad (7.292)$$

Case 6: This is the case with delamination at the top and the middle interfaces

For the interface, "i," i.e., 2, the interface condition can be written as follows:

$$\mathbf{UF}_{2b1^*}\mathbf{A}_{1^*} + \mathbf{UF}_{2b2b}\mathbf{A}_{2b} + \mathbf{UF}_{2b2a}\mathbf{A}_{2a} = \mathbf{UF}_{2b2b^*}\mathbf{A}_{2b^*} + \mathbf{UF}_{2b2a^*}\mathbf{A}_{2b^*} + \mathbf{UF}_{2b3b}\mathbf{A}_{3b} + \mathbf{UF}_{2b3d}\mathbf{A}_{3d} + \mathbf{UF}_{2b3a}\mathbf{A}_{3a} \quad (7.293.1)$$

$$\mathbf{UF}_{2d1^*}\mathbf{A}_{1^*} + \mathbf{UF}_{2d2b}\mathbf{A}_{2b} + \mathbf{UF}_{2d2d}\mathbf{A}_{2d} + \mathbf{UF}_{2d2a}\mathbf{A}_{2a} = 0 \quad (7.293.2)$$

$$\mathbf{UF}_{2a1^*}\mathbf{A}_{1^*} + \mathbf{UF}_{2a2b}\mathbf{A}_{2b} + \mathbf{UF}_{2a2a}\mathbf{A}_{2a} = \mathbf{UF}_{2a2b^*}\mathbf{A}_{2b^*} + \mathbf{UF}_{2a2a^*}\mathbf{A}_{2b^*} + \mathbf{UF}_{2a3b}\mathbf{A}_{3b} + \mathbf{UF}_{2a3d}\mathbf{A}_{3d} + \mathbf{UF}_{2a3a}\mathbf{A}_{3a} \quad (7.293.3)$$

Case 7: This is the case with delamination at the middle and at the bottom interfaces

For the interface, "i," i.e., 2, the interface condition can be written as follows:

$$\mathbf{UF}_{2b1b^*}\mathbf{A}_{1b^*} + \mathbf{UF}_{2b1d^*}\mathbf{A}_{1d^*} + \mathbf{UF}_{2b1a^*}\mathbf{A}_{1a^*} + \mathbf{UF}_{2b2b}\mathbf{A}_{2b} + \mathbf{UF}_{2b2a}\mathbf{A}_{2a} = \mathbf{UF}_{2b2b^*}\mathbf{A}_{2b^*} + \mathbf{UF}_{2b2a^*}\mathbf{A}_{2b^*} + \mathbf{UF}_{2b3}\mathbf{A}_3 \quad (7.294.1)$$

$$\mathbf{UF}_{2d1b^*}\mathbf{A}_{1b^*} + \mathbf{UF}_{2d1d^*}\mathbf{A}_{1d^*} + \mathbf{UF}_{2d1a^*}\mathbf{A}_{1a^*} + \mathbf{UF}_{2d2b}\mathbf{A}_{2b} + \mathbf{UF}_{2d2d}\mathbf{A}_{2d} + \mathbf{UF}_{2d2a}\mathbf{A}_{2a} = 0 \quad (7.294.2)$$

$$\mathbf{UF}_{2a1b^*}\mathbf{A}_{1b^*} + \mathbf{UF}_{2a1d^*}\mathbf{A}_{1d^*} + \mathbf{UF}_{2a1a^*}\mathbf{A}_{1a^*} + \mathbf{UF}_{2a2b}\mathbf{A}_{2b} + \mathbf{UF}_{2a2a}\mathbf{A}_{2a} = \mathbf{UF}_{2a2b^*}\mathbf{A}_{2b^*} + \mathbf{UF}_{2a2a^*}\mathbf{A}_{2b^*} + \mathbf{UF}_{2a3}\mathbf{A}_3 \quad (7.294.3)$$

Case 8: This is the case with delamination at all the interfaces

For the interface, "i," i.e., 2, the interface condition can be written as follows:

$$\mathbf{UF}_{2b1b^*}\mathbf{A}_{1b^*} + \mathbf{UF}_{2b1d^*}\mathbf{A}_{1d^*} + \mathbf{UF}_{2b1a^*}\mathbf{A}_{1a^*} + \mathbf{UF}_{2b2b}\mathbf{A}_{2b} + \mathbf{UF}_{2b2a}\mathbf{A}_{2a} = \mathbf{UF}_{2b2b^*}\mathbf{A}_{2b^*} + \mathbf{UF}_{2b2a^*}\mathbf{A}_{2b^*} + \mathbf{UF}_{2b3b}\mathbf{A}_{3b} + \mathbf{UF}_{2b3d}\mathbf{A}_{3d} + \mathbf{UF}_{2b3a}\mathbf{A}_{3a} \quad (7.295.1)$$

$$\mathbf{UF}_{2d1b^*}\mathbf{A}_{1b^*} + \mathbf{UF}_{2d1d^*}\mathbf{A}_{1d^*} + \mathbf{UF}_{2d1a^*}\mathbf{A}_{1a^*} + \mathbf{UF}_{2d2b}\mathbf{A}_{2b} + \mathbf{UF}_{2d2d}\mathbf{A}_{2d} + \mathbf{UF}_{2d2a}\mathbf{A}_{2a} = 0 \tag{7.295.2}$$

$$\mathbf{UF}_{2a1b^*}\mathbf{A}_{1b^*} + \mathbf{UF}_{2a1d^*}\mathbf{A}_{1d^*} + \mathbf{UF}_{2a1a^*}\mathbf{A}_{1a^*} + \mathbf{UF}_{2a2b}\mathbf{A}_{2b} + \mathbf{UF}_{2a2a}\mathbf{A}_{2a} = \mathbf{UF}_{2a2b^*}\mathbf{A}_{2b^*} + \mathbf{UF}_{2a2a^*}\mathbf{A}_{2b^*} + \mathbf{UF}_{2a3b}\mathbf{A}_{3b} + \mathbf{UF}_{2a3d}\mathbf{A}_{3d} + \mathbf{UF}_{2a3a}\mathbf{A}_{3a} \tag{7.295.3}$$

To populate the DPSM matrix automatically, the user should also visually understand how the DPSM matrix is formed. Here, Case 1 and 2 are showcased for demonstration. However, the filling of the Green's functions matrices in the big DPSM matrix for other cases can be done in a similar manner. DPSM matrix for 2 layered plate without and with delamination are shown in Figs. 7.76 and 7.77. By looking at the above two DPSM matrices (in Figs. 7.76 and 7.74), one can realize that the equations and the segments are divided into before delamination ("b"), during delamination ("d"), and after delamination ("a") when the delamination exists in the interface. In this scenario (Fig. 7.61), interface 2 has delamination hence it was split into 2b, 2d, and 2a as required to satisfy the interface conditions. Whereas, without delamination the interface conditions are satisfied without dividing the components for the interface 2. The first and the last equations are obtained for each transducer. The next four equations from the top and the bottom are to satisfy the fluid solid interface equations. The six set of equations in between without delamination are for the solid-solid interface. But the eighteen set of equations in between in case of delamination are for the solid-solid interface present in Fig. 7.61. With the understanding of these crucial components, the automated population of the DPSM Matrix for any N layered plate with delamination can be implemented.

$$\begin{bmatrix}
VG_{SS} & VG_{S1} & Zeros & Zeros & Zeros & Zeros & Zeros & Zeros \\
P_{1S} & P_{11} & \Sigma33_{11^*} & \Sigma33_{12} & Zeros & Zeros & Zeros & Zeros \\
UFZ_{1S} & UFZ_{11} & -UAZ_{11^*} & -UAZ_{12} & Zeros & Zeros & Zeros & Zeros \\
Zeros & Zeros & \Sigma31_{11^*} & \Sigma31_{12} & Zeros & Zeros & Zeros & Zeros \\
Zeros & Zeros & \Sigma32_{11^*} & \Sigma32_{12} & Zeros & Zeros & Zeros & Zeros \\
Zeros & Zeros & \Sigma31_{21^*} & \Sigma31_{22} & -\Sigma31_{22^*} & -\Sigma31_{23} & Zeros & Zeros \\
Zeros & Zeros & \Sigma31_{21^*} & \Sigma31_{22} & -\Sigma31_{22^*} & -\Sigma31_{23} & Zeros & Zeros \\
Zeros & Zeros & \Sigma31_{21^*} & \Sigma31_{22} & -\Sigma31_{22^*} & -\Sigma31_{23} & Zeros & Zeros \\
Zeros & Zeros & UAX_{21^*} & UAX_{22} & -UAX_{22^*} & -UAX_{23} & Zeros & Zeros \\
Zeros & Zeros & UAY_{21^*} & UAY_{22} & -UAY_{22^*} & -UAY_{23} & Zeros & Zeros \\
Zeros & Zeros & UAZ_{21^*} & UAZ_{22} & -UAZ_{22^*} & -UAZ_{23} & Zeros & Zeros \\
Zeros & Zeros & Zeros & Zeros & \Sigma31_{32^*} & \Sigma31_{33} & Zeros & Zeros \\
Zeros & Zeros & Zeros & Zeros & \Sigma32_{32^*} & \Sigma32_{33} & Zeros & Zeros \\
Zeros & Zeros & Zeros & Zeros & -\Sigma33_{32^*} & -\Sigma33_{33} & P_{33^*} & P_{3R} \\
Zeros & Zeros & Zeros & Zeros & -UAZ_{32^*} & -UAZ_{33} & UFZ_{33^*} & UFZ_{3R} \\
Zeros & Zeros & Zeros & Zeros & Zeros & Zeros & VG_{R3^*} & VG_{RR}
\end{bmatrix}
\begin{Bmatrix} A_S \\ A_1 \\ A_{1^*} \\ A_2 \\ A_{2^*} \\ A_3 \\ A_{3^*} \\ A_R \end{Bmatrix}
=
\begin{Bmatrix} V_{S0} \\ Zeros \\ Zeros \\ Zeros \\ Zeros \\ Zeros \\ Zeros \\ V_{R0} \end{Bmatrix}$$

FIGURE 7.76 Populated DPSM matrix with Green's Function matrices for the CNDE problem in Fig. 7.37 with 2 layers.

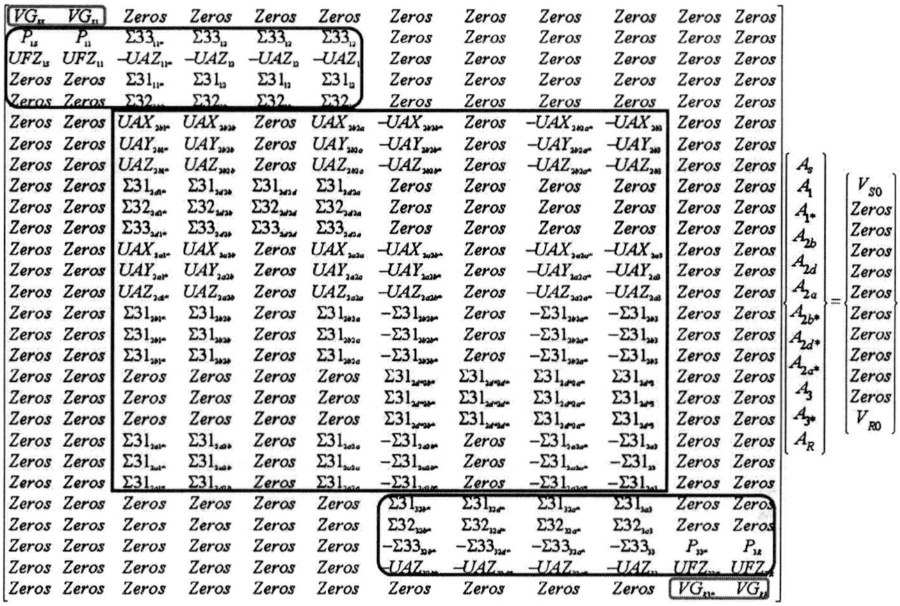

FIGURE 7.77 Populated DPSM matrix with Green's Function matrices for the CNDE problem in Fig. 7.61 with 2 layers.

7.11 IMPLEMENTATION OF PARALLEL COMPUTING FOR DPSM

Basics and fundamental descriptions of parallel computing is presented in Chapter 6. The objective of the implementation of DPSM was to develop an automated and one-stop wave simulation platform for any anisotropic media with damage scenario such as material degradation, delamination, etc. The simulation platform is developed by combining the knowledge of the physics of wave propagation in conjunction to the computational technique, DPSM to model the wave field in accurate and computationally efficient manner [38]. The fundamental framework showcasing how the DPSM computation follows few specific steps are shown in Fig. 7.78. Using the flowchart, user must start with discretization of the problem domain. The user must calculate the Christoffel Solution to obtain the required wave modes. Then Green Function matrices should be initialized and the big DPSM matrix should be formed, and the user should calculate the required components in the matrix for the problem domain, i.e., for both the fluid and solid domain. After that, solving the DPSM matrix source strengths should be computed which are required for the implementation of the DPSM. Next, user could implement the DPSM formulation again to compute the wave field in fluid and solid domain. However, runtime for the execution of the code is very slow. With the help of applications such as profiler and valgrind, one can find that more than 90% of runtime is required for the calculation of Green's function. The instances where Green's function needs to be calculated is circled in the flowchart. Hence, one need to parallelize the instances where calculation of Green's function occurs. To do that, let us first look at the analytical expression of the Green's

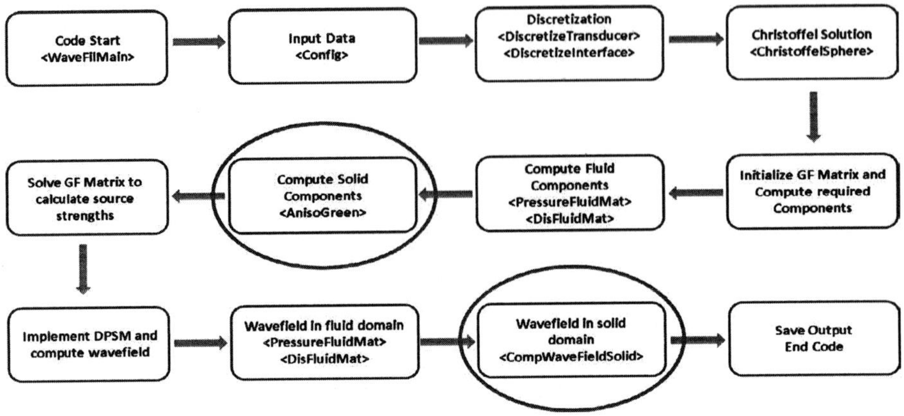

FIGURE 7.78 A flow chart showing the DPSM steps involved in a CNDE code based on DPSM formulation.

function which is expressed in Fig. 7.79 [38, 39]. Looking at the analytical expression, it can be seen that there is a computationally intensive nested loop, hence it is necessary to parallelize this nested loop, such that the computation becomes faster and more efficient. Similarly, there are few other sequential sections that can be implemented in parallel such as discretization, calculation of Green's function components, and calculation of wave field components in fluid and solid domain. The computation in fluid and solid are independent of each other and hence can be executed in parallel. For the parallelization, based on the description presented in Chapter 6 one could use two approaches: CPU parallel computing using OpenMP and GPU parallel computing using CUDA. DPSM code is transformed into CUDA code such that the computation intensive process, i.e., the calculation of Green's function is executed in parallel in the GPU while the main code is executed in CPU. The flow of the execution of the calculation of Green's function is shown in Fig. 7.80. After the implementation of CUDA into the DPSM problem, the comparison test is organized for different configurations such as half-space and 1-Layer plate under same input data settings such that the speed up (e.g., X times) can be calculated in straight-forward manner. The GPU used in this test was p2.xlarge (Tesla K80). Few new libraries such as BLAS, LAPACK, LAPACKE, and cuSOLVE are added for faster implementation.

With the help of CUDA implementation, a significant improvement was evident in the runtime for the DPSM program as shown in Table 7.4.

$$G_{mp}(x_n,\omega) = \frac{i\omega f_p}{2(2\pi)^2\rho}\sum_{z=1}^{3} \boxed{\iiint_{r=0;\,\theta=0;\,\phi=0}^{r=1\,\theta=\pi;\,\phi=2\pi}}\left[(s_z)^3\left(P_{mp}\right)^{(z)} exp\left(i(k_i x_i)^{(z)}\right) sin\theta d\theta d\phi\right]$$
$$+\frac{1}{2(2\pi)^2\rho|x|}\int_0^{2\pi}(s_z)^2\left(P_{mp}\right)^{(z)} d\phi$$

$$G_{mp} = u_m^p \longrightarrow \varepsilon^p{}_{mj} = \frac{1}{2}\left(u^p_{m,j} + u^p_{j.m}\right) \longrightarrow \sigma^p{}_{ik}(\mathrm{x}) = C_{ikmj}\varepsilon^p{}_{mj}(\mathrm{x})$$

FIGURE 7.79 Green's function in Anisotropic Solid and its computationally taxing parts.

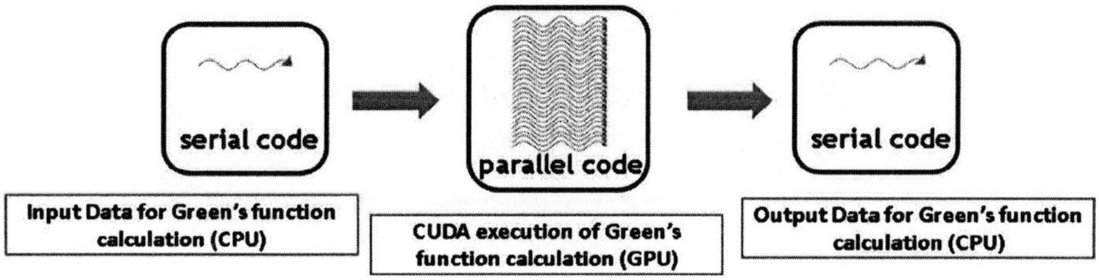

FIGURE 7.80 CUDA enabled GPU parallelization of DPSM code for calculating Green's functions.

7.12 APPENDIX

7.12.1 Effect of Microscale Voids on C_{IJKL} Matrix

To understand the effect of microscale defects, e.g., voids on the effective material properties, a representative volume element (RVE) is considered. An unidirectional composite RVE with length, width, and height of 62 μm is simulated to obtain the pristine constitutive coefficients using the finite element method [40, 41]. The fiber diameter was 7 mm and the fiber volume fraction was 50%. Interfaces between the fiber and the matrix were assumed to be perfectly bonded. The material properties of the fiber and the matrix are tabulated in Table A.7.1. The RVE method utilized in Swaminathan et al. [40, 41] is employed to govern the size of the RVE with void contents, and the size of the RVE is calculated further. The size of the RVE with void contents in a unidirectional composite was found to be 90 μm. The RVE schematics with spherical and ellipsoid void contents and mesh generation for the RVEs are shown in Fig. A7.1. The void measurements such as centers and radiuses are randomly generated [42, 43] by means of normal distribution. While automatic generation of the voids the void sizes were monitored closely through an algorithm, such that the voids does not interfere with the fibers. The process is repeated to attain the desired configuration. Simulation of RVE is not the focus of this book, hence, much details are omitted and reader are requested to refer Refs. [44, 45]

When the percentage of void content is constant and void shape is fixed to a spherical shape, a unit cell with different distributions of voids are simulated to understand the effect of various distribution on the effective material properties.

TABLE 7.4
Comparison Test to find the Speedup after Implementation of Parallelization

	Speed Comparison Test		
	Serial Implementation	**CUDA Implementation**	**Speedup**
Half-space	2946.3 sec	26.9 sec	~110x
1 Layer Plate	355.4 min	4.6 min	~77x

TABLE A.7.1
Material Properties used in RVE Simulation

Material Properties of Individual Constituents, Modulus are in GPa					
T 300 Carbon Fiber				Epoxy	
E_{fl}	E_{fv}	ν_{fl}	ν_{fv}	E	ν
230	15	0.25	0.07	5	0.16

Data were generated through FEM simulation applying periodic boundary conditions and error bounds were estimated within three times the standard deviation which are wings out from the mean value. Detail method is quite intensive and omitted herein and can be found in Ref. [46]. Similar study was conducted with ellipsoid void shape. Thus, both spherical voids between randomly generated fiber locations and ellipsoid void shapes parallel to the fiber direction are modeled into the unit cell to thoroughly understand the effect of void shapes on constitutive coefficients. Next, RVEs with different void contents are also analyzed. The full matrix of effective material properties was found. The perturbation of each constitutive coefficient with respect to increasing void percentage in the RVE is shown in Fig. A7.2. Percentage of degradation is calculated from the third order polynomial trend line within the calculated error bound. The percentage of degradation of each element in the constitutive matrix due to 5% void is calculated [46] and is as presented below:

$$C^{5\%void} = \begin{bmatrix} -8.8\% & -4.2\% & -10.1\% & 0 & 0 & 0 \\ -4.2\% & -8.2\% & -10.1\% & 0 & 0 & 0 \\ -10.1\% & -10.1\% & -11.9\% & 0 & 0 & 0 \\ 0 & 0 & 0 & -8.5\% & 0 & 0 \\ 0 & 0 & 0 & 0 & -8.3\% & 0 \\ 0 & 0 & 0 & 0 & 0 & -8.8\% \end{bmatrix} \tag{A.7.1}$$

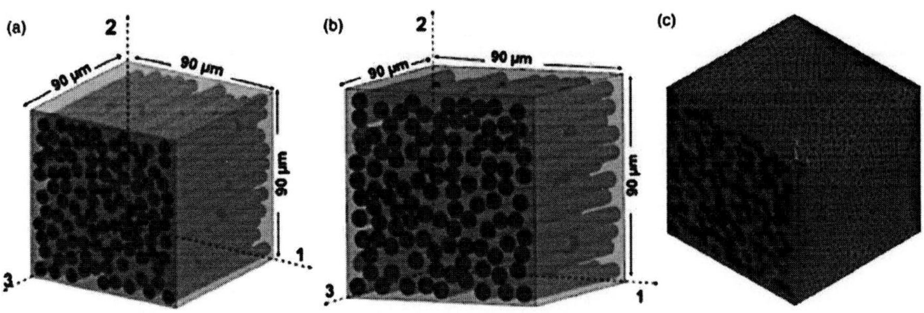

FIGURE A7.1 Schematic of an RVE including (a) Spherical voids (b) Ellipsoid voids and (c) Mesh generation. RVE: representative volume element [3].

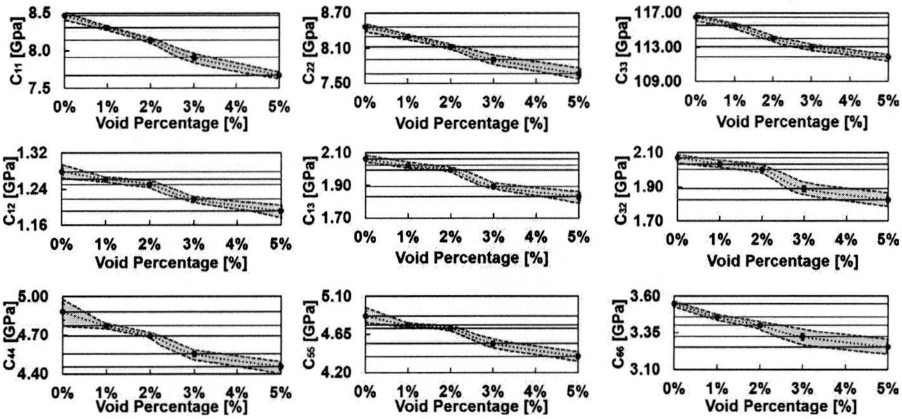

FIGURE A7.2 Perturbation to the constitutive coefficients due to the different percent of void contents fitted by third-order polynomial but bounded by error within 3σ, dots shows the mean values.

7.12.2 Distribution of Point Sources with Convergence

Like any other numerical methods, DPSM also has a specific convergence criterion. In this method, before simulating a CNDE problem, first it is necessary to find the maximum frequency that the experimental set up will experience. In NDE experiments, mostly the central frequency is mentioned as an operating frequency. However, beyond the central frequency a transducer is capable of generating higher frequencies but with lower amplitude compared to the central frequency. Although generates lower amplitudes, higher frequencies are embedded in an ultrasonic signal, if not considered in CNDE, signals generated from the model may not be able to replicate the experimental scenario. Hence, in DPSM, first it is crucial to identify the maximum frequency that CNDE problem is going to see. Let us assume the angular maximum frequency is ω_{max}. Next it is necessary to find the lowest modulus from the set of materials used in an experiment. For example, in an NDE experiment of an aluminum plate submerged under water will have higher wave velocity (P-wave velocity c_{ap}=6.5 km/sec, S-wave velocity c_{as}=~3.13 km/sec as per Table 7.1) in the solid but will have lower wave velocity (c_w=~1.48 km/sec) in fluid, which is water. If wave numbers for respective waves in respective materials are calculated, following three expressions are achieved

$$k_1 = \frac{\omega_{max}}{c_{ap}} \quad ; \quad k_2 = \frac{\omega_{max}}{c_{as}} \quad ; \quad k_3 = \frac{\omega_{max}}{c_w} \tag{A.7.2}$$

As ω_{max} identified in a problem is fixed, from the above expression it is evident that the lowest velocity will result highest wave number which is k_3 in water and k_2 in aluminum. Using Eq. (4.5) from Chapter 4 the wave lengths can be calculated as follows:

$$\lambda_2 = 2\pi / \frac{\omega_{max}}{c_{as}} \quad ; \quad \lambda_3 = 2\pi / \frac{\omega_{max}}{c_{as}} \tag{A.7.3}$$

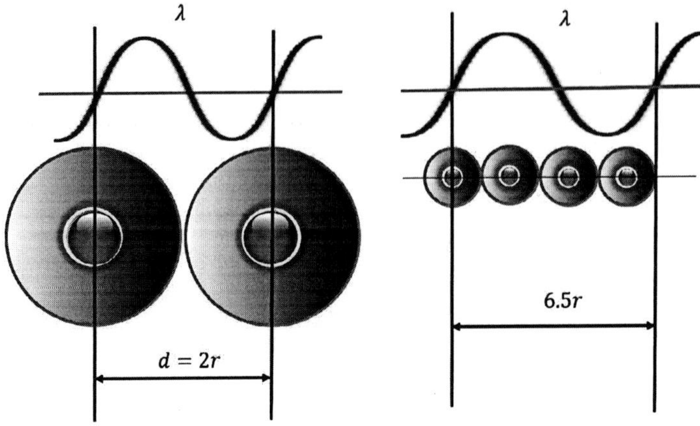

FIGURE A7.3 Calculation of number of point sources for converged DPSM results.

From the above two numbers λ_3 is the lowest with smallest wavelength. In DPSM, this smallest wavelength is used to calculate the distance between two-point sources as shown in Fig. A7.3. The minimum distance between two-point sources should be equal to the diameter of the boundary layer of the source bulbs, i.e., equal to $d = 2r$, twice the radius of the source bulb. Next to simulate the wave correctly, it is necessary to fit minimum one wavelength within that distance d to have at least two source bulbs within one wavelength. Hence, the radius of each bulb should be $r \leq \lambda/2$. This criterion gives satisfactory stable result; however, it was found [11] from simulations that the solutions are not converged with this criterion. In FEM and SEM method for wave simulation, a thumb rule is used to create an element size $a \leq \lambda/10$. Further to see if that criterion is necessary for DPSM, several simulations were conducted. Through many studies [5, 11, 13, 14, 34, 47–54] it was found that the most convenient, computationally nontaxing, stable and converged result can be found using $r \leq \lambda/2\pi$, which is close to $r \leq \lambda/6.5$. Similar criterion was also used to model converged wave field in anisotropic media. Although this criterion is just a recommendation based on author's findings, user is recommended to perform their own convergence study respective to their CNDE problem in hand.

7.12.3 Flow Chart for the DPSM Algorithm

Fig. A7.4 shows the DPSM algorithm flow through a flow chart broken into three parts. In Part 1, user should provide the input data, for example, number and geometry of transducer, geometry of material to be inspected, material properties, arrangement of fluid and solid, and frequency of CNDE virtual experiment. Part 1 shows the process till user need to generate the big DPSM or Green's function matrix. First the memory required for the matrix depending on the size based on CNDE problem has to be allocated. Next in Part 2 the user needs to populate the matrix by filling right the Green's function matrices based on the boundary and interface conditions. Here, parallelization is necessary. After constructing the DPSM matrix, unknown source strengths should be solved. After solving the source strengths in Part 3 user needs to compute the wave field in frequency or time domain at a desired point or over the entire problem domain. Here, again parallelization might be necessary. Fig. A7.4a–c shows the Partl, Part 2, and Part 3 of the DPSM algorithm.

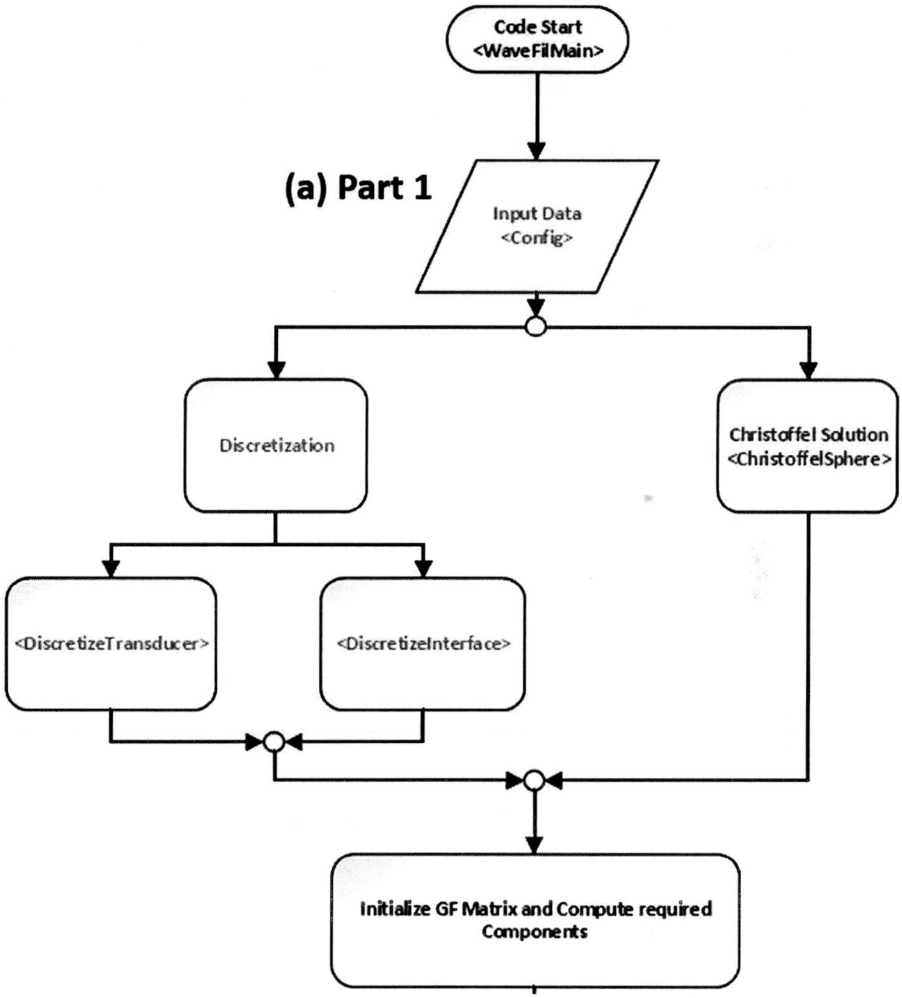

FIGURE A7.4 DPSM Flow Chart a) Part 1, from the input data for a CNDE problem to Initializing Green's Function (GF) or DPSM matrix b) Part 2 from Initialization to the Solution of GF or DPSM matrix to find the source Strength c) Part 3 from finding the source strength to the computation of wave field in a user defined area.

(*Continued*)

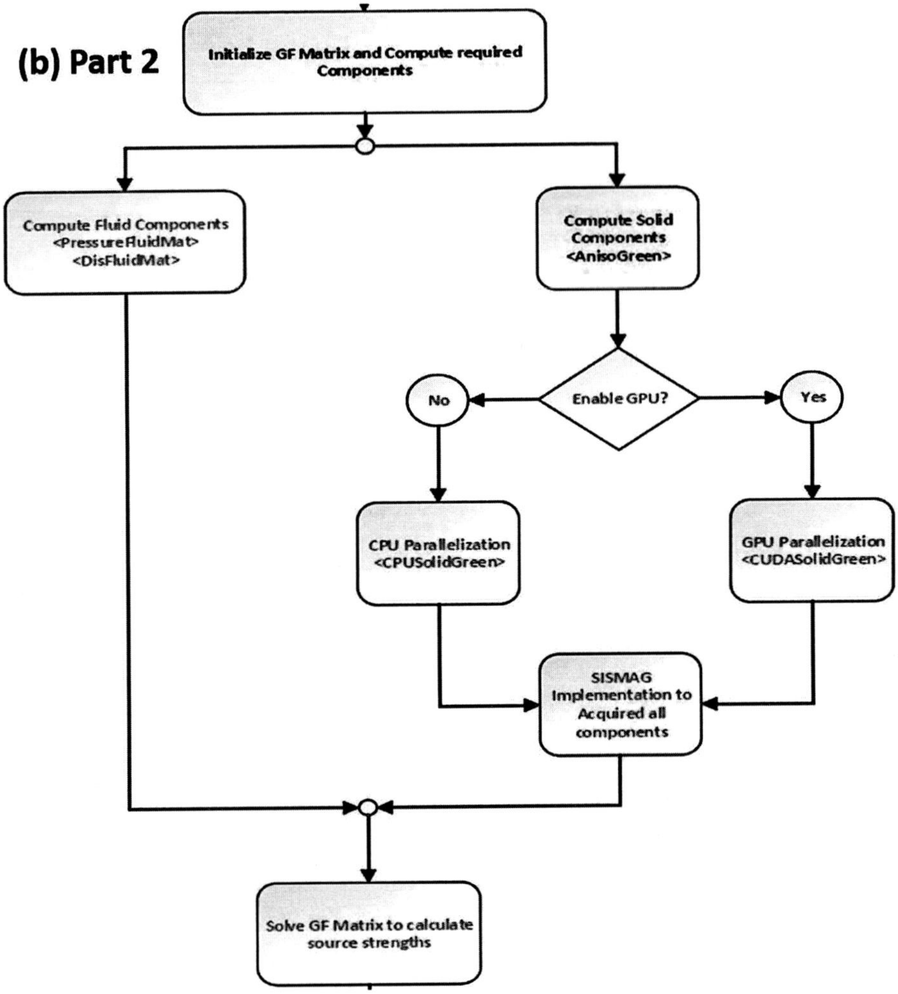

FIGURE A7.4 *(Continued)*

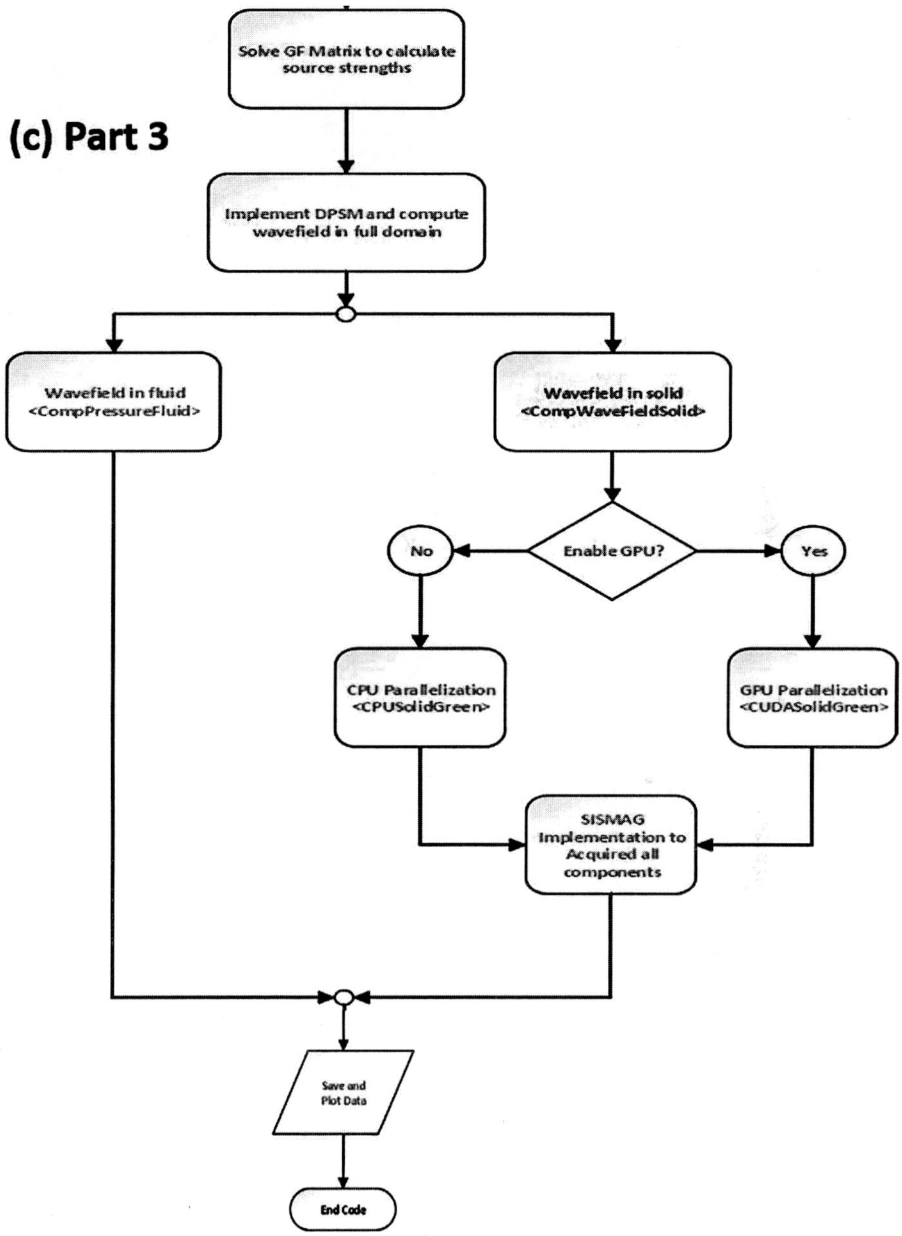

FIGURE A7.4 (*Continued*)

REFERENCES

1. Shaw, R.P., Ch.6 Boundary integral equation methods applied to wave problems, in *Developments in Boundary Element Methods -- 1*, P.K. Banerjee and R. Butterfield, Editors. 1979, London: Applied Science Publishers. pp. 121–153.
2. X. Zhao. and J.L. Rose, Boundary element modeling for defect characterization potential in a wave guide. *International Journal of Solids and Structures*, 2003. **40**(11): p. 2645.
3. Banerjee, S. and T. Kundu, Ultrasonic field modeling in plates immersed in fluid. *International Journal of Solids and Structures,* 2007. **44**(18–19): pp. 6013–6029.
4. Banerjee, S., T. Kundu, and N.A. Alnuaimi, DPSM technique for ultrasonic field modelling near fluid–solid interface. *Ultrasonics,* 2007. **46**(3): pp. 235–250.
5. Banerjee, S., Das, S., Kundu, T., Placko, D., Controlled space radiation concept for mesh-free semi-analytical technique to model wave fields in complex geometries. *Ultrasonics,* 2009. **48**(8): pp. 615–622.
6. Haberman, R., *Elementary Applied Partial Differential Equations With Fourier Series and Boundary Value Problems,* 3rd ed. 1997, New Jersey: Prentice Hall.
7. Auld, B.A., *Acoustic Fields and Waves in Solids.* 1973: Рипол Классик.
8. Graff, K., *Wave Motion in Elastic Solids.* 1991, New York: Dover Publications Inc.
9. Nayfeh, A.H., *Wave Propagation in Layered Anisotropic Media: With Application to Composites*, Vol. **39**. 1995, Amsterdam: Elsevier.
10. Schmerr Jr., L.W., *Fundamentals of Ultrasonic Nondestructive Evaluation: A Modeling Approach.* 1998, New York: Springer.
11. Banerjee, S., Alnuaimi, N., Kundu, T., DPSM technique for ultrasonic field modelling near fluid-solid interface. *Ultrasonics,* 2007. **46**(3): pp. 235–250.
12. Placko, D., Kundu, T., *DPSM for Modeling Engineering Problems,* 1st ed. 2007, Hoboken, NJ: Wiley-Interscience.
13. Banerjee, S. and T. Kundu, Advanced Application of distributed point source method – ultrasonic field modeling in solid media, in *DPSM for Modeling Engineering Problems,* T. Kundu and D. Placko, Editors. 2007, Hoboken, NJ: John & Willey Publication.
14. Banerjee, S., T. Kundu, and D. Placko, Ultrasonic field modeling in multilayered fluid structures using the distributed point source method technique. *ASME Journal of Applied Mechanics,* 2005. **73**(4): pp. 598–609.
15. Mal, A.K., Singh, S. J., *Deformation of Elastic Solids.* 1991, New Jersey: Prentice Hall.
16. Banerjee, S. and T. Kundu, Scattering of ultrasonic waves by internal anomalies in plates. *Optical Engineering Journal,* 2006. **46**(5).
17. Achenbach, J., *Wave Propagation in Elastic Solids.* 1975, Amsterdam: Elsevier Science Publishers B. V.
18. Kraut, E.A., Advances in the theory of anisotropic elastic wave propagation. *Reviews of Geophysics,* 1963. **1**(3): pp. 401–448.
19. Burridge, R., Lamb's problem for an anisotropic half-space. *The Quarterly Journal of Mechanics and Applied Mathematics,* 1971. **24**(1): pp. 81–98.
20. Willis, J.R., Self-similar problems in elastodynamics. *Philosophical Transactions of the Royal Society of London A: Mathematical, Physical and Engineering Sciences,* 1973. **274**(1240): pp. 435–491.
21. Willis, J., Inclusions and cracks in constrained anisotropic media, in *Modern Theory of Anisotropic Elasticity and Applications.* 1991, Philadelphia: SIAM. pp. 87–102.
22. Payton, R., *Elastic Wave Propagation in Transversely Isotropic Media,* Vol. **4**. 2012, Berlin: Springer Science & Business Media.
23. Yeatts, F.R., Elastic radiation from a point force in an anisotropic medium. *Physical Review B,* 1984. **29**: pp. 1674–1684.

24. Wang, C.Y. and J.D. Achenbach, A new look at 2-D time-domain elastodynamic Green's functions for general anisotropic solids. *Wave Motion,* 1992. **16**(4): pp. 389–405.
25. Wang, C.Y. and J.D. Achenbach, A new method to obtain 3-D Green's functions for anisotropic solids. *Wave Motion,* 1993. **18**(3): pp. 273–289.
26. Tverdokhlebov, A., Rose, J., On Green's function for elastic waves in anisotropic media. *Journal of Acoustical Society of America,* 1988. **83**: pp. 844–870.
27. Kim, K.Y., A.G. Every, and W. Sachse, Focusing of fast transverse modes in (001) silicon at ultrasonic frequencies. *Journal of Acoustical Society of America,* 1994. **95**(4): p. 1942.
28. Wang, C.Y., Achenbach, J.D., Elastodynamic fundamental solutions for anisotropic solids. *Geophysical Journal International,* 1994. **118**: pp. 384–392.
29. Pluta, M., et al., Angular spectrum approach for the computation of group and phase velocity surfaces of acoustic waves in anisotropic materials. *Ultrasonics,* 2000. **38**: pp. 232–236.
30. Every, A.G., et al., Phonon focusing caustics in crystals and their diffraction broadening at ultrasonic frequencies. *Transactions of the Royal Society of South Africa,* 2004. **58**(2).
31. Deans, S.R., *The Radon Transform and some of its Applications.* 2007: Courier Corporation.
32. Wang, C.-Y. and J. Achenbach. Three-dimensional time-harmonic elastodynamic Green's functions for anisotropic solids, in *Proceedings of the Royal Society of London A: Mathematical, Physical and Engineering Sciences.* 1995, London: The Royal Society.
33. Huk, W.J., Gademann, G., Magnetic resonance imaging (MRI): method and early clinical experiences in diseases of the central nervous system. *Neurosurgery Review,* 1984. **7**(4): pp. 259–80.
34. Das, S., Dao, C.M., Banerjee, S., Kundu, T., Interaction between bounded acoustic beams and corrugated plates. *IEEE Transaction on Ultrasonics, Ferroelectric and Frequency Control,* 2007. **54**(9): pp. 1860–1872.
35. Wang, C.-Y., Green's tensors for solids of general anisotropy. *WIT Transactions on Modelling and Simulation,* 1970. **10**.
36. Every, A.G. and K.Y. Kim, Determination of elastic constants of anisotropic solids from elastodynamic Green's functions. *Ultrasonics,* 1996. **34**(2–5): pp. 471–472.
37. Shrestha, S., Banerjee, S., Virtual nondestructive evaluation of anisotropic plates using symmetry informed sequential mapping of anisotropic Green's function (SISMAG). *Ultrasonics,* 2018. **88**: pp. 53–61.
38. Shrestha, S. and S. Banerjee, Virtual nondestructive evaluation for anisotropic plates using Symmetry Informed Sequential Mapping of Anisotropic Green's function (SISMAG). *Ultrasonics,* 2018. **88**: pp. 51–63.
39. Shrestha, S. and S. Banerjee. Computational wave modeling of anisotropic plates with implementation of Symmetry Informed Sequential Mapping of Anisotropic Green's function (SISMAG), in *Health Monitoring of Structural and Biological Systems XII.* 2018, International Society for Optics and Photonics.
40. Swaminathan, S., S. Ghosh, and N. Pagano, Statistically equivalent representative volume elements for unidirectional composite microstructures: Part I-Without damage. *Journal of Composite Materials,* 2006. **40**(7): pp. 583–604.
41. Swaminathan, S. and S. Ghosh, Statistically equivalent representative volume elements for unidirectional composite microstructures: part II-with interfacial debonding. *Journal of Composite Materials,* 2006. **40**(7): pp. 605–621.
42. Tavaf, V., et al., Quantification of material degradation and its behavior of elastodynamic Green's function for computational wave field modeling in composites. *Materials Today Communications,* 2018. **17**: pp. 402–412.

43. Tavaf, V., et al., Effect of multiscale precursor damage on wave propagation through modulated constitutive properties of composite materials, in *Health Monitoring of Structural and Biological Systems XII*. 2018, International Society for Optics and Photonics.
44. Tavaf, V., Saadatzi, M.S., Shrestha, S., Banerjee, S., Quantification of material degradation and its behavior of elastodynamic Green's function for computational wave field modeling in composites. *Materials Today Communication* 2018. **17**: pp. 402–412.
45. Tavaf, V., Saadatzi, MS., Banerjee, S., Quantification of degraded constitutive coefficients of composites in the presence of distributed defects. *Journal of Composite Materials,* 2019. **53**(18).
46. Tavaf, V., M. Saadatzi, and S. Banerjee, Quantification of degraded constitutive coefficients of composites in the presence of distributed defects. *Journal of Composite Materials,* 2019: p. 0021998319832351.
47. Banerjee, S., Kundu, T., Symmetric and anti-symmetric Rayleigh-Lamb modes in sinusoidally corrugated waveguides: an Analytical approach. *International Journal of Solids and Structures,* 2006. **43**(6551–6567).
48. Banerjee, S. and T. Kundu, Elastic wave propagation in sinusoidally corrugated waveguides. *Journal of the Acoustical Society of America,* 2006. **119**(4): pp. 2006–2017.
49. Banerjee, S., Kundu, T., Ultrasonic field modeling in plates immersed in fluids. *International Journal of Solids and Structures,* 2007. **44**(18–19): pp. 6013–6029.
50. Banerjee, S., Kundu, T., Semi-analytical modeling of ultrasonic fields in solids with internal anomalies immersed in a fluid. *Wave Motion,* 2008. **45**(5): pp. 581–595.
51. Banerjee, S., Kundu, T., Elastic wave field computation in multilayered non-planar solid structures: a mesh-free semi-analytical approach. *Journal of Acoustical Society of America,* 2008. **123**(3): pp. 1371–1382.
52. Das, S., Banerjee, S., Kundu, T., Elastic wave scattering in solid half space with a circular cylindrical hole using distributed point source method. *International Journal of Solids and Structures,* 2008. **45**: pp. 4498–4508.
53. Dao, C.M., Das, S., Banerjee, S., Kundu. T., Wave propagation in a fluid wedge over a solid half space – mesh free analysis with experimental verification. *International Journal of Solids and Structures,* 2009. **46**(11–12): pp. 2486–2492.
54. Das, S., Banerjee, S., Kundu, T. Transient ultrasonic wave field modeling in an elastic half-space using distributed point source method, in *SPIE Smart Structure NDE*. 2009, San Diego, California: SPIE.

8 Elastodynamic Finite Integration Technique

8.1 INTRODUCTION

Elastodynamic Finite Integration Technique (EFIT), for which this chapter is named, is closely related to the broader numerical method category of finite difference (FD) methods. FD methods can be applied to solve ordinary and partial differential equations, including the equations with relevance to nondestructive evaluation (NDE) (e.g., thermal diffusion for thermographic NDE, elastodynamic and acoustic wave propagation for ultrasonic NDE, electromagnetic wave propagation for methods such as terahertz inspection). While this chapter focuses on the use of EFIT for modeling ultrasonic NDE, details on the use of FD methods to model other NDE techniques, such as thermography and electromagnetic based methods, can be found in References [28, 29, 32, 45].

FD schemes use an approach of approximating derivatives as finitely spaced differences. As such, FD methods represent a class of numerical methods where the equation is solved over the discretized domain space. As with other numerical methods (such as finite element method), FD may be an appropriate approach when an analytical solution cannot be readily derived. As an example in NDE, numerical methods may be an appropriate choice for simulating ultrasonic wave behavior in complex geometry components made of advanced materials (such as anisotropic materials) containing complex geometry defects.

The physics-based methods used for most NDE techniques can be represented by differential equations with spatial and time dependence. The associated FD schemes can be spatially discretized in 1-dimension (1D), 2-dimensions (2D), or 3-dimensions (3D) and can use implicit or explicit methods to move the solution forward in time. Explicit time stepping is mathematically and computationally straightforward. However, explicit schemes can also require strict time step sizes to meet stability requirements. Implicit schemes generally require a more complex mathematical solution compared to explicit schemes; however, a benefit is that they do not have the strict time step size requirements of explicit schemes (thus, with implicit schemes, it may be possible to take larger time steps compared to explicit schemes).

Two commonly used FD approaches are standard staggered grid (SSG) methods and rotated staggered grid (RSG) methods [39]. In staggered grid schemes, the solution converges more quickly than in nonstaggered schemes due to the strategic placement of derivatives centered at half-grid points (i.e., between grid points) [19, 48]. For elastodynamics, this approach results in a grid cell where velocity and stress components are located at different positions on the grid. An example of a 2D

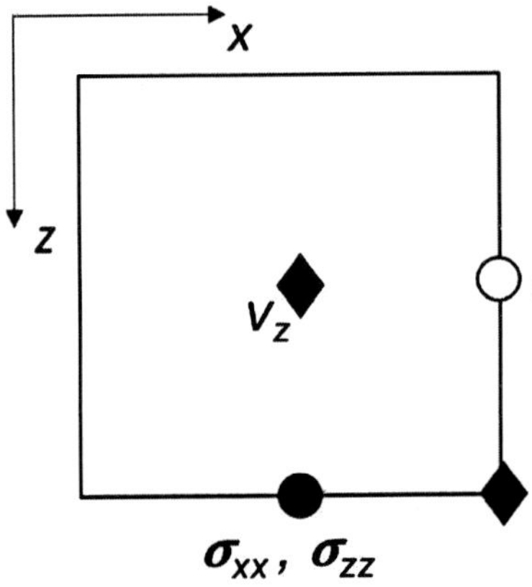

FIGURE 8.1 Example of a 2D Virieux standard-staggered grid [48].

elastodynamic standard staggered grid cell is given in Fig. 8.1. SSG and RSG FD methods (as well as mixed grid methods) have been heavily explored and utilized in the field of geophysics for elastic wave propagation in isotropic and anisotropic materials, but are also highly applicable to NDE [12, 18, 39, 49]. A thorough overview on the basic mathematics of general FD schemes can be found in books on the subject such as in Reference [44].

The remainder of this chapter focuses on a particular SSG type approach called Finite Integration Technique (FIT). When this technique is applied to ultrasonic wave propagation in fluids, the resulting mathematical scheme is called Acoustic Finite Integration Technique (AFIT). Since ultrasonic NDE often involves sending waves into a coupling medium (commonly a fluid medium) before the waves enter the solid medium under inspection, AFIT is discussed with regard to simulating ultrasound in a fluid. When FIT is applied to ultrasound propagation in solid media, the approach is referred to as Elastodynamic Finite Integration Technique (EFIT). Sections 8.2.1 and 8.3 describe the mathematics and computational implementation of these two schemes in 3D. EFIT simulation examples for ultrasonic wave propagation in solid media are described in Section 8.4. It is also noted that in this chapter, the discussion on simulating ultrasonic NDE is limited to simulating waves in inviscid fluids and elastic solids.

8.1.1 Finite Integration Technique

As mentioned in the prior section, FIT is an explicit SSG type of approach [43]. A general overview of FIT, as applied to acoustics, elastodynamics, and electromagnetics, can be found in a prior review paper by Marklein [29]. In this chapter, the equations for Acoustic Finite Integration Technique (AFIT) are briefly discussed, followed by a more in-depth discussion of Elastodynamic Finite Integration Technique (EFIT). Mathematical and computational implementation details of EFIT are described in detail, with the intent to be used a practical aid by readers.

Regarding computational implementation, FD approaches, such as FIT, lead to algebraic equations that are practical to parallelize for use on multicore, manycore, or cluster computing hardware. In recent years, EFIT has been effectively parallelized to run on multiple central processing unit (CPU) cores using tools such as Message Passing Interface (MPI) and OpenMP. The method has also been implemented on manycore accelerator hardware such as graphics processing units (GPUs) and Intel Xeon Phi [9, 40]. Considerations of the computational implementation of EFIT are discussed in Section 8.3.6.

The staggered grid mathematical approaches shown here for both AFIT and EFIT are second-order accurate (though the accuracy can reduce to first order at boundaries depending on the boundary condition that is implemented) [16]. A detailed discussion on the order of error of FIT compared with other methods (such as finite element and spectral element methods) can be found in Reference [46].

8.2 ACOUSTIC FINITE INTEGRATION TECHNIQUE

AFIT can be used to model ultrasonic wave propagation in fluids. For the field of NDE, the AFIT approach is relevant for modeling ultrasonic waves in water or other fluid coupling media before the waves enter a solid medium (i.e., the specimen under inspection). There may be also NDE cases where the material under inspection is fluid-like. In the following section, the mathematical equations of AFIT are outlined in detail, followed by a discussion on step size, stability, initial conditions, and boundary conditions.

8.2.1 MATHEMATICAL EQUATIONS: AFIT

Wave motion in a fluid can be described using the Navier-Stokes equations. In this case, we begin with the linearized Navier-Stokes equations for an inviscid fluid, written in terms of pressure [29, 33] as follows:

$$\dot{p} = -\rho c^2 \nabla \cdot \vec{v} \tag{8.1}$$

$$\rho \dot{\vec{v}} = -\nabla p \tag{8.2}$$

where the dot notation stands for the partial derivative in time (e.g., $\dot{p} = \partial p / \partial t$), p is the pressure, v is the velocity, and ρ is the density of the fluid.

Discretization of the above equations over the Cartesian cubic grid, shown in Fig. 8.2, leads to the four equations below for pressure and velocity:

$$\dot{p} = -\frac{\rho c^2}{\Delta x}\left(v_x^{(n)} - v_x^{(n-\hat{x})} + v_y^{(n)} - v_y^{(n-\hat{y})} + v_z^{(n)} - v_z^{(n-\hat{z})}\right) \tag{8.3}$$

$$\begin{aligned}
\rho \dot{v}_x &= -\frac{1}{\Delta x}\left(p^{(n+\hat{x})} - p^{(n)}\right) \\
\rho \dot{v}_y &= -\frac{1}{\Delta x}\left(p^{(n+\hat{y})} - p^{(n)}\right) \\
\rho \dot{v}_z &= -\frac{1}{\Delta x}\left(p^{(n+\hat{z})} - p^{(n)}\right)
\end{aligned} \tag{8.4}$$

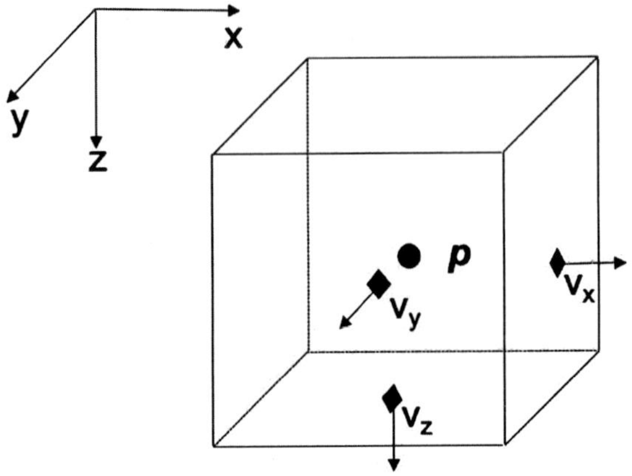

FIGURE 8.2 The diagram above shows the 3D AFIT grid in Cartesian coordinates. Pressure is located at the center of the cell with velocities located on cell faces.

where the cubic spatial grid assumes that $\Delta x = \Delta y = \Delta z$, and the notation $v_x^{(n-\hat{x})}$ represents the v_x velocity component located at one grid cell in the $-\hat{x}$ direction from the nth cell. Central differences in time can be used to yield velocity and pressure equations that are discretized in time and space:

$$v(t) = v(t - \Delta t) + \dot{v}(t - \Delta t/2)\Delta t \tag{8.5}$$

$$p(t + \Delta t/2) = p(t - \Delta t/2) + \dot{p}(t)\Delta t \tag{8.6}$$

here, the notation Δt represents one whole time step.

Thus, the fully discretized velocity equations for an inhomogenous fluid are:

$$\begin{aligned} v_x(t) &= v_x(t-\Delta t) - \frac{2}{\left(\rho^{(n+\hat{x})} + \rho^{(n)}\right)} \frac{\Delta t}{\Delta x}\left(p^{(n+\hat{x})} - p^{(n)}\right) \\ v_y(t) &= v_y(t-\Delta t) - \frac{2}{\left(\rho^{(n+\hat{y})} + \rho^{(n)}\right)} \frac{\Delta t}{\Delta y}\left(p^{(n+\hat{y})} - p^{(n)}\right) \\ v_z(t) &= v_z(t-\Delta t) - \frac{2}{\left(\rho^{(n+\hat{z})} + \rho^{(n)}\right)} \frac{\Delta t}{\Delta z}\left(p^{(n+\hat{z})} - p^{(n)}\right) \end{aligned} \tag{8.7}$$

and the pressure is given by

$$p(t + \Delta t/2) = p(t - \Delta t/2) - \rho c^2 \frac{\Delta t}{\Delta x}\left(v_x^{(n)} - v_x^{(n-\hat{x})} + v_y^{(n)} - v_y^{(n-\hat{y})} + v_z^{(n)} - v_z^{(n-\hat{z})}\right) \tag{8.8}$$

It can be observed that the velocity and pressure equations are staggered in time by $\Delta t/2$, leading to a leap-frog in time scheme as the simulation progresses [29].

8.2.2 Step Size and Stability Conditions

As shown in Fig. 8.2, the AFIT grid is based on a cubic voxelization scheme. In order to accurately capture the physics of wave propagation, a maximum size limit is placed on the spatial step size to ensure enough computational points for the smallest wavelength expected in the simulated scenario. The recommended minimum number of grid points per wavelength reported in the literature varies, ranging from $\lambda/8$ to $\lambda/15$ [33, 43]. In practice, it is not uncommon to use spatial step sizes on the order of $\lambda/40$. Additionally, the Courant-Friedrichs-Lewy condition is used to determine the maximum time step size [4, 33], to ensure stability of the scheme. The spatial and time step size requirements are given by:

$$\begin{aligned} \Delta x & \leq \frac{c_{min}}{8 f_{max}}, \\ & \leq \lambda_{min}/8 \\ \Delta t & \leq \frac{\Delta x}{c_{max}\sqrt{3}}, \end{aligned} \tag{8.9}$$

where it is assumed that $\Delta x = \Delta y = \Delta z$, c_{min} and c_{max} are the minimum and maximum speeds of sound expected in the simulated scenario (which could involve multiple fluid materials), and f_{max} is the maximum frequency.

8.2.3 Initial Conditions and Boundary Conditions

In order to introduce ultrasonic waves in the AFIT simulation, an incident pressure function can be added to Eq. (8.3) as an initial condition at desired spatial grid cell locations (for example, over a circular region representing a transducer face). Alternatively, a function can be added to Eq. (8.4) to create an initial change in displacement (velocity). To model typical NDE ultrasound excitations in a fluid, the initial condition will likely be a time-dependent function with a varying amplitude as the simulation steps forward in time (for example, a tone burst excitation such as a windowed sine wave with a specified center frequency in the kHz or MHz frequency range). Since inviscid fluids (such as water) can support longitudinal (compressional) ultrasonic waves but cannot support shear (transverse) waves, the initial condition will result in the generation of longitudinal waves.

For many NDE simulation applications, the simple approach of adding an initial condition to the AFIT pressure and/or velocity equations to excite ultrasonic waves is sufficient. However, in some applications, one may desire to simulate the electromechanical behavior of a piezoelectric transducer. This route for generating the simulated excitation is far more complex, and details for a FIT approach can be found in the literature [29].

A boundary condition that is relevant for NDE applications is that of rigid body reflection (i.e., total reflection). In particular, such a boundary condition could be used to model reflection from boundaries of the simulation space. The same condition can also be used to model wave scatter from rigid reflectors within the fluid, if desired. The rigid reflector boundary condition is implemented by setting the velocities (v_x, v_y, and v_z) equal to zero at the location of rigid reflectors. For example, for total reflection at minimum and maximum $\hat{x}$ boundaries:

$$\begin{aligned} v_x \,|_{x_{max}} &= 0 \\ v_x \,|_{x_{min}} &= 0 \end{aligned} \tag{8.10}$$

Alternatively, absorbing boundary conditions could be used to damp out reflections from simulation edges. A perfectly matched layer (PML) approach can be used to efficiently reduce reflections from simulation boundaries. A brief discussion on PML conditions is given in Section 8.3.4.

For NDE applications, the treatment of a fluid-solid interface (i.e., acoustic-elastodynamic interface) is highly relevant. Such an interface occurs at the boundary of a fluid coupling medium (such as water) and the solid media specimen under inspection. There are two common approaches for the treatment of such an interface. One is to use a partitioned approach, in which acoustic equations are used in the fluid region and elastodynamic equations are used in the solid region, with the enforcement of appropriate boundary conditions at the interface. The other common approach is a monolithic approach, in which the same equations are used in both regions (in this case, elastodynamic equations, EFIT) and material properties are adjusted to represent the fluid region. In practice, some EFIT researchers have found it adequate to model waves in the fluid by simply letting the shear modulus go to zero, rather than coupling AFIT and EFIT models [6]. A study on the accuracy of several monolithic methods for various numerical methods, including multiple FD schemes, can be found in the geophysics literature [11]. Basabe and Sen reported that the use of monolithic approaches for a fluid-solid interface can result in varying degrees of accuracy, and for some numerical methods can result in spurious modes.

8.3 ELASTODYNAMIC FINITE INTEGRATION TECHNIQUE

FIT was applied to elastodynamics as early as the 1990s by Fellinger and colleagues [13]. Since that time, the technique has been used to simulate ultrasound in solid media for various wave types, including Rayleigh waves, surface acoustic waves, bulk waves, and guided waves. The method has been applied to inhomogeneous, isotropic, anisotropic, and attenuative materials; as well as cases involving nonlinear wave propagation [4, 7, 10, 21, 22, 26, 30, 38, 41, 47]. Application areas for EFIT simulation reported recently in the literature include: ultrasonic inspection of railroad tracks [35], shear wave attenuation in concrete [1], and guided wave simulation for aerospace components [17], among others. For isotropic materials, studies with experimental comparisons have been reported in the literature by several authors, including [27, 38]. Additionally, recent benchmarks against experiment for ultrasonic wave propagation in orthotropic composite laminates were reported by

Leckey et al. [22, 26]. Experimental benchmark studies along with verification against theory can be used by readers to aid in verification and validation of their computational implementations of EFIT.

In this section, the mathematical equations for isotropic and anisotropic EFIT are described in detail. Stability considerations, as well as computational considerations, will also be discussed. Examples of EFIT results for isotropic and anisotropic materials are then presented.

8.3.1 Mathematical Equations: Isotropic EFIT

In this section, the discretized EFIT equations in 3D Cartesian coordinates for isotropic materials and anisotropic materials are described. The anisotropic EFIT equations reduce to the isotropic case if the stiffness matrix of the material(s) being simulated is of an isotropic form. However, in this chapter, the equations for both isotropic and anisotropic materials are listed explicitly for the purpose of aiding the reader in an easier implementation of the target scenario. It is also noted that while the EFIT equations presented here are for Cartesian coordinates, cylindrical coordinate versions of the equations can be found in the literature [38, 42]. Referred to as cylindrical EFIT (CEFIT), the choice of cylindrical coordinates is most appropriate for applications where simulated components are of cylindrical shape (such as ultrasonic inspection of pipes).

As described in detail in Chapter 3, linear motion in the form of elastic wave propagation in isotropic media can be described mathematically using Cauchy's equation for momentum and Hooke's Law:

$$\rho \dot{v}_j = \frac{\partial \sigma_{ij}}{\partial x_i}, \tag{8.11}$$

$$\sigma_{ij} = \mathbb{L} e_{kk} \delta_{ij} + 2\mathrm{m} e_{ij}, \tag{8.12}$$

where ρ is the density, σ_{ij} is the stress tensor, $\mathbb{L}$ is the Lamé's first parameter, m is the shear modulus, and strain, e_{ij}, is defined in terms of displacement, u, by

$$e_{ij} = \frac{1}{2}\left(\frac{\partial u_i}{\partial x_j} + \frac{\partial u_j}{\partial x_i} \right). \tag{8.13}$$

The Lamé parameters can be defined in terms of longitudinal (c_L) and transverse (c_T) speeds of sound:

$$\begin{aligned} c_L &= \sqrt{(\mathbb{L} + 2\mathrm{m})/\rho} \\ c_T &= \sqrt{\mathrm{m}/\rho}\,. \end{aligned} \tag{8.14}$$

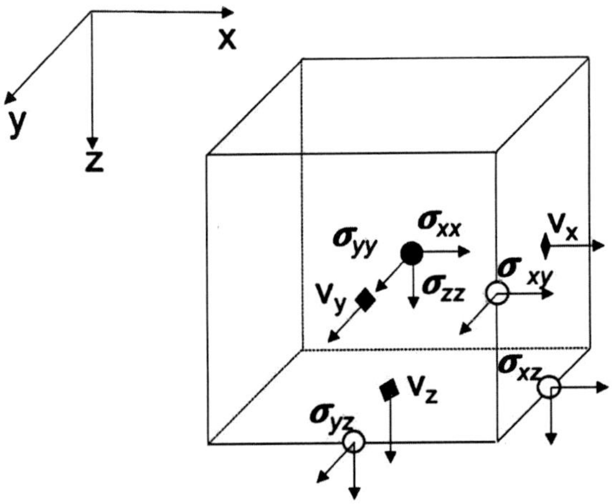

FIGURE 8.3 The diagram above shows the 3D EFIT grid in Cartesian coordinates. Normal stresses are located at the center of the cell with shear stresses located at edges of the cell and velocities located on cell faces.

One can then take the time derivative of Eq. (8.12). In 3D Cartesian coordinates, Eqs. (8.12) and (8.13) result in three velocity equations and six stress equations. Following work of Fellinger and colleagues, the equations can be written on a discrete spatial grid (a cubic grid is shown in this chapter), as shown in Fig. 8.3. Additionally, the material parameters $m, \mathbb{L}$, and ρ are also distributed onto the spatial grid, leading to computationally implementable equations for wave propagation in isotropic inhomogeneous solid media (the material within each cell is assumed to be homogeneous) [14].

Using the grid layout in Fig. 8.3 leads to the nine equations listed below, where x, y, and z are standard Cartesian coordinates, and the notation $(n + / - \hat{i})$ represents a single grid step in the corresponding Cartesian direction [4, 13, 29]:

$$\begin{aligned}
\dot{v}_x^{(n)}(t) &= \frac{1}{\Delta x}\frac{2}{\rho^{(n)}+\rho^{(n+\hat{x})}}\Big[\Big(\sigma_{xx}^{(n+\hat{x})}(t)-\sigma_{xx}^{(n)}(t)\Big)+\Big(\sigma_{xy}^{(n)}(t)-\sigma_{xy}^{(n-\hat{y})}(t)\Big) \\
&\quad +\Big(\sigma_{xz}^{(n)}(t)-\sigma_{xz}^{(n-\hat{z})}(t)\Big)\Big], \\
\dot{v}_y^{(n)}(t) &= \frac{1}{\Delta x}\frac{2}{\rho^{(n)}+\rho^{(n+\hat{y})}}\Big[\Big(\sigma_{xy}^{(n)}(t)-\sigma_{xy}^{(n-\hat{x})}(t)\Big)+\Big(\sigma_{yy}^{(n+\hat{y})}(t)-\sigma_{yy}^{(n)}(t)\Big) \\
&\quad +\Big(\sigma_{yz}^{(n)}(t)-\sigma_{yz}^{(n-\hat{z})}(t)\Big)\Big], \\
\dot{v}_z^{(n)}(t) &= \frac{1}{\Delta x}\frac{2}{\rho^{(n)}+\rho^{(n+\hat{z})}}\Big[\Big(\sigma_{xz}^{(n)}(t)-\sigma_{xz}^{(n-\hat{x})}(t)\Big)+\Big(\sigma_{yz}^{(n)}(t)-\sigma_{yz}^{(n-\hat{y})}(t)\Big) \\
&\quad +\Big(\sigma_{zz}^{(n+\hat{z})}(t)-\sigma_{zz}^{(n)}(t)\Big)\Big],
\end{aligned} \tag{8.15}$$

$$\begin{aligned}
\dot{\sigma}_{xx}^{(n)}(t) &= \frac{1}{\Delta x}\Big[\big(\mathbb{L}^{(n)}+2m^{(n)}\big)\big(v_x^{(n)}(t)-v_x^{(n-\hat{x})}(t)\big)+\mathbb{L}^{(n)}(v_y^{(n)}(t)-v_y^{(n-\hat{y})}(t) \\
&\quad + v_z^{(n)}(t)-v_z^{(n-\hat{z})}(t))\Big], \\
\dot{\sigma}_{yy}^{(n)}(t) &= \frac{1}{\Delta x}\Big[\big(\mathbb{L}^{(n)}+2m^{(n)}\big)\big(v_y^{(n)}(t)-v_y^{(n-\hat{y})}(t)\big)+\mathbb{L}^{(n)}(v_x^{(n)}(t)-v_x^{(n-\hat{x})}(t) \\
&\quad + v_z^{(n)}(t)-v_z^{(n-\hat{z})}(t))\Big], \\
\dot{\sigma}_{zz}^{(n)}(t) &= \frac{1}{\Delta x}\Big[\big(\mathbb{L}^{(n)}+2m^{(n)}\big)\big(v_z^{(n)}(t)-v_z^{(n-\hat{z})}(t)\big)+\mathbb{L}^{(n)}(v_x^{(n)}(t)-v_x^{(n-\hat{x})}(t) \\
&\quad + v_y^{(n)}(t)-v_y^{(n-\hat{y})}(t))\Big], \\
\dot{\sigma}_{xy}^{(n)}(t) &= \frac{1}{\Delta x}\frac{4}{\frac{1}{m^{(n)}}+\frac{1}{m^{(n+\hat{x})}}+\frac{1}{m^{(n+\hat{y})}}+\frac{1}{m^{(n+\hat{x}+\hat{y})}}}\Big[\big(v_x^{(n+\hat{y})}(t)-v_x^{(n)}(t)\big) \\
&\quad +\big(v_y^{(n+\hat{x})}(t)-v_y^{(n)}(t)\big)\Big], \\
\dot{\sigma}_{xz}^{(n)}(t) &= \frac{1}{\Delta x}\frac{4}{\frac{1}{m^{(n)}}+\frac{1}{m^{(n+\hat{x})}}+\frac{1}{m^{(n+\hat{z})}}+\frac{1}{m^{(n+\hat{x}+\hat{z})}}}\Big[\big(v_x^{(n+\hat{z})}(t)-v_x^{(n)}(t)\big) \\
&\quad +\big(v_z^{(n+\hat{x})}(t)-v_z^{(n)}(t)\big)\Big], \\
\dot{\sigma}_{yz}^{(n)}(t) &= \frac{1}{\Delta x}\frac{4}{\frac{1}{m^{(n)}}+\frac{1}{m^{(n+\hat{y})}}+\frac{1}{m^{(n+\hat{z})}}+\frac{1}{m^{(n+\hat{x}+\hat{z})}}}\Big[\big(v_y^{(n+\hat{z})}(t)-v_y^{(n)}(t)\big) \\
&\quad +\big(v_z^{(n+\hat{y})}(t)-v_z^{(n)}(t)\big)\Big],
\end{aligned} \tag{8.16}$$

These nine equations are then discretized in time using central differences.

$$v_i(t) = v_i(t-\Delta t)+\dot{v}_i(t-\Delta t/2)\Delta t, \tag{8.17}$$

$$\sigma_{ij}(t+\Delta t/2) = \sigma_{ij}(t-\Delta t/2)+\dot{\sigma}_{ij}(t)\Delta t, \tag{8.18}$$

where t is the current time step and Δt and $\Delta t/2$ are full and half time steps. The resulting velocity and stress components are therefore staggered in time by $\Delta t/2$.

8.3.2 Mathematical Equations: Anisotropic EFIT

For anisotropic materials, the velocity equations (requiring only density as a material property) remain the same as above. However, for anisotropic cases, Eq. (8.12) becomes

$$\sigma_{ij} = \mathbb{C}_{ijkl}e_{kl} \tag{8.19}$$

where $\mathbb{C}_{ijkl}$ is the stiffness tensor $(i, j, k, l = 1:3)$. Thus, the EFIT stress equations must now incorporate anisotropic material properties. Materials with general anisotropy (triclinic anisotropy) correspond to a fully populated stiffness matrix with 21 unique constants [37]:

$$\mathbb{C} = \begin{bmatrix} \mathbb{C}_{11} & \mathbb{C}_{12} & \mathbb{C}_{13} & \mathbb{C}_{14} & \mathbb{C}_{15} & \mathbb{C}_{16} \\ \mathbb{C}_{21} & \mathbb{C}_{22} & \mathbb{C}_{23} & \mathbb{C}_{24} & \mathbb{C}_{25} & \mathbb{C}_{26} \\ \mathbb{C}_{31} & \mathbb{C}_{32} & \mathbb{C}_{33} & \mathbb{C}_{34} & \mathbb{C}_{35} & \mathbb{C}_{36} \\ \mathbb{C}_{41} & \mathbb{C}_{42} & \mathbb{C}_{43} & \mathbb{C}_{44} & \mathbb{C}_{45} & \mathbb{C}_{46} \\ \mathbb{C}_{51} & \mathbb{C}_{52} & \mathbb{C}_{53} & \mathbb{C}_{54} & \mathbb{C}_{55} & \mathbb{C}_{56} \\ \mathbb{C}_{61} & \mathbb{C}_{62} & \mathbb{C}_{63} & \mathbb{C}_{64} & \mathbb{C}_{65} & \mathbb{C}_{66} \end{bmatrix} \tag{8.20}$$

where Voigt notation is used, and there is symmetry about the main diagonal (e.g., $\mathbb{C}_{12} = \mathbb{C}_{21}$). The resulting form of the normal stress equations is given below:

$$\begin{aligned} \dot{\sigma}_{ii}^{(n)}(t) = \frac{1}{\Delta x}\Big[& \mathbb{C}_{i1}^{(n)}\left(v_x^{(n)} - v_x^{(n-\hat{x})}\right) + \mathbb{C}_{i2}^{(n)}\left(v_y^{(n)} - v_y^{(n-\hat{y})}\right) + \mathbb{C}_{i3}^{(n)}\Big(v_z^{(n)} \\ & - v_z^{(n-\hat{z})}(t)\Big) + (1/4)\mathbb{C}_{i4}^{(n)}\Big[\left(v_y^{(n)} + v_y^{(n-\hat{y})} + v_y^{(n+\hat{z})} + v_y^{(n-\hat{y}+\hat{z})}\right) \\ & - \left(v_y^{(n)} + v_y^{(n-\hat{y})} + v_y^{(n-\hat{z})} + v_y^{(n-\hat{y}-\hat{z})}\right) + \Big(v_z^{(n)} + v_z^{(n-\hat{z})} \\ & + v_z^{(n+\hat{y})} + v_z^{(n-\hat{z}+\hat{y})}\Big) - \left(v_z^{(n)} + v_z^{(n-\hat{z})} + v_z^{(n-\hat{y})} + v_z^{(n-\hat{z}-\hat{y})}\right)\Big] \\ & + (1/4)\mathbb{C}_{i5}^{(n)}\Big[\left(v_x^{(n)} + v_x^{(n-\hat{x})} + v_x^{(n+\hat{z})} + v_x^{(n-\hat{x}+\hat{z})}\right) - \Big(v_x^{(n)} + v_x^{(n-\hat{x})} \\ & + v_x^{(n-\hat{z})} + v_x^{(n-\hat{x}-\hat{z})}\Big) + \left(v_z^{(n)} + v_z^{(n-\hat{z})} + v_z^{(n+\hat{x})} + v_z^{(n+\hat{x}-\hat{z})}\right) \\ & - \left(v_z^{(n)} + v_z^{(n-\hat{z})} + v_z^{(n-\hat{x})} + v_z^{(n-\hat{x}-\hat{z})}\right)\Big] + (1/4)\mathbb{C}_{i6}^{(n)}\Big[\Big(v_x^{(n)} \\ & + v_x^{(n-\hat{x})} + v_x^{(n+\hat{x})} + v_x^{(n+\hat{y}-\hat{x})}\Big) - \left(v_x^{(n)} + v_x^{(n-\hat{x})} + v_x^{(n-\hat{x})} + v_x^{(n-\hat{y}-\hat{x})}\right) \\ & + \left(v_y^{(n)} + v_y^{(n-\hat{y})} + v_y^{(n+\hat{x})} + v_y^{(n+\hat{x}-\hat{y})}\right) - \left(v_y^{(n)} + v_y^{(n-\hat{y})} + v_y^{(n-\hat{x})} + v_y^{(n-\hat{x}-\hat{y})}\right)\Big]\Big] \end{aligned} \tag{8.21}$$

Note that some terms in the above equation cancel, but the full equation form is listed to demonstrate to the reader the symmetries in selecting the required velocity components from the EFIT grid. More details can be found in work by Halkjaer [16].

Due to length, a single anisotropic EFIT shear stress equation is listed below. Again, the full equation form is shown to demonstrate how to select the appropriate velocity grid elements when deriving the other two shear stress equations.

$$\begin{aligned}
\sigma_{yz}^{(n)}(t+\Delta t/2) = \sigma_{yz}^{(n)}(t-\Delta t/2) + \frac{\Delta t}{4\Delta x}\Big[& \bar{\mathbb{C}}_{yz,41}^{(n)}\Big(v_x^{(n)} + v_x^{(n+\hat{z})} + v_x^{(n+\hat{y})} \\
& + v_x^{(n+\hat{y}+\hat{z})} - v_x^{(n-\hat{x})} - v_x^{(n-\hat{x}+\hat{z})} - v_x^{(n-\hat{x}+\hat{y})} - v_x^{(n-\hat{x}+\hat{y}+\hat{z})}\Big) \\
& + \bar{\mathbb{C}}_{yz,42}^{(n)}\Big(v_y^{(n)} + v_y^{(n+\hat{z})} + v_y^{(n+\hat{y})} + v_y^{(n+\hat{y}+\hat{z})} \\
& - v_y^{(n)} - v_y^{(n+\hat{z})} - v_y^{(n-\hat{y})} - v_y^{(n-\hat{y}+\hat{z})}\Big) \\
& + \bar{\mathbb{C}}_{yz,43}^{(n)}\Big(v_z^{(n)} + v_z^{(n+\hat{y})} + v_z^{(n+\hat{z})} + v_z^{(n+\hat{y}+\hat{z})} \\
& - v_z^{(n)} - v_z^{(n+\hat{y})} - v_z^{(n-\hat{z})} - v_z^{(n+\hat{y}-\hat{z})}\Big) \\
& + \bar{\mathbb{C}}_{yz,45}^{(n)}\Big(v_x^{(n+\hat{z})} + v_x^{(n-\hat{x}+\hat{z})} + v_x^{(n+\hat{y}+\hat{z})} + v_x^{(n-\hat{x}+\hat{y}+\hat{z})} \\
& - v_x^{(n)} - v_x^{(n-\hat{x})} - v_x^{(n+\hat{y})} - v_x^{(n-\hat{x}+\hat{y})} \\
& + v_z^{(n)} + v_z^{(n+\hat{y})} + v_z^{(n+\hat{x})} + v_z^{(n+\hat{x}+\hat{y})} \\
& - v_z^{(n)} - v_z^{(n+\hat{y})} - v_z^{(n-\hat{x})} - v_z^{(n-\hat{x}+\hat{y})}\Big) \\
& + \bar{\mathbb{C}}_{yz,46}^{(n)}\Big(v_x^{(n+\hat{y})} + v_x^{(n-\hat{x}+\hat{y})} + v_x^{(n-\hat{x}+\hat{y}+\hat{z})} + v_x^{(n+\hat{y}+\hat{z})} \\
& - v_x^{(n)} - v_x^{(n-\hat{x})} - v_x^{(n+\hat{z})} - v_x^{(n-\hat{x}+\hat{z})} \\
& + v_y^{(n)} + v_y^{(n+\hat{z})} + v_y^{(n+\hat{x})} + v_y^{(n+\hat{x}+\hat{z})} \\
& - v_y^{(n)} - v_y^{(n+\hat{z})} - v_y^{(n-\hat{x})} - v_y^{(n-\hat{x}+\hat{z})}\Big)\Big] \\
& + \frac{\Delta t}{\Delta x}\bar{\mathbb{C}}_{yz,44}^{(n)}\Big(v_y^{(n+\hat{z})} - v_y^{(n)} + v_z^{(n+\hat{y})} - v_z^{(n)}\Big)
\end{aligned} \tag{8.22}$$

In the anisotropic EFIT shear stress equations, the $\bar{C}$ terms are defined in terms of the compliance matrix, $\mathbb{S}$. One example of $\bar{S}_{yz,41}$ is shown below (where other terms can be derived following the pattern shown below):

$$\bar{\mathbb{S}}_{yz} = \frac{1}{4}\Big(\mathbb{S}^{(n)} + \mathbb{S}^{(n+\hat{x})} + \mathbb{S}^{(n+\hat{y})} + \mathbb{S}^{(n+\hat{x}+\hat{y})}\Big) \tag{8.23}$$

Recall that the stiffness matrix is the inverse of the compliance matrix:

$$\bar{\mathbb{C}}_{yz} = (\bar{\mathbb{S}}_{yz})^{-1} \tag{8.24}$$

The equations above reduce significantly for common anisotropy cases occurring in NDE, such as transversely isotropic or orthotropic materials (e.g., composite laminates made of unidirectional prepreg carbon fiber-reinforced polymer material).

In such cases, several stiffness matrix values are equal to zero. The stiffness matrix for an orthotropic material takes the general form:

$$\mathbb{C} = \begin{bmatrix} \mathbb{C}_{11} & \mathbb{C}_{12} & \mathbb{C}_{13} & 0 & 0 & 0 \\ \mathbb{C}_{12} & \mathbb{C}_{22} & \mathbb{C}_{23} & 0 & 0 & 0 \\ \mathbb{C}_{13} & \mathbb{C}_{23} & \mathbb{C}_{33} & 0 & 0 & 0 \\ 0 & 0 & 0 & \mathbb{C}_{44} & 0 & 0 \\ 0 & 0 & 0 & 0 & \mathbb{C}_{55} & 0 \\ 0 & 0 & 0 & 0 & 0 & \mathbb{C}_{66} \end{bmatrix} \tag{8.25}$$

8.3.3 Grid Sizing and Stability Requirements

The cubic voxelization scheme means that the spatial step size is the same in all directions. The reported number of recommended points per wavelength in the literature varies, ranging from $\lambda/8$ to $\lambda/15$ [43, 33]. In practice, a value on the order of $\lambda/40$ is commonly used for EFIT. As with AFIT, stability conditions require setting a limit on the time step size, with the Courant-Friedrichs-Levy condition determining this limit [4, 33]. The spatial and time step size requirements are:

$$\begin{aligned} \Delta x &\leq \frac{c_{min}}{8 f_{max}}, \\ &\leq \lambda_{min}/8 \\ \Delta t &\leq \frac{\Delta x}{c_{max}\sqrt{3}}, \end{aligned} \tag{8.26}$$

f_{max} is the maximum frequency in the simulated ultrasonic field (often determined by the specified incident wave created by initial conditions). c_{min} and c_{max} are defined as the minimum and maximum speeds of sound in the simulated ultrasonic field. To ensure stability, it is not uncommon to multiply the right-hand side of the time step equation by a factor less than 1 (for example, 0.99 or 0.95) rather than setting equality.

Speed of sound values for determining Δt and Δx is not always simply defined by the shear and longitudinal wave speeds. For example, in the case of simulating Lamb wave propagation, the velocity values used for specifying the spatial and time steps would be based on the expected Lamb modes for the simulated scenario (and could be determined using dispersion curves). In this case, the Lamb mode with the lowest phase velocity value would be used to determine λ_{min}, and the mode with the maximum velocity would determine c_{max}. Further, in cases with varying material properties, small-scale defects, or other special considerations – it may be necessary to use an even smaller step size than the recommended values stated above (see Section 8.4.3 for example).

8.3.4 Boundary Conditions

The equations in Sections 8.3.1 and 8.3.2 describe ultrasonic wave propagation within a material. When simulating waves in an NDE specimen, it is frequently necessary to model scattering from component edges (i.e., material-air interfaces). In practice, such boundaries can be adequately modeled using stress-free boundary conditions. For applications in ultrasonic inspection, it is also necessary to simulate wave interaction with flaws defects. Many common defect types (e.g., voids, open cracks, disbond/delamination) can be represented by implementing stress-free boundaries.

Stress-free boundary conditions are implemented in EFIT by setting the stress components equal to zero at the boundaries. The shear components are located on the surface, so one may simply set them equal to zero (e.g., at a maximum $\hat{x}$ boundary one sets $\sigma_{xy} = 0$ and $\sigma_{xz} = 0$). To set the normal stresses equal to zero, one can set $\sigma_{ii}^{(n)} = -\sigma_{ii}^{(n+\hat{x}_i)}$ on minimum boundaries and $\sigma_{ii}^{(n+\hat{x}_i)} = -\sigma_{ii}^{(n)}$ on maximum boundaries [38, 43]. Applying the stress-free boundary conditions at a spatial maximum boundary results in

$$v_i^{(n)}(t) = v_i^{(n)}(t-\Delta t) - \frac{2\Delta t}{\rho^{(n)}\Delta x}\sigma_{ii}^{(n)}(t), \tag{8.27}$$

and at a spatial minimum boundary:

$$v_i^{(n)}(t) = v_i^{(n)}(t-\Delta t) + \frac{2\Delta t}{\rho^{(n+\hat{x}_i)}\Delta x}\sigma_{ii}^{(n+\hat{x}_i)}(t). \tag{8.28}$$

A key limitation of the anisotropic EFIT approach is in the implementation of these stress-free boundary equations for monoclinic or triclinic anisotropy cases, for which it has been observed that numerical instability can occur at the location of stress-free boundaries [34]. Alternative finite difference approaches that have shown initial promise for avoiding such instabilities are the rotated staggered grid and Lebedev methods [15, 34, 39]. Finite element and spectral finite element approaches may also allow for stable implementation of stress-free boundaries for monoclinic and triclinic anisotropy. There are currently few publications in the scientific literature showing validated cases of ultrasound propagation and scattering from stress-free boundaries for monoclinic or triclinic materials.

Realistic defect geometries are straightforward to incorporate into EFIT simulations when using the above stress-free boundary approach [22, 23]. An example of the incorporation of realistic cracking in a thermal-fatigue cycled composite is shown in Fig. 8.4. The defect geometries and locations were taken directly from volumetric x-ray computed tomography data for a real specimen. In the figure, a bulk ultrasonic wave is excited over a circular region on the top surface of the composite plate.

One scenario where it would not be adequate to use stress-free boundaries for NDE specimen edges is in cases where ultrasound leakage into the air (leaky waves [37]) is being studied. Furthermore, accurate modeling of certain defects, such as closed-faced cracks or weak bonds, may require changes to the material properties or the introduction of more complex boundary conditions. In order to represent weak bonding, one approach is to model the bond interface using a spring stiffness parameter at the boundary. This approach is enabled by studies reporting links between interfacial stiffness and a weak bond interface [2, 8].

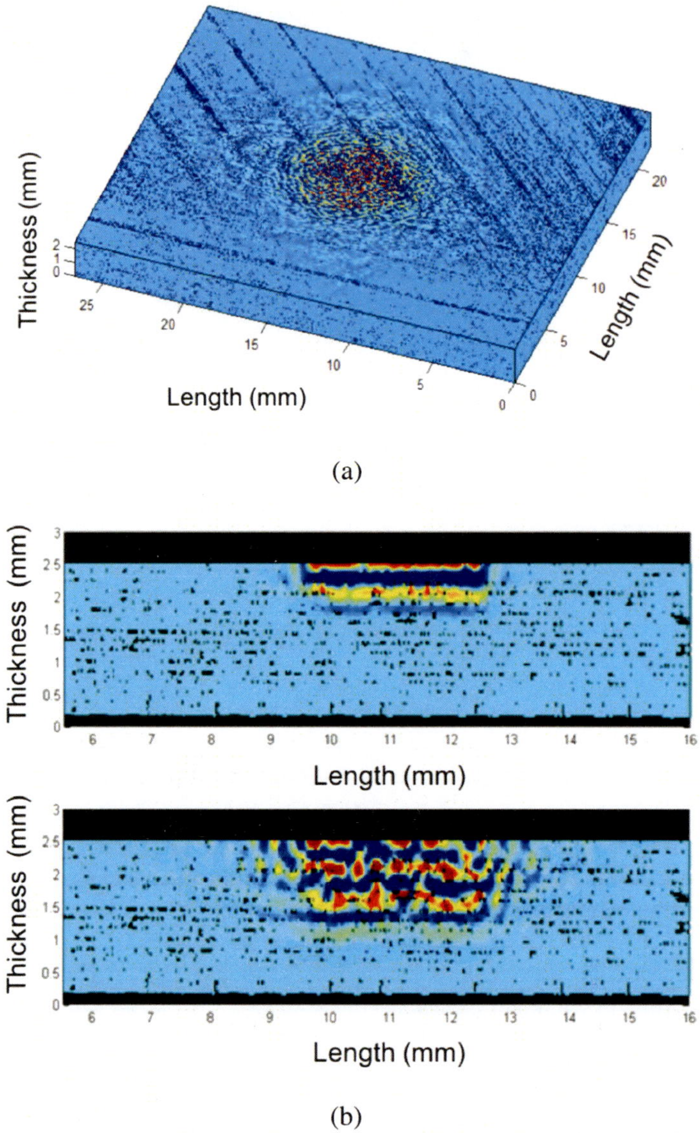

FIGURE 8.4 (a) 3D EFIT bulk wave simulation incorporating realistic cracking in a composite plate (image shows a single snapshot in time). (b) Through-thickness image slices showing wave interaction with realistic crack geometries at two different points in time.

Another boundary condition that may be of interest for NDE modeling is the implementation of the perfectly matched layer (PML) condition. A PML boundary can be used to minimize scattering from simulation boundaries, and hence is desirable in cases where the real component is larger than the simulated domain. Various approaches exist for creating a PML at computational boundaries [3, 36]. A recursive convolution time-domain PML has been successfully implemented for 2D EFIT by Calvo et al. [6]. The time-domain approach results in EFIT stress and velocity equations with an additional term containing a parameter to control the attenuation rate approaching the boundary. The mathematical approach is described in detail in Reference. [6].

8.3.5 Initial Conditions for Ultrasound Excitation

When implementing EFIT, the stress and velocity equations are solved for all 3D spatial points at each time step. A nonzero initial condition is necessary to initiate elastic wave propagation. An initial change in displacement (i.e., velocity) can used to excite ultrasonic waves by simply adding a corresponding function to the velocity equations. Alternatively, an initial forcing function can be added to the stress equations. In either case, the initial condition can be implemented as normal incidence excitation (e.g., adding a term to the v_z equation assuming $\hat{z}$ is the normal direction for the simulated transducer), in-plane excitation (e.g., adding terms to the v_x and v_y equations), or simultaneous normal and shear excitation. To generate the typical time-dependent excitations used in ultrasonic NDE (such as a tone burst), a time-dependent initial condition can be specified (e.g., varying amplitude values as the simulation marches forward in time). Three examples of initial conditions with particular relevance to NDE are discussed below.

8.3.5.1 Normal Incidence Example

One approach to model the mechanical input from a normal incidence contact transducer is to add a function to the normal velocity term. For example, the function can be added to the v_z term, as demonstrated in the equations below and in Fig. 8.5, where the excitation is inserted over a circular region (within the limitation of a cubic grid) representing the transducer. Note that the excitation is added to the boundary equation for v_z in this case, as is appropriate due to transducer placement on the specimen.

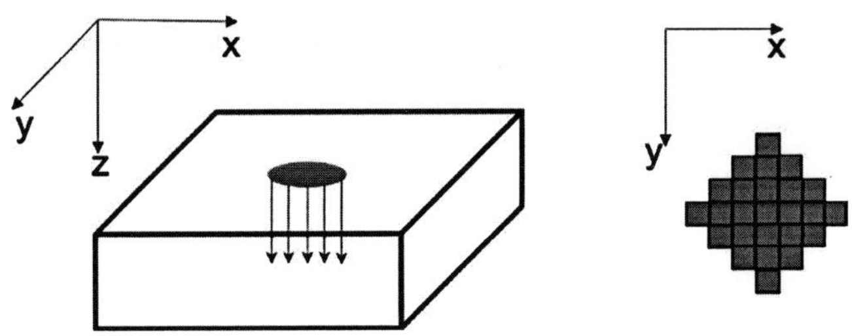

FIGURE 8.5 The figure demonstrates the implementation of a normal incidence sine wave excitation of amplitude A, in the $\hat{z}$ direction for a circular transducer of radius r, located on the minimum z surface of a plate-like specimen. Left: the excitation placement on the specimen (black arrows represent the direction of specified excitation energy); Right: the geometric result of representing a circle with a cubic grid. In this case, the excitation is introduced with the same amplitude across all grid cells within the specified transducer diameter (where the constant gray coloration in the right image represents amplitude).

For this scenario, an incident sine wave can be generated over a circular region of radius r, centered at (x_o, y_o) in the x-y plane and on the maximum z surface:

$$\text{if } (x-x_o)^2+(y-y_o)^2 \le r^2$$
$$v_z^{(n)}(t)=v_z^{(n)}(t-\Delta t)-2\frac{\Delta t}{\Delta x \rho^{(n)}}\sigma_{zz}^{(n)}(t)+\frac{\Delta t}{\rho^{(n)}}A\sin(2\pi ft) \tag{8.29}$$

where A is the specified excitation amplitude.

8.3.5.2 Shear Excitation Example

Another useful initial condition for ultrasonic NDE is a shear excitation representing a piezoelectric wafer active sensor (PWAS) excitation. One model to represent the mechanical input of a PWAS is to add the excitation function to the in-plane velocity terms, with the amplitude of excitation increasing toward the PWAS edges [50]. The equations below are one option to model this type of excitation for EFIT. Fig. 8.6 shows a diagram representing circular PWAS excitation. In the figure, v_x and v_y are the in-plane velocity terms.

$$\begin{aligned}
&\text{if } (x-x_o)^2+(y-y_o)^2 \le r^2\\
&v_x^{(n)}(t)=v_x^{(n)}(t-\Delta t)+\frac{\Delta t}{\Delta x}\frac{2}{\rho^{(n)}+\rho^{(n+\hat{x})}}\Big[\Big(\sigma_{xx}^{(n+\hat{x})}(t)-\sigma_{xx}^{(n)}(t)\Big)\\
&\quad+\Big(\sigma_{xy}^{(n)}(t)-\sigma_{xy}^{(n-\hat{y})}(t)\Big)+\Big(\sigma_{xz}^{(n)}(t)-\sigma_{xz}^{(n-\hat{z})}(t)\Big)\Big]+\frac{\Delta t}{\rho^{(n)}}x_{amp}\sin(2\pi ft),\\
&v_y^{(n)}(t)=v_y^{(n)}(t-\Delta t)+\frac{\Delta t}{\Delta x}\frac{2}{\rho^{(n)}+\rho^{(n+\hat{y})}}\Big[\Big(\sigma_{xy}^{(n)}(t)-\sigma_{xy}^{(n-\hat{x})}(t)\Big)\\
&\quad+\Big(\sigma_{yy}^{(n+\hat{y})}(t)-\sigma_{yy}^{(n)}(t)\Big)+\Big(\sigma_{yz}^{(n)}(t)-\sigma_{yz}^{(n-\hat{z})}(t)\Big)\Big]+\frac{\Delta t}{\rho^{(n)}}y_{amp}\sin(2\pi ft),
\end{aligned} \tag{8.30}$$

where $x_{amp}=(x-x_o)/r$, $y_{amp}=(y-y_o)/r$.

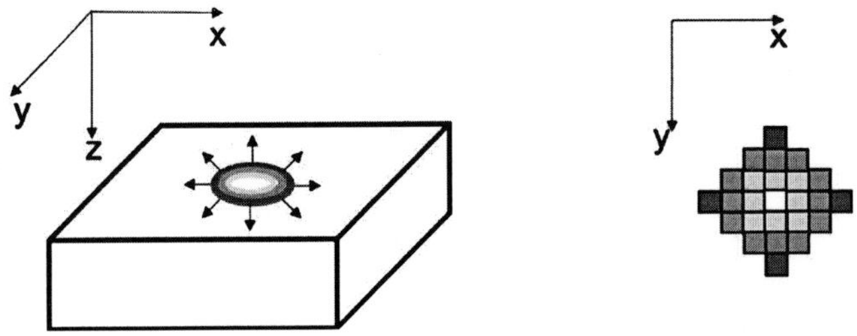

FIGURE 8.6 The diagram shows the implementation of a shear incidence sine wave excitation for a circular PWAS type transducer of radius r, located on the minimum z surface of a plate-like specimen. Left: the excitation placement on the specimen (black arrows indicate direction of specified excitation energy); Right: the excitation is introduced with a growing amplitude towards the PWAS edges (represented by the change in gray coloration in the right image). Maximum amplitude occurs at the outermost grid cells that are within the specified PWAS diameter.

It should be noted that for both of the previous examples, a smaller grid size with respect to transducer radius would result in a more circular excitation, and in practice is often the case since ultrasonic wavelengths require small grid sizes (as discussed in Section 8.3.3.). The figures show a simple case where the circular transducer becomes a diamond shape due to the effect of coarsely discretizing a circle.

8.3.5.3 Angled Incidence Example

Angled incidence or phased array excitations are straightforward to implement in EFIT, and can be modeled by simply adding a spatially dependent time delay to the initial conditions [16]. In practice, this approach is implemented by sending an altered (e.g., delayed) excitation function to each grid cell within the transducer region. The superposition of the waves can then lead to an angled incident ultrasound beam. Fig. 8.7 shows an example for exciting an angled incident wave along a simple linear array transducer, with the delay applied along the $\hat{x}$ direction. The equations associated with this excitation are given below. More complex 2D or 3D excitations can be implemented by following the general concept outlined below, but adding delays along additional spatial directions.

$$
\begin{gathered}
\text{for} \quad x_n = x_{min} \quad \text{to } x_{max} \\
t_{delay}\left(x_n\right) = \frac{\left(n - n_{min}\right)\Delta x}{v_{sp}\Delta t} sin\left(\theta_{sp}\right) \\
v_z^{(n)}(t) = v_z^{(n)}(t - \Delta t) - 2\frac{\Delta t}{\Delta x \rho^{(n)}}\sigma_{zz}^{(n)}(t) + \frac{\Delta t}{\rho^{(n)}} f\left(t_{delay}\right)
\end{gathered}
\tag{8.31}
$$

where f is the excitation function with the appropriate time delay applied along the $\hat{x}$ direction to create the angle θ_{sp} in the specimen medium, which corresponds to velocity v_{sp}.

8.3.6 Computational Implementation

One benefit of the finite integration approach is that it is straightforward to implement computationally. As stated in Sections 8.2.1 and 8.3, the numerical scheme results

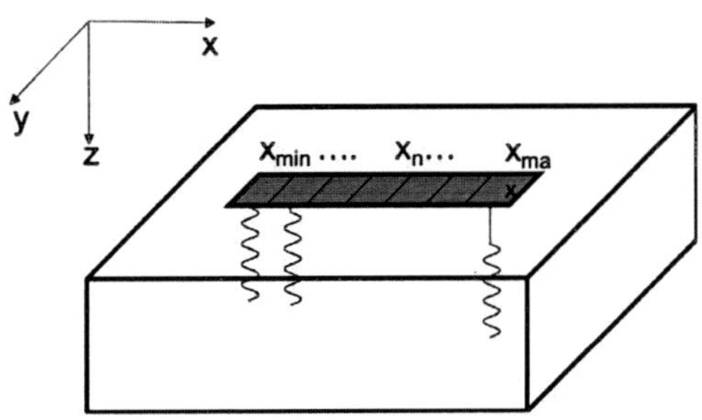

FIGURE 8.7 The diagram represents a time delay applied to consecutive spatial grid cells to excite a wave at angle θ, with the delay applied in the $\hat{x}$ direction only. Associated equations are given in the text.

in a leap-frog in time approach. A detailed (pseudo-code) example of the computational implementation of isotropic EFIT is given below, with a function in the velocity term, v_i (where i can represent x, y, and/or z), acting as the initial excitation.

<u>Isotropic EFIT Implementation:</u>

- Outer loop over time, from $t = t_0$ to t_{end}
  - First iteration of time loop:
    - Use initial condition to determine non-zero $v_i(t_0)$ for spatial locations corresponding to the excitation region, $v_i(t_0) = 0$ at all other spatial locations
    - Implement nested loops over all x, y, and z spatial locations to calculate $\sigma_{ij}(\Delta t / 2)$ using Eqs. (8.16) and (8.18)
  - Second iteration of time loop:
    - Implement nested loops over all x, y, and z spatial locations to calculate $v(\Delta t)$ using $\sigma_{ij}(\Delta t/2)$ and Eqs. (8.15) and (8.17)
    - Implement nested loops over all x, y, and z spatial locations to calculate $\sigma(\Delta t + \Delta t/2)$ using $v(\Delta t)$ and Eqs. (8.16) and (8.18)
- etcetera...marching forward in time

The small step size requirements for ultrasonic wavelengths can lead to large computational domain sizes for NDE simulation scenarios. Fortunately, the form of the EFIT equations is straightforward to implement using parallel computing. A common approach is to break up the spatial domain, calculating portions of the problem on separate CPU or GPU cores. The form of the equations requires that stress and velocity values at the edges of the partitioned domain be passed between neighboring computational cores. As an example, for a 1D partitioning that breaks up the spatial domain in the $\hat{x}$ direction, the form of the isotropic EFIT equations requires the passing of velocities and σ_{xx}, σ_{xy}, and σ_{xz} stresses (as shown in Fig. 8.8). All stresses and velocities are required to be passed to neighboring cores in the case of the anisotropic EFIT equations. Depending on the dimensions of the simulated specimen, 2D or 3D partitioning may be reasonable for effective parallelization. However, one must account for the time required for passing information between neighboring cores. Too much passing can lead to diminishing computational speed returns.

On modern compute hardware, the EFIT algorithm tends to be a bandwidth bound algorithm. This means that the time to perform arithmetic operations is generally not the primary factor determining the overall simulation speed (time to solution). Rather, the available memory bandwidth, memory hierarchy (e.g., cache levels), and network bandwidth (e.g., for data passing between compute nodes) are major factors determining simulation speed. Hence, reasonable approaches for optimizing EFIT code should focus on data locality for cache reuse and overlapping network communication (message passing) with calculations. For example, an EFIT implementation can be improved by ordering the calculations for within the time loop such that grid cells next to compute node boundaries are calculated first, enabling data transfer to

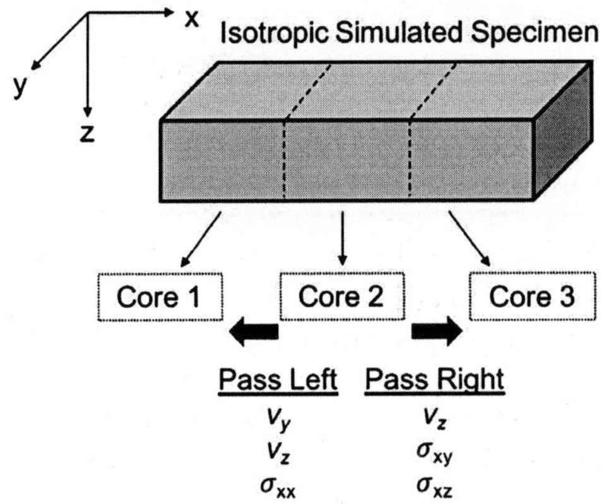

FIGURE 8.8 The diagram shows an example of 1D partitioning of the spatial domain along the x direction for a plate-like isotropic specimen. Variables required by neighboring computer cores are passed according to requirements of the isotropic EFIT equations.

neighboring processors while the interior grid cells are calculated (i.e., overlap halo exchange with calculations). Further, in order to reduce the required memory, it is beneficial to design a computational implementation that does not require the storage of an excessive number of "empty" cells (e.g., avoiding the storage of a large number of air cells surrounding a complex-shaped specimen).

Parallel EFIT implementations have been successfully run on traditional Beowulf style clusters, multicore and manycore CPUs, and GPUs [20, 24, 40]. Message Passing Interface (MPI) can be used to pass the required variables between compute cores that do not physically share memory (such as between nodes on a traditional computing cluster). Tools such as OpenMP or OpenACC (among others) can be used to parallelize the simulation across shared memory cores (CPU or GPU) [5, 31].

8.4 EXAMPLES

This section describes several examples of EFIT simulation for ultrasonic wave propagation. The selected examples include bulk wave and guided wave scenarios, isotropic and anisotropic materials, pristine and flawed specimens, and demonstrate normal, shear, and angled incidence scenarios.

8.4.1 Bulk Wave Angled Incidence with Arbitrary Backwall

The simulation case discussed in this section was chosen to demonstrate the use of EFIT for bulk wave propagation. It also demonstrates a case where more than one material is present in the simulated specimen. Additionally, this case demonstrates angled incident excitation and a nonuniform specimen geometry.

The simulated specimen is a metal block with a nonuniform (arbitrary shaped) back wall, as shown in Fig. 8.9. As shown in the figure, the specimen material is part aluminum and part stainless steel (no bonding layer was simulated in this example case).

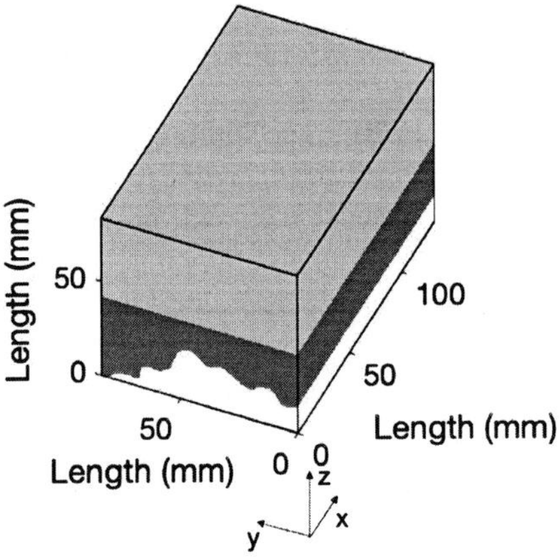

FIGURE 8.9 Diagram showing the simulated specimen, which contains an aluminum-steel interface. The light gray region is aluminum and the dark gray region is stainless steel. The simulated specimen also contains an arbitrary geometry back wall in the y-z plane. Overall block dimension is 132 mm × 83 mm × 83 mm.

The material properties for aluminum were set to $\rho_{Al} = 2780\ kg/m^3$, $\mathbb{m}_{Al} = 27.4$ GPa, $\mathbb{L}_{Al} = 53.3$ GPa, and properties for stainless steel were set to $\rho_{SS} = 7890\ kg/m^3$, $\mathbb{m}_{SS} =$ 82.6 GPa, $\mathbb{L}_{SS}$ = 99.3 GPa. The simulation spatial step size was set to 0.21 *mm*, allowing for 15 spatial steps per wavelength for the minimum wavelength in the simulation (in this case based on the shear wavespeed in aluminum). A 45 degree angled incident excitation was generated using the approach and equation shown in Fig. 8.7. A 1 MHz 5-cycle Hann windowed sine wave excitation was generated over an 12.5 mm square region centered in the x-y plane via v_z, with a time delay creating the oblique incident angle in the x-z plane. In this case, each grid cell acts as an individual source with an associated time delay (i.e., a phased array-type excitation).

Fig. 8.10 shows the resulting angled incident wave that is generated (with v_z plotted). The majority of wave energy is directed towards the specified 45 degree angle.

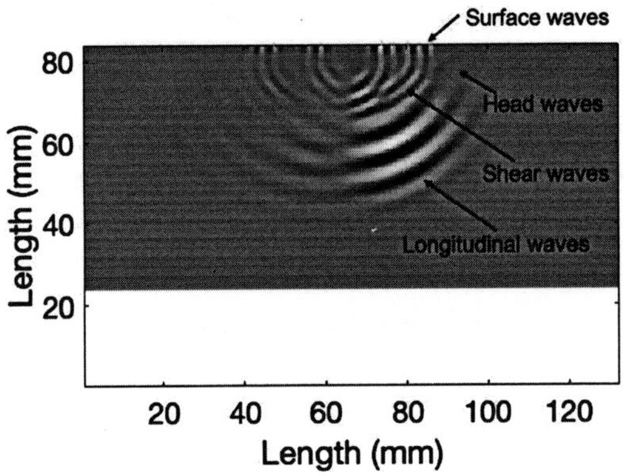

FIGURE 8.10 Isotropic bulk wave example for a case with aluminum steel interface, arbitrary back wall geometry and angled incident wave.

The figure shows a 2D slice through the 3D simulation at a single point in time (9.7 μs after the start of the excitation). A faster longitudinal wave front, slower shear wave front, and both surface and head waves (low amplitude) can be observed in the image (as labeled). Fig. 8.11 shows simulation output at three points in time, as labeled. The simulation output is plotted at the maximum z surface and for a slice

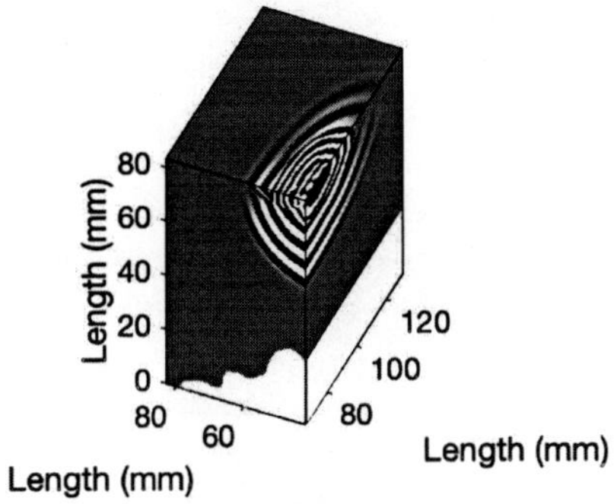

(a) Time = 5.8 μs

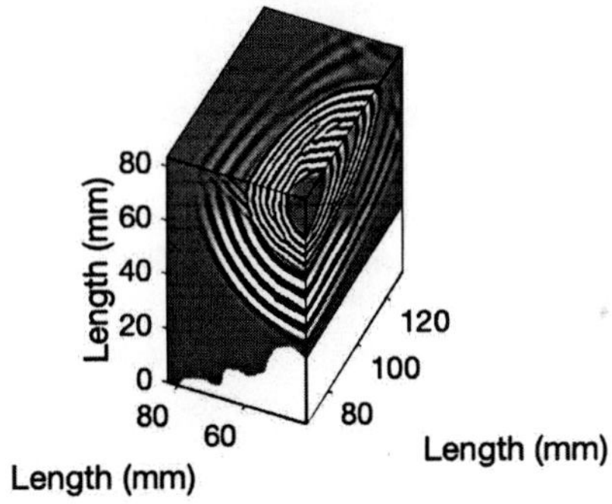

(b) Time = 9.7 μs

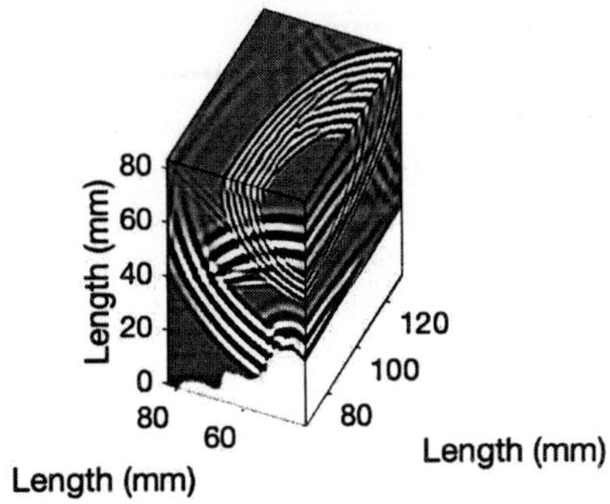

(c) Time = 13.6 μs

FIGURE 8.11 3D EFIT simulation results bulk waves in an isotropic media with an aluminum - stainless steel interface. An angled incident bulk wave is used as the excitation and the backwall has an arbitrary geometry along the x dimension. The images show three points in time (as labeled) and show waves at x = 66 mm, y = 41 mm, z = 83 mm.

halfway through x-z and y-z planes (i.e., in the x-z and x-y planes the images show waves at the center of the block, and center of the excitation).

For cases such as this, where the physics is simple (isotropic media and nondispersive waves), easy verification checks can be performed on simulation output. For example, it is straightforward to track wave propagation versus time and compare to expected wave speeds. In this example, the aluminum-steel interface is located at depth z = 41 mm. As expected, based on the longitudinal wavespeed in aluminum (v_L = 6235 m/s), initial longitudinal wave interaction with the interface can be observed in the simulation data around t = 6.58 μs (not pictured), and leads to strong wave reflections (observable in Fig. 8.11c). Shear wave interaction with the interface occurs around t = 13.1μs, as expected based on a shear wavespeed in aluminum of v_S = 3140 m/s, and also leads to wave reflections from the interface. Wave reflections from the arbitrary geometry back wall, and reflections from edges of the simulated specimen (i.e., stress-free boundaries) can also be observed in Fig. 8.11c.

8.4.2 Lamb Waves in an Aluminum Plate

Fig. 8.12 shows isotropic EFIT simulation results for Lamb wave propagation in an aluminum plate containing numerous through-holes and a through-crack. The in-plane v_x velocity is plotted on the plate surface (2D slice taken from the 3D simulation). The holes and crack were included by implementing stress-free boundaries at the corresponding locations. Stress-free boundaries were also implemented at the plate edges. The simulated plate dimensions are 508 mm x 610 mm x 1 mm (*x* by *y* by *z*). The material properties correspond to AL 2024, with density $\rho = 2780\ kg/m^3$, Lamé's first parameter $\mathbb{L}$ = 53.3 GPa, and shear modulus $\mathbb{m}$ = 27.4 GPa. The spatial step size was set to 0.126 *mm* and the time step size to 0.0017 μs. The selected spatial step size allowed for 8 spatial grid cells through the plate thickness and corresponds to 60 spatial steps for the smallest expected wavelength in the simulation (A_0 mode wavelength). A 3 cycle Hann windowed sine wave with frequency 310 kHz was used as the excitation and was included by adding the sine wave term to the v_x and v_y equations, thus creating an in-plane excitation (Fig. 8.13). The excitation was inserted with center location x = 120 mm, y = 317 mm, over a spatial region corresponding to a 7 mm diameter circle (following the piezoelectric wafer active sensor (PWAS) sensor type of excitation discussed in Section 8.3.5.2 [50]). The fastest, and highest amplitude, mode visible in Fig. 8.12 is the in-plane S_0 guided wave mode. A lower amplitude, and slower, out-of-plane A_0 Lamb mode is also visible. Fig. 8.12c shows the S_0 Lamb wave mode interaction with the crack defect. Comparisons between simulation and experiment are discussed in [50].

8.4.3 Guided Waves in a Cross-ply Composite Plate

This section describes two example cases that use the anisotropic EFIT equations: (1) guided waves in a pristine cross-ply composite laminate, (2) guided waves in a cross-ply laminate containing delaminations. These cases have been selected to demonstrate the decisions that must be made regarding spatial step size when modeling layered media and layered media containing defects. Additionally, the two cases

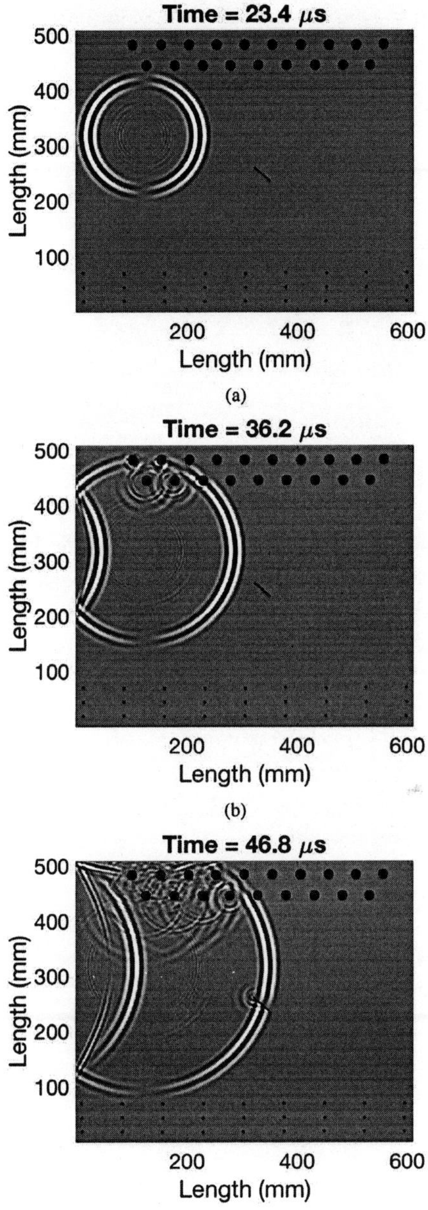

FIGURE 8.12 EFIT simulation results for an aluminum plate with through-holes and an angled crack. The in-plane v_x velocity is plotted at three points in time after the excitation, as labeled.

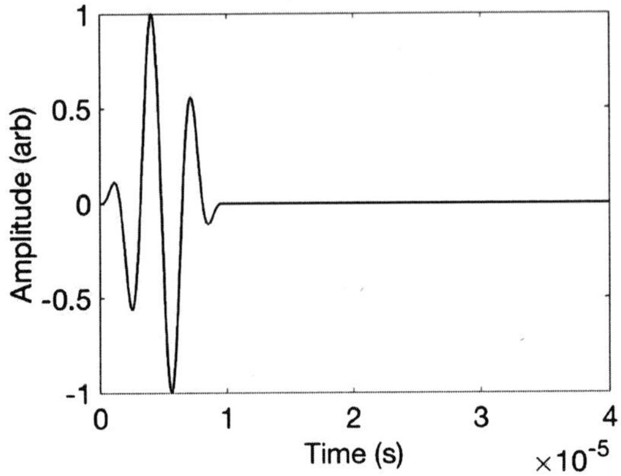

FIGURE 8.13 Excitation used in the aluminum composite example, 3 cycle Hann windowed sine wave (y-axis is arbitrary units).

yield example results for an isotropic wave propagation modeled using the anisotropic EFIT equations.

Figs. 8.14 and 8.15 show anisotropic EFIT simulation results (v_z output) for the case of an 8 ply IM7/8552 carbon fiber reinforced polymer (CFRP) composite plate of a cross-ply layup, [0/90/0/90]s. The simulated plate size was set to 105 mm x 105 mm x 0.916 mm (*x* by *y* by *z*). Material properties for a single unidirectional ply are shown in Table 8.1 [22]. In this example, all ply layers were modeled and the corresponding stiffness matrix was rotated accordingly for each ply layer. The spatial step size was set to allow for 2 steps per ply layer, $57.2\mu m$. This step size also meets the requirements discussed in Section 8.3.3 regarding wavelength for the

TABLE 8.1
IM7/8552 material properties*

Property	IM7/8552
ρ (kg/m^3)	1570
E_1 (GPa)	171.4
E_2 (GPa)	9.08
E_3 (GPa)	9.08
G_{12} (GPa)	5.29
G_{13} (GPa)	5.29
G_{23} (GPa)	2.80
ν_{12}	0.32
ν_{13}	0.32
ν_{23}	0.5

* for a single ply, with E1 in the fiber direction

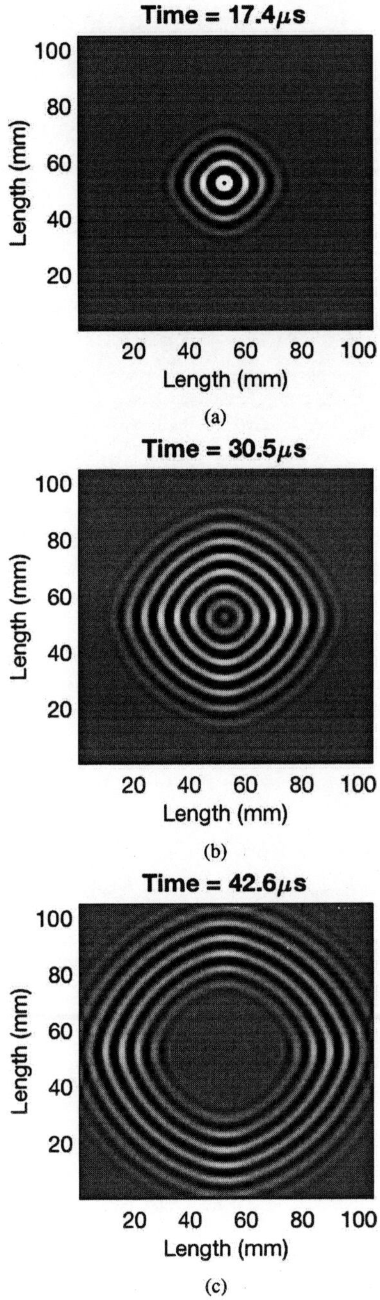

FIGURE 8.14 EFIT simulation results for a [0/90/0/90]s CFRP plate. The figures show out-of-plane v_z velocity on the top plate surface at three different points in time after the start of the excitation, as labeled.

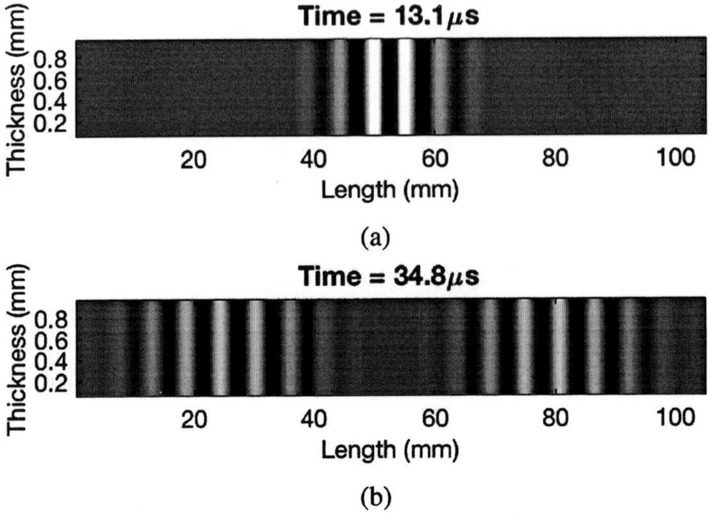

FIGURE 8.15 EFIT simulation results for a [0/90/0/90]s CFRP plate. The figures show out-of-plane v_z velocity through the plate thickness at two different points in time after the start of the excitation, as labeled.

shortest wavelength guided wave mode that exists in the simulation scenario. A 6.5 cycle Hann windowed sine wave at a center frequency of 200 kHz was used as the excitation (Fig. 8.16). The excitation was inserted by adding the sine function to the out-of-plane (v_z) velocity equation over a single grid cell (representing a point source excitation).

Fig. 8.14 represents a 2D slice from the 3D simulation, showing the v_z wavefield at the plate surface for three different points in time. The wavefield plot shows the dominant out-of-plane guided wave mode (A_0) that exists for the excitation frequency and plate thickness. The initial in-plane guided wave mode (S_0) has a significantly

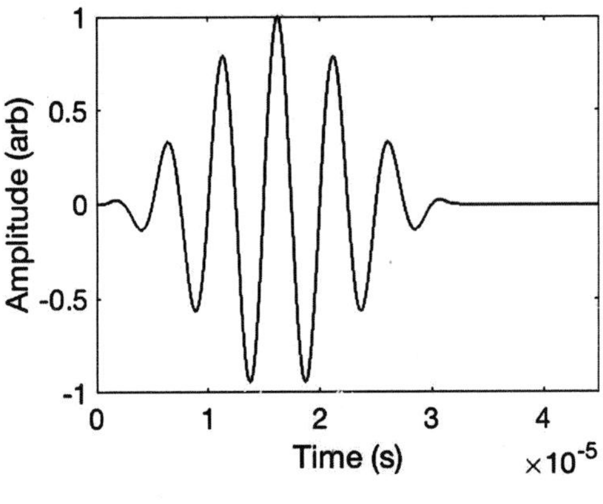

FIGURE 8.16 Excitation used in the cross-ply composite example, 6.5 cycle Hann windowed sine wave (y-axis is arbitrary units).

lower amplitude and is not visible with the color scaling shown in the figured. Initial A_0 mode reflections can be observed in Fig. 8.14c. Fig. 8.15 shows v_z for a through-thickness slice taken at the center of the plate (directly through the center of the excitation). The vertical (thickness) axis in the figure is stretched to make the wave behavior more visible. As with Fig. 8.14, the A_0 wave mode is visible while the much lower S_0 wave mode amplitude is below the selected color scale of the figure. It is noted, that any of the velocity or stress terms could be output and plotted in a similar fashion to that shown in Figs. 8.14 and 8.15. Validation comparisons between this case and experimental data can be found in Reference [22].

Figs. 8.18 and 8.19 show anisotropic EFIT simulation results (v_z output) for the case of an 8 ply IM7/8552 carbon fiber reinforced polymer (CFRP) composite plate of a cross-ply layup, [0/90/0/90]s containing delaminations at three depths [25]. The simulated plate size was set to 70 mm x 60 mm x 0.916 mm (*x* by *y* by *z*). Material properties for a single unidirectional ply are shown in Table 8.1 [22]. In this example, all ply layers were modeled and the corresponding stiffness matrix was rotated accordingly for each ply layer. The delaminations were modeled as square in shape (for simplicity) and were located above the 3rd, 5th and 6th plies and were modeled as regions with stress-free boundaries (i.e., vacuum and complete disbond). The delamination sizes were 15 mm^2, 25 mm^2, 20 mm^2 (above 3rd, 5th, 6th plies, respectively), see Fig. 8.17. All delaminations were centered at 35 mm x 30 mm. A 3 cycle

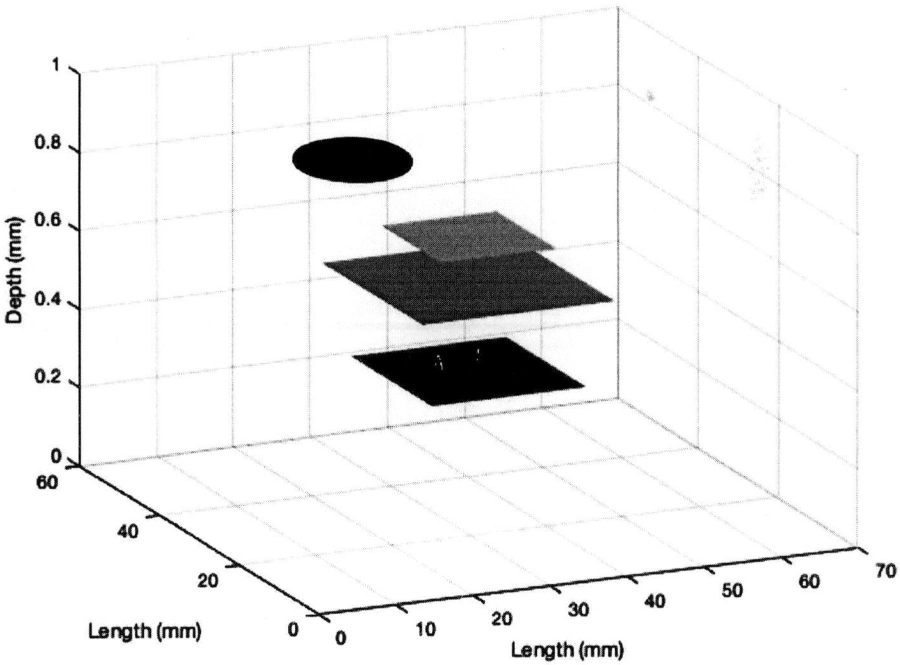

FIGURE 8.17 Diagram showing the location and size of three square delaminations within the cross-ply laminate. The excitation region (located on the maximum z surface at z = 0.92 mm) is noted by the black circle.

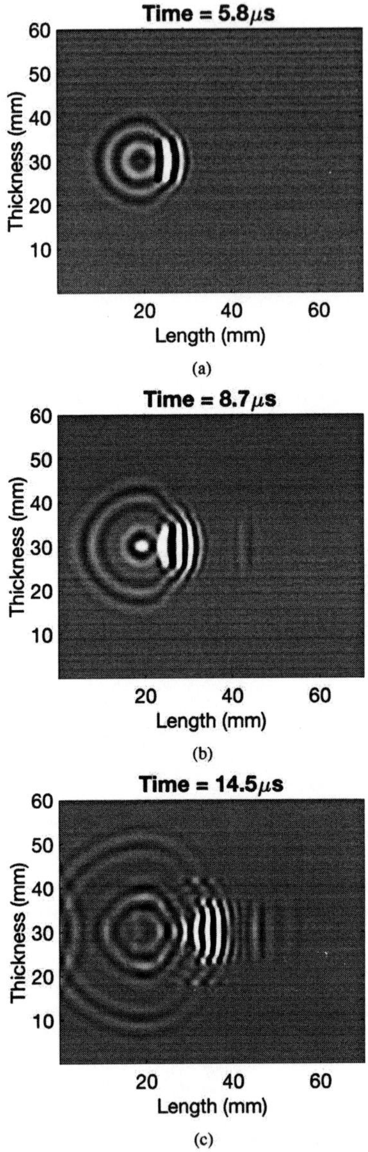

FIGURE 8.18 EFIT simulation results for a [0/90/0/90]s CFRP plate containing three delaminations located between three ply depths. The figures show out-of-plane v_z velocity on the top plate surface at three different points in time after the start of the excitation, as labeled.

Hann windowed sine wave with 300 kHz center frequency inserted over a 12.5 mm diameter circular region (centered at 20 mm x 30 mm) was used for the excitation.

The spatial step size was set to allow for 6 steps per ply layer, 19.1μm. The step size is set much smaller than the prior example for a pristine [0/90/0/90]s laminate in order to capture the shortened wavelengths in the delamination region. As the

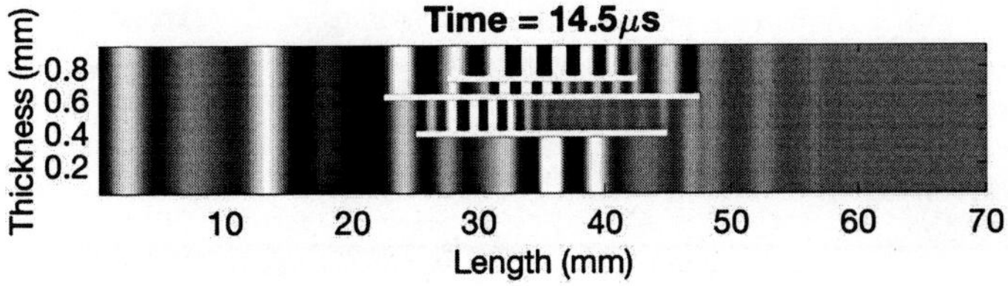

FIGURE 8.19 EFIT simulation results for a [0/90/0/90]s CFRP plate containing three delaminations located between three ply depths. The figures show out-of-plane v_z velocity on through the plate thickness at a single point in time after the start of the excitation, as labeled.

waves pass above and below the delaminations, the wavelength changes according to the new thickness encountered by the wave (between/above delaminations). This example demonstrates the need to understand the shortest wavelength present in a given simulation scenario in order to select an appropriate spatial step size. In this case, dispersion curves can be used to predict the expected wavelength in the delamination region.

Fig. 8.18 shows the excited guided waves immediately interacting with the delamination defects due to their close proximity to the excitation location. Scattering and changes in wavelength (i.e., wavenumber) can be observed as the waves interact with the delamination region. Fig. 8.19 shows the guided waves passing above/below the delaminations.

8.5 SUMMARY

The EFIT method is straightforward may be a good choice when modeling ultrasonic waves in inhomogeneous isotropic or orthotropic materials. The method that can be used to simulate any of the ultrasonic wave types typically used in NDE (e.g., bulk waves, surface waves, guided waves, etc.). The instabilities that have been observed at stress-free boundaries for EFIT when modeling monoclinic and triclinic anisotropic materials, means that other finite difference approaches, such as the Rotated Staggered Grid method, would be a better option in such scenarios [15, 34, 39]. Yet, the ease of implementing a variety of excitation scenarios, as well as the ease of incorporating changes in material properties and for including stress-free boundaries, make the method particularly fitting for many NDE scenarios. The mathematical details and simulation approaches discussed in this chapter can be used as a guide to implement EFIT in a computationally efficient manner.

REFERENCES

1. Asadollahi, Aziz and Lev Khazanovich, Numerical investigation of the effect of heterogeneity on the attenuation of shear waves in concrete. *Ultrasonics*, 2019. **91**: pp. 34–44.
2. Baltazar, A., L. Wang, B. Xie, and S.I. Rokhlin, Inverse ultrasonic determination of imperfect interfaces and bulk properties of a layer between two solids. *The Journal of the Acoustical Society of America*, 2003. **114**(3): pp. 1424–1434.

3. Berenger, J-P., Improved pml for the fdtd solution of wave-structure interaction problems. *IEEE Transactions on Antennas and Propagation*, 1997. **45**(3): pp. 466–473.
4. Bingham, J.P. and M.K. Hinders, Lamb wave characterization of corrosion-thinning in aircraft stringers: Experiment and three-dimensional simulation. *J Acoust Soc Am*, 2009. **126**: pp. 103–113.
5. OpenMP Architecture Review Board. *OpenMP Specifications*, 2019 (accessed June 19, 2019).
6. Calvo, D.C., K.E. Rudd, M. Zampolli, W.M. Sanders, and L.D. Bibee, Simulation of acoustic scattering from an aluminum cylinder near a rough interface using the elastodynamic finite integration technique. *Wave Motion*, 2010. **47**(8): pp. 616–634.
7. Canepa, Gaetano and David Calvo, Elastodynamic finite integration technique model of a sound channel with objects and attenuation. In *Proceedings of Meetings on Acoustics ECUA2012*, 2012. **17**: p. 070049. ASA.
8. Cantrell, John H. Hydrogen bonds, interfacial stiffness moduli, and the interlaminar shear strength of carbon fiber-epoxy matrix composites. *AIP Advances*, 2015. **5**(3): p. 037125.
9. Castro, M., E. Francequini, F. Dupros, H. Aochi, P.O. Navaux, and J. Mehaut, Seismic wave propagation simulations on low-power and performance-centric manycores. *Journal of Computational Physics*, 2016. **54**: pp. 108–120.
10. Chinta, Prashanth K., Klaus Mayer, Karl-Jörg Langenberg, and Jens Prager, Three-dimensional elastic wave modeling in austenitic steel welds using elastodynamic finite integration technique. In *Proceedings of 18th World Conference on Nondestructive Testing*, 2012, *Durban, South Africa*, pp. 16–20. Citeseer.
11. De Basabe, Jonás D. and Mrinal K. Sen, A comparison of finite-difference and spectral-element methods for elastic wave propagation in media with a fluid-solid interface. *Geophysical Journal International*, 2014. **200**(1):pp. 278–298.
12. Faria, E. and P. Stoffa, Finite-difference modeling in transversely isotropic media. *Geophysics*, 1994. **59**(2): pp. 282–289.
13. Fellinger, F. and K.J. Langenberg, Numerical techniques for elastic wave propagation and scattering. In *Elastic waves and ultrasonic nondestructive evaluation*; Proceedings of the IUTAM Symposium on Elastic Wave Propagation and Ultrasonic Evaluation, Boulder, CO, July 30-Aug. 3, 1989 (A91-55476 24-71). Amsterdam and New York, North-Holland, 1990, p. 81–86.
14. Fellinger, F., R. Marklein, K.J. Langenberg, and S. Klaholz, Numerical modeling of elastic wave propagation and scattering with EFIT - elastodynamic finite integration technique. *Wave Motion*, 1995: **21**(1) pp. 47–66.
15. Frankforter, Erik, Cara Leckey, and William Schneck III, Comparison of staggered grid finite difference schemes for ultrasound simulation in curving composites. In *AIP Conference Proceedings*, 2019. **2102**: p. 130002. AIP Publishing.
16. Halkjaer, Soren, *Elastic wave propagation in anisotropic, inhomogeneous materials.* PhD thesis, Technical University of Denmark, 1999.
17. He, Jiaze, Cara A.C. Leckey, Patrick E. Leser, and William P Leser, Multi-mode reverse time migration damage imaging using ultrasonic guided waves. *Ultrasonics*, 2019. **94**: pp. 319–331.
18. Hustedt, B., S. Operto, and J. Virieux, Mixed-grid and staggered-grid finite-difference methods for frequency-domain acoustic wave modelling. *Geophysical Journal International*, 2004. **157**(3): pp. 1269–1296.
19. Igel, Heiner, Bruno Riollet, and Peter Mora, Accuracy of staggered 3-d finite-difference grids for anisotropic wave propagation. In *SEG Technical Program Expanded Abstracts 1992*. 1992, Society of Exploration Geophysicists. pp. 1244–1246.

20. Iturrarán-Viveros, Ursula and Miguel Molero-Armenta, Gpu computing with opencl to model 2d elastic wave propagation: exploring memory usage. *Computational Science & Discovery*, 2015. **8**(1): p. 014006.
21. Kozic, Eva, René Hammer, Jördis Rosc, Bernhard Sartory, Joerg Siegert, Franz Schrank, and Roland Brunner, Metallization defect detection in 3d integrated components using scanning acoustic microscopy and acoustic simulations. *Microelectronics Reliability*, 2018. **88**: pp. 262–266.
22. Leckey, C., M. Rogge, and R. Parker, Guided waves in anisotropic and quasi-isotropic aerospace composites: Three-dimensional simulation and experiment. *Ultrasonics*, 2014. **54**(1): pp. 385–394.
23. Leckey, C.A.C., M.D. Rogge, and F.R. Parker, Microcracking in composite laminates: Simulation of crack-induced ultrasound attenuation. In *AIP Conference Proceedings*, 2013. **1511**: pp. 947–954.
24. Leckey, Cara A.C., Matthew D. Rogge, Corey A. Miller, and Mark K. Hinders, Multiple-mode lamb wave scattering simulations using 3d elastodynamic finite integration technique. *Ultrasonics*, 2012. **52**(2): pp. 193–207.
25. Leckey, Cara A.C. and Jeffrey P. Seebo, Guided wave energy trapping to detect hidden multilayer delamination damage. In *AIP Conference Proceedings*, 2015. **1650**: pp. 1162–1169.
26. Leckey, Cara A.C., Kevin R. Wheeler, Vasyl N. Hafiychuk, Halyna Hafiychuk, and Doğan A. Timuçin, Simulation of guided-wave ultrasound propagation in composite laminates: Benchmark comparisons of numerical codes and experiment. *Ultrasonics*, 2018. **84**: pp. 187–200.
27. Lingyu, Y., C. Leckey, and Z. Tian, Study on crack scattering in aluminum plates with Lamb wave frequency–wavenumber analysis. *Smart Materials and Structures*, 2013. **22**(6): pp. 065019.
28. Maldague, X., *Theory and Pracitce of Infrared Technology for Nondestructive Testing*. 2001, Wiley-Interscience.
29. Marklein, René, The finite integration technique as a general tool to compute acoustic, electromagnetic, elastodynamic, and coupled wave fields. *Review of Radio Science*, 1999. **2002**: pp. 201–244.
30. Nakahata, Kazuyuki, Sohichi Hirose, Frank Schubert, and Bernd Koehler, Image based efit simulation for nondestructive ultrasonic testing of austenitic steel. *Journal of Solid Mechanics and Materials Engineering*, 2009. **3**(12): pp. 1256–1262.
31. OpenACC-Standard.org. OpenACC Specifications, 2019 (accessed June 19, 2019).
32. Ozisik, M.N., H. Orlande, M¿ Colaco, and R. Cotta. *Finite Difference Methods in Heat Transfer*. 2017, CRC Press.
33. Pieffer, A., B. Kohler, and S. Petzold, The acoustic finite integration technique for waves of cylindrical symmetry. *Journal of the Acoustical Socitety of America*, 1997. **104**: pp. 697–706.
34. Quintanilla, F.H. and Leckey, C.A.C., Lebedev scheme for ultrasound simulation in composites. *Ultrasonics*, 2018. **86**: pp. 28–40.
35. Raley, Sean and Eric A. Dieckman, Application of elastodynamic finite integration technique (efit) to three-dimensional wave propagation and scattering in arbitrary geometries. *The Journal of the Acoustical Society of America*, 2019. **145**(3): pp. 1781–1781.
36. Roden, J Alan and Stephen D. Gedney, Convolution pml (cpml): An efficient fdtd implementation of the cfs–pml for arbitrary media. *Microwave and Optical Technology Letters*, 2000. **27**(5): pp. 334–339.
37. Rose, J., *Ultrasonic Waves in Solid Media*. 1999, New York NY: Cambridge University Press.

38. Rudd, Kevin E., Kevin R. Leonard, Jill P. Bingham, and Mark K. Hinders, Simulation of guided waves in complex piping geometries using the elastodynamic finite integration technique. *The Journal of the Acoustical Society of America*, 2007. **121**(3): pp. 1449–1458.
39. Saenger, E. and T. Bohlen, Finite-difference modeling of viscoelastic and anisotropic wave propagation using the rotated staggered grid. *Geophysics*, 2004. **69**(2): pp. 583–591.
40. Schneck, W., E. Gregory, and C. Leckey, Optimization of elastodynamic finite integration technique on intel xeon phi knights landing processors. *Journal of Computational Physics*, 2018. **374**: pp. 550–562.
41. Schubert, F., Numerical tim-domain modeling of linear and nonlinear ultrasonic wave propagation using finite integration techniques - theory and applications. *Ultrasonics*, 2004. **42**: pp. 221–229.
42. Schubert, F., A. Peiffer, B. Köhler, and T. Sanderson, The elastodynamic finite integration technique for waves in cylindrical geometries. *The Journal of the Acoustical Society of America*, 1998. **104**(5): pp. 2604–2614.
43. Schubert, Frank, Alexander Peiffer, Bernd Köhler, and Terry Sanderson, The elastodynamic finite integration technique for waves in cylindrical geometries. *The Journal of the Acoustical Society of America*, 1998. **104**(5): pp. 2604–2614.
44. Strikwerda, J., *Finite Difference Schemes and Partial Differential Schemes*. 1989, Belmont CA: Wadsworth and Brooks.
45. Sullivan, D., *Electromagnetic Simulations Using the FDTD Method*. 2013, Wiley-IEEE Press.
46. Tschöke, Kilian and Hauke Gravenkamp, On the numerical convergence and performance of different spatial discretization techniques for transient elastodynamic wave propagation problems. *Wave Motion*, 2018. **82**: pp. 62–85.
47. Vanaverbeke, S. and K. Abeele, Two-dimensional modeling of wave propagation in materials with hysteretic nonlinearity. *Journal of the Acoustical Society of America*, 2007. **122**(1): pp. 58–72.
48. Virieux, J., P-SV wave propagation in heterogeneous media: Velocity-stress finite difference method. *Geophysics*, 1986. **51**(4): pp. 889–901.
49. Virieux, Jean, Henri Calandra, and R-E. Plessix, A review of the spectral, pseudo-spectral, finite-difference and finite-element modelling techniques for geophysical imaging. *Geophysical Prospecting*, 2011. **59**(5): pp. 794–813.
50. Yu, Lingyu and Cara A.C. Leckey, Lamb wave–based quantitative crack detection using a focusing array algorithm. *Journal of Intelligent Material Systems and Structures*, 2013. **24**(9): pp. 1138–1152.

9 Local Interaction Simulation Approach

9.1 INTRODUCTION

Local Interaction Simulation Approach (LISA) was developed in the 1990s by Delsanto and colleagues for modeling ultrasonic wave propagation in one, two, or three dimensions [1–3]. Similar to traditional finite difference (FD) approaches, the method approximates spatial and temporal derivatives over a discretized domain. LISA replaces the second-order derivatives in the elastodynamic equations of motion with recursive relations [4]. A feature of LISA that differs from traditional FD approaches is the use of sharp interface model (SIM) for treatment of boundaries and discontinuities. The LISA method lends itself to parallel computational implementation, and has been implemented on traditional computing cores and on graphics processing units (GPUs) [5, 6].

Since the 1990s, LISA has been applied to a variety of NDE related scenarios, including guided wave-based structural health monitoring for metals and composites, as well as inspection of railroad track [4, 7–9]. While LISA is a general elastodynamic numerical method that can be used to model any type of ultrasonic wave used in NDE (e.g., bulk waves, surface waves, etc.), prior literature on LISA is primarily focused on guided wave simulation. LISA has been used to study wave propagation in isotropic, anisotropic, and viscoelastic materials [4, 8, 10, 11]. LISA has also been used to model nonlinear wave behavior [8, 12, 13]. In particular, the method has been used to model nonlinear interaction between ultrasonic guided waves and material damage. Shen and Cesnik have shown that LISA may be a more efficient computational approach than traditional finite element method (FEM)-based commercial packages for modeling nonlinear behavior (such as the contact acoustic nonlinearity for a breathing crack) [8]. Several comparisons between LISA simulation and dispersion curves and/or experiment can be found in several publications, including References [4, 8, 11].

Cesnik and colleagues studied hybrid modeling approaches entailing the use of LISA in combination with other methods, such as FEM or global matrix method (GMM) [8, 14, 15]. Such hybrid approaches can improve model accuracy by employing the most accurate method for portions of the model scenario. For example, a hybrid approach with GMM and LISA has been used to effectively represent a circular actuator using GMM and then propagate the resulting waves using LISA. Hybrid methods can also aid computational efficiency by employing models representing only the level of detail needed for portions of the domain.

This chapter outlines LISA with a goal of providing details needed for implementing the method. The mathematical equations for LISA, for both elastic and viscoelastic materials, are given in Section 9.2. Considerations on grid sizing and stability are discussed in Section 9.3, followed by a description of various boundary conditions with relevance to NDE (Section 9.4). Two different approaches for exciting utlrasonic waves in LISA are given in Section 9.5. Finally, a brief discussion on computational implementation is given in Section 9.6, followed by several examples of LISA simulation cases taken from the scientific literature (Section 9.7).

9.2 MATHEMATICAL EQUATIONS: LISA

As with other elastodynamic numerical methods, one begins with the elastic wave equation (previously introduced in Chapters 3 and 8). Following the LISA literature, the equation is written in terms of displacement, u [2, 4, 16]:

$$\begin{aligned}\rho \ddot{u}_i &= \frac{\partial \sigma_{ij}}{\partial x_j} \\ &= \frac{\partial}{\partial x_j}\left(\mathbb{C}_{ijkl}\frac{\partial u_k}{\partial x_l}\right)\end{aligned} \tag{9.1}$$

where ρ is the density and $\mathbb{C}$ is the stiffness tensor. For an orthotropic material, one can write 9.1 as:

$$\rho \ddot{u}_i = \sum_{l=1}^{3}\left(\mathbb{A}_{il}\frac{\partial^2 u_i}{\partial x_l^2} + \mathbb{B}_{il}\frac{\partial^2 u_l}{\partial x_i x_l}\right) \tag{9.2}$$

where $\mathbb{A}_{il}$ and $\mathbb{B}_{il}$ are the corresponding stiffness values (a subset of $\mathbb{C}_{il}$). This equation is discretized with the assumption that the spatial grid is rectangular with spatial grid steps Δx, Δy, and Δz. A FD formalism, as described in Reference [16], is applied to the second-order spatial derivatives and central differences are applied to the time derivatives. The second-order spatial derivatives result in the displacement calculation at point, (n) (as shown in Fig. 9.1), requiring the 18 neighboring grid points. For the case of an inhomogeneous material, it is assumed that within each grid cell the material is homogeneous and that there is an interface between differing materials that corresponds to the interface between grid cells.

Further, a sharp interface model (SIM) is assumed, with the stress and displacement values assumed to be uniform within a cell while all interfaces between cells are assumed to be in perfect contact. Material properties of neighboring cells can be discontinuous and SIM requires that all $\ddot{u}_i$ converge to a common displacement value at the interfaces between cells. SIM has been reported to lead to more accurate results for wave propagation in complex media with complex boundaries, when compared to other FD typical schemes [5].

The details of imposing SIM and deriving discretized equations in a useful form for computation are quite lengthy and can be found in [2, 4, 16]. Here, we show the final discretized equations for the case of orthotropic inhomogeneous materials. These are the form of equations that can be implemented computationally to

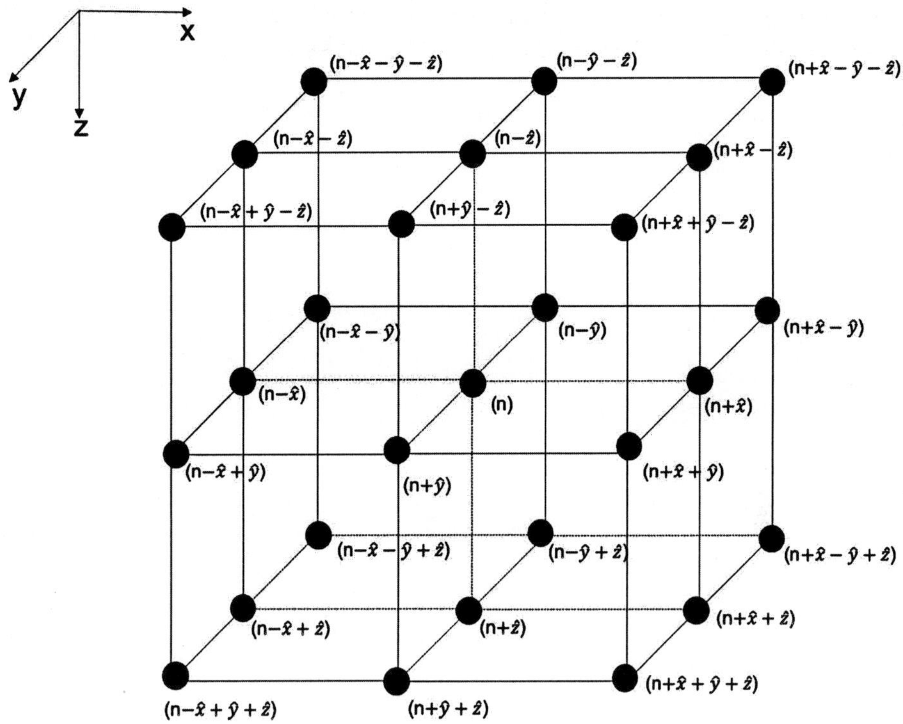

FIGURE 9.1 LISA grid layout showing current computational point (n) and the eighteen nearest neighbors. (Based on [4].)

numerically model ultrasonic waves using LISA. The LISA equations shown below result in a second-order accurate, time-marching scheme. Below, the notation $u^{(n)}$ is shorthand to note the displacement at the current ("nth") grid step (see Fig. 9.1). For consistency with the notation used in Chapter 8, the notation $(n+/-i)$ represents a single grid step in the corresponding Cartesian direction, and where shorthand has been used in the equations below with the slightly altered notation $(n+\boldsymbol{i})$ corresponding to the listed summation with $\boldsymbol{i}$ =+/-1, and Δt is the time step size.

$$u_x^{(n)}(t+\Delta t) = -u_x^{(n)}(t-\Delta t) + 2u_x^{(n)} + \frac{\Delta t^2}{\rho} \sum_{\boldsymbol{x},\boldsymbol{y},\boldsymbol{z}=\pm 1} \left[-\frac{1}{4} u_x^{(n)} \left[\frac{1}{\Delta x^2}\tilde{\mathbb{C}}_{11} + \frac{1}{\Delta y^2}\tilde{\mathbb{C}}_{12} + \frac{1}{\Delta z^2}\tilde{\mathbb{C}}_{13} \right] \right.$$

$$+\frac{1}{4}\left[\frac{1}{\Delta x^2}\tilde{\mathbb{C}}_{11} u_x^{(n+\boldsymbol{x})} + \frac{1}{\Delta y^2}\tilde{\mathbb{C}}_{12} u_x^{(n+\boldsymbol{y})} + \frac{1}{\Delta z^2}\tilde{\mathbb{C}}_{13} u_x^{(n+\boldsymbol{z})} \right]$$

$$+\frac{1}{8}\left[\frac{\boldsymbol{x}\boldsymbol{y}}{\Delta x \Delta y}\Big((\tilde{\mathbb{C}}_{12} + \tilde{\mathbb{C}}_{66})(u_y^{(n+\boldsymbol{x}+\boldsymbol{y})} - u_y^{(n)}) + (\tilde{\mathbb{C}}_{12} - \tilde{\mathbb{C}}_{66})(u_y^{(n+\boldsymbol{y})} - u_y^{(n+\boldsymbol{x})}) \Big) \right]$$

$$+\frac{1}{8}\left[\frac{\boldsymbol{x}\boldsymbol{z}}{\Delta x \Delta z}\Big((\tilde{\mathbb{C}}_{13} + \tilde{\mathbb{C}}_{55})(u_z^{(n+\boldsymbol{x}+\boldsymbol{z})} - u_z^{(n)}) + (\tilde{\mathbb{C}}_{13} - \tilde{\mathbb{C}}_{55})(u_z^{(n+\boldsymbol{z})} - u_z^{(n+\boldsymbol{x})}) \Big) \right]$$

$$-\frac{1}{4}\left[\frac{\hat{x}\hat{y}}{\Delta x\Delta y}\tilde{\mathbb{C}}_{16}\left(u_x^{(n)}-u_x^{(n+\hat{x}+\hat{y})}\right)\right]-\frac{1}{4}u_y^{(n)}\left[\frac{\tilde{\mathbb{C}}_{16}}{\Delta x^2}+\frac{\tilde{\mathbb{C}}_{26}}{\Delta y^2}\right]$$

$$-\frac{1}{4}\left[\frac{1}{\Delta x^2}\tilde{\mathbb{C}}_{16}u_y^{(n+\hat{x})}+\frac{1}{\Delta y^2}\tilde{\mathbb{C}}_{26}u_y^{(n+\hat{y})}\right]-\frac{1}{8}\left[\frac{\hat{y}\hat{z}}{\Delta y\Delta z}(\tilde{\mathbb{C}}_{36}+\tilde{\mathbb{C}}_{45})u_z^{(n)}\right]$$

$$+\frac{1}{8}\left[\frac{\hat{y}\hat{z}}{\Delta y\Delta z}\tilde{\mathbb{C}}_{36}\left(u_z^{(n+\hat{y}+\hat{z})}+u_z^{(n+\hat{z})}-u_z^{(n+\hat{y})}\right)\right]$$

$$+\frac{1}{8}\left[\frac{\hat{y}\hat{z}}{\Delta y\Delta z}\tilde{\mathbb{C}}_{45}\left(u_z^{(n+\hat{y}+\hat{z})}-u_z^{(n+\hat{z})}+u_z^{(n+\hat{y})}\right)\right]+\frac{1}{4}\frac{\tilde{\mathbb{C}}_{45}}{\Delta z^2}\left(\left(u_y^{(n+\hat{z})}-u_y^{(n)}\right)\right)\Bigg] \tag{9.3}$$

$$u_y^{(n)}(t+\Delta t)=-u_y^{(n)}(t-\Delta t)+2u_y^{(n)}$$

$$+\frac{\Delta t^2}{\rho}\sum_{\hat{x},\hat{y},\hat{z}=\pm 1}\left[-\frac{1}{4}u_y^{(n)}\left[\frac{1}{\Delta x^2}\tilde{\mathbb{C}}_{12}+\frac{1}{\Delta y^2}\tilde{\mathbb{C}}_{22}+\frac{1}{\Delta z^2}\tilde{\mathbb{C}}_{23}\right]\right.$$

$$+\frac{1}{4}\left[\frac{1}{\Delta x^2}\tilde{\mathbb{C}}_{12}u_y^{(n+\hat{x})}+\frac{1}{\Delta y^2}\tilde{\mathbb{C}}_{22}u_y^{(n+\hat{y})}+\frac{1}{\Delta z^2}\tilde{\mathbb{C}}_{23}u_y^{(n+\hat{z})}\right]$$

$$+\frac{1}{8}\left[\frac{\hat{x}\hat{y}}{\Delta x\Delta y}\left(\left(\tilde{\mathbb{C}}_{12}+\tilde{\mathbb{C}}_{66}\right)\left(u_x^{(n+\hat{x}+\hat{y})}-u_x^{(n)}\right)+\left(\tilde{\mathbb{C}}_{12}-\tilde{\mathbb{C}}_{66}\right)\left(u_x^{(n+\hat{x})}-u_x^{(n+\hat{y})}\right)\right)\right]$$

$$+\frac{1}{8}\left[\frac{\hat{x}\hat{z}}{\Delta x\Delta z}\left(\left(\tilde{\mathbb{C}}_{23}+\tilde{\mathbb{C}}_{44}\right)\left(u_z^{(n+\hat{y}+\hat{z})}-u_z^{(n)}\right)+\left(\tilde{\mathbb{C}}_{23}-\tilde{\mathbb{C}}_{44}\right)\left(u_z^{(n+\hat{z})}-u_z^{(n+\hat{y})}\right)\right)\right]$$

$$-\frac{1}{4}\left[\frac{\hat{x}\hat{y}}{\Delta x\Delta y}\tilde{\mathbb{C}}_{26}\left(u_y^{(n)}-u_y^{(n+\hat{x}+\hat{y})}\right)\right]-\frac{1}{4}u_x^{(n)}\left[\frac{\tilde{\mathbb{C}}_{16}}{\Delta x^2}+\frac{\tilde{\mathbb{C}}_{26}}{\Delta y^2}\right]$$

$$-\frac{1}{4}\left[\frac{1}{\Delta x^2}\tilde{\mathbb{C}}_{16}u_x^{(n+\hat{x})}+\frac{1}{\Delta y^2}\tilde{\mathbb{C}}_{26}u_x^{(n+\hat{x}+\hat{y})}\right]-\frac{1}{8}\left[\frac{\hat{x}\hat{z}}{\Delta x\Delta z}(\tilde{\mathbb{C}}_{36}+\tilde{\mathbb{C}}_{45})u_z^{(n)}\right]$$

$$+\frac{1}{8}\left[\frac{\hat{x}\hat{z}}{\Delta x\Delta z}\tilde{\mathbb{C}}_{36}\left(u_z^{(n+\hat{x}+\hat{z})}+u_z^{(n+\hat{z})}-u_z^{(n+\hat{x})}\right)\right]$$

$$+\frac{1}{8}\left[\frac{\hat{x}\hat{z}}{\Delta x\Delta z}\tilde{\mathbb{C}}_{45}\left(u_z^{(n+\hat{x}+\hat{z})}-u_z^{(n+\hat{z})}+u_z^{(n+\hat{x})}\right)\right]+\frac{1}{4}\frac{\tilde{\mathbb{C}}_{45}}{\Delta z^2}\left(\left(u_x^{(n+\hat{z})}-u_x^{(n)}\right)\right)\Bigg] \tag{9.4}$$

$$u_z^{(n)}(t+\Delta t)=-u_z^{(n)}(t-\Delta t)+2u_z^{(n)}$$

$$+\frac{\Delta t^2}{\rho}\sum_{\hat{x},\hat{y},\hat{z}=\pm 1}\left[-\frac{1}{4}u_y^{(n)}\left[\frac{1}{\Delta x^2}\tilde{\mathbb{C}}_{13}+\frac{1}{\Delta y^2}\tilde{\mathbb{C}}_{23}+\frac{1}{\Delta z^2}\tilde{\mathbb{C}}_{33}\right]\right.$$

$$+\frac{1}{8}\left[\frac{1}{\Delta x^2}\tilde{\mathbb{C}}_{13}u_z^{(n+\hat{x})}+\frac{1}{\Delta y^2}\tilde{\mathbb{C}}_{23}u_z^{(n+\hat{y})}+\frac{1}{\Delta z^2}\tilde{\mathbb{C}}_{33}u_z^{(n+\hat{z})}\right]$$

$$+\frac{1}{8}\left[\frac{\mathscr{yz}}{\Delta y\Delta z}\left((\tilde{\mathbb{C}}_{23}+\tilde{\mathbb{C}}_{44})(u_y^{(n+\mathscr{y}+\mathscr{z})}-u_y^{(n)})+(\tilde{\mathbb{C}}_{23}-\tilde{\mathbb{C}}_{44})(u_y^{(n+\mathscr{y})}-u_y^{(n+\mathscr{z})})\right)\right]$$

$$+\frac{1}{8}\left[\frac{\mathscr{xz}}{\Delta x\Delta z}\left((\tilde{\mathbb{C}}_{13}+\tilde{\mathbb{C}}_{55})(u_x^{(n+\mathscr{x}+\mathscr{z})}-u_x^{(n)})+(\tilde{\mathbb{C}}_{13}-\tilde{\mathbb{C}}_{55})(u_x^{(n+\mathscr{x})}-u_x^{(n+\mathscr{z})})\right)\right]$$

$$-\frac{1}{8}\left[\frac{\mathscr{yz}}{\Delta y\Delta z}(\tilde{\mathbb{C}}_{36}+\tilde{\mathbb{C}}_{45})(u_x^{(n)}-u_x^{(n+\mathscr{y}+\mathscr{z})})\right]-\frac{1}{8}\left[\frac{\mathscr{xz}}{\Delta x\Delta z}(\tilde{\mathbb{C}}_{36}+\tilde{\mathbb{C}}_{45})(u_y^{(n)}-u_y^{(n+\mathscr{x}+\mathscr{z})})\right]$$

$$-\frac{1}{8}\left[\frac{\mathscr{yz}}{\Delta y\Delta z}(\tilde{\mathbb{C}}_{36}+\tilde{\mathbb{C}}_{45})(u_x^{(n+\mathscr{z})}-u_x^{(n+\mathscr{y})})\right]-\frac{1}{8}\left[\frac{\mathscr{xz}}{\Delta x\Delta z}(\tilde{\mathbb{C}}_{36}+\tilde{\mathbb{C}}_{45})(u_y^{(n+\mathscr{z})}-u_y^{(n+\mathscr{x})})\right]$$

$$+\frac{1}{4}\left[\frac{\mathscr{xy}}{\Delta x\Delta y}\tilde{\mathbb{C}}_{45}(u_z^{(n+\mathscr{x}+\mathscr{y})}-u_z^{(n)})\right]\Bigg] \tag{9.5}$$

Following the notation of Reference [4], $\mathbb{C}_{ij}=\mathbb{C}_{ij}^{(n+\mathscr{x}+\mathscr{y}+\mathscr{z})}$ represents one of the 8 cells surrounding point (n). Further, it is assumed that ρ is the average density of all 8 cells surrounding point (*n*).

The above equations can be implemented to model in 3D ultrasonic waves in an elastic media. As described in Section 9.5.2, there are multiple approaches to creating an excitation source. An initial condition on displacement of an electromechanical model can be added into the LISA framework to explicitly model piezoelectric coupling to the elastodynamic substrate [17, 18]. Further, a Kelvin-Voigt viscoelastic damping model has been included in the LISA framework to capture the attenuation of ultrasonic waves that can occur in some materials (such as composites). The incorporation of viscoelastic damping adds complexity to the above equations, in that the LISA model with damping included requires the prior three time steps to calculate displacement at the current time step [15].

In this section, a brief overview on the approach to include damping in the LISA method is given. Further details of including damping in LISA can be found in References [8, 15]. In order to account for material based damping, stress components are represented as a combination of both elastic and viscous terms:

$$\sigma_i j=\mathbb{C}_{ijkl}e_{kl}+\mathbb{D}_{ijkl}\dot{e}_{kl} \tag{9.6}$$

where $\mathbb{D}$ is the viscosity matrix and *dot* $\boldsymbol{e}$ is the strain rate.

Thus, when viscoelasticity is included in LISA, the resulting equations are simply a linear combination of Eqs. (9.3)–(9.5), plus damping terms:

$$\begin{aligned}
u_x^{(n)}(t+\Delta t) &= u_x^{(n)}(t+\Delta t,\mathbb{C})+u_x^{(n)}(t+\Delta t,\mathbb{D})\\
u_y^{(n)}(t+\Delta t) &= u_y^{(n)}(t+\Delta t,\mathbb{C})+u_y^{(n)}(t+\Delta t,\mathbb{D})\\
u_z^{(n)}(t+\Delta t) &= u_z^{(n)}(t+\Delta t,\mathbb{C})+u_z^{(n)}(t+\Delta t,\mathbb{D})
\end{aligned} \tag{9.7}$$

For the case of a transversely isotropic material, such as a unidirectional composite ply, the viscosity coefficient matrix contains only diagonal terms when the fiber direction is aligned along one of the axes of the computational grid. However, if the unidirectional ply is rotated to an off-axis angle with respect to the global computational grid (i.e., monoclinic matrix form), the viscosity matrix will contain off-diagonal terms:

$$\mathbb{D} = \begin{bmatrix} \mathbb{D}_{11} & \mathbb{D}_{12} & 0 & 0 & 0 & \mathbb{D}_{16} \\ \mathbb{D}_{21} & \mathbb{D}_{22} & 0 & 0 & 0 & \mathbb{D}_{26} \\ 0 & 0 & \mathbb{D}_{33} & 0 & 0 & 0 \\ 0 & 0 & 0 & \mathbb{D}_{44} & \mathbb{D}_{45} & 0 \\ 0 & 0 & 0 & \mathbb{D}_{54} & \mathbb{D}_{55} & 0 \\ \mathbb{D}_{61} & \mathbb{D}_{62} & 0 & 0 & 0 & \mathbb{D}_{66} \end{bmatrix} \tag{9.8}$$

Using the above form of the viscosity coefficient matrix, the resulting discretized LISA terms for $u_i^{(n)}(t+\Delta t, \mathbb{D})$ are listed below. These additional terms can be added to the corresponding Eqs. (9.3)–(9.5), as shown in Eq. (9.7) to model waves in a viscoelastic media.

$$\begin{aligned} u_x^{(n)}\quad (t+\Delta t,\mathbb{D}) = \frac{\Delta t^2}{\rho} \sum_{\hat{x},\hat{y},\hat{z}=\pm 1} \Bigg[& -\frac{1}{4}\dot{u}_x^{(n)}\left[\frac{1}{\Delta x^2}\tilde{\mathbb{D}}_{11} + \frac{1}{\Delta y^2}\tilde{\mathbb{D}}_{66} + \frac{1}{\Delta z^2}\tilde{\mathbb{D}}_{55}\right] \\ & + \frac{1}{4}\left[\frac{1}{\Delta x^2}\tilde{\mathbb{D}}_{11}\dot{u}_x^{(n+\hat{x})} + \frac{1}{\Delta y^2}\tilde{\mathbb{D}}_{66}\dot{u}_x^{(n+\hat{y})} + \frac{1}{\Delta z^2}\tilde{\mathbb{D}}_{55}\dot{u}_x^{(n+\hat{z})}\right] \\ & + \frac{1}{8}\left[\frac{\hat{x}\hat{y}}{\Delta x \Delta y}\left(\left(\tilde{\mathbb{D}}_{12} + \tilde{\mathbb{D}}_{66}\right)\left(\dot{u}_y^{(n+\hat{x}+\hat{y})} - \dot{u}_y^{(n)}\right) + \left(\tilde{\mathbb{D}}_{12} - \tilde{\mathbb{D}}_{66}\right)\left(\dot{u}_y^{(n+\hat{y})} - \dot{u}_y^{(n+\hat{x})}\right)\right)\right] \\ & + \frac{1}{8}\left[\frac{\hat{x}\hat{z}}{\Delta x \Delta z}\left(\left(\tilde{\mathbb{D}}_{55}\left(\dot{u}_z^{(n+\hat{x}+\hat{z})} - \dot{u}_z^{(n)}\right) - \left(\dot{u}_z^{(n+\hat{z})} + \dot{u}_z^{(n+\hat{x})}\right)\right)\right)\right] \\ & - \frac{1}{4}\left[\frac{\hat{x}\hat{y}}{\Delta x \Delta y}\tilde{\mathbb{D}}_{16}\left(\dot{u}_x^{(n)} - \dot{u}_x^{(n+\hat{x}+\hat{y})}\right)\right] - \frac{1}{4}\dot{u}_y^{(n)}\left[\frac{\tilde{\mathbb{D}}_{16}}{\Delta x^2} + \frac{\tilde{\mathbb{D}}_{26}}{\Delta y^2}\right] \\ & + \frac{1}{4}\left[\left(\frac{1}{\Delta x^2}\tilde{\mathbb{D}}_{16}\dot{u}_y^{(n+\hat{x})} + \frac{1}{\Delta y^2}\tilde{\mathbb{D}}_{26}\right)\dot{u}_y^{(n+\hat{y})}\right] - \frac{1}{8}\left[\frac{\hat{y}\hat{z}}{\Delta y \Delta z}\left(\tilde{\mathbb{D}}_{45}\right)\dot{u}_z^{(n)}\right] \\ & + \frac{1}{8}\left[\frac{\hat{y}\hat{z}}{\Delta y \Delta z}\tilde{\mathbb{D}}_{45}\left(\dot{u}_z^{(n+\hat{y}+\hat{z})} + \dot{u}_z^{(n+\hat{z})} - \dot{u}_z^{(n+\hat{y})}\right)\right] \\ & + \frac{1}{4}\frac{\tilde{\mathbb{D}}_{45}}{\Delta z^2}\left(\dot{u}_y^{(n+\hat{z})} - \dot{u}_y^{(n)}\right)\Bigg] \end{aligned} \tag{9.9}$$

$$
\begin{aligned}
u_y^{(n)}(t+\Delta t,\mathbb{D}) = \frac{\Delta t^2}{\rho}\sum_{\hat{x},\hat{y},\hat{z}=\pm 1} & \Bigg[-\frac{1}{4}\dot{u}_y^{(n)}\left[\frac{1}{\Delta x^2}\tilde{\mathbb{D}}_{66}+\frac{1}{\Delta y^2}\tilde{\mathbb{D}}_{22}+\frac{1}{\Delta z^2}\tilde{\mathbb{D}}_{44}\right] \\
& +\frac{1}{4}\left[\frac{1}{\Delta x^2}\tilde{\mathbb{D}}_{66}\dot{u}_y^{(n+\hat{x})}+\frac{1}{\Delta y^2}\tilde{\mathbb{D}}_{22}\dot{u}_y^{(n+\hat{y})}+\frac{1}{\Delta z^2}\tilde{\mathbb{D}}_{44}\dot{u}_y^{(n+\hat{z})}\right] \\
& +\frac{1}{8}\left[\frac{\hat{x}\hat{y}}{\Delta x\Delta y}\left(\left(\tilde{\mathbb{D}}_{12}+\tilde{\mathbb{D}}_{66}\right)\left(\dot{u}_x^{(n+\hat{x}+\hat{y})}-\dot{u}_x^{(n)}\right)+\left(\tilde{\mathbb{D}}_{12}-\tilde{\mathbb{D}}_{66}\right)\left(\dot{u}_x^{(n+\hat{x})}-\dot{u}_x^{(n+\hat{y})}\right)\right)\right] \\
& +\frac{1}{8}\left[\frac{\hat{x}\hat{z}}{\Delta x\Delta z}\left(\tilde{\mathbb{D}}_{44}\left(\dot{u}_z^{(n+\hat{y}+\hat{z})}-\dot{u}_z^{(n)}-\dot{u}_z^{(n+\hat{z})}+\dot{u}_z^{(n+\hat{y})}\right)\right)\right] \\
& -\frac{1}{4}\left[\frac{\hat{x}\hat{y}}{\Delta x\Delta y}\tilde{\mathbb{D}}_{26}\left(\dot{u}_y^{(n)}-\dot{u}_y^{(n+\hat{x}+\hat{y})}\right)\right]-\frac{1}{4}\dot{u}_x^{(n)}\left[\frac{\tilde{\mathbb{D}}_{16}}{\Delta x^2}+\frac{\tilde{\mathbb{D}}_{26}}{\Delta y^2}\right] \\
& -\frac{1}{4}\left[\frac{1}{\Delta x^2}\tilde{\mathbb{D}}_{16}\dot{u}_x^{(n+\hat{x})}+\frac{1}{\Delta y^2}\tilde{\mathbb{D}}_{26}\dot{u}_x^{(n+\hat{x}+\hat{y})}\right]-\frac{1}{8}\left[\frac{\hat{x}\hat{z}}{\Delta x\Delta z}\tilde{\mathbb{D}}_{45}\dot{u}_z^{(n)}\right] \\
& +\frac{1}{8}\left[\frac{\hat{x}\hat{z}}{\Delta x\Delta z}\tilde{\mathbb{D}}_{45}\left(\dot{u}_z^{(n+\hat{x}+\hat{z})}-\dot{u}_z^{(n+\hat{z})}+\dot{u}_z^{(n+\hat{x})}\right)\right]+\frac{1}{4}\frac{\tilde{\mathbb{D}}_{45}}{\Delta z^2}\left(\left(\dot{u}_x^{(n+\hat{z})}-\dot{u}_x^{(n)}\right)\right)\Bigg]
\end{aligned}
\tag{9.10}
$$

$$
\begin{aligned}
u_z^{(n)}(t+\Delta t,\mathbb{D}) = \frac{\Delta t^2}{\rho}\sum_{\hat{x},\hat{y},\hat{z}=\pm 1} & \Bigg[-\frac{1}{4}\dot{u}_z^{(n)}\left[\frac{1}{\Delta x^2}\tilde{\mathbb{D}}_{55}+\frac{1}{\Delta y^2}\tilde{\mathbb{D}}_{44}+\frac{1}{\Delta z^2}\tilde{\mathbb{D}}_{33}\right] \\
& +\frac{1}{4}\left[\frac{1}{\Delta x^2}\tilde{\mathbb{D}}_{55}\dot{u}_z^{(n+\hat{x})}+\frac{1}{\Delta y^2}\tilde{\mathbb{D}}_{44}\dot{u}_z^{(n+\hat{y})}+\frac{1}{\Delta z^2}\tilde{\mathbb{D}}_{33}\dot{u}_z^{(n+\hat{z})}\right] \\
& +\frac{1}{8}\left[\frac{\hat{y}\hat{z}}{\Delta y\Delta z}\left(\tilde{\mathbb{D}}_{44}\left(\dot{u}_y^{(n+\hat{y}+\hat{z})}-\dot{u}_y^{(n)}-\dot{u}_y^{(n+\hat{y})}+\dot{u}_y^{(n+\hat{z})}\right)\right)\right] \\
& +\frac{1}{8}\left[\frac{\hat{x}\hat{z}}{\Delta x\Delta z}\left(\tilde{\mathbb{D}}_{55}\left(\dot{u}_x^{(n+\hat{x}+\hat{z})}-\dot{u}_x^{(n)}-\dot{u}_x^{(n+\hat{x})}+\dot{u}_x^{(n+\hat{z})}\right)\right)\right] \\
& +\frac{1}{8}\left[\frac{\hat{y}\hat{z}}{\Delta y\Delta z}\tilde{\mathbb{D}}_{45}\left(\dot{u}_x^{(n+\hat{y}+\hat{z})}-\dot{u}_x^{(n)}+\dot{u}_x^{(n+\hat{z})}-\dot{u}_x^{(n+\hat{y})}\right)\right] \\
& +\frac{1}{8}\left[\frac{\hat{x}\hat{z}}{\Delta x\Delta z}\tilde{\mathbb{D}}_{45}\left(\dot{u}_y^{(n+\hat{x}+\hat{z})}-\dot{u}_y^{(n)}+\dot{u}_y^{(n+\hat{z})}-\dot{u}_y^{(n+\hat{x})}\right)\right] \\
& +\frac{1}{4}\left[\frac{\hat{x}\hat{y}}{\Delta x\Delta y}\tilde{\mathbb{D}}_{45}\left(\dot{u}_z^{(n+\hat{x}+\hat{y})}-\dot{u}_z^{(n)}\right)\right]\Bigg]
\end{aligned}
\tag{9.11}
$$

where $\tilde{\mathbb{D}}_{ij} = \mathbb{D}_{ij}^{(n+\hat{x}+\hat{y}+\hat{z})}$ represents the viscosity coefficient in one of the 8 cells surrounding point (n), (where $\hat{x}, \hat{y}, \hat{z}$ = +/- 1). As with the earlier LISA equations, ρ is the average density of all 8 cells surrounding point (*n*). Unlike Eqs. (9.3)–(9.5), the discretized viscoelastic equations above contain the time derivative of the displacement. $\dot{u}_i$ can be defined as:

$$\dot{u}_i^{(n)} = \frac{3u_i^{(n)} - 4u_i^{(n)}(t - \Delta t) + u_i^{(n)}(t - 2\Delta t)}{2\Delta t} \tag{9.12}$$

where a second-order finite difference definition was used. Thus, as stated earlier, the viscoelastic terms require values from the three prior time steps.

9.3 GRID SIZING AND STABILITY REQUIREMENTS

The Courant-Friedrichs-Lewy (CFL) condition determines the required time step size to ensure stability. Additionally, the spatial grid size should be set to have 8 points per wavelength at a minimum. For a rectangular Cartesian grid, these conditions result in:

$$\begin{aligned} \Delta x &\leq \frac{c_{min}}{8 f_{max}}, \\ &\leq \lambda_{min} / 8 \\ \Delta t &\leq \frac{1}{c_{max}} \frac{1}{\sqrt{1/\Delta x^2 + 1/\Delta y^2 + 1/\Delta z^2}}. \end{aligned} \tag{9.13}$$

As discussed in detail in Chapter 8, speed of sound values for determining Δt and Δx is not always simply defined by the shear and longitudinal wave speeds. In particular, for the case of guided wave propagation, the speeds of relevant wave modes must be considered when determining the appropriate spatial and time step sizes. Additionally, for thin plate-guided wave cases, enough grid points are needed through the thickness to accurately capture mode shape [8].

9.4 BOUNDARY CONDITIONS

Several boundary conditions are highly relevant for ultrasonic NDE modeling. A common requirement for an NDE setup is to model scattering from free edges of the component/specimen. This situation may arise from a structural component containing portions surrounded by air (such as a stiffener in an aircraft panel), or from inspection of a subcomponent of a larger structure (as is often the case for laboratory tests, calibration studies, probability of detection studies, etc.). For such situations, stress-free boundaries can be implemented by setting the stresses beyond the material boundary to zero using a computational grid layer beyond the simulated specimen edges where the stresses are prescribed to be zero [16].

However, in practice, to represent stress-free boundaries in LISA, multiple researchers report padding the grid with a layer of cells that correspond to the material properties for air [9, 11]. For example, Nadella and Cesnik report setting material properties in the "air" padded cells to a density of 1.3 kg/m^3 and stiffness to 10,000 times less than the solid elastic media in the simulation [11].

Nonreflecting and/or absorbing boundaries are also highly relevant for NDE applications since they can be used to decrease scattering from simulation domain edges (for example, when a subsection of a component is being modeled and waves in the real component would propagate farther than the simulated domain). Zhang and Cesnik have reported the use of both nonreflective boundary conditions and absorbing boundary layers with LISA [6].

To implement nonreflective boundary conditions, the authors apply a series of viscous boundary conditions to the model boundaries, following the approach by Lysmer and Kuhlemeyer [19]. For an isotropic material, the approach leads to boundary conditions of the form shown below:

$$\begin{aligned} \sigma_{ii} &= a\rho V_l \dot{u}_i \\ \sigma_{ij} &= b\rho V_s \dot{u}_j \qquad (i \neq j) \end{aligned} \tag{9.14}$$

where a and b are fitting constants that can be taken as $a = b = 1$ for a standard viscous boundary, and V_l and V_s are phase velocities for the longitudinal and shear waves, respectively [6]. For an orthotropic material case, these equations must account for directional dependent velocities:

$$\sigma_{ij} = a_{ij}\rho V_{ij}\dot{u}_j \tag{9.15}$$

where V_{ij} represents phase velocities in the respective directions. Details of the resulting LISA equations can be found in Reference [6].

An alternative approach to reduce boundary reflections is the application of absorbing layers with increasing damping (ALID) [6]. This approach requires adding an absorbing region beyond the edges of the simulated material/structure. The added computational grid cells (beyond the modeled specimen) have a prescribed increase in damping towards the computational domain edges. A general recommendation for absorbing layers, is that the length of the absorbing region, L_a, be set to a minimum value of $L_a = 2\lambda_{max}$, where λ_{max} is the longest wavelength in the simulation. Further, it is recommended that a cubic polynomial be used to define the increase in damping from the minimum (for absorbing cells neighboring the simulated material cells) to maximum (for absorbing cells at the computational domain boundary) damping value. Hybrid methods combining ALID and the L-K method have been studies by Zhang and Cesnik. An example in Section 9.7.1 shows results for wave propagation in an isotropic plate using the above methods to decrease wave scattering from boundaries of the computational domain.

9.5 INITIAL CONDITIONS FOR ULTRASOUND EXCITATION

When implementing LISA, an excitation source is necessary to initiate ultrasonic wave propagation. For NDE applications, a common excitation would be a kHz or MHz frequency tone burst or sinusoidal wave (see Fig. 9.2 as an example). A simple approach to excite ultrasonic waves is to prescribe an incident displacement function. A more complex, but more rigorous approach is to explicitly model piezoelectric behavior. The sections below describe both approaches.

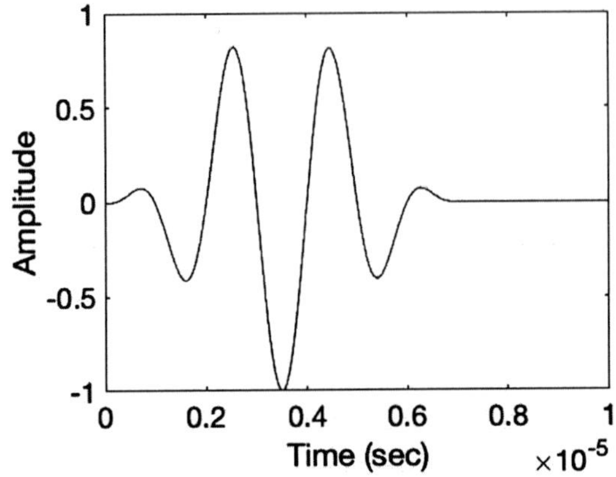

FIGURE 9.2 The figure shows an example excitation function, a 3.5 cycle 0.5 MHz Hann windowed sine wave. This signal is an example of a common ultrasonic excitation with relevance to NDE.

9.5.1 Displacement Excitation

A simple model an ultrasonic excitation can be implemented by specifying an initial displacement for the spatial region corresponding to the modeled actuator (i.e., transducer) [4]. Using this approach, both the excitation amplitude and the time dependent functional form of the initial displacement can be set on a per-node basis. It is therefore possible to define the displacement pattern across the actuator region, such as a 2D Gaussian. It is also possible to use wave superposition to define angled incidence, similar to that discussed in Chapter 8. Thus, a nonuniform excitation across the actuation region can easily be represented using the simple displacement based approach. As an example, in the case of modeling a uniform normal displacement in a plate-like component, with the z direction taken as the through-thickness direction, one would prescribe u_z with uniform amplitude across the excitation region. Alternatively, an in-plane excitation can be specified by adding an initial condition for u_x and u_y. Fig. 9.3 shows an example based on [4], where the magnitude of the excitation increases toward the edges of the actuator region.

9.5.2 Electromechanical Model of Actuation

Coupling the electromechanical behavior of a piezoelectric actuator to the elastodynamic LISA model is a second, and likely more accurate, approach to represent ultrasound excitation from a piezoelectric transducer. This excitation method involves explicitly discretizing and modeling an actuator on the specimen surface (see Fig. 9.4). The approach has been explored in detail be Nadella and Cesnik [17, 18], who found significant differences in the excitation results when explicitly modeling the electromechanical behavior of the actuator compared to using the simple

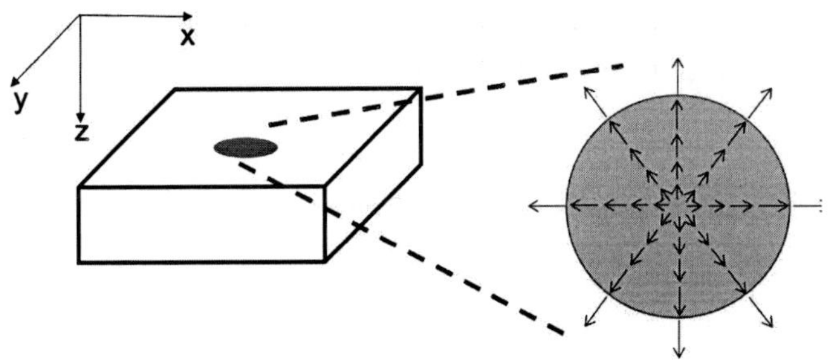

FIGURE 9.3 A diagram showing in-plane excitation in the LISA model where an initial condition on displacement is used to specify the excitation. The magnitude of the specified incident excitation grows towards the excitation region's edges (as represented by the size of the black arrows). In the case shown, the z direction is taken as the through-thickness direction, and initial conditions on u_x and u_y specify the in-plane excitation.

displacement excitation method (discussed in Section 9.5.1). Further, they found that the coupled approach led to excitation sources matching more closely with experiment.

To add a model of the electromechanical behavior into LISA, the charge equations for electrostatics, along with constitutive equations for mechanical-piezoelectric coupling, must be considered in addition to Eq. (9.1):

$$\begin{aligned} \frac{\partial D_j}{\partial x_j} &= 0 \\ \sigma_{ij} &= C^E_{ikjl} e_{kl} - \varepsilon_{ijk} E_k \\ D_i &= \varepsilon_{ijk} e_{jk} + \kappa^S_{ij} E_j \end{aligned} \tag{9.16}$$

where D is electric displacement, C^E is the stiffness tensor at constant electric field, E is the electric field, e_{jk} is the strain tensor, κ^S is the permittivity at constant strain, ε is the tensor of piezoelectric stress constants, and indices $i, j, k, l = 1,2,3$.

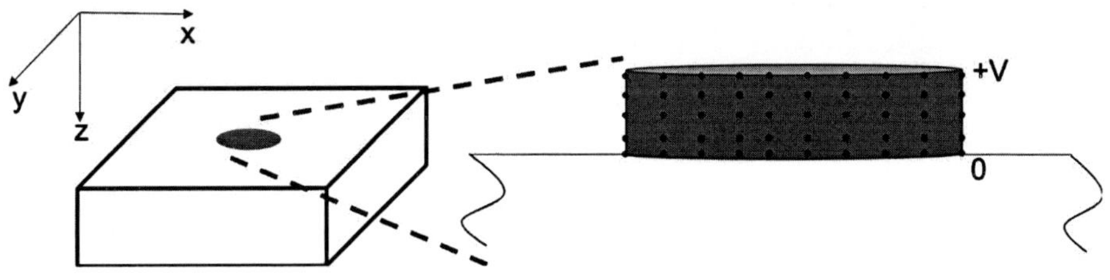

FIGURE 9.4 Diagram showing a discretized actuator in the coupled LISA method. While not shown in the figure, the substrate is also discretized. The top actuator nodes are prescribed a time varying potential, while the bottom nodes have a potential set to zero. (Image is based on [18].)

It is further required to define the strain-mechanical displacement and the electric field-electric potential, following Nadella and Cesnik [17]:

$$e_{ij} = \frac{1}{2}\left(\frac{\partial u_i}{\partial x_j} + \frac{\partial u_j}{\partial x_i}\right)$$
$$E_j = -\frac{\partial \psi}{\partial x_j} \tag{9.17}$$

where Ψ is the electric potential. With electromechanical behavior now being modeled, the SIM must be extended to include continuity of electric displacement and electric potential across the interface between cells. Discretization of the equations in 9.16 and application of the updated SIM results in a coupled electromechanical LISA model that is derived and listed in detail in [17].

Using this approach, the mechanical and piezoelectric coupling between the actuator and substrate can be explicitly modeled. A time-dependent electric potential can be applied to the discretized actuator, resulting in a more rigorous representation of a transducer compared to the simple displacement approach. Fig. 9.5 gives an example taken from Reference [18] showing results for using the simple displacement excitation approach versus explicitly modeling the actuator behavior. These plots from Nadella and Cesnik clearly show that there are significant differences between the displacement and electromechanical coupled approaches discussed in this section. More details on the cases shown in Fig. 9.5, including comparisons between experiment and simulation, can be found in Reference [18].

9.6 COMPUTATIONAL IMPLEMENTATION

As stated earlier, LISA is an explicit time domain method. Therefore, the computational implementation explicitly marches forward in time. At each time step, displacements are calculated at all spatial grid points. Similar to other FD techniques, the calculation at a single grid point requires information from neighboring grid points. For computational implementation, this requirement means that parallelization onto compute nodes/cores that do not have a shared memory space must involve data passing. As with other elastodynamic FD-based methods, LISA can be efficiently parallelized to run on multicore, cluster, or accelerator (such as GPU) hardware using tools such as Message Passing Interface (MPI), OpenMP, and OpenACC [16, 20, 21].

GPU implementation of techniques such as LISA requires careful planning on memory usage in order to reduce transfers between the host (CPU) and device (GPU), as excessive transfers will have a significant impact on increasing computational time. Paćko et al. describe the implementation of LISA on GPU hardware. In order to reduce memory footprint, only material property values per cell were stored on the GPU device (e.g., density and Lamé parameters for an isotropic case), with any other required values left to be calculated in real-time on the GPU. The memory requirements of LISA can also be reduced by not storing an excessive number of

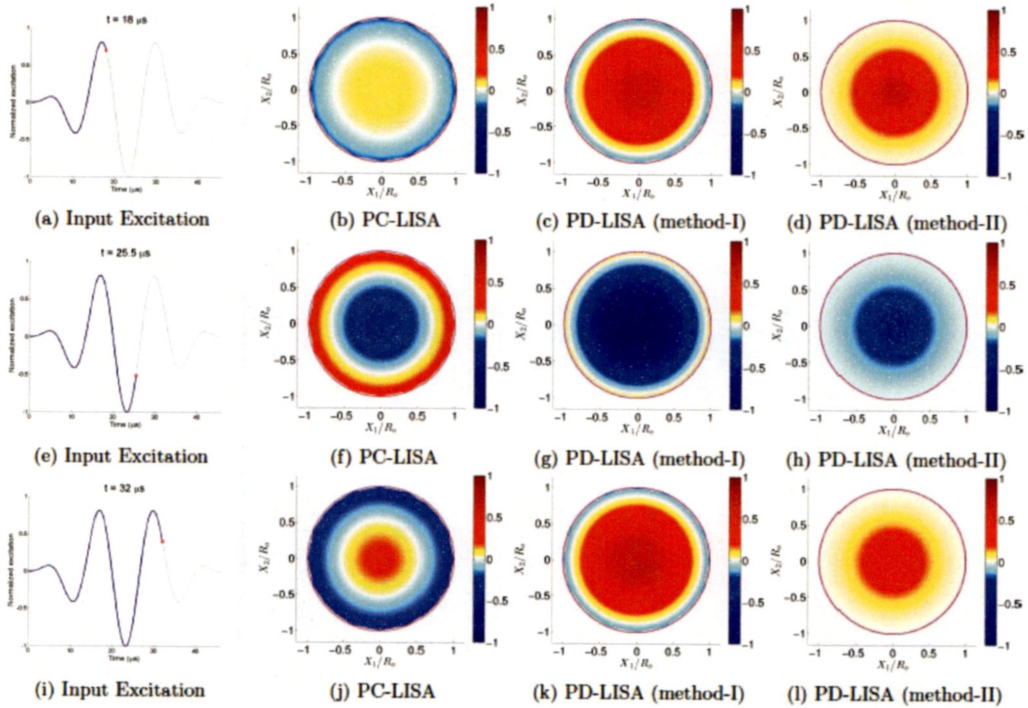

FIGURE 9.5 The figure shows out-of-plane displacement patterns for actuation on an isotropic substrate. Each rows corresponds to a different point in time in the excitation (corresponding to the left-most column (a),(e),(i)). (b), (f), (j) show the resulting excitation pattern when the actuator is modeled using coupled LISA method described in this section. (c),(g),(k) show the pattern that results from using the prescribed displacement (PD) method discussed in Section 9.5.1 for a case PD-LISA (method I) with uniformly spaced radially increasing in-plane displacement. The right-most column, (d), (h), (l), shows the result for a case PD-LISA(method II) using out-of-plane displacements and a Gaussian profile enforced across the actuator face. (Image from Nadella, K. and Cesnik, C. 2014. Effect of piezoelectric actuator modeling for wave generation in LISA. Proc. SPIE 9064, Health Monitoring of Structural and Biological Systems, reprinted with permission from SPIE and the author [18].)

"empty" (e.g., air) cells. For example, Paćko and team stored only one to two air cells around material cells.

9.7 EXAMPLES

This section describes several examples of LISA simulation cases for ultrasonic wave propagation. The examples are taken from the scientific literature and are used with permission of the authors. As mentioned in Section 9.1, prior reported work using LISA is focused on guided wave scenarios. Accordingly, the examples here are focused on guided wave simulation cases.

9.7.1 Guided Waves in an Isotropic Plate

Due to the widespread use of metallic plate-like components across various industries (such as aerospace, and shipbuilding), guided wave-based defect detection methods for isotropic plates is an area of extensive study in the scientific literature ([22–24], as a few examples). Zhang and Cesnik report the use of LISA to model guided wave propagation in an isotropic steel plate. Specifically, the team used the simple isotropic plate model to study the boundary methods discussed in Section 9.4 for decreasing wave scattering from simulation domain edges.

Results from their work are given in Fig. 9.6, and show guided waves interaction with boundaries for four different conditions placed on the simulation edges (as noted in the figure). In the steel plate case, a tone burst 200 kHz excitation is generated on the top and bottom plate surfaces to enable separate excitation of symmetric and antisymmetric guided wave modes. The materials properties for the steel plate are set to E=200 GPa, v=0.30, ρ=7850 kg/m^3. The plate thickness is 2 mm, and therefore only two guided waves exist at the corresponding frequency-thickness. The modeled region includes an extension to incorporate absorbing layers, and corresponds to the region beyond the black box. As stated earlier, general guidance is to set the absorbing layer region twice the longest wavelength in the simulation. In this case, the region size is set to twice the S0 mode wavelength (λ_{S0}=26.41 mm).

Since out-of-plane displacement is plotted, the A0 mode is the dominant mode that is observable in the figure. A spatial step size of 1 mm is used in x and y directions, with a spatial step size of 0.5 mm in the z (through-thickness) direction. The time step size was 69.6 ns. Fig. 9.6 shows a point 62.6 mus after the start of the wave excitation. The figure with stress-free edges shows symmetric scattering, as expected. As shown in Fig. 9.6d, wave scattering from edges can be significantly decreased using the hybrid L-k/ALID method.

9.7.2 Guided Waves in Composite Plates

The LISA equations listed in Section 9.2 can be used for modeling waves in materials containing orthotropic anisotropy (or a lower degree of anisotropy). Several authors have reported the use of LISA for modeling guided waves in composites,

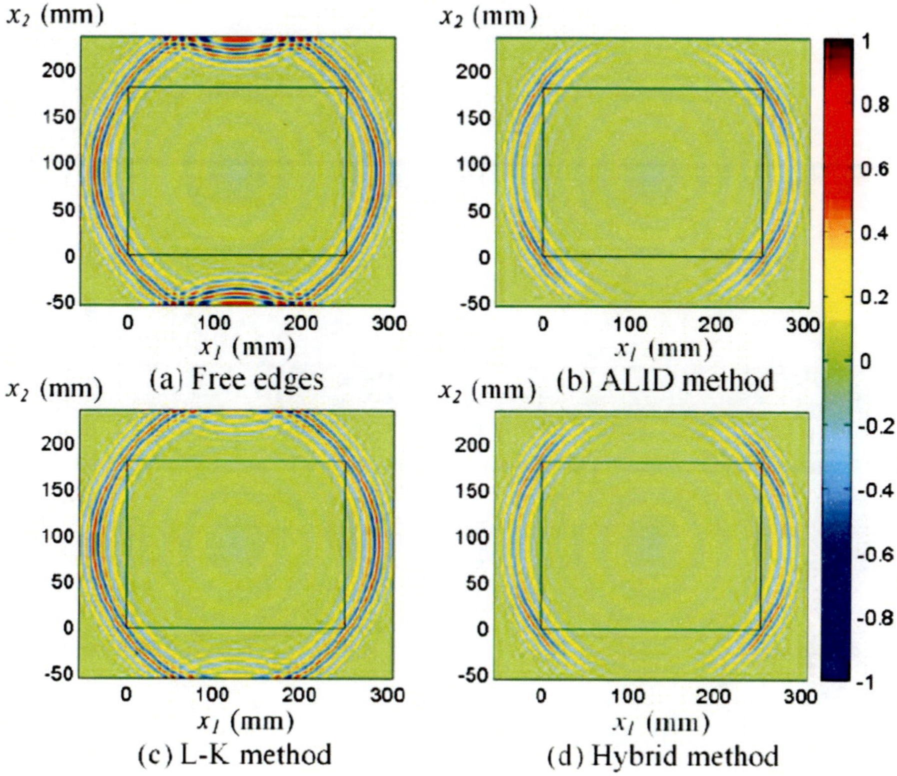

FIGURE 9.6 LISA out-of-plane displacement simulation results for guided waves in an isotropic steel plate at 62.6 μs after the initial excitation with (a) stress-free boundaries at simulation edges, (b) damping at simulation edges using ALID method, (c) non-reflective boundaries using the L-K method, and (d) decreased edge reflections using a hybrid L-K/ALID method (see Section 9.4 for details). The region beyond the black box corresponds to the added damping region. (Image from Zhang, H. and Cesnik, C. 2016. A hybrid non-reflective boundary technique for efficient simulation of guided waves using local interaction simulation approach. Proc. SPIE 9805, Health Monitoring of Structural and Biological Systems 2016, 98050U, reprinted with permission from SPIE and the author [27].)

including References [4, 8, 9, 15, 25]. Material-induced attenuation of ultrasonic waves can occur in composite materials, with the degree of attenuation depending on direction of propagation. As such, the incorporation of a directional damping model into LISA can allow for more realistic amplitude calculations for waves propagating in composites (see Section 9.2).

An example of LISA simulation results from Shen and Cesnik for guided wave propagation in unidirectional and cross-ply carbon fiber composite laminates are shown in Fig. 9.7 [8]. Fig. 9.7 also shows comparisons between the LISA simulation results and data acquired using scanning laser Doppler vibrometry (SLDV). The cases shown in the figure correspond to wave propagation in two IM7/977-3 composite laminates with layups $[0]_{12T}$ and $[0/90]_{3S}$ and thicknesses of 1.5 mm (material

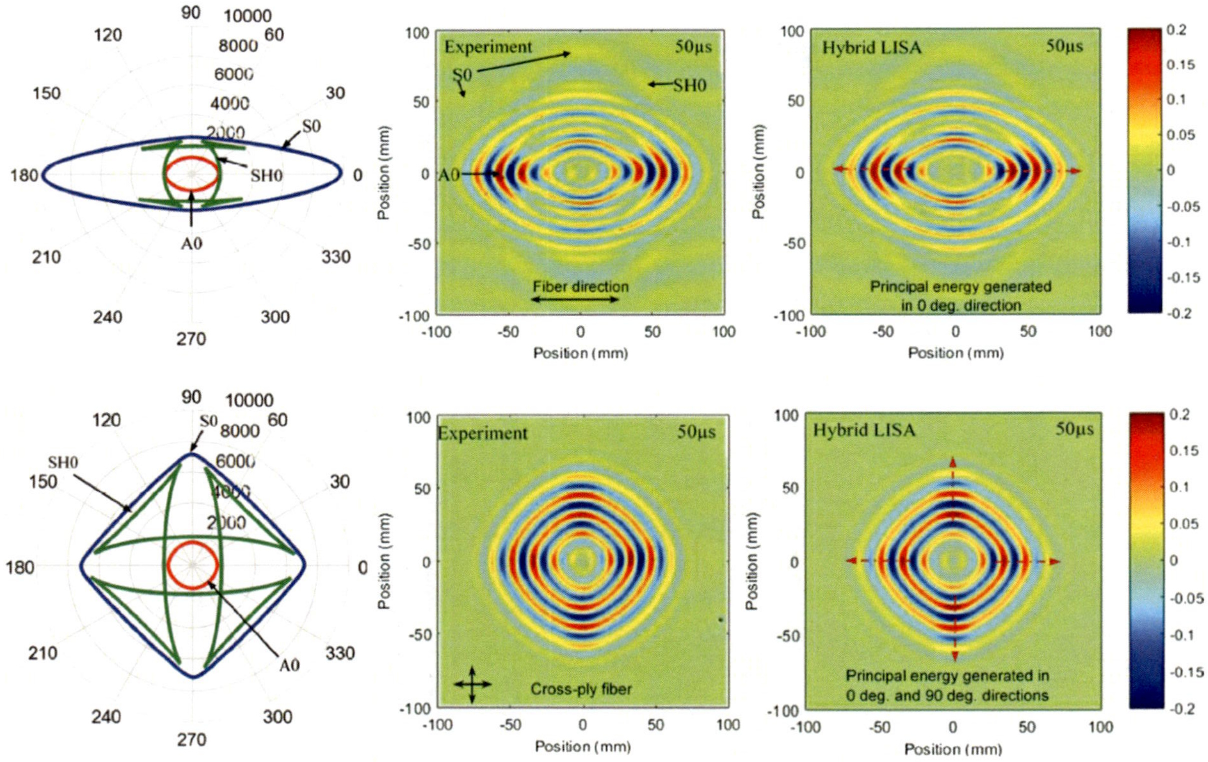

FIGURE 9.7 Left column: group velocity curves for the unidirectional (top) and cross-play (bottom) composite laminate cases, Middle: SLDV experimental results for the unidirectional (top) and cross-ply (bottom) cases, Right: LISA simulation results for unidirectional (top) and cross-ply (bottom) cases. (Shen, Y. and Cesnik, C. 2018. Local interaction simulation approach for efficient modeling of linear and nonlinear ultrasonic guided wave active sensing of complex structures. Journal of Nondestructive Evaluation, Diagnostics and Prognostics of Engineering Systems 1.1 (2018): 011008, reprinted with permission from ASME [8].)

TABLE 9.1
IM7/977-3 Material Properties*

Property	IM7/977-3
ρ (kg/m³)	1558
E_1 (GPa)	147.0
E_2 (GPa)	9.80
E_3 (GPa)	9.80
G_{12} (GPa)	5.3
G_{13} (GPa)	5.3
G_{23} (GPa)	3.31
ν_{12}	0.41
ν_{13}	0.41
ν_{23}	0.69

Properties from: [22]

properties are shown in Table 9.1) [8]. The example includes the damping model outlined earlier in the chapter to account for material induced attenuation, and viscosity coefficients are shown in Table 9.2. For both cases a 75 kHz excitation was created using frequency domain FEM to generate the guided waves. This hybrid approach, using local FEM to generate waves in LISA, is outlined further in [8, 15].

It can be observed in the unidirectional laminate case that the LISA results show A0 mode elongation along the fiber direction, as expected based on the expected group velocity (left column in Fig. 9.7). The primary energy direction is also observed to be in the 0 degree (fiber) direction, as expected. For the cross-ply case, the expected wavefield shape is also observed in the LISA results (compared to the group velocity curves and experiment). LISA results for both cases visually compare well with experiment. More details can be found in Reference [8].

TABLE 9.2
IM7/977-3 Viscosity Coefficients

Viscosity Coeff. (Pa s)	IM7/977-3
D_{11}	328.6
D_{22}	525.7
D_{33}	525.7
D_{44}	722.8
D_{55}	394.3
D_{66}	394.3

Properties from: [8]

9.7.3 Guided Waves in Rail Track

In terms of the NDE application, guided waves have been extensively studied and used for detection of cracking and material property assessment in rail track [26, 27]. Shen and Cesnik report the use of LISA to model guided wave propagation in a rail track [8]. Additionally, the simulated rail track contains a fatigue crack modeled in a manner to include contact acoustic nonlinearity [8]. In order to appropriately capture the crack opening/closing behavior, the LISA equations are updated to account for contact interaction and include a penalty function to restrict motion when crack faces are in contact. Further, the ability to add air grid cells between crack faces during crack opening was included. The updated LISA approach allowed the team to represent the nonlinear behavior involved in contact interaction of the crack interfaces due to the ultrasonic waves. The mathematical details of adding nonlinear behavior to LISA can be found in Reference [7].

The simulation setup used two line sources to excite a 100 kHz 10-cycle tone burst via surface traction forces. The rail track geometry and excitation location is shown in Fig. 9.8. Results of the LISA guided wave simulation are shown in Fig. 9.9. As described in the paper, a 1 mm spatial mesh size was used for the rail cross-section and a 2 mm mesh size was used along the track length, with a 110.33 ns time step size. The model size corresponded to over 10 million degrees of freedom, and was implemented using LISA code that was parallelized to run on GPUs. The team also implemented absorbing layers at the ends of the simulation domain along the track length. Beyond the inclusion of nonlinearity, this case demonstrates the ability of LISA to model waves in complex geometries and the use of absorbing layers at simulation domain edges.

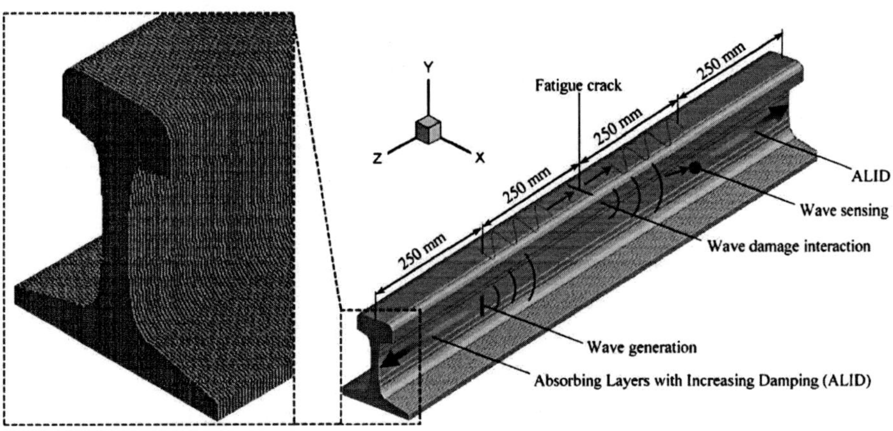

FIGURE 9.8 Model of rail track geometry implemented using LISA. (Image from Shen, Y. and Cesnik, C. 2018. Local interaction simulation approach for efficient modeling of linear and nonlinear ultrasonic guided wave active sensing of complex structures. Journal of Nondestructive Evaluation, Diagnostics and Prognostics of Engineering Systems 1.1 (2018): 011008, reprinted with permission from ASME [8].)

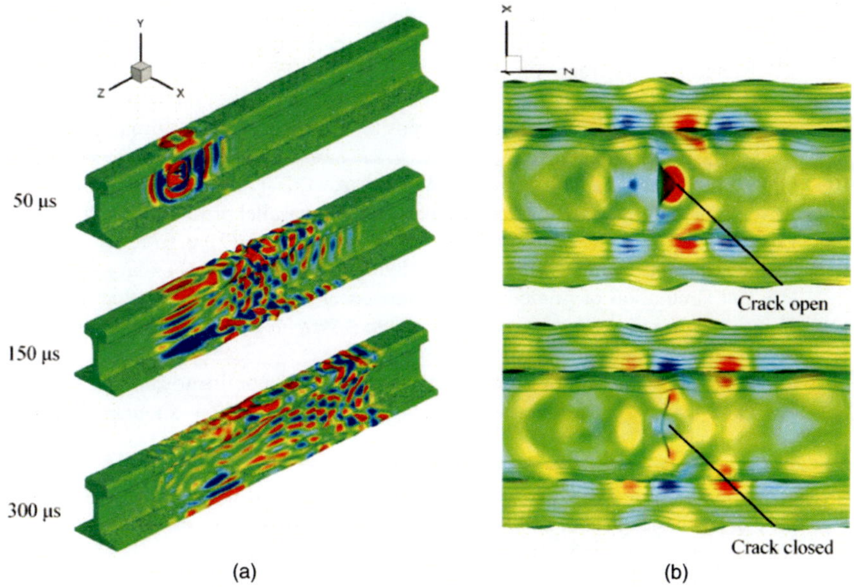

FIGURE 9.9 Left: LISA simulation results at three points in time (as noted) of guided wave propagation in a rail track. Right: View of the fatigue crack region as guided waves interact with the open and closed fatigue crack. (Shen, Y. and Cesnik, C. 2018. Local interaction simulation approach for efficient modeling of linear and nonlinear ultrasonic guided wave active sensing of complex structures. Journal of Nondestructive Evaluation, Diagnostics and Prognostics of Engineering Systems 1.1 (2018): 011008, reprinted with permission from ASME [8].)

9.8 SUMMARY

This chapter outlined LISA as a method for simulating ultrasonic waves in elastic and viscoelastic materials. The method involves implementation of the fairly straightforward equations shown in Section 9.2, and lends itself to parallelization for computational efficiency. The SIM approach used with LISA makes it well-suited for modeling waves in inhomogeneous, complex media with complex boundaries. While prior uses reported in the literature focus on guided ultrasonic waves, fundamentally LISA is a method for numerically calculating elastodynamic wave propagation and can be used to simulate any of the ultrasonic wave types typically used in NDE (e.g., bulk waves, surface waves, guided waves, etc.). At this time, the method has only been demonstrated for anisotropy up to monoclinic (i.e., triclinic anisotropy in LISA has not been demonstrated).

REFERENCES

1. Delsanto, P.P., R.S. Schechter, H.H. Chaskelis, R.B. Mignogna, and R. Kline, Connection machine simulation of ultrasonic wave propagation in materials. II: The two-dimensional case. *Wave Motion*, 1994. **20**(4): pp. 295–314.
2. Delsanto, P.P., R.S. Schechter, and R.B. Mignogna, Connection machine simulation of ultrasonic wave propagation in materials. III: The three-dimensional case. *Wave Motion*, 1997. **26**(4): pp. 329–339.

3. Delsanto, P.P., T. Whitcombe, H.H. Chaskelis, and R.B. Mignogna, Connection machine simulation of ultrasonic wave propagation in materials. I: The one-dimensional case. *Wave motion*, 1992. **16**(1): pp. 65–80.
4. Nadella, Kalyan S. and Carlos E.S. Cesnik, Local interaction simulation approach for modeling wave propagation in composite structures. *CEAS Aeronautical Journal*, 2013. **4**(1): pp. 35–48.
5. Paćko, P., T. Bielak, A.B. Spencer, W.J. Staszewski, T. Uhl, and K. Worden, Lamb wave propagation modelling and simulation using parallel processing architecture and graphical cards. *Smart Materials and Structures*, 2012. **21**(7): p. 075001.
6. Zhang, Hui and Carlos E.S. Cesnik, A hybrid non-reflective boundary technique for efficient simulation of guided waves using local interaction simulation approach, In *Health Monitoring of Structural and Biological Systems 2016*. 2016, **9805**, International Society for Optics and Photonics. p. 98050U.
7. Shen, Yanfeng and Carlos E.S. Cesnik, Modeling of nonlinear interactions between guided waves and fatigue cracks using local interaction simulation approach. *Ultrasonics*, 2017. **74**: pp. 106–123.
8. Shen, Yanfeng and Carlos E.S. Cesnik, Local interaction simulation approach for efficient modeling of linear and nonlinear ultrasonic guided wave active sensing of complex structures. *Journal of Nondestructive Evaluation, Diagnostics and Prognostics of Engineering Systems*, 2018. **1**(1): p. 011008.
9. Sundararaman, Shankar and Douglas E. Adams, Modeling guided waves for damage identification in isotropic and orthotropic plates using a local interaction simulation approach. *Journal of vibration and acoustics*, 2008. **130**(4): p. 041009.
10. Lee, B.C., and W.J. Staszewski, Modelling of lamb waves for damage detection in metallic structures: Part II. Wave interactions with damage. *Smart Materials and Structures*, 2003. **12**(5): pp. 815.
11. Nadella, Kalyan, and Carlos Cesnik, Simulation of guided wave propagation in isotropic and composite structures using LISA. In *53rd AIAA/ASME/ASCE/AHS/ASC Structures, Structural Dynamics and Materials Conference 20th AIAA/ASME/AHS Adaptive Structures Conference 14th AIAA*, 2012. p. 1387.
12. Radecki, Rafal, Zhongqing Su, Li Cheng, Pawel Packo, and Wieslaw J. Staszewski, Modelling nonlinearity of guided ultrasonic waves in fatigued materials using a nonlinear local interaction simulation approach and a spring model. *Ultrasonics*, 2018. **84**: pp. 272–289.
13. Scalerandi, Marco, Valentina Agostini, Pier Paolo Delsanto, Koen Van Den Abeele, and Paul A. Johnson, Local interaction simulation approach to modelling nonclassical, nonlinear elastic behavior in solids. *The Journal of the Acoustical Society of America*, 2003. **113**(6): pp. 3049–3059.
14. Obenchain, Matthew B., Kalyan S. Nadella, and Carlos E.S. Cesnik, Hybrid global matrix/local interaction simulation approach for wave propagation in composites. *AIAA Journal*, 2014. **53**(2): pp. 379–393.
15. Shen, Yanfeng and Carlos E.S. Cesnik, Hybrid local fem/global LISA modeling of damped guided wave propagation in complex composite structures. *Smart Materials and Structures*, 2016. **25**(9): p. 095021.
16. Sinor, Milan, Numerical modelling and visualisation of elastic wave propagation in arbitrary complex media. In *Proceedings of the Eighth Workshop on Multimedia in Physics Teaching and Learning of the European Physical Society*, 2004. pp. 9–11.
17. Nadella, Kalyan S. and Carlos E.S. Cesnik, Piezoelectric coupled LISA for guided wave generation and propagation. In *Health Monitoring of Structural and Biological Systems 2013*, 2013. **8695**, p. 86951R. International Society for Optics and Photonics.

18. Nadella, Kalyan S. and Carlos E.S. Cesnik, Effect of piezoelectric actuator modeling for wave generation in LISA. In *Health Monitoring of Structural and Biological Systems 2014*, 2014. **9064**, p. 90640Z. International Society for Optics and Photonics.
19. Lysmer, John, and Roger L. Kuhlemeyer, Finite dynamic model for infinite media. *Journal of the Engineering Mechanics Division*, 1969. **95**(4): pp. 859–878.
20. OpenMP Architecture Review Board. *OpenMP Specifications*, 2019 (accessed June 19, 2019).
21. OpenACC-Standard.org. *OpenACC Specifications*, 2019 (accessed June 19, 2019).
22. Ren, Baiyang and Cliff J. Lissenden, Modeling guided wave excitation in plates with surface mounted piezoelectric elements: coupled physics and normal mode expansion. *Smart Materials and Structures*, 2018. **27**(4): p. 045014.
23. Wilcox, Paul, Mike Lowe, and Peter Cawley, Omnidirectional guided wave inspection of large metallic plate structures using an EMAT array. *IEEE Transactions on Ultrasonics, Ferroelectrics, and Frequency Control*, 2005. **52**(4): pp. 653–665.
24. Yu, Lingyu, Zhenhua Tian, and Cara A.C. Leckey, Crack imaging and quantification in aluminum plates with guided wave wavenumber analysis methods. *Ultrasonics*, 2015. **62**: pp. 203–212.
25. Li, Guoyi, Rajesh Kumar Neerukatti, and Aditi Chattopadhyay, Ultrasonic guided wave propagation in composites including damage using high-fidelity local interaction simulation. *Journal of Intelligent Material Systems and Structures*, 2018. **29**(5): pp. 969–985.
26. Loveday, Philip W., Rebecca M.C. Taylor, Craig S. Long, and Dineo A. Ramatlo, Monitoring the reflection from an artificial defect in rail track using guided wave ultrasound. In *AIP Conference Proceedings*, 2018. **1949**: p. 090003. AIP Publishing.
27. Setshedi, Isaac I., Philip W. Loveday, Craig S. Long, and Daniel N. Wilke, Estimation of rail properties using semi-analytical finite element models and guided wave ultrasound measurements. *Ultrasonics*, 2019. **96**: pp. 240–252.

10 Spectral Element Method for CNDE

10.1 INTRODUCTION

The objective of this chapter is to solve CNDE problem using spectral element method (SEM). First, a comparison between finite element method (FEM) and SEM is presented in detail. Then a mathematical formulation to solve the wave propagation problem in a generalized material system (isotropic and laminated composite materials) is derived. Specifically, simulation of wave interaction in composite structure using pulse-echo (PE) ultrasonic transducer as depicted in Chapter 1 is discussed in this chapter. The PE ultrasonic signals have many features that are not known because actual interaction of the wave with different material layers in isotropic and ply and ply thickness in composites are poorly understood. The anisotropy of composite material plays a critical role in affecting the signal pattern, which is not apparent in isotropic materials. Many (FEM) based modeling approaches have been proposed to simulate the wave interactions without any satisfactory results at high frequencies in the Mega Hertz (MHz) range. Therefore, many alternate methods have been discussed in this book and SEM is one of them and probably a very strong competitor with the other methods presented in this book. Mathematical details of SEM method and its implementation in computer code is presented in this book to solve NDE problem numerically.

SEM is relatively a new addition to the history of computational NDE techniques. This technique basically combines the advantages of two different numerical techniques which is the spectral methods and FEMs. In structural analysis, these two techniques had been utilized widely for their accuracy and ease of use. SEMs are based on high-order weighted-residual techniques for solving partial differential equations that exploit both the common foundations and competitive advantages of h-type FEMs and p-type spectral techniques. Therefore, in short, it can be said that SEM is a higher order finite element technique that combines the geometric flexibility of finite elements with the high accuracy of spectral method. As proposed by Patera in 1984 [1], SEM is capable of simulating propagation of probing energy into a material system although proposed for fluid dynamics. Similarly, SEM should be able to simulate elastic waves in structures of varying geometries to investigate the structural state or state of the material and facilitate the understanding of wave-damage interactions as proposed by many researchers [2–4]. The basic idea and working principle of SEM is very similar to the FEM with some exception in approximating specific functions. Hence, in foregoing discussion several FEM related terminologies are used assuming the knowledge and familiarity of the readers with the FEM method. The very first difference between SEM and FEM is that the location of the interpolation nodes in an element used in SEM. In SEM, the elemental interpolation nodes are chosen at the points corresponding to the zeros of an appropriate family of orthogonal polynomial functions such as Lobatto or Legendre or Chebyshev polynomials [5, 6]. A set of local shape functions consists of Lagrange polynomials spanned on these nodal points are constructed. Due to this approximation and the use of Gauss-Lobatto-Legendre integration rule [7] a diagonal form of the mass matrix can be obtained. Adopting this technique, the computational cost is reduced, and the method becomes less expensive than that of a classic finite element approach. The second difference with the FEM is related to the degree of approximation of the approximating polynomials. In FEM, linear or quadratic shape functions are widely used in commercial applications with C_0, C_1, or C_2 continuity [8]; however, in SEM, higher order polynomials are commonly used for higher order continuity. Introduction of the diagonal mass matrix and the higher order polynomial decreases the numerical errors, faster than any power of 1/p where the p is the order of the applied polynomial. CNDE problems using SEM are associated with the simulation of wave propagation, transmission, reflection, refraction, scattering, interference, and diffraction of the ultrasonic elastic waves in continuous media (solid and fluid) [9].

10.1.1 A Comparative Analysis of FEM and SEM

As a popular and widely used method, the FEM is employed to solve complex problems from multi-disciplinary physical sciences that can be expressed by partial differential equations or integral equations. Examples of such problems usually can be found in fluid mechanics, solid mechanics, thermodynamics, electro and magneto statistics and dynamics, and many others. The working principle of FEM involves the discretization of the analyzed area into a finite number of smaller subdivisions normally termed as finite elements, within which solutions are determined by approximating suitable polynomials over evenly spaced nodal points. As the number of evenly spaced interpolation points or nodes increases, a so-called oscillation near the ends of the interpolation domain arises, which can be described as Runge's phenomenon. Unless a low-order polynomial approximation is employed, an even distribution of interpolation nodes is detrimental to the accuracy of the interpolation. To demonstrate the potential effect of the interpolation, let us approximate the Runge's function as $f(\xi)=\frac{1}{1+25\xi^2}$ in a normalized interval [-1, 1] with 6th order interpolating polynomial that passes through 6+1 evenly distributed data points. As the number of data points increases the polynomial order increases, and the interpolation worsens near the two ends of the interpolation domain which is referred as Runge's phenomenon. This is an obvious effect because the interpolated polynomial is reconstructed from evenly spaced data that contributes equally to the middle and towards the two ends of the interpolation domain. Insufficient information is provided beyond the boundaries of the interpolation domain. As a result, a wrong approximation is observed near the element boundaries as shown in Fig. 10.1a. Therefore, as the polynomial order increases, the results obtained from evenly spaced interpolation nodes in an element may become sensitive to the numerical error and may even fail to converge with erroneous results causing spurious reflections at the element boundaries, which is the case for the FEM. In order to avoid potential failure of a CNDE simulation or incorrect assessment of the wave fields, a judicial placement

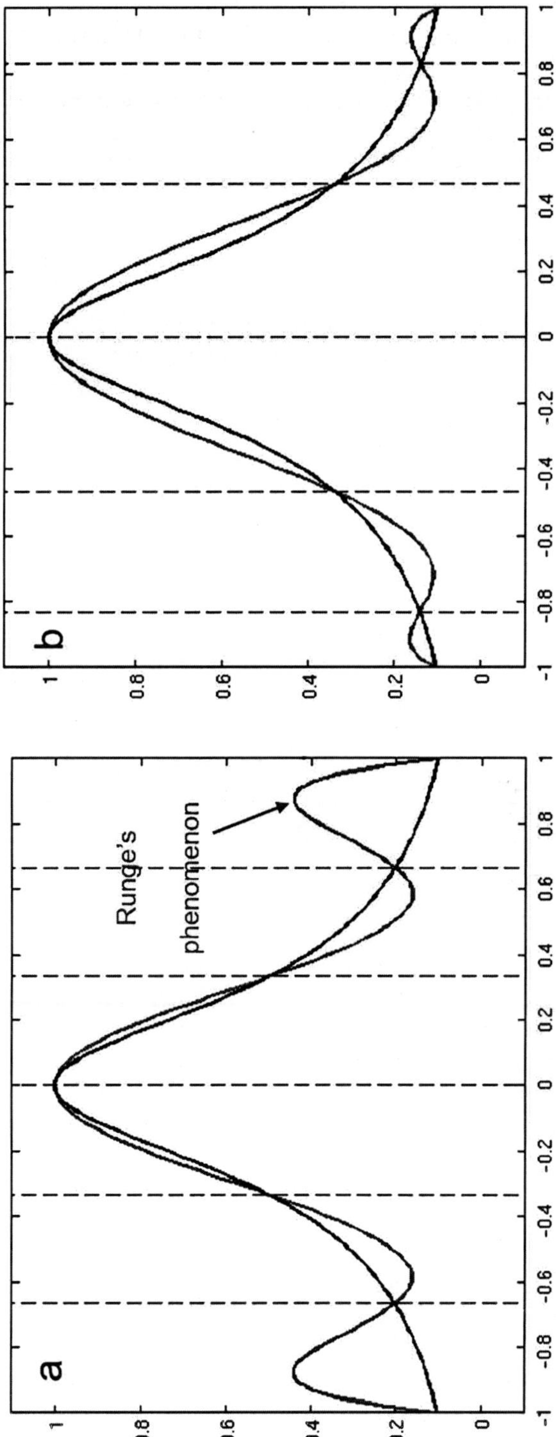

FIGURE 10.1 Plot of (a) 7-node interpolation functions corresponding to polynomial order 6 with evenly spaced nodal points, (b) 7-node interpolation functions corresponding to polynomial order 6 with nodal points at the roots of Gauss-Lobatto-Legendre polynomials.

of the interpolating nodes over an element to positions corresponding to the zeros of orthogonal polynomials can be utilized. In Fig. 10.1b, the nodes are located at the zeros of the Gauss-Lobatto-Legendre polynomial. As the nodes are not evenly spaced, the Runge's effect is reduced at the two ends of the normalized element interval between [-1,1].

10.1.2 Classification of SEM

SEM method can be essentially classified in to two types. Type I – SEM is a Frequency domain SEM [2–4] where the problem is solved at specific individual frequency, we call FSEM. A loading function is essentially transformed in to the frequency domain using Fourier transformation law, and the problem of interest is solved in all discrete frequencies at specific frequency interval governed by the sampling rate of time in the Fourier domain. Next, the response functions from the problem of interest at all the frequencies are collected, and an inverse Fourier transform (IFFT) is conducted to retrieve the time domain signal. The process in this method will be similar to the DPSM method presented in Chapter 7. DPSM is also a frequency domain CNDE solution method. In Type I – SEM the wave functions obtained from the solution of the wave propagation equation are used as the element shape functions. Whereas, Type II SEM is a transient or direct time domain simulation method [10] where the problem is solved in time domain utilizing the specific polynomial functions to express the quantity of interest inside the element and the nodal points are distributed at the zero points of that polynomial. We call Type II – SEM a TSEM. Hence, the TSEM is essentially a type of FEM with specific element types that are designed based on the spectral wave mode shapes. In this chapter we will primarily discuss TSEM as one of the frequency domain CNDE method is already discussed in this book in Chapter 7.

10.2 MATHEMATICAL FORMULATION OF SEM

In wave propagation and structural mechanics problems, the fundamental objective is to find the displacement amplitudes at each material point. Hence, the fundamental philosophy is in line with the fact that the material body is discretized in to multiple elements with multiple nodal points. The fundamental differential equations with primary unknowns are applied to the nodal points with approximate distribution of the primary unknowns between the nodal points along an imaginary line. The approximation of the distribution of the primary unknowns could be linear or quadratic or following higher order polynomials. Thus, in both FEM and SEM, the material body is discretized into a finite number of elements. The profile of the displacements which is the primary unknown in wave propagation solution is assumed in a form to obtain approximate element displacements from element level equations. The discretized equations obtained from each element are then assembled together with the adjoining elements to form global finite or spectral element equations for the whole material body. Equations are created to solve the displacement field in the entire material domain as a system of simultaneous linear algebraic equations. The problem of interest herein is CNDE or wave propagation in solid material.

The primary unknown variables in the governing differential equation are the displacement functions in three directions in the Cartesian coordinate system (u_1, u_2, and u_3). The secondary variables subsequently are the strain (e_{ij}) and stress variables (σ_{ij}). The strong form of the problem requires a strong continuity of the unknown variables across the whole domain of the body. Since obtaining the exact solution with a strong form of the equations is usually difficult for the wave propagation problems, a weak form of the equations is required to be derived which is suitable for obtaining an approximate solution of the primary unknowns. The weak form can be derived using variational principles. Following two forms of variational principles are used in general discussed below:

A. Energy principles such as Hamiltonian principles (discussed in Section 3.10.4 in Chapter 3): this can be categorized as a special form of the variational principle which is particularly applicable for problems related to mechanics of solids and structures, which here is wave propagation in solids. Minimum potential energy law is limited to the static equilibrium of solids, but Hamilton's principle is a generalization principle for solids under both static and dynamic conditions.
B. Weighted residual methods such as Ritz method and Galerkin method [6, 8]: This is a more general mathematical tool applicable, in principle, for solving different kinds of partial differential equations.

10.2.1 Application of Hamiltonian Principle

Application of Hamilton's principle guarantees solution for the system that is governed by strong form of equations. When the material is discretized, Hamilton principle still holds for the discrete material system. The primary unknowns, which are the displacement field in wave propagation problems, are assumed in a form distributed over the discretized elements with unknown values at the nodal points. The set of assumed displacements must satisfy the following three admissible displacement conditions:

1. The compatibility conditions (discussed in Chapter 3)
2. The essential or the kinematic boundary conditions
3. The conditions at initial and final time (after and before deformation)

The equation of motion of the system can be derived directly from the Lagrange equation as it was discussed in Section 4.5.2 in Chapter 4. According to the Hamiltonian principle of a system where wave propagation is sought, the governing equation can be written in bit different form compared to Eq. (A4.14) as

$$\frac{d}{dt}\left\{\frac{\partial \mathcal{L}}{\partial \dot{\mathbf{u}}}\right\} - \left\{\frac{\partial \mathcal{L}}{\partial \mathbf{u}}\right\} + \left\{\frac{\partial R}{\partial \dot{\mathbf{u}}}\right\} = 0 \tag{10.1}$$

where the problem of variation is nonconservative and the dissipative force is artificially introduced into the principle of least action. In the above equation, $\mathcal{L}$ is the Lagrangian functional of the dynamic system. Lagrangian of the system is the difference between the kinetic energy ($\mathcal{K}$) and the potential energy (Π_p) or elastic strain energy ($\mathcal{E}$), i.e., $\mathcal{L} = \mathcal{K} - \Pi_p$ like discussed in Chapter 4 (Section 4.5.2). The Lagrangian is assumed to be independent of the displacement gradients but only the function of displacements ($\mathbf{u}$) and velocities ($\dot{\mathbf{u}}$). $\mathbf{u}$ is the vector of dependent variables; here, u_1, u_2, and u_3 ($\mathbf{u} = [\, u_1 \; u_2 \; u_3 \,]^T$) are the displacements in three orthogonal directions in Cartesian coordinate system $\dot{\mathbf{u}}$ is the vector of time derivative of the dependent variables or the velocity; here, $\dot{u}_1 = v_1$, $\dot{u}_2 = v_2$, and $\dot{u}_3 = v_3$ are the displacements in three orthogonal directions in Cartesian coordinate system. The kinetic energy ($\mathcal{K}$), the potential energy (Π_p), and the dissipation of energy (R) in the system can be written as

$$\mathcal{K} = \frac{1}{2}\int_{\mathrm{V}} \rho \dot{\mathbf{u}}^T \dot{\mathbf{u}} dV \tag{10.2.1}$$

$$\Pi_p = \frac{1}{2}\int_{\mathrm{V}} \mathbf{e}^T \sigma d\mathrm{v} - \int_{\mathrm{V}} \mathbf{u}^T \psi_\Omega d\mathrm{v} - \int_{\mathcal{S}} \mathbf{u}^T \psi_\Gamma d\Gamma \tag{10.2.2}$$

$$R = \frac{1}{2}\int_{\mathrm{V}} \mu \dot{\mathbf{u}}^T \dot{\mathbf{u}} d\mathrm{v} \tag{10.2.3}$$

where ρ and μ represent the density and damping coefficient, respectively. Also, ψ_Ω and ψ_Γ denote the volume and surface force vectors, respectively. Ω represents the inside of the body or the volume of the body and Γ represents the surface of the body. In wave propagation problem, it can be easily realized that these force vectors are the functions of time. When a material domain is discretized using the spectral elements, the unknown displacement function inside the element can be approximated by the shape functions, and the nodal displacements at the i-th node point of the element can be written as U_i^{u1}, U_i^{u2}, and U_i^{u3} in the three mutually perpendicular directions, x_1, x_2, and x_3. Hence, the displacement at any point (x_j) inside the element is the superposition of the shape function multiplied by the nodal displacements. The shape functions can be visualized as the basis functions and the nodal displacements as the contribution factor. Displacement at any point in the domain can be

$$u_1(x_1, x_2, x_3) = \sum_{i=1}^{m} N_i(x_1, x_2, x_3) U_i^{u_1} \tag{10.3.1}$$

$$u_2(x_1, x_2, x_3) = \sum_{i=1}^{m} N_i(x_1, x_2, x_3) U_i^{u_2} \tag{10.3.2}$$

$$u_3(x_1, x_2, x_3) = \sum_{i=1}^{m} N_i(x_1, x_2, x_3) U_i^{u_3} \tag{10.3.3}$$

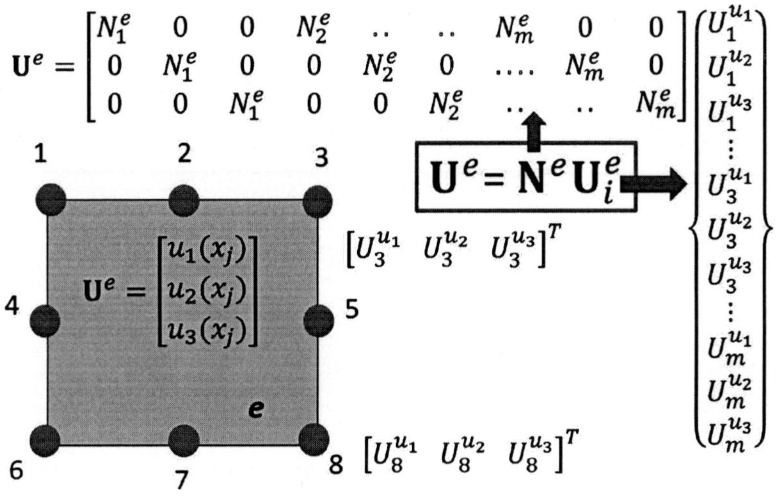

FIGURE 10.2 Discretization of displacement.

In Eq. (10.3.i), the shape function has m nodal points and thus there are m shape functions in the system. Fig. 10.2 can explain the further discussion with the formation of vectors filled with unknown displacement functions and shape functions. Let us assume an arbitrary element e with m nodal points (8 in Fig. 10.2) as a representative spectral element. Displacements in three orthogonal directions of a k-th nodal point can be written in a vector form $[U_k^{u_1}\ U_k^{u_2}\ U_k^{u_3}]^T$, like it is expressed in the third and eighth nodal points in Fig. 10.2. The displacements u_1, u_2, and u_3 at any point x_j inside the element can be expressed in a vector form $\mathbf{U}^e = [\,u_1\ u_2\ u_3\,]^T$ as shown in Fig. 10.2. As the displacement in x_1 direction, which is u_1, is only contributed by the $U_k^{u_1}$ displacement values at the respective nodal points, the displacement vector for the element $\mathbf{U}^e$ can be collectively written using a shape function matrix with zeros in proper respective places as shown in Fig. 10.2 and written in Eq. (10.4).

$$\mathbf{N}^e = \begin{bmatrix} N_1^e & 0 & 0 & N_2^e & \cdots\cdots & N_m^e & 0 & 0 \\ 0 & N_1^e & 0 & 0 & N_2^e & 0 & N_m^e & 0 \\ 0 & 0 & N_1^e & 0 & 0 & N_2^e & \cdots\cdots & N_m^e \end{bmatrix} \tag{10.4}$$

The nodal displacement vector is composed of $[U_k^{u_1}\ U_k^{u_2}\ U_k^{u_3}]^T$ vector sequentially placed for each respective nodal point in a column vector also shown in Fig. 10.2. Shape functions are not the function of time, and hence, the time derivatives of the displacement functions, i.e., the velocity and accelerations will have the same shape functions as written below. The relationships could be written in the following simpler form if the displacement vector of and element is written as

$$\mathbf{U}^e = \begin{bmatrix} u_1(x_j) \\ u_2(x_j) \\ u_3(x_j) \end{bmatrix} \tag{10.5}$$

$$\mathbf{U}^e{}_i = [U_1^{u_1},\ U_1^{u_2},\ U_1^{u_3},\ \ldots .U_m^{u_1},\ U_m^{u_2},\ U_m^{u_3}]^T \tag{10.6}$$

$$\mathbf{U}^e = \mathbf{N}^e\mathbf{U}_i^e;\ \dot{\mathbf{U}}^e = \mathbf{N}^e\dot{\mathbf{U}}_i^e;\ \ddot{\mathbf{U}}^e = \mathbf{N}^e\ddot{\mathbf{U}}_i^e \tag{10.7}$$

where $\mathbf{N}^e$ is the shape function matrix in Eq. (10.4). The superscript e signifies that the relation is valid only in one element with designation e (Fig. 10.2). Similarly, for that same element, the strain ($\mathbf{e}^e$) and the stress (σ^e) equations can be written as follows: The strain function inside the element e will be divided into two parts, linear (refer Eq. (A.3.2.i) in Chapter 3) and nonlinear strains (refer Eq. (T3.4) in Table 3 in Chapter 3). Considering both the linear and the nonlinear parts of the strain components, the total strain inside the element can be written as

$$\{\mathbf{e}^e\} = \mathbf{S}_l\mathbf{U}^e + \mathbf{S}_n\mathbf{U}^e \tag{10.8}$$

where

$$\{\mathbf{e}^e\} = \begin{Bmatrix} \mathbf{e}_{11} \\ \mathbf{e}_{22} \\ \mathbf{e}_{33} \\ \mathbf{e}_{12} \\ \mathbf{e}_{23} \\ \mathbf{e}_{31} \end{Bmatrix} \tag{10.9}$$

$$\mathbf{S}_l = \begin{bmatrix} \frac{\partial}{\partial x_1} & 0 & 0 \\ 0 & \frac{\partial}{\partial x_2} & 0 \\ 0 & 0 & \frac{\partial}{\partial x_3} \\ \frac{\partial}{\partial x_2} & \frac{\partial}{\partial x_1} & 0 \\ 0 & \frac{\partial}{\partial x_3} & \frac{\partial}{\partial x_2} \\ \frac{\partial}{\partial x_3} & 0 & \frac{\partial}{\partial x_1} \end{bmatrix},\ \mathbf{S}_n = \mathbf{S}_n(\mathbf{U}) = \frac{1}{2}\begin{bmatrix} \left(\frac{\partial}{\partial x_1}\mathbf{U}\right)^T \frac{\partial}{\partial x_1} \\ \left(\frac{\partial}{\partial x_2}\mathbf{U}\right)^T \frac{\partial}{\partial x_2} \\ \left(\frac{\partial}{\partial x_3}\mathbf{U}\right)^T \frac{\partial}{\partial x_3} \\ \left(\frac{\partial}{\partial x_1}\mathbf{U}\right)^T \frac{\partial}{\partial x_2} + \left(\frac{\partial}{\partial x_2}\mathbf{U}\right)^T \frac{\partial}{\partial x_1} \\ \left(\frac{\partial}{\partial x_2}\mathbf{U}\right)^T \frac{\partial}{\partial x_3} + \left(\frac{\partial}{\partial x_3}\mathbf{U}\right)^T \frac{\partial}{\partial x_2} \\ \left(\frac{\partial}{\partial x_3}\mathbf{U}\right)^T \frac{\partial}{\partial x_1} + \left(\frac{\partial}{\partial x_1}\mathbf{U}\right)^T \frac{\partial}{\partial x_3} \end{bmatrix} \tag{10.10}$$

Substituting Eq. (10.7) into Eq. (10.8), we get

$$\boldsymbol{\varepsilon}^e = \mathbf{S}_l \mathbf{N}^e \mathbf{U}_i^e + \mathbf{S}_n \mathbf{N}^e \mathbf{U}_i^e = \mathbf{B}_l^e \mathbf{U}_i^e + \mathbf{B}_n^e \boldsymbol{I}_i^e = \mathbf{B}^e \mathbf{U}_i^e \tag{10.11}$$

On the other hand, the constitutive equation gives the relationship between the stresses and strains in the material of a solid, which is governed by the Hooke's law. Therefore, the stress equation for an element can be reiterated similar to Eq. (3.75) but in vector form as

$$\sigma^e = \mathbb{C}^e \boldsymbol{\varepsilon}^e \tag{10.12}$$

Substituting Eq. (10.11) in Eq. (10.12), we get

$$\sigma^e = \mathbb{C}^e \mathbf{B}^e \mathbf{U}_i^e \tag{10.13}$$

where $\mathbb{C}^e$ is the representative constitutive property matrix of the element. Hence, it is apparent that it is not necessary to have same material properties for every element in the model. At every point in the material, one could provide different material properties based on the local constituent material. This is especially convenient for fiber composite materials where constituent material may change point to point in a material body. For example, change in fiber orientation may manifest different material properties and thus different $\mathbb{C}^e$ should be provided at different points as appropriate. Similarly, localized degraded or damaged material properties (like the degraded properties are calculated in Appendix section of Chapter 7) could also be used for certain element to simulate the effect of material degradation on the ultrasonic wave signal obtained from the NDE of isotropic or anisotropic composite materials. Once the stresses and strains are defined in terms of displacements for an element, substituting their expression into Eq. (10.2.i) will be.

$$\mathcal{K}^e = \frac{1}{2}(\dot{\mathbf{U}}_i^e)^T \left[\int_{\mathrm{V}^e} \rho (\mathbf{N}^e)^T \mathbf{N}^e d\mathrm{v}^e \right] \dot{\mathbf{U}}_i^e \tag{10.14.1}$$

$$\begin{aligned} \Pi_p^e = {} & \frac{1}{2}(\mathbf{U}_i^e)^T \left[\int_{\mathrm{V}^e} (\mathbf{B}_l^e + \mathbf{B}_n^e)^T \mathbb{C}^e (\mathbf{B}_l^e + \mathbf{B}_n^e) d\mathrm{v}^e \right] \mathbf{U}_i^e \\ & - (\mathbf{U}_i^e)^T \left[\int_{V^e} (\mathbf{N}^e)^T \psi_\Omega d\mathrm{v}^e + \int_{A^e} (\mathbf{N}^e)^T \psi_\Gamma d\Gamma^e + \mathbf{f}_c^e \right] \end{aligned} \tag{10.14.2}$$

$$R^e = \frac{1}{2}(\dot{\mathbf{U}}_i^e)^T \left[\int_{\mathrm{V}^e} \mu (\mathbf{N}^e)^T \mathbf{N}^e d\mathrm{v}^e \right] \dot{\mathbf{U}}_i^e \tag{10.14.3}$$

where $\mathbf{f}_c^e$ in the potential energy term, is a vector of point forces at the nodal points, if any. This force vector is also a function of time. The integrals in Eq. (10.14.i), represents the characteristics matrices and vectors obtained from spectral element discretization. The characteristics mass matrix (**M**), stiffness matrix (**K**), damping matrix (**D**), and the force vectors (surface loads ($\mathbf{f}_\Gamma$) and body force or volume force ($\mathbf{f}_\Omega$)) from Eq. (10.14.i) could be written as

$$\mathbf{M}^e = \int_{v^e} \rho(\mathbf{N}^e)^T \mathbf{N}^e dv^e \tag{10.15}$$

$$\mathbf{K}^e = \int_{v^e} (\mathbf{B}_l^e + \mathbf{B}_n^e)^T \mathbb{C}^e (\mathbf{B}_l^e + \mathbf{B}_n^e) dv^e \tag{10.16}$$

$$\mathbf{D}^e = \int_{v^e} \mu(\mathbf{N}^e)^T \mathbf{N}^e dv^e \tag{10.17}$$

$$\mathbf{f}_\Omega^e = \int_{v^e} (\mathbf{N}^e)^T \psi_\Omega dv^e \tag{10.18}$$

$$\mathbf{f}_\Gamma^e = \int_{\Gamma^e} (\mathbf{N}^e)^T \psi_\Gamma d\Gamma^e \tag{10.19}$$

With the newly written matrices in Eqs. (10.15) through (10.19) the kinetic energy, potential energy, and the dissipation functions can be rewritten in the following form. Please note that all the equations are written for a single spectral element.

$$\mathcal{K}^e = \frac{1}{2}(\dot{\mathbf{U}}_i^e)^T \mathbf{M}^e \dot{\mathbf{U}}_i^e \tag{10.20.1}$$

$$\Pi_p^e = \frac{1}{2}(\mathbf{U}_i^e)^T \mathbf{K}^e \mathbf{U}_i^e - (\mathbf{U}_i^e)^T \left[\mathbf{f}_\Omega^e + \mathbf{f}_\Gamma^e + \mathbf{f}_c^e \right] \tag{10.20.2}$$

$$R^e = \frac{1}{2}(\dot{\mathbf{U}}_i^e)^T \mathbb{C}^e \dot{\mathbf{U}}_i^e \tag{10.20.3}$$

Next, substituting Eq. (10.20.i) into Eq. (10.1), the governing dynamic equation with a spectral element discretization can be written as

$$\mathbf{M}^e \ddot{\mathbf{U}}_i^e + \mathbf{D}^e \dot{\mathbf{U}}_i^e + \mathbf{K}^e \mathbf{U}_i^e = \mathbf{f}_i^e(t) \tag{10.21}$$

where $\mathbf{f}_i^e(t) = \mathbf{f}_\Omega^e(t) + \mathbf{f}_\Gamma^e(t) + \mathbf{f}_c^e(t)$.

In a material body if there are multiple elements (say S number spectral of elements) and which is always the case, after assembling the element terms for a global problem, the spectrally discretized global governing equation of the NDE problem can be

$$\mathbf{M}\ddot{\mathbf{U}}_i + \mathbf{D}\dot{\mathbf{U}}_i + \mathbf{K}\mathbf{U}_i = \mathbf{f}_i(t) \tag{10.22}$$

The global mass matrix (**M**), damping matrix (**D**), stiffness matrix (**K**), and the global displacement vector ($\mathbf{U}_i$) of all material points considered in the problem, and global velocity ($\dot{\mathbf{U}}_i$) and global acceleration ($\ddot{\mathbf{U}}_i$) vectors, and the force vector in Eq. (10.22) can be further written as

$$\mathbf{M} = \sum_{e=1}^{S} (\mathbf{P}^e)^T \mathbf{M}^e \mathbf{P}^e, \quad \mathbf{D} = \sum_{e=1}^{S} (\mathbf{P}^e)^T \mathbf{D}^e \mathbf{P}^e, \quad \mathbf{K} = \sum_{e=1}^{S} (\mathbf{P}^e)^T \mathbf{K}^e \mathbf{P}^e \tag{10.23}$$

$$\ddot{\mathbf{U}}_i = \sum_{e=1}^{S} (\mathbf{P}^e)^T \ddot{\mathbf{U}}_i^e, \quad \mathbf{U}_i = \sum_{e=1}^{S} (\mathbf{P}^e)^T \dot{\mathbf{U}}_i^e \tag{10.24}$$

$$\mathbf{U}_i = \sum_{e=1}^{S} (\mathbf{P}^e)^T \mathbf{U}_n^e, \quad \mathbf{f}_i = \sum_{e=1}^{S} (\mathbf{P}^e)^T \mathbf{f}_i^e \tag{10.25}$$

where S is the total number of spectral elements used in the formulation. The **P** matrix is composed of zeros and ones to assemble the elements for the global representation of the contribution of each element based on the topology of the orientation of the spectral elements. The assembly procedure is explained later in this chapter for a CNDE problem used in the case study. Further, the boundary conditions are the necessary part to implement in the global model. Here in this problem, the time domain displacement is provided on the material directly to simulate real time NDE problem. If a multilayered structure is solved, the internal layers will use stresses and displacements continuity conditions. Eq. (10.21) can be solved using standard FEM explicit method time integration method as discussed in Chapter 6.

10.2.2 Application of Weighted Residual Method

The weighted residual method does not require the knowledge of the principles of virtual displacements or total minimum potential energy but only needs the governing differential equations of the problem. In case of plane elastic wave propagation in a structure, the three-dimensional equations of motion are the governing differential equations, which are the strong form of the problem.

The equations of motion can be reiterated from Eqs. (3.71) and (A.3.1.i) and can be written as

$$\frac{\partial \sigma_{11}}{\partial x_1} + \frac{\partial \sigma_{12}}{\partial x_2} + \frac{\partial \sigma_{13}}{\partial x_3} + f_1 = \rho \frac{\partial^2 u_1}{\partial t^2} \tag{10.26.1}$$

$$\frac{\partial \sigma_{21}}{\partial x_1} + \frac{\partial \sigma_{22}}{\partial x_2} + \frac{\partial \sigma_{23}}{\partial x_3} + f_2 = \rho \frac{\partial^2 u_2}{\partial t^2} \tag{10.26.2}$$

$$\frac{\partial \sigma_{31}}{\partial x_1} + \frac{\partial \sigma_{32}}{\partial x_2} + \frac{\partial \sigma_{33}}{\partial x_3} + f_3 = \rho \frac{\partial^2 u_3}{\partial t^2} \tag{10.26.3}$$

Due to stress symmetry, it can be stated that $\sigma_{12} = \sigma_{21}$; $\sigma_{13} = \sigma_{31}$; $\sigma_{23} = \sigma_{32}$. Determining the stress parameters, generalized Hook's law or the constitutive relation can be utilized which are previously written in Eq. (3.79) for a generalized anisotropic material as follows:

$$\begin{Bmatrix} \sigma_{11} \\ \sigma_{22} \\ \sigma_{33} \\ \sigma_{23} \\ \sigma_{31} \\ \sigma_{12} \end{Bmatrix} = \begin{bmatrix} \mathbb{C}_{11} & \mathbb{C}_{12} & \mathbb{C}_{13} & \mathbb{C}_{14} & \mathbb{C}_{15} & \mathbb{C}_{16} \\ \mathbb{C}_{21} & \mathbb{C}_{22} & \mathbb{C}_{23} & \mathbb{C}_{24} & \mathbb{C}_{25} & \mathbb{C}_{26} \\ \mathbb{C}_{31} & \mathbb{C}_{32} & \mathbb{C}_{33} & \mathbb{C}_{34} & \mathbb{C}_{35} & \mathbb{C}_{36} \\ \mathbb{C}_{41} & \mathbb{C}_{42} & \mathbb{C}_{43} & \mathbb{C}_{44} & \mathbb{C}_{45} & \mathbb{C}_{46} \\ \mathbb{C}_{51} & \mathbb{C}_{52} & \mathbb{C}_{53} & \mathbb{C}_{54} & \mathbb{C}_{55} & \mathbb{C}_{56} \\ \mathbb{C}_{61} & \mathbb{C}_{62} & \mathbb{C}_{63} & \mathbb{C}_{64} & \mathbb{C}_{65} & \mathbb{C}_{66} \end{bmatrix} \begin{Bmatrix} \dfrac{\partial u_1}{\partial x_1} \\ \dfrac{\partial u_2}{\partial x_2} \\ \dfrac{\partial u_3}{\partial x_3} \\ \left(\dfrac{\partial u_2}{\partial x_3} + \dfrac{\partial u_3}{\partial x_2}\right) \\ \left(\dfrac{\partial u_3}{\partial x_1} + \dfrac{\partial u_1}{\partial x_3}\right) \\ \left(\dfrac{\partial u_1}{\partial x_2} + \dfrac{\partial u_2}{\partial x_1}\right) \end{Bmatrix} \tag{10.27}$$

In most cases, NDE of generalized material system is rare. Common NDE applications are generally with isotropic materials, hybrid, or composite materials or materials that have up to two planes of symmetry. Materials with two planes of symmetry are called the orthotropic materials, and commonly occurs in NDE problems. Hence, without complications, further discussions will be based on an orthotropic material where an isotropic material will be a subset of orthotropic material case as previously discussed in Chapter 3. Generalized constitutive relation in orthotropic material can be written as (refer Eq. (3.82)):

$$\begin{Bmatrix} \sigma_{11} \\ \sigma_{22} \\ \sigma_{33} \\ \sigma_{23} \\ \sigma_{31} \\ \sigma_{12} \end{Bmatrix} = \begin{bmatrix} \mathbb{C}_{11} & \mathbb{C}_{12} & \mathbb{C}_{13} & 0 & 0 & 0 \\ & \mathbb{C}_{22} & \mathbb{C}_{23} & 0 & 0 & 0 \\ & & \mathbb{C}_{33} & 0 & 0 & 0 \\ & & & \mathbb{C}_{44} & 0 & 0 \\ & Sym & & & \mathbb{C}_{55} & 0 \\ & & & & & \mathbb{C}_{66} \end{bmatrix} \begin{Bmatrix} \dfrac{\partial u_1}{\partial x_1} \\ \dfrac{\partial u_2}{\partial x_2} \\ \dfrac{\partial u_3}{\partial x_3} \\ \left(\dfrac{\partial u_2}{\partial x_3} + \dfrac{\partial u_3}{\partial x_2}\right) \\ \left(\dfrac{\partial u_3}{\partial x_1} + \dfrac{\partial u_1}{\partial x_3}\right) \\ \left(\dfrac{\partial u_1}{\partial x_2} + \dfrac{\partial u_2}{\partial x_1}\right) \end{Bmatrix} \tag{10.28}$$

Expanding each stress parameters in Eq. (10.28) one can get,

$$\sigma_{11} = \mathbb{C}_{11}\frac{\partial u_1}{\partial x_1} + \mathbb{C}_{12}\frac{\partial u_2}{\partial x_2} + \mathbb{C}_{13}\frac{\partial u_3}{\partial x_3} \tag{10.29.1}$$

$$\sigma_{22} = \mathbb{C}_{21}\frac{\partial u_1}{\partial x_1} + \mathbb{C}_{22}\frac{\partial u_2}{\partial x_2} + \mathbb{C}_{23}\frac{\partial u_3}{\partial x_3} \tag{10.29.2}$$

$$\sigma_{33} = \mathbb{C}_{31}\frac{\partial u_1}{\partial x_1} + \mathbb{C}_{32}\frac{\partial u_2}{\partial x_2} + \mathbb{C}_{33}\frac{\partial u_3}{\partial x_3} \tag{10.29.3}$$

$$\sigma_{23} = \mathbb{C}_{44}\left(\frac{\partial u_2}{\partial x_3} + \frac{\partial u_3}{\partial x_2}\right) = \sigma_{32} \tag{10.29.4}$$

$$\sigma_{31} = \mathbb{C}_{55}\left(\frac{\partial u_3}{\partial x_1} + \frac{\partial u_1}{\partial x_3}\right) = \sigma_{13} \tag{10.29.5}$$

$$\sigma_{12} = \mathbb{C}_{66}\left(\frac{\partial u_1}{\partial x_2} + \frac{\partial u_2}{\partial x_1}\right) = \sigma_{21} \tag{10.29.6}$$

Differentiating the normal and shear stresses like they appear in Eq. (10.26.i) the derivatives can be written as:

$$\frac{\partial \sigma_{11}}{\partial x_1} = \frac{\partial}{\partial x_1}\left(\mathbb{C}_{11}\frac{\partial u_1}{\partial x_1} + \mathbb{C}_{12}\frac{\partial u_2}{\partial x_2} + \mathbb{C}_{13}\frac{\partial u_3}{\partial x_3}\right) \tag{10.30.1}$$

$$\frac{\partial \sigma_{22}}{\partial x_2} = \frac{\partial}{\partial x_2}\left(\mathbb{C}_{21}\frac{\partial u_1}{\partial x_1} + \mathbb{C}_{22}\frac{\partial u_2}{\partial x_2} + \mathbb{C}_{23}\frac{\partial u_3}{\partial x_3}\right) \tag{10.30.2}$$

$$\frac{\partial \sigma_{33}}{\partial x_3} = \frac{\partial}{\partial x_3}\left(\mathbb{C}_{31}\frac{\partial u_1}{\partial x_1} + \mathbb{C}_{32}\frac{\partial u_2}{\partial x_2} + \mathbb{C}_{33}\frac{\partial u_3}{\partial x_3}\right) \tag{10.30.3}$$

$$\frac{\partial \sigma_{12}}{\partial x_2} = \frac{\partial}{\partial x_2}\mathbb{C}_{66}\left(\frac{\partial u_1}{\partial x_2} + \frac{\partial u_2}{\partial x_1}\right) \tag{10.30.4}$$

$$\frac{\partial \sigma_{21}}{\partial x_1} = \frac{\partial}{\partial x_1}\mathbb{C}_{66}\left(\frac{\partial u_1}{\partial x_2} + \frac{\partial u_2}{\partial x_1}\right) \tag{10.30.5}$$

$$\frac{\partial \sigma_{31}}{\partial x_1} = \frac{\partial}{\partial x_1}\mathbb{C}_{55}\left(\frac{\partial u_3}{\partial x_1} + \frac{\partial u_1}{\partial x_3}\right) \tag{10.30.6}$$

$$\frac{\partial \sigma_{13}}{\partial x_3} = \frac{\partial}{\partial x_3}\mathbb{C}_{55}\left(\frac{\partial u_3}{\partial x_1} + \frac{\partial u_1}{\partial x_3}\right) \tag{10.30.7}$$

$$\frac{\partial \sigma_{23}}{\partial x_3} = \frac{\partial}{\partial x_3}\mathbb{C}_{44}\left(\frac{\partial u_2}{\partial x_3} + \frac{\partial u_3}{\partial x_2}\right) \tag{10.30.8}$$

$$\frac{\partial \sigma_{32}}{\partial x_2} = \frac{\partial}{\partial x_2}\mathbb{C}_{44}\left(\frac{\partial u_2}{\partial x_3} + \frac{\partial u_3}{\partial x_2}\right) \tag{10.30.9}$$

Therefore, substituting these values into Eq. (10.26.i), one can write:

$$\mathbb{C}_{11}\frac{\partial^2 u_1}{\partial x_1^2}+\mathbb{C}_{12}\frac{\partial^2 u_2}{\partial x_1\,\partial x_2}+\mathbb{C}_{13}\frac{\partial^2 u_3}{\partial x_1\,\partial x_3}+\frac{1}{2}\mathbb{C}_{66}\left(\frac{\partial^2 u_1}{\partial x_2^2}+\frac{\partial^2 u_2}{\partial x_1\,\partial x_2}\right)+\frac{1}{2}\mathbb{C}_{55}\left(\frac{\partial^2 u_3}{\partial x_1\,\partial x_3}+\frac{\partial^2 u_1}{\partial x_3^2}\right)=\rho\ddot{u}_1-f_1 \tag{10.31.1}$$

$$\mathbb{C}_{21}\frac{\partial^2 u_1}{\partial x_2\,\partial x_1}+\mathbb{C}_{22}\frac{\partial^2 u_2}{\partial x_2^2}+\mathbb{C}_{23}\frac{\partial^2 u_3}{\partial x_2\,\partial x_3}+\frac{1}{2}\mathbb{C}_{66}\left(\frac{\partial^2 u_1}{\partial x_2\,\partial x_1}+\frac{\partial^2 u_2}{\partial x_1^2}\right)+\frac{1}{2}\mathbb{C}_{44}\left(\frac{\partial^2 u_2}{\partial x_3^2}+\frac{\partial^2 u_3}{\partial x_2\,\partial x_3}\right)=\rho\ddot{u}_2-f_2 \tag{10.31.2}$$

$$\mathbb{C}_{31}\frac{\partial^2 u_1}{\partial x_3\,\partial x_1}+\mathbb{C}_{32}\frac{\partial^2 u_2}{\partial x_2\,\partial x_3}+\mathbb{C}_{33}\frac{\partial^2 u_3}{\partial x_3^2}+\frac{1}{2}\mathbb{C}_{55}\left(\frac{\partial^2 u_3}{\partial x_1^2}+\frac{\partial^2 u_1}{\partial x_3\,\partial x_1}\right)+\frac{1}{2}\mathbb{C}_{44}\left(\frac{\partial^2 u_2}{\partial x_2\,\partial x_3}+\frac{\partial^2 u_3}{\partial x_2^2}\right)=\rho\ddot{u}_3-f_3 \tag{10.31.3}$$

Eq. (10.31.i) is the strong form of the governing equations of an elastic body undergoing small deformations. At this point, a weak formulation of these equations is performed as follows: First, Eq. (10.31.1) is multiplied with a weight function w_1, which is assumed to be differentiable once with respect to directions 1, 2, and 3, and then integrations over the element domain Ω_e are performed. This gives

$$\int w_1\left[\begin{array}{l}-\dfrac{\partial}{\partial x_1}\left(\mathbb{C}_{11}\dfrac{\partial u_1}{\partial x_1}+\mathbb{C}_{12}\dfrac{\partial u_2}{\partial x_2}+\mathbb{C}_{13}\dfrac{\partial u_3}{\partial x_3}\right)-\dfrac{\partial}{\partial x_2}\mathbb{C}_{66}\left(\dfrac{\partial u_1}{\partial x_2}+\dfrac{\partial u_2}{\partial x_1}\right)\\ -\dfrac{\partial}{\partial x_3}\mathbb{C}_{55}\left(\dfrac{\partial u_3}{\partial x_1}+\dfrac{\partial u_1}{\partial x_3}\right)+\rho\ddot{u}_1-f_1\end{array}\right]dv=0 \tag{10.32}$$

Let us assume

$$F_1=\mathbb{C}_{11}\frac{\partial u_1}{\partial x_1}+\mathbb{C}_{12}\frac{\partial u_2}{\partial x_2}+\mathbb{C}_{13}\frac{\partial u_3}{\partial x_3} \tag{10.33}$$

$$F_2=\mathbb{C}_{66}\frac{\partial u_1}{\partial x_2}+\mathbb{C}_{66}\frac{\partial u_2}{\partial x_1}\ \ and\ \ F_3=\mathbb{C}_{55}\frac{\partial u_3}{\partial x_1}+\mathbb{C}_{55}\frac{\partial u_1}{\partial x_3} \tag{10.34}$$

From (10.32.1), it can be written as

$$\int w_1\left[-\frac{\partial}{\partial x_1}(F_1)-\frac{\partial}{\partial x_2}(F_2)-\frac{\partial}{\partial x_3}(F_3)+\rho\ddot{u}_1-f_1\right]dv=0 \tag{10.35}$$

Now, product of differentiation rule gives us

$$\frac{\partial}{\partial x_1}(w_1F_1) = \frac{\partial w_1}{\partial x_1}F_1 + w_1\frac{\partial F_1}{\partial x_1} \Rightarrow -w_1\frac{\partial F_1}{\partial x_1} = \frac{\partial w_1}{\partial x_1}F_1 - \frac{\partial}{\partial x_1}(w_1F_1) \quad (10.36)$$

$$\frac{\partial}{\partial x_2}(w_1F_2) = \frac{\partial w_1}{\partial x_2}F_2 + w_1\frac{\partial F_2}{\partial x_2} \Rightarrow -w_1\frac{\partial F_2}{\partial x_2} = \frac{\partial w_1}{\partial x_2}F_2 - \frac{\partial}{\partial x_2}(w_1F_2) \quad (10.37)$$

$$\frac{\partial}{\partial x_3}(w_1F_3) = \frac{\partial w_1}{\partial x_3}F_3 + w_1\frac{\partial F_3}{\partial x_3} \Rightarrow -w_1\frac{\partial F_3}{\partial x_3} = \frac{\partial w_1}{\partial x_3}F_3 - \frac{\partial}{\partial x_3}(w_1F_3) \quad (10.38)$$

Using divergence theorem (discussed in Chapter 2), one can get

$$\int \frac{\partial}{\partial x_1}(w_1F_1)dx_1dx_2dx_3 = \oint w_1F_1n_1\, d\Gamma \quad (10.39)$$

$$\int \frac{\partial}{\partial x_2}(w_1F_2)dx_1dx_2dx_3 = \oint w_1F_2n_2\, d\Gamma \quad (10.40)$$

$$\int \frac{\partial}{\partial x_3}(w_1F_3)dx_1dx_2dx_3 = \oint w_1F_3n_3\, d\Gamma \quad (10.41)$$

where $dv = dx_1dx_2dx_3$, n_j is the unit vector along the j-th direction and the direction cosines. Therefore,

$$-w_1\frac{\partial F_1}{\partial x_1} = \frac{\partial w_1}{\partial x_1}F_1 - \oint w_1F_1n_1\, d\Gamma \quad (10.42)$$

$$-w_1\frac{\partial F_2}{\partial x_2} = \frac{\partial w_1}{\partial x_2}F_2 - \oint w_1F_2n_2\, d\Gamma \quad (10.43)$$

$$-w_1\frac{\partial F_3}{\partial x_3} = \frac{\partial w_1}{\partial x_3}F_3 - \oint w_1F_3n_3\, d\Gamma \quad (10.44)$$

Putting these values in Eq. (10.35), be

$$\begin{aligned}&\frac{\partial w_1}{\partial x_1}F_1dv - \oint w_1F_1n_1\, d\Gamma + \frac{\partial w_1}{\partial x_2}F_2dv - \oint w_1F_2n_2\, d\Gamma + \frac{\partial w_1}{\partial x_3}F_3dv \\ &- \oint w_1F_3n_3\, d\Gamma + \int w_1\rho\ddot{u}\, dv - \int w_1f_1\, dv = 0\end{aligned} \quad (10.45)$$

Substituting the expressions in Eqs. (10.33) and (10.44) into Eq. (10.45), be

$$
\begin{aligned}
&\int [\frac{\partial w_1}{\partial x_1}\left(\mathbb{C}_{11}\frac{\partial u_1}{\partial x_1}+\mathbb{C}_{12}\frac{\partial u_2}{\partial x_2}+\mathbb{C}_{13}\frac{\partial u_3}{\partial x_3}\right)+\frac{\partial w_1}{\partial x_2}\left(\mathbb{C}_{66}\left(\frac{\partial u_1}{\partial x_2}+\frac{\partial u_2}{\partial x_1}\right)\right)\\
&+\frac{\partial w_1}{\partial x_3}\left(\mathbb{C}_{55}\left(\frac{\partial u_3}{\partial x_1}+\frac{\partial u_1}{\partial x_3}\right)\right)+w_1\rho\ddot{u}_1-w_1\ f_1]\ dv\\
&-\oint w_1\left[n_1\left(\mathbb{C}_{11}\frac{\partial u_1}{\partial x_1}+\mathbb{C}_{12}\frac{\partial u_2}{\partial x_2}+\mathbb{C}_{13}\frac{\partial u_3}{\partial x_3}\right)\right.\\
&\left.+n_2\left(\mathbb{C}_{66}\frac{\partial u_1}{\partial x_2}+\mathbb{C}_{66}\frac{\partial u_2}{\partial x_1}\right)+n_3\left(\mathbb{C}_{55}\frac{\partial u_3}{\partial x_1}+\mathbb{C}_{55}\frac{\partial u_1}{\partial x_3}\right)\right]d\Gamma=0
\end{aligned}
\tag{10.46}
$$

or

$$
\begin{aligned}
&\int\left[\left(\mathbb{C}_{11}\frac{\partial w_1}{\partial x_1}\frac{\partial u_1}{\partial x_1}+\mathbb{C}_{12}\frac{\partial w_1}{\partial x_1}\frac{\partial u_2}{\partial x_2}+\mathbb{C}_{13}\frac{\partial w_1}{\partial x_1}\frac{\partial u_3}{\partial x_3}\right)+\left(\mathbb{C}_{66}\frac{\partial w_1}{\partial x_2}\frac{\partial u_1}{\partial x_2}+\mathbb{C}_{66}\frac{\partial w_1}{\partial x_2}\frac{\partial u_2}{\partial x_1}\right)\right.\\
&\left.+\left(\mathbb{C}_{55}\frac{\partial w_1}{\partial x_3}\frac{\partial u_3}{\partial x_1}+\mathbb{C}_{55}\frac{\partial w_1}{\partial x_3}\frac{\partial u_1}{\partial x_3}\right)+w_1\rho\ddot{u}_1-w_1\ f_1\right]dv\\
&-\oint w_1\left[n_1\left(\mathbb{C}_{11}\frac{\partial u_1}{\partial x_1}+\mathbb{C}_{12}\frac{\partial u_2}{\partial x_2}+\mathbb{C}_{13}\frac{\partial u_3}{\partial x_3}\right)+n_2\left(\mathbb{C}_{66}\frac{\partial u_1}{\partial x_2}+\mathbb{C}_{66}\frac{\partial u_2}{\partial x_1}\right)\right.\\
&\left.+\ n_3\left(\mathbb{C}_{55}\frac{\partial u_3}{\partial x_1}+\mathbb{C}_{55}\frac{\partial u_1}{\partial x_3}\right)\right]d\Gamma=0
\end{aligned}
\tag{10.47}
$$

Which can be further simplified to

$$
\begin{aligned}
&\int\left[\mathbb{C}_{11}\frac{\partial w_1}{\partial x_1}\frac{\partial u_1}{\partial x_1}+\mathbb{C}_{12}\frac{\partial w_1}{\partial x_1}\frac{\partial u_2}{\partial x_2}+\mathbb{C}_{13}\frac{\partial w_1}{\partial x_1}\frac{\partial u_3}{\partial x_3}+\mathbb{C}_{66}\frac{\partial w_1}{\partial x_2}\frac{\partial u_1}{\partial x_2}+\mathbb{C}_{66}\frac{\partial w_1}{\partial x_2}\frac{\partial u_2}{\partial x_1}\right.\\
&\left.+\mathbb{C}_{55}\frac{\partial w_1}{\partial x_3}\frac{\partial u_3}{\partial x_1}+\mathbb{C}_{55}\frac{\partial w_1}{\partial x_3}\frac{\partial u_1}{\partial x_3}+w_1\rho\ddot{u}_1-\ w_1 f_1\right]dv-\oint w_1 t_1\,d\Gamma=0
\end{aligned}
\tag{10.48}
$$

where

$$
\begin{aligned}
t_1=&\ n_1\left(\mathbb{C}_{11}\frac{\partial u_1}{\partial x_1}+\mathbb{C}_{12}\frac{\partial u_2}{\partial x_2}+\mathbb{C}_{13}\frac{\partial u_3}{\partial x_3}\right)+n_2\left(\mathbb{C}_{66}\frac{\partial u_1}{\partial x_2}+\mathbb{C}_{66}\frac{\partial u_2}{\partial x_1}\right)\\
&+n_3\left(\mathbb{C}_{55}\frac{\partial u_3}{\partial x_1}+\mathbb{C}_{55}\frac{\partial u_1}{\partial x_3}\right)
\end{aligned}
\tag{10.49}
$$

Similarly, from Eqs. (10.31.2) and (10.31.3), it can be written as

$$\int\left[\mathbb{C}_{21}\frac{\partial w_2}{\partial x_2}\frac{\partial u_1}{\partial x_1}+\mathbb{C}_{22}\frac{\partial w_2}{\partial x_2}\frac{\partial u_2}{\partial x_2}+\mathbb{C}_{23}\frac{\partial w_2}{\partial x_2}\frac{\partial u_3}{\partial x_3}+\mathbb{C}_{66}\frac{\partial w_2}{\partial x_1}\frac{\partial u_1}{\partial x_2}+\mathbb{C}_{66}\frac{\partial w_2}{\partial x_1}\frac{\partial u_2}{\partial x_1}\right.$$
$$\left.+\mathbb{C}_{44}\frac{\partial w_2}{\partial x_3}\frac{\partial u_2}{\partial x_3}+\mathbb{C}_{44}\frac{\partial w_2}{\partial x_3}\frac{\partial u_3}{\partial x_2}+w_2\rho\ddot{u}_2-w_2 f_2\right]dv-\oint w_2 t_2\,d\Gamma=0 \tag{10.50}$$

and

$$\int\left[\mathbb{C}_{31}\frac{\partial w_3}{\partial x_3}\frac{\partial u_1}{\partial x_1}+\mathbb{C}_{32}\frac{\partial w_3}{\partial x_3}\frac{\partial u_2}{\partial x_2}+\mathbb{C}_{33}\frac{\partial w_3}{\partial x_3}\frac{\partial u_3}{\partial x_3}+\mathbb{C}_{55}\frac{\partial w_3}{\partial x_1}\frac{\partial u_3}{\partial x_1}+\mathbb{C}_{55}\frac{\partial w_3}{\partial x_1}\frac{\partial u_1}{\partial x_3}\right.$$
$$\left.+\mathbb{C}_{44}\frac{\partial w_3}{\partial x_2}\frac{\partial u_2}{\partial x_3}+\mathbb{C}_{44}\frac{\partial w_3}{\partial x_2}\frac{\partial u_3}{\partial x_2}+w_3\rho\ddot{u}_3-w_3 f_3\right]dv-\oint w_3 t_3\,d\Gamma=0 \tag{10.51}$$

Eqs. (10.48), (10.50), and (10.51) give the weak form of the governing equations. At this stage, a spectral element approximation is introduced which is a standard procedure similar to the FEM. The displacement at the element level can be approximated like previously discussed in Section 0.6.1, as

$$\begin{aligned}
u_1^e &= \sum U_j^{u_1}(t)\ \psi_j^e(x_k)\\
u_2^e &= \sum V_j^{u_2}(t)\psi_j^e(x_k)\\
u_3^e &= \sum W_j^{u_3}(t)\psi_j^e(x_k)
\end{aligned} \tag{10.52}$$

where ψ_j^e is the shape function for the spectral element, like the N_i^e shape functions introduced in the previous section. Reason for using different nomenclature is just to distinguish two different formulations in this book, however, they are interchangeable. Hence, the elemental displacement function anywhere in the element can be expressed like the expressions written in Eqs. (10.5) and (10.6) for a m nodded element as follows:

$$\mathbf{U}^e=\begin{Bmatrix}u_1^e\\u_2^e\\u_3^e\end{Bmatrix}=\begin{bmatrix}\psi_1^e & 0 & 0 & \psi_2^e & .. & .. & \psi_m^e & 0 & 0\\ 0 & \psi_1^e & 0 & 0 & \psi_2^e & 0 & .... & \psi_m^e & 0\\ 0 & 0 & \psi_1^e & 0 & 0 & \psi_2^e & .. & .. & \psi_m^e\end{bmatrix}\begin{Bmatrix}U_1^{u_1}\\V_1^{u_2}\\W_1^{u_3}\\U_2^{u_1}\\V_2^{u_2}\\W_2^{u_3}\\\vdots\\\vdots\\U_m^{u_1}\\V_m^{u_2}\\W_m^{u_3}\end{Bmatrix} \tag{10.53}$$

Next, using Galerkin approximation [8], where the weighted function is assumed to be the shape functions, one can set $w_1 = \psi_i$, $w_2 = \psi_i$, and $w_3 = \psi_i$ in Eqs. (10.48), (10.50), and (10.51). The ith equation over an element can be written as follows:

$$\int\Big[\left(\mathbb{C}_{11}\frac{\partial\psi_i}{\partial x_1}\frac{\partial\psi_j}{\partial x_1}+\mathbb{C}_{66}\frac{\partial\psi_i}{\partial x_2}\frac{\partial\psi_j}{\partial x_2}+\mathbb{C}_{55}\frac{\partial\psi_i}{\partial x_3}\frac{\partial\psi_j}{\partial x_3}\right)U_j^{u1}$$
$$+\left(\mathbb{C}_{12}\frac{\partial\psi_i}{\partial x_1}\frac{\partial\psi_j}{\partial x_2}+\mathbb{C}_{66}\frac{\partial\psi_i}{\partial x_2}\frac{\partial\psi_j}{\partial x_1}\right)V_j^{u2}+\left(\mathbb{C}_{13}\frac{\partial\psi_i}{\partial x_1}\frac{\partial\psi_j}{\partial x_3}+\mathbb{C}_{55}\frac{\partial\psi_i}{\partial x_3}\frac{\partial\psi_j}{\partial x_1}\right)W_j^{u3} \quad (10.54)$$
$$+\ \psi_i\rho\ddot{u}_1-\psi_i f_1\Big]\,d\text{v}-\oint\psi_i t_1\,d\Gamma=0$$

$$\int\Big[\left(\mathbb{C}_{21}\frac{\partial\psi_i}{\partial x_2}\frac{\partial\psi_j}{\partial x_1}+\mathbb{C}_{66}\frac{\partial\psi_i}{\partial x_1}\frac{\partial\psi_j}{\partial x_2}\right)U_j^{u1}$$
$$+\left(\mathbb{C}_{66}\frac{\partial\psi_i}{\partial x_1}\frac{\partial\psi_j}{\partial x_1}+\mathbb{C}_{22}\frac{\partial\psi_i}{\partial x_2}\frac{\partial\psi_j}{\partial x_2}+\mathbb{C}_{44}\frac{\partial\psi_i}{\partial z}\frac{\partial\psi_j}{\partial x_3}\right)V_j^{u2} \quad (10.55)$$
$$+\left(\mathbb{C}_{23}\frac{\partial\psi_i}{\partial x_2}\frac{\partial\psi_j}{\partial x_3}+\mathbb{C}_{44}\frac{\partial\psi_i}{\partial x_3}\frac{\partial\psi_j}{\partial x_2}\right)W_j^{u3}+\psi_i\rho\ddot{u}_2-\psi_i f_2\Big]\,d\text{v}-\oint\psi_i t_2\,d\Gamma=0$$

$$\int\Big[\left(\mathbb{C}_{31}\frac{\partial\psi_i}{\partial x_3}\frac{\partial\psi_j}{\partial x_1}+\mathbb{C}_{55}\frac{\partial\psi_i}{\partial x_1}\frac{\partial\psi_j}{\partial x_3}\right)U_j^{u1}+\left(\mathbb{C}_{32}\frac{\partial\psi_i}{\partial x_3}\frac{\partial\psi_j}{\partial x_2}+\mathbb{C}_{44}\frac{\partial\psi_i}{\partial x_2}\frac{\partial\psi_j}{\partial x_3}\right)V_j^{u2}$$
$$+\left(\mathbb{C}_{55}\frac{\partial\psi_i}{\partial x_1}\frac{\partial\psi_j}{\partial x_1}+\mathbb{C}_{44}\frac{\partial\psi_i}{\partial x_2}\frac{\partial\psi_j}{\partial x_2}+\mathbb{C}_{33}\frac{\partial\psi_i}{\partial x_3}\frac{\partial\psi_j}{\partial x_3}\right)W_j^{u3} \quad (10.56)$$
$$+\psi_i\rho\ddot{u}_3-\psi_i f_3\Big]\,d\text{v}-\oint\psi_i t_3\,d\Gamma=0$$

In these three equations, let us assume

$$K_{ij}^{11}=\mathbb{C}_{11}\frac{\partial\psi_i}{\partial x_1}\frac{\partial\psi_j}{\partial x_1}+\mathbb{C}_{66}\frac{\partial\psi_i}{\partial x_2}\frac{\partial\psi_j}{\partial x_2}+\mathbb{C}_{55}\frac{\partial\psi_i}{\partial x_3}\frac{\partial\psi_j}{\partial x_3} \quad (10.57.1)$$

$$K_{ij}^{12}=\mathbb{C}_{12}\frac{\partial\psi_i}{\partial x_1}\frac{\partial\psi_j}{\partial x_2}+\mathbb{C}_{66}\frac{\partial\psi_i}{\partial x_2}\frac{\partial\psi_j}{\partial x_1} \quad (10.57.2)$$

$$K_{ij}^{13}=\mathbb{C}_{13}\frac{\partial\psi_i}{\partial x_1}\frac{\partial\psi_j}{\partial x_3}+\mathbb{C}_{55}\frac{\partial\psi_i}{\partial x_3}\frac{\partial\psi_j}{\partial x_1} \quad (10.57.3)$$

$$K_{ij}^{21}=\mathbb{C}_{21}\frac{\partial\psi_i}{\partial x_2}\frac{\partial\psi_j}{\partial x_1}+\mathbb{C}_{66}\frac{\partial\psi_i}{\partial x_1}\frac{\partial\psi_j}{\partial x_2} \quad (10.57.4)$$

$$K_{ij}^{22} = \mathbb{C}_{66}\frac{\partial\psi_i}{\partial x_1}\frac{\partial\psi_j}{\partial x_1} + \mathbb{C}_{22}\frac{\partial\psi_i}{\partial x_2}\frac{\partial\psi_j}{\partial x_2} + \mathbb{C}_{44}\frac{\partial\psi_i}{\partial z}\frac{\partial\psi_j}{\partial x_3} \quad (10.57.5)$$

$$K_{ij}^{23} = \mathbb{C}_{23}\frac{\partial\psi_i}{\partial x_2}\frac{\partial\psi_j}{\partial x_3} + \mathbb{C}_{44}\frac{\partial\psi_i}{\partial x_3}\frac{\partial\psi_j}{\partial x_2} \quad (10.57.6)$$

$$K_{ij}^{31} = \mathbb{C}\frac{\partial\psi_i}{\partial x_3}\frac{\partial\psi_j}{\partial x_1} + \mathbb{C}_{55}\frac{\partial\psi_i}{\partial x_1}\frac{\partial\psi_j}{\partial x_3} \quad (10.57.7)$$

$$K_{ij}^{32} = \mathbb{C}_{32}\frac{\partial\psi_i}{\partial x_3}\frac{\partial\psi_j}{\partial x_2} + \mathbb{C}_{44}\frac{\partial\psi_i}{\partial x_2}\frac{\partial\psi_j}{\partial x_3} \quad (10.57.8)$$

$$K_{ij}^{33} = \mathbb{C}_{55}\frac{\partial\psi_i}{\partial x_1}\frac{\partial\psi_j}{\partial x_1} + \mathbb{C}_{44}\frac{\partial\psi_i}{\partial x_2}\frac{\partial\psi_j}{\partial x_2} + \mathbb{C}_{33}\frac{\partial\psi_i}{\partial x_3}\frac{\partial\psi_j}{\partial x_3} \quad (10.57.9)$$

$$M_{ij} = \int \rho\ \psi_i\psi_j dv \quad (10.57.10)$$

$$F_i = \int \psi_i dv \quad (10.57.11)$$

$$Q_i = \oint \psi_i\, d\Gamma \quad (10.57.12)$$

Therefore, by substituting them, it can be written as

$$\begin{bmatrix} K_{ij}^{11} & K_{ij}^{12} & K_{ij}^{13} \\ K_{ij}^{21} & K_{ij}^{22} & K_{ij}^{23} \\ K_{ij}^{31} & K_{ij}^{32} & K_{ij}^{33} \end{bmatrix} \begin{Bmatrix} \{U\} \\ \{V\} \\ \{W\} \end{Bmatrix} + \begin{bmatrix} M_{ij} & 0 & 0 \\ 0 & M_{ij} & 0 \\ 0 & 0 & M_{ij} \end{bmatrix} \begin{Bmatrix} \{\ddot{U}\} \\ \{\ddot{V}\} \\ \{\ddot{W}\} \end{Bmatrix} = F_i \begin{Bmatrix} f_1 \\ f_2 \\ f_3 \end{Bmatrix} + Q_i \begin{Bmatrix} t_1 \\ t_2 \\ t_3 \end{Bmatrix} \quad (10.58)$$

In matrix form,

$$[\mathbf{K}^e]\{\mathbf{U}^e\} + [\mathbf{M}^e]\{\ddot{\mathbf{U}}^e\} = \{\mathbf{F}^e\} + \{\mathbf{Q}^e\} \quad (10.59)$$

Once the equations are formulated for an element, a standard finite difference explicit solution method can be utilized after assembling them into global equations as discussed and implemented in a computer code.

10.2.3 Spectral Shape Function

Assumption of spectrally guided shape functions with appropriate order of polynomials make the SEM different from FEM. As a generic comment, it can be said that in all numerical methods, the dependent variables (primary unknowns) in the governing differential equations are always unknown. But it is necessary to assume the pattern of the solution and plug or enforce the solution into the governing equation to solve for the unknown coefficients by satisfying the boundary and interface conditions. In FEM, the solution is enforced at the element level. When the solution of an unknown function say $f(x)$ is assumed, then the function is expressed in terms of superposition of few basis functions with specific properties when the basis functions contributes to the solution with their respective contribution factors. The most important property of the basis function is the property of orthogonality with respect to a suitable weight function. As an example, a function $f(x)$ can be written as a superposition of n basis functions (b_i) and their respective contribution factors (f_i).

$$f(x) = b_1 f_1 + b_2 f_2 + b_2 f_2 + b_3 f_3 \ldots\ldots + b_n f_n \tag{10.60}$$

This concept of function is very similar to the concept of vectors presented in Chapter 2. In Eqs. (2.1) through (2.3) a few vectors are presented in a basis system governed by unit vectors $\mathbf{e}_i$, where all the unit vectors are orthogonal to each other. In Chapter 2 the vectors are presented with respective magnitudes along with their respective unit vectors in an orthogonal basis system (Cartesian coordinate). Similar concept is presented in Chapter 6 for function. Here, similarly, a function $f(x)$ is assumed to be in a system of n basis (b_i, where i takes values 1 to n). f_i is the magnitude of the function along the respective basis.

Similarly, using a Taylor series expansion of n-th order of the same function within a certain zone in the neighborhood of a, one can write,

$$f(x) = f(a) + \frac{x-a}{1!} f^{(1)}(a) + \frac{(f-a)^2}{2!} f^2(a) + \cdots + \frac{(x-a)^n}{n!} f^{(n)}(a) \tag{10.61}$$

Comparing Eqs. (10.60) and (10.61), it can be visualized that they are spinned off from the similar concept. Taylor series expansion helps any unknown function to expand into a polynomial function, and thus the polynomial orders can be assumed to be the basis functions and the derivatives of the unknown function at the pivotal point "a" can be assumed to be the contribution factors or the magnitude of the function along the polynomial basis. Hence, the n-th basis function and Lagrange residual in the Taylor series expansion can be written as,

$$b_n = \frac{(x-a)^n}{(n)!}; \quad \frac{db_n}{\mathrm{dx}} = \frac{(x-a)^{n+1}}{(n+1)!} f^{(n+1)}(a) \tag{10.62}$$

The basis functions can also be the trigonometric functions. If sine and cosine functions are used as basis functions, the expansion of the series will be the Fourier series [5]. Irrespective of the method, the polynomial expansions are imposed at the element level in SEM or FEM and unknown functions inside the element are approximated. In most of the polynomial expansions, the nodal points where the unknown function values are zero are always equally spaced between the element boundaries. However, as per the discussion in the previous section to avoid the Runge's phenomenon, it is necessary to use polynomial functions where the nodal points are not uniformly distributed between the element boundaries. Hence, the basis function must be defined such a way that the above requirements are satisfied. Lobatto, Laguerre, and Chebyshev polynomials are of that kind and are frequently used in SEM are discussed in the following sections. Further in this work, however, orthogonal Lobatto polynomials are utilized.

10.2.3.1 Lobatto Polynomials

The basis functions in the Lobatto polynomial are described as the first-order derivative of the Legendre polynomial. Using Rodrigues's formula, the Legendre polynomial can be written as

$$P_n(\xi) = \frac{1}{2^n n!}\frac{d^n}{d\xi^n}(\xi^2 - 1)^n, \quad n = 0, 1, 2, \ldots k \ldots .\, n \tag{10.63}$$

Hence, the Lobatto polynomial can be written as

$$L_n(\xi) = \frac{d}{d\xi}P_{n+1}(\xi), \quad n = 0,1, 2, \ldots . k \ldots . n \tag{10.64}$$

The k-th order Lobatto polynomial will be

$$L_k(\xi) = \frac{1}{2^{k+1}(k+1)!}\frac{d^{k+2}}{d\xi^{k+2}}(\xi^2 - 1)^{k+1} \tag{10.65}$$

As discussed in the previous section, the specific property of the basis functions is the orthogonality of the function with suitable weight function. Hence, it was found that the Lobatto polynomials are orthogonal with respect to the weight function $(1-\xi^2)$ in the domain [−1,+1].

$$\int_{-1}^{1} L_i(\xi)L_j(\xi)(1-\xi^2)d\xi = \frac{2(i+1)(i+2)}{(2i+3)}\delta_{ij}, \quad i, j = 0, 1, 2 \tag{10.66}$$

where δ_{ij} is the well-known Kronecker delta function. In an n-th order spectral element, the internal node points are distributed at the roots of the $(n-2)$-th order of Lobatto polynomial $(L_{n-2}(\xi))$ because other two end points are placed at −1 and at +1, respectively. The coordinates of the spectral element nodes are solved from the following equation

$$(1-\xi^2)L_{n-2}(\xi) = 0 \quad n = 2,3,\ldots \tag{10.67}$$

10.2.3.2 Laguerre Polynomials

Laguerre polynomials can be written as

$$g_n(\xi) = \frac{e^{\xi}}{n!}\frac{d^n}{d\xi^n}\left(e^{-\xi}\xi^n\right) \quad n = 0, 1, 2, 3 \ldots k \ldots n \tag{10.68}$$

Laguerre polynomials are orthogonal with respect to the weight function $e^{-\xi}$ in the domain $[0,+\infty]$.

$$\int_0^\infty g_i(\xi)g_j(\xi)e^{-\xi}d\xi = \delta_{ij}, \quad i, j = 0,1,2 \tag{10.69}$$

In an n-th order spectral element, the internal node points are distributed at the roots of the $(n-1)$-th order of Laguerre polynomial $(g_{n-1}(\xi))$ because other end point is placed at 0. The coordinate of the spectral element nodes is solved from the following equation

$$(\xi)g_{n-1}(\xi) = 0 \quad n = 1,2, \ldots \tag{10.70}$$

10.2.3.3 Chebyshev Polynomials

In SEM, the Chebyshev polynomial of second-type is used and can be written as

$$\begin{gathered} C_0(\xi) = 1 \\ C_1(\xi) = 2\xi \\ \ldots \\ C_{n+1}(\xi) = 2\xi C_n(\xi) - C_{n-1}(\xi) \quad n = 0,1,2 \ldots \end{gathered} \tag{10.71}$$

Chebyshev polynomials are orthogonal with respect to the weight function $(\sqrt{1-\xi^2})$ in the domain $[-1,+1]$.

$$\int_{-1}^{+1} C_i(\xi)C_j(\xi)\sqrt{1-\xi^2}d\xi = \frac{\pi}{2}\delta_{ij}, \quad i, j = 0,1,2 \tag{10.72}$$

In an n-th order spectral element, the internal node points are distributed at the roots of the $(n-2)$-th order of Chebyshev polynomial $(C_{n-2}(\xi))$ because other end point is placed at -1 and $+1$, respectively. The coordinate of the spectral element nodes are solved from the following equation

$$(1-\xi^2)C_{n-2}(\xi) = 0 \quad n = 1,2, \ldots \tag{10.73}$$

results $\xi_j = cos\left(\frac{\pi i}{n-1}\right)$ where $i = 0, 1, 2, \ldots., n-1$

10.2.4 Lobatto Integration Quadrature

In practice, the integration terms stated in the mass matrix and stiffness matrix in equations (10.22) or (10.59) are computed by numerical methods. An appropriate quadrature rule is thus important to get the most effective results from the integration and overall solution of the problem. If Lobatto polynomials are used, the Lobatto quadrature rule should be used, whereas for Chebyshev and Laguerre polynomials, the Gauss quadrature and Gauss-Laguerre quadrature rules are used, respectively. Here, in this chapter, Lobatto polynomials are used for CNDE and Lobatto quadrature rule will be used. Gauss quadrature and Gauss-Laguerre quadrature rules can be found in many other books on numerical method [11]. Since numerical implementations are performed in element level before assembling them in a global equation, an element coordinate system is preferable to implement any of the integration quadrature. Hence, the shape function matrices are mapped to the normalized coordinate system and the Jacobian matrix is introduced. Say, for example, the integral of any function $\mathbf{H}(x, y, z)$ over the element volume can be transformed to a nondimensional coordinate system of ξ, η, β each defined between $[-1,+1]$ and the result of the integral will be identical. We can write this case for any arbitrary function $\mathbf{H}(x_j)$ as follows:

$$I^e = \int_{\Omega} \mathbf{H}(x_j) d\mathrm{v}^e = \int_{-1}^{+1}\int_{-1}^{+1}\int_{-1}^{+1} \mathbf{H}(\xi,\eta,\beta)\det(\mathbf{J})d\xi\, d\eta\, d\beta \tag{10.74}$$

where $\mathbf{J}$ is the Jacobi matrix discussed under contravariant vectors in Section 2.5.2 and calculated as follows:

$$\mathbf{J} = \begin{bmatrix} \frac{\partial x_1}{\partial \xi} & \frac{\partial x_2}{\partial \xi} & \frac{\partial x_3}{\partial \xi} \\ \frac{\partial x_1}{\partial \eta} & \frac{\partial x_2}{\partial \eta} & \frac{\partial x_3}{\partial \eta} \\ \frac{\partial x_1}{\partial \beta} & \frac{\partial x_2}{\partial \beta} & \frac{\partial x_3}{\partial \beta} \end{bmatrix} \tag{10.75}$$

which was previously written extensively in Eq. (2.48). The derivative of the spectral shape functions of an m-noded spectral element can be calculated as follows:

$$\begin{Bmatrix} \frac{\partial N_i}{\partial x_1} \\ \frac{\partial N_i}{\partial x_2} \\ \frac{\partial N_i}{\partial x_3} \end{Bmatrix} = \mathbf{J}^{-1} \begin{Bmatrix} \frac{\partial N_i}{\partial \xi} \\ \frac{\partial N_i}{\partial \eta} \\ \frac{\partial N_i}{\partial \beta} \end{Bmatrix} \quad i = 1, \ldots \ldots ., m \tag{10.76}$$

The quadrature rule to calculate the integral in Eq. (10.74) can be written in a generalized form as follows:

$$I^e = \int_{-1}^{+1}\int_{-1}^{+1}\int_{-1}^{+1} \mathbf{H}(\xi,\eta,\beta)\det(\mathbf{J})d\xi\, d\eta\, d\beta = \sum_{i=1}^{n_1}\sum_{j=1}^{n_2}\sum_{k=1}^{n_3} \omega_i\omega_j\omega_k \mathbf{H}(\phi_i, \phi_j, \phi_k)\det(\mathbf{J}) \tag{10.77}$$

where ω_i, ω_j, *and* ω_k are the weight factors and ϕ_i, ϕ_j, ϕ_k are the abscissae where the function values are obtained. The values depend on the number of quadrature points (n_1, n_2, n_3) used in the quadrature rule. The numerical integration in Lobatto quadrature uses the following formula [6]:

$$\int_{-1}^{+1} H(\xi)d\xi = \frac{2}{n(n-1)}\left[H(-1)+H(+1)\right]+\sum_{i=1}^{n-2}\omega_i H(\phi_i)+E_r \tag{10.78}$$

where n is the number of quadrature points. Next, the weight factors can be calculated as follows:

$$\omega_i = \frac{2}{q(q-1)P_{q-1}(a_i)^2}, \quad i=1,2,\ldots\ \ldots,n \tag{10.79}$$

The coordinate of the abscissae can be calculated from the root of the equation

$$\left(1-\phi_i^2\right)\frac{d}{d\xi}P_{n-1}(\phi_i)=0, \quad i=1,2,\ldots\ \ldots\ \ldots,n \tag{10.80}$$

where $P_{n-1}(\xi)$ is the Legendre polynomial of order $n-1$. The error in Eq. (10.78) is

$$E_r = \frac{n(n-1)^2\,2^{2n-1}\left((n-2)!\right)^4}{(2n-1)\left((2n-2)!\right)^3}H^{(2n-2)}(\eta), \quad \eta \in [-1,+1] \tag{10.81}$$

10.7 MODELING PIEZOELECTRIC EFFECT USING SEM

This particular topic has not been discussed under any other methods that are introduced in this book. Introducing the model of piezoelectric effect using SEM is timely herein because the SEM is the most effective method for modeling piezoelectricity that can be exploited in CNDE problems. Of course, this statement is based on author's different trials and experience in this field of NDE and SHM. In the forgoing discussion, readers' basic understanding of smart materials and piezoelectricity is assumed. For more detailed understanding of piezoelectricity, books listed in Ref. [6] can be referred.

In a piezoelectric material, the coupled (mechanical and electrical coupling) constitutive equation can be written as follows:

$$\sigma = \mathbb{C}e - e^T\mathbf{E} \tag{10.82.1}$$

$$\mathbf{D} = ee + \mathbf{g}\mathbf{E} \tag{10.82.2}$$

where e is the strain vector (6 × 1) as defined in Chapter 3, $\mathbb{C}$ is the elastic constitutive matrix (6 × 6) with elastic material constants of the piezoelectric material, **D** is the piezoelectric displacement vector (3 × 1), σ is the stress vector (6 × 1), matrix (3 × 6) of piezoelectric coupling constant is e, **g** is the matrix of dielectric constant of piezoelectric material, and **E** is the vector (3 × 1) electric field. Eqs. (10.82.1) and (10.82.2), respectively, are used for modeling piezoelectric material in actuator mode where the electric field is given to the material and material respond with output stress and piezoelectric material in sensing mode where resultant strain in the piezoelectric material cause the piezoelectric displacement. Hence, when the piezoelectric material is attached to a host structure, Eq. (10.82.1) is used for actuating a piezoelectric material to induce the elastic energy to cause wave propagation in the host material and Eq. (10.82.2) is used for sensing the propagating elastic waves from a host structure. Using electromagnetic phenomena, the electric field can be generated from an electric potential field, which is a scalar potential function. The electric field vector can be written as the gradient of the scalar potential function as follows:

$$\mathbf{E} = -\nabla\phi = -\phi_{,i} \tag{10.83}$$

While modeling piezoelectricity, the scalar potential input function is first discretized using spectral element. However, to have conformity, the order of discretization of the potential function should be the same in the order of mechanical discretization. Shape functions in three orthogonal directions can be assumed as $N^e(\xi,\eta,\beta) = N_i^e(\xi)N_j^e(\eta)N_k^e(\beta)$. Hence, after discretization, Eq. (10.83) will be visualize as

$$\mathbf{E} = -\nabla N^e(\xi,\eta,\beta)\varphi^e = -\mathbf{B}_\phi\varphi^e \tag{10.84}$$

where φ^e is a vector contains nodal values of potentials at the spectral element nodes. $\mathbf{B}_\phi$ is the gradient matrix with elements that are derivatives of the shape functions.

$$\mathbf{B}_\phi = \begin{bmatrix} \dfrac{\partial}{\partial x_1} & \dfrac{\partial}{\partial x_2} & \dfrac{\partial}{\partial x_3} \end{bmatrix}^T N_i^e(\xi)N_j^e(\eta)N_k^e(\beta) \tag{10.85}$$

Let us assume the SEM element has n, m, and p nodes in three respective orthogonal directions. The potential function in a discretized form in a $(n+m+p)$ nodded SEM element can be written as

$$\phi = N^e(\xi,\eta,\beta)\varphi^e = \sum_{i=1}^{n}\sum_{j=1}^{m}\sum_{k}^{p} N_i^e(\xi)N_j^e(\eta)N_k^e(\beta)\varphi^e(\xi_i,\eta_j,\beta_k) \tag{10.86}$$

To simulate wave propagation excited by a piezoelectric material, the equation of motion in a piezoelectric material should be derived using the same Hamilton's principle. Although the detailed derivation is omitted herein, Eq. (10.12) or (10.27) can be replaced by Eq. (10.82.1) and one can follow the steps using both methods presented in Sections 10.2.1 and 10.2.2, respectively, to obtain the following equation of motion for an element in a matrix form.

$$\mathbf{M}^e\ddot{\mathbf{U}}_i^e + \mathbf{D}^e\dot{\mathbf{U}}_i^e + \mathbf{K}^e{}_{uu}\mathbf{U}_i^e - \mathbf{K}^e{}_{u\varphi}\varphi^e = \mathbf{f}_i^e(t) \tag{10.87.1}$$

$$\mathbf{K}^e{}_{\varphi u}\mathbf{U}_i^e - \mathbf{K}^e{}_{\varphi\varphi}\varphi^e = \mathbf{q}_i^e(t) \tag{10.87.2}$$

where $\mathbf{K}^e{}_{u\varphi}$ and $\mathbf{K}^e{}_{\varphi u}$ are the matrix originated from piezoelectric coupling $\boldsymbol{e}$ in Eq. (10.82.1). $\mathbf{K}^e{}_{\varphi\varphi}$ is the matrix of dielectric constants originated from $\mathbf{g}$ in Eq. (10.82.2), $\mathbf{q}_i^e(t)$ is the vector of external electric charges defines the boundary condition given to the nodes of the piezoelectric SEM element. These new matrices can be expressed in a mathematical form as follows

$$\mathbf{K}^e{}_{\varphi\varphi} = \int_{\Omega_e} \left(\mathbf{B}_u^e\right)^T \mathbf{g}^e \mathbf{B}_\varphi^e d\mathrm{v}_e \tag{10.88.1}$$

$$\mathbf{K}^e{}_{u\varphi} = \left(\mathbf{K}^e{}_{\varphi u}\right)^T = -\int_{\Omega_e} \left(\mathbf{B}_u^e\right)^T e^e \mathbf{B}_\varphi^e d\mathrm{v}_e \tag{10.88.2}$$

$$\mathbf{q}_i^e = \int_{\Omega_e} \left(N^e\right)^T \mathbf{Q}_0 d\mathrm{v}_e \tag{10.88.3}$$

where $\mathbf{g}^e$, e^e, and $\mathbf{Q}_0$ are the dielectric matrix, piezoelectric coupling matrix, and electric charge distribution vector, respectively. A numerical instability occurs when Eqs. (10.87.1) and (10.87.2), are implemented in a SEM code. Primary reasons for such instability are the bad conditioning with the order of the magnitudes of $\mathbf{K}^e{}_{\varphi\varphi}$ ($\sim 10^{-8}$) and $\mathbf{K}^e{}_{uu}$ ($\sim 10^{11}$) matrices. They differ by almost $\sim 10^{19}$, which is very significant [6]. To circumvent this problem, Eq. (10.87.i) is statically condensed to separate into two parts, one to express the governing equation to simulate the actuator part and another part to simulate the sensor. Eq. (10.87.i) can be rewritten after static condensation as follows:

$$\mathbf{M}^e\ddot{\mathbf{U}}_i^e + \mathbf{D}^e\dot{\mathbf{U}}_i^e + \left(\mathbf{K}^e{}_{uu} + \mathbf{K}^e{}_I\right)\mathbf{U}_i^e = \mathbf{f}_i^e(t) + \mathbf{f}_A^e \tag{10.87.1}$$

$$\mathbf{K}^e{}_I = \left(\mathbf{K}^e{}_{\varphi u}\right)^T \left(\mathbf{K}^e{}_{\varphi\varphi}\right)^{-1} \mathbf{K}^e{}_{\varphi u} \tag{10.87.2}$$

where $\mathbf{K}^e{}_I$ is the stiffness developed by electromechanical coupling, which is dependent on the electrical boundary condition, e.g., open circuit, closed circuit, or piezoelectric material actuates as a transducer. $\mathbf{f}_A^e$ is the force generated by the SEM nodes in the piezoelectric element as actuator. Next, how the electrical boundary conditions should be given to simulate different circuit conditions is discussed in Appendix section.

10.8 IMPLEMENTATION OF SEM IN CNDE COMPUTATION

To solve CNDE problem, the above mathematical formulation is translated in to a computer code. The code has been developed using MATLAB to solve wave propagation problem. The algorithm and the list of variables are described below. First, wave propagation in multilayered composite plate is considered.

10.8.1 Setting up Initial Parameters

The purpose of computer implementation is to demonstrate the effectiveness of SEM to evaluate wave propagation in multilayer structure. Therefore, a suitable type of elements is chosen as the building block in formulating the SEM algorithm. All initial parameters are computed in three-dimensional space. The geometry and the material properties of the composite structure are declared in this part. In addition, the number of elements in three-dimensional space is optimized based on available computational resources. To facilitate master spectral element, the code is formulated based on N^{th}-order approximate polynomials in each of the available dimensions. For example, 3rd-, 5th-, and 7th-order polynomials can be approximated in directions 1, 2, and 3, (x_1, x_2, and x_3), respectively. This information is required to discretize the domain, determining the shape functions (Eq. (10.53)) and formulating stiffness and mass matrices (Eq. (10.58)).

10.8.2 Discretization of the Problem Domain

The first step of the SEM is to discretize the problem domain depending on respective CNDE problem into three-dimensional elements. Each of these elements is composed of group of geometrical nodes. The combination of these element nodes constitutes a larger set of unique global nodes. In SEM, several types of elements can be assumed. The element types can be tetrahedral, pyramidal, prismatic, and hexahedral (brick). Further, hexahedral brick elements are chosen considering simple geometry of the problem domain. As stated earlier, in SEM, the node points are located at the roots of Gauss-Lobatto-Legendre (GLL) polynomials; unlike FEM, these node points are not spaced equally as shown in Fig. 10.3 [12].

In SEM discretization, first a master brick element is mapped to a parent domain (ξ, η, ζ) with standard intervals [-1, 1], [-1, 1], and [-1, 1]. A subroutine is used to generate the element mapping and the discretization points. The global coordinates of

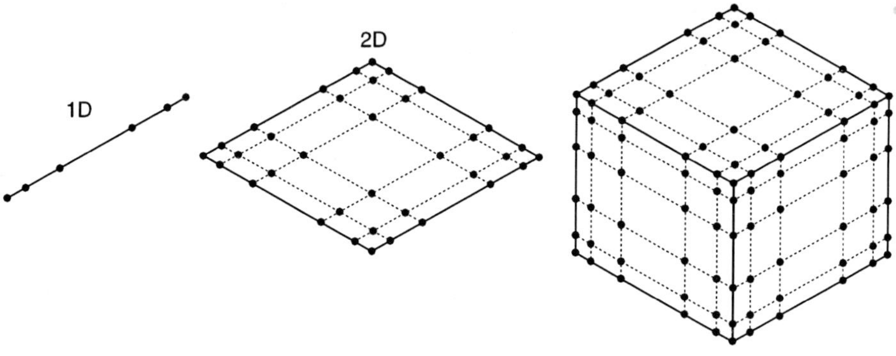

FIGURE 10.3 Node distribution in Spectral Element methods which are roots of 5th order GLL polynomials for 1D, 2D and 3D cases.

the node points are calculated based on the master element local node points. While calculating the global node points, the overlapping faces of the adjoining faces are assigned to the sharing elements and thus C^0 continuity is ensured. This step is much like FEM. A typical example of a discretized domain with $2 \times 2 \times 2$ elements having $4 \times 3 \times 2$ GLL points in each element is shown in Fig. 10.4. It can be observed that a total of eight elements are present in the problem domain. Out of these eight elements, four elements are located on the front row (two in front top and two in front bottom) and rest four are located on the back row (two in back top and two in back bottom). The origin of this domain is considered at node 1 of element 1. To maintain C^0 continuity, during discretization, specific node points of each element are designated to be shared by the node points of adjoining elements. For example, in Fig. 10.4, $x_2 - x_3$ face of the element 1 that contains nodes 4, 8, 12, 16, 20, and 24 is shared by $x_2 - x_3$ face of element 2. Similarly, $x_1 - x_3$ face of the element 2 is shared by $x_1 - x_3$ face of element 6. Thus, element 1 is connected with elements 2, 3, 4, 5, and 6 either by face or by edge. Therefore, it can be seen that a pair of elements share a face to each other. In a global sense, all the front faces rectangles are in one $x_2 - x_3$ plane and similarly back rectangles of each element of Fig. 10.4 are in the other $x_2 - x_3$ plane. While discretizing, a variable is formulated that maps the local numbering of the computational nodes to their global (nonredundant) numbering. Based on this map, the three-dimensional coordinates of global node points

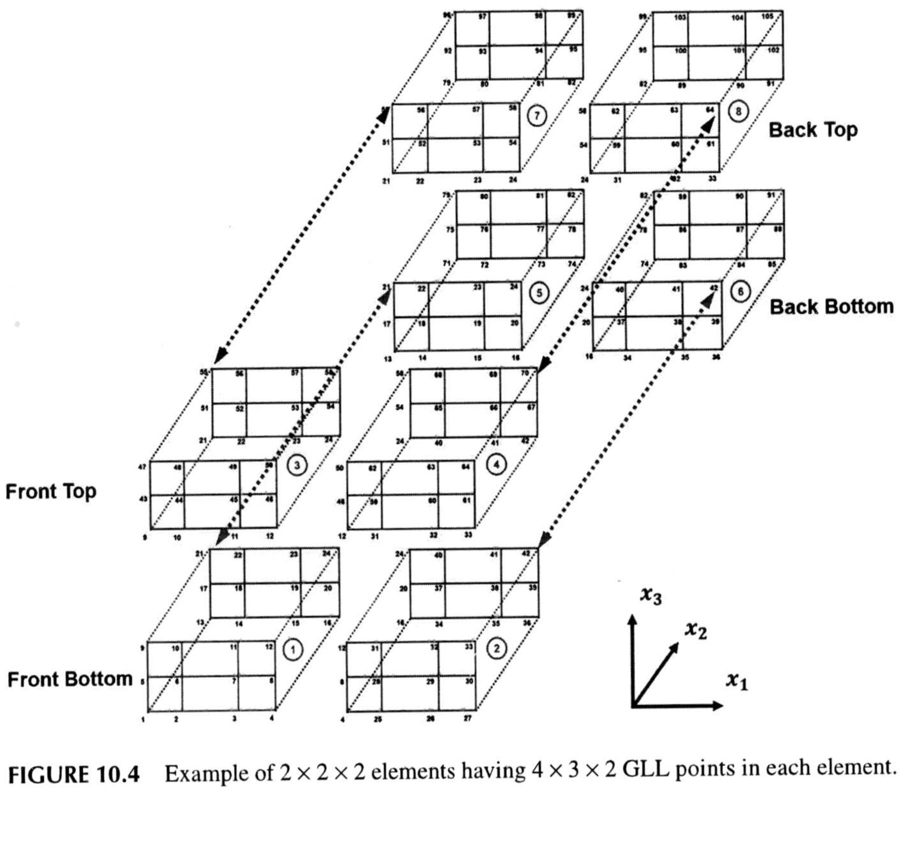

FIGURE 10.4 Example of $2 \times 2 \times 2$ elements having $4 \times 3 \times 2$ GLL points in each element.

are determined. Formulation of this map is helpful to identify each points of the global domain and assign any properties as required by the problem statement.

10.8.3 Determination Global Mass and Stiffness Matrix

This is relatively a long step where local components of the element stiffness and mass matrices are determined. Determination of local mass and stiffness matrices allows us to formulate element equation applicable to each type of elements used in the discretization process. In this section, the implementation of computer application is based on the element equation formulated for master hexahedral element. The concluding equation, Eq. (10.59), describes the general element level SEM equation. Each of the components of this equation is required to be determined and formulated for the SEM computer implementation.

10.8.3.1 Local Stiffness Matrix

The element stiffness matrix, denoted by $\left[\mathbf{K}^e\right]$, has nine components. For brevity, evaluation of one component is described in this article. The first component is

$$K_{ij}^{11} = \int \left[\mathbb{C}_{11}\frac{\partial\psi_i}{\partial x_1}\frac{\partial\psi_j}{\partial x_1} + \mathbb{C}_{66}\frac{\partial\psi_i}{\partial x_2}\frac{\partial\psi_j}{\partial x_2} + \mathbb{C}_{55}\frac{\partial\psi_i}{\partial x_3}\frac{\partial\psi_j}{\partial x_3}\right]dv \tag{10.88}$$

which has an integral form on its right-hand side. This integral is computed by numerical methods using Lobatto integration quadrature. Therefore, in numerical method, the value of K_{ij}^{11} can be calculated as,

$$\begin{aligned}\int \left[\mathbb{C}_{11}\frac{\partial\psi_i}{\partial x_1}\frac{\partial\psi_j}{\partial x_1} + \mathbb{C}_{66}\frac{\partial\psi_i}{\partial x_2}\frac{\partial\psi_j}{\partial x_2} + \mathbb{C}_{55}\frac{\partial\psi_i}{\partial x_3}\frac{\partial\psi_j}{\partial x_3}\right]dv = \iiint_{-1}^{1}\left[\mathbb{C}_{11}\frac{\partial\psi_i}{\partial x_1}\frac{\partial\psi_j}{\partial x_1}\right.\\ \left. + \mathbb{C}_{66}\frac{\partial\psi_i}{\partial x_2}\frac{\partial\psi_j}{\partial x_2} + \mathbb{C}_{55}\frac{\partial\psi_i}{\partial x_3}\frac{\partial\psi_j}{\partial x_3}\right]dx_1dx_2dx_3 = \sum_{p=1}^{m}\sum_{p=1}^{n}\sum_{p=1}^{k}\left[\mathbb{C}_{11}\frac{\partial\psi_i}{\partial\xi}\frac{\partial\psi_j}{\partial\xi}\omega_{\xi_p}\omega_{\eta_p}\omega_{\zeta_p}J_{x_1x_1}\right.\\ \left. + \mathbb{C}_{66}\frac{\partial\psi_i}{\partial\eta}\frac{\partial\psi_j}{\partial\eta}\omega_{\xi_p}\omega_{\eta_p}\omega_{\zeta_p}J_{x_2x_2} + \mathbb{C}_{55}\frac{\partial\psi_i}{\partial\zeta}\frac{\partial\psi_j}{\partial\zeta}\omega_{\xi_p}\omega_{\eta_p}\omega_{\zeta_p}J_{x_3x_3}\right]\end{aligned} \tag{10.89}$$

The mass matrix M_{ij}^{11} can be calculated as

$$M_{ij}^{11} = \int \rho\ \psi_i\psi_j dv = \iiint_{-1}^{1}\rho\ \psi_i\psi_j dx_1dx_2dx_3 = \sum_{p=1}^{m}\sum_{p=1}^{n}\sum_{p=1}^{k}\rho\psi_i\psi_j\omega_{\xi_p}\omega_{\eta_p}\omega_{\zeta_p}J_{x_1x_1} \tag{10.90}$$

where $\mathbb{C}_{11}$, $\mathbb{C}_{66}$, and $\mathbb{C}_{55}$ are components of the elastic properties and ρ is the density of the material properties defined in initial parameter steps; ψ_i and ψ_j are the three-dimensional shape functions; $\frac{\partial\psi_i}{\partial x_1}$ and $\frac{\partial\psi_j}{\partial x_1}$ are the first derivatives of the three-dimensional shape functions calculated at each of the node points in each of the element; ω_{ξ_p}, ω_{η_p}, and ω_{ζ_p} are the integration weights at each of the normalized axis; $J_{x_1x_1}$ is the determinant of the Jacobian matrix to transfer coordinates from x_1, x_2, and x_3 to ξ, η, and β as described in Eq. (10.75). In the next subsection, calculation of the above parameters is discussed.

10.8.3.2 Material Properties

The components of the elasticity matrix are calculated at each element nodes. Since the material properties of an anisotropic material depend on the direction of the coordinate axis, an element wise evaluation of the elasticity matrix is a pivotal approach to determine its components. Therefore, the nodal map determined in discretization step is used to determine the material properties at each node point. However, this does not hinder the implementation of SEM for isotropic material. In case of isotropic material, Eq. (3.84) in Chapter 3 is used to replace the respective anisotropic material parameters by the isotropic parameters. Before evaluating local stiffness matrix, a local variable is created by collecting the node points that have similar material properties. For example, in case of a transversely isotropic material, the material properties are prescribed in each layer. If the layers are stacked in x_3 direction, thickness of each layer is defined by the distance between two lines where the material properties remain same. In case of isotropic material, no such demarcation is necessary. Similarly, the density of the material can be determined at each node point following the same approach.

10.8.3.3 Shape Functions

The shape functions are determined using the Lagrange interpolation functions of appropriate order. The Lagrange interpolation functions associated with hexahedral elements can be obtained from the corresponding one-dimensional Lagrange interpolation functions and taking the tensor products of the x_1 direction interpolation functions with the tensor product of the x_2 and x_3 directional interpolation functions. The Nth-order Lagrange interpolation functions can be associated with the given abscissas ξ. The Lagrange interpolation functions associated with the i^{-th} abscissa can be defined in terms of the abscissas of the data points, ξ, where $j = 1, 2, 3, \ldots\ldots, N+1$ and denoted by $L_{N,i}(\xi)$. In line with the properties of the shape functions, $L_{N,i}(\xi)$ is equal to zero at all the data points except at the i-th data point where it becomes one, which is same to the Kronecker delta, δ_{ij} function. Therefore, the Lagrange interpolation functions can be written as follows:

$$L_{N,i}(\xi) = \frac{(\xi-\xi_1)(\xi-\xi_2)\ldots\ldots(\xi-\xi_{i-1})(\xi-\xi_{i+1})\ldots\ldots(\xi-\xi_N)(\xi-\xi_{N+1})}{(\xi_i-\xi_1)(\xi_i-\xi_2)\ldots\ldots(\xi_i-\xi_{i-1})(\xi_i-\xi_{i+1})\ldots\ldots(\xi_i-\xi_N)(\xi_i-\xi_{N+1})} \tag{10.91}$$

If one-dimensional shape functions in three coordinate axes are denoted as $\psi_{\xi,m}$, $\psi_{\eta,n}$, and $\psi_{\zeta,k}$, the tensor product of three vectors can be performed as follows:

$$\psi_{2d,(m,n)} = \psi_{\xi,m} \otimes \psi_{\eta,n}^{T} \tag{10.92}$$

which is a matrix and ⊗ denotes the tensor product. This needs to be reshaped as a vector before performing next tensor product, which will yield three-dimensional shape functions.

$$\psi_{2d,(mn,1)} = reshape\ (\psi_{2d,\,mxn}) \tag{10.93}$$

$$\psi_{3d} = \psi_{2d,(mn,1)} \otimes \psi_{\zeta,k}{}^{T} \tag{10.94}$$

10.8.3.4 First Derivate of the Shape Functions

The first derivative of the one-dimensional shape functions can be determined using the following expressions [13]:

$$\frac{d\psi_{1d,ij}}{d\xi} = \frac{1}{\xi_j - \xi_i}\,\frac{(\xi_j - \xi_1)(\xi_j - \xi_2)\ldots\ldots(\xi_j - \xi_{j-1})(\xi_j - \xi_{j+1})\ldots\ldots(\xi_j - \xi_{m+1})}{(\xi_i - \xi_1)(\xi_i - \xi_2)\ldots\ldots(\xi_i - \xi_{i-1})(\xi_j - \xi_{i+1})\ldots\ldots(\xi_i - \xi_{m+1})} \tag{10.95}$$

For $i \neq j$, and

$$\frac{d\psi_{1d,ii}}{d\xi} = \frac{1}{\xi_i - \xi_1} + \ldots + \frac{1}{\xi_i - \xi_{i-1}} + \frac{1}{\xi_i - \xi_{i+1}} + \ldots + \frac{1}{\xi_i - \xi_{m+1}} \tag{10.96}$$

In case of first derivative of three-dimensional shape functions, it can be calculated as

$$\frac{d\psi(\xi,\eta,\zeta)}{d\xi} = \frac{d\psi(\xi)}{d\xi} \otimes \psi(\eta)^{T} \otimes \psi(\zeta)^{T} \tag{10.97}$$

Similarly, the first derivative of other two three-dimensional shape functions can be determined as follows:

$$\frac{d\psi(\xi,\eta,\zeta)}{d\eta} = \psi(\xi) \otimes \frac{d\psi(\eta)}{d\eta}^{T} \otimes \psi(\zeta)^{T} \tag{10.98}$$

$$\frac{d\psi(\xi,\eta,\zeta)}{d\zeta} = \psi(\xi) \otimes \psi(\eta)^{T} \otimes \frac{d\psi(\zeta)}{d\zeta}^{T} \tag{10.99}$$

10.8.3.5 Weighting Function

Integration weights are determined to perform Lobatto integration quadrature. Therefore, the following relationships are utilized to calculate the integration weights for a m-th order one-dimensional shape function

$$\omega_m = \frac{2}{k(k+1)}\,\frac{1}{L_k^2(\xi)} \tag{10.100}$$

TABLE 10.1
Zeros of GLL Points and the Corresponding Integration Quadrature

Polynomial Order	Zeros of GLL Points	Integration Weights
1	-1	1
	1	1
2	-1	0.3333
	0	1.3333
	1	0.3333
3	-1	0.1667
	-0.4472	0.8333
	0.4472	0.8333
	1	0.1667
4	-1	0.1
	-0.6547	0.5444
	0	0.7111
	0.6547	0.5444
	1	0.1
5	-1	0.0667
	-0.7651	0.3785
	-0.2852	0.5549
	0.2852	0.5549
	0.7651	0.3785
	1	0.0667
6	-1	0.0476
	-0.8302	0.2768
	-0.4688	0.4317
	0	0.4876
	0.4688	0.4317
	0.8302	0.2768
	1	0.0476

where $m = 2, \ldots \ldots, k$ and L_k is a Legendre polynomial. The numerical values in ξ axis for all ξ_i points are shown in Table 10.1.

10.8.3.6 Coordinate Transformation

The node points of a hexahedral element are expressed in normalized (ξ, η, ζ) coordinates. The sole objective of utilizing these coordinates is to evaluate the integrands of Eq. (10.59) numerically. This helps to evaluate this integrand for an arbitrary master element, which can be adjusted based on the location of any element in the entire domain in global coordinates. However, evaluation of the integrand for a master element should be such that there exist no spurious gaps between elements and no element overlaps with each other. The three-dimensional shape function, $\psi_i^e(\xi, \eta, \zeta)$

can be expressed in terms of local coordinates ξ,η, and ζ and by the chain rule of differentiation can be written as

$$\frac{d\psi_i^e}{d\xi}=\frac{d\psi_i^e}{dx_1}\frac{dx_1}{d\xi}+\frac{d\psi_i^e}{dx_2}\frac{dx_2}{d\xi}+\frac{d\psi_i^e}{dx_3}\frac{dx_3}{d\xi} \tag{10.101.1}$$

$$\frac{d\psi_i^e}{d\eta}=\frac{d\psi_i^e}{dx_1}\frac{dx_1}{d\eta}+\frac{d\psi_i^e}{dx_2}\frac{dx_2}{d\eta}+\frac{d\psi_i^e}{dx_3}\frac{dx_3}{d\eta} \tag{10.101.2}$$

$$\frac{d\psi_i^e}{d\zeta}=\frac{d\psi_i^e}{dx_1}\frac{dx_1}{d\zeta}+\frac{d\psi_i^e}{dx_2}\frac{dx_2}{d\zeta}+\frac{d\psi_i^e}{dx_3}\frac{dx_3}{d\zeta} \tag{10.101.3}$$

which can be written in a matrix form as follows

$$\begin{Bmatrix}\dfrac{d\psi_i^e}{d\xi}\\ \dfrac{d\psi_i^e}{d\eta}\\ \dfrac{d\psi_i^e}{d\zeta}\end{Bmatrix}=\begin{bmatrix}\dfrac{dx_1}{d\xi} & \dfrac{dx_2}{d\xi} & \dfrac{dx_3}{d\xi}\\ \dfrac{dx_1}{d\eta} & \dfrac{dx_2}{d\eta} & \dfrac{dx_3}{d\eta}\\ \dfrac{dx_1}{d\zeta} & \dfrac{dx_2}{d\zeta} & \dfrac{dx_3}{d\zeta}\end{bmatrix}\begin{Bmatrix}\dfrac{d\psi_i^e}{dx_1}\\ \dfrac{d\psi_i^e}{dx_2}\\ \dfrac{d\psi_i^e}{dx_3}\end{Bmatrix} \tag{10.102}$$

This gives the relationship between the derivatives of three-dimensional shape function ψ_i^e with respect to the global and local coordinates. Here, the matrix can be defined as the Jacobian matrix of the transformation like it is previously defined in Chapter 2 and in Eq. (10.75).

$$[J]=\begin{bmatrix}\dfrac{dx_1}{d\xi} & \dfrac{dx_2}{d\xi} & \dfrac{dx_3}{d\xi}\\ \dfrac{dx_1}{d\eta} & \dfrac{dx_2}{d\eta} & \dfrac{dx_3}{d\eta}\\ \dfrac{dx_1}{d\zeta} & \dfrac{dx_2}{d\zeta} & \dfrac{dx_3}{d\zeta}\end{bmatrix} \tag{10.103}$$

The determinant of the Jacobian matrix can be calculated as

$$Det[J]=\frac{dx_1}{d\xi}\frac{dx_2}{d\eta}\frac{dx_3}{d\zeta} \tag{10.104}$$

$$\frac{dx_2}{d\xi}=0,\ \frac{dx_3}{d\xi}=0,\ \frac{dx_1}{d\eta}=0,\ etc.$$

An example of determining, $J_{x_1x_1}$, which transforms the first derivative of three-dimensional shape function with respect to x_1, x_2, and x_3 can be written as follows:

$$\frac{\partial\psi_i}{\partial x_1}\frac{\partial\psi_j}{\partial x_1}dx_1dx_2dx_3 = \frac{\partial\psi_i}{\partial\xi}\frac{\partial\xi}{\partial x_1}\frac{\partial\psi_j}{\partial\xi}\frac{\partial\xi}{\partial x_1}\frac{\partial x_1}{\partial\xi}\partial\xi\frac{\partial x_2}{\partial\eta}\partial\eta\frac{\partial x_3}{\partial\zeta}\partial\zeta$$

$$= \frac{\partial\psi_i}{\partial\xi}\frac{\partial\psi_j}{\partial\xi}\partial\xi\partial\eta\partial\zeta\frac{\partial\xi}{\partial x_1}\frac{\partial\xi}{\partial x_1}\frac{\partial x_1}{\partial\xi}\frac{\partial x_2}{\partial\eta}\frac{\partial x_3}{\partial\zeta} \quad (10.105)$$

$$= \frac{\partial\psi_i}{\partial\xi}\frac{\partial\psi_j}{\partial\xi}\partial\xi\partial\eta\partial\zeta\frac{\left(\frac{\partial x_1}{\partial\xi}\frac{\partial x_2}{\partial\eta}\frac{\partial x_3}{\partial\zeta}\right)}{\frac{\partial x_1}{\partial\xi}\frac{\partial x_1}{\partial\xi}} = \frac{\partial\psi_i}{\partial\xi}\frac{\partial\psi_j}{\partial\xi}\partial\xi\partial\eta\partial\zeta\, J_{x_1x_1}$$

Therefore, $J_{x_1x_1} = \frac{\left(\frac{\partial x1}{\partial\xi}\frac{\partial x2}{\partial\eta}\frac{\partial x3}{\partial\zeta}\right)}{\frac{\partial x1}{\partial\xi}\frac{\partial x1}{\partial\xi}}$

Similarly, $J_{x_1x_2}$ can be calculated as $J_{x_1x_2} = \frac{\left(\frac{\partial x1}{\partial\xi}\frac{\partial x2}{\partial\eta}\frac{\partial x3}{\partial\zeta}\right)}{\frac{\partial x1}{\partial\xi}\frac{\partial x2}{\partial\eta}}$

10.8.3.7 Assembly of Local Stiffness Matrix into a Global Stiffness Matrix

Assembly of local stiffness matrix in SEM into a global stiffness matrix is like the FEM. Here, assembly of local stiffness matrix with two hexahedral elements is briefly described for reader who wants to implement SEM from the beginning. Once the local stiffness matrix is determined using Eq. (10.59), each component of the local stiffness matrix is assembled in the global stiffness matrix following two basic principles:

1. Continuity of primary variables
2. Balance of secondary variables.

In the next step, the global equations are solved using a suitable time integration solution method like they are described in Chapter 6. The global equation takes the form

$$[\mathbf{K}]\{\mathbf{U}\} + [\mathbf{M}]\{\ddot{\mathbf{U}}\} = \{\mathbf{F}\} \quad (10.106)$$

where

$[\mathbf{K}]$ = Global Stiffness matrix
$[\mathbf{M}]$ = Global mass matrix
$\{\mathbf{U}\}$ = Primary variable which is displacement in wave propagation problem
$\{\ddot{\mathbf{U}}\}$ = Second derivative of primary variable with respect to time
$\{\mathbf{F}\}$ = Secondary variable which is applied force with frequency range

In three-dimensional problem, the primary variable has three degree of freedom which, therefore, consists of three mutually perpendicular values at each geometric node points. It means that, for each nodal point, three equations are needed to determine the primary variable. Therefore, at each node point, $U = f(u_1, u_2, u_3)$, where u_1, u_2, and u_3 are the displacements along x_1, x_2, x_3 directions, respectively. For more clarity, here a change in nomenclature would be easier and the displacements in three orthogonal directions can be substituted as $u_1 = u$, $u_2 = v$, $u_3 = w$ such that displacement vectors for all nodes in all elements could be written together. In fact, in a computer code, similar nomenclature should be used. u_i^j, v_i^j, and w_i^j are further used to designate the respective displacements at the i-th node in j-th element. In this chapter, the global displacement vector, which contains displacements of all nodes, is arranged in the following fashion

$$\{\mathbf{U}\} = \left\{ u_1^1 \ v_1^1 \ w_1^1 \ \ldots \ u_m^1 \ v_m^1 \ w_m^1 \ \ldots\ldots\ldots \ u_1^e \ v_1^e \ w_1^e \ \ldots \ u_m^e \ v_m^e \ w_m^e \right\}^T \quad (10.107)$$

where $u_1^e,.., w_m^e$ denote the displacements of eth elements for all its node points. It can be noted that the nodal map contains element wise nodal positions while the global displacement vector assumed in Eq. (10.107) contains the displacements in global sequence. Fulfilment of global displacement requirements necessitates the global stiffness and mass matrices to be assembled in global nodal sequence. An example of this assembly method is described as follows:

Consider two hexahedral elements having two node points in each axis direction. Therefore, each of the elements has eight (8) node points as shown in Fig. 10.5. Since each of the node point has three degree of freedom, the global displacement vector consists of $12 \times 3 = 24$ displacement values. Therefore, the global stiffness matrix should consist of 24 rows and 24 columns. This 24×24 square matrix gets its data values from its local counterparts. The local stiffness matrix of element 1 can be computed using the method described in "Local Stiffness matrix" section.

As observed in Fig. 10.5, the element 1 has 8 nodes labeled {1, 2, 3, 4, 5, 6, 7, 8} and element 2 has 8 nodes labeled {2, 9, 4, 10, 6, 11, 8, 12}. In global SEM equations, there exist 3×12 or 36 values of the displacement vectors. The assigned indexes of the global displacement vector are described in Fig. 10.6. Note that, first, second, and third indices of each node are assigned with directions 1, 2, and 3 respectively.

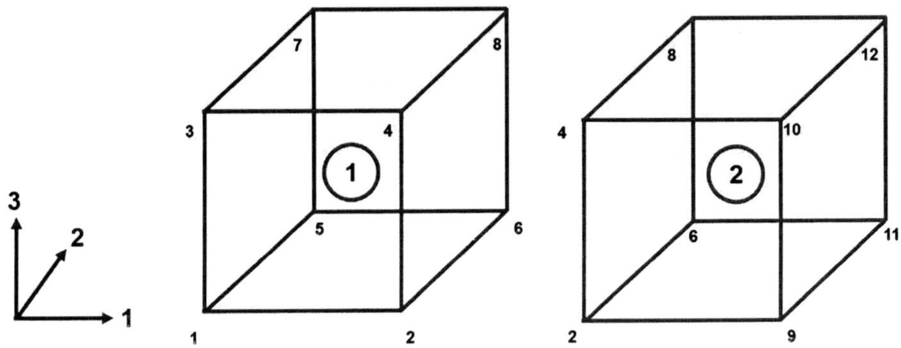

FIGURE 10.5 Global node points of two adjoining elements having two nodes at each axis direction.

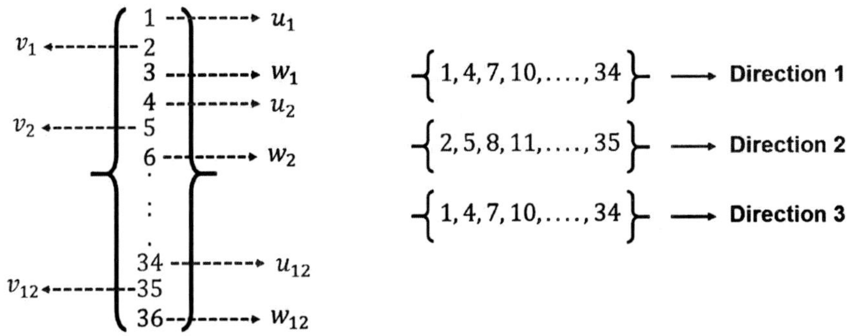

FIGURE 10.6 Global displacement vector having values in direction 1, 2 and 3.

Therefore, the global stiffness matrix contains 36 rows and 36 columns, which need to be filled in by the local stiffness matrices of the elements. Since each element has 8 nodes, the local stiffness matrix contains 24 rows and 24 columns due to three degrees of freedom in each node. Therefore, the components of the local stiffness matrix for element 1 can be written as

$$\left[K_{11}^{1}\right]=\begin{bmatrix} K_{11}^{1,11} & K_{11}^{1,12} & K_{11}^{1,13} & K_{11}^{1,14} & K_{11}^{1,15} & K_{11}^{1,16} & K_{11}^{1,17} & K_{11}^{1,18} \\ K_{11}^{1,21} & K_{11}^{1,22} & K_{11}^{1,23} & K_{11}^{1,24} & K_{11}^{1,25} & K_{11}^{1,26} & K_{11}^{1,27} & K_{11}^{1,28} \\ \cdot & \cdot & \cdot & \cdot & \cdot & \cdot & \cdot & \cdot \\ \cdot & \cdot & \cdot & \cdot & \cdot & \cdot & \cdot & \cdot \\ \cdot & \cdot & \cdot & \cdot & \cdot & \cdot & \cdot & \cdot \\ \cdot & \cdot & \cdot & \cdot & \cdot & \cdot & \cdot & \cdot \\ \cdot & \cdot & \cdot & \cdot & \cdot & \cdot & \cdot & \cdot \\ K_{11}^{1,81} & K_{11}^{1,82} & K_{11}^{1,83} & K_{11}^{1,84} & K_{11}^{1,85} & K_{11}^{1,86} & K_{11}^{1,87} & K_{11}^{1,88} \end{bmatrix} \tag{10.108}$$

$$\left[K_{12}^{1}\right]=\begin{bmatrix} K_{12}^{1,11} & K_{12}^{1,12} & K_{12}^{1,13} & K_{12}^{1,14} & K_{12}^{1,15} & K_{12}^{1,16} & K_{12}^{1,17} & K_{12}^{1,18} \\ K_{12}^{1,21} & K_{12}^{1,22} & K_{12}^{1,23} & K_{12}^{1,24} & K_{12}^{1,25} & K_{12}^{1,26} & K_{12}^{1,27} & K_{12}^{1,28} \\ \cdot & \cdot & \cdot & \cdot & \cdot & \cdot & \cdot & \cdot \\ \cdot & \cdot & \cdot & \cdot & \cdot & \cdot & \cdot & \cdot \\ \cdot & \cdot & \cdot & \cdot & \cdot & \cdot & \cdot & \cdot \\ \cdot & \cdot & \cdot & \cdot & \cdot & \cdot & \cdot & \cdot \\ \cdot & \cdot & \cdot & \cdot & \cdot & \cdot & \cdot & \cdot \\ K_{12}^{1,81} & K_{12}^{1,82} & K_{12}^{1,83} & K_{12}^{1,84} & K_{12}^{1,85} & K_{12}^{1,86} & K_{12}^{1,87} & K_{12}^{1,88} \end{bmatrix} \tag{10.109}$$

$$\cdot \quad \cdot \quad \cdot \quad \cdot \quad \cdot \quad \cdot \quad \cdot \quad \cdot \quad \cdot \quad \cdot \quad \cdot$$

$$\cdot \quad \cdot \quad \cdot \quad \cdot \quad \cdot \quad \cdot \quad \cdot \quad \cdot \quad \cdot \quad \cdot \quad \cdot$$

$$\left[K_{33}^{1}\right] = \begin{bmatrix} K_{33}^{1,11} & K_{33}^{1,12} & K_{33}^{1,13} & K_{33}^{1,14} & K_{33}^{1,15} & K_{33}^{1,16} & K_{33}^{1,17} & K_{33}^{1,18} \\ K_{33}^{1,21} & K_{33}^{1,22} & K_{33}^{1,23} & K_{33}^{1,24} & K_{33}^{1,25} & K_{33}^{1,26} & K_{33}^{1,27} & K_{33}^{1,28} \\ \cdot & \cdot & \cdot & \cdot & \cdot & \cdot & \cdot & \cdot \\ \cdot & \cdot & \cdot & \cdot & \cdot & \cdot & \cdot & \cdot \\ \cdot & \cdot & \cdot & \cdot & \cdot & \cdot & \cdot & \cdot \\ \cdot & \cdot & \cdot & \cdot & \cdot & \cdot & \cdot & \cdot \\ \cdot & \cdot & \cdot & \cdot & \cdot & \cdot & \cdot & \cdot \\ K_{33}^{1,81} & K_{33}^{1,82} & K_{33}^{1,83} & K_{33}^{1,84} & K_{33}^{1,85} & K_{33}^{1,86} & K_{33}^{1,87} & K_{33}^{1,88} \end{bmatrix} \tag{10.110}$$

Similarly, components of the local stiffness matrix of element 2 can be derived. At this stage, the components of the local stiffness matrix need to be assembled in a matrix that contains 24 rows and 24 columns where local contributions of the components are evaluated. Fig. 10.7 shows the location convention of the local stiffness matrix. While assembling the local stiffness matrix from its components, the contributions of a node to itself and contributions of other nodes to the element need to be considered. This concept is illustrated in Fig. 10.8a and an example of

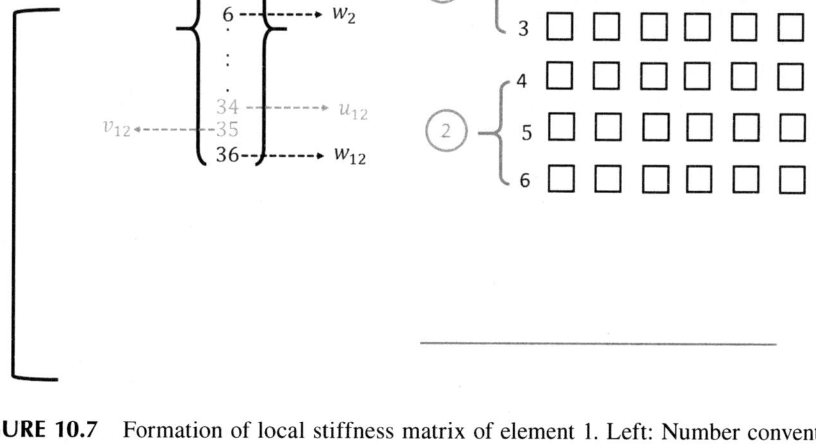

FIGURE 10.7 Formation of local stiffness matrix of element 1. Left: Number convention of a member of a component matrix. Right: Location convention of a local stiffness matrix.

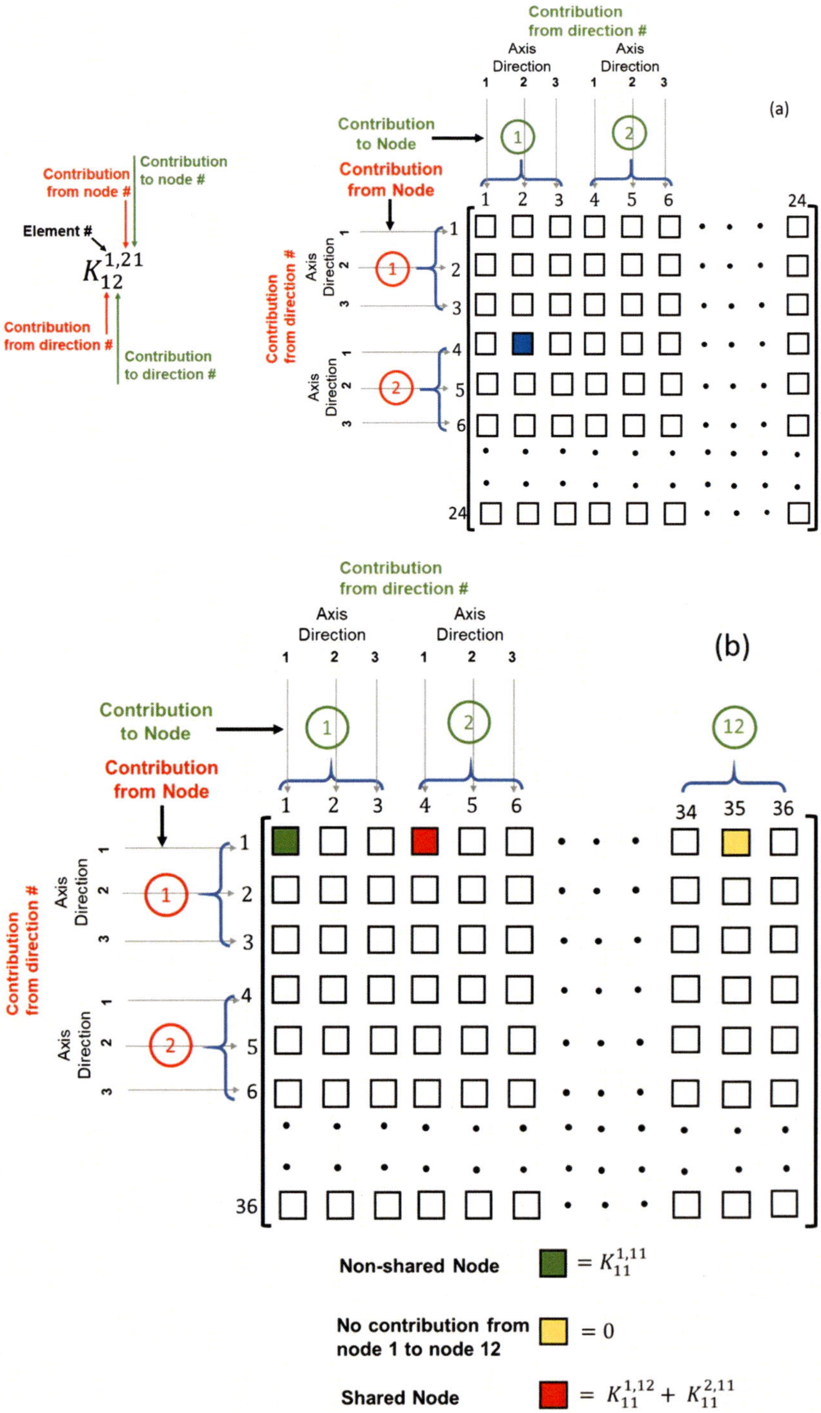

FIGURE 10.8 Evaluation and assembly of global stiffness matrix from its local stiffness matrix.

TABLE 10.2
Assembly of Element Local Stiffness Matrix from its Components

Component	Contribution from Node	Contribution to Node	Contribution from Direction	Contribution to Direction	Placement in Local Matrix
$K_{11}^{1,11}$	1	1	1	1	[1, 1]
$K_{11}^{1,12}$	1	2	1	1	[1, 4]
$K_{11}^{1,13}$	1	3	1	1	[1, 7]
$K_{11}^{1,18}$	1	8	1	1	[1, 22]
$K_{12}^{1,11}$	1	1	1	2	[1, 2]
$K_{32}^{1,11}$	1	1	3	2	[3, 2]
$K_{32}^{1,17}$	1	7	3	2	[3, 20]

implementation is described in Table 10.2. Based on this contribution, each member of the components is placed in the local stiffness matrix. The notation $K_{12}^{1,21}$ in Fig. 10.8 indicates that it has a contribution from node 2 and it is contributing to node 1. Therefore, it can be located at possible nine positions in the local stiffness matrix, which are [4, 1], [4, 2], [4, 3], [5, 1], [5, 2], [5, 3], [6, 1], [6, 2], and [6, 3]. However, the contribution from direction 1 and contribution to direction 2 determines the exact location in local stiffness matrix, which is [4, 2].

Similarly, the local stiffness matrix of element 2 can be derived from its stiffness components. Once the local stiffness matrices are determined, the global stiffness matrix can be evaluated using the local stiffness parts. The nodes which are not shared by any elements, for them, the local stiffness values of those nodes are directly placed in to the global stiffness matrix using the convention protocol of the matrices stated above. Moreover, if there is no contribution from one node to another node, the stiffness value is zero. However, nodes that are shared by the elements, values of global stiffness matrix are evaluated by adding the local stiffness components of shared nodes. In this example, global nodes [2, 4, 6, 8] are shared by both the elements. Therefore, respective components for the nodes [2, 4, 6, 8] need to be summed up. For example, first component of node 2 (which is a global node 2) of element 1 is located at [1, 4] of the local stiffness matrix of element 1. On the other hand, first local node of element 2 is global node 2. Therefore, first nodal component of local stiffness matrix of element 2 needs to be added with the first component of node 2 of element 1 in global stiffness matrix. Therefore

$$K\left[1,\,4\right]= K_{11}^{1,12} + K_{11}^{2,11} \tag{10.111}$$

$$K\left[1,\,5\right]= K_{11}^{1,13} + K_{11}^{2,12} \tag{10.112}$$

and so on. Evaluation and assembly of global stiffness matrix are illustrated in Fig. 10.8b.

10.8.4 Necessary Variables and Flowchart

The complete process for SEM described above could be difficult to implement into a computer code without the visualization of the flow of information in the program. Hence, Fig. 10.9 shows a flowchart of a generalized simple SEM problem where multiple variables are implemented. A list of variables with their respective meaning and their information identifier marked in Fig. 10.9 is presented in Table 10.3.

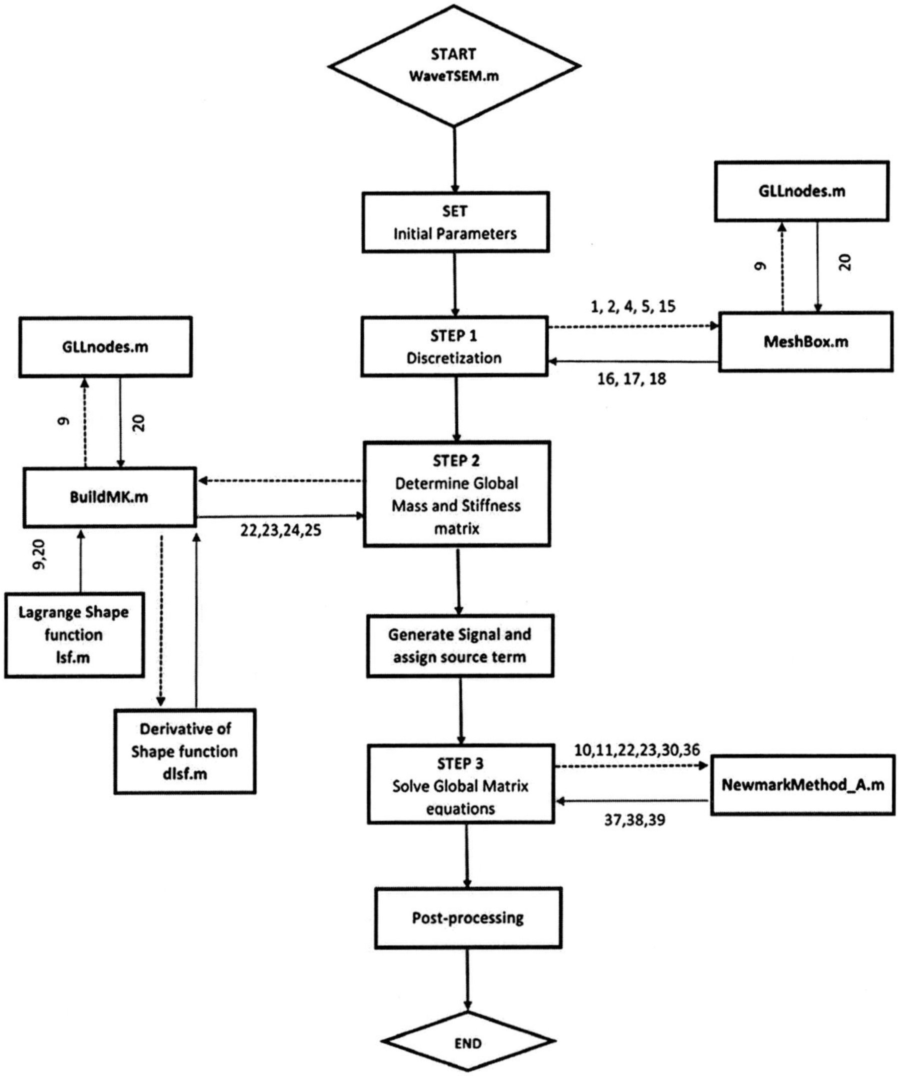

FIGURE 10.9 Algorithm for solving wave propagation problem using SEM.

TABLE 10.3
List of Variables Used to Write SEM Code

Parameters/Variables Description	Name	Identifier
Geometry Length in *x*-direction (in m)	LX	1
Geometry Length in *y*-direction (in m)	LY	2
Geometry Length in *z*-direction (in m)	LZ	3
Number of elements in *x*-direction	NELX	4
Number of elements in *y*-direction	NELY	5
Density (kg/m^3)	rho	6
Density × *z*-direction length	mu	7
Stiffness (GPa)	C	8
Degree of polynomial	P	9
Time step (nano-second)	dt	10
Total number of time step	NT	11
STEP 1: Spectral Element Mesh Generation		
Length per element in *x*-direction	dxe	12
Length per element in *y*-direction	dye	13
Total Number of Element	NEL	14
Number of GLL nodes per element	NGLL	15
X coordinate of Global geometry	x	16
Y coordinate of Global geometry	y	17
Local to Global node map	iglob	18
Number of global nodes	nglob	19
Node location in normalized coordinate [xi or eta]	xgll	20
Integration weights using Lobatto quadrature	wgll	21
STEP 2: Determination of Global Mass and Stiffness matrix		
Global Mass matrix	M	22
Global Stiffness matrix	K	23
Global *x*-coordinates in ascending order	x1	24
Global *y*-coordinates in ascending order	y1	25
Values of shape function in xi-coordinate	Sx	26
Values of shape function in eta-coordinate	Sy	27
Derivative of shape function in xi-coordinate	DSx	28
Derivative of shape function in eta-coordinate	DSy	29
Damping matrix in the system	Damping	30
Signal generation and source term assignment		
Central frequency	CentFreq	31
Number of tones burst cycle	NumCycles	32
Total time	TotTim	33
X-coordinate of force location	Fx	34
Y-coordinate of force location	Fy	35
Applied force	F	36
STEP 3: Implicit Solver – Newmark Beta Method		
Displacement	disp	37
Velocity	vel	38
Acceleration	accl	39

10.9 CNDE CASE STUDIES AT LOW FREQUENCIES (<~1 MHz)

In all foregoing discussions, x_1, x_2, and x_3 axes' notations are synonymously used with x, y, and z.

10.9.1 Propagation of Elastic Waves in an Angle Bar [6]

SEM simulation setup: In this simulation, an elastic wave is induced to propagate through an Angle bar made of aluminum alloy with Young's modulus of 72.7 GPa, Poisson's ratio 0.33, and density 2700 kg/m^3. The length, width, and thickness of each bar are 1000 mm, 250 mm, and 10 mm, respectively, as shown in Fig. 10.10. The carrier frequency f_c and modulation frequency f_m of the excitation signal $F(t)$ with an amplitude of 1 N are assumed as 75 kHz and 18.75 kHz, respectively. In total, 1800 spectral elements following the six-mode theory of shells are considered and the Lobatto distribution of nodes within each spectral element is utilized. In this time-domain simulation, the calculation time t_T is divided into 8000 time steps to accommodate 0.5 ms, which allows for multiple reflections of the propagating waves from both ends of the bar.

The results of SEM numerical simulation for the case of the aluminum bar element in the form of three-dimensional snapshots of the displacement with amplitudes $\sqrt{u_x^2 + u_y^2 + u_z^2}$ are presented in Fig. 10.11. In this simulation, mode conversion phenomena at geometrical discontinuity in the form of the middle edge of the angle bar are evidenced. The snapshots at multiple time stamps are presented in Fig. 10.11. Clearly, the initial wave propagating as the A_0 wave mode transmits and reflects from the middle edge of the bar as different wave modes. Due to the geometry of the angle

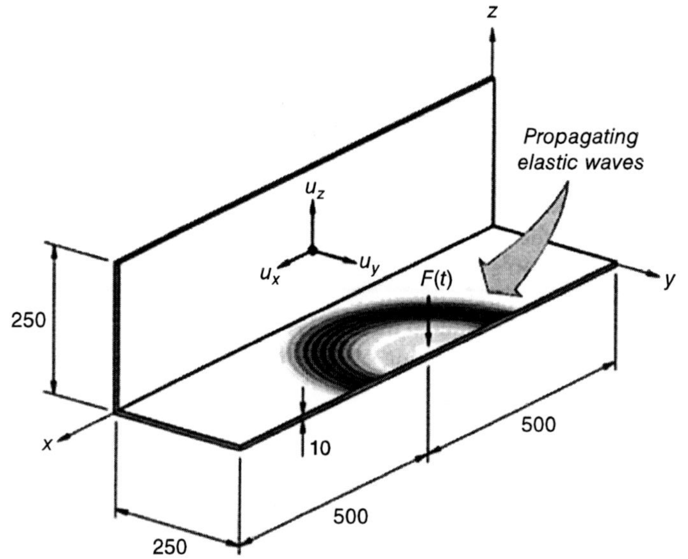

FIGURE 10.10 Geometry of an Aluminium angle bar [Directly from Ref. 6].

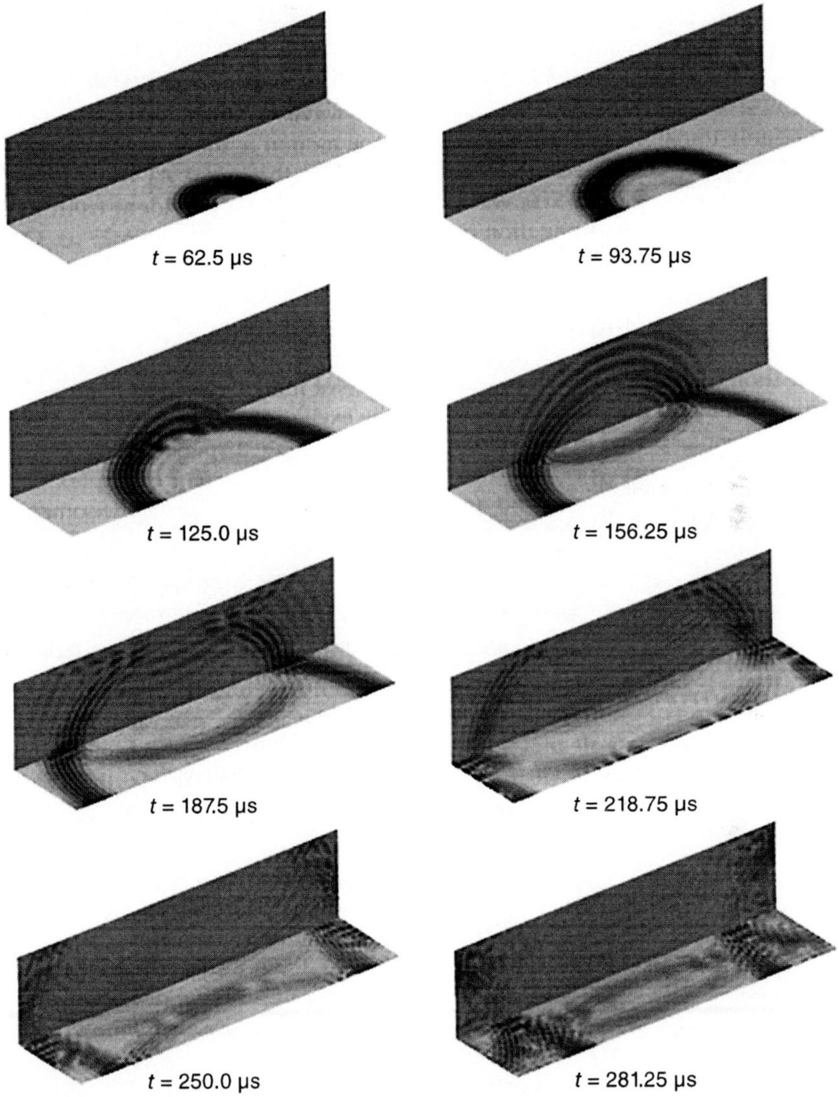

FIGURE 10.11 SEM simulation results to show the snapshots of wave propagation patterns in an Aluminium angle bar at multiple time stamps according to the six-mode theory of shells. Based on Ostachowicz et al. (2012).

bar, the wave motion at its middle edge is clearly transformed from the out-of-plane motion to the in-plane motion, which results in a wave mode conversion from the initial A_0 wave mode to the resulting S_0 and SH_0 wave modes. This can be seen from the snapshots taken at 125 μs and 156.25 μs. Moreover, the initial A_0 wave mode is back-reflected without any change in its motion type, as two out-of-plane wave modes A_0 and SH_1. For the understanding of typical guided wave modes in isotropic plate like structures, please refer Chapters 5 and 11. This can be seen from the snapshots taken at 187.5 μs and 218.75 μs. Therefore, it should be emphasized that the presence of mode conversion phenomena depends on the frequency range of the excitation signal, which is always interpreted in conjunction with appropriate dispersion curves.

10.9.2 Propagation of Elastic Waves in a Half-pipe, Aluminum Shell Structure [6]

In this simulation, the elastic wave is propagated through a half-pipe made of aluminum alloy with Young's modulus of 72.7 GPa, Poisson's ratio 0.33, and density 2700 kg/m³. The length, circumferential width, and thickness of the half-pipe are 1000 mm, 500 mm, and 10 mm, as shown in Fig. 10.12. A total of 16 actuators and 16 sensors are placed on the two opposing ends of the half-pipe and these transducers are made of PZT-5H, which are located in the circumferential direction. The dimensions of each transducer are 16.12 mm in circumferential direction, 16.67 mm in x-direction, and 1 mm in thickness direction. Spectral element modeling has been utilized with three nodes in the thickness direction for both main and piezoelectric structures. Additionally, six nodes are used in each of the remaining directions, spaced according to the Lobatto node distribution. Thus, each three-dimensional spectral element consists of nine nodes. Circumferentially originated, a crosswise, through-thickness, open crack with a length of 15 mm is placed at the midsection of the half-pipe. This crack is modeled by disconnecting the appropriate nodes in adjacent solid spectral elements. All the actuators are excited by applying a voltage $V(t)$ to the surfaces with a central frequency f_c and modulation frequency f_m of 75 kHz and 18.75 kHz, respectively, with an amplitude of 100 V. These values are equivalent to the range of 0.375 MHz.mm to 1.125 MHz.mm in the dispersion curve for the velocity ratio of c_g/c_p. In this time domain simulation, the calculation time $t_T = 0.5$ ms

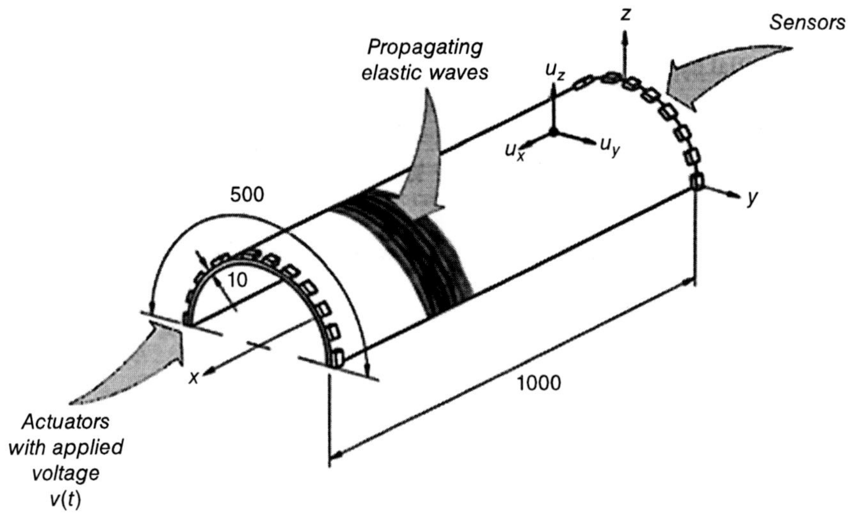

FIGURE 10.12 Geometry of a three-dimensional Aluminum half-pipe with piezoelectric transducers. Based on Ostachowicz et.al. (2012).

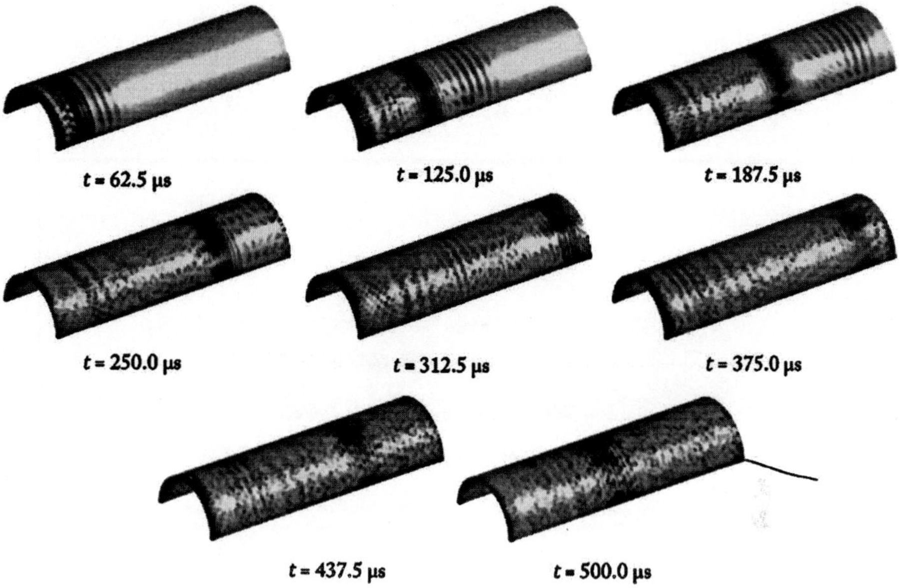

FIGURE 10.13 Results of wave propagation patterns in a cracked Aluminium half-plate shell at multiple time stamps according to nine-mode theory of solids. Based on Ostachowicz et al. (2012).

is divided into 8000 time steps, which allows multiple reflections of the propagating waves to be simulated from all edges of the three-dimensional shell.

The snapshots of numerical simulations shown in Fig. 10.13 are the representations of the three-dimensional frames of displacement amplitudes $\sqrt{u_x^2 + u_y^2 + u_z^2}$. It is clearly evidenced that the piezoelectric transducers induce a continuous front of the wave mode S_0, with the wave mode A_0 propagating behind. The induced wave front is consistent with a phased-array transducer system. The continuity of the S_0 wave mode front guarantees that the distance between the actuator centers is less than the wavelength. However, the analogous condition is not met for the A_0 wave mode, which demonstrates the waves reflecting among actuators and causing additional interactions behind the main wave front. As for simulating cracks, each time an incoming wave encounters with any kind of geometrical discontinuity of the pipe, its reflections are multimodal in character. It can be noted that multiple wave reflections among piezoelectric transducers make mode conversions difficult to identify and thus wave reflections from the damage sites are less noticeable.

10.10 CNDE CASE STUDIES AT HIGH FREQUENCIES (>~1 MHz)

High-frequency ultrasonic is used in traditional ultrasonic inspection with pulse-echo setup. Similar setup is further considered to be implemented using SEM. It is realized from the discussions in previous sections that SEM to solve wave propagation problem requires setting up the initial parameters based on the material properties and applied force with specific frequency range. In this study, beyond the capability of solving wave propagation in isotropic material, wave propagation simulation in anisotropic composite

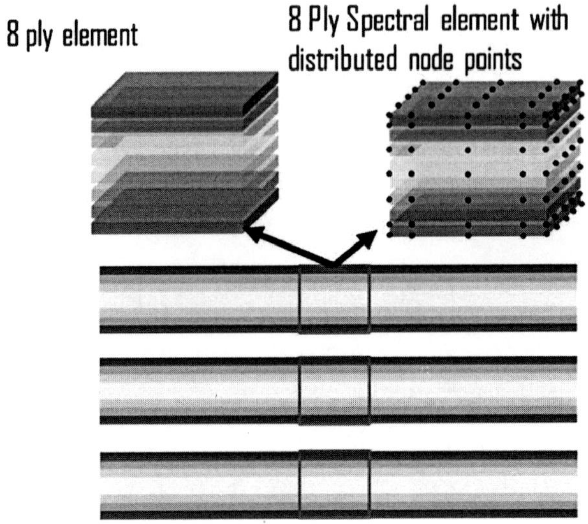

FIGURE 10.14 8-ply symmetric laminated composite is illustrated for simulation and discretization purpose.

material is demonstrated. The geometric dimension of the problem is assumed as 8-ply composite structure (Fig. 10.14) having a dimension of 50 mm × 20 mm × 10 mm. The material properties are written below (reiterating the material properties of a transversely isotropic materials written in Eq. (4.102)) with density: 1560 kg/m^3. Here, the fiber direction of the composite is assumed to be along x_1 axis or x axis.

$$\text{Elasticity matrix: } \mathbb{C} = \begin{bmatrix} 143.8 & 6.2 & 6.2 & 0 & 0 & 0 \\ & 13.3 & 6.5 & 0 & 0 & 0 \\ & & 13.3 & 0 & 0 & 0 \\ & & & 3.4 & 0 & 0 \\ & Sym & & & 5.7 & 0 \\ & & & & & 5.7 \end{bmatrix} Gpa$$

Further other simulation parameters are set as follows:

Order of the Lobatto polynomial: 5 in all directions
Order of the shape function: 5 in all directions
Location of point source: 25 mm × 10 mm × 10 mm
Type of loading: Tone burst signal with central frequency 1 MHz, 5 MHz and 7.5 MHz with 5 cycles and unit amplitude. Fig. 10.15 shows a typical 1 MHz tone burst signal.

10.10.1 Pulse-Echo Simulation at 1 MHz

Discretized area using SEM mesh and the point of signal application are shown in Fig. 10.16.

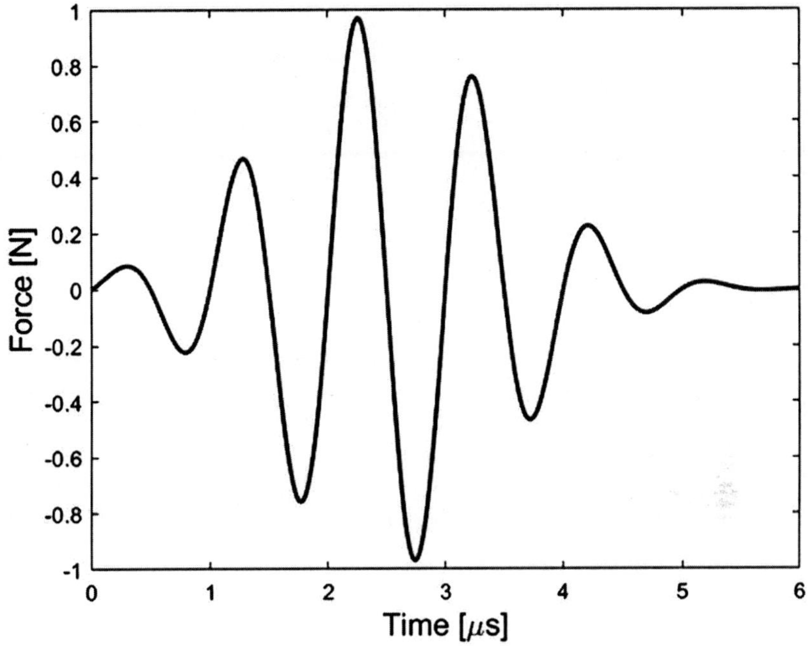

FIGURE 10.15 5 count tone burst with central frequency of 1 MHz.

After successful translation of SEM concept into computer codes, various simulations were performed by changing the geometry and simulation parameters. Time history signals are collected at multiple configurations and presented herein. First, simulation results of a two-layered composite plate while excited with a 1 MHz tone burst signal are presented. Afterward, layers are added to the composite plate and results with four-layered and eight-layered composite plates are presented. Figs. 10.17, 10.18, and 10.19 show the time domain signals obtained from 0-90 2 layers plate, $(0/90)_2$ 4 layers plate, $(0/90)_4$ 8 layers plate. Please note, here layers do not refer the

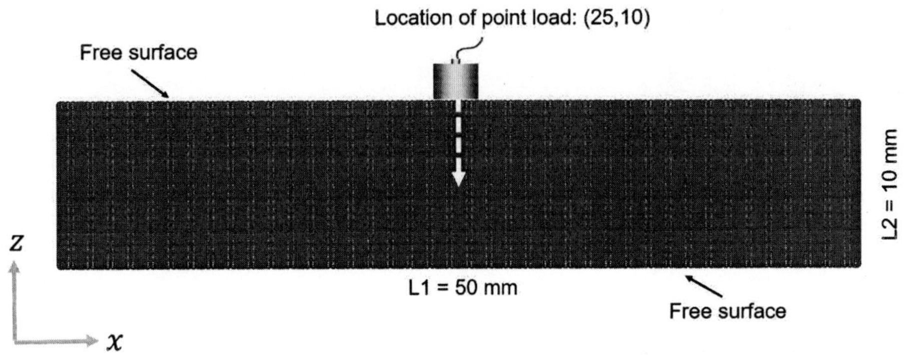

FIGURE 10.16 Discretized SEM domain with the location of applied point force.

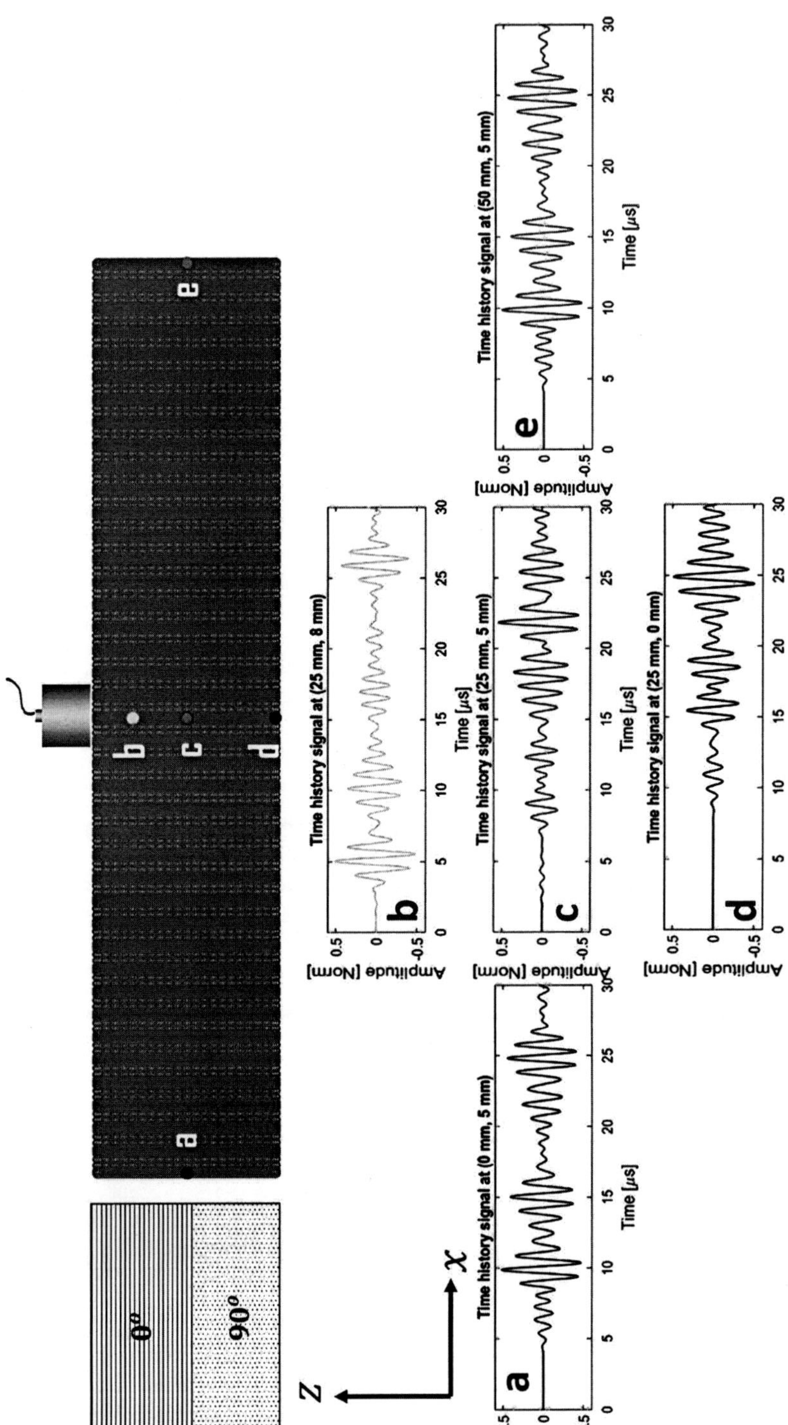

FIGURE 10.17 (top) Discretized domain of two-layered [0, 90] composite plate with five sensing points. (bottom five) Time history signals of respective sensing points.

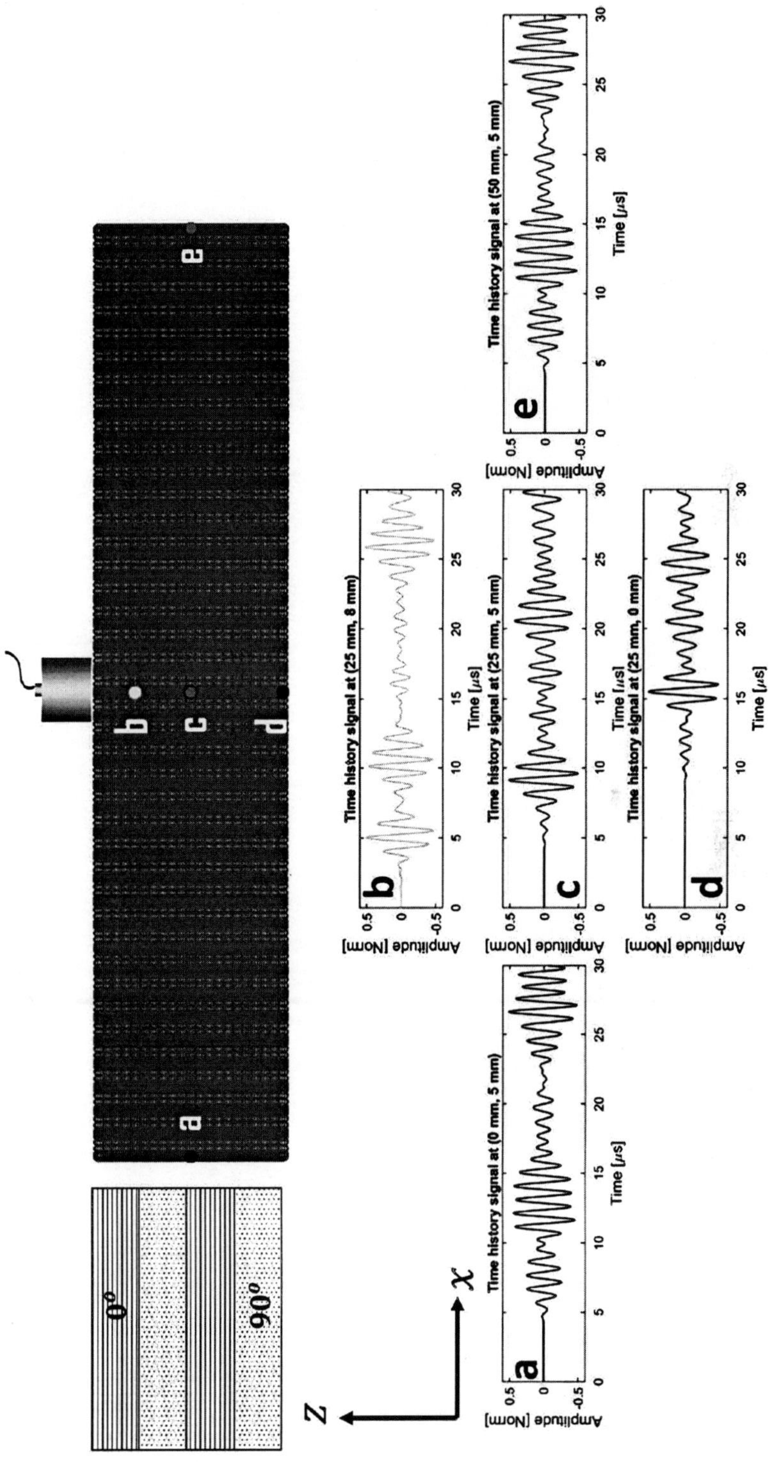

FIGURE 10.18 Time history signals at the sensing points of a four-layered [0, 90, 0, 90] composite specimen.

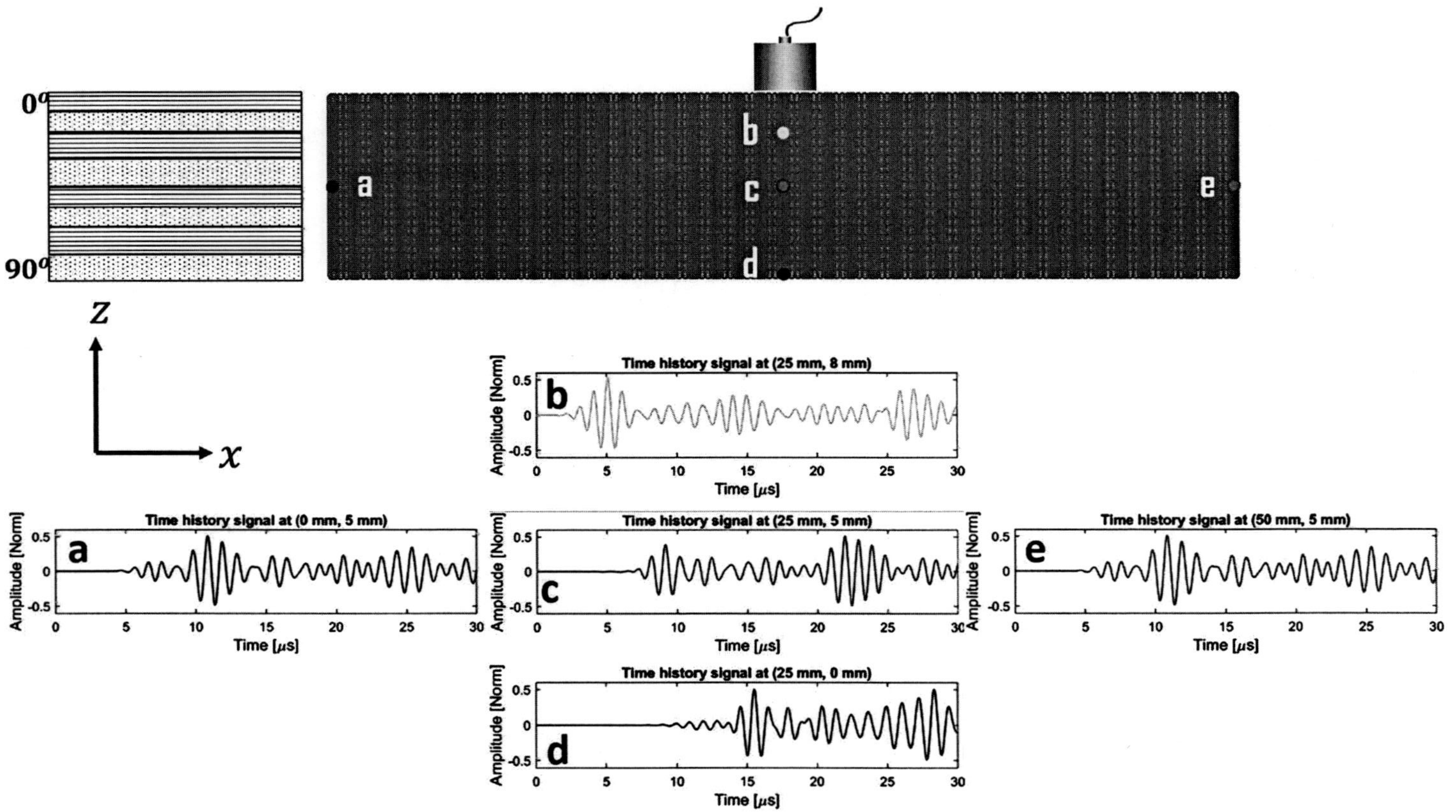

FIGURE 10.19 Time history signals at the sensing points of eight-layered $[0, 90]_4$ composite specimen.

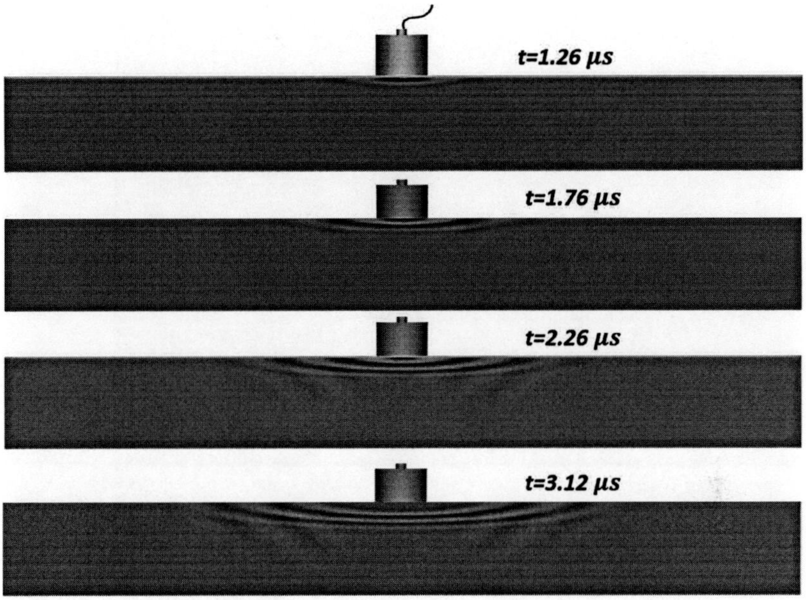

FIGURE 10.20 Full wavefield view of a pulse-echo NDE experiment at 1 MHz on a composite specimen from SEM simulation.

lamina but a cluster of laminae with similar fiber orientation. To check if SEM correctly captured the physics of wave propagation and simulated the obvious phenomena in time domain signals, the time of arrival of the wave from the incident point to the points, a, b, c, d, and e are evaluated and comparted. It is found that the time of arrival of the wave at b, c, and d gradually increased in all material type. This is obvious because the wave has traveled longer distance from the point of actuation to b and d. As the distance between the actuation point and the points a and e is equal, as they are symmetrically placed on either side of the center line, the wave should reach at the same time point. Which is in fact true from the signals received through SEM simulations for all the material types. Further, a 100 mm × 10 mm thick plate is simulated to see the full view of the propagating wavefield v in a 1-layer composite plate. Total 160 SEM elements were used along x and y directions and 40 elements were used along the z direction. Like before, 5th-order polynomial was used in the elements. Total 5.84 μs was simulated. Cross section of the wavefield on a central $x - z$ plane is presented in Fig. 10.20 at time 1.26 μs, 1.76 μs, 2.26 μs, 3.12 μs.

10.10.2 Pulse-echo Simulation at 5 MHz

A similar problem shown in Fig. 10.16 is solved with 5 MHz actuation and the full wavefield from SEM solution is presented in Fig. 10.21 after time 0.51 μs, 0.76 μs, 1.01 μs, 1.26 μs, 1.51 μs, and 1.76 μs, respectively. This clearly shows that the CNDE method SEM is applicable for simulating high-frequency traditional pulse-echo NDE experiments.

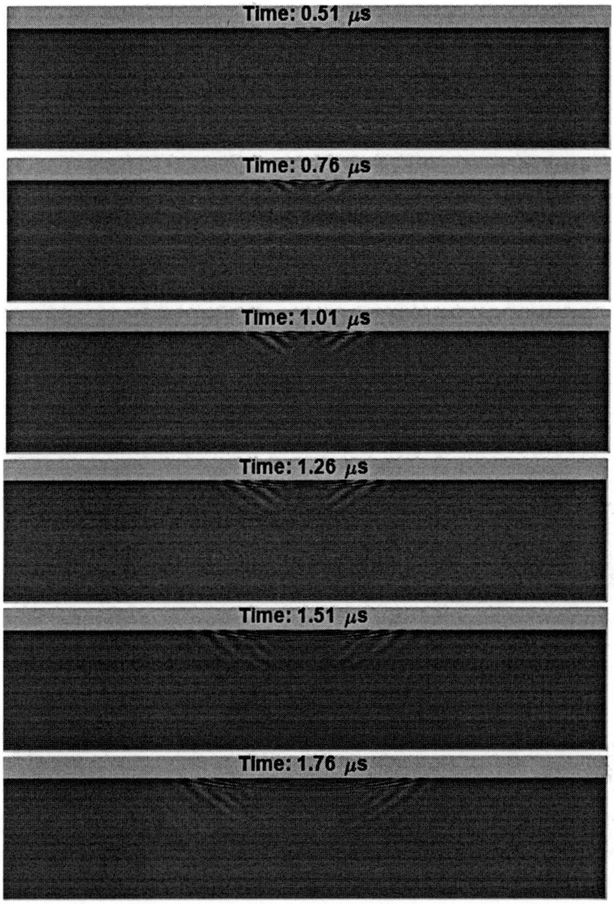

FIGURE 10.21 Full wavefield view of a pulse-echo NDE experiment at 5 MHz on a composite specimen from SEM simulation.

10.11 EXPERIMENTAL VALIDATION

The results obtained from CNDE using SEM simulation require experimental validation to prove its effectiveness. An NDE experiment was performed on a 1.7 mm composite plate made in the laboratory from pre-preg. The orientation of the carbon fibers was $[0, 90]_4$. This plate was manufactured from pre-preg using hot-press machine available at McNair center of University of South Carolina. The dimension of this plate was 12 in × 12 in. A center point was identified, which was excited by a 5-count tone burst signal with a central frequency of 1 MHz. The experimental setup is shown in Fig. 10.22. The input amplitude of the tone burst signal was set to 100 VPP. The through thickness transmitted signal was obtained from a 1 MHz contact transducer shown in Fig. 10.22. The input and sensing transducers have 25 mm element diameter. In order to validate the effectivity of SEM formulation, a simulation

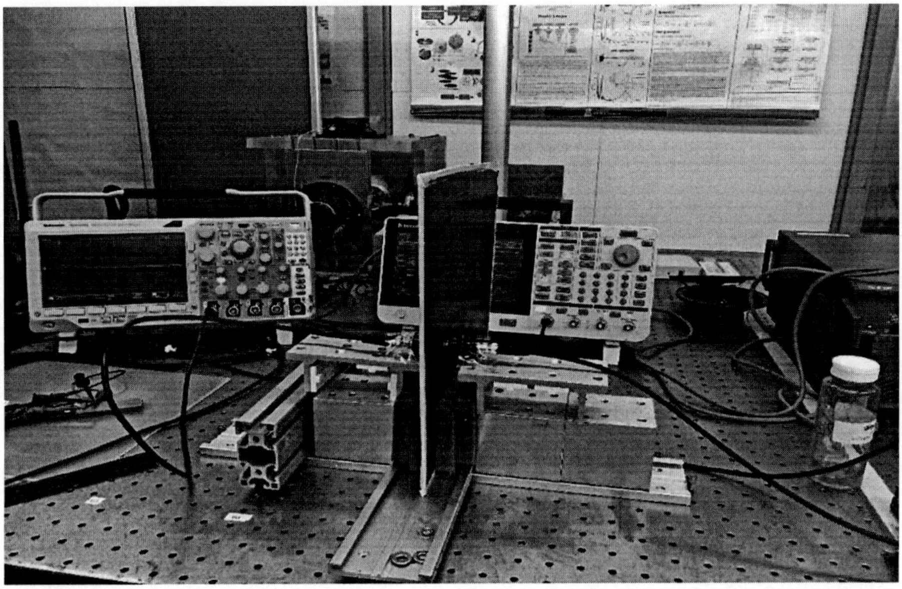

FIGURE 10.22 Experimental setup to acquire NDE signal through a 1.7 mm thick composite plate using through transmission mode.

was performed in line with the specification followed for experimental setup. In this simulation, the dimension of the plate was considered as 100 mm × 1.7 mm × 10 mm. The excitation signal of 5 count tone burst with a central frequency of 1 MHz was applied over a line of 12.5 mm. Force boundary condition was provided only to the nodal points that are present within the area under the 25 mm diameter transducer. The output or sensing signal was collected from a single point from the bottom of the plate "d" point as shown in Fig. 10.19. The simulation and experimental signal were then superimposed to each other as shown in Fig. 10.23. It can be noted that the SEM simulation result partially followed the experimental results. The first arrival of the wave packet is matched when normalized experimental and simulated results are compared. However, later part of the signal did not match accurately and can be eliminated by improving the model in the future versions. The possible reasons for the partial matching are as follows:

- SEM simulation does not consider damping in it, which is evident from the SEM results and multiple vibrations of the sensing point are observed. On the other hand, the experimental signal contains damping in it, which damped out the later part of the signal. The part of the signal, which has large damping effect, is colored gray in Fig. 10.23.
- In SEM simulation, reflecting boundary conditions are not assumed. For this reason, the reflections from the boundary are evident form the SEM signal. However, the experimental plate was large enough to avoid the boundary reflections.

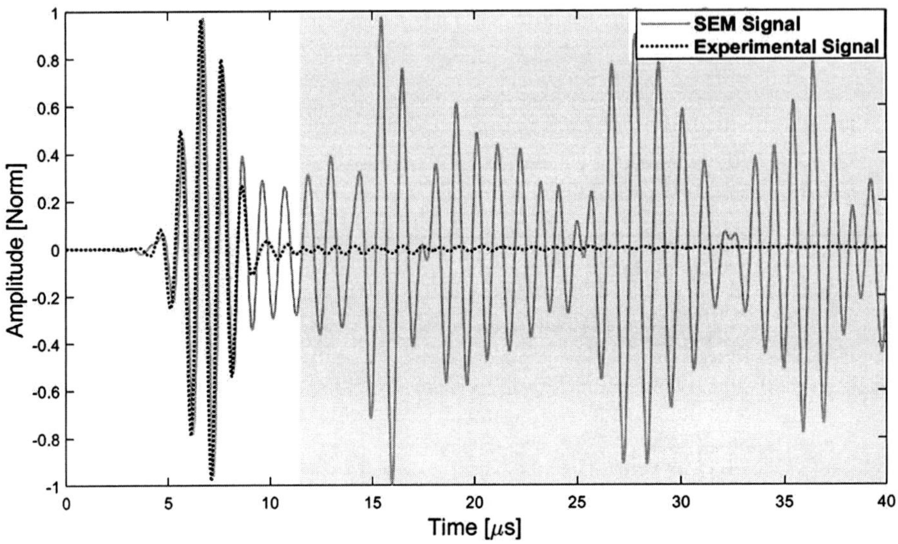

FIGURE 10.23 Comparison of experimental and SEM simulation results excited with a 5-count tone burst signal with a central frequency of 1 MHz.

- The finite excitation regions in SEM simulations were half of the experimental length of the simulation. Moreover, in experimental setup, the excitation was of a circular shape, which was absent in SEM configurations. All these contributions could be the reasons for the partial mismatch of the SEM results with the experimental data.
- Currently, the SEM code is being developed to adopt the experimental configurations, which will simulate the composite structures with considerable accuracy. However, such code needs high computing capability. However, if solved, an adequate extension of the computer code with parallelization would be the key to develop highly accurate CNDE predictive tool.

10.12 APPENDIX

10.12.1 Electrical Boundary Conditions for Piezoelectric Crystal

The basic equations of motion of a piezoelectric element need mechanical and electrical boundary conditions. The reason to impose electrical boundary conditions is that the matrix of dielectric constants $K_{\emptyset\emptyset}$ is not positively definite. Depending on the type of electrical boundary conditions in static condensation, the representative submatrices of matrices $K_{\emptyset\emptyset}$ and $K_{\emptyset u}$ are taken into account. In the case of wave propagation in solids, a piezoelectric transducer is assumed to be attached to the structure surface and electrodes to be located on its upper and lower surfaces. Based on this configuration, three different types of electrical boundary conditions could be considered [6].

10.12.1.1 Condition 1: Piezoelectric Sensor in Closed Circuit

When a piezoelectric material experiences deformation, an electric charge gathers on its electrodes. Therefore, closing the electric circuit causes the charge to flow away from the top and bottom surfaces of the piezoelectric transducer. At the same time, it is assumed that there is no free electric charge inside the piezoelectric material. In other words, the electric potential in nodes on the upper and lower surfaces of a solid spectral finite element can be assumed to be zero (grounding), while in the other nodes of the element, the electric potential is induced. Therefore, the charge equation for a sensor in a closed circuit can be written as

$$\begin{bmatrix} K_{\varnothing u}^{00} & K_{\varnothing u}^{0i} & K_{\varnothing u}^{0n} \\ K_{\varnothing u}^{i0} & K_{\varnothing u}^{ii} & K_{\varnothing u}^{in} \\ K_{\varnothing u}^{n0} & K_{\varnothing u}^{nn} & K_{\varnothing u}^{ni} \end{bmatrix} \begin{Bmatrix} \hat{u}^0 \\ \hat{u}^i \\ \hat{u}^n \end{Bmatrix} + \begin{bmatrix} K_{\varnothing\varnothing}^{00} & K_{\varnothing\varnothing}^{0i} & K_{\varnothing\varnothing}^{0n} \\ K_{\varnothing\varnothing}^{i0} & K_{\varnothing\varnothing}^{ii} & K_{\varnothing\varnothing}^{in} \\ K_{\varnothing\varnothing}^{n0} & K_{\varnothing\varnothing}^{nn} & K_{\varnothing\varnothing}^{ni} \end{bmatrix} \begin{Bmatrix} 0 \\ \hat{\varnothing}_c^i \\ 0 \end{Bmatrix} = \begin{Bmatrix} \hat{q}^0 \\ 0 \\ \hat{q}^n \end{Bmatrix} \tag{A.10.1}$$

where the upper indices denote degrees of freedom of the nodes of a solid spectral element such that *0*, *i*, and *n* indicate bottom, middle, and top layers, respectively. It can be noted that the matrices and vectors of above equation have been formulated for a spectral element and therefore an index "*e*" has been omitted. The induced potential can be calculated by solving this equation and we can get:

$$\left\{\hat{\varnothing}_c^i\right\} = -\left[K_{\varnothing\varnothing}^{ii}\right]^{-1}\left[\begin{matrix} K_{\varnothing u}^{i0} & K_{\varnothing u}^{ii} & K_{\varnothing u}^{in} \end{matrix}\right] \begin{Bmatrix} \hat{u}^0 \\ \hat{u}^i \\ \hat{u}^n \end{Bmatrix} \tag{A.10.2}$$

According to static condensation equations, the electrically induced stiffness of a sensor in the closed circuit K_I^c can be written as

$$K_I^c = \begin{bmatrix} K_{\varnothing u}^{i0} \\ K_{\varnothing u}^{ii} \\ K_{\varnothing u}^{in} \end{bmatrix} \left[K_{\varnothing\varnothing}^{ii}\right]^{-1}\left[\begin{matrix} K_{\varnothing u}^{i0} & K_{\varnothing u}^{ii} & K_{\varnothing u}^{in} \end{matrix}\right] \tag{A.10.3}$$

10.12.1.2 Condition 2: Piezoelectric Sensor in an Open Circuit

It has been noted that an electrical charge gathers on electrodes of a piezoelectric material as a result of its deformations. On the other hand, the electric potential on the bottom surface of a sensor is assumed to be zero. Due to this grounding, the electric potential is induced in the other nodes of a solid spectral finite element. According to basic equations of motion, the charge equation for a sensor working in an open circuit can be written as

$$\begin{bmatrix} K_{\varnothing u}^{00} & K_{\varnothing u}^{0i} & K_{\varnothing u}^{0n} \\ K_{\varnothing u}^{i0} & K_{\varnothing u}^{ii} & K_{\varnothing u}^{in} \\ K_{\varnothing u}^{n0} & K_{\varnothing u}^{nn} & K_{\varnothing u}^{ni} \end{bmatrix} \begin{Bmatrix} \hat{u}^0 \\ \hat{u}^i \\ \hat{u}^n \end{Bmatrix} + \begin{bmatrix} K_{\varnothing\varnothing}^{00} & K_{\varnothing\varnothing}^{0i} & K_{\varnothing\varnothing}^{0n} \\ K_{\varnothing\varnothing}^{i0} & K_{\varnothing\varnothing}^{ii} & K_{\varnothing\varnothing}^{in} \\ K_{\varnothing\varnothing}^{n0} & K_{\varnothing\varnothing}^{nn} & K_{\varnothing\varnothing}^{ni} \end{bmatrix} \begin{Bmatrix} 0 \\ \hat{\varnothing}_c^i \\ \hat{\varnothing}_0^n \end{Bmatrix} = \begin{Bmatrix} \hat{q}^0 \\ 0 \\ 0 \end{Bmatrix} \tag{A.10.4}$$

Therefore, the induced potential can be obtained by solving the below equation:

$$\begin{Bmatrix} \hat{\varnothing}_0^i \\ \hat{\varnothing}_0^n \end{Bmatrix} = -\begin{bmatrix} K_{\varnothing\varnothing}^{ii} & K_{\varnothing\varnothing}^{in} \\ K_{\varnothing\varnothing}^{ni} & K_{\varnothing\varnothing}^{nn} \end{bmatrix}^{-1} \begin{bmatrix} K_{\varnothing u}^{i0} & K_{\varnothing u}^{ii} & K_{\varnothing u}^{in} \\ K_{\varnothing u}^{n0} & K_{\varnothing u}^{ni} & K_{\varnothing u}^{nn} \end{bmatrix} \begin{Bmatrix} \hat{u}^0 \\ \hat{u}^i \\ \hat{u}^n \end{Bmatrix} \tag{A.10.5}$$

According to static condensation equations, the electrically induced stiffness of a sensor in the open circuit K_I^0 can be written as

$$K_I^0 = \begin{bmatrix} K_{\varnothing u}^{i0} & K_{\varnothing u}^{n0} \\ K_{\varnothing u}^{ii} & K_{\varnothing u}^{ni} \\ K_{\varnothing u}^{in} & K_{\varnothing u}^{nn} \end{bmatrix} \begin{bmatrix} K_{\varnothing\varnothing}^{ii} & K_{\varnothing\varnothing}^{in} \\ K_{\varnothing\varnothing}^{ni} & K_{\varnothing\varnothing}^{nn} \end{bmatrix}^{-1} \begin{bmatrix} K_{\varnothing u}^{i0} & K_{\varnothing u}^{ii} & K_{\varnothing u}^{in} \\ K_{\varnothing u}^{n0} & K_{\varnothing u}^{ni} & K_{\varnothing u}^{nn} \end{bmatrix} \tag{A.10.6}$$

10.12.1.3 Condition 3: Actuator

In case of a piezoelectric actuator, the electrical energy is being converted to mechanical energy. Therefore, the electric charge is applied to the electrodes of the piezoelectric actuator. In such case, the voltage is assumed to be supplied to the top electrode, while the electric potential on the bottom surface of the piezoelectric transducer is assumed to be zero. Thus, the electric potential is induced in nodes of a solid spectral finite element between the top and bottom layers of nodes. According to basic equations of motion, the charge equation for the actuator can be written as:

$$\begin{bmatrix} K_{\varnothing u}^{00} & K_{\varnothing u}^{0i} & K_{\varnothing u}^{0n} \\ K_{\varnothing u}^{i0} & K_{\varnothing u}^{ii} & K_{\varnothing u}^{in} \\ K_{\varnothing u}^{n0} & K_{\varnothing u}^{nn} & K_{\varnothing u}^{ni} \end{bmatrix} \begin{Bmatrix} \hat{u}^0 \\ \hat{u}^i \\ \hat{u}^n \end{Bmatrix} + \begin{bmatrix} K_{\varnothing\varnothing}^{00} & K_{\varnothing\varnothing}^{0i} & K_{\varnothing\varnothing}^{0n} \\ K_{\varnothing\varnothing}^{i0} & K_{\varnothing\varnothing}^{ii} & K_{\varnothing\varnothing}^{in} \\ K_{\varnothing\varnothing}^{n0} & K_{\varnothing\varnothing}^{nn} & K_{\varnothing\varnothing}^{ni} \end{bmatrix} \begin{Bmatrix} 0 \\ \hat{\varnothing}_A^i \\ \hat{V}_A^n \end{Bmatrix} = \begin{Bmatrix} \hat{q}^0 \\ 0 \\ \hat{q}^n \end{Bmatrix} \tag{A.10.7}$$

where $\hat{V}_A^n$ denotes the voltage vector in nodes corresponding to the top surface of a solid spectral finite element. Therefore, the induced potential can be calculated by solving the below equation:

$$\left\{\hat{\varnothing}_A^i\right\} = -\left[K_{\varnothing\varnothing}^{ii}\right]^{-1} \left[K_{\varnothing u}^{i0} \; K_{\varnothing u}^{ii} \; K_{\varnothing u}^{in} \right] \begin{Bmatrix} \hat{u}^0 \\ \hat{u}^i \\ \hat{u}^n \end{Bmatrix} - \left[K_{\varnothing\varnothing}^{ii}\right]^{-1} \left[K_{\varnothing\varnothing}^{in}\right] \left\{\hat{V}_A^n\right\} \tag{A.10.8}$$

According to static condensation equations, the electrically induced stiffness of the actuator K_I^A can be written as

$$K_I^A = \begin{bmatrix} K_{\varnothing u}^{i0} \\ K_{\varnothing u}^{ii} \\ K_{\varnothing u}^{in} \end{bmatrix} \left[K_{\varnothing\varnothing}^{ii}\right]^{-1} \left[K_{\varnothing u}^{i0} \; K_{\varnothing u}^{ii} \; K_{\varnothing u}^{in} \right] \tag{A.10.9}$$

Considering the electrical boundary conditions, the static condensation equations can be solved in the same fashion as described earlier.

10.13 SUMMARY

In this chapter, classification of spectral method is presented in the context of CNDE problems. Two types of frequency domain and time-domain SEMs are discussed. Predominantly time-domain formulation for wave propagation in any generalized material system is presented. CNDE problem can span at low frequencies (below 750 kHz) representing SHM problems and at high frequencies (> ~1 MHz) representing traditional pulse echo, through transmission or phase-array NDE scenarios. Both low- and high-frequency CNDE problems are presented through time-domain SEM formulation in this chapter.

REFERENCES

1. Patera, A.T., A spectral element method for fluid dynamics – Laminar flow in a channel expansion. *Journal of Computational Physics*, 1984. **54**: pp. 468–488.
2. Chakraborty A., and Gopalakrishnan, S., A spectral finite element model for wave propagation analysis in laminated composite plate. *Journal of Vibration and Acoustics-Transactions of the ASME*, 2006. **128**(4): pp. 477–488.
3. Doyle, J.F., *Wave Propagation in Structures: An FFT Based Spectral Analysis Methodology*. 1989, New York: Springer.
4. Park, I., Kim, T., Lee, U, Frequency domain spectral element model for the vibration analysis of a thin plate with arbitrary boundary conditions. *Mathematical Problems in Engineering*, 2016. **2016**: p. 20.
5. Boyd, J.P., *Chebyshev and Fourier Spectral Methods*, 2nd ed. 2001, Mineola, NY: Dover Publications Inc.
6. Ostachowicz, W.M., Kudela, P., Krawczuk, M., Zak, A, *Guided Waves in Structures for SHM: The Time-Domain Spectral Element Method*. 2012, New York: Wiley.
7. Wang, L.-L., Guo, B., Interpolation approximations based on Gauss–Lobatto–Legendre–Birkhoff quadrature. *Journal of Approximation Theory*, 2009. **161**(1): pp. 142–173.
8. Krishnamoorthy, C.S., *Finite Element Analysis: Theory and Programming*. 2000, New Delhi, India: Tata McGraw Hill.
9. Ostachowicz, W., et al., *Guided Waves in Structures for SHM The time domain Spectral Element Method*. 2012, New York: A John Wiley & Sons. pp. 46–92.
10. Ostachowicz, W.M., Damage detection of structures using spectral finite element method. *Computers and Structures*, 2008. **86**(3–5): pp. 454–462.
11. Chapra, S.C., Canale, R. P., *Numerical Methods for Engineers*, 7th ed. 2010, New York: McGraw-Hill Education.
12. Ostachowicz, W., et al., *Guided Waves in Structures for SHM: The Time-Domain Spectral Element Method*. 2011, New York: John Wiley & Sons.
13. Pozrikidis, C., *Introduction to Finite and Spectral Element Methods using MATLAB*. 2014, New York: Chapman and Hall/CRC.

11 Perielastodynamic Simulation Method for CNDE

11.1 INTRODUCTION

In Chapter 1, Figs. 1.3 through 1.8 several modes of NDE and SHM problems are presented. In Chapters 7 and 10, bulk wave NDE problems are simulated for CNDE to visualize traditional NDE mode with a transducer. Here in this chapter only the SHM problems are simulated replicating the situations presented in Figs. 1.5 and 1.6. In these modes of NDE/SHM guided Lamb wave modes are used for material inspection, damage detection, and material state awareness [1, 2]. This chapter is directly written using the materials presented in Refs. [3, 4] published by the corresponding author. High-frequency ultrasonic actuators and sensors are strategically mounted on the plate-like structure to detect, localize, and characterize the damages [5] as discussed in Chapters 1 and 5. First symmetric (S_0) and antisymmetric (A_0) Rayleigh-Lamb wave modes discussed in Fig. 5.5 in Chapter 5 are predominantly used in guided wave-based NDE/SHM. While traveling through a structure these wave modes interact with the structure boundaries, discontinuities, inclusions, etc. [6]. Sometimes due to these multiple interaction the guided wave modes are subjected to mode conversion. The guided wave signals that are received at the sensor location play a critical role in quantifying the damages, if any in the structures. Considering the most practical scenario, the damage states in the materials are unknown, and the sensor signals are the only observables. Therefore, as discussed before, there could be infinite possibilities of damage states in the material and it is impossible to experimentally obtain the understanding of the sensor signals due to the varying damage states. This was the primary motivation for CNDE as discussed in Chapter 1. Thus recently, computational NDE and SHM [7–9] have gained enormous popularity. Like it is discussed in this book in several chapters, a number of numerical techniques such as, finite element method (FEM) [10], boundary element method (BEM) [11], mass-spring lattice model (MSLM) [12], finite difference method (FDM) [13], finite strip method [14, 15], cellular automata [16, 17], and elastodynamic finite integration technique (EFIT) [18] (Chapter 8) were developed. While these techniques can predict the sensor signals with a considerable accuracy, fine discretization in spatial and time domains makes them computationally expensive. To overcome this issue, a few semi-analytical techniques, such as distributed point source method (DPSM) [19] (Chapter 7), local interaction simulation approach (LISA) [8, 20, 21] (Chapter 9), spectral finite element method (SEM) [22–24] (Chapter 10), and semi-analytical finite element (SAFE) [25] (Chapter 5) methods were developed to reduce computational burden while increased accuracy. DPSM as discussed in Chapter 7 is a meshless semi-analytical method which requires displacement and stress Green's functions in the problem formulation. In this chapter, a newly formulated technique, Perielastodynamics is presented as an alternative approach to simulate the wave propagation in three dimensional (3D). The reason for incorporating Perielastodynamics as one of the selected methods in this book, is that by changing the boundary conditions virtual wave propagation can be studied while the material is still under operation or loaded. Complementary to the existing methods like DPSM, EFIT, LISA, and SEM, presented in previous chapters, Perielastodynamics can be used to predict both the damage growth as well as the wave propagation signals, simultaneously. In Perielastodynamics, damage detection and characterization can be performed while the damage is still growing, without altering any meshing or discretization keeping the same parent model. This would not only be impossible using FEM but would be equally impossible by the newer models like DPSM, EFIT, LISA, and SEM. Hence, Perielastodynamics could be one of the top candidates to be considered for wave simulations and CNDE. To visualize the wave damage interaction of a growing or continuously changing damage state inside the material, in perielastodynamic simulation, it is only necessary to modify the damage matrix. Damage propagation in metallic and composite structures [26, 27] was successfully presented by the earlier researchers using peridynamics. Similarly, two dimensional (2D) in plane wave propagation was also simulated [15, 16, 28, 29] using peridynamic theory. However, CNDE of a complete NDE/SHM problem using peridynamic is rare. Simulation of 3D Lamb wave modes in a plate-like structures that are frequently used for NDE/SHM are recently been simulated using perielastodynamic [3]. So far, the simulations are performed using bond-based peridynamic models and thus simulations were only possible for metallic structures. CNDE of composites or wave simulation in any generalized anisotropic media using perielastodynamic requires state-based peridynamic model which is still an open research problem in NDE/SHM.

Classical theory of continuum mechanics is used successfully to solve problems in solid mechanics. The underlying assumption of classical mechanics is that the material body remains continuous before and after deformation. Although, the classical approaches presented in Chapter 3 are successfully used to solve mechanics and elastodynamic problems at macro-scale, it encounters difficulty in solving problems with discontinuities and sharp edges. This is because the classical approach described in Chapter 3 uses the partial differential equation with spatial derivatives of stresses (Eq. (3.49)) which become undefined at the crack locations (gradient of stress-tensor become indeterminate). To overcome these limitations other approaches like linear elastic fracture mechanics [30] and peridynamic approaches were developed. For CNDE peridynamic is more appropriate and the theory is briefly described below.

Peridynamic theory belongs to a class of nonlocal mechanics-based formulation. Peridynamic theory was first developed by Silling et al. [26, 27, 31–34] with Sandia National Laboratory. Since inception, this new theory is being used successfully to understand material behavior at different length scales. The word "peridynamic" was derived from two Greek words which are "Peri" and "Dynamic". In the Greek Language "Peri" means near and "Dynamic" means force. It was proposed that modeling of a continuous body or a material body with discontinuities (e.g., cracks, delaminations, etc.) can be achieved within a single framework of peridynamic theory.

In contrast to the partial differential equation (Eq. (3.49)) derived from classical continuum theory, peridynamic theory utilizes integro-differential equation, which makes the approach suitable for solving crack propagation problem and here we show that based on peridynamic theory newly formulated perielastodynamic method is also suitable for elastodynamic problems. Integro-differential equations from peridynamic approach are inherently valid at the crack surface. In peridynamics, damage parameters are included in the constitutive equations which make the peridynamic approach beneficial over others for modeling crack initiation and crack branching in the material while simulating the wave propagation using perielastodynamic. Hence, in a nutshell it can be said that the integral approach in peridynamic makes it suitable to simulate the damage propagation problem [34] without altering the mesh. Ha and Bobaru [35] studied dynamic crack propagation and crack branching in glass under dynamic loading. Madenci and Oterkus [36] employed peridynamic theory to predict the crack propagation in a composite and metallic plate. Above works are promising but does not present the solution of elastodynamic problem using peridynamic which is necessary to be discussed for CNDE. Although in small numbers, a few articles can be found where solution of wave propagation problems is attempted using peridynamics. Nishawala et al. [16] used bond-based peridynamic theory to simulate Rayleigh wave propagation in a 2D isotropic (CR-39) plate. A short time pulse of a ramp loading was given to generate the surface skimming Rayleigh wave. Hafezi et al. [28, 37, 38] employed peridynamic theory named peri-ultrasound to simulate in plane longitudinal ultrasonic wave in an aluminum plate [15, 29]. In these works, wave propagation was simulated in a one layer of the material points where the out of plane deformations is ignored. Although promising none of these methods could be used for CNDE. In this chapter from the fundamentals of peridynamic theory to the development of perielastodynamic modeling method is described with several technical requirements to properly simulate waves in NDE/SHM scenarios for CNDE problems, which was first discussed in Ref. [3].

11.2 FUNDAMENTAL OF PERIDYNAMIC APPROACH

In peridynamic theory, a material body (in Fig. 11.1) is discretized into a number of material points where each point has finite volume. Interaction between the material points takes place within a finite internal length, which is called the Horizon. Interaction between two material points depends on the material properties, internal length, and relative distance between the particles. Internal length parameter in peridynamic approach is selected based on the nonlocality of the problem. In continuum mechanics, the internal length scale approaches to zero which signifies no interactions of a point with its immediate neighbor. Whereas, in atomistic simulation internal length is selected as interatomic distance. Thus, peridynamic approach can be used to analyze the material behavior across different length scales [33]. Following section introduces the peridynamic theory in more detail. In the following sections vector notations and index notations will be used simultaneously based on the introduction presented in Chapter 2 and readers understanding is assumed herein.

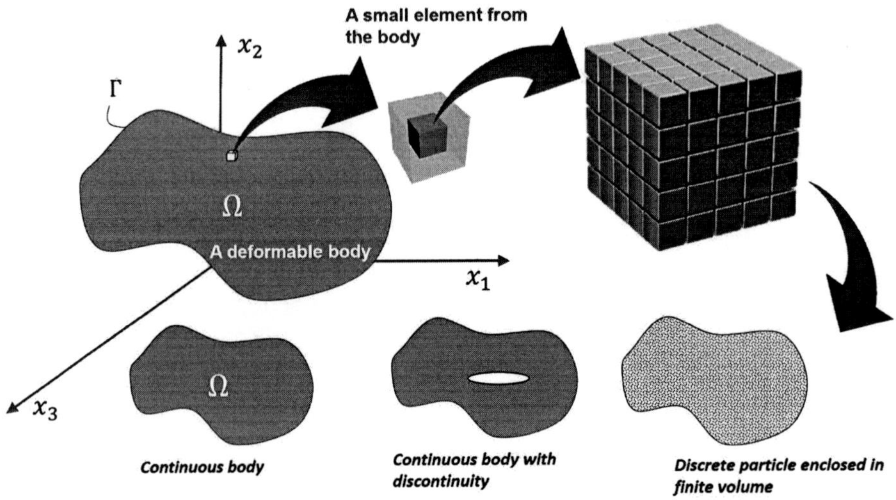

FIGURE 11.1 A continuum body and its peridynamic representation.

11.2.1 Fundamentals of Bond-based Peridynamic Theory

In this section only necessary mathematical formulas for simulating wave propagation are pointed out, whereas detailed discussion on peridynamic can be found in Refs. [26, 27, 31–34]. The equation of motion using bond-based peridynamic theory at point **x** in reference configuration can be written as [27, 39],

$$\rho\ddot{u}(\mathbf{x},t) = \int_H \mathbb{F}(\mathbf{u}'(x_j',t) - \mathbf{u}(x_j,t), \mathbf{P}' - \mathbf{P})dv_{\mathbf{x}'} + b(\mathbf{x},t) \tag{11.1}$$

where H denotes the internal length scale, ρ is the density of the material in reference configuration, **u** is the particle displacement, b is the body force density, $v_{\mathbf{x}'}$ is volume of each material point, and $\mathbb{F}$ is pair wise force function. A pair wise force at any instant can be described by the force exerted by a material point $\mathbf{P}'$ located at $\mathbf{x}'$ on the material point **P** located at **x**. Here it is necessary to refer the discussion presented in Section 3.1 in Chapter 3. In Lagrangian coordinate system (Fig. 3.4a), the deformed material points are described by the reference coordinate system and this description is suitable for peridynamic approach. Based on similar concept presented in Fig. 3.4a a new representative Fig. 11.2 in depicts $d\mathbf{P} = \xi$ and $d\mathbf{p} = \eta$. Relative distance between two material points before and after deformation are expressed by,

$$\xi = \mathbf{P}' - \mathbf{P} \tag{11.2}$$

$$\eta = \mathbf{u}'(x_j',t) - \mathbf{u}(x_j,t) \tag{11.3}$$

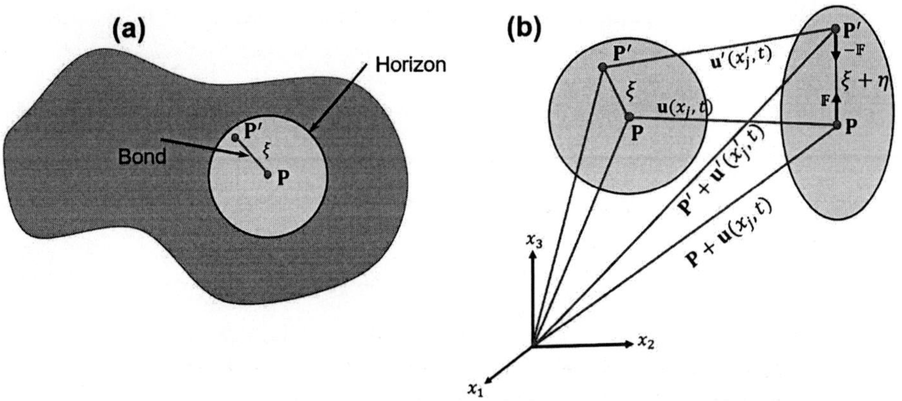

FIGURE 11.2 A schematic showing the force in bond based peridynamic approach and its horizon.

Relative displacement between the two material points in Lagrangian coordinate system is expressed by,

$$(\xi + \eta) = \left(\mathbf{P}' + \mathbf{u}'\left(x_j', t\right)\right) - \left(\mathbf{P} + \mathbf{u}(x_j, t)\right) \tag{11.4}$$

The interaction between two material points $\mathbf{P}'$ located at $\mathbf{x}'$ and $\mathbf{P}$ located at $\mathbf{x}$ happens through a straight line connecting the two particles. A bond is imagined between these two material points. The line connecting these two points is called the bond line, along which they exert force on each other. Somehow this concept is synonymous to a spring mass system where two material particles with two local lumped masses and finite volume are attached with a spring element. Hence, it can be imagined that the length between two material particles can get bigger and bigger for two distant particles located further away from each other. All material particles in a body cannot be influenced by all particles in the body. Hence, a restriction should be devised. A question to be answered is, how far a particle will be influenced by a particle of interest through their imaginary attached spring along the bond line. For example, a house of interest in a community with several hundred houses, if suffers a local calamity or damage due to a local event like, fire, gas leakage, water leakage, etc., the immediate neighboring houses get affected by the event. However, the houses that are further away from the house of interest are not affected by the event. Hence, there would emerge an event horizon inside which the houses are affected, but beyond the horizon the houses are unaffected. This is purely a nonlocal effect with a horizon of influence. Similarly, in peridynamic, a certain radius centering the material particle of interest called Horizon H is considered, beyond which the material particles do not interact. Distance between two material points if exceeds this defined Horizon H, the pairwise interaction force function becomes zero, such that [27, 39],

$$\mathbb{F}(\xi, \eta) = 0 \ \forall \eta \ \textit{if} \ |\xi| > H \tag{11.5}$$

The pair wise force function is required to satisfy the conservation of the linear momentum as follows [27, 39],

$$\mathbb{F}(-\xi,-\eta)=-\mathbb{F}(\xi,\eta) \quad \forall\xi,\eta \tag{11.6}$$

Eq. (11.6) also called the linear admissibility condition. To satisfy the conservation of the angular momentum following equation must hold [27, 39],

$$\mathbb{F}(\xi,\eta)\times(\eta+\xi)=0 \quad \forall\xi,\eta \tag{11.7}$$

Eq. (11.7) is called the angular admissibility condition. The pair-wise force vector between two particles acts opposite to each other and is parallel to the current relative position in order to hold the defined conditions depicted in Eqs. (11.6) and (11.7). In peridynamic theory, pair-wise force function is derived from a scalar micro-potential function ϖ, [27, 39] viz.

$$\mathbb{F}(\xi,\eta)=\frac{\partial\varpi}{\partial\eta}(\xi,\eta) \tag{11.8}$$

Micro-potential of the bond is nothing but the strain energy of a single bond. A peridynamic body is said to be microelastic if Eq. (11.8) is satisfied. Total strain energy at a point can be defined by the following equation [27, 39],

$$W_E=\frac{1}{2}\int_H \varpi(\xi,\eta)dV_\xi \tag{11.9}$$

The factor of half is included in the equation because each endpoint of the bond shares half of the strain energy of the bond.

11.2.2 Peridynamic Constitutive Model

In peridynamic and similarly in perielastodynamic, a constitutive equation is necessary along each bond line. The equation will define how these two nonlocal particles will interact between each other. As discussed before, in peridynamic solid in simplest form, two particles will exert force on each other. Hence, generally speaking, force should be equal to the elongation, multiplied with the modulus of the attachment along the bond line. Hence, in peridynamic theory, force should be expressed by stretch of the bond multiplied with the modulus of the bond. Higher the stretch, higher the force generated. However, this expansion cannot be infinite. For example, no spring mass system can take infinite force. If the stretch is higher than a certain defined value, the bond should be considered broken. Thus force in a perielastodynamic model is similarly considered following the previous descriptions in [27, 39] as follows

$$\mathbb{F}(\xi,\eta,t)=\mathbb{C}(\xi)s(\xi,\eta)\frac{\xi+\eta}{|\xi+\eta|}\mu(t,\xi,\eta) \tag{11.10}$$

where $\mathbb{C}(\xi)$ is a bond constant, $s(\xi,\eta)$ is stretch of the bond, and $\mu(t,\xi,\eta)$ is a history-depended binary function. Value of $\mu(t,\xi,\eta)$ either 0 or 1.

$$\begin{aligned} &\mu(t,\xi,\eta)=1 && \text{if } s(t',\xi,\eta)<s_0 \text{ for } 0\le t'\le t \\ &\mu(t,\xi,\eta)=0 && \text{otherwise} \end{aligned} \tag{11.11}$$

s_0 is the critical stretch of the bond beyond which the bond is assumed broken. Bond stretch is defined by,

$$s(\xi,\eta)=\frac{|\xi+\eta|-|\xi|}{|\xi|} \tag{11.12}$$

Local damage index at a material point is defined by [27, 39],

$$\psi(x,t)=1-\frac{\int_H \mu(t,\xi,\eta)dv_\xi}{\int_H dv_\xi} \tag{11.13}$$

11.2.3 Bond Constant Estimation in Isotropic Material

Peridynamic bond constant in an isotropic material can be determined considering a large homogenous body under isotropic expansion as shown in Fig. 11.3. A uniform stretch s is applied to all the material points for uniform expansion of the body. Length of a bond before and after the deformation can be expressed as ξ and $(1+s)\xi$, respectively. Applying Eq. (11.12)

$$\xi+\eta=(1+s)\xi, \qquad \text{where } \eta=s\xi \tag{11.14}$$

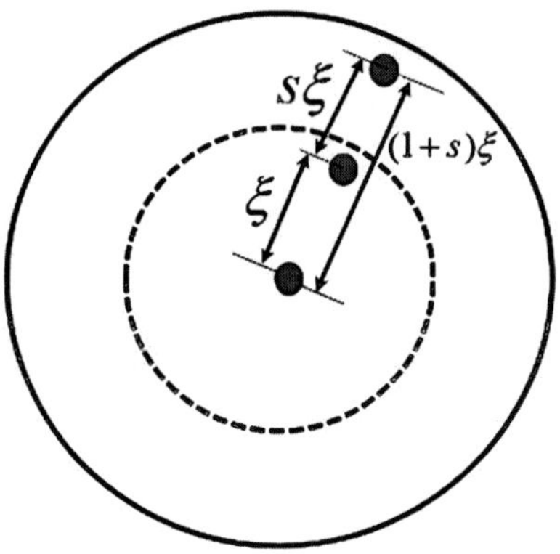

FIGURE 11.3 Homogenous expansion of isotropic material.

Hence, force density vector is expressed as,

$$\mathbb{F} = \mathbb{C}(\xi)s = \frac{\mathbb{C}(\xi)\eta}{\xi} \tag{11.15}$$

Additionally, a micropotential is introduced in Eq. (11.8). Perfroming the integral in Eq. (11.15) the micropotential can takes the follwing from

$$\varpi = \frac{\mathbb{C}(\xi)\eta^2}{2\xi} = \mathbb{C}(\xi)s^2\xi/2 \tag{11.16}$$

Hence, from peridynamic approach, using Eq. (11.9) the starin energy stored at a point can be expressed by,

$$W_E = \frac{1}{2}\int_H \varpi dv_\xi = \frac{1}{2}\int_0^{\mathcal{D}} \frac{\mathbb{C}(\xi)s^2\xi}{2} \int_0^{2\pi}\int_0^{\pi} \xi^2 \sin\theta \, d\theta d\phi d\xi = \frac{\mathbb{C}s^2\pi\mathcal{D}^4}{4} \tag{11.17}$$

In the above expression dv_ξ is an elemental volume of a sphere of radius ξ expressed in Fig. A.3.2 in Chapter 3. The modulus function $\mathbb{C}(\xi)$ in Eq. (11.17) is assumed to be constant and is assumed to be independent of pair wise distance. $\mathcal{D}$ is assumed to be the radius of the sphere that defines the Horizon. Inside which the particle interactions are considered active but beyond the Horizon no particles interact with the particle of interest at the center of the sphere. On the other hand, using simple continuum approach the potential energy without any dissipation stored in the particle should be equal to the strain energy density for the same uniform expansion of the material body and can be written as

$$W_E = \frac{9}{2}\mathbb{K}s^2 \tag{11.18}$$

Next equating the peridynamic strain energy density and the strain energy density obtained from continuum approach the peridynamic modulus can viz.

$$\mathbb{C} = \frac{18\mathbb{K}}{\pi\mathcal{D}^4} \tag{11.19}$$

where $\mathbb{K}$ is the bulk modulus of the material. Following similar equality of the strain energy densities obtained from peridynamic approach and continuum approach, peridynamic modulus or the bond constant in any material body can be found. For example, if a material body is considered thin with three dimensional represented by a finite thickness ($2h$), then the peridynamic modulus or the bond constant in a representative two-dimensional body can be presented as [36].

$$\mathbb{C} = \frac{9\mathbb{E}}{(2\pi h\mathcal{D}^3)} \tag{11.20}$$

For guided wave propagation most of the structures are considered thin with finite thickness and hence, in following perielastodynamic simulation of wave propagation presented in this chapter mostly uses the formula presented in Eq. (11.20).

11.3 FUNDAMENTALS OF PERIELASTODYNAMIC SIMULATIONS

11.3.1 Perielastodynamic Spatial and Temporal Discretization

Numerical simulation of wave propagation in solids comes with specific requirement of discretization of the problem domain. Moreover, perielastodynamic is a temporal computational method and hence, there is additional restriction on the temporal discretization must be enforced. Physical restrictions on spatial and temporal discretization to achieve converged wave simulation results are discussed in Chapters 8 and 9 in detail. Proper spatial and temporal discretization are the critical parameters for the convergence of the solution in wave propagation simulation in perielastodynamic. Resonating with similar restrictions discussed in earlier chapters, briefly the perielastodynamic discretization is discussed herein. Maximum spatial discretization (ΔS), which governs the distance between two nearest material particles in perielastodynamic must meet the criterion below [40],

$$\lambda_{min} = \frac{c_{min}}{2\pi f}\ ;\ \Delta S < \lambda_{min}\,/\,10 \tag{11.21}$$

where λ_{min} is the minimum wavelength appears in the targeted NDE system with different combination of materials. For example, in an NDE problem if a material system is made of two materials with phase wave velocities c_1 and c_2 ($c_1 > c_2$) then $c_{min} = c_2$ should be used to compute the minimum wave length that will appear in the system and may propagate at a maximum frequency f in the system. Hence, c_{min} is the minimum phase wave velocity of the simulated wave modes at the maximum excitation frequency f. For guided waves, the dispersion of phase wave velocities of propagating wave modes as a function of excitation frequency can be easily obtained from the theoretical dispersion curves discussed in Chapter 5. Phase wave velocity in a bulk material can be found using their respective material properties discussed in Chapter 4. If the minimum phase velocity that may appear in an NDE system is known and the minimum wave length in the system can be found with maximum excitation frequency, then the discretization of the problem domain should be such that the distance between two perielastodynamic material points should be less than one-tenth of the minimum wave length as written in Eq. (11.21). Once the spatial discretization is finalized, the temporal discretization should follow the following rule which is also discussed in Chapter 8.

$$\Delta t = \frac{\Delta S}{c_{max}\sqrt{3}} \tag{11.22}$$

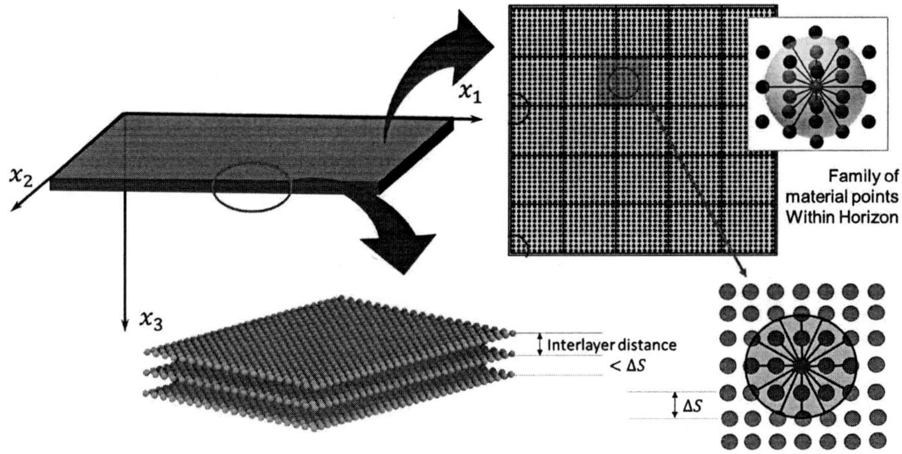

FIGURE 11.4 Discretization of a flat plate and defining the horizon for converged solution.

where c_{max} is the maximum phase wave velocity of the propagating wave mode in the NDE system in consideration. Once the spatial and temporal discretization are finalized, the NDE problem domain must be divided into several layers where, at each layer a set of perielastodynamic material points are placed following the rule in Eq. (11.21). In a three-dimensional material body as shown in Fig. 11.4, each layer of perielastodynamic material points are placed keeping the minimum interlayer distance governed by Eq. (11.21). As shown in Fig. 11.4 each layer has two-dimensional distributions of perielastodynamic material points following the similar rule of 1/10th discussed above.

11.3.2 Numerical Time Integration

In the perielastodynamics approach, the material is spatially discretized into a finite number of material points. Each material point has finite volume in the reference configuration. Material volume with 3D uniform discretization of the grid is calculated by $v_f = dl^3$, where $dl = \Delta S$ is the element length [36, 41]. By replacing the integration in Eq. (11.1) with a finite summation over all the material points inside the Horizon shown in Fig. 11.4, the equation of motion at material point i after the time step n can be expressed as,

$$\rho \ddot{\mathbf{u}}_i^n = \sum_{N_f} \mathcal{F}(\mathbf{u}_f^n - \mathbf{u}_i^n, \mathbf{x}_f - \mathbf{x}_i) v_f + b_i^n \tag{11.23}$$

The net force ($\mathcal{F}$) acting on a material point is calculated by summing the peridynamic forces on the parent material point due to all the neighboring points inside its

Horizon. N_f represents the number of material points within the Horizon enclosing the parent material point i. b_i^n is the body force at the i-th material point after n-th time step. $\ddot{\mathbf{u}}_i^n$ is the acceleration or the second-order derivative of displacements of the i-th material point after n-th time step. Introducing velocity as the first-order derivative of displacement, $\ddot{\mathbf{v}}_i^n$ is the velocity or the second-order derivative of displacements of the i-th material point after n-th time step. To obtain the displacements at each material point after n-th time step a numerical time integration method should be employed as discussed in Chapter 6. Various time integration methods are available such as Newmark β method, Wilson θ method, etc. which are second-order method and can be very useful. However, here in perielastodynamic simulation Velocity-Verlet integration [31] scheme is employed to calculate the velocity and subsequently the displacements in the time domain [3, 4] for given boundary and initial conditions. There is a specific reason for using Velocity-Verlet method. In wave simulation it is necessary to find the position of the particle and the velocity of the particle with respect to current configuration where no future information is available. Velocity-Verlet method is perfect in that sense to initiate the algorithm. Moreover, unlike Newmark method in Velocity-Verlet method there is no need for the assumption of linear or average acceleration between two-time steps. The time integration equations used in second-order Velocity-Verlet method are as follows,

$$\mathbf{v}_i^{n+1/2} = \mathbf{v}_i^n + \frac{\Delta t}{2\rho_i}\mathcal{F}_i^n \tag{11.24.1}$$

$$\mathbf{u}_i^{n+1/2} = \mathbf{u}_i^n + \frac{\Delta t}{2\rho_i}\mathbf{v}_i^{n+1/2} \tag{11.24.2}$$

$$\mathbf{v}_i^{n+1} = \mathbf{v}_i^{n+1/2} + \frac{\Delta t}{2\rho_i}\mathcal{F}_i^{n+1} \tag{11.24.3}$$

Stability of the numerical solution can be obtained for a small-time step Δt and a spatial discretization step ΔS. To have a convergence of the displacements, the detailed procedure to select the time step (Δt) and spatial discretization (ΔS) is discussed in Section 11.3.1.

11.4 CNDE CASE STUDY: MODELING GUIDED WAVES IN ISOTROPIC PLATE

11.4.1 Problem Statement

In this section a representative structural computational NDE problem is solved which frequently occurs in structural health monitoring (SHM) problems with surface mounted sensors as depicted in Fig. 1.5 in Chapter 1. In SHM, piezoelectric (PZT) actuator is attached to the host structure and in this problem as square PZT is used as shown in Fig. 11.5. In this problem a 300 mm × 200 mm, 2 mm

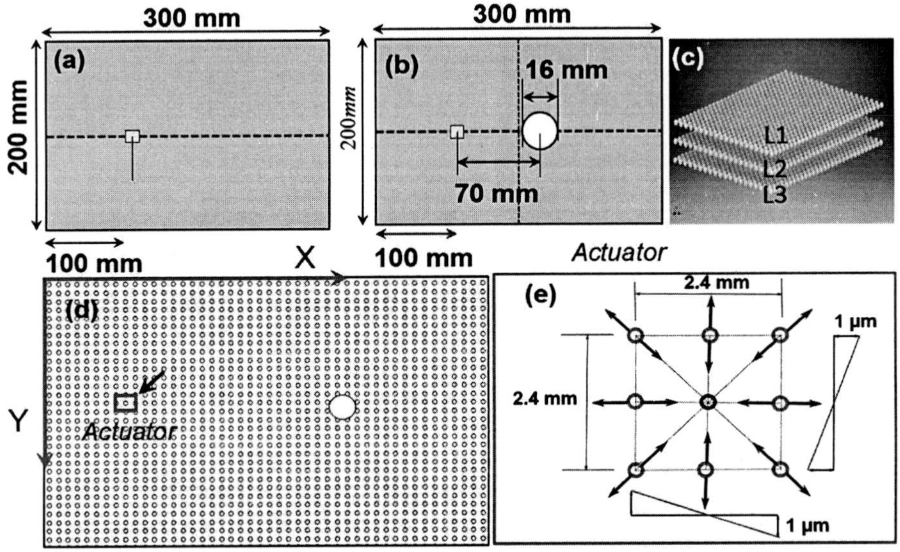

FIGURE 11.5 The schematics showing the geometry of an Aluminum 6061-T6 plate used in the simulation: (**a**) Pristine plate with PZT mounted on the top surface, (**b**) Plate with a through-thickness hole and a PZT patch, (**c**) Discretization of the plate and material layers (top, middle and bottom surfaces, L1, L2 and L3, respectively), (**d**) Schematics of Peri-elastodynamics discretization of the plate showing PZT and the hole (**e**) Boundary condition: Particle displacement due to the PZT excitation.

thick isotropic aluminum plate (Fig. 11.5a) Aluminum 6061-T6 is used to demonstrate the perielastodynamic method if can correctly simulate the guided wave propagation. An Aluminum 6061-T6 plate with a hole is also considered to study the wave damage interactions (Fig. 11.5b). Guided Lamb waves are generated in the plate by applying a standard tone-burst voltage signal to the PZT actuators [5]. Except the interface between the PZT actuator and the plate, all other boundaries of the plate are considered stress-free in the simulation. The voltage signal actuates the PZT and through electromechanical transduction the given energy is transformed into in plane mechanical strain. The in-plane strain causes the rapid localized displacement in the host structure, which results in the Lamb wave propagation in the plate. The Lamb wave modes create out of plane displacements and hence, to accommodate such deformation three layers of material points are used in the modeling (Fig. 11.5c) as previously depicted in Fig. 11.4. Fig. 11.5d shows the discretization used in this study satisfying the criteria described in Section 11.3. Displacement in the structure varies linearly along the length of the PZT and attains a maximum value at the boundaries [42] as shown in Fig. 11.5e. In this study, a square PZT with a dimension of 2.4 mm × 2.4 mm is modeled by applying a maximum 1 μm in plane radial displacement to the circumferential material points, shown in Fig. 11.5e. The displacements of the material points at the center of the PZT were enforced to zero. Next the boundary-initial condition, the voltage is given to the PZT to generate a tone burst signal. In this study electromechanical transduction is not model explicitly, hence, an in-plane displacement function due to the application of a tone burst signal is used to the provide the displacement boundary-initial condition. The equation for the displacement variation is expressed as.

$$\mathbf{u}(\mathbf{x},\mathrm{t}) = \mathbf{U}(\mathrm{x})e^{-pt^2/2}\sin(\omega_c t) \tag{11.25}$$

where ω_c and $\mathbf{U}(\mathrm{x})$ are the central frequency and the maximum displacement amplitude of the excitation signal given to different material points, respectively. Please note that the displacement ($\mathbf{u}(\mathbf{x},t=0) \equiv u_1(x_j)$, $u_2(x_j)$ and $u_3(x_j)$) boundary-initial conditions are provided to the each material point in each direction following the displacement pattern shown in Fig. 11.5. For example, both $u_1(x_j)$, $u_2(x_j) = \pm 1\cos 45°$ μm are prescribed at the corner points of the PZT shown in Fig. 11.5 depending on their location. The parameter p in Eq. (11.25) is expressed as

$$p = (2kh\omega_c/N_c)^2 \tag{11.26}$$

where k, h, and N_c are the signal shape factor, the half thickness of the specimen, and the number of cycles of the actuation signal, respectively. In Eq. (11.26) h is 1 mm for a 2 mm thick plate used in this problem. Let's assume a 3.5 count tone burst signal is to be used for actuation of the wave in the plate and hence, $N_c = 3.5$. In Eq. (11.26) ω_c, the central frequency of actuation is a user defined choice to generate different frequencies in the plate. However, to generate guided wave in the plate, one should use the dispersion behavior of the guided wave modes in the plate, which is previously discussed in Chapter 5.

11.4.2 Dispersion Behavior and Wave Tuning

Dispersion curves of various Lamb wave modes are used to predict the existence of various modes at a particular excitation frequency [5]. Two types of Lamb wave modes exist in a plate based on the particle motion, named symmetric modes (i.e., S_0, S_1, S_2...) and antisymmetric modes (i.e., A_0, A_1, A_2...). Generation of the Lamb wave modes in a plate depends on the frequency of excitation, the thickness of the plate and the material properties (Density, Young's modulus or Shear Modulus and Poisson's ratio) of the material. Dispersion of various Lamb wave modes is obtained by solving Rayleigh-Lamb wave equations discussed in Section 5.2.2 in Chapter 5. In Eq. (5.31) a generalized Rayleigh-Lamb equation is presented, however, solution of Eq. (5.31) in Fig. 5.6 generated two sets of modes, symmetric and antisymmetric (refer Fig. 5.5) modes. Although not discussed in Chapter 5 the Rayleigh-Lamb wave equation in Eq. (5.31) can be further divided into two sets of equations [43]. One for symmetric Lamb wave modes and another for antisymmetric Rayleigh-Lamb modes. The symmetric Rayleigh-Lamb modes are generated solely from the following equation

$$\frac{tan(qh)}{tan(ph)} = -\frac{4k^2qp}{(k^2 - q^2)^2} \tag{11.27}$$

Similarly, antisymmetric modes are generated solely from the following equation,

$$\frac{tan(qh)}{tan(ph)} = -\frac{(k^2 - q^2)^2}{4k^2qp} \tag{11.28}$$

Parameters in the above equations are expressed as,

$$p^2 = \frac{\omega^2}{c_L^2} - k^2 \; ; \; q^2 = \frac{\omega^2}{c_S^2} - k^2 \; ; \; c_L = \sqrt{\frac{2\mathrm{m}(1-\nu)}{\rho(1-2\nu)}} \; ; \; c_s = \sqrt{\frac{\mathrm{m}}{\rho}} \tag{11.29}$$

where ω, k, c_s, c_L, ν, m, ρ, and h are the angular frequency, wavenumber, the shear wave velocity, the longitudinal wave velocity, the poison's ratio, the shear modulus, the density, and the half thickness of the plate, respectively. In this work, the dispersion curves for various Lamb wave modes in an Aluminum 6061-T6 plate were calculated using the commercially available "Disperse" software [44], designed by the Imperial College, London, UK, as shown in Fig. 11.6a. The plate thickness was set to 2 mm and the material properties were set to the values listed in Table 11.1.

Next to select the desired excitation frequency, the tuning curves for S_0 and A_0 modes in the Aluminum 6061-T6 were obtained from an open source software "Waveform Revealer" [45], developed by the LAMSS laboratory at the University of South Carolina (USC). As shown in Fig. 11.6b, the central frequency of 150 kHz is chosen to make sure that only the S_0 and A_0 modes are excited when the modal amplitudes are comparable but nonequal. In the present study, a 3.5 count tone-burst signal with the central frequency (ω_c) of 150 kHz is used. Fig. 11.6c and d shows the time domain signal and its frequency transform, respectively.

TABLE 11.1
Material Properties

Aluminum 6061-T6 Material Properties	
Density, ρ	2700 kg/m^3
Young's modulus, E	69 GPa
Poisson ratio, ν	0.33

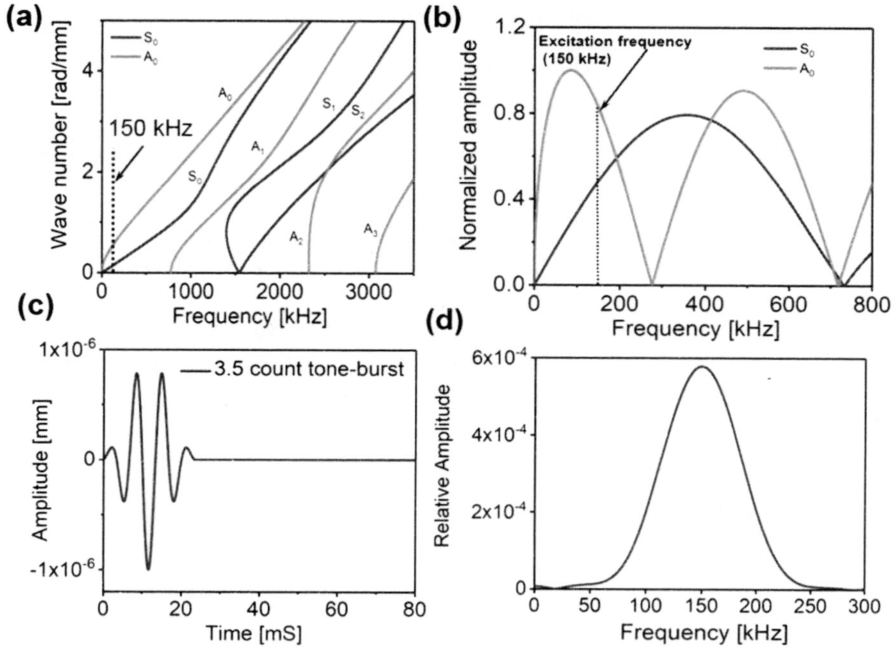

FIGURE 11.6 (**a**) Dispersion curves for 2 mm thick Aluminum 6061-T6 plate, (**b**) Tuning curve of an Aluminum 6061-T6 plate (2 mm thickness) with a standard 7 mm PZT, (**c**) 3.5 count tone burst signal (displacement input signal) with 150 kHz central frequency shown in time domain, (**d**) Frequency domain representation of the excitation signal.

11.4.3 Discretization of Perielastodynamic Problem Domain

In this work, the phase velocity of the S_0 mode at 150 kHz is the c_{max} in Eq. (11.22). Similarly, the phase velocity of the A_0 mode at 150 kHz is used as c_{min} in Eq. (11.21). The Courant-Friedrichs-Levy condition is used to obtain a numerically stable time step (Δt) [40]. Thus, to obtain a converging solution, the spatial and the temporal step sizes are chosen to be 1.2 mm and 0.01 µS, respectively, satisfying Eqs. (11.21) and (11.22). Material points are chosen in a grid fashion with a spacing of ΔS to model the plate with a layer spacing of 1 mm between each layer L1, L2, and L3 (Fig. 11.5c). 41,750 material points were used in each layer in the pristine plate. A total of 125,250 material points was used in the simulation including all the three layers L1, L2, and L3. In case of the damaged plate (with hole), there was 41,610 material points in each layer and a total of 124,830 material points were used for the perielastodynamic simulation. Each parent point is assigned with a family based on its Horizon and bonds were established between each pair of material points within the family. A 3.015ΔS was used as the Horizon size in the simulation. To model a plate with a through-thickness hole, a pristine discretization is performed and then the material points are removed from the geometry to produce the hole.

11.4.4 Numerical Computation and Results

11.4.4.1 Displacement Filed Presentation

The perielastodynamic simulation was performed on a workstation with two Intel Xeon (R) CPU E5-2650 V3 2.30 GHz processors with total 128 GB RAM, in a single core. The simulation was run on MATLAB-18a. One simulation was completed within a reasonable time of ~24 h. Note that this problem is highly parallelizable and can be implemented with distributed clusters, GPUs, multiple threads, or a combination of these methods. Preliminary translation of this MATLAB program into multithreaded C++ resulted in a 20 times speedup.

In this CNDE simulation, fundamental Lamb wave modes (S_0 and A_0) are simulated which are widely used in the damage detection with ultrasonic SHM. Fig. 11.7

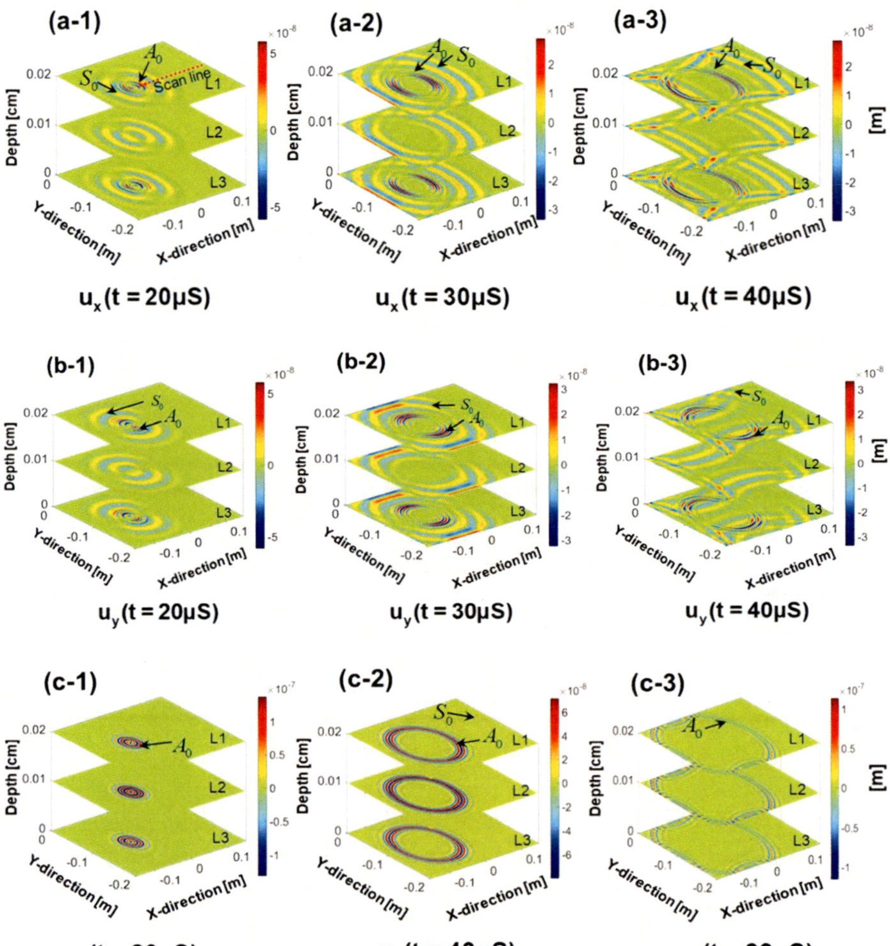

FIGURE 11.7 Time domain in plane and out of plane displacement waveform: (a) at t = 20, 30 and 40, (b) at t = 20, 40 and 60, (c) at t = 20, 40 and 60.

shows the in plane displacements $u_1(x_1,x_2,t)$ and $u_2(x_1,x_2,t)$ and out of plane displacement $u_3(x_1,x_2,t)$, for the Lamb wave propagation plotted after $t = 20$, 40, and 60 μS, respectively. Like before in earlier chapter the axes 1, 2, and 3 (x_1,x_2, and x_3) are synonymous to the axes x, y, and z in the figures presented in this chapter. Each stack of the figure is plotted with the respective displacement pattern in the layers L1, L2, and L3. The S_0 mode travels faster than the A_0 mode and confirms the dispersion curves. In wave propagation simulation, most inaccuracy comes from the boundary reflections if the results are not converged. The best approach to judge if a CNDE simulation of the wave propagation is converged is to evaluate the boundary reflections. In the present simulation, the reflected modes from the plate boundaries are clearly visible in the figures. Usually, there is a possibility of divergence of the solution at the boundaries due to the nonconvergence of the solution. However, with the specific steps within the framework of perielastodynamic process described in this chapter, the results are converged, and bounded boundary reflections are achieved. In the top (L1) and the bottom (L3) layers in Fig. 11.7 (a-1–a-3) and (b-1–b-3), both symmetric and antisymmetric modes (S_0 and A_0) are visible in the wave fields composed of in plane displacements. The contribution of the A_0 mode in the in-plane motion at the middle layer (L2) of the plate is negligible but the out of plane motion is dominant. As shown in Fig. 11.7 (c-1–c-3), the out of plane displacement of the A_0 mode is visible (i.e., contribution from $u_3(x_1,x_2,t)$) while the S_0 mode is barely noticeable. This is because the displacements of the in-plane particles of the S_0 mode dominate over the out of plane motion of the particles. Also, the contribution of the S_0 mode in the displacement of the middle layer (L2) of the plate is negligible, because, in S_0 mode, the middle layer (L2) remains undisturbed. Top view of the respective displacements after 10 μs, 24 μs, and 38 μs are shown in Fig. 11.8. Symmetric and antisymmetric wave modes are visible from the top view of the propagating guided waves in isotropic plate.

Next, along a centerline of the plate, space-time representation of the in plane ($u_1(\mathbf{x},t)$) and the out of plane displacement ($u_3(\mathbf{x},t)$) are presented in Fig. 11.9, where, displacement fields are presented. Displacements at the top, middle, and the bottom surfaces (L1, L2, and L3), are presented to investigate the existence of the different Lamb wave modes and their contribution to the displacement in each layer. In Fig. 11.9 (a-1–a-3), it is observed that the S_0 mode contributes to the in-plane displacement in all layers and the A_0 mode contributes only at the top and the bottom layers. Space-time representation of the out of plane displacement is shown in Fig. 11.9 (b-1–b-3), which shows that the A_0 mode had a higher amplitude than the S_0 mode. A_0 mode contributed to the displacement of all the layers (L1, L2, and L3), whereas, S_0 mode contributed only to the top (L1) and the bottom (L3) layers.

11.4.4.2 Vector Field Representation of the Guided Wave Modes

To prove the accuracy of the perielastodynamic simulation, the characteristics of the Lamb wave modes (S_0 and A_0), in plane and out of plane particle motion

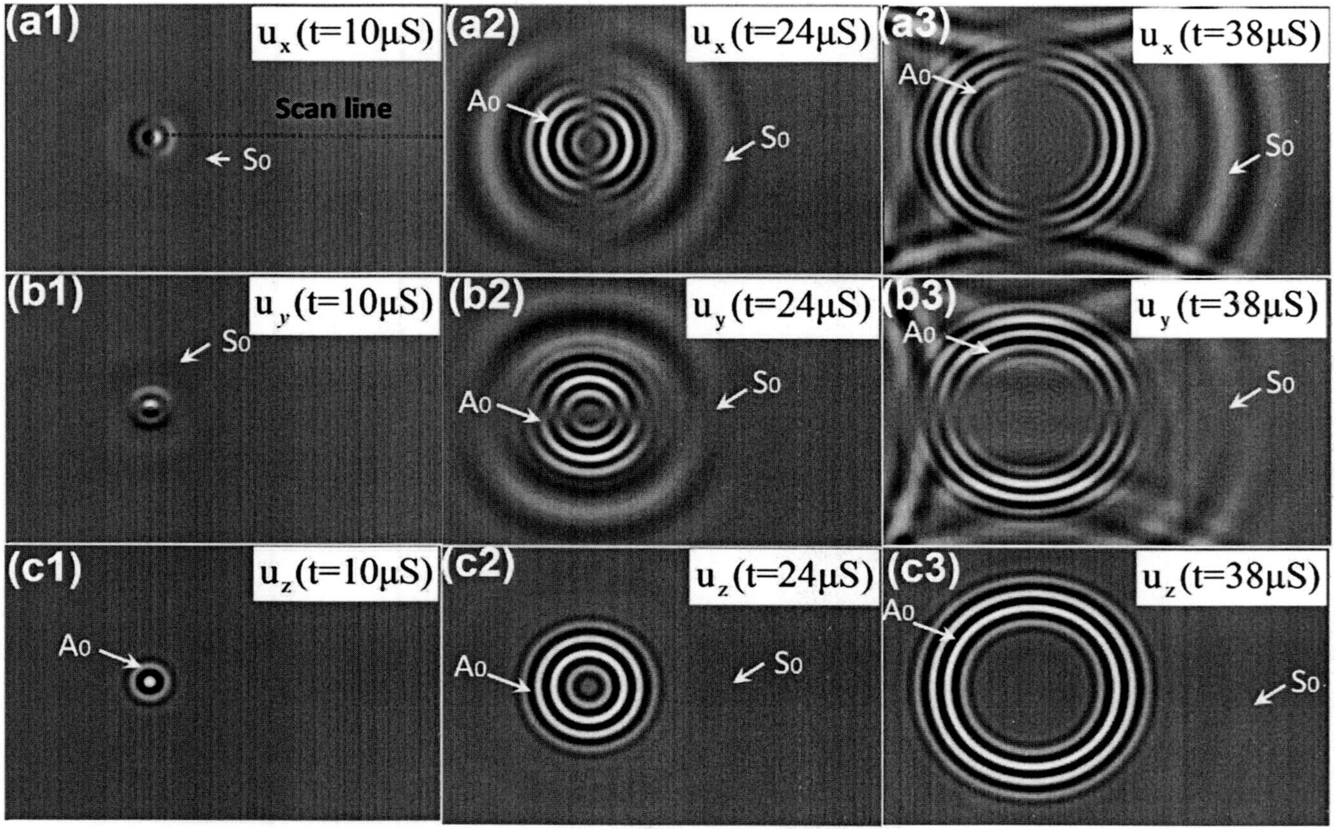

FIGURE 11.8 Time domain in-plane and out-of-plane displacement waveform. (a) at t = 10, 24, and 38, (b) at t = 10, 24, and 38, (c) at t = 10, 24, and 38.

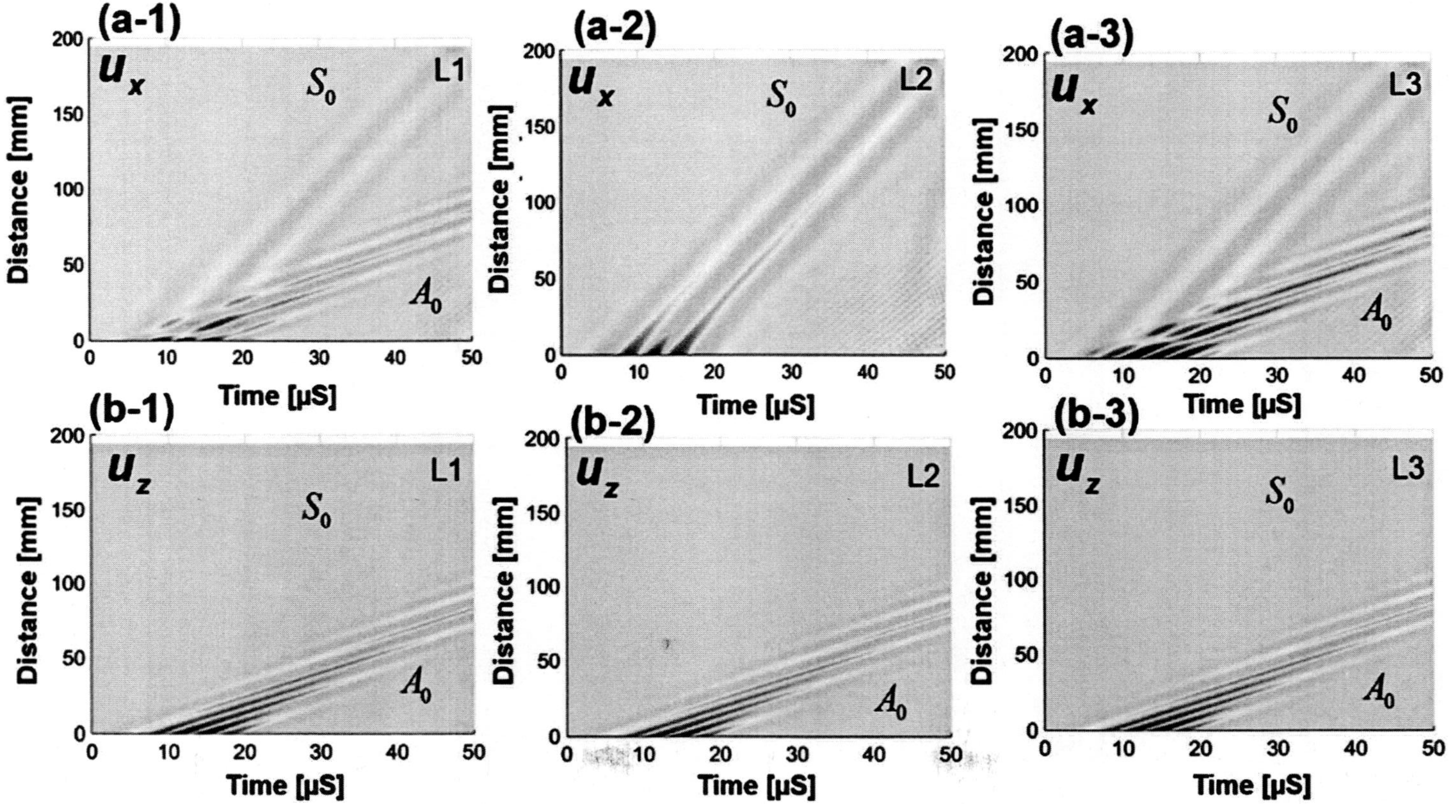

FIGURE 11.9 Space-time in plane and out of plane displacement fields: (a-1) $u_x(x,t)$ at the top (L1), (a-2) $u_x(x,t)$ at the middle layer (L2), (a-3) $u_x(x,t)$ at the bottom layer (L3), (b-1) $u_z(x,t)$ at the top layer (L1), (b-2) $u_z(x,t)$ at the middle layer (L2), (b-3) $u_z(x,t)$ at the bottom layer (L3).

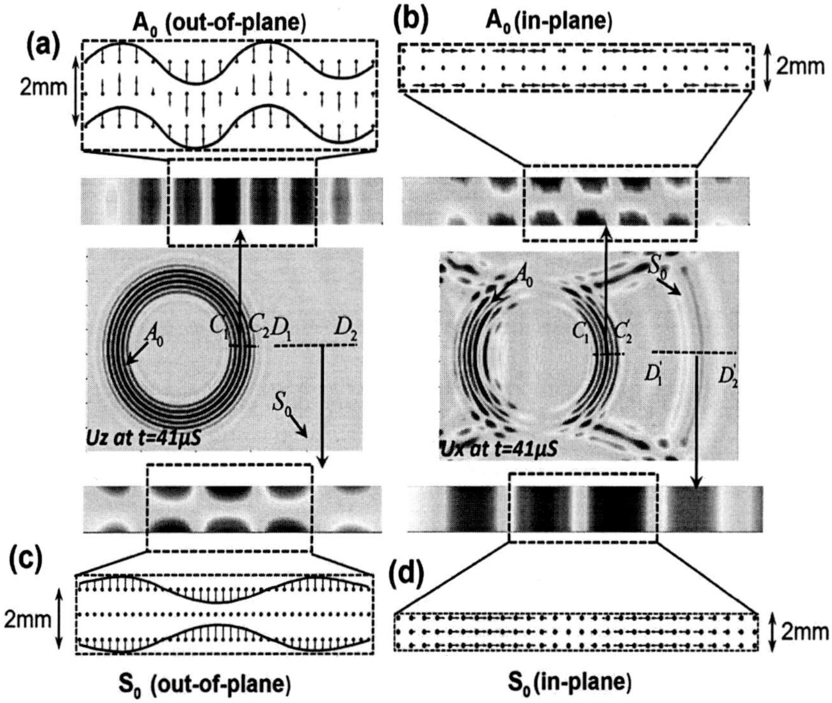

FIGURE 11.10 Perielastodynamics simulation, vector field and displacement distribution of the S_0 and A_0 modes across the thickness of the plate: (a) Vector field of the A_0 mode for out of plane motion, (b) Vector field of the A_0 mode for in plane motion, (c) Vector field of the S_0 mode for out of plane motion, (d) Vector field of the S_0 mode for in plane motion.

across-the-thickness of the plate are plotted in Fig. 11.10, after 41 μs. Out of plane ($u_3(\mathbf{x},t)$) and in plane ($u_1(\mathbf{x},t)$) displacement field of A_0 mode are extracted along the cross-sections $C_1 - C_2$ and $C_1' - C_2'$, on the plate. Similarly, the out of plane and the in-plane displacement distribution of S_0 mode is plotted along the cross-section lines $D_1 - D_2$ and $D_1' - D_2'$, respectively. Vector fields and displacement distributions are shown in Fig. 11.10a–d. It is observed in Fig. 11.10a that all particles moved either upwards (+) or downwards (–) with variable amplitudes (like bending motion) due to the generation of the A_0 mode. In Fig. 10.11c, the top and the bottom layers are symmetrically displaced with respect to the mid-plane and the displacement of the mid-plane is almost zero due to the generation of the S_0 mode. In plane particle motion in A_0 and S_0 modes are also shown in Fig. 11.10b and d. In Fig. 11.10b, the particles at the top and the bottom layers are moved in the opposite directions along the in-plane direction and the displacement of the mid-plane is zero due to the generation of the A_0 mode. In Fig. 11.10c, the particle displacements are constant across the thickness due to the S_0 mode. Vector fields and the mode shapes in Fig. 11.10 indicate that the Perielastodynamics can simulate the guided Lamb wave modes accurately.

11.4.4.3 Fourier Analysis of the Sensor Signals

Fundamental understanding of Fourier transform and related mathematical equations are described in Chapter 6. Fourier transform can be one dimensional, two dimensional and could be multi-dimensional. A space-time data set if transformed through the two-dimensional Fourier transform, it will result wavenumber-frequency domain plots. Multidimensional Fourier transform is widely used to separate the different Lamb wave modes [46] in NDE and SHM applications. Two-dimensional and three-dimensional Fourier transforms (2D or 3D FFT) are performed on space and time domain data. The equation that transforms the space-time wave field data into the wavenumber-frequency representation of the wave field can be expressed by [46],

$$u_p\left(k_j,\omega\right)=\int_{-\infty}^{\infty}\int_{-\infty}^{\infty}\int_{-\infty}^{\infty}u_p\left(x_j,t\right)e^{-2\pi i(\mathbf{k}.\mathbf{x})}e^{-i\omega t}dx_1dx_2dt \qquad (11.30)$$

where j take the values, 1 and 2, p take values 1, 2, and 3, $u_p(x_j,t)$ designates the p-th displacement after time t at a point x_j located on the 2D x_1-x_2 plane. $u_p\left(k_j,\omega\right)$ designates the p-th displacement at frequency ω in reciprocal space of wavenumbers at the point k_j located on the 2D reciprocal k_1-k_2 plane. Here index, 1, 2, and 3 synonymous to the axes x, y, and z. In multi-modal wave propagation analysis, it is not only difficult but impossible to distinguish the different wave modes from a time domain signal, especially on a small plate where the wave modes tend to overlap. In this chapter, wavenumber-frequency analyses are presented to visualize and identify the different modes separately. This is also verified by comparing the simulated wave modes with the theoretical dispersion curves. For this purpose, 2D and 3D fast Fourier transforms (FFT) were performed on the simulated displacement wave field to obtain the wavenumber-frequency representations.

To perform the 2D-FFT, out of plane ($u_3(x_j,t)$) and in plane ($u_1(x_j,t)$), displacement data are obtained across-the-thickness of the plate along the selected scan line shown in Fig. 11.8(a1). 163 spatial points with a resolution of 1.2 mm along the line shown in Fig. 11.8(a1) were used in the analysis. Matrix size used to store the displacements wave field was 163×8000, where 8000 data point along the time domain was used for the analysis. The 2D FFT was performed on the displacement vectors obtained from all the three material layers (L1, L2, and L3).

Wavenumber-frequency domain representation of the in-plane displacement ($u_1(x_j,t)$) is presented in Fig. 11.11(a-1–a-3), respectively. Both the S_0 and A_0 modes are identified at the top (L1) and the bottom (L3) material layers. This is because they both significantly contribute to the energy of the in-plane wave motion. Whereas middle layer (L2) only identified the symmetric wave mode. The amplitudes of the A_0 mode are slightly greater than S_0 mode. A similar phenomenon is predicted from the tuning curve of the plate at 150 kHz. The contribution of the in-plane motion of the A_0 mode to the energy of the middle layer is almost zero. Similarly, Fig. 11.11 (b-1–b-3) is obtained from the wavenumber-frequency domain representation of the out of plane ($u_3(x_j,t)$) displacements at the top (L1), middle (L2), and the bottom

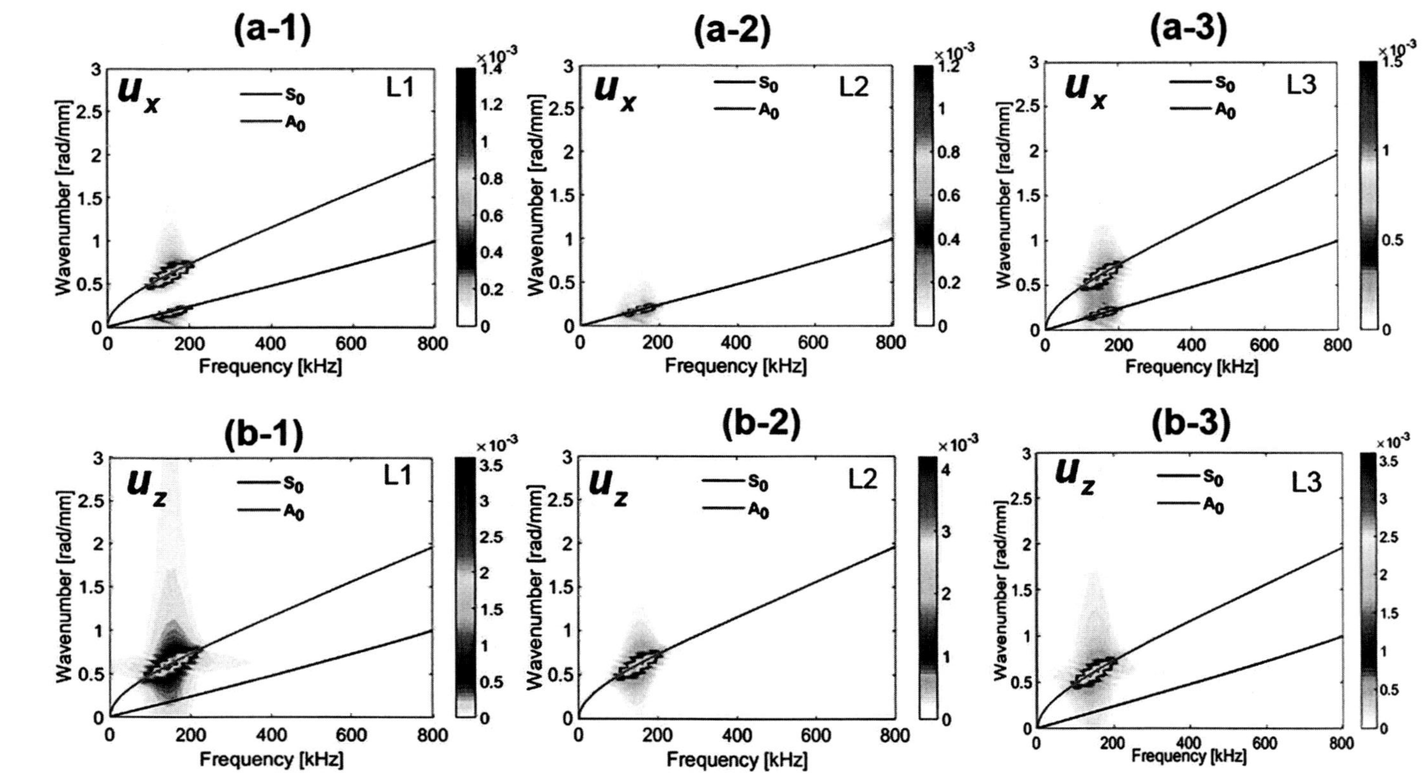

FIGURE 11.11 Frequency-wavenumber (FW) representation of the displacement field at the pristine state: (a-1) FW of the in plane displacement at the top surface (L1), (a-2) FW of the in plane displacement at the mid-surface (L2), (a-3) FW of the in plane displacement at the bottom surface (L3), (b-1) FW of the out of plane displacement at the top surface (L1), (b-2) FW of the out of plane displacement at the mid-surface (L2), (b-3) FW of the out of plane displacement at the bottom surface (L3).

layer (L3), respectively. The energy distribution of the A_0 mode is higher than the S_0 mode in all the material layers. The S_0 mode is visible only at the top and the bottom layers with very low amplitude. This is because the out of plane motion of the particles in S_0 mode is very low and the displacement at the midplane is almost zero.

Next, the 3D-FFT is employed to transform the 3D displacement data ($u_1(x_j,t)$, $u_2(x_j,t)$, and $u_3(x_j,t)$) into the wavenumber-frequency domain ($u_1(k_1,k_2,\omega)$, $u_2(k_1,k_2,\omega)$, and $u_3(k_1,k_2,\omega)$) and are shown in Fig. 11.12a–c, respectively. In this problem, frequency transformation is performed only on the data obtained from the top surface (L1). The size of the matrix used to store the 3D displacement data was $163 \times 250 \times 8000$. Fig. 11.12 shows the 3D Fourier transform of the in plane ($u_1(k_1,k_2,\omega)$, $u_2(k_1,k_2,\omega)$) and the out of plane ($u_3(k_1,k_2,\omega)$) displacement at the top surface (L1). Wavenumber domain displacement (a) u_1 at 110 kHz, 150 kHz, 185 kHz, and 225 kHz, (b) u_2 at 110 kHz, 150 kHz, 185 kHz, and 225 kHz, (c) u_3 at 110 kHz, 150 kHz, 185 kHz, and 225 kHz are presented by 3D stacking the 2D representations of two-dimensional wavenumber domain plots. It can be seen that both the S_0 and the A_0 modes appeared in the form of two concentric circular rings in Fig. 11.12. The radius of the circles corresponds to the wavenumbers at the respective frequencies. Wavenumbers of the S_0 and A_0 modes at the 150 kHz are obtained from the perielastodynamic simulation are 0.56 rad/mm and 0.187 rad/mm, respectively. A smaller circle corresponds to the S_0 mode while the larger circle corresponds to the A_0 mode. It is also observed that the energy of the modes at the frequencies 110 kHz, 185 kHz, and 225 kHz are lower compared to that at the 150 kHz. This is because most of the energy of the modes is concentrated around the excitation frequency (150 kHz) selected from the tuning curve presented in Fig. 11.6.

To verify the directional dependency of the Lamb wave propagation, 2D wavenumber plots ($u_x(k_x,k_y)$, $u_y(k_x,k_y)$ and $u_z(k_x,k_y)$) at $\omega_c = 150$ kHz, are obtained from the 3D FFT and were compared with those obtained from the theoretical predictions using Disperse software at 150 kHz. Theoretical wavenumber plot is superimposed on the numerically obtained wavenumber plots in Fig. 11.13a–c. Good agreements between the numerical and analytical results are obtained.

11.5 CNDE CASE STUDY: WAVE-DAMAGE INTERACTION IN ISOTROPIC PLATE

11.5.1 CNDE of a Plate with Hole: Comment on the Sensor Placement

The study of wave-damage interaction is an important part of wave propagation and CNDE simulations. Thus, to demonstrate the feasibility, in this chapter, several simulations are performed on a plate with a circular hole (Fig. 11.14) and cracks. Here in this section the simulation results are presented for a plate with hole. The center of a 16 mm diameter circular hole is located in the plate at a 70 mm distance from the center of the actuator. The in-plane displacements in the simulated wave field demonstrate the reflection and transmission of the respective wave modes in three

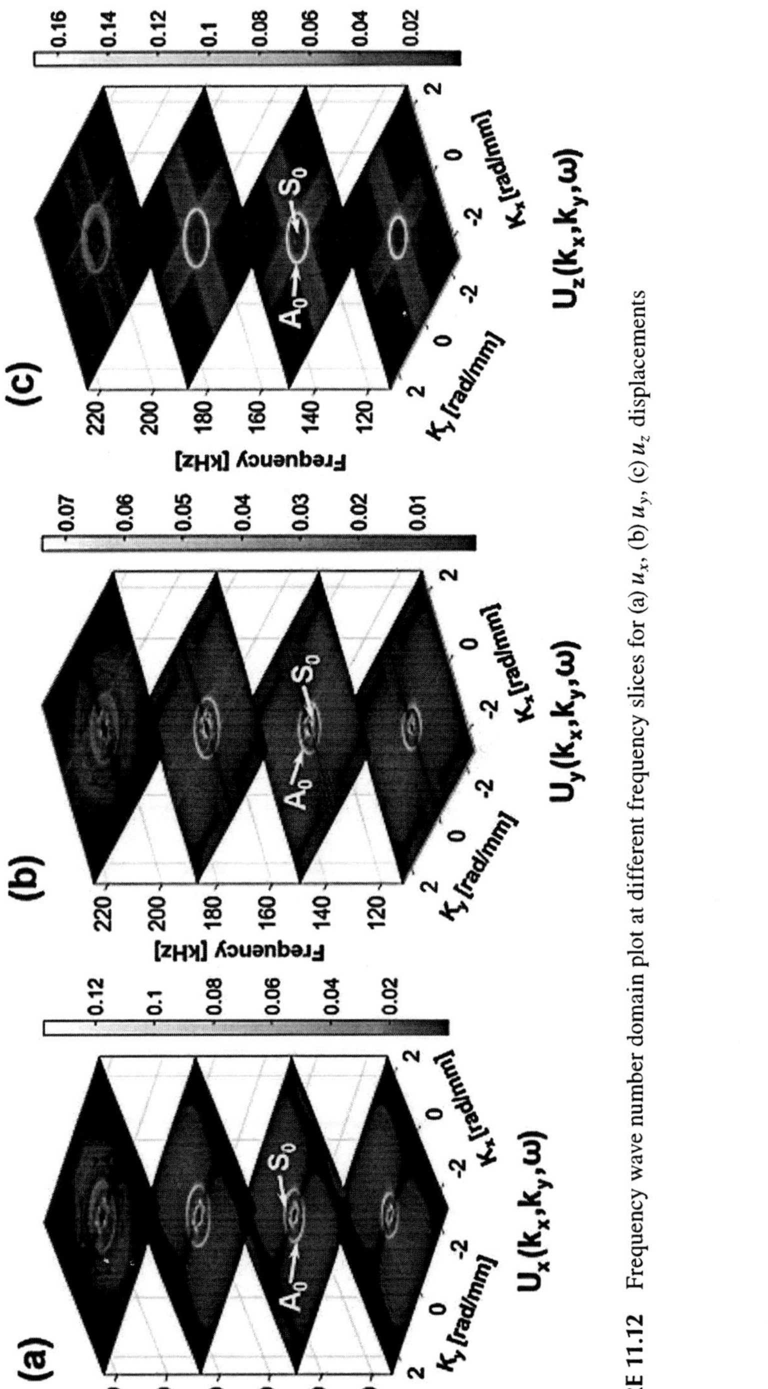

FIGURE 11.12 Frequency wave number domain plot at different frequency slices for (a) u_x, (b) u_y, (c) u_z displacements

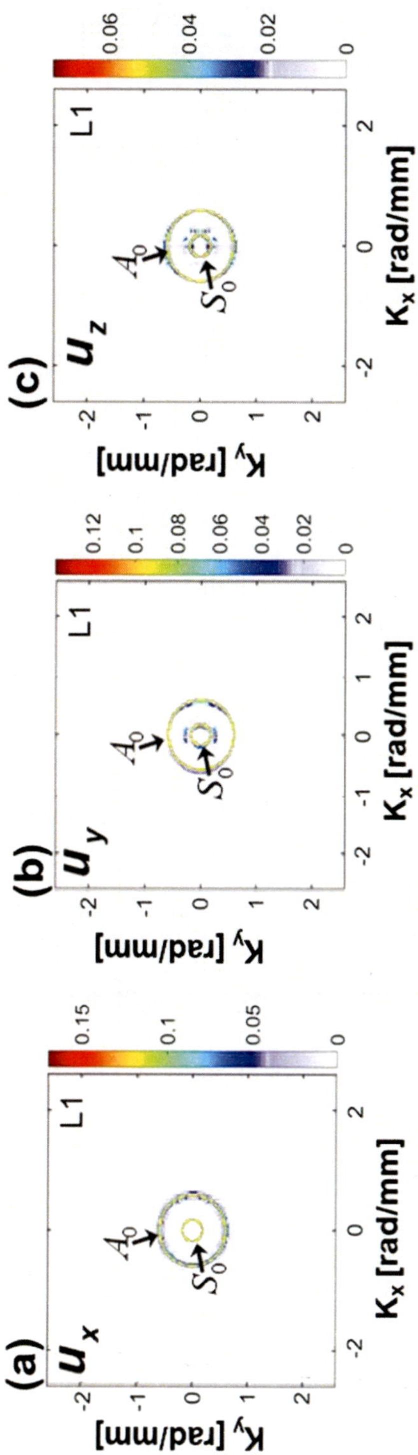

FIGURE 11.13 Comparison of theoretical and numerical (Perielastodynamics) wavenumber domain at 150 kHz: (a) u_x at 150 kHz, (b) u_y at 150 kHz, (c) u_z at 150 kHz.

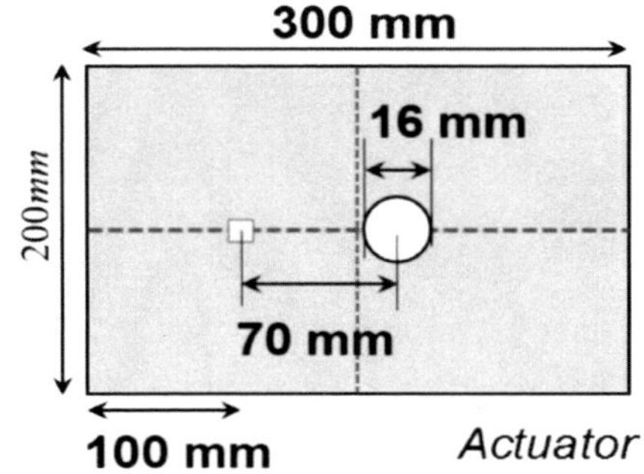

FIGURE 11.14 Geometry of an aluminum 6061-T6 plate with hole for a SHM problem to be solved using perielastodynamic

different layers across the depth. Fig. 11.15a–c shows the displacement wave fields in the plate with a circular hole after 48 μs. Reflection from the hole in all three layers is manifested only in the vertical displacement ($u_3(x_j,t)$) wave fields. However, the reflections from the hole are apparent only in the in-plane displacement filed plotted for the top and the bottom layers. Fig. 11.16a and b shows the 2D-FFT analysis of the $u_1(x_j,t)$ and $u_3(x_j,t)$ displacements of the top surface of the plate, respectively. Reflected and transmitted S_0 and A_0 wave modes are clearly visible in both displacement wave fields, but they much stronger in $u_1(x_j,t)$ displacement field in Fig. 11.16a and weaker in $u_3(x_j,t)$ displacement field in Fig. 11.16b. Boundary reflection of the same modes is also clearly visible. Based on the simulation results, a PZT sensor at a distance of 45 mm from the center of the circular hole along any direction will be suitable to detect the damage in the PZT mounted plate-like structure. If the PZT sensor is placed toward the positive x_1 axis from the hole, the transmitted wave signals will be received by the sensor, but if the PZT sensor is placed toward the negative x_1 axis from the hole, the reflected wave signal will be received. Based on these simulation results presented in Fig. 11.16, it can be said that if a PZT with 1-3 and 1-1 polarization [47] is used, any location is suitable to detect the hole with both transmitted and reflected wave fronts. Because the superposition of the transmitted and reflected wave front will contain both the in plane $u_1(x_j,t)$and $u_2(x_j,t)$ and the out of plane deformation $u_3(x_j,t)$ of the PZT attached to the plate.

11.5.2 CNDE of a Plate with Crack with Experimental Validation

In this section wave-damage interaction is studied for through thickness cracks in a plate like structure. A through-thickness crack of length 16 mm x 2.4 mm located at a 70 mm distance from the PZT actuator (Fig. 11.17) is considered in this problem. A crack on the centerline (dotted line in Fig. 11.17a) and a crack offset from the centerline are considered as shown in Fig. 11.17a and b, respectively. Satisfying earlier discretization requirement for perielastodynamic simulation, a total of 41,750 material points was used to discretize each layer. Each parent point is assigned with a family based on its Horizon and bonds were established between each pair of material points within the family. A 3.015ΔS was used as the Horizon size in the

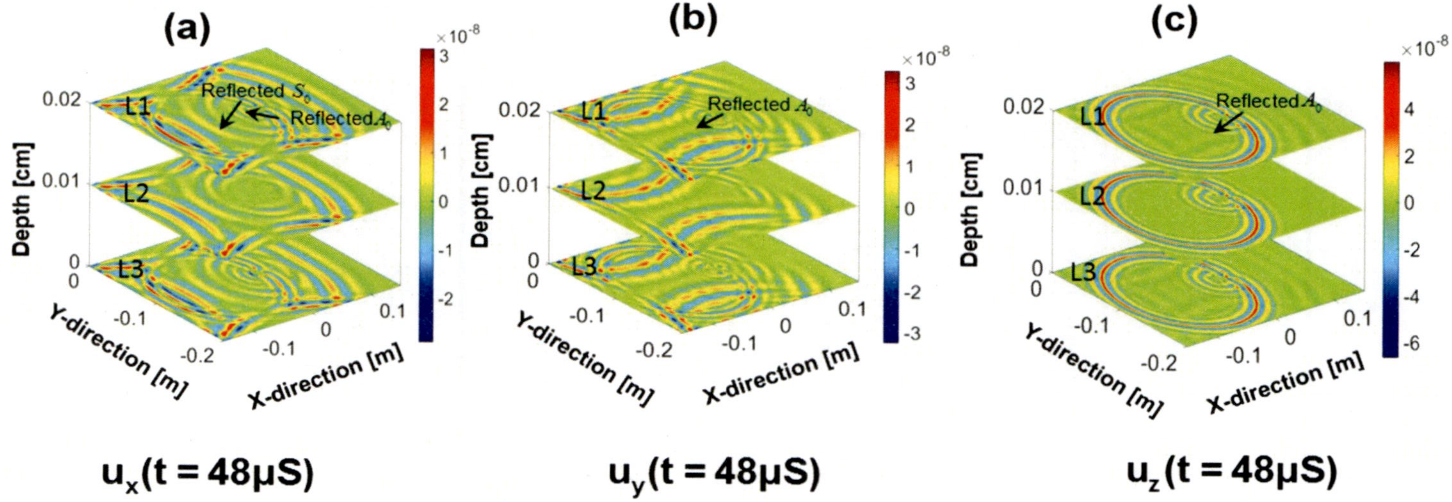

FIGURE 11.15 Time domain displacement waveform in a plate with a circular hole. (**a**) at t = 48, (**b**) at t = 48, (**c**) at t = 48.

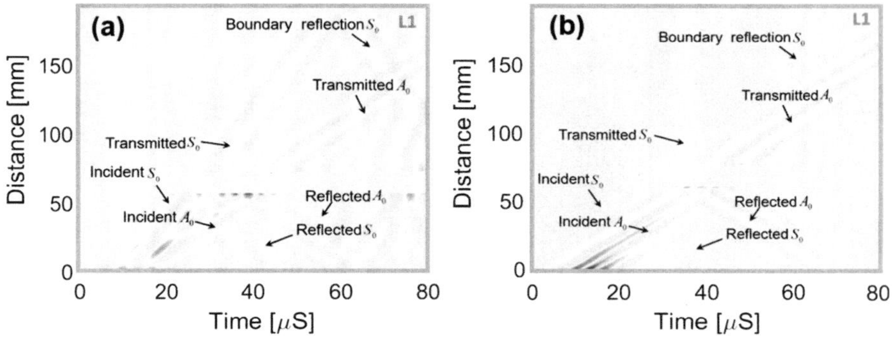

FIGURE 11.16 Space-time wavefield representations for top surface of the plate: (**a**) $u_x(x, y, t)$ for plate with a through-thickness hole (**b**) $u_z(x, y, t)$ for plate with a through-thickness hole.

simulation like it was used for pristine plate discussed in Section 11.4. To model a plate with a through-thickness crack, first a pristine discretization is performed and then crack is modeled by assigning bond constant zero to bonds those pass through the cracked zone. Ultrasonic wave fields ($u_1(x_j, t)$ and $u_3(x_j, t)$) in a plate with central and offset crack obtained from the perielastodynamic simulations are shown in Figs. 11.18 and 11.19, respectively.

To observe reflections and transmissions of the respective wave modes from the damage location, in-plane displacement wave fields, after three different time steps (40 μs, 50 μs, and 60 μs), are shown in Fig. 11.18 (a1–a3). It can be seen that, while the reflected S_0 mode from the crack location is observed after time 40 μs, the same mode disappeared after 50 μs and 60 μs due to the interference with the parent mode and with the reflected boundary mode. Alternatively, reflected A_0 mode is identifiable even after 50 μs and 60 μs time steps. In case of out-of-plane displacements, reflected S_0 mode is not quite visible in any of the time steps (Fig. 11.18 (b1–b3)) due

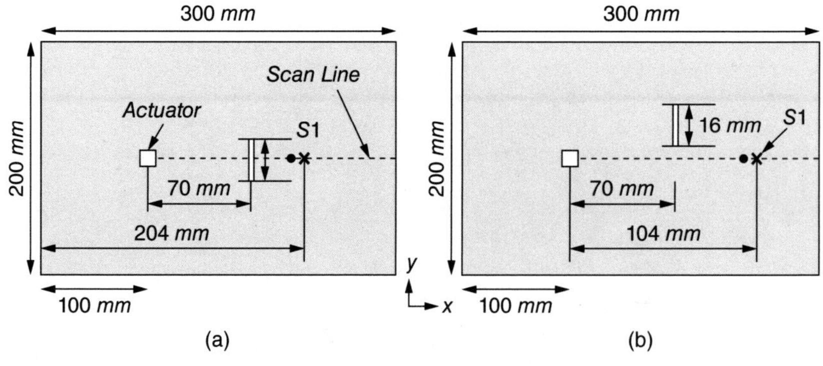

FIGURE 11.17 The geometry of aluminum 6061-T6 plate with crack: (a) central-crack, (b) offset-crack.

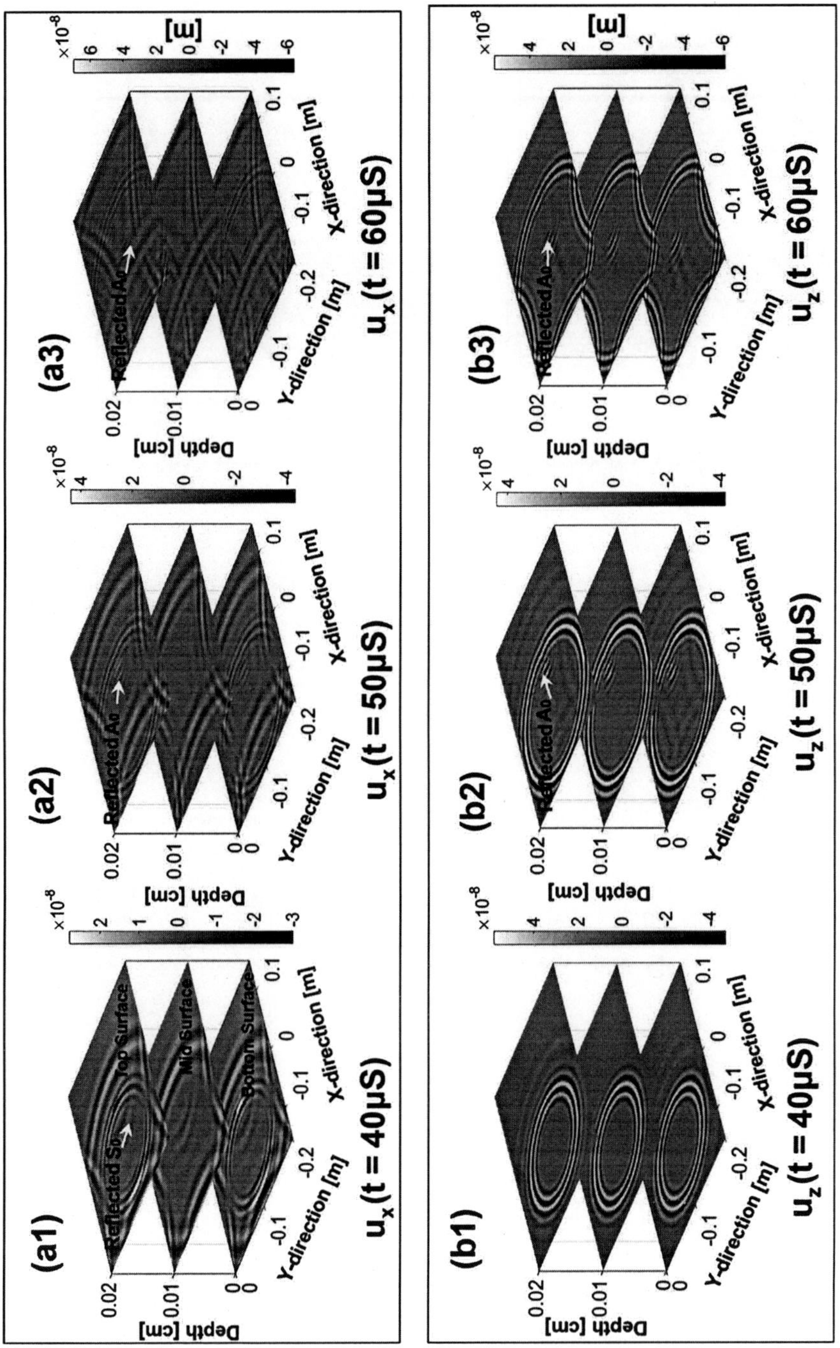

FIGURE 11.18 Time-domain displacement waveform in a plate with a central-crack: (a1) at t = 40, (a2) at t = 50, (a3) at t = 60 ,(b1) at t = 40, (b2) at t = 50, (b3) at t = 60.

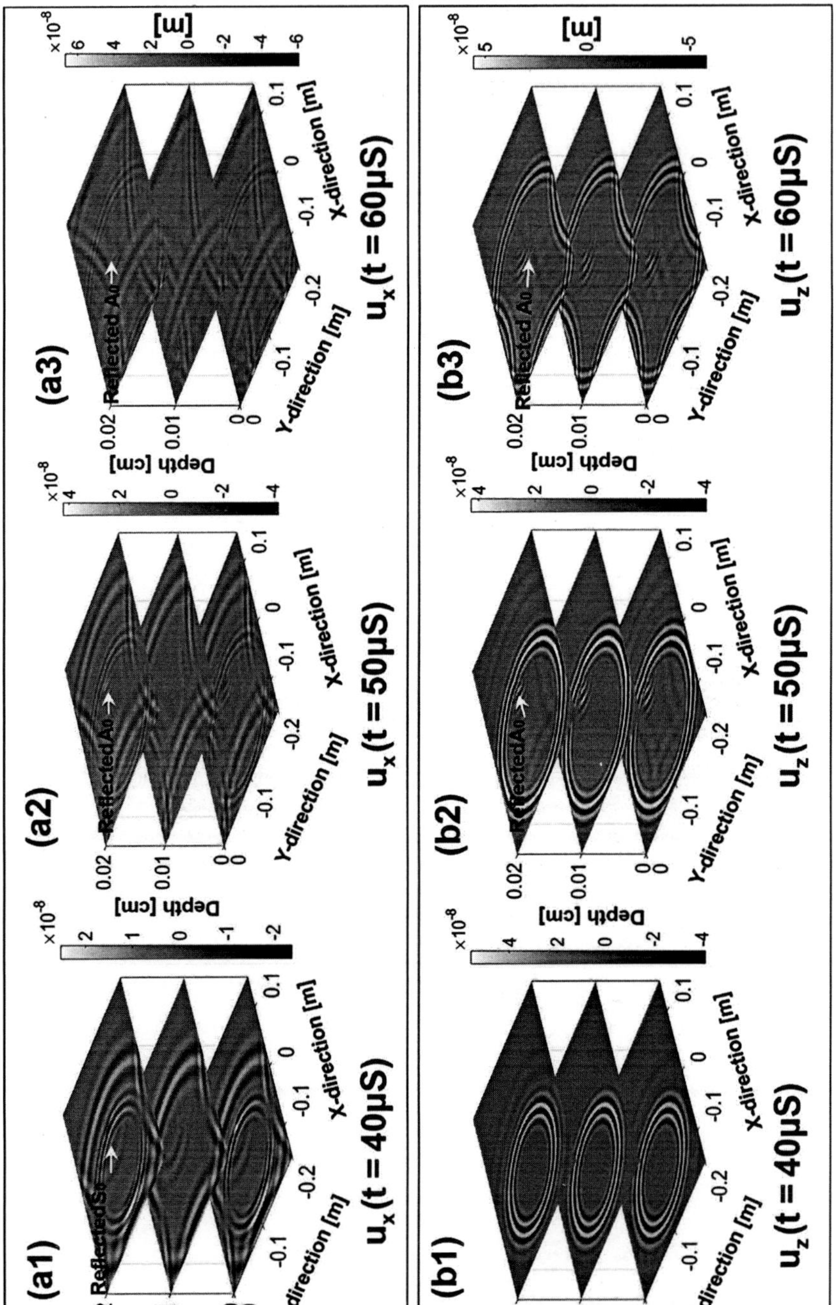

FIGURE 11.19 Time-domain displacement waveform in a plate with a offset-crack: (a1) at t = 40, (a2) at t = 50, (a3) at t = 60, (b1) at t = 40, (b2) at t = 50, (b3) at t = 60.

to its minor contribution to the out-of-plane wave propagation. However, reflected A_0 mode is observed after both 50 μs and 60 μs time steps.

In-plane displacement wave fields with offset-crack (Fig. 11.17b) after times 40 μs, 50 μs, and 60 μs are shown in Fig. 11.19 (a1–a3). Both reflection and transmission of the fundamental Lamb wave modes are observed. Like the center-crack scenario, in case of an offset-crack, the reflected S_0 mode is observed only after the time 40 μs. Reflection of the S_0 mode after other two-time steps is barely noticeable due to the interference with the boundary reflections. However, the reflected A_0 mode is visible after 50 μs and 60 μs. Evaluating Fig. 11.19 (b1–b3), in comparison to the central-crack, a similar argument can also be made for the offset-crack in relation to the A_0 and S_0 modes existed at different time steps.

11.5.2.1 Experimental Design for the Validation of Perielastodynamic

To validate the perielastodynamic results, a pitch-catch experiment was designed on a pristine Aluminum 6061-T6 plates as shown in Fig. 11.20a and b. Fig. 11.20a shows the complete experimental setup consists of, (1) the plate structure Aluminum 6061-T6, (2) PZT sensors and actuators (type PZT 5A, STEMiNC, Florida, 7 mm in diamater and 0.5 mm in thickness) attached to the structure with Hysol 9340 adhesive, (3) a Tektronix AFG3021C (25 MHz, 1-Ch Arbitrary Function Generator, Tektronix Inc.), (4) a Tektronix MDO3024 (200 MHz, 4-Ch Mixed Domain Oscilloscope, Tektronix Inc.), (5) nF HSA series Power Amplifier, and (6) A laptop computer to control all devices through MATLAB. In this study, the plates with a crack along the center line and with a crack offset from the center line were considered as shown in Fig. 11.20c and d. Dimensons of the plate are the similar to the dimensions used in perielastodynamic simulation presented above. Distance between two PZT's was kept 104 mm, measured from their center. One of the PZT was used as an actuator while the other was used as a sensor. A 3.5 count tone-burst, with a central-frequency of ~150 KHz and ~20 V amplitude (~10 V peak-to-peak) was used to excite the actuator to generate the Lamb wave propagation in the plate. Tektronix AFG3021C was used to generate the tone-burst actuation at the interval of 1 ms and Tektronix MDO3024 was used to record the signals from the sensor. Sensor signal was recorded after averaging a total of 500 signals (to improve signal-to-noise ratio). Sampling rate and signal length were set to 50 MS/s and 10,000, respectively.

11.5.2.2 Other Computational Method for Verification of Perielastodynamic

Commercially available finite element (FE)-based computational tool COMSOL Multi-physics was used to simulate the structural mechanics problem coupled with the piezoelectric actuator and sensor. To model the ultrasonic guided wave propagation in the aluminum plate, solid mechanics and electrostatics modules, a multiphysics approach in COMSOL, was employed. COMSOL utilizes implicit scheme

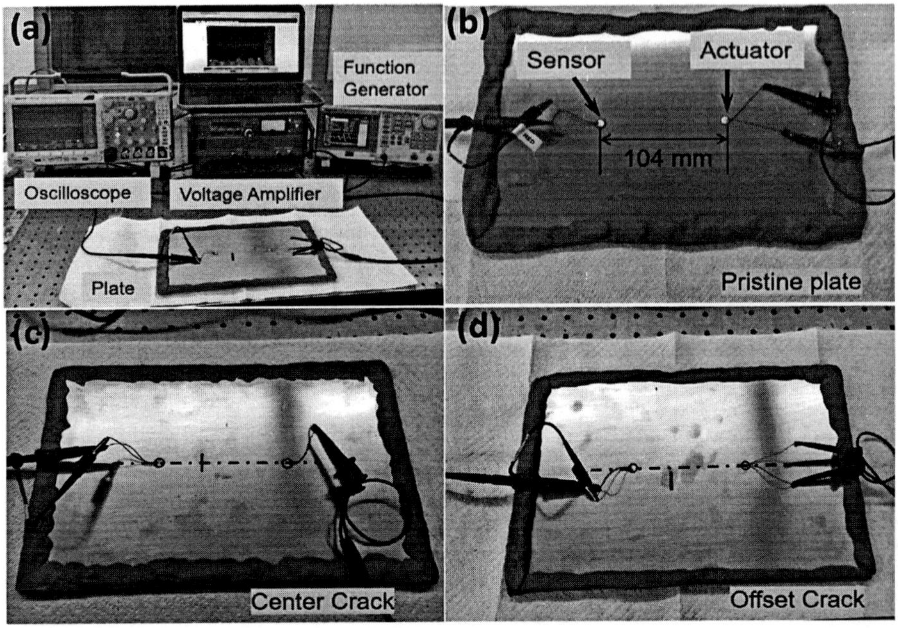

FIGURE 11.20 (a) Experimental set-up of pitch-catch experiments, (b) Pristine plate, (c) Plate with a center-crack, (d) Plate with a offset-crack.

to solve the transient problems. In this study, the direct solver MUMPS was chosen over an iterative solver for its robustness. All the direct solvers in COMSOL require significant amounts of RAM where MUMPS can store the solution out-of-core, i.e., on to the hard disk. Moreover, MUMPS is substantially faster than iterative solvers. The absolute tolerance of the time-dependent solver used a global method of scaling with a specified tolerance of 0.001. The setting for time steps was set to generalized-alpha method with intermediate time steps, a linear predictor and a maximum time step of 50 ns. The mechanical and electrical properties of the aluminum plate and the piezoelectric components are like the properties used in the experimental design. To excite the PZT actuator, a 20 V 3.5 count tone-burst signal (Fig. 11.6c) was applied at the electrical terminal. The signal response was collected from the PZT sensor for the entire duration of the simulation which was 80 μs. Free tetrahedral (tets) meshes generated by COMSOL multi-physics were utilized to generate mesh for the entire plate domain where the minimum mesh size was varied from 0.1 mm to 1.2 mm as shown in Fig. 11.21b and c. On the other hand, the minimum mesh size for the PZT actuator and sensor was varied from 0.01 mm to 0.2 mm as shown in Fig. 11.21a. A mesh convergence study was performed starting from the maximum mesh size of 2 mm to a minimum mesh size of 1.2 mm. As the mesh size was decreased the accuracy of the simulation improved at the expense of increased computational time. A total of 24 CPU cores with a maximum memory of 80 GB were utilized to solve this problem in 45 hours.

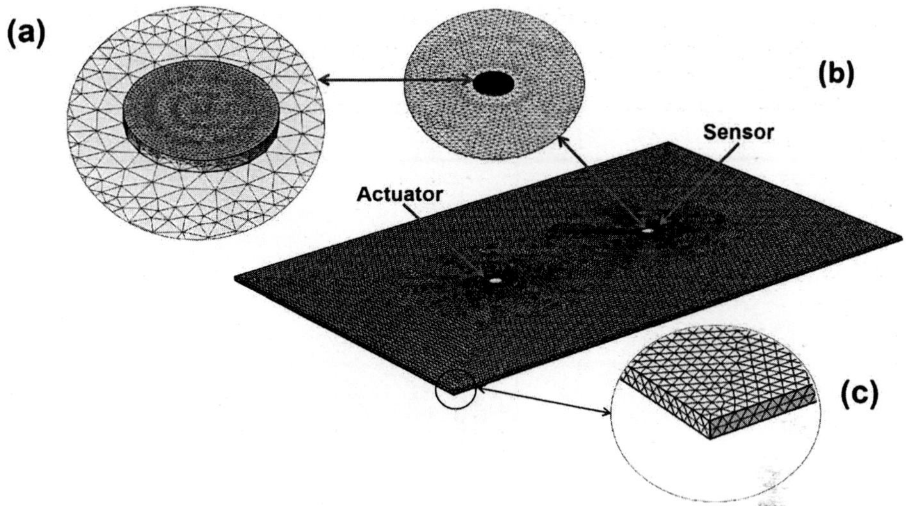

FIGURE 11.21 Three-dimensional FE discretization of the aluminum plate and PZT: (a) Discretization of the PZT, (b) Discretization of the plate and PZT, (c) Discretization of the plate.

11.5.2.3 Verification and Validation of Perielastodynamic Simulation

In this section, accuracy and efficiency of the perielastodynamic technique to simulate Lamb wave propagation is discussed. Time domain signal (at sensor location S1, in Fig. 11.17) obtained from perielastodynamic was compared with the numerically (COMSOL) and analytically (WaveFormRevealer [45, 48] (WFR)) obtained sensor signals and the signals acquired from the pitch-catch experiments on a pristine aluminum plate. Note that, in this study a square PZT was employed for perielastodynamic simulation for convenience, whereas circular PZT was used for COMSOL and the experiment. However, due to isotropic nature of the plate, at the far-field wave fronts were circular and the guided wave mode when fully developed, the effect of PZT size was thus ignored. In plane displacement ($u_1(x_j,t)$) from perielastodynamic was used to compare with the output voltage obtained from COMSOL, WFR, and Experiment. Since the sensors on the plate are located on the center line along the x_1-axis or the x-axis shown in Fig. 11.22b, c, and d, the output voltage at the sensor location is contributed primarily by the in-plane $u_1(x_j,t)$ displacement. To capture sensor signal from the perielastodynamic simulation, time-domain signal was collected from a material point located at 97 mm away from the PZT edge along the center line from the actuator ($L_{center} = 104$ mm and $L_{eff} = 97$ mm). Normalized amplitude of the sensor signals obtained from Experiment, COMSOL, WFR, and Perielastodynamics, respectively, are plotted in time-domain in Fig. 11.22. Sensor signals obtained from Perielastodynamics, COMSOL, and WFR were also compared with the experimental results to check the accuracy of those techniques. Good agreement between perielastodynamic and the experiment was observed for

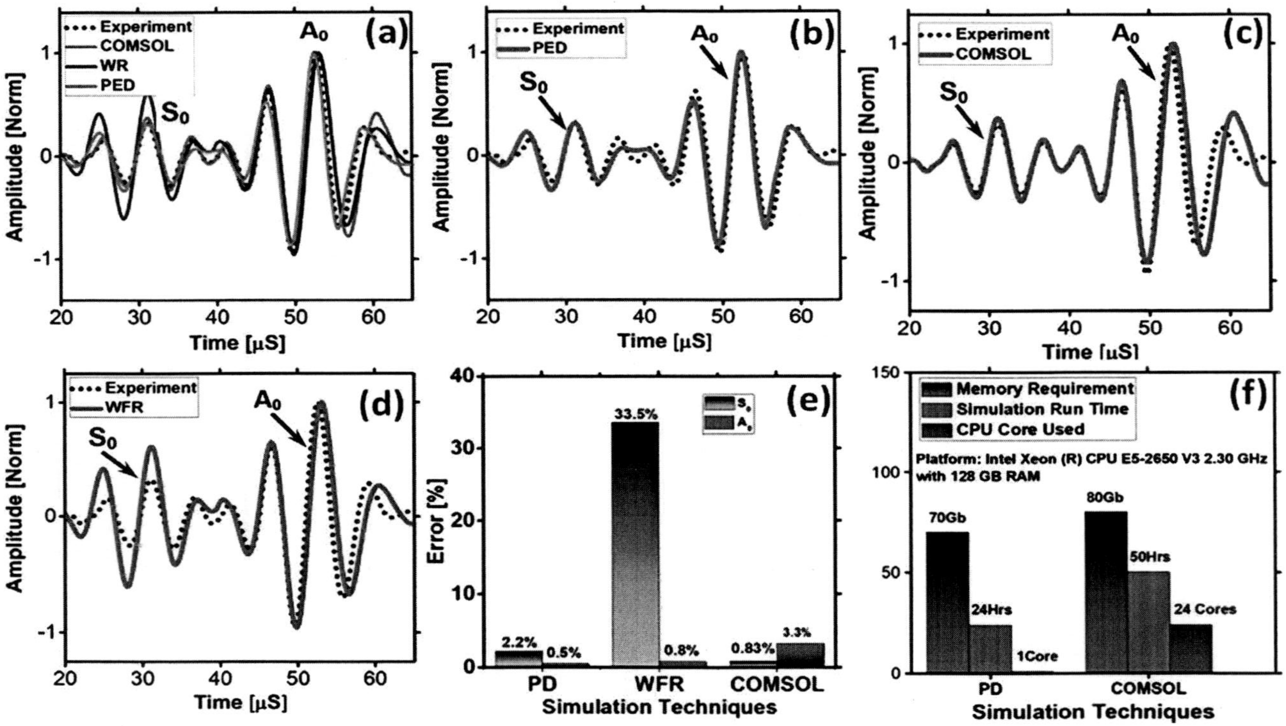

FIGURE 11.22 Time-domian comaprison of sensor signal: (a) Experiment, COMSOL, WFR and PED, (b) PED and Experiment, (c) Comsol and Experiment, (d) WFR and Experiment, (d) Error of simulated symmteric and anti-symmteric modes with respect to experimental results, (d) Memory requirement and simulation run time of PED and COMSOL simulation.

both symmetric and antisymmetric modes as shown in Fig. 11.22b. The symmetric mode generated by COMSOL simulation is in good agreement with the experiment whereas the antisymmetric mode slightly deviates (Fig. 11.22c). WFR predicted the antisymmetric mode well but overestimated the symmetric mode (Fig. 11.22d). WFR in based on the analytical method discussed in Chapter 5, and hence, material property of PZT could not be provided explicitly. To verify the accuracy of perielastodynamic technique and to compare with existing simulation techniques, error in simulating symmetric and antisymmetric modes is calculated by comparing them with experimental results using the following equation

$$Error = \frac{\int_{t_i}^{t_f \int} (A_{sim}^{env}(t) - A_{exp}^{env^2}}{\int_{t_i}^{t_f \int} A_{exp}^{env} \times 100} \tag{11.31}$$

where $A_{sim}^{env}(t) = |H(A_{sim}(t)|$; $A_{exp}^{env}(t) = |H(A_{exp}\,\|| \; A_{sim}^{env}(t)$ and $A_{exp}^{env}(t)$ are the envelope of the experimental and simulated sensor signals. A similar approach was used to calculate the error from COMSOL and WFR predicted sensor signals. Error analysis through a bar chart is shown in Fig. 11.22e. Error associated with the symmetric and antisymmetric modes from perielastodynamic simulation was 2.2% and 0.5%, respectively. Significant error in predicting symmetric mode from WFR was observed. Error percentage for COMSOL simulation for symmetric and antisymmetric modes is 0.83% and 3.3%, respectively. Considering both the modes, perielastodynamic simulation provided a better accuracy for wave propagation simulation then COMSOL and WFR.

To investigate the efficiency of perielastodynamic in simulating wave propagation problem with respect to the state-of-the-art finite element tool, memory requirement, simulation run time, and CPU core used are compared in Fig. 11.22f. Note that, element size was kept same. While COMSOL can use multiple cores (i.e., 24 cores) for running the simulation, perielastodynamic used only one core. Memory consumption and simulation run time for perielastodynamic is smaller than that of COMSOL simulation. However, parallelization of the perielastodynamic code can improve simulation run time significantly.

11.5.2.4 Wave Field Computation with Cracks and Comparisons

Perielastodynamic simulation results in a plate with a central-crack and an offset-crack are presented in Figs. 11.18 and 11.19, respectively. Next, to distinguish the amplitude of the reflected and transmitted Lamb wave modes, space-time representation of the in-plane and out-of-plane displacements for both the scenarios with center-crack and the offset-crack are analyzed. Displacement wave fields are computed along a dotted line shown in Fig. 11.20c and d. In case of central-crack scenario, both in-plane ($u_1(x_j,t)$) reflection and the transmission of the incident waves

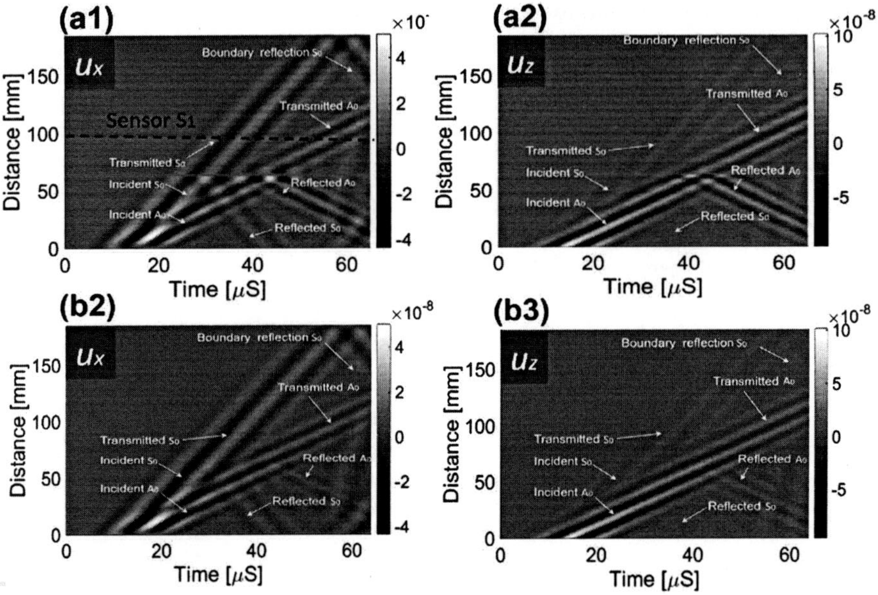

FIGURE 11.23 Space-time wavefield representations for the top surface of the plate with a through-thickness crack: (a-1) $u_x(x,t)$ for a plate with a central-crack, (a-2) $u_z(x,t)$ for a plate with a with a central-crack, (b-1) $u_x(x,t)$ for a plate with offset-crack, (b-2) $u_z(x,t)$ for a plate with offset- crack.

are clearly visible in Fig. 11.23(a1). A similar phenomenon can also be noticed from Fig. 11.23(a2) where out-of-plane displacement ($u_3(x_j,t)$) is shown. Boundary reflection of the modes can be observed both in Fig. 11.23(a1 & a2). Likewise, considering both the in-plane and the out-of-plane displacement modes, in case of offset-crack scenario, reflection and transmission from offset-crack are also visible in the space-time wave field in Fig. 11.23(b1 & b2).

Further the time history signals from the damage free pristine plate at sensor location S1 marked in Fig. 11.17a are compared with the signals when the central crack and the offset-crack are present in the plate. Fig. 11.24a compares the output response (at S1) obtained from the experiments in pristine state, with central-crack and with offset-crack, respectively. Fig. 11.24b compares the output response (at S1) obtained from perielastodynamic at pristine state, with central-crack and with offset crack, respectively. The first arrival of the symmetric and antisymmetric wave modes at the sensor location S1 obtained from both the experiment and perielastodynamic simulation are slightly delayed due to the presence of the crack. The delay is comparatively higher at the S1 location due to the central-crack, compared to the offset-crack. Due to an offset-crack edge, the reflected wave energy reflects at an angle and the senor S1 on the centerline have less effect compared to a crack present along the centerline. In Fig. 11.24c, sensor signal with central crack obatined from both CNDE perielastodynamic simulation and experiments are compared. Error in predicting symmetric and antisymmetric wave modes by

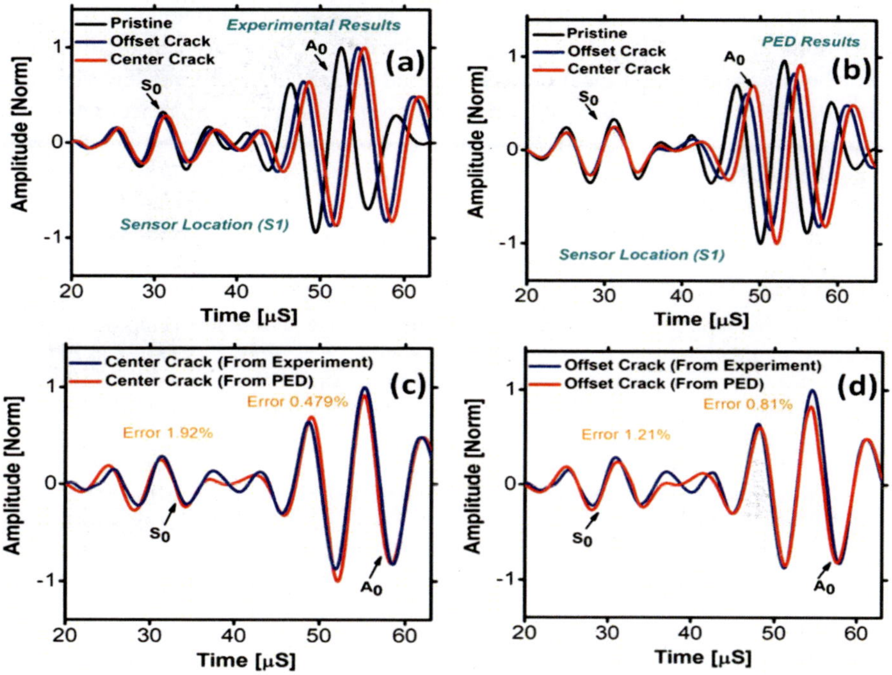

FIGURE 11.24 Comparison of time dependent signals obtained from PED and experiment at sensor location S1, in a pristine plate, plate with a crack along centerline and a plate with a off-set crack a) sensor signals at location S1 obtained from experiment, b) sensor signals at location S1 obtained from PED, c) sensor signals for centerline crack obtained from PED and experiment, d) sensor signals for offset crack obtained from PED and experiment.

perielastodynamic was 1.92% and 0.479%, respectively. In Fig. 11.24d, sensor signals for offset-crack obatined from both the perielastodynamic simulation and experiment are compared. Error in predicting symmetric and antisymmetric wave modes by perielastodynamic was 1.21% and 0.81%, respectively. Hence, it can be concluded that perielastodynamic method is a suitable wave simulation tool for CNDE and Computational SHM.

11.6 SUMMARY

A numerical wave field computational tool called Perielastodynamics is presented to simulate the guided waves in a plate-like structure with surface mounted PZT. Feasibility of the method is proved by simulating an SHM problem with PZT induced Lamb wave propagation in an isotropic aluminum plate. Fundamental symmetric (S_0) and antisymmetric (A_0) Lamb wave modes were generated. Further, their characteristics were investigated and compared with the theoretical predictions. Particle displacements due to S_0 and A_0 mode propagation were visualized through the vector-field plots across-the-thickness of the plate. Lamb wave modes simulated by the numerical technique were presented in the frequency-wavenumber domain and compared with those obtained from the analytical predictions. It can be concluded that if the process described in this book is adopted meticulously, the Perielastodynamics can simulate the Lamb wave propagation accurately. The computational time for the SHM problem can be easily accelerated by implementing the parallel computing.

REFERENCES

1. Kessler, S.S., S.M. Spearing, and C. Soutis, Damage detection in composite materials using Lamb wave methods. *Smart Materials and Structures*, 2002. **11**(2): p. 269.
2. Patra, S. and S. Banerjee, Material state awareness for composites Part I: Precursor damage analysis using ultrasonic guided coda wave interferometry (CWI). *Materials*, 2017. **10**(12): p. 1436.
3. Patra, S., Ahmed, H., Banerjee, S., Peri-elastodynamic simulations of guided ultrasonic waves in plate-like structure with surface mounted PZT. *Sensors*, 2018. **18**(1).
4. Patra, S., Ahmed, H., Saadatzi, MS., Banerjee, S., Experimental verification and validation of nonlocal peridynamic approach for simulating guided Lamb wave propagation and damage interaction. *Structural Health Monitoring Journal*, 2019. **18**(5–6).
5. Giurgiutiu, V., *Structural Health Monitoring: With Piezoelectric Wafer Active Sensors*. 2007, New York: Academic Press.
6. Giurgiutiu, V., Lamb wave generation with piezoelectric wafer active sensors for structural health monitoring. *Smart Structures and Materials*, 2003. **5056**: pp. 111–122.
7. Lee, B. and W. Staszewski, Lamb wave propagation modelling for damage detection: I. Two-dimensional analysis. *Smart Materials and Structures*, 2007. **16**(2): p. 249.
8. Paćko, P., et al., Lamb wave propagation modelling and simulation using parallel processing architecture and graphical cards. *Smart Materials and Structures*, 2012. **21**(7): p. 075001.
9. Raghavan, A. and C.E. Cesnik, Finite-dimensional piezoelectric transducer modeling for guided wave based structural health monitoring. *Smart Materials and Structures*, 2005. **14**(6): p. 1448.
10. Moser, F., L.J. Jacobs, and J. Qu, Modeling elastic wave propagation in waveguides with the finite element method. *NDT&E International*, 1999. **32**(4): pp. 225–234.
11. Cho, Y. and J.L. Rose, A boundary element solution for a mode conversion study on the edge reflection of Lamb waves. *The Journal of the Acoustical Society of America*, 1996. **99**(4): pp. 2097–2109.
12. Yim, H. and Y. Sohn, Numerical simulation and visualization of elastic waves using mass-spring lattice model. *IEEE Transactions on Ultrasonics, Ferroelectrics, and Frequency Control*, 2000. **47**(3): pp. 549–558.
13. Balasubramanyam, R., et al., A finite-difference simulation of ultrasonic Lamb waves in metal sheets with experimental verification. *Journal of Physics D: Applied Physics*, 1996. **29**(1): p. 147.
14. Bergamini, A. and F. Biondini, Finite strip modeling for optimal design of prestressed folded plate structures. *Engineering Structures*, 2004. **26**(8): pp. 1043–1054.
15. Diehl, P. and M.A. Schweitzer, Simulation of wave propagation and impact damage in brittle materials using peridynamics, in *Recent Trends in Computational Engineering-CE2014*. 2015, Berlin: Springer. pp. 251–265.
16. Nishawala, V.V., et al., Simulation of elastic wave propagation using cellular automata and peridynamics, and comparison with experiments. *Wave Motion*, 2016. **60**: pp. 73–83.

17. Kluska, P., et al., Cellular automata for Lamb wave propagation modelling in smart structures. *Smart Materials and Structures*, 2013. **22**(8): pp. 085022.
18. Leckey, C.A., et al., Multiple-mode Lamb wave scattering simulations using 3D elastodynamic finite integration technique. *Ultrasonics*, 2012. **52**(2): pp. 193–207.
19. Banerjee, S., T. Kundu, and N.A. Alnuaimi, DPSM technique for ultrasonic field modelling near fluid–solid interface. *Ultrasonics*, 2007. **46**(3): pp. 235–250.
20. Kijanka, P., et al., GPU-based local interaction simulation approach for simplified temperature effect modelling in Lamb wave propagation used for damage detection. *Smart Materials and Structures*, 2013. **22**(3): p. 035014.
21. Shen, Y. and C.E. Cesnik, Local interaction simulation approach for efficient modeling of linear and nonlinear ultrasonic guided wave active sensing of complex structures. *Journal of Nondestructive Evaluation, Diagnostics and Prognostics of Engineering Systems*, 2018. **1**(1): p. 011008.
22. Ge, L., X. Wang, and F. Wang, Accurate modeling of PZT-induced Lamb wave propagation in structures by using a novel spectral finite element method. *Smart Materials and Structures*, 2014. **23**(9): p. 095018.
23. Ha, S. and F.-K. Chang, Optimizing a spectral element for modeling PZT-induced Lamb wave propagation in thin plates. *Smart Materials and Structures*, 2009. **19**(1): p. 015015.
24. Zou, F. and M. Aliabadi, On modelling three-dimensional piezoelectric smart structures with boundary spectral element method. *Smart Materials and Structures*, 2017. **26**(5): p. 055015.
25. Marzani, A., et al., A semi-analytical finite element formulation for modeling stress wave propagation in axisymmetric damped waveguides. *Journal of Sound and Vibration*, 2008. **318**(3): pp. 488–505.
26. Silling, S.A., Reformulation of elasticity theory for discontinuities and long-range forces. *Journal of the Mechanics and Physics of Solids*, 2000. **48**(1): pp. 175–209.
27. Silling, S.A. and E. Askari, A meshfree method based on the peridynamic model of solid mechanics. *Computers & Structures*, 2005. **83**(17): pp. 1526–1535.
28. Hafezi, M.H., R. Alebrahim, and T. Kundu, Peri-ultrasound for modeling linear and nonlinear ultrasonic response. *Ultrasonics*, 2017. **80**: pp. 47–57.
29. Martowicz, A., et al. Non-local modeling and simulation of wave propagation and crack growth. in *AIP Conference Proceedings*. 2014. AIP.
30. Rice, J., Elastic fracture mechanics concepts for interfacial cracks. *Journal of Applied Mechanics*, 1988. **55**(1): pp. 98–103.
31. Macek, R.W. and S.A. Silling, Peridynamics via finite element analysis. *Finite Elements in Analysis and Design*, 2007. **43**(15): pp. 1169–1178.
32. Silling, S.A., et al., Peridynamic states and constitutive modeling. *Journal of Elasticity*, 2007. **88**(2): pp. 151–184.
33. Askari, E., et al. Peridynamics for multiscale materials modeling. *Journal of Physics: Conference Series*, 2008. **125**: p. 012078.
34. Bobaru, F., et al., *Handbook of Peridynamic Modeling*. 2016, New York: CRC Press.
35. Ha, Y.D. and F. Bobaru, Studies of dynamic crack propagation and crack branching with peridynamics. *International Journal of Fracture*, 2010. **162**(1): pp. 229–244.
36. Madenci, E. and E. Oterkus, *Peridynamic Theory and its Applications*. Vol. **17**. 2014, Berlin: Springer.
37. Hafezi, M.H. and T. Kundu, Peri-ultrasound modeling for surface wave propagation. *Ultrasonics*, 2018. **84**: pp. 162–171.
38. Hafezi, M.H. and T. Kundu, Peri-ultrasound modeling of dynamic response of an interface crack showing wave scattering and crack propagation. *Journal of Nondestructive Evaluation, Diagnostics and Prognostics of Engineering Systems*, 2018. **1**(1): p. 011003.

39. Silling, S. and R. Lehoucq, Peridynamic theory of solid mechanics. *Advances in Applied Mechanics*, 2010. **44**: pp. 73–168.
40. Leckey, C.A.C., M.D. Rogge, and F. Raymond Parker, Guided waves in anisotropic and quasi-isotropic aerospace composites: three-dimensional simulation and experiment. *Ultrasonics*, 2014. **54**(1): pp. 385–394.
41. Parks, M.L., et al., *Peridynamics with LAMMPS: A User Guide, v0. 3 Beta*. Sandia Report (2011–8253), 2011.
42. Giurgiutiu, V., Tuned lamb wave excitation and detection with piezoelectric wafer active sensors for structural health monitoring. *Journal of Intelligent Material Systems and Structures*, 2005. **16**(4): pp. 291–305.
43. Kundu, T., Mechanics of elastic waves and ultrasonic nondestructive evaluation, in *Ultrasonic Nondestructive Evaluation*, T. Kundu, Editor. 2004, New York: CRC Press. pp. 1–142.
44. Pavlakovic, B., et al., Disperse: a general purpose program for creating dispersion curves, in *Review of Progress in Quantitative Nondestructive Evaluation: Volume 16A*, D.O. Thompson and D.E. Chimenti, Editors. 1997, Boston, MA: Springer US. pp. 185–192.
45. Shen, Y. and V. Giurgiutiu. *WFR-2D:* an analytical model for PWAS-generated 2D ultrasonic guided wave propagation. in *SPIE Smart Structures and Materials + Nondestructive Evaluation and Health Monitoring*. 2014. New York: SPIE.
46. Michaels, T.E., J.E. Michaels, and M. Ruzzene, Frequency–wavenumber domain analysis of guided wavefields. *Ultrasonics*, 2011. **51**(4): pp. 452–466.
47. Giurgiutiu, V., *Structural Health Monitoring with Piezoelectric Wafer Active Sensors*, 2nd Ed. 2014, New York: Elsevier Academic Press.
48. Shen, Y. and V. Giurgiutiu, WaveForm Revealer-An analytical predictive tool for the simulation of multi-mode guided waves interaction with damage. *AIAA SDM*, 2013.

Index

A

B

E

T